Biology of Conidial Fungi

Volume 2

Biology of Conidial Fungi

Volume 2

Edited by

Garry T. Cole

Department of Botany
University of Texas at Austin
Austin, Texas

Bryce Kendrick

Department of Biology
University of Waterloo
Waterloo, Ontario, Canada

ACADEMIC PRESS   1981
A Subsidiary of Harcourt Brace Jovanovich, Publishers
New York   London   Toronto   Sydney   San Francisco

UNIVERSITY LIBRARY
2 5 MAY 1982
LANCASTER

ACADEMIC PRESS, INC.
111 Fifth Avenue, New York, New York 10003

United Kingdom Edition published by
ACADEMIC PRESS, INC. (LONDON) LTD.
24/28 Oval Road, London NW1   7DX

Library of Congress Cataloging in Publication Data
Main entry under title:

The Biology of conidial fungi.

    Includes bibliographies and index.
    1. Fungi. 2. Conidia. I. Cole, Garry T., Date.
II. Kendrick, Bryce. [DNLM: 1. Fungi. 2. Mycoses.
QW 180 B614]
QK603.B5   589.2'3   80-1679
ISBN 0-12-179502-0 (v. 2)

PRINTED IN THE UNITED STATES OF AMERICA

81 82 83 84   9 8 7 6 5 4 3 2 1

81 009819

To Heather and Allison

Contents

IV   CONIDIAL FUNGI AND MAN

15   Clinical Aspects of Medically Important Conidial Fungi
J. W. Rippon

16   Mycotoxin Production by Conidial Fungi
Philip B. Mislivec

17   Development of Parasitic Conidial Fungi in Plants
James R. Aist

18   Food Spoilage and Biodeterioration
John I. Pitt

19   Use of Conidial Fungi in Biological Control
T. E. Freeman

20   Predators and Parasites of Microscopic Animals
G. L. Barron

21   Entomogenous Fungi
Donald W. Roberts and Richard A. Humber

22   Food Technology and Industrial Mycology
William D. Gray

V   ULTRASTRUCTURE, DEVELOPMENT, PHYSIOLOGY, AND BIOCHEMISTRY

23   Conidiogenesis and Conidiomatal Ontogeny
Garry T. Cole

24   Biochemistry of Microcycle Conidiation
J. E. Smith, J. G. Anderson, S. G. Deans, and D. R. Berry

25   Nuclear Behavior in Conidial Fungi
C. F. Robinow

26   Viruses of Conidial Fungi
Paul A. Lemke

**31   Techniques for Examining Developmental and Ultrastructural
       Aspects of Conidial Fungi**
       Garry T. Cole

List of Contributors

Numbers in parentheses indicate the pages on which the authors' contributions begin.

James R. Aist (75), Department of Plant Pathology, Cornell University, Ithaca, New York 14853

J. G. Anderson (329), Department of Applied Microbiology, University of Strathclyde, Glasgow IG IXW, Scotland

Jerome M. Aronson (459), Botany and Microbiology Department, Arizona State University, Tempe, Arizona 85281

G. L. Barron (167), Department of Environmental Biology, University of Guelph, Guelph, Ontario N1G 2W1, Canada

D. R. Berry (329), Department of Applied Microbiology, University of Strathclyde, Glasgow IG IXW, Scotland

Garry T. Cole (271, 577), Department of Botany, University of Texas at Austin, Austin, Texas 78712

S. G. Deans (329), Department of Microbiology, West of Scotland Agricultural College, Auchincruive, Ayr, Scotland

T. E. Freeman (143), Plant Pathology Department, University of Florida, Gainesville, Florida 32611

William D. Gray* (237), Department of Biological Sciences, Northern Illinois University, DeKalb, Illinois 60115

Robert Hall (417), Department of Environmental Biology, University of Guelph, Guelph, Ontario N1G 2W1, Canada

A. C. Hastie (511), Department of Biological Sciences, The University of Dundee, Dundee DD1 4HN, Scotland

Richard A. Humber (201), Insect Pathology Research Unit, Boyce Thompson Institute for Plant Research, Cornell University, Ithaca, New York 14853

S. C. Jong (551), Mycology Department, American Type Culture Collection, Rockville, Maryland 20852

*Present address: 512 N. High Street, Apartment E, Lancaster, Ohio 43130.

Paul A. Lemke (395), Department of Botany, Plant Pathology, and Microbiology, Auburn University, Auburn, Alabama 36849

Philip B. Mislivec (37), Food Microbiology Branch, Division of Microbiology, Food and Drug Administration, Washington, D.C. 20204

John I. Pitt (111), Division of Food Research, C.S.I.R.O., North Ryde, New South Wales 2113, Australia

J. W. Rippon (3), Department of Medicine, Section of Dermatology, Pritzker School of Medicine, University of Chicago, Chicago, Illinois 60637

Donald W. Roberts (201), Insect Pathology Research Unit, Boyce Thompson Institute for Plant Research, Cornell University, Ithaca, New York 14853

C. F. Robinow (357), Department of Microbiology and Immunology, University of Western Ontario, London, Ontario, Canada

J. E. Smith (329), Department of Applied Microbiology, University of Strathclyde, Glasgow G1 1XW, Scotland

Foreword

With the publication of these volumes it can be truly said that the conidial fungi—long the stepchildren of mycologists—have finally been legitimized, and the editors are to be congratulated on putting together these authoritative volumes.

No important phase of the biology of these fungi has been neglected and the editors were fortunate to have each topic treated by a recognized authority. Although the discussions vary greatly in length and in detail, the editors have succeeded admirably in intertwining them into a unified whole.

From systematics to biochemistry, from saprobes to human pathogens, from mycotoxins to mycoviruses, every topic that has been explored by mycologists in the last two centuries is included and beautifully summarized. Congratulations to the editors and the authors for a job well done.

Constantine J. Alexopoulos
University of Texas at Austin

Preface

The term Fungi Imperfecti could be broadly defined as including any and all anamorphic (asexual) expressions of true fungi that occur separately in time and/or space from their respective teleomorphs (sexual phases), or are not known to have a teleomorph. This interpretation embraces such phenomena as the sporangial forms of many Zygomycetes, several of the spore forms of the Uredinales, and the Mycelia Sterilia, which do not produce spores. These volumes exclude the Mycelia Sterilia. The conidial fungi, for the purposes of this work, are anamorphic fungi of presumed ascomycetous or basidiomycetous origin (the higher fungi or Dikaryomycota). Sporangial morphs of Zygomycetes, etc., are specifically excluded, though their ecology and biology, but not their phylogeny, closely parallel those of the dikaryomycotan anamorphs. Why, then, are they excluded? The reasons are largely historical. While separate systems of binomials and taxonomic schemes have grown up around dikaryomycotan anamorphs and teleomorphs, respectively, no such dichotomy has occurred in the Zygomycetes. Even if only the sporangial anamorph is known, the basic zygomycotan nature of the organism is assumed by inference, and a teleomorphic, and thus holomorphic, generic name is invariably applied to such anamorphs. Treatments of Zygomycetes have an integrated approach that has been lacking in our dealings with the higher fungi and their various life forms. The history and rationale of this schism between anamorphic and teleomorphic classifications is carefully expounded by Weresub and Pirozynski (1979).* There are other reasons. While several hundred Zygomycetes are known, the Dikaryomycota comprise about 90% of all true fungi, and many thousands of confirmed or presumed dikaryomycotan anamorphs confront mycologists wherever they turn their attention. Such anamorphs affect our lives in many ways. They cause such superficial mycoses as the various kinds of ringworm, and deep-seated mycoses like aspergillosis and histoplasmosis. They cause the majority of important fungal plant

*Weresub, L. K., and Pirozynski, K. A. (1979). Pleomorphism of fungi as treated in the history of mycology and nomenclature. *In* The Whole Fungus (B. Kendrick, ed.), Vol. I, pp. 17–26. Nat. Mus., Canada.

diseases, such as southern blight of corn, Panama disease of bananas, and a plethora of wilts, anthracnoses, leaf spots, soft rots, etc. They cause enormous losses of stored processed food (anything from bread to jam), and produce insidious and dangerous toxins such as aflatoxin, vomitoxin, zearalenone, and sporidesmins. They cause deterioration of manufactured natural products including paper, leather, cottons, and even paint.

On the positive side, dikaryomycotan anamorphs produce extremely valuable secondary metabolites, such as antibiotics and organic acids, which we have been able to exploit. They are used in processing gourmet cheeses (camembert, brie, roquefort, gorgonzola, stilton, danish blue); they have potential in biological control of insects, and possibly of nematodes and pathogenic fungi. And perhaps most important of all, they are principally responsible for the conditioning or mineralization of the enormous quantities of plant litter that fall to the ground each year, they comprise the greater part of the microbial biomass in the soil and in many ponds and streams, and they are thus vitally involved in energy flow in many ecosystems. It is probably fair to say that dikaryomycotan anamorphs are the commonest fungal manifestations of all. And yet the only volumes so far that treat them exclusively are concerned with their systematics. We felt it was time that other aspects of their existence were given some concentrated attention. Hence these two volumes. We have tried to bring together in one place detailed considerations of many facets of conidial fungi—some of which have been almost entirely though undeservedly neglected in previous literature. We hope that these volumes will fill some lacunae in our knowledge of anamorphs, and serve as a useful reference to the advanced student who probably encounters many such fungi, but has tended to regard them simply as weeds or contaminants. If we can convey to students our belief that dikaryomycotan anamorphs are among the most successful and versatile fungi of all, and the reasons for that belief, we shall be well satisfied.

Garry T. Cole
Bryce Kendrick

Contents of Volume 1

IV

CONIDIAL FUNGI
AND MAN

15

Clinical Aspects of
Medically Important Conidial Fungi

J.W. Rippon

I. INTRODUCTION

With the exception of a few dermatophytes, fungi do not depend for survival on their ability to cause infection in man. The production of human disease by fungi is an accidental phenomenon. Pathogenesis appears to be a peculiar trait of a particular organism—a trait not usually shared by taxonomically related

Biology of Conidial Fungi, Vol. 2
Copyright © 1981 by Academic Press, Inc.
All rights of reproduction in any form reserved.
ISBN 0-12-179502-0

species. In most cases it is the ability to adapt to and survive in an unfavorable environment which differentiates pathogenic from nonpathogenic species. However, under circumstances where the host's normal defenses against infection are altered, species of fungi not usually considered virulent may invade and cause disease, thus earning the term "pathogenic."

Clinical infections of man caused by fungi may be divided into four categories (Table I). Each category is based primarily on the location of the infection, and each involves a group of organisms peculiar to it. However, within the disease category there is usually a consistent pattern of mechanism of infection, host

TABLE I

Clinical Types of Fungal Infections

Type	Disease	Causative organism
Superficial infections	Pityriasis versicolor	*Malassezia furfur*
	Piedra	*Trichosporon beigelii* (white), *Piedraia hortae* (black)
	Tinea nigra	*Exophiala werneckii*
Cutaneous infections	Ringworm of scalp, body, feet, groin, and so on; the tineas	Dermatophytes (*Microsporum, Trichophyton, Epidermophyton* sp.)
	Candidiasis of skin, mucus membranes, and nails	*Candida albicans* and related species
Subcutaneous infections	Chromoblastomycosis	*Fonsecaea pedrosoi, Wangiella dermatitidis, Phialophora verrucosa*
	Mycotic mycetoma	*Petriellidium boydii, Madurella mycetomatis, Exophiala jeanselmei*
	Entomophthoromycosis	*Basidiobolus ranaram Conidiobolus coronata*
	Rhinosporidiosis	*Rhinosporidium seeberii*
	Lobomycosis	*Loboa loboi*
	Sporotrichosis	*Sporothrix schenckii*
Systemic infections	Pathogenic fungus infections	
	Histoplasmosis	*Histoplasma capsulatum*
	Coccidioidomycosis	*Coccidioides immitis*
	Paracoccidioidomycosis	*Paracoccidioides brasiliensis*
	Blastomycosis	*Blastomyces dermatitidis*
	Opportunistic fungus infections	
	Aspergillosis	*Aspergillus fumigatus,* and so on
	Candidiasis, systemic	*Candida albicans* and related species
	Mucormycosis	*Mucor* sp., *Absidia* sp., *Rhizopus* sp.
	Cryptococcosis	*Cryptococcus neoformans*

response, type of disease produced, and natural course of infection. The first category, superficial infections, involves colonization of the surface of the host without the production of significant injury. The second category, cutaneous disease, involves dermatophytes. In this type of infection a gamut of host responses may be present, from mild scaling and itching to severe, inflammatory denuding lesions of the epithelium. The etiological agents are a closely related group of fungi, dermatophytes, some of which show an evolutionary trend toward dependency on infection for survival. The subcutaneous mycoses are a disparate group of clinical entities caused by a diverse group of unrelated organisms. The trait they have in common is that infection is usually the result of traumatic implantation. Also, the subcutaneous disease usually remains localized. Systemic infections, on the other hand, are of two types. The first is a homogeneous group of clinical diseases caused by closely related species termed true pathogenic fungi, and the second is a diverse group of infections caused by opportunistic fungi.

II. SUPERFICIAL INFECTIONS

Superficial fungal infections are due to colonization by a fungus of the outer surface of some cutaneous structure such as skin or hair. They are usually symptomless and are brought to the attention of the patient only because of a cosmetic change. In some infections this is due to an increase in the quantity of a species that is part of the normal flora, such as *Malassezia furfur* in pityriasis versicolor and *Trichosporon beigelii* in white piedra. In other diseases, the infection is caused by a soil-inhabiting species such as *Piedraia hortae* in black piedra and *Exophiala werneckii* in tinea nigra.

A. Pityriasis Versicolor

Pityriasis versicolor, a common superficial mycosis, involves areas of overgrowth by *Malassezia furfur*. These areas, characterized by branny, variably discolored, furfuraceous scales on the outer layers of the stratum corneum give rise to the name of the disease. In white or light-complected people, fawn or yellow-brown to dark-brown patches appear. Sometimes the lesion consists of one continuous sheet involving the entire chest, trunk, or abdomen. If the patient has been exposed to the sun, the patches will appear as light areas, as substances in the organism filter out the ultraviolet rays. In dark-skinned people, the patches are mottled light and darker than the normal pigmentation. *Malassezia furfur*, a lipophilic yeast, is part of the normal flora of the scalp and skin. Overgrowth and disease appear to be related to the rate of epidermal turnover. A slowing of the rate due to either environment (high temperatures) or drugs (patients on steroids) may lead to the condition. Because of favorable climatic conditions the disease is almost universal in the tropics. In skin scales the organisms appear as small, budding yeasts and short, branching hyphal elements.

B. Piedra

White piedra is an aggregation around the hair shaft of *Trichosporon beigelii*, a yeast which is a member of the normal flora of the skin. The organism appears to take up residence in a damaged area on the cuticle of the hair shaft some time after it emerges above the skin. The organism then replicates, forming a soft, granular mass consisting of arthroconidia and budding yeasts as well as various other species of fungi, protozoa, and bacteria. There is no damage to the patient.

Black piedra consists of a hard, knobby, black incrustation of the fungus *Piedraia hortae* around a hair shaft. The nodule consists of a stroma in which locules containing asci and ascospores are formed. It is the only sexual form of a fungus to be found in human disease. The organism can be found infecting the hair shafts of any member of the primates, although closely related species of Asterineae form hard masses of similar morphology on banana leaves and other tropical trees.

C. Tinea Nigra

Tinea nigra is a colonization of the thick stratum corneum of the palm of the hand, the sole of the foot or, rarely, some other area of the skin by *Exophiala werneckii* and related species. The fungi are found in the soil and gain entrance through an abrasion of the skin surface. They are pigmented and cause a brownish discoloration of the area. There are no other symptoms. *In situ*, the fungi grow as highly pleomorphic yeastlike cells and short mycelial strands. The yeastlike cells, which are actually reduced annellidic conidiogenous cells, are found in culture along with compact, septate hyphae. The lesion, which is sometimes mistaken for malignant melanoma, can lead to mutilative surgery.

III. CUTANEOUS INFECTIONS

Cutaneous disease is the result of colonization of the stratum corneum by a fungus that incites an allergic or toxic inflammatory response. This causes various degrees of discomfort or debility to the host. There are two large categories of these infections. The first involves a dermatophyte. The second represents colonization of the skin surface by a member of the normal flora of the alimentary tract, *Candida albicans*.

A. Ringworm or Dermatophytosis

Dermatophytes are specialized ectodermal parasites ultimately derived from the soil. They represent a homogeneous group of closely related species, a single species of which can induce a wide variety of disease processes. They constitute one of the most complicated and controversial aspects of medical mycology. The morphology of the various species is similar both in infection and in culture, whether the organism grows as a conidium-producing hyphomycete or forms its ascigerous teleomorph.

Dermatophytes include approximately 40 species of fungi grouped in three anamorph-genera (Table II). The genera are differentiated by the type of macroconidium, a nondeciduous phragmospore produced directly on the hypha or from the ballooning out of a short pedicle. In the genus *Microsporum* the macroconidia are spindle-shaped and lenticular or broad in the middle, with tapering ends. Various echinulations are formed on the walls of the spores, which are thick in some species (*M. canis*) and thin (*M. racemosum*) in others. The walls of the macroconidia of the genus *Trichophyton* are thin and without echinulations (Fig. 1), and the conidia are pencil-shaped or cigar-shaped. In both genera, single-celled microconidia of variable morphology are produced. The genus *Epidermophyton* produces macroconidia with a blunt, expanded distal end which results in a club or beaver tail shape (Fig. 2). The walls are thin and may be echinulated. No microconidia are produced.

Most dermatophytic species for which a teleomorph is known are heterothallic and of bipolar incompatibility. The ascoma is a loosely woven gymnothecium (Fig. 3) in which evanescent asci containing eight ascospores are produced. The form of the peridial hyphae defines the two teleomorphic genera: *Arthroderma*, representing the ascigerous state of the genus *Trichophyton*, and *Nannizzia*, that of the genus *Microsporum* (Table III). Note in Table III that a single anamorph-species, e.g., what is known as *T. terrestre*, may have several teleomorphs. *Epidermophyton* has no known teleomorph. The two teleomorphic genera are included in the family Gymnoascaceae of the Eurotiales. This family also contains the teleomorphs of the true pathogenic fungi *Histoplasma capsulatum* and *Blastomyces dermatitidis*.

Soil abounds with other closely related genera that have as a common characteristic the ability to utilize keratin. However, only the dermatophytes of the anamorph-genera *Trichophyton*, *Microsporum*, and *Epidermophyton* are regularly isolated from human and animal infections.

In examining the ecology, reservoirs, and distribution of the various species listed in Table IV, several patterns emerge. The first, mentioned earlier, is the evolutionary trend toward dependence on the infection for the maintenance and dissemination of the species. Thus the species can be categorized as to their principal reservoir. The first group includes anthropophilic species for which

TABLE II

Currently Recognized Dermatophytes: Anamorph Genera and Species

Epidermophyton Sabouraud 1910
 E. floccosum (Harz 1870) Langeron & Milochevitch 1930
 E. stockdaleae Prochacki & Engelhard Zasuda 1974
Microsporum Gruby 1843
 M. amazonicum Moraes, Borelli, & Feo, 1967
 M. audouinii Gruby 1943[a]
 M. boullardii Dominik & Majchrowicz 1965
 M. canis Bodin 1902[a,b]
 M. cookei Ajello 1959
 M. distortum DiMenna & Marples 1954[b]
 M. equinum (Delacroix & Bodin 1896) Gueguen 1904[b]
 M. ferrugineum Ota 1921[a]
 M. fulvum Uriburu 1909[a]
 M. gallinae (Megnin 1881) Grigorakis 1929
 M. gypseum (Bodin 1907) Guiart & Grigorakis 1928[a]
 M. nanum Fuentes 1956[b]
 M. persicolor (Sabouraud 1910) Guiart & Grigorakis 1928
 M. praecox Rivalier 1954
 M. racemosum Borelli 1965
 M. ripariae Hubalek & Rush-Monroe 1973
 M. vanbreuseghemii Georg, Ajello, Friedman, & Brinkman 1962
Trichophyton Malmsten 1845
 T. ajelloi (Vanbreuseghem 1952) Ajello 1968
 T. concentricum Blanchard 1895[a]
 T. equinum (Matruchot & Dassonville 1898) Gedoelst 1902
 T. flavescens Padhye & Carmichael 1971
 T. georgiae Varsavsky & Ajello 1964
 T. gloriae Ajello 1967
 T. gourvilii Catanei 1933[a]
 T. longifusum (Florian & Galgoczy 1964) Ajello 1968
 T. megninii Blanchard 1896[a]
 T. mentagrophytes (Robin 1853) Blanchard 1896 var. *mentagrophytes*[a,b]
 T. mentagrophytes (Robin 1853) Blanchard 1896 var. *interdigitale*
 T. mentagrophytes (Robin 1853) Blanchard 1896 var. *erinacei*
 T. mentagrophytes (Robin 1853) Blanchard 1896 var. *quinckeanum*
 T. phaseoliforme Borelli & Feo 1966
 T. rubrum (Castellani 1910) Sabouraud 1911[b]
 T. schoenleinii (Lebert 1845) Langeron & Milochevitch 1930[a]
 T. simii (Pinoy 1912) Stockdale, Mackenzie, & Austwick 1965[b]
 T. soudanense Joyeux 1912[a]
 T. terrestre Durie & Frey 1957
 T. tonsurans Malmsten 1845
 T. vanbreuseghemii Rioux, Tarry, & Tuminer 1964
 T. verrucosum Bodin 1902[a]
 T. violaceum Bodin 1902[a]
 T. yaoundei Cochet & Doby Dubois 1957[a]

[a] Commonly isolated from human infection.
[b] Commonly isolated from animal infection.

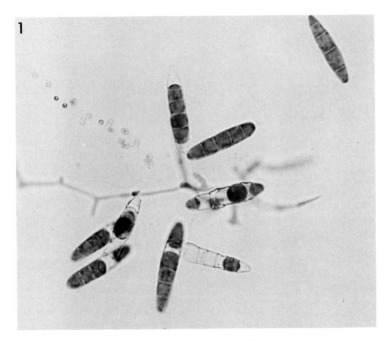

Fig. 1. Macroconidia of *Trichophyton simii*. The thin-walled conidia are blunt at both ends and without echinulations. In this species endoarthrospores are sometimes seen developing within the macroconidia.

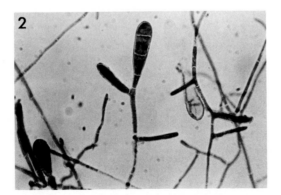

Fig. 2. Macroconidia of *Epidermophyton floccosum*.

Fig. 3. Gymnothecium of *Arthroderma benhamiae,* teleomorph of *Trichophyton mentagrophytes* (500 μm in diameter), growing on soil-baited hair (photograph from G. Rebell).

TABLE III

Ascigerous Genera and Species of Dermatophytes

Teleomorph	Anamorph
Arthroderma	*Trichophyton*
A. benhamiae Ajello & Cheng 1967+	T. mentagrophytes
A. ciferrii Varsavsky & Ajello 1964	T. georgiae
A. flavescens Padhye & Carmichael 1971	T. flavescens
A. gertlerii Bohme 1967	T. vanbreuseghemii
A. gloriae Ajello 1967	T. gloriae
A. insingulare Padhye & Carmichael 1972	T. terrestre
A. lenticularum Pore, Tsao, & Plunkett 1965	T. terrestre
A. quadrifidum Dawson & Gentles 1961	T. terrestre
A. simii Stockdale, Mackenzie, & Austwick 1965	T. simii
A. uncinatum Dawson & Gentles 1959	T. ajelloi
A. vanbreuseghemii Takashio 1973	T. mentagrophytes
Nannizzia	*Microsporum*
N. borellii Morales, Padhye, & Ajello 1975	M. amazonicum
N. cajetani Ajello 1961	M. cookei
N. fulva Stockdale 1963	M. fulvum
N. grubia Georg, Ajello, Friedman, & Brinkman 1967	M. vanbreuseghemii
N. gypsea Stockdale 1963	M. gypseum
N. incurvata Stockdale 1961	M. gypseum
N. obtusa Dawson & Gentles 1961	M. nanum
N. otae Hasegawa & Usui 1974	M. canis
N. persicolor Stockdale 1967	M. persicolor
N. racemosa Rush-Monro, Smith, & Borelli 1970	M. racemosum

TABLE IV

Ecology of Human Dermatophytic Species

Anthropophilic	Zoophilic	Geophilic
Cosmopolitan species		
E. floccosum	*M. canis*	*M. gypseum*
M. audouinii	*M. gallinae*	*M. fulvum*
T. mentagrophytes var. *interdigitale*	*T. mentagrophytes* var. *mentagrophytes*	*T. ajelloi*
T. rubrum	*T. verrucosum*	*T. terrestre*
T. schoenleinii	*T. equinum*	
T. tonsurans	*M. nanum*	
T. violaceum		
Rare and geographically limited species		
M. ferrugineum	*M. distortum*	*M. racemosum*
T. concentricum	*T. erinacei*	
T. gourvillii	*T. simii*	
T. megninii	*M. persicolor*	
T. soudanense	*M. ripariae*	
T. yaoundei		

man is the chief reservoir. Infections are spread by fomites, articles such as clothing or blankets that have recently been used by the infected person. Another category is zoophilic dermatophytes, which grow preferentially on lower animals rather than man. Organisms like *T. mentagrophytes*, which infect animal species such as rodents, cows, pigs, horses, dogs, and cats are found in this group. Some zoophilic organisms are specialized to a single host, e.g., *T. equinum* on horses and *M. nanum* on pigs. A third category includes species regularly isolated from human and animal diseases whose reservoir appears to be the soil. These are geophilic species.

It is evident from examining the list of dermatophytes that almost half are anthropophilic. These could be considered obligate, antagonistic symbionts. They are specialized for infection only on the human skin or its appendages, although they will digest other sources of keratin saprophytically. Like other dermatophytes, they invade only nonliving keratinous appendages, but they invoke a less inflammatory reaction than geophilic or zoophilic species. They incite just enough irritation, perhaps, to increase the epithelial turnover rate. This in turn increases the keratin substrate and increases shedding of infectious material, thus spreading the organism to new hosts. In some species (*E. floccosum*) chronic infections are rare. There has to be rapid transmission from host to host in order to preserve the species. In others, chronic infections of the foot and nails may occur, allowing the infected individual to spread infectious material in his or her environment over a period of many years. *Trichophyton rubrum* infection is spread in this way (Fig. 4).

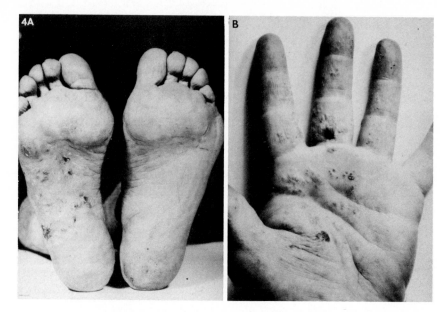

Fig. 4.   Chronic scaling of the foot (A) caused by *Trichophyton rubrum*. In this case, there is a sympathetic "id" or allergic reaction seen on the palm of the hand (B). This type of chronic infection usually involves a nail as well and is essentially impossible to cure.

Keratin or some other factor of an individual human host often dictates who will and who will not become infected. Almost everyone is susceptible to transient infection due to *E. floccosum,* but a smaller number are infected by *T. rubrum* on a long-term basis (Fig. 4). The latter group may often have life-long infections, while cohabiting partners remain infection-free. In some chronic infections of the foot, e.g., occult tinea pedis, the organism *T. mentagrophytes* var. *interdigitale* may be present for years in the absence of clinical symptoms. There appears to be some specialization to the keratin of particular races as well. *Trichophyton concentricum* incites a clinical entity called tinea imbricata or tokelau (Fig. 5) in Indonesians and Polynesians, yet neither cohabiting caucasians nor blacks are infected.

In considering the list in Table V another set of patterns can be seen. There is evident specialization among some species as to the site and type of keratin on the body. *Trichophyton rubrum* can infect any area of skin or dermal appendage, whereas *Microsporum audouinii* infects mainly hair, and *E. floccosum* mainly the groin and feet. The basic infection mechanism is the same regardless of the anatomic site. Infectious material lands on the stratum corneum, and the dermatophyte grows radially from this site. Though there is great clinical variation depending on the species and strain of etiological agent and the response of an individual host, the basic pathology is the same. The initial reaction induced is

Fig. 5. Tinea imbricata (tokelau). Concentric rings of scales on the body of a Polynesian. *Trichophyton concentricum* is the etiological agent of this chronic family-centered infection.

toxic dermatitis. The same response is mounted for any skin irritant. The second is allergic dermatitis and inflammation incited by the enzymes and metabolites elaborated by the fungus and absorbed through the skin. If the organism growing from the initial site encounters a hair shaft, it may invade and grow downward to the hair bulb (tinea capitis). If it encounters a nail, it may invade beneath the nail plate either from the side or the distal end (tinea unguium). If the infection is on glabrous (smooth) skin (tinea corporis), growth is rapid at first, quiescent for a while, and then rapid again. This type of infection may produce concentric rings. Each of the various appendages and anatomic sites has a subtly unique spectrum of pathological responses, duration of infection, and morphology of the lesion.

TABLE V

Dermatophytic Infections—Clinical Diseases and Common Etiologies

Disease	Dermatophyte involved
Tinea capitis	*Microsporum*, any species; *Trichophyton*, any species except *T. concentricum*
Tinea favosa	*T. schoenleinii, T. violaceum* (rare), *M. gypseum* (rare)
Tinea barbae	*T. mentagrophytes, T. rubrum, T. violaceum, T. verrucosum, T. megninii, M. canis*
Tinea corporis	*T. rubrum, T. mentagrophytes, M. audouinii, M. canis*; almost any dermatophyte (*Candida albicans*)
Tinea imbricata	*T. concentricum*
Tinea cruris	*E. floccosum, T. rubrum, T. mentagrophytes* (*Candida albicans*)
Tinea pedis	*T. rubrum, T. mentagrophytes, E. floccosum* (*Candida albicans*)
Tinea manuum	*T. rubrum, E. floccosum, T. mentagrophytes*
Tinea unguium	*T. rubrum, T. mentagrophytes, T. violaceum* (rare), *T. schoenleinii* (rare), *T. concentricum* (rare), *T. tonsurans* (rare) (*C. albicans* and other fungi are involved in a similar clinical disease—onychomycosis)

An even more interesting evolutionary trend appears in examining the etiological agents and pathogenesis of tinea capitis. Among some agents not only has specialization for type of keratin evolved but also a single, set life cycle on a single host. In addition, some species produce a special spore type for survival and dissemination to a new host. In its complexity, the life history of *Microsporum audouinii* approaches that of corn smut or wheat rust. Infectious material lands on the scalp of a child, grows radially in the stratum corneum, and produces a finite-sized colony. All hair shafts in the area are invaded. The organism grows downward, approaching but not invading the keratinizing cells. The equilibrium of upward hair and downward fungal growth is called Adamson's fringe. There is little irritation to the child, but enough to cause a little itching and induce spread of the infection to other areas by autoinoculation. Patches of various sizes can be observed on the scalp. All are gray and scaly with little abnormal reddening (erythema). The usual host is caucasian and the infection benign; in blacks there is often more of an inflammatory response. As the hair shaft grows above the scalp line, the mycelium within grows out and surrounds the cuticle. This then fragments into masses of arthroconidia, which are produced only on growing hair. In time rings cease to form, and the mycelium in the stratum corneum dies and is shed. Perhaps host immunity then plays a role, as reinfection is unknown. The infected hairs remain, conidia being rubbed off on a variety of fomites. As the normal telagen–anagen cycle of hair occurs, the infected hairs are shed and the infection spontaneously resolves. Once the cycle has been completed, the host is immune to another infection by that particular species. The clinical name for this disease is small spore ectothrix tinea capitis of *Microsporum audouinii* (Fig. 6). The disease appears to come in epidemics that occur in 50-yr cycles.

Many other agents of tinea capitis occur, each with a unique cycle, a preferred race, a certain geographical distribution, and a distinct pathogenesis. In addition, the so-called immunity to reinfection varies from host to host and species to species of etiological agent. Among the most interesting clinical forms of disease is the long, family centered exposure required for the development of favus, a disease caused by *Trichophyton schoenleinii* (Fig. 7).

In culture the anthropophilic organisms tend to grow slowly and produce few asexual conidia. As yet, no one has discovered a teleomorph among them. Perhaps under evolutionary pressure one mating type had a selective advantage over the other, and it alone survived.

The zoophilic species offer a wide variety of disease types, life cycles, specialization or nonspecialization for hosts, and duration of infection. Some, such as *M. nanum* of pigs, are very rarely encountered in human infection, whereas *T. verrucosum* of cows is a frequently encountered human infection in rural areas. In man it induces the most severe clinical disease of all the der-

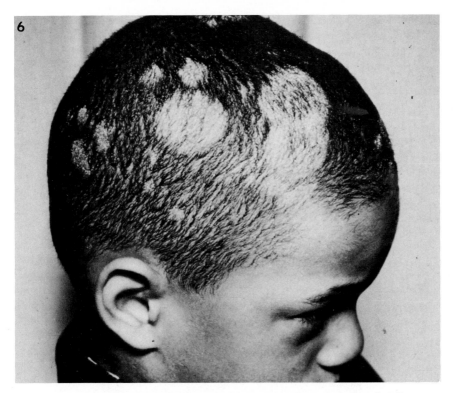

Fig. 6.   Small-spore gray patch (ectothrix tinea capitis) of *Microsporum audouinii*.

Fig. 7.   *Trichophyton schoenleinii* (favus). Few spores are produced in culture by this obligate anthropophilic species, but these "elk horn" or "chandelier form" mycelial growths are characteristic.

matophytes. The infection often leads to permanent baldness, pitting, epidermal denudement, and scar formation on the scalp, face, chin, or other areas (Figs. 8 and 9).

Trichophyton mentagrophytes var. *mentagrophytes* may be a complex of various species. It represents the emergence of soil dermatophytes toward close association and finally dependence on an animal host. The fungus can be found in the soil but is more commonly discovered in association with rodents or other animals. It may live on such animals without inducing symptoms. However, if by chance it encounters species such as dog or man, it may induce a severe, short-term, highly inflammatory infection.

The third category of etiological agent is geophilic. Here the species is not usually associated with benign colonization of an animal but with the soil. In culture it grows rapidly, producing masses of asexual conidia as it would in the

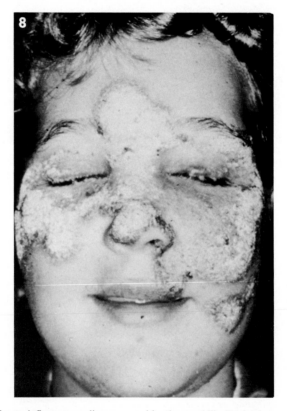

Fig. 8.   Severe inflammatory disease caused by the zoophilic *Trichophyton verrucosum*.

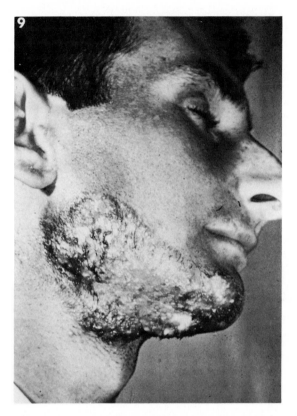

Fig. 9.   Severe inflammatory tinea barbae (sycosis barbae) caused by *Trichophyton verrucosum.*

soil. When infection of an animal or human occurs, it is usually short-term and highly inflammatory (Fig. 10).

In studying the various species adapted for human, animal, or soil existence other biological properties are evident. In culture the types of enzymes and metabolites elaborated are many and varied. They all produce numbers of proteases, peptidases, some elastases, and so on, each of which is irritating and toxic. However, when growing on its special host, the organism does not appear to produce the same irritants in the same quantities found in culture. It has been speculated that some entity of the host represses the production of offensive agents, thus tending to increase the duration of infection. Many aspects of this equilibrium along with other details of dermatophytoses (e.g., the problems of natural and acquired immunity, and pathogenic mechanisms) are in need of careful investigative research.

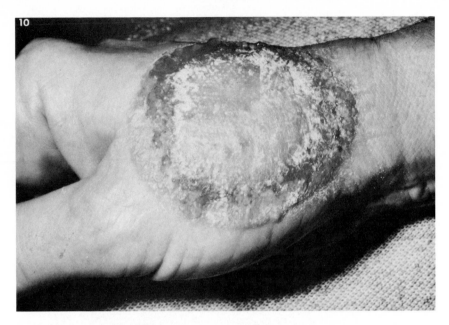

Fig. 10.   Inflammatory lesion on the hand produced by geophilic *Microsporum racemosum.* (From Rippon and Andrews, 1980.)

B. Cutaneous Candidiasis

In contrast to the dermatophytoses caused by soil-derived ectodermal parasites, cutaneous candidiasis is caused by a component of the normal flora. The organism involved is almost always *Candida albicans.* This species is part of the normal flora of the alimentary tract of mammals and birds. The organism is a potential pathogen par excellence, causing the widest variety of clinical disease of any fungus. It is an opportunist that may overgrow, colonize, and cause disease in response to a slight change in normal bacterial flora, normal physiology, or immune state. Simple conditions such as excessive moisture, occlusive clothing, or frequent immersion of hands in water can lead to colonization. In patients severely debilitated by disease or immunosuppressed by cytotoxins and steroids, colonization can lead to the direct invasion of tissue and systemic, sometimes fatal, disease. In cutaneous infections the disease is limited to the epidermal surface, but the lesion is often painful and inflamed.

Infection of mucocutaneous tissue by *Candida albicans* is frequently an overgrowth of normal flora. This may be due to the lack of other antagonistic flora, as in oral thrush, or a change in physiology and secretions, as in vaginitis. Cutaneous disease is the result of excessive moisture and occlusion predisposing to colonization. The organisms are derived from the anus or mouth. Such clinical manifestations as diaper rash, paronychia (infection around the fingernails) and onychomycosis (infection of the fingernails), and intertrigenous disease [between the fingers, under the mammaries, intraaxillary (armpits), or intercrural (crotch)] may ensue. The pathology is similar in all types. Powerful water-soluble toxins elaborated by the organisms appear to be responsible for the initial erythema. This is followed by colonization of the epidermis, and a pruritic (itchy), highly inflamed lesion is produced. Punctate satellite lesions with ragged borders are a hallmark of the infection. The disease is easily treated by topical nystatin. Proper hygiene and clothing prevent recurrences. Chronic disease manifested by chronic mucocutaneous candidiasis is associated with an underlying abnormality. The possible abnormalities include various leukocyte dysfunctions, thymus abnormalities, hypoparathyroidism and polyendocrinopathies. The *Candida* infection often precedes clinical symptoms of the underlying disease.

IV. SUBCUTANEOUS INFECTIONS

Subcutaneous mycoses are induced by a variety of unrelated soil-, water-, and plant-infecting fungi. They include chromoblastomycosis, mycetoma, sporotrichosis, lobomycosis, entomophthoromycosis, and rhinosporidiosis. These diseases share at least three characteristics: The infection is usually the result of traumatic implantation of the organisms; the anatomic site is cutaneous or subcutaneous tissue; and the infection tends to remain localized or involves adjacent areas by extension. The etiological agents are of low inherent virulence capacity. The organisms differ from related species in being able to adapt to and finally proliferate in the abnormal environment of human tissue. In so doing they usually show some morphological or physiological alteration. In sporotrichosis there is a transition from a mycelial to a yeast phase, as seen in the thermal dimorphism exhibited by the agents of systemic mycoses. In mycetoma the infecting agent forms compact microcolonies called grains. In chromoblastomycosis (verrucous dermatitis) the various etiological agents develop into planate, dividing "sclerotic" cells. In other diseases grouped in the category of subcutaneous mycoses the organism is unknown in culture or the adaptation obscure.

It is to be emphasized that these are diseases of the normal host. In general the pathology seen in these infections consists of the formation of foreign-body granulomas mixed with microabscesses. This is in contrast to the tuberculoid granulomas of systemic mycoses. Some aspect of the pathological response may be exaggerated in a particular disease, such as excessive fibrosis in mycetoma, keloid formation (excessive production of collagen) in lobomycosis, and dermal hyperplasia in chromoblastomycosis. The same agents may also be involved in opportunistic infection of a debilitated host. In such cases the morphology of the organism and the pathology elicited may be quite unusual.

A. Chromoblastomycosis

In the field of medical mycology most of the names for diseases are derived from the etiological agent e.g., histoplasmosis, blastomycosis. In subcutaneous mycoses, however, there is no consistency or widely accepted classificatory scheme. For example, "mycetoma" is a clinical term for a disease process that can be caused by about two dozen completely unrelated organisms. About half of these organisms are bacteria. In other diseases, such as sporotrichosis, a single agent causes a wide spectrum of clinical manifestations. Probably the most confusing, complicated, and controversial subject is the group of diseases called chromoblastomycosis and related diseases. Their only common denominator is the brown pigment in the fungal cell walls when seen in host tissue. The agents involved are diverse, and the type of infections produced are quite distinct one from the other. Many names and schemes of classification have been proposed for this group of diverse infections. Their popularity waxes and wanes over the years, and a completely satisfactory system still is lacking. One recent suggestion is to resurrect the old term "chromoblastomycosis" for the distinct entity of verrucous dermatitis and call all other infections phaeohyphomycoses. There is some merit to this suggestion.

In discussing these infections a traditional approach to nomenclature will be used. The diseases are grouped with respect to the type of pathology induced, the morphology of the agent in tissue, and the anatomic site of the infection. The diseases are chromoblastomycosis (verrucous dermatitis), phaeomycotic cyst and phaeohyphomycosis (cerebral and systemic).

Chromoblastomycosis or verrucous dermatitis is a distinct clinical entity. As the name implies, there are warty growths on the skin. The infection is common throughout the tropics in rural and non-shoe-wearing populations. The etiological agent, a soil fungus, is implanted in the skin by trauma—thorns, slivers, or sticks act as vectors. The lesions are usually on the feet or legs. In tissue the fungus may remain dormant for some time. Eventually it begins to replicate as brown-walled, planate-dividing sclerotic cells. The tissue response is that of pseudoepitheliomatous hyperplasia or extreme acanthosis (epidermal overgrowth). This hyperplasia gives rise to warty, stalked lesions that resemble the florets of cauliflower. Crops of lesions may occur either by autoinoculation or lymphatic spread. The more common etiological agents are *Phialophora verrucosa*, which produces phialospores, *Cladosporium carrioni*, which produces branched acropetal chains of conidia, and *Fonsecaea pedrosoi*, a fungus that demonstrates polymorphic conidiation. In rare cases dissemination to the brain occurs. The morphology of the organism in tissue is that of distorted mycelial strands.

Phaeomycotic cyst, which usually results from a puncture wound, consists of single or multiple deep cysts in which brown-pigmented fungi of varied morphology are seen. About 50 cases have been noted. The most common etiological agent is *Exophiala jeanselmii,* an organism more often associated with black-grain mycetoma. In this disease the organism appears as distorted hyphal strands, yeastlike bodies, and moniliform hyphae in addition to a few sclerotic cells.

About two dozen cases of a mysterious disease called cerebral phaeohyphomycosis have been reported. The infection consists of single or multiple encapsulated abscesses in various parts of the brain. The lesions are filled with masses of brown hyphal strands. Concurrently, lesions may also be seen in the skin or lung, suggesting a portal of entry. In most cases the responsible agent has been *Cladosporium bantianum.* Often patients were on immunosuppressive steroid therapy prior to infection.

A large number of dematiaceous fungi have been isolated a few times from a variety of clinical diseases. In most cases there were opportunistic infections of debilitated patients. In others, however, no predisposing factors could be found. The more commonly isolated fungi include *Wangiella dermatitidis, Phialophora parasitica, P. richardsiae, P. repens,* and *Exophiala spinifera.* The disease may involve the skin or subcutaneous tissue or be systemic in all organs. On histological examination brown-walled hyphal units are seen. Because of the paucity of such isolation and the great number of genera and species involved, the suggestion of Libero Ajello that these infections be grouped under the term ''phaeohyphomycosis'' has been accepted.

B. Mycotic Mycetoma

Mycotic mycetoma is a clinical entity consisting of tumefaction (a swelling), sinus tracts, and grains. Like verrucous dermatitis, it is a disease of non-shoe-wearing populations of tropical areas. The responsible agent is implanted beneath the skin of the feet, legs, chest, or arms. There it forms a microcolony and begins to digest its way through the subcutaneous tissue, muscles, fascia, and bones. This induces hyperplasia, and a tumorlike growth occurs. Within this tissue sinus tracts containing the grains are scattered. They enlarge, break apart, and grow again. When the tracts rupture the surface of the skin, large grains (2000–4000 μm in diameter) can be extruded.

The etiological agents of this disease are extremely diverse. Actinomycotic mycetoma is caused by bacteria such as *Nocardia brasiliensis, Actinomadura madurae,* and *Streptomyces somaliensis.* True fungal agents include *Madurella mycetomi, Petriellidium boydii, Leptosphaeria senegalensis,* and *Exophiala jeanselmi.* The grains may be white, black, or red, depending on the species.

C. Entomophthoromycosis

There are two subcutaneous mycoses involving members of the Entomophthorales. The first, often referred to as subcutaneous phycomycosis, is more correctly termed entomophthoromycosis basidiobolae. The etiological agent is *Basidiobolus ranarum*, a ubiquitous species that occurs in decaying vegetation and the gastrointestinal tracts of many reptiles and amphibians. About 100 cases have been recorded. These have occurred mainly in Uganda, Kenya, Nigeria, and Indonesia. The mode of transmission is unknown but may be by insect bites. The infection begins as a subcutaneous nodule which gradually enlarges. It is of firm consistency, well circumscribed, and painless. The palpable mass is attached to the overlying skin but not the underlying fascia. As the mass continues to grow, it may involve the whole shoulder, chest, trunk, or leg. On histological examination, broad, nonseptate hyphae 4–10 μm in diameter are seen to course through the tissue. Upon examination of hematoxylin–eosin stain preparations, these seem to have a sleeve of eosinophilic precipitate representing an antigen–antibody complex. Many cases of the disease were previously ascribed to worm infestations. The disease usually responds to potassium iodide therapy.

The second of the entomophthoraceous diseases is even more rare. In the literature it is called rhinoentomophthoromycosis, but the correct designation is entomophthoromycosis conidiobolae. Fourteen cases have been reported, most from the tropical rain forests of Africa. A few have occurred in the New World, namely, in Puerto Rico, Grand Cayman, Brazil, and Columbia. The etiological agent is *Conidiobolus coronatus,* a soil organism pathogenic to spiders and termites. The mode of transmission is unknown. The symptoms of the disease are similar to those of entomophthoromycosis basidiobolae, except that in all cases only the nose was involved. A swelling begins on the interior turbinates. As this enlarges, it may involve the entire side of the face and forehead. Unlike entomophthoromycosis basidiobolae, these lesions may resolve spontaneously. Scrapings from the nose reveal broad, branching, nonseptate hyphae.

D. Rhinosporidiomycosis

Rhinosporidiosis and lobomycosis, discussed in the following paragraph, remain complete mysteries of medical mycology. Neither of the causative agents has been grown in culture. Nothing about the organisms in nature is known. Even whether they are fungi or not is uncertain. Rhinosporidiosis is an infection of the mucocutaneous tissue by an entity termed *Rhinosporidium seeberii*. Approximately 2000 cases have been reported, the vast majority in India and Ceylon. Scattered cases have occurred in Brazil, Argentina, South Africa, Mexico, Tennessee (one of the earliest cases), North Carolina, and Texas (15 cases).

The lesions consist of stalked, polyplike growths. They usually involve the nose (Fig. 11) but have also occurred on the conjunctiva, uvula, epiglottus, pharynx, labia of the vagina, anus, and meatus (opening) of the penis. The tissue contains spherules of various sizes ranging up to 350 μm. At maturity they contain up to 16,000 endospores. As these are released through a pore at the

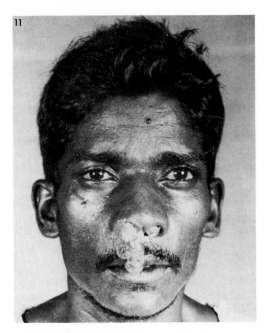

Fig. 11. Rhinosporidiosis. Stalked lesion of the nose.

mucosal surface, they reinfect the adjacent mucosa. It has been speculated that the etiological agent may be a member of the Chytridiaceae or Olpidiaceae. An association with stagnant fresh water has been postulated. The disease occurs in horses, cows, mules, sheep, and dogs, as well as in man.

E. Lobomycosis

Lobomycosis is a strange disease of which about 100 human cases have been reported. Approximately 70 of these were from the Matto Grosso of Brasil and 15 from the Saramaccaner River region of Surinam. The other cases occurred in northern South and Central America. The disease is also encountered frequently in the bottle-nosed dolphin *Tursiops truncatus* off the east and west coasts of Florida and in dolphin off the west coast of Africa. The symptoms, which are the same in human and dolphin, include hard, knobby, painless growths in the subepidermis. They occur singly or in clusters of dozens. An exaggerated fibroblastic response is elicited that gives rise to keloid formation. Within the tissue numerous chains of yeastlike cells, uniform in size (10 μm in diameter), are observed. The knobby growth, which may occur anywhere on the body, remains stationary for years. The organisms have not been grown in culture but are referred to as *Loboa loboi*.

F. Sporotrichosis

The etiological agent of sporotrichosis is the thermal dimorphic fungus *Sporothrix schenckii*. Of worldwide distribution, it is the most virulent of the subcutaneous infecting fungi. Following traumatic implantation from contaminated plant material, the organisms transform into elongate budding yeast cells. Several disease types may be produced. They are lymphocutaneous, fixed cutaneous, extracutaneous and systemic, and pulmonary.

The lymphocutaneous type is most frequently encountered in temperate climates. A raised bubo develops at the site of inoculation. This may ulcerate and drain (sporotrichotic chancre). This lesion usually heals. Later other subcutaneous nodules develop along the draining lymph channels of the primary lesion. In time lesions may be scattered over wide areas of the skin surface and become gummateous. Early lesions are similar in histology to tuberculoid granulomas, and the organisms are almost never seen. Culture is usually positive.

The fixed cutaneous disease occurs in semiarid climates where the organism is abundant and exposure is constant. A hypersensitivity develops in the exposed population, which alters the course of the infection. After abrasion and inoculation the lesions become ulcerative, superficial, and verrucous. The lesion remains fixed and does not spread. A few satellite lesions may be seen around the central ulcer. Such lesions are very common among certain populations in Mexico and northern South America. The organisms are rarely seen on histological examination, and culture is almost always positive.

Lymphocutaneous disease (and, more rarely, the fixed cutaneous type) may ultimately develop into systemic mycosis. The synovial membranes of joints are the first extracutaneous areas to become involved; then osseous and finally other internal organs are infected by hematogenous spread.

Primary pulmonary infection is an infrequently reported form of sporotrichosis. About 50 cases are found in the literature. This could be termed the "falling-down-drunk-in-a-vacant-lot-syndrome." Often there is a history of a derelict "sleeping it off" in a grassy area to recover from alcoholic intoxication. Apparently the patient inhales a number of conidia, and his or her defenses are sufficiently abrogated so that a primary infection may ensue. The course of the disease is usually fulminant but resolves spontaneously. Chronic pulmonary infections also have been recorded.

V. SYSTEMIC INFECTIONS

Systemic mycoses fall into two very distinct categories delineated by the interaction of two factors: inherent virulence of the fungus and constitutional adequacy of the host. The first category includes infections caused by the true pathogenic fungi: histoplasmosis, blastomycosis, coccidioidomycosis, and

TABLE VI

Systemic Mycoses

Characteristic	True pathogenic fungal infections	Opportunistic fungal infections
Diseases	Histoplasmosis, blastomycosis,[a] paracoccidioidomycosis, coccidioidomycosis	Aspergillosis, candidiasis (systemic), mucormycosis, cryptococcosis[a]
Host	Normal	Abrogated
Portal of entry	Primary infection, pulmonary	Various
Prognosis	99% of cases resolve spontaneously	Recovery depends on severity of host defense impairment
Immunity	Resolution imparts strong specific immunity	No specific resistance to reinfection
Host response	Tuberculoid granuloma, mixed pyogenic	Depends on degree of impairment— necrosis to pyogenic to granulomatous
Morphology	All agents show dimorphism to a tissue form	No change in morphology
Distribution	Geographically restricted	Ubiquitous

[a] This disease has significant exceptions to the usual pattern.

paracoccidioidomycosis. True pathogenic fungi are defined as species whose conidia are capable of producing an infectious process in a normal host. The second category of systemic infections consists of opportunistic fungal infections. There are four common types of these infections: aspergillosis, systemic candidiasis, mucormycosis, and cryptococcosis. The etiological agents do not have sufficient virulence capacity to infect a normal host but take advantage of one whose immune defense mechanisms are abrogated (Table VI).

A. Pathogenic Fungal Infections

The four etiological agents of the systemic pathogenic fungal infections, described below, share many attributes. Probably the most important as far as disease production is concerned is dimorphism. This appears to be an adaptation of the organism to the abnormal environment of elevated temperature and host tissue. As infection of a mammalian host is a dead end, this adaptability seems to be a physiological accident in the fungus. In three of the four fungi some type of budding yeastlike cell is formed in tissue or in culture at elevated temperatures. In the fourth, a spherule containing endospores is produced. Though the details differ with the species, all can grow in the reduced oxidation–reduction potential of living tissue, and the special morphology is induced by a combination of temperature, pH, carbon dioxide tension, and free sulfhydryl groups. As the metabolic rate escalates in the parasitic phase, the production of cell wall and other metabolites increases. In some of the organisms there is an increase in chitin and a shift from β- to α-glucans in the cell wall.

Human infection by these organisms is common in their endemic areas. Approximately 40 million people have, or have had, histoplasmosis in the Ohio and Mississippi valleys of the United States. The number infected by *Coccidioides* is almost as great. However, most infections are completely symptomless, and fully 99% resolve spontaneously. One is left with a strong specific immunity to reinfection and is skin test-positive. Frequently a small, calcified lesion is left in the lung and/or spleen. Of the few patients who develop serious progressive pulmonary or systemic disease, the great majority are males in the third to fifth decade of life. All the agents have a restricted geographical distribution. A few, e.g., *Histoplasma,* are found sporadically throughout the world, but the major endemic focus remains in the New World. The one exception to these general patterns is blastomycosis. The etiological agent is not regularly encountered in soil. Subclinical, self-resolving disease is debatable, and there are no serological or immunological parameters to determine the incidence of exposure or endemicity. Geographical distribution and thermal morphogenesis follow the patterns of the other agents, however.

Histoplasmosis is a very common granulomatous disease of worldwide distribution caused by the dimorphic fungus *Histoplasma capsulatum.* Infection, initiated by the inhalation of conidia, results in a variety of clinical manifestations. Approximately 95% or more of cases are inapparent, subclinical, or completely benign. Following resolution the patient will have a positive histoplasmin skin test and a calcified lesion in the lungs and will appear to be resistant to reinfection.

The areas of highest endemicity in the world are found in the Ohio and Mississippi valleys. The organism grows abundantly in high-nitrogen substances such as chicken manure, bat dung (cave disease), and starling droppings. The starling arrived in the Midwest in the late 1890s and proliferated in great numbers. It may have filled the vacuum left by the demise of the passenger pigeon. The starling congregates in large groups, leaving piles of guano that provide a growth substrate for the fungus. Perhaps the organism was much less common before the arrival of the bird.

Following human inhalation, the conidia convert to small yeast cells 1–3 μm in diameter. Though first encountered in pulmonary macrophages, which distribute them throughout the body, the yeasts become intracellular parasites of histiocytes (phagocytic cells). In tissue section the organisms are found in tuberculoid granulomas consisting of lymphocytes, plasma cells, Langham's giant cells, and fibroblasts, with an epithelioid cell layer around the periphery. If the infection does not resolve spontaneously, two general types of disease may develop. The first is a slowly progressive lung disease that consists of a single enlarging lesion (histoplasmoma) or multiple foci that extend to involve more lung tissue. The second is an acute or chronic progressive systemic infection. All organs may be involved. If untreated, this disease is fatal. The organism produces a dermatophyte-like colony in culture, which forms microconidia (1 μm in diameter) and tuberculate macroconidia (5 μm in diameter). Its teleomorph is *Ajellomyces capsulatus*. The ascoma resembles the gymnothecia of *Nannizzia* and *Arthroderma* spp. In soil isolations the ratio of plus to minus mating type is 50:50. In clinical material, however, the minus mating type predominates.

Coccidioidomycosis is a common, though benign, infection in its endemic regions. As in the pattern seen for histoplasmosis, skin reactivity, a few calcified lesions of the lung, and apparent resistance to reinfection follow resolution of this disease. The organism, *Coccidioides immitis,* is highly selective in its environmental requirements. It does not appear to be able to compete with other fungi in rich, moist cultivated soil; its habitat is restricted to high-salt, high-nitrogen, semiarid areas. These areas are referred to as the lower Sonoran life zone. The fungus is found in the southwestern United States, northern Mexico, and sporadically in Central and South America. It does not exist in Europe, Asia, or Africa.

The types of disease produced, clinical manifestations, pathology elicited, and general history are similar to those outlined for histoplasmosis. In chronic disease, however, cutaneous lesions are more common in coccidiomycosis than in histoplasmosis. Marked acanthosis and pseudoepitheliomatous hyperplasia give the lesions a verrucous appearance. Granulomas that resemble those of tuberculosis are formed in lung and other tissue. Unlike *Histoplasma,* the organism is not intracellular. and has the form of a spherule not a yeast. As the spherule grows, a granulomatous reaction surrounds it. When it matures and breaks, releasing endospores, there is an infiltrate of neutrophils. The taxonomy of the etiological agent is obscure. In culture it produces only chains of alternate arthroconidia. No other type of conidiation has been observed. The teleomorph is unknown. In tissue or in conidia media at 37°–40°C under high carbon dioxide tension, the organisms form a spherule approximately 60 μm in diameter containing variable numbers of ''endospores.'' When released, these spores develop into new spherules. In culture or in experimentally infected mice, the diameter of the spherule may reach 200 μm.

The third mycosis caused by a pathogenic fungus is called paracoccidioidomycosis. Its etiological agent is *Paracoccidioides brasiliensis*. Endemic foci are scattered from central Mexico and Central America to most of South America (except Chile and southern Argentina). It is not found in any other part of the world. The fungus is associated with the topography and climate known as the Holridge tropical mountain forest. Infection appears to be common in the endemic regions, but significant disease is infrequent. As with the other mycoses, the male/female ratio of skin test reactors, indicating resolved infection, is approximately 50:50. However, 80% of those with significant disease are males.

The primary infection is pulmonary and follows inhalation of spores. About 95% of patients have inapparent infections that resolve completely. In those that do not resolve, a variety of clinical conditions may ensue. The most common manifestations of secondary paracoccidioidomycosis are mucocutaneous lesions of the oral, buccal, and nasal regions. X-ray examination of the lungs reveals complete resolution of infection with few or no residual effects. In regional lymph nodes of the head and neck and in the mucosa of the mouth, gingiva, tongue, pharynx, or the nasal turbinates, however, the organism proliferates. *Paracoccidioicles brasiliensis* is a significant cause of tooth loss in endemic regions. The granulomatous lesions develop very slowly and eventually erode into adjacent cutaneous tissue. At this point they very closely resemble secondary cutaneous leishmaniasis (a chronic ulcerative granuloma caused by the protozoan *Leishmania brasiliensis*). In fact, both diseases may coexist in the same patient.

Chronic progressive pulmonary disease is less common than the mucocutaneous form but is not infrequently seen. Development of the lesion may be gradual, over a period of years, or extremely rapid. The morphology and development of these lesions resemble that seen in tuberculosis or the other mycoses of this category, but calcification is less common. Progressive disease may lead to systemic involvement of the liver, spleen, kidney, bone, and particularly the gastrointestinal tract. In tissue section, the lesions appear granulomatous, with some areas of pyogenic reaction (i.e., producing pus). The organisms are usually free but may be found in giant cells of the Langham's type. The fungus, in the form of a large, multiple budding yeast, has a mother cell that ranges up to 60 μm in diameter. One or more large buds develop, giving the cell a "Mickey Mouse cap" appearance, or numerous small buds are formed and the cell is called a "pilot's wheel." The organism in culture is not fastidious. It grows at 25°C as a fluffy fungus of variable colony morphology. Conidia of the *Chrysosporium* type are produced. Endoarthroconidia and chlamydoconidia are also seen. The teleomorph has not as yet been described. At 37°C the fungus grows as a yeasty, mealy, whitish colony consisting of multiple budding yeast cells.

Not all of the general patterns seen in the preceding systemic mycoses apply to blastomycosis. The etiological agent, *Blastomyces dermatitidis,* appears to be a rarely encountered fungus. The organism has been recovered from soil only a few times. The endemic regions include the eastern United States and Canada to the mouth of the St. Lawrence River. It is not found in South America, Europe, or Asia, although autochthonous (indigenous) cases of the disease are regularly encountered in northern and central East Africa and Israel. The incidence of subclinical *Blastomyces* infections or self-resolving blastomycosis is not known. A reliable skin test to detect the rate of exposure in a given population is lacking. Other standard serological procedures—complement fixation, immunodiffusion, and a tube precipitin test—are also lacking.

According to case histories, the initial infection is often acquired during cool times of the year and is associated with disturbing dead vegetation. Epidemics have occurred among hunters and their dogs and people cleaning up brush piles around buildings. In its yeast form it has been found in composting pigeon manure fertilizer, yet the yeast cell lyses when mixed with regular soil. Some authors consider the fungus to be rare in nature and infection in man to be progressive, with little or no spontaneous resolution observed. Like the other mycoses of this section, the disease is much more frequent in men than in women.

The primary site of infection is the lungs. Slowly enlarging granulomas of the pulmonary parenchyma are observed. These grow to involve adjacent areas and often resemble carcinomas on x-ray examination. They do not tend to cavitate or calcify. In tissue section the lesions are seen to be granulomas that contain scattered microabscesses. The organisms are generally found in areas of pyogenic reactions, often within giant cells. The single budding yeast cell is 8–20 μm in diameter. It has a thick cell wall, is multinucleate, and has a characteristic broad base. As in paracoccidioidomycosis, the pulmonary lesions may resolve and the organism reappears in another organ system. In blastomycosis the most common form of secondary involvement is cutaneous (Figs. 12 and 13). The lesions are

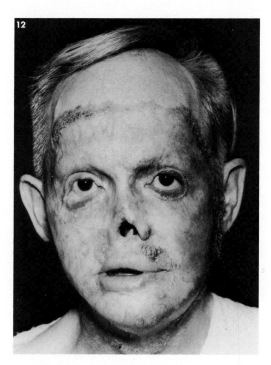

Fig. 12. Blastomycosis, secondary cutaneous form. The lesions started on the face, spread, and have become confluent. Only the advancing border contains the organism. Behind this border denuded skin is replaced by scar tissue. Part of the nose has also been destroyed.

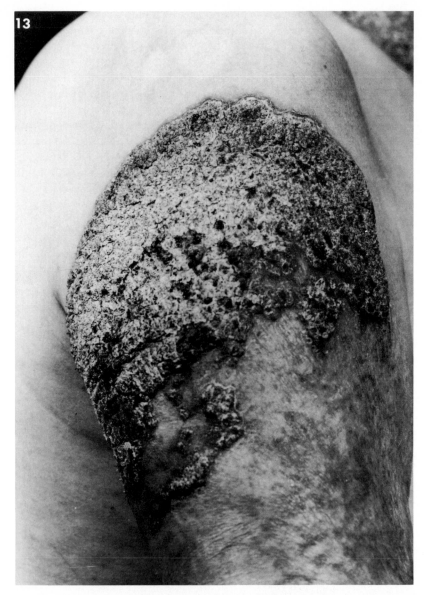

Fig. 13. Advancing verrucous lesion of the shoulders. With time the lesion has spread to the fingertips.

most often found on the sun-exposed areas of the face and neck. They appear as verrucous, granulomatous growths and are characterized by extreme acanthosis and pseudoepitheliomatous hyperplasia. The center of the lesion heals with fibrosis, while the active serpigenous (creeping) border is studded with pustules. The pustules contain the fungus, which can be demonstrated by potassium hydroxide examination of aspirated vesicular contents. If the cutaneous lesions were primary sites of infection, the morphology would be different. Primary fungus infection of the skin, as seen in sporotrichosis, is characterized by an eroding chancreform or ulcerated lesion and regional adenositis (enlargement of the lymph nodes). Both of these aspects are lacking in cutaneous blastomycosis.
are lacking in cutaneous blastomycosis.

Another common manifestation of secondary disease is the involvement of osseous tissue. This involvement may be occult, and the disease may remain stationary for long periods. Most infections become progressive in time and involve many other organs, particularly the prostate and meninges. The kidney and gastrointestinal tract are usually spared.

On culture the fungus grows as a fluffy or glabrous colony of variable morphology. All the agents of systemic mycoses and dermatophytes produce colonies that are usually some shade of beige or whitish tan. None are black or green. The conidia, of the *Chrysosporium* type, are not distinctive. They are 5–7 μm in diameter, smooth, and often dumbbell-shaped. At 37°C a mealy, whitish-tan colony is produced. This contains yeasts with single, broad-based buds. The cells are 8–20 μm in diameter. When mated with a strain of the opposite mating type, the fungus produces ascospores in an evanescent ascus contained in a gymnothecium. The ascoma is typical of the Gymnoascaceae. The teleomorph is *Ajellomyces dermatitidis*. The sex ratio of fungi isolated from clinical disease is about 50:50.

The mode of treatment for all the systemic mycoses is essentially the same. This consists of intravenous amphotericin B at the rate of 0.65 mg/kg body weight. One to three grams is the usual course dose. Hydroxystilbamidine has also been found useful in some forms of blastomycosis but appears ineffective in the other diseases.

B. Opportunistic Infections

Unlike the four etiological agents of the true pathogenic fungus infections, the species involved in opportunistic infection share few attributes. Of the taxa involved—*Aspergillus, Candida, Cryptococcus* and the Mucorales—only one or two species are involved in such infections; closely related organisms are benign. One of the few similarities among the agents of opportunistic infection is thermotolerance. All species involved grow well or preferentially at 37°C. Yet this characteristic alone cannot explain the ability to cause disease, as many other thermotolerant organisms are innocuous. Extensive research has sought to delineate what the other pathogenic factors are. The agents appear to have a greater capacity to adapt to a tissue environment than related species, and a greater potential virulence. That is, the total pathogenic potential of these fungi falls just short of the capacity to induce spontaneous infection in a normal host. Thus some abrogation of host defenses is necessary for successful invasion. In sufficient numbers, however, these organisms can invade a healthy individual.

To a certain extent, the kind and type of host debilitation will determine the form and agent of the infection. For example, the diabetic in a state of ketoacidosis is a candidate for infection by a *Rhizopus* species that has a ketoreductase. Vaginitis in a pregnant female is predisposed by changes in the glycogen content of secretions that promote the growth of *Candida albicans*. In the case of aspergilloma, the organism simply colonizes a preformed cavity that is available for homesteading.

Aspergillosis is a collection of distinct clinical entities that have in common an *Aspergillus* sp. as etiological agent. The diseases include allergic bronchopulmonary aspergillosis, aspergilloma, invasive aspergillosis, and colonization and invasion of debilitated tissue such as burned skin and injured eyes. Allergic bronchopulmonary aspergillosis occurs in the atopic patient with severe asthma. Bronchiectasis is present along with cylindric dilatation of subsegmental bronchi of the apical regions of the lung. Mucus plugs are often coughed up. Upon examination they reveal dichotomously branched, septate mycelium growing within the material. In addition to IgE antibodies, the patients have IgG antibodies to *Aspergillus* and often show pulmonary eosinophilia. *In situ,* the aspergilli growing in the mucus material within the lumen of the bronchi produce showers of antigenic material. This material reacts with antibody at the bronchial surface, producing a continuous arthus reaction. Afflicted patients demonstrate precipitin bands to *Aspergillus* in the immunodiffusion test and show immediate and delayed skin test sensitivity. The most common etiological agent is *Aspergillus fumigatus*. *Petriellidium boydii* has also caused the disease.

Aspergilloma is colonization of a preformed cavity without tissue invasion. Most commonly it involves cavities in the lung that result from resolved tuberculosis or sarcoidosis. The fungus forms a ball that consists of concentric rings of growth. On x ray this opacity within the cavity is seen to move with movement of the patient. Enlargement of the cavity may occur with the erosion of vessels. Thus bouts of hemoptysis (expectoration of blood) are characteristic of such patients. Sometimes the balls deliquesce and disappear. Others require surgical intervention. *Aspergillus fumigatus, A. niger,* and *Petriellidium boydii* are frequent invaders. The sinus cavities of the skull and the nose may also be involved.

Invasive aspergillosis is the least common and most severe form of aspergillosis. The disease occurs in patients who are severely debilitated as a result of disease or medication. Patients with leukemia or lymphoma, or taking cytotoxic drugs and steroids, and organ transplant patients on immunosuppressive therapy, are especially susceptible. The onset of the disease is insidious; there are no serological or successful cultural procedures available. The diagnosis is often not made until autopsy. The fungus grows radially from the point of inoculation within pulmonary tissue. It is characterized by septate, dichotomously branching hyphal units. These invade vessels and are spread to other organs as septic emboli. *Aspergillus fumigatus* is most frequently involved. Aspergilli also invade the lacerated cornea of the eye, causing mycotic keratitis which is usually manifested as a corneal ulcer. Invasion of burned cutaneous tissue, decubitus ulcers (bed sores), and intravenous therapy sites have been recorded.

As discussed in Section III, B, *Candida albicans* is a member of the normal flora of the buccal cavity, the gastrointestinal tract, and the human vagina. Under certain conditions, however, *Candida albicans* is the quintessential opportunist. The clinical conditions are myriad. In general, the predisposing factors fall into one of three categories: extreme youth, physiological change, and general debility.

In the neonate or newborn, the normal bacterial flora is often not established sufficiently quickly to check the growth of *Candida*. It appears to be the lactic acid-producing bacteria that are responsible for curtailing the proliferation of *C. albicans*. If growth is not inhibited, colonization may occur, and thrush, stomatitis (inflammation of the oral mucosa), and glossitis (inflammation of the tongue) may occur. A similar event may be the result of antibiotic therapy in later life. In such cases serious and sometimes fatal invasive disease of the gastrointestinal tract has occurred. Pregnancy, which appears to affect the carbohydrate content of the vagina, can lead to an overgrowth of *C. albicans*. The overgrowth may be sufficient to elicit clinically apparent vaginitis. The administration of steroids to males or females may also lead to proliferation of *Candida*. Much more serious physiological disorders lead to a clinical entity called chronic mucocutaneous candidiasis. Various polyendocrinopathies, hypoparathyroidism, thymic dysfunctions, leukocyte abnormalities, and so on, may predispose to this disease. Often the *Candida* infection is discovered before the underlying condition is diagnosed.

The debilitations that predispose patients to colonizing by *Candida* are numerous. The degree and severity of the *Candida* infection depends on the degree and severity of the debility. In the diabetic patient who wears occlusive clothing *Candida* often colonizes the skin, causing a "scalded skin syndrome" similar to staphlococcal disease. If the patient is leukemic, colonization, invasion, and fatal systemic disease may result.

"Mucormycosis" is a term used to cover various disease types in which the infecting agent is a member of the order Mucorales. Although these are not conidial fungi, they do represent anamorphic expressions of zygomycetous fungi. The commonly isolated agents are *Rhizopus oryzae* and *R. arrhizus* from rhinocerebral and pulmonary infection, and *Absidia corymbifera* from gastrointestinal and animal disease. Isolates of other genera such as *Mucor* (especially *M. pusillus*), *Mortierella, Cunninghamella,* and *Saksenaea* have accounted for a few infections. All species involved are thermotolerant. However, many closely related thermotolerant species have never been isolated. It must be remembered, however, that pure cultures have been obtained in only 10% of the cases. In tissue, all organisms appear as broad, distorted, nonseptate, irregularly branching hyphae. Four quite different sets of predisposing factors have given rise to four quite distinct clinical diseases. The clinical types of infection are rhinocerebral, thoracic, gastrointestinal, and cutaneous.

Rhinocerebral disease progresses in a rapid and devastating manner in the acidotic diabetic. The patient is in ketoacidosis and has high blood sugar and some impairment of normal leukocyte function. The organism usually colonizes the upper turbinates or paranasal sinuses. It then grows outward to involve the orbit of the eye, inward into the cranial cavity, and finally into the brain. The patient usually has eye symptoms including proptosis (bulging), ophthalmoplegia (paralysis of the eye muscles), ptosis (drooping of the upper eyelid), and localized anesthesia. If cranial nerves have been affected, some degree of facial palsy may be present. The fungus invades and occludes arteries. This is followed by necrosis and sloughing. Although the course is rapid and usually fatal, some cures and spontaneous recoveries (following control of the diabetes) have occurred. Both common agents, *R. oryzae* and *R. arrhizus,* have a ketoreductase system and are saccharolytic, so the high-sugar–high-ketone body content of a patient's serum is stimulatory to growth.

Thoracic mucormycosis is most prevalent among patients with leukemia or lymphoma. The symptoms are those of progressive nonspecific bronchitis and pneumonia complicated by signs of thrombosis and infarction. Large areas are involved, which may necrose and cavitate. The infection is usually fatal. Pulmonary disease has complicated organ transplant patients on steroids, vein graft recipients, and patients on dialysis, among others.

Gastrointestinal disease is encountered in areas of abject poverty. The victims are almost always children with kwashiorkor. The etiological agent of this disease, *Absidia corymbifera,* is also responsible for a similar disease in cattle, horses, and other animals. For this reason, it is speculated that it comes from the eating of moldy grain. The walls of the stomach and intestine are pierced, and the arteries are invaded and occluded. This causes necrosis and perforations followed rapidly by death.

Cutaneous mucormycosis results from the colonization of burned skin or skin grafts. In the burn patient the infection may be rapid and fatal. The Mucorales have been found to colonize any injured tissue and may or may not evoke significant pathology.

Cryptococcosis is a particularly perplexing disease. In only 50% of patients can an underlying defect or predisposing factor be determined. Patients with leukemia, lymphomas, organ transplants (with accompanying immunosuppression therapy), and various other debilitating conditions or treatments are prone to cryptococcosis. Their disease is usually pulmonary initially, progresses with some rapidity, and may develop into a fatal systemic infection. Evidence from routine autopsies indicates that many people have had small pulmonary cryptococcomas at some time in their life. These lesions resolve or remain quiescent. In others they enlarge to the point that they cause impairment of an adjacent structure and must be surgically removed. There are no sequelae.

Far more serious and problematic are the cases of cryptoccal meningitis. The patients, otherwise healthy, complain of headaches of increasing severity. Eventually other meningitic signs occur, and on spinal tap *Cryptococcus neoformans* is seen in the fluid. Although the lungs must be the site of primary infection, they are usually clear. Meningitis is usually treatable using amphotericin B, unless far advanced. Treatment failure rate is 25%. Occult lesions in bone, skin, viscera, and eye have been found.

The organism, *Cryptococcus neoformans,* is found at a variety of natural sites, particularly pigeon dung, and all humans have an almost constant exposure to yeast-containing dust. *Cryptococcus neoformans,* like all cryptococci, is encapsulated. It can grow at 37°C. In tissue it is of the same morphology as in nature: an encapsulated, budding yeast. It evokes little inflammatory reaction in infected tissue. This anamorphic fungus has a basidiomycetous teleomorph in the order Ustilaginales, *Cryptococcus neoformans* serotype A and D are *Filobasidiella neoformans* and serotype B and C are *Filobasidiella bacillispora.*

REFERENCES

Emmons, C. W., Binford, C. H., Utz, J. P., and Kwon-Chung, K. J (1977). "Medical Mycology." Lea & Febiger, Philadelphia, Pennsylvania.
Rebell, G., and Taplin, D. (1964). "The Dermatophytes." Univ. of Miami Press, Miami, Florida.
Rippon, J. W. (1974). "Medical Mycology." Saunders, Philadelphia, Pennsylvania.
Rippon, J. W., and Andrews, T. (1980). *Microsporum racemosum.* Second clinical isolation in the United States and Chicago area. *Mycopathologia* (in press).

16

Mycotoxin Production by Conidial Fungi

Philip B. Mislivec

Biology of Conidial Fungi, Vol. 2
Copyright © 1981 by Academic Press, Inc.
All rights of reproduction in any form reserved.
ISBN 0-12-179502-0

I. INTRODUCTION

This chapter deals primarily with toxic metabolites produced by conidial fungi. Toxic fungal metabolites have been termed mycotoxins. Discomforts resulting from the ingestion of, or other involvement with, mycotoxins are called mycotoxicoses. According to Detroy *et al.* (1971) mycotoxicoses have the following clinical symptoms: (1) the disease is noncommunicable; (2) drugs and antibiotics are ineffective; (3) outbreaks usually are associated with a specific food or feed; and (4) examination of implicated foodstuffs reveals active fungal growth.

II. BRIEF HISTORY OF MYCOTOXICOSES

A. Stachybotryotoxicosis

Perhaps the first conidial fungus implicated in a mycotoxicosis was *Stachybotrys alternans*. The disease, which primarily affected horses, was first recorded in the Ukraine in the early 1930s and was associated with the consumption of moldy hay. The disorder involved irritations of the mouth, lips, nose, throat, and sinuses, plus necrotic lesions in the respiratory and digestive tracts (Forgacs, 1972; Hintikka, 1978). Similar outbreaks were subsequently reported from other Eastern European countries. According to the reports of Forgacs (1972) and Eppley (1977), stachybotryotoxicosis is undoubtedly a mycotoxicosis.

B. Alimentary Toxic Aleukia

During and after the war years (1942–1947), a serious disease occurred in Russia caused by the ingestion of moldy grain that had overwintered in the field. The disease, termed alimentary toxic aleukia (ATA), affected humans and animals of all age groups, reaching its peak in 1944 when more than 10% of the residents of Russia's Orenburg District contracted the often fatal illness (Joffe, 1965). Other Russian districts were similarly affected. The disease, which involves blood-forming tissue, resulted in leukopenia, severe hemorrhage, agranulocytosis (leukocyte reduction), necrotic angina, and bone marrow exhaustion, as well as external skin necrosis (Joffe, 1965, 1978). Reports of earlier outbreaks of ATA had been recorded, but none was as severe and widespread as that of 1942–1947. Joffe (1965) isolated molds from more than 950 overwintered grain samples and screened several thousand pure culture isolates for toxicity to rabbits and guinea pigs. Virtually all isolates examined were conidial fungi, primarily of the genera *Aspergillus, Fusarium, Alternaria, Cladosporium,* and

Penicillium. Of the isolates tested, *Fusarium poae, F. sporotrichioides,* and *Cladosporium epiphyllum* were extremely toxic, causing animal death in 5–21 days. Isolates of other species of the above genera were toxic to lesser degrees. Joffe (1978) has further reported on the toxicity of isolates of 18 *Fusarium* species and species varieties isolated from overwintered and summer-harvested Russian cereals. Of 107 isolates found to be toxic, 44 were *F. poae* and 42 *F. sporotrichioides,* both species producing the clinical symptoms of ATA. Thus the excellent and extensive efforts of Joffe confirmed that the ATA outbreaks were mycotoxicoses and strongly suggested that the two *Fusarium* species were the causative agents.

C. Toxic Moldy Rice

That mycotoxicoses are caused by the consumption of moldy rice has been suspected in the Orient for nearly a century. In 1891 (Kinosita and Shikata, 1965) it was thought that periodic outbreaks of beriberi in Japan were due to the consumption of moldy rice. During the 1894–1895 Sino-Japanese war, the incidence of beriberi rose alarmingly among the Japanese military. Since rice was the staple of their diet, studies were initiated to determine the cause of the apparently rice-related polyneuritis. But it was not until 1940 that a culture of *Penicillium citreo-viride,* isolated from toxic moldy rice, produced polyneuritis symptoms in laboratory animals, thus adding credence to the suspicion that a conidial fungus was associated with the illness (Kinosita and Shikata, 1965).

After World War II, incidences of serious illness due to the consumption of moldy rice again increased in Japan, prompting a number of studies to determine if mold species were responsible. Several conidial fungi, primarily aspergilli and penicillia, were screened. Three *Penicillium* species, *P. citreo-viride* (related to nervous disorders), *P. citrinum* (related to kidney disorders), and *P. islandicum* (related to liver disorders), were found to be particularly toxic. Saito *et al.* (1971) have extensively reviewed the literature on moldy rice mycotoxicoses.

D. Moldy Corn Toxicosis

In 1952 a severe, often fatal, disease of swine occurred in parts of Florida and Georgia. The disease was noncommunicable and, although the precise cause was not known, the outbreaks seemed to involve the consumption of moldy corn in the field. Subsequently, Burnside *et al.* (1957) isolated 13 mold species from the corn implicated in the disease. Of these, *Aspergillus flavus* and *Penicillium rubrum* were found to be toxic in experimental studies, causing severe hemorrhage in several body tissues. Forgacs (1965), who reviewed the pathology of moldy corn toxicosis, considered the disease a mycotoxicosis.

E. Hemorrhagic Syndrome in Poultry

Detroy *et al*. (1971) discussed a noncommunicable disease involving hemor-rhaging among poultry, which occurred in certain parts of the United States during the 1950s. The disease was severe, mortality rates reaching 40% in some affected flocks. Several possible causes were suggested, including the consumption of moldy feed. Molds were subsequently isolated from the feed and experimentally fed to poultry. Results showed that the ingestion of certain isolates of *Aspergillus* and *Penicillium* by poultry resulted in hemorrhages. In the absence of any stronger evidence regarding the cause of the disease, it was designated a mycotoxicosis. However, no specific mycotoxins have been isolated or characterized.

F. Turkey "X" Disease

Prior to 1960 the interest in toxic metabolites elaborated by fungi, and the resultant mycotoxicoses, was quite low in the scientific community. Mycotoxin research efforts were sporadic and often inconclusive, usually being initiated as a result of isolated outbreaks of apparently noncommunicable diseases of livestock feeding on moldy diets. In addition, scientific efforts were usually restricted to isolated geographical regions, arousing little or no interest in other apparently unaffected areas of the world. Few of these earlier efforts were interdisciplinary in nature, involving contributions of mycologists, chemists, toxicologists, and pathologists, all of which are now known to be necessary in order to define the nature, seriousness, and possible cure and control of mycotoxin-induced diseases. However, in 1960 a noncontagious disease resulting in the death of more than 100,000 turkey poults occurred in England (Blount, 1961). The disease was initially called the turkey "X" disease, since its cause was a mystery. In any case, events and reports stemming from this disease were responsible for the birth of modern mycotoxicology, in that coordinated interdisciplinary efforts finally were established in attempts to solve the problem. The investigation of turkey X disease is a truly fascinating story. Below is a brief account.

Following the 1960 disease outbreaks among turkeys, reports quickly appeared describing similar disorders of other types of poultry (Asplin and Carnaghan, 1961). The only common factor in the diets of the affected animals was Brazilian groundnut (peanut) meal. Later in 1960, several outbreaks of disease in swine and calves were reported in England (Allcroft and Carnaghan, 1963). Their diets also included rations of Brazilian groundnut meal. Simultaneously, a turkey X-like disease occurred in Kenya involving ducklings whose diets included East African-grown groundnut meal. What followed was an intensive investigation by English and Dutch scientists of various disciplines. Stoloff (1977) acknowledges the researchers involved by name and affiliation. Initial findings in 1960 were that methanol extracts of implicated groundnut meal were hepatotoxic to day-old ducklings, the extracts having a bright blue fluorescence under long-wave ultraviolet light. During 1961–1962, methanol extracts of groundnut meal samples from 13 different countries were found to be toxic, with typical turkey X symptoms. Extracts of other groundnut meal samples from the same countries were nontoxic and did not fluoresce. Thus a fluorescent substance was suspected. During the same period, laboratory studies with rats showed that a diet of 20% toxic groundnut meal produced a high incidence of liver carcinoma within 30 days (Lancaster *et al.*, 1961). Turkey X disease took on additional serious dimensions.

Since toxic groundnut samples were found to be free of toxic plant alkaloids and pesticides (Blount, 1961), microbiological involvements were considered. Sargeant *et al.* (1961) confirmed this when they found that a single isolate of *Aspergillus flavus*, subcultured from Ugandan groundnuts implicated in duckling disorders in Kenya, was hepatotoxic. Methanol extracts of the fungus fluoresced bright blue. Further studies confirmed that turkey X disease was a mycotoxicosis and that the mycotoxin involved was a metabolite of *Aspergillus flavus*, which was given the trivial name "aflatoxin" (van der Zidjen *et al.*, 1962). Characterization of the fluorescent material revealed that four distinct but closely related chemical compounds were present, two with blue fluorescence (called aflatoxins B_1 and B_2) and two with greenish-blue fluorescence (called aflatoxins G_1 and G_2).

At about the same time as the turkey X outbreak in England, serious outbreaks of trout hepatomas in U.S. fish hatcheries and rat hepatomas in experimental rat colonies in France occurred. In both cases the diet was suspected, but it did not contain groundnut meal. This time the common ingredient was cottonseed meal. Following the groundnut meal studies, scientists successfully isolated aflatoxins from the cottonseed meal and concluded that the two outbreaks were aflatoxicoses. Halver (1965) presented a thorough review of the trout hepatoma outbreaks.

III. MYCOTOXIN-PRODUCING CONIDIAL FUNGI

A. Initial Considerations

In the past 20 yr, numerous conidial fungi have been implicated in the production of mycotoxins and mycotoxicoses. Species of *Aspergillus, Penicillium,* and *Fusarium* have been involved most frequently, but several other genera have also been identified as toxin producers. For instance, Christensen *et al.* (1968) screened 943 molds isolated from grains and peanuts for animal toxicity. Of the 439 isolates found to be toxic, no fewer than 413 (94%) were from 18 anamorph-genera. Joffe (1978) found toxic isolates in 16 anamorph-genera. Butler and Crisan (1977), in their excellently illustrated key to toxic fungi, include more than 50 anamorph-genera with at least some history of toxicogenicity. Conidial fungi are thus seen to be the most important group causing potential or actual hazards to human and animal health through mycotoxin contamination of foods.

Although the list of toxic species is large, in order to evaluate properly the hazard potential of a single species, many points must be considered. How frequently is the species encountered, and in what types of foods? Is its presence due to surface contamination or to actual internal colonization? Is its presence restricted to selected foods or does it contaminate a wide variety of foods? How much of the food is consumed and by what human or animal age group? How many toxins does the fungus under consideration produce? How toxic are the compounds? What proportion of isolates of the species produce the toxin(s), and at what levels? Do the toxins occur naturally in foods? How stable are the toxins, and are their breakdown products also toxic? What are the environmental requirements for toxin production? Has the fact that the substance(s) actually causes human mycotoxicoses been established? Ideally answers to the above questions should be available for all toxicogenic conidial fungi. In reality, with the possible exception of *Aspergillus flavus, A. parasiticus,* and aflatoxin production, most of this information has not been satisfactorily established. Thus proper evaluation of the hazard potential of many individual species is still a difficult task ultimately requiring additional scientific information. For lack of necessary information, the hazard potential of many of the species and genera discussed below has not yet been conclusively established. However, an attempt has been made to place each species in proper perspective using the available data. The mycological point of view is stressed.

B. The Anamorph-Genus *Penicillium*

This genus consists of at least 137 species and varieties (Raper and Thom, 1949), and many of these produce mycotoxins. Scott (1977) reported on more than 50 toxicogenic species. Because the penicillia are seemingly always involved when foodstuffs—in the field, in storage, in raw materials, and in finished products—are undergoing deterioration and spoilage, serious consideration of their hazard potential due to toxin production is in order. The species discussed below are the most commonly encountered toxic penicillia in foodstuffs, at least in the United States. The brief species descriptions are based on Raper and Thom (1949) and Mislivec (1977b). For more detailed information, consult these authors.

1. *Penicillium cyclopium*

This fungus is perhaps the most frequently encountered toxic *Pencillium* in foods and feeds, especially in storage. It can grow at or below 2°C and has relatively low moisture requirements (Mislivec *et al.,* 1975a). Colonies on

Czapek agar are bright blue to blue-gray. The colony surface is velvety, but sometimes granular to tufted near the margin. Clear to faint-yellow surface droplets are usually produced. The reverse of the colony becomes brownish purple with time. Virtually all isolates produce a strong, penetrating, musty odor. Conidiophores are compact and asymmetric, consisting of branches, metulae, and flask-shaped phialides which produce smooth-walled, globose phialoconidia. Branches and metulae are usually rough-walled.

The mycotoxins most frequently associated with *P. cyclopium* are penicillic acid, cyclopiazonic acid, and two toxic tremorgenic compounds. Penicillic acid is toxic to several laboratory animals, producing liver, kidney, and thyroid disorders (Scott, 1977). This compound may be the most important *P. cyclopium* toxin, since it is produced by most isolates (Mislivec, 1977a) and has been implicated as a carcinogen following subcutaneous injection in mice and rats (Dickens and Jones, 1965). At least 10 additional *Penicillium* species and 5 species of the *Aspergillus ochraceus* group also produce the toxin (Scott, 1977), and it occurs naturally in dried beans (Thorpe, 1974), high-moisture corn (Kurtzman and Ciegler, 1970), and cigarette tobacco (Snow *et al.*, 1972). To date, penicillic acid has not been implicated in field outbreaks of disease, and its effect on humans is unknown. Cyclopiazonic acid, also produced by *Aspergillus versicolor* (Scott, 1977), causes convulsions in rats upon interperitoneal injection, and liver and kidney damage upon oral administration. Its importance as a mycotoxin remains uncertain, since it has not yet been detected as a natural food contaminant nor has it been directly implicated in any animal disease outbreak. The two tremorgens produced by *P. cyclopium* (penetrem A and penetrem B) are neuropathic to rats, mice, guinea pigs, poultry, calves, swine, and sheep. Although they have not been found as naturally occurring food contaminants, they are considered important because of their extreme toxicity. Cysewski (1977) has presented a review of these and six other tremorgens produced by conidial fungi.

2. *Penicillium expansum*

An important pathogen of apples in transit, in storage, and in the home, this fungus is responsible for the rapid and destructive "blue rot." Periodically it contaminates other foods. Colonies produced on Czapek agar are dull blue-gray with a granular to tufted surface. Distinct fascicles may be produced near the margin. Surface droplets are rare. The colony reverse is light to deep walnut brown. A strong sourish odor, suggestive of rotting apples, is produced by most isolates. A few produce a distinct acetic acid odor. Conidiophores are asymmetric, consisting of smooth-walled branches and metulae and flask-shaped phialides giving rise to small (3 μm), ellipsoid, smooth-walled conidia. All isolates invariably cause rapid rot when inoculated into Golden Delicious apples.

The mycotoxin most frequently associated with *P. expansum* is patulin. The metabolite has strong antibiotic properties and was actually used for therapeutic purposes during the 1940s. However, toxic side effects (dermal irritation, upset stomach, vomiting) were found to be too severe to permit its continued use (Ciegler, 1977). Patulin has subsequently been shown to be toxic to mice, rats, poultry, and rabbits, the toxicoses involving the kidney, liver, lungs, brain, and spleen (Ciegler, 1977; Scott, 1977). There is one report of patulin being a carcinogen (Dickens and Jones, 1965). Ciegler (1977) and Scott (1977) have reported in detail on patulin toxicity. The importance of this compound as a mycotoxin is still unclear. However, it has been found in commercially available apple juice in both the United States (Ware *et al.*, 1974) and Canada (Scott and Kennedy, 1973), thus providing at least one avenue for human exposure, and has been experimentally produced in meats, other fruits, and bread (Ciegler, 1977).

3. *Penicillium urticae*

This fungus is periodically encountered as a contaminant of a number of foods, particularly alimentary paste products (Mislivec, 1977a). It grows, though poorly, at refrigerator temperatures and at relative humidities as low as 81% (Mislivec *et al.*, 1975a). Colonies on Czapek agar are light gray, with granular to fasciculate surfaces; the fascicles are often produced in concentric circles near the margin. Dull-red surface droplets are usually produced. The colony reverse is orange-brown. A distinct, fruity, strawberry-like odor is produced. Conidiophores are asymmetric, with smooth-walled branches, metulae, and unusually small (less than 6 μm) flask-shaped phialides which produce smooth, globose conidia.

Penicillium urticae produces at least two mycotoxins, patulin and griseofulvin (Pohland and Mislivec, 1976). Griseofulvin is a powerful antimycotic agent in the treatment of dermatophyte infections. However, numerous toxic manifestations of this metabolite have been documented in humans, including skin disorders, hematopoietic disturbances, neurological disorders involving blurred vision and vertigo, and gastrointestinal stress (Wilson, 1971). No reports exist of griseofulvin as a natural contaminant of any food product. Although *P. urticae* is encountered only periodically, every isolate I have personally examined produced both patulin and griseofulvin in large amounts (Mislivec, 1977a), and it may therefore be an important toxicogenic fungus.

4. *Penicillium islandicum*

This fungus has been reported as a regular contaminant of soybeans (Mislivec and Bruce, 1977), rice (Miyaki and Saito, 1965), and white peppercorns (Mislivec *et al.*, 1972). It is sporadically encountered in other foods, particularly small grains (personal observations). Colonies grown on Czapek agar are bright orange-brown in mycelial areas and metallic blue in areas of sporulation. The colony surface is velvety, and the reverse is orange-brown. Colonies are odorless, and surface droplets are rare. Conidiophores are symmetric, having only closely packed, smooth-walled metulae and lanceolate, finely tapered phialides producing smooth-walled, fusiform conidia.

Perhaps no other *Penicillium* species has been as thoroughly studied for tox-icogenicity as *P. islandicum,* partly because of its involvement in the moldy rice toxicosis outbreaks in Japan (Miyaki and Saito, 1965). Studies show that isolates produce three potent hepatotoxins: luteoskyrin, cyclochlorotine, and is-landitoxin. Luteoskyrin and cyclochlorotine produce cirrhosis and hepatomas upon oral administration to mice. Islanditoxin causes cirrhosis, but its car-cinogenicity is still in doubt. Additional, less toxic, hepatotoxins produced by *P. islandicum* are rugulosin and erythroskyrine. The *P. islandicum* toxins have been reviewed by Saito *et al.* (1971) and Scott (1977). Considering that this species produces at least five hepatotoxins, two of them carcinogens, and is involved in the moldy rice toxicoses in Japan, its hazard potential cannot be overlooked.

5. *Penicillium viridicatum*

This species is second only to *P. cyclopium* in importance in *Penicillium* invasion of food in storage. It will grow at or below 2°C and at a relative humidity of 81% (Mislivec and Tuite, 1970; Mislivec *et al.,* 1975a). Colonies grown on Czapek agar are indistinguishable from those of *P. cyclopium,* with the following exceptions. Colonies are bright green rather than blue. Conidia are rough-walled. Not all isolates produce a strong, musty odor.

Because of their frequent occurrence on stored food, isolates of *P. viridicatum* have been studied for toxicity since the mid-1960s and have been found to produce several toxic metabolites, primarily ochratoxin A, citrinin, xanthomeg-nin, and viomellein. Carlton and Tuite (1977) have reviewed the *P. viridicatum* toxins and resultant pathologies in detail.

Ochratoxin A and citrinin have been considered the most important of the *P. viridicatum* toxins. Like ochratoxin A (see Section III, C, 2), citrinin is a neph-rotoxin, although apparently less toxic than ochratoxin A (Scott, 1977). Rats, mice, guinea pigs, rabbits, swine, and dogs develop sometimes lethal kidney damage in response to citrinin (Carlton and Tuite, 1977). The metabolite has been isolated from naturally contaminated wheat, rye, oats, and barley in Canada (Scott *et al.,* 1972), and from barley and oats in Denmark (Krogh *et al.,* 1973). The Danish report involved a serious mycotoxicosis of swine termed Balkan porcine nephropathy. The feeds associated with the outbreak were highly con-taminated with *P. viridicatum* and contained both citrinin and ochratoxin A. Studies indicated that ochratoxin A was primarily responsible for the disease but that citrinin was also involved (Krogh *et al.,* 1973; Scott, 1977).

A number of reports, reviewed by Stack *et al.* (1977), showed that several cultures of *P. viridicatum* isolated from Indiana corn were hepatotoxic or neph-rotoxic to laboratory animals. The Indiana isolates did not produce ochratoxin A or citrinin. Rather, they produced xanthomegnin and viomellein which, when administered to mice, produced symptoms identical to those produced by feeds experimentally inoculated with the Indiana isolates. Thus xanthomegnin and viomellein were implicated as the mycotoxins. All the Indiana isolates produced a strong, musty odor. This point is made because a recent report on nine *P. viridicatum* isolates (Stack and Mislivec, 1978) showed that only three of the isolates had a musty odor, and that all three produced xanthomegnin and viomellein but no ochratoxin A or citrinin. The six odorless isolates produced ochratoxin A or citrinin, but no xanthomegnin or viomellein. Earlier, Ciegler *et al.* (1973) concluded that only odorless or slightly fragrant *P. viridicatum* isolates produced ochratoxin A or citrinin. These authors did not report on xanthomegnin or viomellein, since at the time the toxins were not known to be metabolites of *P. viridicatum*.

Since *P. viridicatum* is a frequent contaminant of foods and feeds, produces at least four toxic metabolites, and has been implicated in serious natural outbreaks of animal illness and death, it may be the most important mycotoxin-producing *Penicillium*.

6. *Penicillium citrinum*

This fungus is widespread in nature, especially in subtropical and tropical climates (Raper and Thom, 1949). It has been detected, though usually at low levels, as a contaminant of several foods including corn (Mislivec and Tuite, 1970), dried beans (Mislivec *et al.,* 1975b), soybeans, (Mislivec and Bruce, 1977), rice (Miyaki and Saito, 1965), wheat flour (Mislivec *et al.,* 1979), pecans (Huang and Hanlin, 1975), and dairy products (Raper and Thom, 1949). Because of its unusually slow growth rate, it may often go undetected when mold floras are determined. Colonies grown on Czapek agar are dull blue in areas of sporulation and faint yellow elsewhere. Clear to yellow surface droplets are abundantly produced. Colonies are odorless. The colony reverse is flesh color to slightly pink. Growth is quite restricted. Conidiophores are symmetric and unique to *P. citrinum,* consisting of only widely divergent metulae and flask-shaped phialides from which long chains of phialoconidia are produced. The latter resemble bluish-green sticks and are readily detected using low-power magnification.

To date, *P. citrinum* is known to produce only one mycotoxin—citrinin. However, virtually every isolate produces this metabolite in large quantities. In fact, Raper and Thom (1949) considered citrinin production to be of diagnostic value in the identification of this species. Because the slow growth rate of the fungus may result in its being overlooked in mold flora surveys, this species could be more extensively involved in mycotoxicoses than heretofore considered.

7. *Penicillium roquefortii*

This fungus has long been used in the production of Roquefort-type cheeses. Colonies grown on Czapek agar are bluish gray and uniformly velvety. Colony margins are very thin and arachnoid (feathery and spiderweb-like). Neither surface droplets nor readily detectable odors are produced. The colony reverse becomes almost black with time. (I know of no other *Penicillium* species with such reverse coloration.) Conidiophores are asymmetric, consisting of branches, metulae, and flask-shaped phialides from which unusually large, smooth-walled conidia (up to 7–8 μm in diameter) are produced. Branches and metulae bear warty encrustations.

Penicillium roqueforti has been implicated in the production of at least four toxic metabolites. The first described toxin, given the trivial name "PR toxin," was isolated from silage and moldy grains suggested to be involved in bovine abortion and placental retention (Still *et al.*, 1972). The compound has yet to be characterized and its toxicogenic properties established. A second toxic metabolite, roquefortine, was reported by Scott *et al.* (1976). The compound is neurotoxic to mice upon peritoneal injection and has been found as a natural contaminant of commercial blue cheese (Scott and Kennedy, 1976). Recently, Olivigni and Bullerman (1978) isolated both penicillic acid (see *P. cyclopium*) and patulin (see Section III, B, 2) from atypical isolates of *P. roqueforti* but were unable to detect these toxins in commercial strains used in manufacture of cheeses. The isolate in question was examined by Fennell and Mislivec, both of whom concluded independently that it was an atypical isolate of *P. roqueforti*.

Although this species has been traditionally assumed to be a harmless mold, recent reports of its toxin production, and the fact that it is directly and deliberately consumed when Roquefort-type cheeses are eaten, strongly suggest that additional studies are needed to clarify the question of its toxicogenicity.

C. The Anamorph-Genus *Aspergillus*

Species of *Aspergillus* are virtually always encountered when foods or feed are undergoing deterioration and spoilage. According to Raper and Fennell (1965), the genus consists of 132 species and varieties which have been placed in 18 groups. Samson (1979) has accepted an additional 38 taxa. Forty-four species have been associated with their teleomorphs. A limited number of toxic species or groups are discussed below. Brief species descriptions are based on Raper and Fennell (1965) and Tuite (1977).

1. *Aspergillus flavus* and *Aspergillus parasiticus*

These are the only confirmed aflatoxin-producing molds. Although other species have been implicated in aflatoxin production, initial reports have not been confirmed by other researchers (Mislivec *et al.*, 1968). *Aspergillus flavus*, initially considered a storage organism (Christensen and Kaufman, 1965), has, since the discovery of aflatoxin, been isolated from a number of field crops, primarily in semitropical and tropical climates (Tuite, 1977). The fungus is relatively xerophytic, capable of initiating growth at 81% relative humidity (Mislivec *et al.*, 1975a). It prefers temperatures of 25°–40°C, and grows poorly, if at all, below 12°C. Below ambient temperatures, higher moisture levels are required for growth. *Aspergillus parasiticus* has the same general geographical range and temperature-moisture characteristics as *A. flavus* but is not as frequently encountered. Colonies of *A. flavus* grown on Czapek agar are bright yellow, then yellow-green, and finally deep green. The reverse is colorless, and surface droplets are rare. Some isolates produce hard, pseudoparenchymatous sclerotia, white at first and becoming reddish black with time. Conidiophores are biseriate with colorless, rough-walled stalks. Conidia are globose, spiny, and relatively large. *Aspergillus parasiticus* differs from *A. flavus* only in that sclerotia are not produced and conidiophores are uniseriate with usually smooth-walled stalks.

a. New Aflatoxins. Since the discovery of aflatoxins B_1, B_2, G_1, and G_2 in the early 1960s several additional aflatoxins have been isolated and characterized. Jones (1977) listed 14 aflatoxins or aflatoxin-like compounds, while Stoloff (1977) recognized 13. Of the newly discovered aflatoxins, M_1 and M_2 may be the most important. They are most frequently associated with dairy products, since lactating animals, upon ingestion of foods contaminated with the B aflatoxins, secrete the M toxins in their milk. The M toxins are also produced by isolates of *A. parasiticus* and less frequently by *A. flavus* isolates.

b. Natural Occurrence of Aflatoxins in Foodstuffs. Jones (1977) has reviewed the literature, and reports of the natural occurrence of aflatoxins in foodstuffs are numerous. Foods most often implicated include groundnuts, cottonseed and other oilseeds, Brazil nuts, pistachios, corn, small grains, and dairy products. Aflatoxin residues have been detected at very low levels in beef, pork, chicken meat, and eggs. In the United States natural contamination has involved mainly peanuts, cottonseed, and corn, primarily in southeastern and southwestern states.

c. Aflatoxin Toxicity in Animals. Patterson's (1977) review showed aflatoxin B_1 to be hepatotoxic to 16 animal species. Based on the LD_{50}, the rabbit is most susceptible and the rat the least. Depending on the animal species and the aflatoxin dose levels, liver pathology may include necrosis, hemorrhage, fibrosis, cell enlargement, and bile duct proliferation. According to Stoloff (1977), aflatoxin B_1 is hepatocarcinogenic to ducklings, swine, rainbow trout, guinea pigs, monkeys, rats, ferrets and, possibly, sheep. Aflatoxins G_1 and M_1 may also be carcinogenic.

d. Aflatoxin Toxicity to Humans. According to Stoloff (1977), only one documented case of human acute aflatoxicosis has been reported, a case of severe hepatitis in India, resulting from the consumption of corn highly contaminated with aflatoxin B_1. Its chronic effect on humans is still uncertain, although epidemiological studies from Africa and parts of Asia, reviewed by Shank (1978), show that in areas where the proportion of aflatoxin-contaminated foodstuffs is high, the incidence of liver disease is elevated. Although there were certain weaknesses in the design of these epidemiological studies (Stoloff, 1977), they suggest that the hazard potential of aflatoxin B_1 to humans may be serious.

2. The *Aspergillus ochraceus* Group

Included in this group are nine species (Raper and Fennell, 1965), most of which are commonly isolated from grains and other seeds (Tuite, 1977). All species produce bright-yellow to ochre colonies on Czapek agar, and eight produce hard, pseudoparenchymatous sclerotia. Size and color (black, brown, pink, yellow, white) of sclerotia are important in diagnosing individual species. Conidiophores are biseriate, their stalks being either smooth or rough and often yellow. Conidia are globose to pyriform, usually with smooth walls.

Seven species produce one or more of a series of seven closely related mycotoxins, ochratoxins. In this series ochratoxin A is the most toxic and most frequently encountered, as a naturally occurring contaminant of small grains, dried beans, peanuts, green coffee beans, and hay, as well as poultry and pork meat (Krogh *et al.,* 1973; Krogh, 1977). Ochratoxin A is a powerful nephrotoxin as well as a hepatotoxin, affecting swine, poultry, rats, mice, and dogs (Carlton and Tuite, 1977). To date, ochratoxin A has been implicated as the cause of only one serious mycotoxicosis outbreak, Balkan porcine nephropathy in Denmark (Krogh *et al.,* 1973). However, the species producing the ochratoxin was *Penicillium viridicatum*.

Members of the *A. ochraceus* group produce at least three other important toxins: Six produce penicillic acid, and at least three produce xanthomegnin and viomellein (Stack and Mislivec, 1978). Thus the importance of this group as agents of mycotoxicoses may be significant, especially since they are widespread in nature and certain of the species produce up to four mycotoxins.

3. *Aspergillus versicolor*

This fungus occurs infrequently in several foods, including small grains, dried beans, soybeans, rice, pecans, peppercorns, and smoked meats (Tuite, 1977; Mislivec, 1977a). The fungus can initiate growth at 81% relative humidity but does not grow below 8°C (Mislivec *et al.,* 1975a). Colonies grown on Czapek agar are bright yellow to orange to red to greenish purple and often demonstrate a mixture of colors in the same colony (hence the epithet "versicolor). Clear to deep-red surface droplets are abundant. The colony reverse is red to purple, the pigment often diffusing throughout the agar. Conidiophores are biseriate, with colorless, smooth-walled stalks. Conidia are globose and spiny. Many isolates produce thick-walled, circular to variously coiled cells in the mycelium called hulle cells.

Aspergillus versicolor produces a single mycotoxin—sterigmatocystin. Virtually every isolate produces this compound in large amounts (Schindler and Abadie, 1973; Mislivec, 1977a). According to Detroy *et al.* (1971), sterigmatocystin is very similar chemically to aflatoxin B_1 and may even be an aflatoxin precursor. Mycotoxicosis caused by this substance involves the liver and kidney. It is carcinogenic, although not nearly as potent as aflatoxin B_1. The

toxin has not been implicated in any field outbreaks. The importance of sterigmatocystin as a mycotoxin is questionable. However, considering its carcinogenic properties and its chemical similarity to aflatoxin B_1, this compound
and the fungus producing it require further study.

4. Other Toxic *Aspergillus* Species

Most other aspergilli with histories of mycotoxin production are patulin producers. Only three are periodically encountered in foods: *A. candidus, A. terreus,*
and *A. clavatus. Aspergillus terreus* also produces citrinin. Briefly, *A clavatus*
produces blue colonies and unusually large club-shaped uniseriate conidiophores. The only other *Aspergillus* that produces this type of conidiophore is
the rarely encountered *A. giganteus* which also produces patulin (Tuite, 1977).
There are only two white *Aspergillus* species: *A. candidus* and *A. niveus.* These
species are distinguished by the fact that *A. candidus* heads are large and radiate,
while those of *A. niveus* are short and columnar. The latter species is rare but
produces citrinin (Tuite, 1977). Colonies of *A. terreus* are beige to cinnamon.
Conidiophores are small and biseriate and produce long, compact columns of
conidia. When colonies are viewed under low magnification, these compact
chains resemble brownish-tan sticks, a characteristic unique to *A. terreus.*

D. The Anamorph-Genus *Fusarium*

Fusaria, widely distributed in nature, include a number of species and varieties
that are active plant pathogens. *Fusarium* taxonomy is at present in a slightly
chaotic state because of the number of monographs available and the number of
legitimate species and varieties recognized in individual monographs (Toussoun
and Nelson, 1975). In my opinion, the most useful *Fusarium* monograph is that
of Joffe (1974), who recognizes 33 species and 14 varieties. Joffe (1977) also
presents an excellent review of toxicogenic fusaria in which only 12 species are
described.

Fusaria have been implicated in a number of field outbreaks of mycotoxicoses,
primarily in Russia but also in the United States and other countries. The following species are believed to be those most often involved.

1. *Fusarium graminearum* (= *F. roseum*)

This fungus, whose teleomorph is *Gibberella zeae,* is primarily a pathogen of
corn. It is responsible for an estrogenic mycotoxicosis of swine called hyperes-
trogenism or vulvovaginitis. The mycotoxin involved is zearalenone, sometimes
referred to as F-2 toxin, which affects the genital system. In females it causes
vaginal enlargement, vaginal prolapse, and atrophy of the ovaries, while in males
it produces atrophy of the testes and enlargement of the mammary glands. Although swine are most susceptible, poultry and dairy cattle are also affected.

Mirocha *et al.* (1977) have recently reviewed the chemistry and toxicity of zearalenone. Other fusaria reported to produce zearalenone are *F. oxysporum, F. moniliforme,* and *F. sporotrichioides* var. *tricinctum.*

2. The *Fusarium sporotrichiella* Section

This section consists of two species and two varieties: *F. poae, F. sporotrichioides, F. sporotrichioides* var. *tricinctum,* and *F. sporotrichioides* var. *chlamydosporum* (Joffe, 1974). These have been associated with a number of serious mycotoxicoses involving both animals and humans. According to Joffe (1965), *F. poae* and *F. sporotrichioides* were chiefly involved in the alimentary toxic aleukia outbreaks in Russia in the 1940s. Certain of these species have also been implicated in other mycotoxicosis outbreaks (Mirocha *et al.,* 1977). Several toxic metabolites produced by these fungi have been isolated and characterized. The metabolites are very similar chemically and are collectively referred to as trichothecenes. To date, at least 37 different trichothecenes have been characterized, although not all are produced by members of the *F. sporotrichiella* section (Mirocha *et al.,* 1977). Trichothecenes are severe dermal irritants that, upon ingestion, cause hemorrhaging in the gastrointestinal tract and other organs. Mirocha *et al.* (1977), in a review of the chemistry and toxicity of trichothecenes, has reported that the following conidial fungi produce one or more trichothecenes: *Fusarium solani, F. equiseti, F. nivale, F. concolor, F. avenaceum, F. sambucinum, F. culmorum,* and *Stachybotrys alternans* (also stachybotryotoxicosis), as well as species of *Myrothecium, Trichothecium, Trichoderma,* and *Cephalosporium* (*Acremonium*).

E. The Anamorph-Genus *Alternaria*

Alternaria is a frequently encountered contaminant of foodstuffs. Considered mainly a field invader (Christensen and Kaufman, 1965; Christensen *et al.,* 1968), it occurs regularly on small grains and other field crops. Certain species are important pathogens of fleshy fruits and vegetable crops, such as *A. solani* (potato, eggplant, tomato rot), *A. citri* (black spot of citrus fruits), *A. brassicicola* (cabbage and cauliflower rot), and *A. radicina* (black rot of carrot). Although these and other fruit rot pathogens usually attack in the field, they are active decomposers during transit and in storage (Ellis, 1971). Many grow at low temperatures and are periodically encountered in home refrigerators, causing food spoilage.

The precise number of legitimate species and varieties in this genus is at present uncertain. Ellis (1971, 1976) has described 41 species. No varieties are described. His descriptions are restricted to active plant-pathogenic species.

Although there have been no confirmed outbreaks of disease directly related to consumption of *Alternaria*-contaminated foodstuffs, several laboratory studies show that such products are severely toxic to animals. Joffe (1960, 1965) isolated *A. humicola* and *A. alternata* from overwintered Russian grains implicated in human and animal toxicities, and several of the isolates were lethal to rabbits in feeding studies. Christensen *et al.* (1968) found that 53 of 60 *Alternaria* isolates were lethal to rats on contaminated corn diets. Doupnic and Sobers (1968) reported on the lethality to chickens of 31 of 96 isolates of *A. longipes* (*A. alternata*) grown on cracked corn. Hamilton *et al.* (1969) found that 30 of 38 *Alternaria* isolates, primarily *A. tenuis* (=*A. alternata*) were toxic and sometimes lethal upon interperitoneal injection. In the above reports, no mycotoxins were isolated. However, Meronouk *et al.* (1972), reporting on 57 isolates of *A. alternata* that were lethal to mice on molded corn–rice diets, isolated and characterized the toxic metabolite tenuazonic acid from 20 of the isolates. Since this metabolite was not detected in the other toxic isolates, it was concluded that *A. alternata* produced additional mycotoxins. In subsequent studies, metabolites were isolated and characterized from *A. alternata* and other *Alternaria* species, including tentoxin, alternaric acid, altenene, altertoxins I and II, altenuene, altenuic acid, zinniol, and so on (Templeton, 1972; Harvan and Pero, 1976). Although animal toxicities of these compounds have not been extensively studied, many are extremely phytotoxic (Templeton, 1972). At present, the most toxic *Alternaria* metabolite is believed to be tenuazonic acid. The chemistry and toxicity of *Alternaria* metabolites have been reviewed by Harvan and Pero (1976).

Considering the high frequency with which *Alternaria* occurs in foodstuffs, and the number of potentially toxic metabolites so far reported, additional research efforts to clarify the hazard potential of *Alternaria* species because of mycotoxin production are clearly warranted.

F. Miscellaneous Toxic Conidial Fungi

1. *Diplodia maydis*

An important corn pathogen, this fungus causes both stalk and ear rot and has been implicated in sporadic disease outbreaks of cattle and sheep in South Africa. The disease is called diplodiasis or cornstalk disease. The coelomycete is recognized by its brownish-black pycnidia in which two types of conidia are produced: elliptical, two-celled, olive to brown conidia and long, narrow, threadlike to filiform conidia which are one-celled and hyaline. Both conidial types may be produced in the same pycnidium, or they may develop in separate pycnidia.

Diplodiasis outbreaks occur sporadically when sheep and cattle graze on stalks and stubble in harvested cornfields. The main symptom is a nervous disorder involving loss of muscle coordination and occasionally paralysis. Although no mycotoxins have been isolated and characterized, the disease is clearly a mycotoxicosis, since pure cultures of *D. maydis*, grown on sterilized corn, reproduce the disease. Although *Diplodia* stalk rot occurs worldwide wherever corn is grown, the only reports of diplodiasis to date have been from South Africa. Marasas (1977b) has reviewed the taxonomy and toxicogenicity of *Diplodia maydis*.

2. *Pithomyces chartarum*

This fungus is a saprophyte encountered mainly on dead grazeland grasses. It is recognized by its distinctive barrel-shaped, rough-walled, brown dictyospores produced on short, simple conidiophores. The spores resemble hand grenades when viewed in slide mounts. DiMenna *et al.* (1977a) provided a good description of the morphology and distinguishing characteristics of this organism.

This hyphomycete is the cause of a disease of cattle and sheep in New Zealand and parts of Australia and South Africa. The disease, which can be fatal, is called pithomycotoxicosis, or facial eczema, and occurs among grazing herds. Affected animals develop severe facial lesions which are highly photoreactive, quickly causing the animals to become photosensitized. Internally, the disease causes severe necrosis, adrenal enlargement, and disruption of the urinary system. The mycotoxins responsible for the disease are a group of chemically related metabolites called sporidesmins which to date are known to be produced only by *P. chartarum*. Sporidesmins and their toxicity have been reviewed by White *et al.* (1977a). Outbreaks of this disease have not been confirmed in the United States, although there is a report of a series of sporidesmin-like outbreaks in Oklahoma which caused hepatogenous photosensitivity in cattle (Monlux *et al.*, 1963).

3. *Rhizoctonia leguminicola*

This species is a frequent pathogen of red clover, causing black, sooty blotches on the leaves. A member of the Mycelia Sterilia, the fungus does not produce conidia. Colonies grown on potato dextrose agar are feathery in consistency, with uniform patches of black to gray mycelium radiating from the center to the margin. The growth pattern resembles the spokes of a wagon wheel. As in *Rhizoctonia* species, hard, bony, black, pseudoparenchymatous microsclerotia of indefinite shape are regularly produced. The most distinguishing characteristic of *R. leguminicola* is its affinity for red clover. It is particularly virulent during periods of wet weather and high humidity.

This fungus is responsible for outbreaks of a disorder among cattle and other farm animals, including sheep, goats, and horses, which is characterized by excessive salivation after ingestion of second-cutting red clover forage. This "slobber factor" disease also causes lacrimation, frequent urination, and diarrhea. In severe cases, animals may become bloated, the lungs, liver, and kidneys becoming enlarged and extremely congested, and death may result. The mycotoxin responsible for slobber factor has been isolated, characterized, and called slaframine. Disease outbreaks, rarely severe, occur periodically in the midwestern and eastern United States and are always associated with red clover ingestion. Smalley (1978) has recently reviewed this mycotoxicosis.

4. *Phoma herbarum* var. *medicaginis*

This pycnidial anamorph is a frequent pathogen of alfalfa, causing a mycotoxicosis symptomatically and pathologically very similar to that produced by *Rhizoctonia leguminicola*. Cattle and sheep are particularly susceptible. The fungus has been described in detail by DiMenna *et al.* (1977b). Although the disease is similar to the slobber factor syndrome, slaframine is apparently not the mycotoxin involved. Instead, brefeldin A and cytochalasin B, both recently characterized, appear to be responsible (White *et al.*, 1977b). However, neither toxin has been isolated from contaminated alfalfa, nor are there any reports of field outbreaks of this disease. The importance of this fungus in mycotoxicoses is at present uncertain.

5. *Phomopsis leptostromiformis*

A parasite of lupine, this pycnidial anamorph has been implicated in a number of field outbreaks of a disease called lupinosis, involving sheep, cattle, horses, and pigs which graze in lupine fields. Reports of outbreaks have come from Germany, Poland, Australia, New Zealand, and South Africa. The organism has been described by Marasas (1977a).

The disease causes a liver ailment, although other organs may be affected, and is characterized by massive lesions, hemorrhage, jaundice, and ultimate liver enlargement and cirrhosis. Hepatomas have not been reported. Although laboratory studies confirm that lupinosis is a mycotoxicosis, to date the responsible mycotoxin(s) have not been isolated and characterized. Marasas (1978) has reviewed our current knowledge of lupinosis.

6. The Genus *Cladosporium*

Personal experience indicates that, except for *Alternaria* spp., members of *Cladosporium* are the most frequently encountered darkly pigmented anamorphs in foodstuffs, both in the field and in storage. Although no mycotoxins have been isolated from *Cladosporium* spp., reports of its toxicogenicity exist. Christensen *et al.* (1968) found that 19 of 41 *Cladosporium* isolates from foodstuffs were lethal to rats in less than 7 days. Joffe (1965) found that, of the fungi isolated from overwintered grains and demonstrated in animal studies to be associated with alimentary toxic aleukia, isolates of *Cladosporium* (e.g., *C. epiphyllum* and *C. fagi*) were second only to those of *Fusarium* in toxicogenicity. In view of the frequency of isolation of cladosporia from foodstuffs, they may play a more important role in mycotoxicoses than is presently suggested.

7. Other Toxic Conidial Fungi

On the basis of reports of Christensen *et al.* (1968) and Joffe (1965), members of the following anamorph-genera also produce toxic compounds: *Colletotrichum, Curvularia, Gliocladium, Myrothecium, Papulaspora, Sclerotium, Scopulariopsis, Thielaviopsis, Trichoderma,* and *Verticillium*. However, with the exception of limited and as yet inconclusive studies on *Myrothecium (M. verrucaria, M. roridum, M. leucotrichum;* DiMenna *et al.*, 1977c), no investigations have been undertaken to clarify the significance of species of these genera in mycotoxin production and mycotoxicoses. A summary of the major mycotoxins produced by conidial fungi, and their structural formulas, natural substrates, and toxic syndromes, is presented in Table I.

IV. CONTROL OF MYCOTOXIN PRODUCTION

A. Initial Considerations

Toxin production by any mold depends upon three necessary conditions: presence of the generating organism, a food substrate, and an environment suitable for growth. Since most toxic conidial fungi are widely distributed in nature and are not usually substrate-specific, the best means of preventing mycotoxin contamination of foodstuffs is to maintain them under environmental conditions which will retard or completely prevent mold growth. Although environmental control is virtually impossible in the field, knowledge of the physical requirements of toxicogenic fungi may serve as an important aid in predicting mycotoxin contamination in the field. For instance, in the United States conditions of excessive moisture and lower than normal temperatures at harvest time are forewarnings of probable zearalenone contamination of field corn, especially in the "corn belt" states, while periods of drought followed by periods of excessive moisture are important predisposing factors for aflatoxin contamination of field corn in the southeastern states. Environmental control can be used with limited success in storage.

Although an environment suitable for mold growth involves a number of interrelated factors, three are of primary importance: suitable temperature, moisture, and oxygen supply. Substrate pH is of limited importance, since most mold species commonly associated with foods have a broad pH tolerance and, in the absence of a strongly buffered substrate, can adjust the pH to a range more suitable for growth.

TABLE I

Summary of Major Mycotoxins Produced by Conidial Fungi

Mycotoxin	Fungi	Natural substrates	Toxic syndrome
Penicillic acid	*Penicillium cyclopium, Aspergillus ochraceus* P. expansum P. viridicatum	Cereals (Moreau and Moss, 1979), corn	Hepatotoxin, nephrotoxin, thyroid disorders, implicated as carcinogen
Cyclopiazonic acid	*P. cyclopium, A. versicolor*	Unknown	Convulsions upon intraperitoneal injection, hepatotoxin, nephrotoxin
Penetrem A, penetrem B	*P. cyclopium*	Unknown	Neurotoxin
Patulin	*P. expansum, P. urticae, A. candidus, A. terreus, A. clavatus, A. giganteus*	Malt feed in Japan (Kurata, 1978), germinating seed (Moreau and Moss, 1979)	Implicated as carcinogen in one report (kidney, liver, lungs, brain, and spleen)

56

Griseofulvin

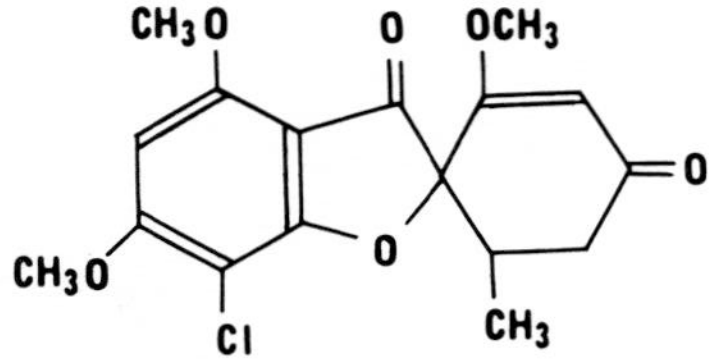

P. urticae

Unknown

Hematopoietic disturbances, neurological disorders, gastrointestinal
stress; also antimycotic agent used
in treatment of, dermatophytic
(skin) disorders

Luteoskyrin

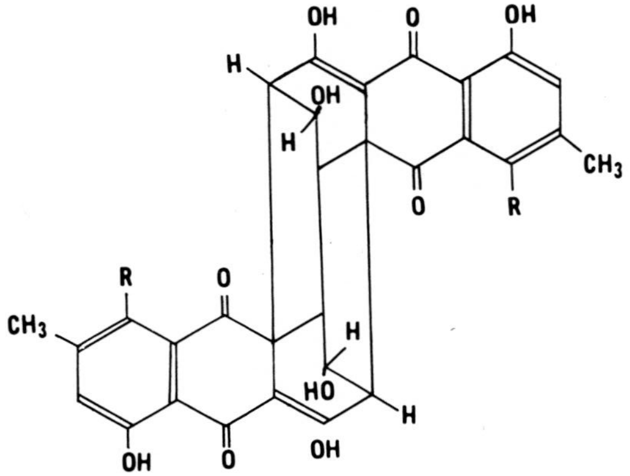

P. islandicum

Rice, sorghum, white peppercorns,
soybeans, millet, and barley[a]

Cirrhosis, hepatomas (carcinogenic)

(continued)

57

TABLE I (*Continued*)

Mycotoxin	Fungi	Natural substrates	Toxic syndrome
Cyclochlorotine	*P. islandicum*	Rice, sorghum, white peppercorns, soybeans, millet, and barley[a]	Cirrhosis, hepatomas (carcinogenic)
Islanditoxin (same structural formula as cyclochlorotine; see Yamazaki, 1978)	*P. islandicum*	Rice, sorghum, white peppercorns, soybeans, millet, and barley[a]	Cirrhosis
Rugulosin	*P. islandicum*	Rice, sorghum, white peppercorns, soybeans, millet, and barley[a]	Hepatotoxin

Erythroskyine

P. islandicum

Rice, sorghum, white peppercorns, soybeans, millet, and barley[a]

Hepatotoxin

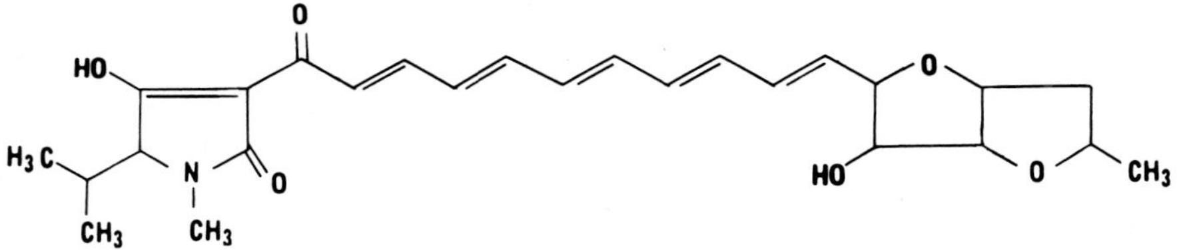

CH₃

HO

O

O

O

HO

N — CH₃

H₃C

CH₃

Ochratoxin A

A. ochraceus, P. viridicatum

Grain, corn, rice, groundnuts, peanuts, green coffee beans, hay, and so on[a]

Nephrotoxin, hepatotoxin

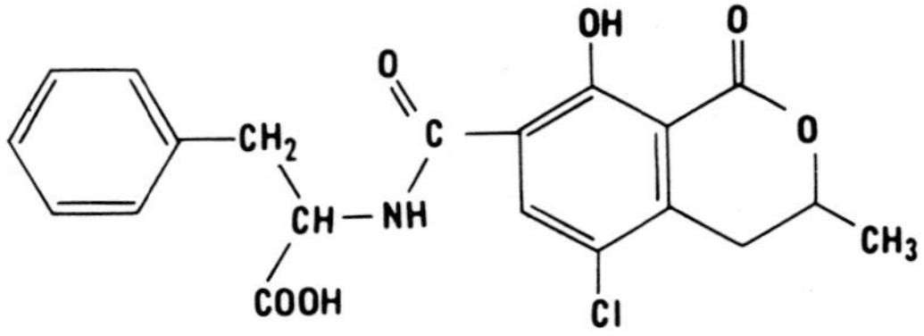

O

O

CH₃

OH

Cl

O=C

NH

CH₂

CH — COOH

Citrinin

P. citrinum, P. viridicatum, A. terreus, A. niveus

Soybeans, wheat flour, pecans, dairy products, groundnuts, rice, cereals, corn, barley, wheat, ryegrass[a]

Nephrotoxin

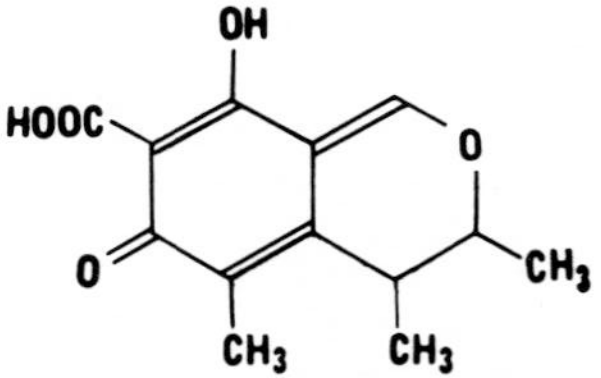

O

CH₃

CH₃

OH

CH₃

HOOC

O

(*continued*)

59

TABLE I (*Continued*)

Mycotoxin	Fungi	Natural substrates	Toxic syndrome
Xanthomegnin	*P. viridicatum, A. ochraceus*	Corn	Hepatotoxin, nephrotoxin
Viomellein	*P. viridicatum, A. ochraceus*	Corn	Hepatotoxin, nephrotoxin

PR toxin

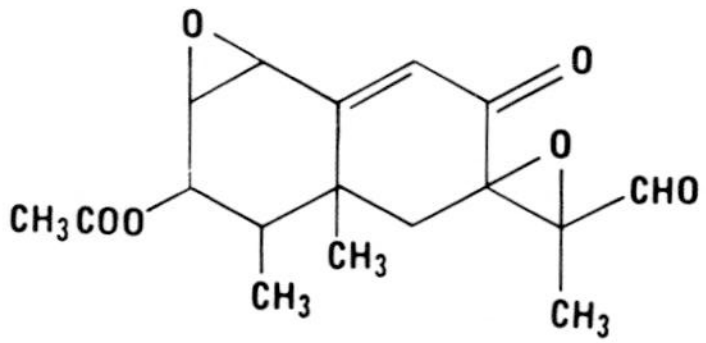

P. roquefortii

Silage and moldy grains

Bovine abortion, placental retention

Roquefortine

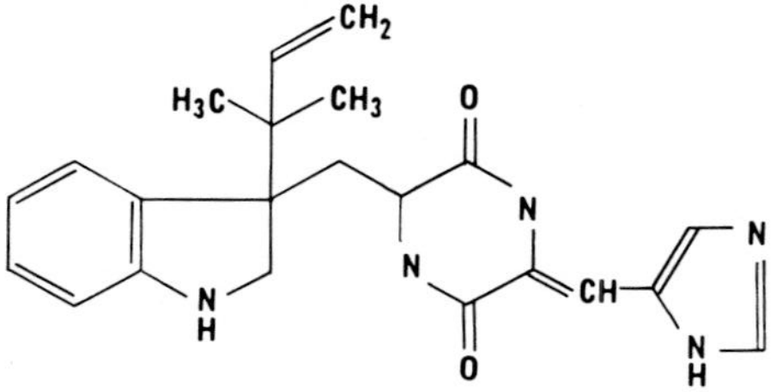

P. roquefortii

Blue cheese

Neurotoxin

Aflatoxin B$_1$

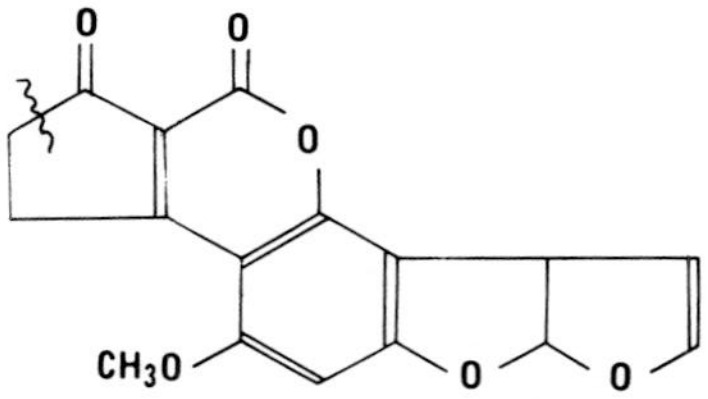

A. flavus, A. parasiticus

Cottonseed and other oil seeds, groundnuts, Brazil nuts, pistachios, corn, small grains, dairy products, flour, and so on (Ciegler, 1978)

B$_1$: hepatotoxin, hepatocarcinogenic

G$_1$ and M$_1$: possibly carcinogenic

(*continued*)

TABLE I (*Continued*)

Mycotoxin	Fungi	Natural substrates	Toxic syndrome

Aflatoxin B$_2$

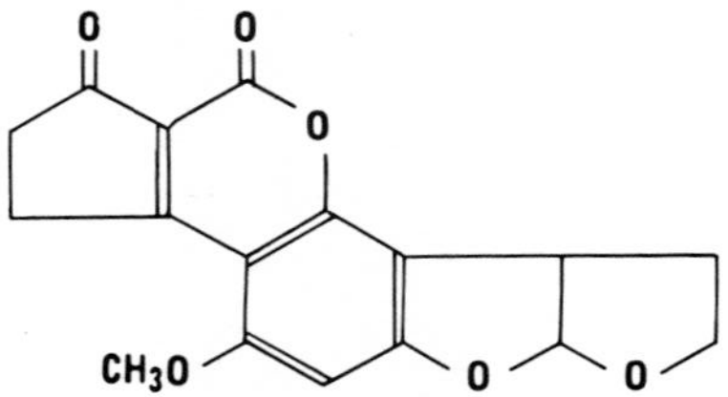

Aflatoxin G$_1$

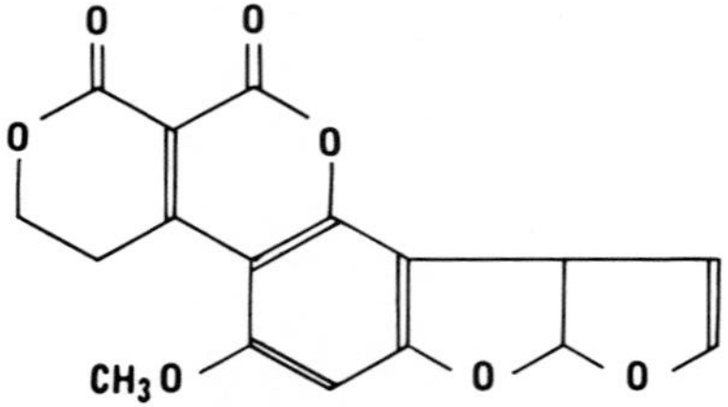

Aflatoxin G$_2$

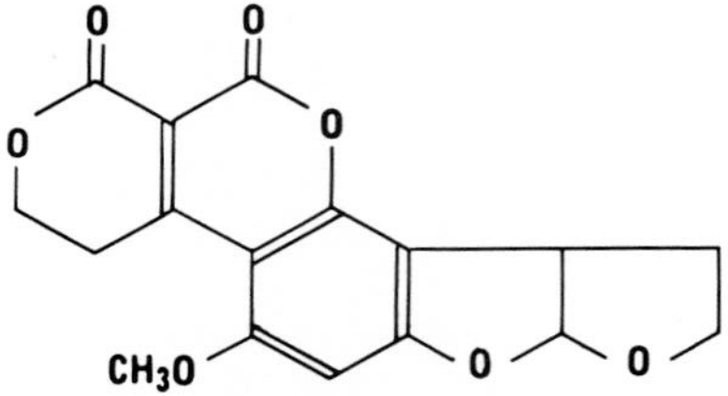

Aflatoxin M_1 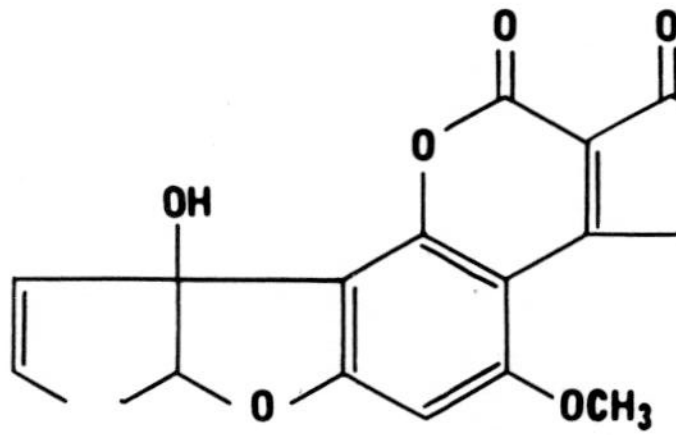			
Aflatoxin M_2 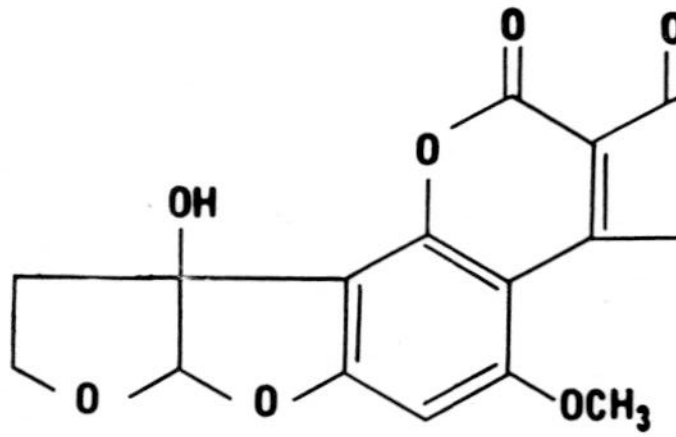			
Sterigmatocystin (R=H) 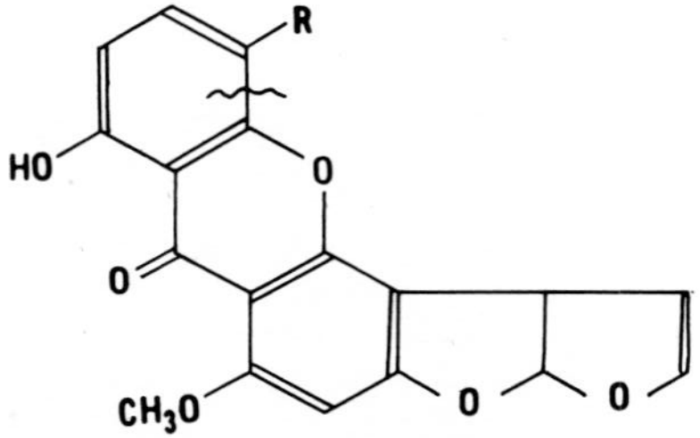	*A. versicolor*	Grain and oilseed products (Moreau and Moss, 1979)	Hepatotoxin, nephrotoxin, carcinogen (possible aflatoxin precursor)

(*continued*)

TABLE I (*Continued*)

Mycotoxin	Fungi	Natural substrates	Toxic syndrome
Zearalenone (F-2 toxin)	*Fusarium graminearum* (=*F. roseum*), *F. oxysporum*, *F. moniliforme*, *F. sporotrichioides* var. *tricinctum*	Stored corn	Female: vaginal enlargement, vaginal prolapse, atrophy of ovaries Male: atrophy of testes, enlargement of mammary glands
Trichothecenes (37 types)	*F. poae*, *F. sporotrichioides*, *F. sporotrichioides* var. *tricinctum*, *F. sporotrichioides* var. *chlamydosporum*, *F. solani*, *F. equiseti*, *F. nivale*, *F. concolor*, *F. avenaceum*, *F. sambucinum*, *F. culmorum*, *Stachybotrys alternans*, *Myrothecium* sp., *Trichothecium* sp., *Trichoderma* sp., *Cephalosporium* sp.	Grains and animal feed	Severe dermal irritation, hemorrhage of gastrointestinal and other organs
No toxin isolated	*Diplodia maydis*	Corn	Diplodiasis: nervous disorder of sheep and cattle, loss of muscle coordination, paralysis

Name	Structure	Fungus	Source	Effect
Sporidesmins	(chemical structure: Cl, OH, OH, OCH₃, H_3CO, OCH_3, N—CH_3, N, S—S, N—CH_3, CH_3, O, O)	*Pithomyces chartarum*	Pasture plants (Moreau and Moss, 1979)	Pithomycotoxicosis: facial eczema
Slaframine	(chemical structure: $OCOCH_3$, H_2N)	*Rhizoctonia leguminicola*	Red clover	Slobber factor disease of cattle and farm animals: bloating, congestion, and enlargement of lungs, liver, and kidneys
Brefeldin A		*Phoma herbarum* var. *medicaginis*	Alfalfa	Slobber factor syndrome
Cytochalasin B		*Phoma herbarum* var. *medicaginis*	Alfalfa	Slobber factor syndrome

(continued)

65

TABLE I (*Continued*)

Mycotoxin	Fungi	Natural substrates	Toxic syndrome
No toxin isolated	*Phomopsis leptostromiformis*	Lupine	Lupinosis: liver ailment characterized by massive lesions, hemorrhage, jaundice, liver enlargement, cirrhosis
No toxin isolated	*Cladosporium* spp., e.g., *C. epiphyllum*, *C. fagi*	Animal feed	Alimentary toxic aleukia
Tentoxin Alternaric acid Altenene Altenuene Altertoxin I Altertoxin II Altenuic acid Zinniol Tenuazonic acid	*A. humicola, A. alternata*	Stored grain	Animal toxicity not extensively studied; many extremely phytotoxic

[a] Common growth substrates of the fungus(i), not substrates from which respective mycotoxin(s) have been detected.

B. Temperature Control

Growth of most food-contaminating conidial fungi is inhibited at 4°C and below. Personal experience indicates that toxicogenic *Aspergillus* species do not grow under these conditions. However, several toxicogenic species of *Fusarium* (Joffe, 1965), *Penicillium* (Mislivec *et al.,* 1975a), and possibly *Alternaria* (Joffe, 1965) grow and elaborate toxins at 4°C and below.

C. Moisture Control

Keeping foods in a dry environment is an excellent means of inhibiting the growth of virtually all known toxicogenic conidial fungi. In general, foods that generate an equilibrium relative humidity (ERH) of no higher than 75% are at a safe moisture level. In the case of small grains the ERH is 13–15% moisture (wet weight basis) (Christensen and Kaufman, 1965). However, for a number of foodstuffs, such as fleshy fruits, storage at an ERH of 75% or below is impossible and they are subject to fungal attack.

D. Oxygen Control

Since all conidial fungi are obligate aerobes, foods maintained in an oxygen-free atmosphere are not subject to invasion by these microbes. Theoretically, oxygen deprivation appears to be the most effective means of controlling mold growth and mycotoxin production. In reality, for economic and other reasons, this prophylactic technique has limited application.

E. Alternative Controls

The two federal agencies most involved in mycotoxin problems in foods, the U.S. Department of Agriculture and the U.S. Food and Drug Administration (with substantial input from academia and industry), have been working hard to develop new approaches for mycotoxin control. Rodricks (1978) has reviewed and summarized these activities, which involve not only means of preventing actual mycotoxin elaboration in foodstuffs but also procedures for preventing already contaminated foodstuffs from reaching the consumer. Research studies include development of more sensitive, reliable, and rapid chemical screening methods for mycotoxin detection, more refined agricultural practices (including the search for mold-resistant plant varieties), detoxification of contaminated products, and removal of contaminated portions of a given food lot from noncontaminated portions. These and other approaches to mycotoxin control are worth pursuing.

V. METHODS FOR DETECTION AND PROPER
HANDLING OF MYCOTOXINS

A. Mycotoxin Standards

When investigating any foodstuff for mycotoxin contamination, or a mold isolate for its ability to elaborate mycotoxins, it is necessary that authentic analytic standards of the mycotoxin(s) under investigation be utilized. Analytic standards of most of the important *Aspergillus* and *Penicillium* toxins and a number of trichothecene toxins are available either commercially or from researchers active in a given field. However, standards for some of the less studied toxins (e.g., the *Alternaria* toxins) are still unavailable, severely hampering research efforts in these areas.

B. Mycotoxin Production by Mold Isolates

If standards are available, the screening of selected fungi for ability to elaborate toxins is not difficult. Most toxic species of *Aspergillus, Penicillium,* and *Fusarium* produce toxins in detectable amounts on a polished rice and tap water substrate, the toxin being detectable via thin-layer chromatography. Briefly, autoclaved 500-ml erlenmeyer flasks or 1000-ml Roux bottles containing 50 g polished rice and 50 ml tap water are inoculated using a spore suspension from pure culture isolates. The rice cultures are incubated at 23°–26°C for 10–14 days. In the case of fusaria, incubation should be at 12°C for up to 30 days. The flask contents are then extracted with chloroform, and the filtrate is concentrated to ca. 5 ml. With analytic standards, filtrates may be analyzed for mycotoxins by thin-layer chromatography utilizing appropriate developing solvent systems. A number of useful solvent systems have been described by Mislivec *et al.* (1968, 1975a). Although certain species may produce their toxins in greater amounts on other substrates (e.g., corn, wheat, peanuts), polished rice is preferred in this rather simple screening procedure, since it is relatively free of highly fluorescent chloroform-extractable substances and gives a relatively "clean" filtrate.

C. Mycotoxin Detection in Foods

Assay procedures for the detection of mycotoxins in foods are rather complicated and, depending on the particular foodstuff and mycotoxin(s) in question, often time-consuming. The crude chloroform or other organic solvent extracts of foods usually have to undergo "clean-up" steps involving column chromatography to separate the mycotoxins from other substances. Since quantitative as well as qualitative information is desired, analysis using thin-layer chromatography may not suffice. Instead, gas–liquid or high-pressure liquid chromatography may be required. Finally, the identity of the suspected mycotoxin must be confirmed, which usually involves preparation of chemical derivatives.

Since numerous analytic methods for the detection of and confirmation of specific mycotoxins from a variety of foodstuffs are available, none will be described. An important reference source is the *Book of Official Methods of Analysis of the Association of Official Analytical Chemists* (AOAC, 1975). Other references are listed by Rodricks and Lovett (1976).

D. Laboratory Safety

As in work with other potentially hazardous compounds, research on mycotoxins demands certain precautions. The following suggestions are presented. Extractions should always be performed in a chemical hood. Analysts should wear proper attire to prevent direct contact with mycotoxin solutions, since some compounds are severe contact irritants. Pure crystalline mycotoxins should be handled in a glove box free of air currents. Toxic extracts should be kept in well-sealed containers and refrigerated.

E. Detoxification

Since most mycotoxins are heat-stable, normal autoclaving procedures do not always suffice to destroy them. Thus certain prior detoxification measures are necessary. A 5% sodium hypochlorite solution is recommended. In the case of preextracted mycotoxin-contaminated foodstuffs, the solution should be used carefully and in a well-vented chemical hood in case of possible unexpected chemical reactions. The treated foodstuffs should then remain undisturbed for 24–48 h and then be autoclaved or, preferably, incinerated. Laboratory equipment (glassware, spatulas, hoods, balances, even bench tops, and so on) should be washed with the 5% sodium hypochlorite solution, followed by soapy water and clear water rinses.

VI. SUMMARY

From the staggering amount of literature available, there is no question that conidial fungi are the most important fungi involved in mycotoxin contamination of foodstuffs. Their positive role in animal discomfort is unquestionable. Their role in human disorders is still inconclusive. However, epidemiological and other circumstantial evidence strongly indicates that human mycotoxicoses caused by conidial fungi are a reality. Positive proof will be provided by future studies.

REFERENCES

Allcroft, R., and Carnaghan, R. (1963). Toxic products in groundnuts. *Chem. Ind. (London)* pp. 50–53.

Asplin, F., and Carnaghan, R. (1961). The toxicity of groundnut meals for poultry with special reference to their effect on ducklings and chickens. *Vet Rec.* **73,** 1215.

Association of Official Analytical Chemists (1975). "Official Methods of Analysis." AOAC.

Blount, W. P. (1961). Turkey "X" disease. *Turkeys* **9,** 52–61.

Burnside, J. D., Sippel, W., Forgacs, J., Carll, W., Atwood, M., and Doll, E. (1957). A disease of swine and cattle caused by eating moldy corn. *Am J. Vet. Res.* **18,** 817.

Butler, E., and Crisan, E. (1977). A key to the genera and selected species of mycotoxin-producing fungi. *In* "Mycotoxic Fungi, Mycotoxins, and Mycotoxicoses" (T. Wyllie and L. Morehouse, eds.), Vol. I, pp. 1–20. Dekker, New York.

Carlton, W. W., and Tuite, J. (1977). Metabolites of *P. viridicatum* toxicology. *In* "Mycotoxins in Human and Animal Health" (J. Rodricks, C. W. Hesseltine, and M. Mehlman, eds.), pp. 526–541. Pathotox Publ., Park Forest, Illinois.

Christensen, C. J., Nelson, G. H., Mirocha, C. J., and Bates, F. (1968). Toxicity to experimental animals of 943 isolates of fungi. *Cancer Res.* **28,** 2293–2295.

Christensen, C. M., and Kaufman, H. H. (1965). Deterioration of stored grains by fungi. *Annu. Rev. Phytopathol.* **3,** 69–84.

Ciegler, A. (1977). Patulin. *In* "Mycotoxins in Human and Animal Health" (J. Rodricks, C. W. Hesseltine, and M. Mehlman, eds.), pp. 609–624. Pathotox Publ., Park Forest, Illinois.

Ciegler, A. (1978). Fungi that produce mycotoxins: conditions and occurrence. *Mycopathologia* **65,** 5–11.

Ciegler, A., Fennell, D., Sansing, G., Detroy, R., and Bennett, G. (1973). Mycotoxin producing strains of *Penicillium viridicatum:* Classification into subgroups. *Appl. Microbiol.* **26,** 271–278.

Cysewski, S. J. (1977). Chemistry of the tremorgenic mycotoxins. *In* "Mycotoxic Fungi, Mycotoxins, and Mycotoxicoses (T. Wyllie and L. Morehouse, eds.), Vol. I, pp. 357–364. Dekker, New York.

Detroy, R. W., Lillihoj, E. B., and Ciegler, A. (1971). Aflatoxins and related compounds. *In* "Microbial Toxins" (A. Ciegler, S. Kadis, and S. J. Ajl, eds.), Vol. 6, pp. 1–178. Academic Press, New York.

Dickens, F., and Jones H. E. H. (1965). Further studies on the carcinogenic action of certain lactones and related substances in the rat and mouse. *Br. J. Cancer* **19,** 392–403.

DiMenna, M. E., Mortimer, P., and White, E. P. (1977a). The genus *Pithomyces. In* "Mycotoxic Fungi, Mycotoxins and Mycotoxicoses" (T. Wyllie and L. Morehouse, eds.), Vol. I, pp. 99–103. Dekker, New York.

DiMenna, M. E., Mortimer, P., and White, E. P. (1977b). The genus *Phoma. In* "Mycotoxic Fungi, Mycotoxins, and Mycotoxicoses" (T. Wyllie and L. Morehouse, eds.), Vol. I, pp. 105–106. Dekker, New York.

DiMenna, M. E., Mortimer, P., and White, E. P. (1977c). The genus *Myrothecium. In* "Mycotoxic Fungi, Mycotoxins and Mycotoxicoses" (T. Wyllie, L. Morehouse, eds.), Vol. I, pp. 107–110. Dekker, New York.

Doupnic, B., and Sobers, E. K. (1968). Mycotoxicoses: Toxicity to chicks of *Alternaria longipes* isolated from tobacco. *Appl. Microbiol.* **16,** 1596–1597.

Ellis, M. B. (1971). Dematiaceous hyphomycetes. *Mycol. Pap.* pp. 464–497.

Ellis, M. B. (1976). More dematiaceous hyphomycetes. *Mycol. Pap.* pp. 411–427.

Eppley, R. (1977). Chemistry of stachybotryotoxicosis. *In* "Mycotoxins in Human and Animal

Health'' (J. Rodricks, C. W. Hesseltine, and M. Mehlman, eds.), pp. 285–293. Pathotox Publ., Park Forest, Illinois.

Forgacs, J. (1972). Stachybotryotoxicosis. *In* ''Microbial Toxins'' (S. Kadis, A. Ciegler, and S. J. Ajl, eds.), Vol. 8, pp. 95–126. Academic Press, New York.

Halver, J. E. (1965). Aflatoxicosis and rainbow trout hepatoma. *In* ''Mycotoxins in Foodstuffs'' (G. N. Wogan, ed.), pp. 209–234. MIT Press, Cambridge, Massachusetts.

Hamilton, P., Lucas, G., and Weltz, R. (1969). Mouse toxicity of fungi of tobacco. *Appl. Microbiol.* **18**, 570–574.

Harvan, D. J., and Pero, R. W. (1976). The structure and toxicity of the *Alternaria* metabolites. *Adv. Chem. Ser.* **149,** 344–355.

Hintikka, E. (1978). Human stachybotryotoxicosis. *In* ''Mycotoxic Fungi, Mycotoxins, and Mycotoxicoses'' (T. Wyllie and L. Morehouse, eds.), Vol III, pp. 87–89. Dekker, New York.

Huang, L., and Hanlin, R. (1975). Fungi occurring in freshly harvested and in-market pecans. *Mycologia* **67,** 689–700.

Joffe, A. Z. (1960). The mycoflora of overwintered cereals and its toxicity. *Bull. Res. Counc. Isr., Sect. D* **9,** 101–126.

Joffe, A. Z. (1965). Toxin production by cereal fungi causing toxic alimentary aleukia in man. *In* ''Mycotoxins in Foodstuffs'' (G. N. Wogan, ed.), pp. 77–85. MIT Press, Cambridge, Massachusetts.

Joffe, A. Z. (1974). A modern system of *Fusarium* taxonomy. *Mycopathol. Mycol. Appl.* **53,** 201–228.

Joffe, A. Z. (1977). The genus *Fusarium. In* ''Mycotoxic Fungi, Mycotoxins, and Mycotoxicoses'' (T. Wyllie and L. Morehouse, eds.), Vol. I, pp. 59–82. Dekker, New, York.

Joffe, A. Z. (1978). *Fusarium poae* and *F. sporotrichioides* as principal causal agents of alimentary toxic aleukia. *In* ''Mycotoxic Fungi, Mycotoxins, and Mycotoxicoses'' (T. Wyllie and L. Morehouse, eds.), Vol. III, pp. 21–86. Dekker, New York.

Jones, B. D. (1977). Aflatoxins and related compounds: Occurrence in foods and feeds. *In* ''Mycotoxic Fungi, Mycotoxins, and Mycotoxicoses'' (T. Wyllie and L. Morehouse, eds.), Vol. I, pp. 190–200. Dekker, New York.

Kinosita, R., and Shikata, T. (1965). On toxic moldy rice. *In* ''Mycotoxins in Foodstuffs'' (G. H. Wogan, ed.), pp. 111–132. MIT Press, Cambridge, Massachusetts.

Krogh, P. (1977). Ochratoxins. *In* ''Mycotoxins in Human and Animal Health'' (J. Rodricks, C. W. Hesseltine, and M. Mehlman, eds.), pp. 489–498. Pathotox Publ., Park Forest, Illinois.

Krogh, P., Hald, B., and Pederson, E. (1973). Occurrence of ochratoxin A and citrinin in cereals associated with mycotoxic porcine nephropathy. *Acta Pathol. Microbiol. Scand.* **81,** 689–695.

Kurata, H. (1978). Current scope of mycotoxin research from the viewpoint of food mycology. *In* ''Toxicology: Biochemistry and Pathology of Mycotoxins'' (K. Uraguchi and M. Yamazaki, eds.), pp. 13–64. Wiley, New York.

Kurtzman, C. P., and Ciegler, A. (1970). Mycotoxin from blue-eye mold of corn. *Appl. Microbiol.* **20,** 204–207.

Lancaster, M., Jenkins, F., and Philip, J. (1961). Toxicity associated with certain samples of groundnuts. *Nature (London)* **192,** 1095–1096.

Marasas, W. F. (1977a). The genus *Phomopsis. In* ''Mycotoxic Fungi, Mycotoxins, and Mycotoxicoses'' (T. Wyllie and L. Morehouse, eds.), Vol. I, pp. 111–118. Dekker, New York.

Marasas, W. F. (1977b). The genus *Diplodia. In* ''Mycotoxic Fungi, Mycotoxins, and Mycotoxicoses'' (T. Wyllie and L. Morehouse, eds.), Vol. I, pp. 119–128. Dekker, New York.

Marasas, W. F. (1978). Lupinosis in sheep. *In* ''Mycotoxic Fungi, Mycotoxins, and Mycotoxicoses'' (T. Wyllie and L. Morehouse, eds.), Vol. II, pp. 213–217. Dekker, New York.

Meronouk, R., Steele, J., Mirocha, C., and Christensen, C. M. (1972). Tenuazonic acid, a toxin produced by *Alternaria alternata*. *Appl. Microbiol.* **23**, 613–617.

Mirocha, C., Pathre, S., and Christensen, C. M. (1977). Chemistry of *Fusarium* and *Stachybotrys* mycotoxins. *In* ''Mycotoxic Fungi, Mycotoxins and Mycotoxicoses'' (T. Wyllie and L. Morehouse, eds.), Vol. I, pp. 365–420. Dekker, New York.

Mislivec, P. (1977a). Toxigenic fungi in foods. *In* ''Mycotoxins in Human and Animal Health'' (J. Rodricks, C. W. Hasseltine, and M. Mehlman, eds.), pp. 469–477. Pathotox Publ., Park Forest, Illinois.

Mislivec, P. (1977b). The genus *Penicillium*. *In* ''Mycotoxic Fungi, Mycotoxins, and Mycotoxicoses'' (T. Wyllie and L. Morehouse, eds.), Vol. I, pp. 41–57. Dekker, New York.

Mislivec, P., and Bruce, V. R. (1977). Incidence of toxic and other mold species and genera in soybeans. *J. Food Protection* **40**, 309–312.

Mislivec, P., and Tuite, J. (1970). Species of *Penicillium* occurring in freshly harvested and in stored dent corn kernels. *Mycologia* **62**, 67–74.

Mislivec, P., Hunter, J., and Tuite, J. (1968). Assay for aflatoxin production by the genera *Aspergillus* and *Penicillium*. *Appl. Microbiol.* **16**, 1053–1055.

Mislivec, P., Douglas, R., and Kautter, D. (1972). Mycotoxic molds in black and white peppercorns. *Abstr. Annu. Meet. Am. Soc. Microbiol.* p. 27.

Mislivec, P., Dieter, C., and Bruce, V. R. (1975a). Effect of temperature and relative humidity on spore germination of mycotoxic species of *Aspergillus* and *Penicillium*. *Mycologia* **67**, 1187–1189.

Mislivec, P., Dieter, C., and Bruce, V. R. (1975b). Mycotoxin producing potential of mold flora of dried beans. *Appl. Microbiol.* **29**, 522–526.

Mislivec, P., Bruce, V. R., and Andrews, W. H. (1979). Mycological survey of selected health foods. *Appl Environ. Microbiol.* **37**, 567–571.

Miyaki, M., and Saito, M. (1965). Liver injury and liver tumors induced by toxins of *Pencillium islandicum* Sopp growing on yellowed rice. *In* ''Mycotoxins in Foodstuffs'' (G. N. Wogan, ed.), pp. 133–146. MIT Press, Cambridge, Massachusetts.

Monlux, A. W., Glenn, B., Panciera, R., and Corcoran, J. (1963). Bovine hepatogenous photosensitivity associated with the feeding of alfalfa hay. *J. Am. Vet. Med. Assoc.* **142**, 989–994.

Moreau, C., and Moss, M. (1979). ''Moulds, Toxins, and Food.'' Wiley, New York.

Olvigni, F., and Bullerman, L. (1978). Production of penicillic acid and patulin by an atypical *Penicillium roqueforti* isolate. *Appl. Environ. Microbiol.* **35**, 435–438.

Palti, J. (1978). Toxigenic Fusaria, their distribution and significance as causes of disease in animal and man. *Acta Phytomedica* **6**, 1–110.

Patterson, D. (1977). Aflatoxin and related compounds: Toxin producing fungi and susceptible animal species. *In* ''Mycotoxic Fungi, Mycotoxins, and Mycotoxicoses'' (T. Wyllie and L. Morehouse, eds.), Vol. I, pp. 156–158. Dekker, New York.

Pohland, A., and Mislivec, P. (1976). Metabolites of various *Penicillium* species encountered in foods. *Adv. Chem. Ser.* **149**, 110–143.

Raper, K., and Fennell, D. (1965). ''The Genus *Aspergillus*.'' Williams & Wilkins, Baltimore, Maryland.

Raper, K., and Thom, C. (1949). ''A Manual of the Penicillia.'' Williams & Wilkins, Baltimore, Maryland.

Rodricks, J. (1978). Regulatory aspects of the mycotoxin problem in the United States. *In* ''Mycotoxic Fungi, Mycotoxins, and Mycotoxicoses'' (T. Wyllie and L. Morehouse, eds.), Vol. III, pp. 159–171. Dekker, New York.

Rodricks, J., and Lovett, J. (1976). Toxigenic fungi. *In* ''Compendium of Methods for Microbiological Examination of Foods'' (M. L. Speck, ed.), pp. 484–503. Am. Public Health Assoc., Washington D.C.

Saito, M., Enomoto, M., and Tatsumo, T. (1971). Yellow rice toxins: Luteoskyrin and related compounds and citrinin. *In* "Microbial Toxins" (A. Ciegler, S. Kadis, and S. J. Ajl, eds.), Vol. 6, pp. 299–380. Academic Press, New York.

Samson, R. A. (1979). Compilation of the Aspergilli described since 1965. *Studies Mycol.* **18**, 1–38.

Sargeant, K. Sheridan, A., O'Kelly, J., and Carnaghan, R. (1961). Toxicity associated with certain samples of groundnuts. *Nature (London)* **192**, 1096–1097.

Schindler, A. F., and Abadie, A. (1973). Sterigmatocystin production by *Aspergillus versicolor* on semi-synthetic and natural substrates: A comparison of 28 isolates and studies on the effect of time. *Proc. Int. Congr. Plant Pathol., 2nd 1973* Abstract 0472.

Scott, P. M. (1977). *Penicillium* mycotoxins. *In* "Mycotoxic Fungi, Mycotoxins, and Mycotoxicoses" (T. Wyllie, and L. Morehouse, eds.), Vol. I, pp. 283–356. Dekker, New York.

Scott, P. M., and Kennedy, B. (1973). Improved method for the thin layer chromatographic determination of patulin in apple juice. *J. Assoc. Off. Anal. Chem.* **58**, 813–816.

Scott, P. M., and Kennedy, B. (1976). Analysis of blue cheese for roquefortine and other alkaloids from *Penicillium roqueforti*. *J. Agric. Food Chem.* **24**, 865–868.

Scott, P. M., van Walbeek, W., Kennedy, B., and Anyeti, D. (1972). Mycotoxins and toxigenic fungi in grains and other argicultural products. *J. Agric. Food Chem.* **20**, 1103–1109.

Scott, P. M., Merrien, M., and Polonsky, J. (1976). Roquefortine and isofumigaclavine A, metabolites from *Penicillium roqueforti*. *Experientia* **32**, 140–142.

Shank, R. C. (1978). Human aflatoxicoses. *In* "Mycotoxic Fungi, Mycotoxins, and Mycotoxicoses" (T. Wyllie and L. Morehouse, eds.), Vol. III, pp. 6–19. Dekker, New York.

Smalley, E. B. (1978). Salivary syndrome in cattle. *In* "Mycotoxic Fungi, Mycotoxins, and Mycotoxicoses" (T. Wyllie and L. Morehouse, eds.), Vol. II, pp. 111–120. Dekker, New York.

Snow, J. P., Lucas, G. B., Harvan, D., Pero, R., and Owens, R. (1972). Analysis of tobacco and smoke condensate for penicillic acid. *Appl. Microbiol.* **24**, 34–36.

Stack, M., and Mislivec, P. (1978). Production of xanthomegnin and viomellein by isolates of *Aspergillus ochraceus, Penicillium cyclopium,* and *Penicillium viridicatum*. *Appl. Environ. Microbiol.* **36**, 552–554.

Stack, M., Eppley, R., and Pohland, A. (1977). Metabolites of *Penicillium viridicatum*. *In* "Mycotoxins in Human and Animal Health" (J. Rodricks, C. W. Hesseltine, and M. Mehlman, eds.), pp. 543–555. Pathotox Publ., Park Forest, Illinois.

Still, P., Wei, R., Smalley, E. B., and Strong, F. (1972). A mycotoxin from *Penicillium roqueforti* isolated from toxic cattle feed. *Fed. Proc., Fed. Am. Soc. Exp. Biol.* **31**, 733–735.

Stoloff, L. (1977). Aflatoxins—An overview. *In* "Mycotoxins in Human and Animal Health" (J. Rodricks, C. W. Hesseltine, and M. Mehlman, eds.), pp. 7–28. Pathotox Publ., Park Forest, Illinois.

Templeton, G. E. (1972). *Alternaria* toxins related to pathogenesis in plants. *In* "Microbial Toxins (S. Kadis, A. Ciegler, and S. J. Ajl, eds.), Vol. 8, pp. 169–192. Academic Press, New York.

Thorpe, C. (1974). Analysis of penicillic acid by gas-liquid chromatography. *J. Assoc. Off. Anal. Chem.* **57**, 861–865.

Toussoun, T. A., and Nelson, P. E. (1975). Variation and speciation in the fusaria. *Annu. Rev. Phytopathol.* **13**, 71–82.

Tuite, J. (1977). The genus *Aspergillus. In* "Mycotoxic Fungi, Mycotoxins, and Mycotoxicoses" (T. Wyllie and L. Morehouse, eds.), Vol. I, pp. 21–39. Dekker, New York.

van der Zidjen, A., Koelensmid, J., Boldingh, J., Barrett, C., Ord, W., and Philp, J. (1962). Toxic metabolites of *Aspergillus flavus*. *Nature (London)* **195**, 1062–1063.

Ware, G., Thorpe, C., and Pohland, A. (1974). Liquid chromatographic method for determination of patulin in apple juice. *J. Assoc. Off. Anal. Chem.* **57**, 1111–1113.

White, E. P., Mortimer, P., and DiMenna, M. (1977a). Chemistry of the sporidesmins. *In* "Mycotoxic Fungi, Mycotoxins, and Mycotoxicoses" (T. Wyllie and L. Morehouse, eds.), Vol. I, pp. 427–447. Dekker, New York.

White, E. P., Mortimer, P., and DiMenna, M. (1977b). Toxins of *Phoma herbarum* var. *medicaginis*. *In* "Mycotoxic Fungi, Mycotoxins, and Mycotoxicoses" (T. Wyllie and L. Morehouse, eds.), Vol. I, pp. 459–463. Dekker, New York.

Wilson, B. (1971). Miscellaneous *Penicillium* toxins. *In* "Microbial Toxins" (A. Ciegler, S. Kadis, and S. J. Ajl, eds.), Vol. 6, pp. 459–521. Academic Press, New York.

Yamazaki, M. (1978). Chemistry of mycotoxins. *In* "Toxicology: Biochemisty and Pathology of Mycotoxins" (K. Uraguchi and M. Yamazaki, eds.), pp. 65–106. Wiley, New York.

17

Development of Parasitic Conidial Fungi in Plants

James R. Aist

I. INTRODUCTION

Among several biotic agents that can parasitize plants, fungi are the most prevalent. A vast number of common and destructive plant diseases, such as rusts, powdery mildews, leaf spots, root rots, and vascular wilts, are caused by conidial fungi. From the viewpoint of the fungus, the benefit derived from associating with a plant depends on the degree to which it can use the plant as a food base to support its own growth and reproduction. Yet most fungi are not equipped to parasitize plants; they lack one or more of the necessary

Biology of Conidial Fungi, Vol. 2
Copyright © 1981 by Academic Press, Inc.
All rights of reproduction in any form reserved.
ISBN 0-12-179502-0

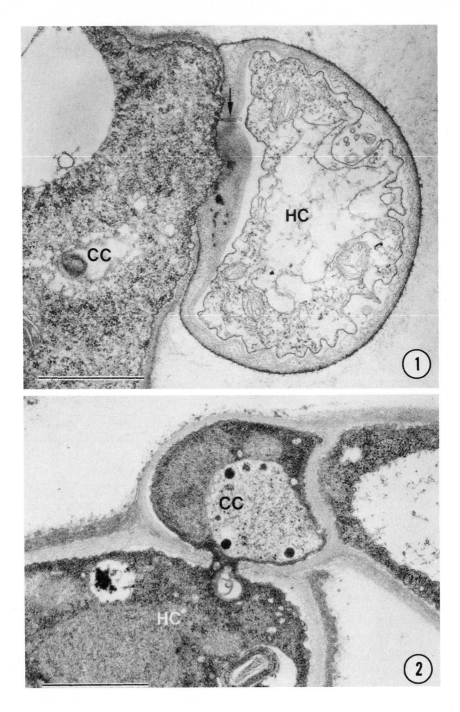

Figs. 1 and 2. Electron micrographs depicting cytoplasmic bridges that connect host and parasite cytoplasms at specialized interfaces.

morphological or biochemical attacking mechanisms, or they have not yet evolved methods to cope with the various resistance mechanisms of a potential host plant.

Fungi that do parasitize plants may be categorized according to their overall nutritional relationship with the host plant (Lewis, 1973). Necrotrophic fungi obtain their nutrients primarily from dead host cells and tend to have broad host ranges. Biotrophic fungi obtain nutrients from living host cells and are usually confined to a small group of taxonomically related plants. This fundamental distinction forms the underlying theme of Section III.

This chapter was written from the viewpoint of the parasite. It describes the broad developmental, morphological, and nutritional relationships of conidial fungi in plants. Although the examples used to illustrate the various concepts and principles are concerned exclusively with conidial fungi, the concepts and principles themselves apply as well to relationships between plants and most other biotic disease agents. Finally, it should be stressed that different fungi react with plants in a variety of different ways. Fungi commonly have more than one way in which they can gain ingress, obtain nutrients, ramify, and gain egress. The prevailing mode of each process may vary with the specific fungal isolate or host plant, the environmental conditions, and the particular plant organ or tissue being attacked.

II. INGRESS

A. Modes of Ingress

To obtain the nutrients for their growth and reproduction, parasitic fungi invariably enter the plant body. Ingress may occur through a wound (Pearson, 1931; Abawi and Lorbeer, 1971; Spencer, 1973), through a natural opening, or directly through the epidermis. The following is a discussion of some of the many approaches conidial fungi have evolved which accomplish this task.

1. Cytoplasmic Bridges

Some mycoparasites attach to host cells by means of specialized contact cells through which it is assumed that nutrients pass into the parasite thallus. Hoch (1977a,b, 1978) has recently discovered that contact cells have direct cytoplasmic connections with host cells (Figs. 1 and 2). The pores established upon

---

Fig. 1.   A plasmodesma (arrow) connects the cytoplasms of a contact cell (CC) of the parasite, *Gonatobotrys simplex,* and a hyphal cell (HC) of the host, *Alternaria tenuis.* ×28,750. Scale bar = 1 μm. (From Hoch, 1977a.)

Fig. 2.   A wide pore has developed between the contact cell (CC) of the parasite, *Calcarisporium parasiticum,* and a hyphal cell (HC) of the host, *Physalospora obtusa.* ×28,750. Scale bar = 1 μm. (From Hoch, 1977b. Reproduced by permission of the National Research Council of Canada from the *Canadian Journal of Botany.*)

contact of *Gonatobotrys* and *Stephanoma* with host cells are plasmodesmata (Fig. 1) and are occluded on the host side by electron-dense plugs. Pores between contact cells of *Calcarisporium* and its host's cells are much larger (Fig. 2) and permit organelle exchange in both directions. These unique forms of entry represent the least obtrusive modes of cell penetration yet documented for conidial fungi. Yet they are also the most intimate, requiring a high degree of cytoplasmic compatibility and providing conduits for direct flow of nutrients from host to parasite. The role of the contact cells in modifying and translocating nutrients obtained from the host needs further research.

2. Germ Tubes and Hyphae

Germ tubes and hyphae sometimes grow directly into the plant epidermis. Ingress may be through the cuticle and into the outer epidermal wall (McKeen, 1974), between epidermal cells (Pearson, 1931; Abawi and Lorbeer, 1971; Paus and Raa, 1973; Wheeler, 1977), or directly into the lumina of epidermal cells (Abawi and Lorbeer, 1971). These fungi, or their near relatives, are known to produce cuticle- and cell wall-degrading enzymes (Bateman and Basham, 1976; Baker and Bateman, 1978; Duddridge and Sargent, 1978), and it is thought that action of these enzymes during entry softens the host cuticle and cell wall sufficiently to allow penetration without the aid of appressoria.

Stomata may also be penetrated by germ tubes. Penetration of host leaves by *Cladosporium fulvum* (Lazarovits and Higgins, 1976a), *Scirrhia acicola* (Patton and Spear, 1978), and *Stemphylium* spp. (Higgins, 1966) occurs typically by growth of germ tubes through stomatal openings. H. Kunoh and co-workers (personal communication) have recently documented a unique mode of ingress by the anamorphic phase of the powdery mildew fungi *Leveillula taurica* and *Phyllactinia moricola;* germ tubes differentiate one or more multilobed, miniature adhesion bodies on the leaf epidermis before growing through stomatal pores. The entry of germ tubes through stomata is a convenient avenue of ingress, obviating the need for either appressoria or cell wall-degrading enzymes, provided the stomata are open when the germ tubes arrive. Moreover, certain epidermal resistance mechanisms may be avoided.

An interesting variation on this general theme was reported by Simmonds (1928) for the penetration of oat by *Fusarium culmorum*. When masses of conidia were deposited in a heavy inoculum, the germ tubes aggregated and grew side by side in a hyphal mass over the epidermis. Penetrations by hyphae into and between epidermal cells occurred along the way. The host cell wall was greatly altered beneath the creeping mass of germ tubes, probably having been softened by fungal enzyme secretions. A related observation was reported by Wheeler (1977) for *Helminthosporium maydis* on corn leaves. Subcuticular hyphae were traced to a compact knot of hyphae from which, presumably, ingress had originated. A great deal of host wall degradation was apparent.

3. Simple Appressoria

Many plant parasitic fungi can penetrate aerial plant parts by means of appressoria (Figs. 6 and 8), swollen apexes of germ tubes, or hyphae that adhere to host surfaces and effect penetration. The subject has been reviewed by Emmett and Parbery (1975). Appressoria appear to aid in the penetration of plants by (1) increasing adhesion to the plant surface, (2) developing, concentrating, and applying force, and (3) providing a means whereby host cell wall degradation may be confined to the immediate wall region to be penetrated.

Direct penetration through intact plant surfaces is the most commonly reported avenue of ingress used by appressoria. The initial entry ends just below the cuticle in more instances than is commonly recognized. Clear examples of such cuticular penetrations have been reported for *Venturia* (Nusbaum and Keitt, 1938; Maeda, 1970), *Diplocarpon* (Aronescu, 1934), *Rhynchosporium* (Jones and Ayers, 1974), *Helminthosporium* (Murray and Maxwell, 1975), *Botrytis* (Clark and Lorbeer, 1976), *Pleiochaeta* (Harvey, 1977), and *Gloeocercospora* (Myers and Fry, 1978). A period of subcuticular ramification follows this kind of entry.

Ingress by appressoria directly into epidermal cell lumina has also been commonly reported and leads to early ramifications in the epidermis. *Colletotrichum* spp. (Mercer *et al.*, 1971; Politis and Wheeler, 1973) and *Pyricularia oryzae* (Hashioka *et al.*, 1967) produce subspherical appressoria with thick, melanized walls (Figs. 6–8). In *Colletotrichum*, a unique funnel-shaped structure usually develops at the base of the appressorium prior to entry, and from it the walls of the penetration peg grow through the host wall (Brown, 1977). This structure and the thickened appressorial wall probably play key roles in force generation during entry. Appressorial penetration into epidermal cells also occurs in such diverse fungi as *Ravenelia* (Hunt, 1968), *Erysiphe* (McKeen and Rimmer, 1973), *Bremia* (Sargent *et al.*, 1973), and certain *Helminthosporium* spp. (Knox-Davies, 1974; Huang and Tinline, 1976).

Appressoria of some fungi routinely send penetration pegs between epidermal cells (Paddock, 1953; Hunt, 1968), initiating early intercellular ramification. Appressorial entry via stomata (Wynn, 1976; Mendgen, 1978) is typical of rust anamorphs. Other well-documented examples include *Cercospora beticola* (Rathaiah, 1976), *Ramularia areola* (Rathaiah, 1977a), and *Dothistroma pini* (Peterson and Walla, 1978). It may seem unnecessary for a fungus to use an appressorium to traverse a natural opening in the plant surface; but appressoria may be vital to ingress when the stomata are closed, to force apart the stomatal lips (Rathaiah, 1976). Appressoria located over stomata, as opposed to the cuticle, could achieve ingress without a battery of cuticle- and wall-degrading enzymes and may avoid triggering certain epidermal resistance responses.

Hunt discussed the possible advantages of direct versus stomatal penetration by appressoria. Since a shorter germ tube is required (the germ tube need not "find" a stomate), less energy and time would be spent in achieving entry. Thus the likelihood of energy depletion or damage from environmental fluctuations would be lessened.

4. Compound Appressoria

Conidial fungi have evolved several different ways to develop contiguous appressoria from which multiple ingress occurs. The simplest way is by the production of branched or septate appressoria (Emmett and Parbery, 1975; Rathaiah, 1976, 1977a; Harvey, 1977). Groups of contiguous appressoria may also arise when individual germ tubes produce appressoria at adjacent sites on the plant surface (Mangin, 1899; Rathaiah, 1977a; Myers and Fry, 1978). In the case of *Leptosphaeria* and *Ophiobolus* (Mangin, 1899), several different germ tubes or hyphae may be involved in production of the compound appressorium, but a great deal of subdivision and branching is also evident.

Multiple ingress may also occur from large, dome-shaped hyphal aggregates called infection cushions (Emmett and Parbery, 1975). Those produced by *Cochliobolus sativus* on wheat and barley apparently originate by repeated branching at the tip of a germ tube, resulting in an initial that consists of several contiguous appressoria (Huang and Tinline, 1976). A similar process was described for *Botrytis cinerea,* but in this case germ tubes from numerous conidia participated in the formation of a single cushion (Sharman and Heale, 1977). The mature infection cushion is compact. In *Cochliobolus* it is composed of a central pseudoparenchymatous zone and an outer layer of elongate runner hyphae. Ingress is achieved by multiple penetration pegs sent into the epidermis from appressoria or hyphal tips at the bases of the cushions.

One obvious advantage of such mass action is to concentrate pathogen attacking mechanisms, such as wall-degrading enzymes and toxins. This concentration of effort enables the fungus to soften the wall, release nutrients that could be used in the penetration process, and ward off host resistance responses by debilitating the host cells at an early stage of entry. Perhaps in this way the pathogen enters the plant through barriers it could not otherwise overcome.

B. External Influences

The effects of external factors on germ tube growth and appressorium formation have been reviewed by Carlile (1966), Wood (1967), Emmett and Parbery (1975), and Allen (1976). Here I will focus on influences affecting the course of germ tubes as they extend over the plant surface that may help determine where appressoria are formed.

1. Preferred Sites

Because the plant epidermis is a highly differentiated tissue that contains several specialized cell types and surface features (Cutter, 1976), one might expect that certain sites on its surface would be more penetrable than others. If this is so, fungi should have evolved special sensory mechanisms whereby the more suitable sites may be found. Although conidia may occupy random sites on a plant surface, their germ tubes or appressoria often do in fact attack at nonrandom locations. One obvious target is stomata, which may be selectively invaded by either germ tubes (Lazarovits and Higgins, 1976a; Rathaiah, 1977b; Patton and Spear, 1978) or appressoria (Wynn, 1976; Rathaiah, 1976, 1977a; Peterson and Walla, 1978). More commonly, germ tubes and appressoria show a preference for anticlinal wall junctures (Paddock, 1953; Murray and Maxwell, 1975; Clark and Lorbeer, 1976; Luttrell, 1977; Wheeler, 1977; Myers and Fry, 1978). Other preferred sites include motor cells and trichomes (Wood, 1967). Penetration of leaf trichomes by anamorphic *Gibberidea* (Mason, 1973) is especially interesting; conidial germ tubes grow across the leaf surface, up the trichome, and to the tip cell before forming appressoria.

2. Tropisms

Such nonrandom behavior indicates that these fungi have evolved mechanisms that respond to directional stimuli and to patterns of the plant surface. *In vitro* studies with conidial fungi have demonstrated that germ tubes can exhibit negative or positive tropisms in response to chemicals (Carlile, 1966; Barnett and Binder, 1973) and can exhibit tropic responses to light (Carlile, 1966) and contact stimuli (Dickinson, 1971; Wynn, 1976). On the plant surface specific patterns and potential sources of tropic signals abound (Cutter, 1976). Regular topographical patterns are associated with parallel epidermal cells and subsidiary cells of stomata (Lewis and Day, 1972; Wynn, 1976). Pectic substances are concentrated in anticlinal wall junctures (Jones and Ayers, 1974), and the junctures may be more permeable to nutrients than are other areas of the epidermis (Clark and Lorbeer, 1976). Stomata release gases such as water and oxygen which presumably establish concentration gradients; certain stomatal tropisms of germ tubes appear to operate on this basis (Rathaiah, 1977b; Peterson and Walla, 1978).

Germ tubes also respond tropically to each other or to their own metabolic products (Carlile, 1966), a phenomenon called autotropism. Negative autotropism of germ tubes could help ensure a maximum amount of tissue invasion by distributing penetration sites more evenly over the plant surface, while positive autotropism could be the basis for formation of compound appressoria.

3. Triggers

Determination of the signals that trigger germ tubes to produce appressoria has proven difficult. Very high percentages of appressoria can be formed *in vitro* either on cellulose membranes floated on water (De Waard, 1971) or in suspension (Dunkle and Allen, 1971). Moreover, a mild heat shock to conidia can substitute *in vitro* for host factors that can stimulate the differentiation of appressoria (Allen, 1976). Obviously, specific plant stimuli, either chemical or physical, are not an absolute requirement.

Nonetheless, appressoria are often produced perferentially at very specific sites on plant surfaces, and it is important to understand why this is so. Plant leachates can strongly stimulate appressorium formation (Grover, 1971; Swinburne, 1976). Thus the diffusion of chemicals in solution, or of gases, from anticlinal wall junctures and stomata, respectively, could establish critical concentrations of these materials at their sources. On the other hand, Wynn (1976) has provided compelling evidence that a contact stimulus provided by topographical features of surfaces triggers the formation of appressoria by germ tubes of *Uromyces phaseoli*. Apparently, for some fungi the *in vivo* trigger is primarily chemical in nature, whereas for others it is mainly physical. The further likelihood that for some fungi both physical and chemical prerequisites are involved should be borne in mind.

III. RAMIFICATION

Ramification represents the principal parasitic phase of the parasite in the host plant. Plant tissues become extensively colonized as the fungal thallus expands and derives from the plant many of the nutrients subsequently packaged in conidia.

A. Modes of Ramification

1. Epicuticular Ramification

Powdery mildew fungi ramify almost exclusively on the outer surface of the plant cuticle (Day and Scott, 1973; Locci and Quaroni, 1974), after the appressorium has produced a primary haustorium in an epidermal cell. During the ramification phase secondary haustoria are formed in other epidermal cells from short branches off the ramifying hyphae (Hirata, 1967; Bracker, 1968). Since only the epidermis is contacted by these fungi, all nutrients must pass from or through the epidermal cells to the mycelium. Bushnell and Gay (1978) discuss this nutritional relationship in detail.

Some fungi may ramify extensively over the plant cuticle, even though their principal development is endophytic (Luttrell, 1976). This epicuticular phase may be preliminary to sporulation (Salmon, 1906; Locci and Bisiach, 1971; Carpenter, 1976). Nutrients are probably translocated to the epicuticular hyphae via direct connections with endophytic hyphae, but direct transfer from the plant cuticle is also possible.

One advantage of epicuticular ramification is the ease and speed with which conidia can be initiated and liberated into the air. However, there are several obvious disadvantages to the epicuticular habit. First, the mycelium is fully exposed to extreme and potentially harmful environmental fluctuations. Second, the fungus must cope with antagonistic effects of a myriad of other microorganisms which commonly inhabit the plant surface (Dickinson and Preece, 1976). Finally, direct contact is made only with epidermal cells that have a very low photosynthetic capacity and sparse cytoplasm and are therefore not the best choice for a food base. The relative infrequency of epicuticular ramification among fungal parasites probably reflects these disadvantages.

2. Subcuticular Ramification

Ramification of conidial fungi just beneath the plant cuticle or within the outer epidermal wall (Fig. 4) is common. Usually, subcuticular ramification occurs concurrently with intra- or intercellular ramification or is significant only during very early stages of infection. This relationship has been illustrated for *Helminthosporium* spp. (Akai *et al.*, 1971; Knox-Davies, 1974; Wheeler, 1977), *Colletotrichum* (Brown, 1977), *Botrytis* (McKeen, 1974), and other conidial fungi (Hashioka and Nakai, 1974; Luttrell, 1977; Mason and Wilson, 1978; Myers and Fry, 1978).

Several fungi exhibit a much more extensive subcuticular phase that lasts for several days and precedes any appreciable ramification below the level of the outer epidermal wall (Heath and Wood, 1969; Jones and Ayers, 1974; Murray and Maxwell, 1975; Clark and Lorbeer, 1976). *Rhynchosporium* leaf blotch of barley is a particularly striking example of this type of behavior (Jones and Ayers, 1974). An extensive subcuticular mycelium develops during the first 10 days after inoculation. Then the epidermis collapses, and hyphae grow into the mesophyll. Other notable examples include *Venturia inaequalis* (Nusbaum and Keitt, 1938; Maeda, 1970) and *Diplocarpon rosae* (Aronescu, 1934), which develop almost exclusively in the subcuticular zone during their entire parasitic phases. Both these fungi ramify by an interesting series of distinct steps. Penetration pegs of *Venturia* enlarge beneath the host cuticle and produce a broad, subglobose primary hypha that further divides to form a primary stroma of short, coarse cells (Nusbaum and Keitt, 1938). From these stromal cells, filamentous runner hyphae ramify outward and produce secondary stromata of subglobose cells along the way. Ramification of *Diplocarpon* occurs in four stages (Aronescu, 1934): (1) The penetration peg produces a primary infection hypha that grows in the outer epidermal cell wall; (2) secondary infection hyphae then develop as branches from the primary infection hypha near the penetration site, and they also grow in the outer epidermal cell wall; (3) the secondary infection hyphae next produce parallel hyphae that grow in groups of two to seven in parallel fashion beneath the cuticle; and (4) haustorium mother cells then develop as outgrowths from the parallel hyphae. Haustoria are sent into host epidermal cells by primary infection hyphae, parallel hyphae, and haustorium mother cells.

Nusbaum and Keitt (1938) considered subcuticular development as exhibited by *V. inaequalis* and *D. rosae* to represent a unique nutritional relationship. These biotrophic fungi, like the powdery mildew fungi, are dependent on the epidermis for nutrition as long as they remain subcuticular. The subcuticular location should negate any effect of the cuticle in restricting the passage of water-soluble nutrients which would be directly accessible to the fungus. Moreover, subcuticular hyphae are in a unique position to degrade the host cuticle and cell wall, releasing nutrients for their own use. Many fungi that can develop subcuticularly are capable of producing cell wall-degrading enzymes (Bateman and Basham, 1976) as well as cutinase (Baker and Bateman, 1978; Duddridge and Sargent, 1978), and ultrastructural studies have provided evidence that cuticle degradation and extensive cell wall hydrolysis (Fig. 4) may occur *in vivo* (Maeda, 1970: Akai *et al.,* 1971; Locci and Quaroni, 1971; McKeen, 1974; Murray and Maxwell, 1975; Wheeler, 1977).

The subcuticular phase may be advantageous for reasons other than derived nutrients. Since the cuticle is usually not ruptured by the hyphae, but only stretched or buckled, it still affords them some degree of protection from dehydration. During this phase, the fungus may be able to build up its battery of enzymes, toxins, and food reserves, enabling it subsequently to intrude into underlying tissues that might otherwise be able to repel the attack. For example, subcuticular ramification of *Helminthosporium carbonum* results in death of the underlying host cells before deeper tissue ramification begins (Murray and Maxwell, 1975). More extensive physiological studies of the subcuticular phase should help clarify its significance.

3. Subepidermal Ramification

a. Intercellular. Ramification of some conidial fungi occurs almost entirely between the cells of invaded host tissues (Fig. 3). Classic examples are rusts (Mendgen, 1978; Plotnikova *et al.,* 1979) and downy mildew fungi (Ingram *et al.,* 1976). Also included in this group are *Cymadothea* (Camp and Whittingham, 1972), *Cladosporium* (Lazarovits and Higgins, 1976b), *Claviceps* (Luttrell, 1977), and *Elsinöe* (Mason and Wilson, 1978). *Elsinöe* ramifies in the plant stem by means of cables of closely appressed hyphae that grow intercellularly for long distances, parallel to the long axis of the stem (Mason and Backus, 1969). These hyphal cables are reminiscent of the parallel hyphae described in Section III,A for *Diplocarpon rosae,* but they do not produce haustoria and are located in the pith. *Leveillula taurica* (Salmon, 1906; H. Kunoh, personal communication) and *Phyllactinia* spp. (Smith, 1900; H. Kunoh, personal communication) diverge from the usual epicuticular habit of the powdery mildew fungi and ramify intercellularly.

All the fungi mentioned above are clearly biotrophic, and this nutritional relationship is probably related to their exclusively intercellular ramification; hyphal invasion of living cells usually leads to their rapid death. The rusts and

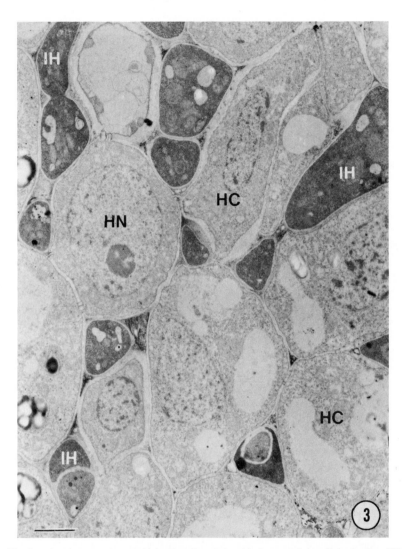

Fig. 3. An electron micrograph showing the relationship between intercellular hyphae (IH) of *Claviceps paspali* and host cells (HC) in the ovary wall of *Paspalum dilatatum* (Gramineae). Host nuclei (HN) can be seen. ×5840. Scale bar = 2 μm. (From Luttrell, 1977.)

mildews in this group produce haustoria in leaf mesophyll cells during the ramification phase (Littlefield and Bracker, 1972; Coffey, 1975; H. Kunoh, personal communication), and it is assumed that the haustoria aid in absorbing nutrients from the host (Bushnell, 1972).

b. Intracellular. It is unusual for biotrophic fungi to ramify through living cells by means of intracellular hyphae (Mason, 1973), presumably because the fungal enzymes released during intracellular development are usually too disruptive. Yet at least two biotrophic conidial fungi, anamorphic *Ravenelia* (Hunt, 1968) and *Gibberidea* (Mason, 1973), ramify intracellularly. These fungi presumably produce cell wall-degrading enzymes that aid host wall traversal during ramification, but probably only small amounts are produced. The enzymes may be bound to the fungal plasmalemma or wall (Berg and Pettersson, 1977) at the tip of the ramification peg, rather than released (see discussion by M. A. Stahman, in Williams *et al.*, 1973). According to this concept, bound enzymes cause highly localized wall degradation and minimal damage to the host protoplasm.

Several principally necrotrophic fungi ramify extensively through host cells (Hashioka *et al.*, 1967; Hess, 1969; Delon *et al.*, 1977; Harvey, 1977). They usually produce copious amounts of wall-degrading enzymes that alter the strength and ultrastructural appearance of the cell walls, as illustrated by Hess (1969) for *Pyrenochaeta terrestris* in onion roots. In lupin leaves infected by *Pleiochaeta setosa*, affected walls may appear swollen or, at localized areas, may seem to have been completely disgested away (Harvey, 1977). This wall softening and solubilization not only aids the process of wall traversal but also releases sugars that can be used by the fungus. These fungi typically kill host cells far in advance of their expanding thalli, which affords the affected host cells little time to muster active defense mechanisms.

Passage of hyphae through interior plant cell walls is usually by way of a thin hyphal strand (Heath and Wood, 1969; Hess, 1969; MacDonald and McNabb, 1970), here termed the ramification peg. Formation of ramification pegs may reflect a degree of physical resistance of the wall. On the other hand, ramifying hyphae of some fungi frequently penetrate host cell walls without an appreciable reduction in diameter (Luttrell, 1976; Harvey, 1977). Occasionally, a hyphal tip swelling or internal appressorium may precede passage of a ramification peg through the wall (Pearson, 1931; Paddock, 1953; Mercer *et al.*, 1975). Whether the walls penetrated by internal appressoria are so unusually resistant that an appressorium is needed to develop adequate penetration force, or whether the hyphal tips do not produce adequate amounts of enzymes to soften the wall, is uncertain.

A special type of intracellular ramification is characteristic of vascular wilt pathogens such as certain *Fusarium* and *Verticillium* spp. (Talboys, 1972). These fungi sporulate in xylem elements (Phipps and Stipes, 1976), and the spores move upward, apparently carried by the transpiration stream (Elgersma *et al.*, 1972). Such transported spores are caught at vessel end walls, germinate there, and again sporulate (Beckman and Keller, 1977). As this cycle is repeated, the fungus ramifies stepwise up the plant stem. This mode of ramification is apparently quite important in enabling the fungus to outdistance host resistance responses.

B. Intermediate Cells

Several fungi that ramify inside the host plant produce a special type of cell immediately upon entry. These cells are produced at the ends of penetration pegs and, once the parasite protoplast has migrated through them, they are cut off from the ramifying mycelium by a septum (Aronescu, 1934; McKeen, 1974; Clark and Lorbeer, 1976). Somewhat analogous, but highly swollen, intermediate cells are produced by many stomatal invaders [e.g., substomatal vesicles (Rathaiah, 1977a; Mendgen, 1978); primary hyphae (Higgins and Millar, 1968)], *Bremia lactucae* [primary vesicles (Ingram *et al.*, 1976)], and *Pleiochaeta setosa* [swollen subcuticular infection hyphae (Harvey, 1977)]. Since these intermediate cells are the first segments of the fungal thallus formed following entry and later occupy an intercalary position between the fungal cells on the exterior and those on the interior of the plant, they could play a key role in establishment of the infection.

An obvious function of some intermediate cells is to serve as an early branch point from which infection hyphae can ramify simultaneously in several directions (Higgins and Millar, 1968). Ellingboe (1968) raised the interesting possibility that they may do more; perhaps they initiate important biochemical interactions with the host, preparatory to extensive colonization. This speculation is in agreement with that of Ingram and co-workers (Sargent *et al.*, 1973; Ingram *et al.*, 1976), who have adopted the view that the primary vesicle of *B. lactucae* receives little or no nutrition from the host and that a nutritional relationship begins only after the parasite protoplast has entered a subsequently formed secondary vesicle. On the other hand, Tani *et al.* (1975) showed that induction of resistance by an incompatible race of the oat crown rust fungus occurs at the substomatal vesicle stage, an interesting result which also falls within the realm of Ellingboe's speculation. Aronescu (1934) found that the cross-wall separating the intermediate cell of *Diplocarpon rosae* from ramifying hyphae was unusually thick and impervious to cotton blue. She supposed that this specialized cell protected the young thallus from adverse weather conditions, particularly dry weather, at the more vulnerable early stages of development. These potentially pivotal cells certainly deserve closer attention in future studies.

C. Nutritional Shifts

Nutritional relationships between a fungus and its host plant are not always as clear-cut as those alluded to in the preceding sections. In many instances, individual host cells at the invasion front are alive during early stages of attack by ramifying hyphae but die shortly thereafter (Griffiths, 1973; Huang and Tinline, 1976; Luttrell, 1976). That these cells are alive during the early stages of direct contact with the fungus is evidenced by their ability to produce wall appositions.

It seems likely that some nourishment is obtained from the host cells at the invasion front and that the bulk of nutrients is eventually obtained from killed cells.

Abrupt switches in the nutritional relationship may not be confined to the invasion front. For example, *Colletotrichum lindemuthianum* ramifies intracellularly in bean leaves for 4–5 days, while the invaded host cells appear healthy (Mercer *et al.*, 1975). Then the biotrophic relationship suddenly becomes necrotrophic; host cells become necrotic, and the lesion expands more rapidly as extensive cell wall degradation occurs. An abrupt switch apparently occurs also in limited lesions caused by *Mycosphaerella pinodes* (Heath and Wood, 1969), but this switch is from necrotrophic to apparently biotrophic; the fungus ramifies far beyond the necrotic zone without causing visible symptoms. It seems that a kind of self-induced compatibility (Ouchi *et al.*, 1976a) may be involved (cf. Section V,A).

IV. EGRESS

Following ramification, parasitic fungi usually grow out of the host plant prior to or during sporulation. This phase of development on the host plant is therefore of utmost importance in the biology of these fungi, yet it has received little attention in morphological studies on host–parasite interactions. It is hoped that the following discussions of the different modes of egress exhibited by different fungi will illustrate the interesting biology represented and stimulate microscopists to focus more attention on this subject. I have categorized the modes of egress on the basis of the type of anatomical feature of the host plant that is breached.

A. Modes of Egress

1. Natural Openings

Many conidial fungi use natural breaks or weak points in the plant epidermis for their exit (Barnett and Hunter, 1972). These include some of the most notorious plant pathogens, such as *Alternaria* (Tsuneda and Skoropad, 1978), *Botrytis* (Clark and Lorbeer, 1976), *Cladosporium* (Lazarovits and Higgins, 1976a), and *Helminthosporium* (Luttrell, 1964), all of which send conidiophores out through stomata. Scanning electron microscopy is particularly well-suited for conveying the three-dimensional aspects of such egress in surface views (Locci and Quaroni, 1971; Guggenheim and Harr, 1978; Tsuneda and Skoropad, 1978). Hashioka and Nakai (1974) used transmission electron microscopy to document some of the ultrastructural aspects of stomatal egress of *Pyricularia* spp. Substomatal hyphae first produce conidiophore mother cells in the substomatal chamber. From the mother cells, conidiophores grow out through the stomatal pore.

Lenticels may also serve as avenues of egress for conidial fungi, as illustrated by Phipps and Stipes (1976) with *Fusarium oxysporum* on *Mimosa*. The fungal hyphae grew into the lenticels where masses of macroconidia were produced on the host surface.

Egress through natural openings may have several advantages over other modes considered here. Natural exits probably demand the least energy. Further, stomatal egress can probably take place relatively quickly in response to favorable environmental conditions, since natural host barriers need not be breached. A tropic growth response (Carlile, 1966) by the fungus is probably involved.

2. Cuticular

Fungi that exhibit an appreciable amount of subcuticular ramification will very often sporulate from the same position. In early stages of sporulation, individual subcuticular hyphae of the *Fusicladium* anamorph of *Venturia* send single conidiophores through the cuticle at separate locations, producing only small openings in the cuticle where egress occurs (Nusbaum and Keitt, 1938; Hughes, 1953). More typically, the development of masses of conidiophores from the surface of an expanding subcuticular stroma produces large ruptures in the cuticle (erumpent stromata), thus effecting egress, as illustrated by *Venturia* (Barnett and Hunter, 1972), its *Spilocaea* anamorph (Hughes, 1953), *Claviceps* (Luttrell, 1977), and *Dothistroma* (Peterson and Walla, 1978).

Cuticular egress by single conidiophores could be facilitated by cutinases, and *Venturia* has been reported to possess an esterase activity (Nicholson *et al.,* 1972). On the other hand, Nusbaum and Keitt (1938) clearly illustrated a bulging out of the cuticle by the conidiophore just prior to emergence, suggesting that mechanical force is a significant component of the process. Massive rupture of the cuticle by an expanding stroma also seems to result from mechanical pressure. The magnitude of force required to breach the cuticle may not be especially large to start with (Martin, 1964), and it could be reduced by even limited cutinase activity. Some very fruitful work could be done on the underlying mechanisms of this process.

3. Epidermal

Conidial fungi such as *Helminthosporium* spp. and *Pyricularia* may begin their exit from hyphae within the lumina of epidermal cells. When such an exit involves individual fungal cells in the plant epidermis, specialized hyphal structures resembling both penetration pegs (Fig. 8) that occur during entry and ramification pegs characteristic of wall traversal during intracellular ramification are sent out through the epidermal cell wall and cuticle (Figs. 4 and 5). These

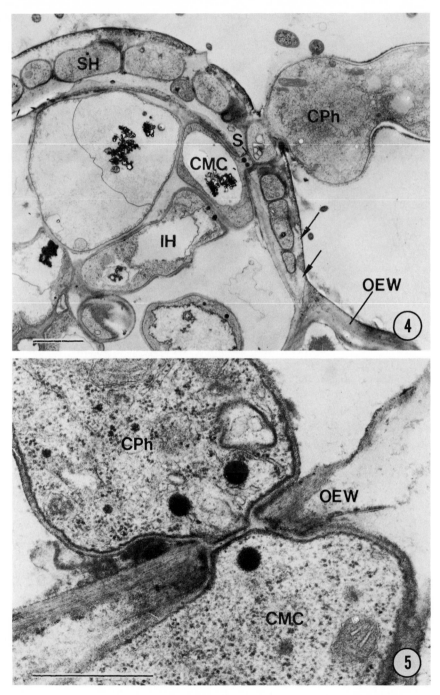

Figs. 4 and 5. Electron micrographs illustrating epidermal egress of *Pyricularia oryzae* from rice leaves.

structures have been variously referred to as "exit tubes" and "exit pegs" (Carpenter, 1976), "fine hyphae" (Haskioka and Nakai, 1973), "hyphal pegs" (Luttrell, 1963), and "egress tubes" (Paddock, 1953). The term "egress peg" seems preferable, because it implies the obvious function of the structure, is not used in a different context ["exit tube" is used to refer to the conduit by which certain fungal zoospores exit from zoosporangia (Alexopoulos, 1962)], and reserves the term "peg" for hyphal segments of reduced diameter which traverse the host wall. Egress pegs often originate from the swollen apexes of intraepidermal hyphae (Paddock, 1953; Luttrell, 1963; Hashioka and Nakai, 1974) that can be considered internal appressoria. The egress peg clearly curves the edges of the epidermal cell wall and cuticle outward in some cases (Hughes, 1953; Paddock, 1953; Hashioka and Nakai, 1974), suggesting that mechanical pressure is exerted during passage through these structures.

Hashioka and Nakai (1974) investigated epidermal egress by *Pyricularia* spp., using transmission electron microscopy. A septum formed in the middle of the egress peg (Fig. 5), which sometimes was swollen distally and had a constriction at the level of the plant cuticle (Fig. 4). Interestingly, subcuticular hyphae, although numerous (Fig. 4), never produced egress pegs. A variation of this mode of egress occurs in *Pollaccia radiosa* (Hughes, 1953) and *Collecephalus hemerocallidis* (Spencer, 1973) in which stromata arise in the host epidermal cells and produce conidiophores as the epidermal walls and cuticle are thrust open (erumpent stromata).

Hashioka and Nakai (1974) considered epidermal egress by *Pyricularia* spp. to be essentially the reverse of ingress, insofar as morphological aspects are concerned. Whether or not the basic mechanisms involved (mechanical force, enzymatic hydrolysis, hydrostatic pressure) are the same is unknown, but in many cases appropriate comparisons could be made.

4. Subepidermal

Most conidial pathogens initiate egress from virtually any location beneath the epidermis. One of the least disruptive modes, and one that requires relatively little energy, is the emergence of conidiophores between epidermal cells. This mode occurs in host plants infected by certain *Helminthosporium* spp. (Luttrell, 1963, 1964). Locci and Quaroni (1971) have shown by scanning electron microscopy that such conidiophores do not cause extensive tearing of the cuticle

---

Fig. 4. An intracellular hypha (IH) in the epidermal cell has produced a conidiophore mother cell (CMC) from which a narrow egress peg has grown into the outer epidermal wall (OEW), enlarged somewhat beyond the septum (S), broken through the cuticle, and expanded to form the conidiophore (CPh). The outer epidermal wall is densely packed with subcuticular hyphae (SH) which have apparently eroded the host wall in places (arrows). ×7000. Scale bar = 2 μm.

Fig. 5. A septum in the egress peg at the level of the outer epidermal wall (OEW) separates the conidiophore mother cell (CMC) from the conidiophore (CPh). ×33,000. Scale bar = 1 μm. (From Hashioka and Nakai, 1974.)

when they emerge singly. Emergence between epidermal cells is apparently uncommon.

Much more common is the production of erumpent, subepidermal stromata (acervuli, sori) or pycnidia. This form of exit is typical of many well-known conidial pathogens, including anamorphic phases of rust fungi (e.g., *Puccinia:* Allen, 1923) and *Ascochyta, Cercospora, Colletotrichum, Phoma,* and *Phyllosticta* (Barnett and Hunter, 1972). Egress is achieved when the expanding stroma or pycnidium ruptures the epidermal layer(s) and pushes it aside. The process obviously involves considerable pressure from the expanding fungal mass, but it could be facilitated by the release of cell wall-degrading enzymes at the fungus–plant interface. The compact, pseudoparenchymatous fungal tissue involved is well designed to withstand its own growth pressure. Pycnidia apparently serve to prevent desiccation of the conidiophores and conidia during periods of dry weather.

B. Via Hyphae

Sometimes the emerging fungal structure is a hypha rather than a conidiophore or stroma. These hyphae may quickly produce conidiophores (Locci and Bisiach, 1971; Luttrell, 1976), or an extensive mycelium (Salmon, 1906) or stroma (Carpenter, 1976) may develop in preparation for an external reproductive phase. A fascinating observation concerning hyphal egress was made by Heath and Wood (1969). They found that the coelomycetous fungus *Ascochyta pisi* typically sent out "exploratory hyphae" from stomata within limited lesions on pear leaves. These hyphae grew along the leaf surface for up to 1 mm beyond the edge of the lesion and formed appressoria on the cuticle over uninvaded leaf areas. Such appressoria rarely caused secondary lesions, however, presumably because the primary infection had induced resistance in the host cells bordering the lesions. This behavior may be a remarkable adaptation of the fungus in trying to escape the host defenses that limit ramification from the initial site of ingress.

V. HOST RESPONSES

The development of conidial fungi on plants is often greatly affected by active responses of the plants to what the fungi are doing. A plant's response may be favorable or unfavorable to the fungus, or it may not affect the fungus at all. This section will examine host responses that are of widespread occurrence and seem to have an effect on the biology of conidial fungi.

A. Favorable Responses

A universal feature of plant diseases caused by fungi is eventual lysis of host organelles, cells, or tissues. This process surely provides the bulk of nutrients for necrotrophic pathogens and a significant proportion of those for biotrophs, especially during sporulation (Aronescu, 1934; Coffey, 1975; Lazarovits and Higgins, 1976b; Luttrell, 1977). Historically, lost lysis has been attributed to the lytic activity of pathogen enzymes, but Pitt and Galpin (1973) and Wilson (1973) have drawn attention to the fact that plant cells have a potent lysosomal system of their own which functions normally in the recycling of molecules during plant growth and development. This largely constitutive autolytic system seems to play a role, along with lytic enzymes of pathogen origin, in lytic processes associated with disease and thus may contribute to nourishment of the pathogen. The idea has been little studied, but it is worthy of continued investigation.

Favorable host responses to biotrophic fungi are initially of a more constructive, or at least less destructive, nature than those described above. Perhaps the most obvious and dramatic of these is the positive chemotropic response of host fungi to certain mycoparasites (Barnett and Binder, 1973). The tropic response leads to the formation of specialized contact interfaces between the host and parasite.

Most of the research on biotrophic relationships has centered on rust and powdery mildew diseases of higher plants and has been reviewed by several authors (Wood, 1967; Coffey, 1975; Daly, 1976; Kosuge, 1978; Bushnell and Gay, 1978). Only a brief outline of the main events will be presented here. Photosynthesis is reduced in virtually all plant diseases caused by fungi, but in diseases caused by biotrophs the reduction is either very small or occurs very slowly. Thus the host is allowed to continue to fix carbon dioxide at a nearly normal rate during ramification. A universal plant response to infection is an increase in the metabolic rate. Daly (1976) suggests that for necrotrophic fungi this increase is coincident with necrosis and is related to resistance responses, whereas for biotrophic fungi the increase is generally greater and provides a favorable environment for their development.

This apparently favorable environment will now be examined in more detail. Various nutrients, such as inorganic ions, sugars, amino acids, organic acids, and alcohols, accumulate to unusually high levels in the infection court and adjacent host tissue. In some cases, an increase in host synthesis is implicated and is probably related to a concomitant enrichment of cytoplasm in host cells (Aist, 1976a). Translocation of nutrients from distant host tissue is clearly involved, as well as decreased translocation from or through the infection court (Daly, 1976; Bushnell and Gay, 1978). The movement of sugars to the infection courts has been attributed to conversion by the fungus of normal host sugars to polyols and trehalose, which are metabolized and stored only in the fungus. These sugar conversions keep the soluble sugar concentrations low near the parasite, thus promoting movement of host sugars along a concentration gradient (Daly, 1976; Bushnell and Gay, 1978). The accumulation and retention of sugars in host cells is commonly manifested as an increase in the number of starch grains, especially in the period just prior to sporulation (Coffey, 1975). This host response could contribute to the concentration gradient mentioned above. In rust infections, the accumulated starch grains are hydrolyzed during sporulation, and it is interesting that at least one rust fungus produces a chemical that can activate host amylase (Coffey, 1975). A very important concomitant of these host responses is an increase in membrane permeability of the infected host tissue (Coffey, 1975; Bushnell and Gay, 1978), which should facilitate nutrient transfer to the fungus.

Ouchi *et al.* (1976a) have recently documented an interesting and possibly pertinent phenomenon in barley powdery mildew–enhanced susceptibility. Ingress and ramification by a compatible race of the fungus were much higher near infection sites established by a prior inoculation with the same race than on previously uninoculated plants. If this phenomenon also occurs in the field, it could increase the rate of development of epidemics.

B. Unfavorable Responses

1. Disease Resistance: General Considerations

The development of conidial fungi on plants involves a continuum of morphological and biochemical changes that can, for convenience, be divided into a sequence of steps: germination, adhesion to the host surface, ingress, ramification, and egress. When a population of fungal units (conidia) is inoculated onto a plant surface, many will not complete the sequence; a certain proportion of them will cease development at each step (Stanbridge *et al.*, 1971; Ellingboe, 1972, 1978; Heath, 1974, 1977). These effects on fungal development, when attributable to factors in the plant, are the result of disease resistance. Similar results are obtained regardless of whether the plants are considered resistant or susceptible to the fungus, but on resistant plants fewer fungal units are able to complete the sequence.

The proportion of fungal units that passes a given developmental step may be governed by the particular set of resistance genes present in a given host (Hooker, 1967; Stanbridge *et al.*, 1971; Ellingboe, 1972, 1978; Heath, 1974, 1977). Thus, if one inoculates a number of different plants, varying in degree or type of resistance, with a genetically uniform population of pathogen conidia, different proportions of the fungal units will be unable to complete each developmental step on most or all of the plants. Some degree of correlation exists between the taxonomic level (e.g., host family, species, cultivar) at which this host–parasite specificity occurs and the developmental step at which the fungus is stopped. At the species level and above, virtually any step from germination through ramification may be affected, depending on the plant–fungus combination (Staub *et al.*, 1974; Heath, 1974, 1977; Hashioka and Ando, 1975; Hashioka and Kusadome, 1975; Johnson *et al.*, 1978), whereas below the species level the steps preceding ingress are usually not affected significantly (Ellingboe, 1972; Skipp and Deverall, 1972; Heath, 1974; Murray and Maxwell, 1975; Wheeler, 1977; Mendgen, 1978).

Finally, it should be mentioned that the degree of development of a fungus on a plant may vary with the age of the plant (Yarwood, 1959; Hooker, 1967), the organ that is attacked (Hooker, 1967), and the kind of host tissue involved (Dickinson, 1960; Wood, 1960). Moreover, various predisposing factors (e.g., seasonal and diurnal cycles, environmental parameters, wounds, and applied chemicals) can modify the plant before the fungus arrives, making the plant a more, or less, congenial host (Yarwood, 1959, 1976).

2. Correlates of Disease Resistance

The precise mechanisms in plants that restrict fungal development are elusive (Wood, 1972; Deverall, 1977). Rather than try to evaluate the evidence for or against the various mechanisms that have been proposed, I will describe some of the more commonly encountered host responses associated with resistance, suggest some developmental steps at which the fungus may be confronted with them, and discuss why the responses might be effective.

a. Wall Appositions. Living plant cells commonly react to an attacking fungus by redirecting cyclosis so that a cytoplasmic aggregate appears at the point of attack (Aist, 1976b). From the cytoplasmic aggregate a heterogeneous material is deposited in the paramural space—between the plasmalemma and cell wall (Fig. 6). This material is referred to as a papilla (Aist, 1976b) or wall apposition (Bracker and Littlefield, 1973).

In many cases of attempted penetration of plant cells by fungi, there is a definite coincidence of wall appositions and penetration failure (Aist, 1976b), suggesting that the appositions may somehow prevent ingress or restrict ramification. Recent experimental work has provided significant new correlations supporting this suggestion. Waterman *et al.* (1978) restricted papilla formation to one end of barley coleoptile cells by centrifugation and showed that a host cytoplasmic response, such as enhanced papilla formation, prevented penetration by *Erysiphe graminis*. Aist *et al.* (1979) found that ingress into barley coleoptile cells by *E. graminis* occurred much less frequently from appressoria located over preformed papillae than from those located over papilla-free sites on some coleoptiles. Vance and Sherwood (1977) used cycloheximide treatments and induced resistance to generate new correlations between papilla formation and resistance to penetration of reed canary grass by *Helminthosporium avenae*.

Exactly how wall appositions impede penetration is not known (Aist, 1976b), but possibilities include (1) presentation to the fungus of a physical barrier that is too strong to grow through and is resistant to enzymatic degradation, (2) prevention of molecular exchange between host and parasite at a critical time and place, and (3) retention of antifungal toxins, such as phenols and phytoalexins.

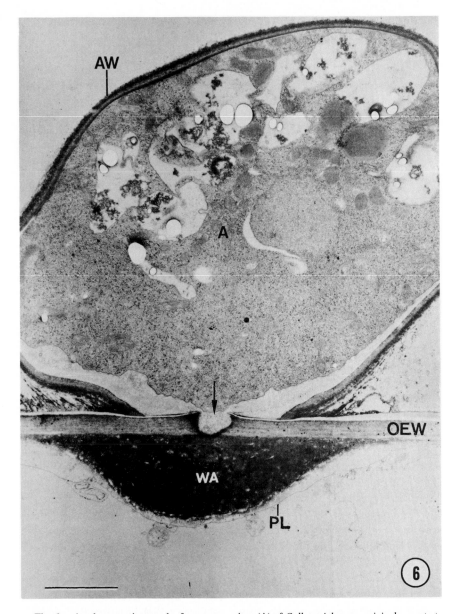

Fig. 6.   An electron micrograph of an appressorium (A) of *Colletotrichum graminicola* penetrating the outer epidermal wall (OEW) of an oat leaf. A wall apposition (WA) has been deposited between the host plasmalemma (PL) and the cell wall, beneath the young penetration peg (arrow). Note the thick appressorial wall (AW). ×20,000. Scale bar = 1 μm. (From Politis, 1976.)

b. Hypersensitive Responses.   Another widespread plant response to fungal attack is the rapid death of one or more cells in the immediate area of the fungus. This so-called hypersensitive response is further characterized by granulation of host cytoplasm, loss of membrane integrity, build-up of variously sized vacuolar inclusions, and host cell collapse (Maeda, 1970; Skipp and Deverall, 1972; Mercer et al., 1974; Politis, 1976; Brown, 1978). These rapid changes seem to result largely from decompartmentalization of the plant's lysosomal system (Wilson, 1973; Pitt and Galpin, 1973).

Although it may be triggered by the initial entry of the fungus, the hypersensitive response is usually associated with ramification, which is severely slowed and eventually stopped (Deverall, 1977). A myriad of biochemical changes, including phytoalexin production, are set in motion during this response, and which of these, if any, have an inhibitory effect on the fungus *in vivo* is not known. Wood (1967) pointed out that starvation would be an adequate explanation of the inhibitory effect of the hypersensitive response on obligate parasites, whereas the production and accumulation of phytoalexins may be more important in restricting the ramification of other fungi.

c. Phytoalexins.   These are antifungal compounds produced by plants in response to infection (Deverall, 1976). Several comprehensive reviews on phytoalexins are available (Kuć, 1976; Friend and Threlfall, 1976; Deverall, 1977). Chemically, these compounds are quite diverse and include terpenoids, isoflavonoids, naphthaldehydes, and polyacetylenes (Deverall, 1976). Most have low solubility in water and remain localized near their site of production in the plant. Several different phytoalexins are often produced by a plant in response to attack by a single fungus.

Phytoalexins inhibit conidial germination and growth of germ tubes and hyphae *in vitro*. *In vivo* their accumulation usually begins during early ramification and reaches a peak one or a few days after ingress; then a decline in concentration sets in (Kuć, 1976). Thus phytoalexins are generally suspected to inhibit ramification. They could also be implicated during ingress as a constituent of wall appositions, since many phytoalexins are present in healthy plants at low concentrations and could be preferentially packaged in papillae as a rapid host response. In addition, catechins produced by infected cotton plants are strong antisporulants (Bell and Stipanovic, 1978). The biochemical bases of the antifungal activity of phytoalexins are generally not known, but some apparently react with and denature proteins, thus altering membrane permeability and inhibiting enzyme activity (Bell and Stipanovic, 1978).

d. Cicatrices.   Akai (1959) cites numerous examples of conidial fungi whose ramification in plant tissues induces a layer of plant cells in advance of the fungus to begin dividing and produce several new layers of flattened cells that surround the infection court. These new cells typically become suberized and/or lignified, and ramification usually does not proceed beyond them. Similar

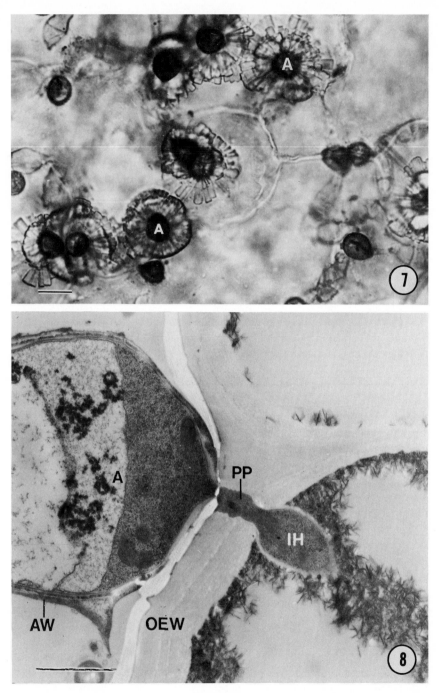

Figs. 7 and 8. Striate inclusions produced in nonhost epidermal cells in response to an attack by *Pyricularia oryzae*.

responses occur when plant tissues are mechanically wounded (Akai, 1959), and such modified surfaces are often highly resistant to infection (Wood, 1967).

Suberin, a complex mixture of polymerized fatty acids, is highly impermeable to polar substances. The low permeability of suberized layers to such substances may restrict not only the translocation of nutrients to the infection court but also the outward movement of metabolic by-products of the fungus—most conidial fungi are negatively autotropic (Wood, 1967).

Lignin is a highly polymerized amorphous material composed largely of methylated and hydroxylated propyl–benzene units. It gives mechanical strength to plant cell walls. Presumably, lignification of plant cell walls adds sufficient strength to make their traversal a most difficult task. This property of lignin is complemented by the fact that lignified plant walls are highly resistant to enzymatic attack. Friend (1976) suggests two possibilities to account for the latter effect: Lignin may form a physical barrier that prevents the enzymes from contacting their substrates, or it may bind chemically to cell wall polysaccharides, rendering them unsuitable as enzyme substrates.

e. Striate Inclusions. I draw attention to this host response because it represents a little known, but potentially significant, resistance response. Hashioka and Kusadome (1975) have reported that the epidermal cells of several nonhost plants react to attack by *Pyricularia oryzae* by producing large aggregates of striate or crystalline intracellular inclusions. These inclusions are localized primarily beneath appressoria in some cases (Fig. 7), or surrounding intracellular infection hyphae in others (Fig. 8). They were detected 2–5 days after inoculation, when ingress and ramification were noticeably inhibited. Huang and Tinline (1976) observed similar structures in wheat cells attacked by *Cochliobolus sativus*. Although papillae produced in response to the fungus were apparently grown through, penetration pegs never emerged from papillae that were ensheathed by striate inclusions. The possibility that striate inclusions reflect extremely high concentrations of antifungal compounds that are localized in the immediate vicinity of the fungi is intriguing and should be investigated.

f. Induced Resistance. An interaction between a plant and a fungus with which it is incompatible very commonly induces a change in the plant that makes it resistant to fungi which are normally able to colonize it (Matta, 1971; Yarwood, 1976). This response may even be induced by a compatible fungus (Kuć *et al.*, 1975), heat-killed conidia (Bell and Presley, 1969), and germination fluids

---

Fig. 7. Light micrograph showing inclusions clustered at sites of attachment of the dark, melanized appressoria (A) to epidermal cells of *Oxalis corniculata* (Oxalidaceae). Ca. ×1000. Scale bar = ca. 10 μm. (From Hashioka and Kusadome, 1975.)

Fig. 8. Electron micrograph illustrating fine, needlelike inclusions coating the intracellular hypha (IH) and the inner surface of the epidermal cell wall of *Sisyrinchium atlanticum* (Iridaceae). Note the thick wall (AW) of the appressorium (A) and the penetration peg (PP) which has traversed the outer epidermal wall (OEW). ×11,000. Scale bar = 2 μm. (From Hashioka and Ando, 1975.)

(Sinha and Das, 1972). Prominent examples of conidial fungi shown to induce resistance include anamorphic phases of both rust (Johnson and Allen, 1975; Tani *et al.*, 1975) and powdery mildew fungi (Ouchi *et al.*, 1976b), *Colletotrichum* (Kuć *et al.*, 1975), *Helminthosporium* (Sinha and Das, 1972), and *Verticillium* (Bell and Presley, 1969). In most cases, induced resistance is limited to the plant surface area inoculated by the inducing fungus, or its immediate vicinity, and is short-lived, lasting only a few days. Induced resistance may be expressed at ingress (Vance and Sherwood, 1977; Richmond and Kuć, 1978) or ramification (Skipp and Deverall, 1973; Ouchi *et al.*, 1976b), with subsequent reduction, delay, or elimination of sporulation (Johnson and Allen, 1975). The mechanisms of induced resistance are not clearly defined as yet, but there is evidence that phytoalexins (Bell and Presley, 1969) or wall appositions (Vance and Sherwood, 1977) may be involved. Richmond and Kuć (1978) inoculated both control and induced plants after removing their epidermal layers and obtained evidence that the induced resistance in cucumber to *Colletotrichum lagenarium* resided in the epidermis. Vance and Sherwood (1977) came to the same conclusion concerning resistance induced in reed canary grass by *Botrytis cinerea*.

Studies on induced resistance may provide some insight into basic disease resistance mechanisms and may eventually prove useful in controlling plant diseases in the field. So far, the most promising candidate among diseases caused by conidial fungi is cucurbit anthracnose, because in this case induced resistance is systemic and lasts at least 10 weeks (Kuć *et al.*, 1975). Caruso and Kuć (1977) have reported that cucumber and watermelon can be systemically protected against *C. lagenarium* in the field by prior inoculation with conidia of the same fungus. The utility of this approach as a practical control measure awaits yield trials on protected plants in the field.

This demonstration of induced resistance under field conditions raises the further possibility that natural infections in the field could operate to reduce the rate of development of epidemics by inducing resistance of the infected plants to subsequent attacks by pathogenic fungi. Johnson and Allen (1975) suggested that induced resistance could significantly reduce the rate of development of natural rust epidemics on resistant multiline varieties by delaying and reducing sporulation. Perhaps artificial inoculations of resistant multilines with avirulent or weakly virulent pathogen races would further enhance such an effect in the field.

Resistance of a plant to a fungus is generally associated with a number of host factors, both constitutive and inducible, that are expressed at various stages of fungal development. Many of these factors have an obvious potential for inhibiting the fungus, and it would be surprising if in most plant–fungus encounters the failure of the fungus to develop fully were not due to more than one host factor.

Resistance of plants to true vascular wilt fungi such as *Fusarium* and *Verticillium* spp. is a good illustration of these points. Talboys (1972) recognized two phases of resistance. The first phase occurs before the fungus reaches the vascular system and may be associated with a hypersensitive response in the epidermis (Tjamos and Smith, 1975), formation of wall appositions by root hairs (Griffiths and Lim, 1964) or cortical cells, or refractiveness of the endodermis to penetration. The second phase of resistance occurs in the vascular system and involves various combinations of factors that will be considered individually.

Vessel end walls serve as primary trapping sites for conidia that are swept up through the vascular system by the transpiration stream (Beckman *et al.*, 1976; Beckman and Keller, 1977). This delaying action is thought to provide valuable time for the host to activate resistance responses in advance of the invader.

One such response is the production of *vascular gels* (Vandermolen *et al.*, 1977) that trap second- and third-generation conidia (Figs. 9 and 10), preventing their further spread upward in the vessels. These gels arise by the swelling of perforation plates, end walls, and pit membranes (Fig. 10) and, at least in banana, are composed of pectins, calcium pectate, hemicellulose, and traces of protein (Beckman and Zaroogian, 1967).

Tyloses are wall-bounded protrusions of parenchyma cell protoplasts into vessel elements. They often form in abundance in advance of the ramifying fungus, sometimes completely occluding the vessel lumina. Their rate and extent of development is correlated with the resistance of tomato cultivars to *Fusarium* and *Verticillium* (Beckman *et al.*, 1972; Tjamos and Smith, 1975) and of cotton to *Verticillium* (Mace, 1978). Tyloses theoretically play a dual role here by restricting transport of conidia by mass flow and creating a stagnant environment within the vessel where phytoalexin accumulation can be facilitated (Tjamos and Smith, 1975).

Both cotton and tomato accumulate *phytoalexins* in response to vascular invasions. These compounds are believed to restrict ramification by inhibiting both growth and sporulation of the inducing fungi (Tjamos and Smith, 1975; Bell and Stipanovic, 1978; Mace, 1978).

Another biochemical response is *phenol infusion* (Mace, 1963; Mace *et al.*, 1972), in which phenolic compounds are released from scattered phenol-storing cells, diffuse into the xylem vessels, and become oxidized and polymerized. Gels and tyloses also become infused by this process. Besides the direct toxic effects phenols and their oxidation products can have on fungi (Schönbeck and Schlösser, 1976), phenol infusion may have a further resistive effect by rendering cell walls more rigid and less susceptible to enzymatic degradation (Beckman *et al.*, 1974). Resistance of tomato plants to *Fusarium* can be enhanced by controlling their natural production of phenols (Carrasco *et al.*, 1978).

All five of the host factors discussed above are found in various combinations in susceptible as well as resistant plants. How, then, could they be involved in limiting the ramification of vascular wilt fungi in resistant plants? The key apparently lies in the timing of the host responses, since hyphal growth and initial

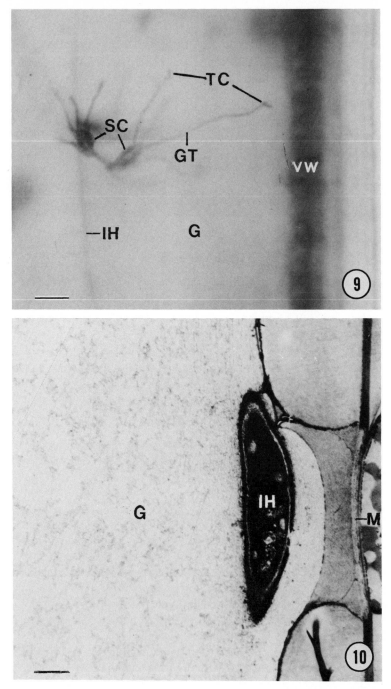

Figs. 9 and 10. Vascular gels in resistant banana roots infected by *Fusarium oxysporum*.

spread to the primary trapping sites may be similar in both resistant and suscepti-ble plants (Beckman *et al.*, 1972, 1976). Resistant cultivars generally respond faster than susceptible ones and thus are able to prepare apparent chemical and physical barriers in advance of the invader (Beckman *et al.*, 1972; Bell and Stipanovic, 1978).

ACKNOWLEDGMENT

I sincerely appreciate the advice and opinions of Drs. W. E. Fry, H. C. Hoch, and R. P. Korf during the preparation of this chapter.

REFERENCES

Abawi, G. S., and Lorbeer, J. W. (1971). Pathological histology of four onion cultivars infected by *Fusarium oxysporum* f.sp. *cepae*. *Phytopathology* **61**, 1164–1169.

Aist, J. R. (1976a). Cytology of penetration and infection—Fungi. *In* "Physiological Plant Pathol-ogy" (R. Heitefuss and P. H. Williams, eds.), pp. 197–221. Springer-Verlag, Berlin and New York.

Aist, J. R. (1976b). Papillae and related wound plugs of plant cells. *Annu. Rev. Phytopathol.* **14**, 145–163.

Aist, J. R., Kunoh, H., and Israel, H. W. (1979). Challenge appressoria of *Erysiphe graminis* fail to breach preformed papillae of a compatible barley cultivar. *Phytopathology* **69**, 1245–1250.

Akai, S. (1959). Histology of defense in plants. *In* "Plant Pathology: An Advanced Treatise" (J. G. Horsfall and A. E. Dimond, eds.), Vol. I, pp. 391–434. Academic Press, New York.

Akai, S., Horino, O., Fukutomi, M., Nakata, A., Kunoh, H., and Shiraishi, M. (1971). Cell wall reaction to infection and resulting change in cell organelles. *In* "Morphological and Biochemi-cal Events in Plant-Parasite Interaction" (S. Akai and S. Ouchi, eds.), pp. 329–347. Phytopathological Society of Japan, Tokyo.

Alexopoulos, C. J. (1962). "Introductory Mycology." Wiley, New York.

Allen, P. J. (1976). Control of spore germination and infection structure formation in the fungi. *In* "Physiological Plant Pathology" (R. Heitefuss and P. H. Williams, eds.), pp. 51–85. Springer-Verlag, Berlin and New York.

Allen, R. F. (1923). Cytological studies of infection of Baart, Kanred, and Mindum wheat by *Puccinia graminis tritici* forms III and XIX. *J. Agri. Res.* **26**, 571–604.

Fig. 9. Light micrograph showing the lumen of a vessel above a trapping site (vessel end wall). The lumen is filled with a gel (G) that has retained second-generation conidia (SC) produced from an intracellular hypha (IH) which grew up from the trapping site. The second-generation conidia germi-nated in place and began to produce third-generation conidia (TC) at the tips of their germ tubes (GT). The vessel wall (VW) is at the right. ×900. Scale bar = 10 μm.

Fig. 10. An electron micrograph illustrating the fine structure of the vascular gel (G) which arises from erosion of vessel wall regions, including the pit membranes (M). An intracellular hypha (IH) is embedded in the gel. Note the two gel layers of different electron density between the hypha and the pit membrane. ×4600. Scale bar = 2 μm. (From Beckman, 1979.)

Aronescu, A. (1934). *Diplocarpon rosae:* From spore germination to haustorium formation. *Bull. Torry Bot. Club* **61,** 291–329.

Baker, C. J., and Bateman, D. F. (1978). Cutin degradation by plant pathogenic fungi. *Phytopathology* **68,** 1577–1584.

Barnett, H. L., and Binder, F. L. (1973). The fungal host-parasite relationship. *Annu. Rev. Phytopathol.* **11,** 273–292.

Barnett, H. L., and Hunter, B. B. (1972). "Illustrated Genera of Imperfect Fungi." Burgess, Minneapolis, Minnesota.

Bateman, D. F., and Basham, H. G. (1976). Degradation of plant cell walls and membranes by microbial enzymes. *In* "Physiological Plant Pathology" (R. Heitefuss and P. H. Williams, eds.), pp. 316–355. Springer-Verlag, Berlin and New York.

Beckman, C. H. (1979). Physical defenses triggered by the invader. *In* "Plant Disease: An Advanced Treatise" (J. G. Horsfall and E. B. Cowling, eds.), Vol. 5, Chapter 13. Academic Press, New York.

Beckman, C. H., and Keller, J. L. (1977). Vessels do end! *Phytopathology* **67,** 954–956.

Beckman, C. H., and Zaroogian, G. E. (1967). Origin and composition of vascular gel in infected banana roots. *Phytopathology* **57,** 11–13.

Beckman, C. H., Elgersma, D. M., and MacHardy, W. E. (1972). The localization of *Fusarium* infections in the vascular tissue of single-dominant-gene resistant tomatoes. *Phytopathology* **62,** 1256–1260.

Beckman, C. H., Mueller, W. C., and Mace, M. E. (1974). The stabilization of artificial and natural cell wall membranes by phenolic infusion and its relation to wilt disease resistance. *Phytopathology* **64,** 1214–1220.

Beckman, C. H., Vandermolen, G. E., Mueller, W. C., and Mace, M. E. (1976). Vascular structure and distribution of vascular pathogens in cotton. *Physiol. Plant Pathol.* **9,** 87–94.

Bell, A. A., and Presley, J. T. (1969). Heat-inhibited or heat-killed conidia of *Verticillium albo-atrum* induce disease resistance and phytoalexin synthesis in cotton. *Phytopathology* **59,** 1147–1151.

Bell, A. A., and Stipanovic, R. D. (1978). Biochemistry of disease and pest resistance in cotton. *Mycopathologia* **65,** 91–106.

Berg, B., and Pettersson, G. (1977). Location and formation of cellulases in *Trichoderma viride. J. Appl. Bacteriol.* **42,** 65–75.

Bracker, C. E. (1968). Ultrastructure of the haustorial apparatus of *Erysiphe graminis* and its relationship to the epidermal cell of barley. *Phytopathology* **58,** 12–30.

Bracker, C. E., and Littlefield, L. J. (1973). Structural concepts of host-pathogen interfaces. *In* "Fungal Pathogenicity and the Plant's Response" (R. J. W. Byrde and C. V. Cutting, eds.), pp. 59–318. Academic Press, New York.

Brown, G. E. (1977). Ultrastructure of penetration of ethylene-degreened Robinson tangerines by *Colletotrichum gloeosporioides. Phytopathology* **67,** 315–320.

Brown, G. E. (1978). Hypersensitive response of orange-colored Robinson tangerines to *Colletotrichum gloeosporioides* after ethylene treatment. *Phytopathology* **68,** 700–706.

Bushnell, W. R. (1972). Physiology of fungal haustoria. *Annu. Rev. Phytopathol.* **10,** 151–176.

Bushnell, W. R., and Gay, J. (1978). Accumulation of solutes in relation to the structure and function of haustoria in powdery mildews. *In* "The Powdery Mildews" (D. M. Spencer, ed.), pp. 183–235. Academic Press, New York.

Camp, R. R., and Whittingham, W. F. (1972). Host-parasite relationships in sooty blotch disease of white clover. *Am. J. Bot.* **59,** 1057–1067.

Carlile, M. J. (1966). The orientation of zoospores and germ-tubes. *In* "The Fungus Spore" (M. F. Madelin, ed.), pp. 175–186. Butterworth, London.

Carpenter, S. E. (1976). Taxonomy, morphology and ontogeny of *Gelatinodiscus flavidus.* Mycotaxon **3,** 209–232.

Carrasco, A., Boudet, A. M., and Marigo, G. (1978). Enhanced resistance of tomato plants to *Fusarium* by controlled stimulation of their natural phenolic production. *Physiol. Plant Pathol.* **12**, 225–232.

Caruso, F. L., and Kuć, J. (1977). Field protection of cucumber, watermelon, and muskmelon against *Colletotrichum lagenarium* by *Colletotrichum lagenarium*. *Phytopathology* **67**, 1290–1292.

Clark, C. A., and Lorbeer, J. W. (1976). Comparative histopathology of *Botrytis squamosa* and *B. cinerea* on onion leaves. *Phytopathology* **66**, 1279–1289.

Coffey, M. D. (1975). Obligate parasites of higher plants, particularly rust fungi. *Symp. Soc. Exp. Biol.* **29**, 297–323.

Cutter, E. G. (1976). Aspects of the structure and development of the aerial surfaces of higher plants. *In* "Microbiology of Aerial Plant Surfaces" (C. H. Dickinson and T. F. Preece, eds.), pp. 1–40. Academic Press, New York.

Daly, J. M. (1976). The carbon balance of diseased plants: Changes in respiration, photosynthesis and translocation. *In* "Physiological Plant Pathology" (R. Heitefuss and P. H. Williams, eds.), pp. 450–479. Springer-Verlag, Berlin and New York.

Day, P. R., and Scott, K. J. (1973). Scanning electron microscopy of fresh material of *Erysiphe graminis* f.sp. *hordei*. *Physiol. Plant Pathol.* **3**, 433–435.

Delon, R., Kiffer, E., and Mangenot, F. (1977). Ultrastructural study of host-parasite interactions. II. Decay of lettuce caused by *Botrytis cinerea* and phyllosphere bacteria. *Can. J. Bot.* **55**, 2463–2470.

Deverall, B. J. (1976). Current perspectives in research on phytoalexins. *In* "Biochemical Aspects of Plant-Parasite Relationships" (J. Friend and D. R. Threlfall, eds.), pp. 207–223. Academic Press, New York.

Deverall, B. J. (1977). "Defense Mechanisms of Plants." Cambridge Univ. Press, London and New York.

De Waard, M. A. (1971). Germination of powdery mildew conidia *in vitro* on cellulose membranes. *Neth. J. Plant Pathol.* **77**, 6–13.

Dickinson, C. H., and Preece, T. F., eds. (1976). "Microbiology of Aerial Plant Surfaces." Academic Press, New York.

Dickinson, S. (1960). The mechanical ability to breach the host barriers. *In* "Plant Pathology: An Advanced Treatise" (J. G. Horsfall and A. E. Dimond, eds.), Vol. 2, pp. 203–232. Academic Press, New York.

Dickinson, S. (1971). Studies in the physiology of obligate parasitism. VIII. An analysis of fungal responses to thigmotropic stimuli. *Phytopathol. Z.* **70**, 62–70.

Duddridge, J. A., and Sargent, J. A. (1978). A cytochemical study of lipolytic activity in *Bremia lactucae* Regel. during germination of the conidium and penetration of the host. *Physiol. Plant Pathol.* **12**, 289–296.

Dunkle, L. D., and Allen, P. J. (1971). Infection structure differentiation by wheat stem rust uredospores in suspension. *Phytopathology* **61**, 649–652.

Elgersma, D. M., MacHardy, W. E., and Beckman, C. H. (1972). Growth and distribution of *Fusarium oxysporum* f.sp. *lycopersici* in near-isogenic lines of tomato resistant or susceptible to wilt. *Phytopathology* **62**, 1232–1237.

Ellingboe, A. H. (1968). Inoculum production and infection by foliage pathogens. *Annu Rev. Phytopathol.* **6**, 317–330.

Ellingboe, A. H. (1972). Genetics and physiology of primary infection by *Erysiphe graminis*. *Phytopathology* **62**, 401–406.

Ellingboe, A. H. (1978). A genetic analysis of host-parasite interactions. *In* "The Powdery Mildews" (D. M. Spencer, ed.), pp. 159–181. Academic Press, New York.

Emmett, R. W., and Parbery, D. G. (1975). Appressoria. *Annu. Rev. Phytopathol.* **13**, 147–167.

Friend, J. (1976). Lignification in infected tissue. *In* "Biochemical Aspects of Plant-Parasite Rela-tionships" (J. Friend and D. R. Threlfall, eds.), pp. 391–403. Academic Press, New York.

Friend, J., and Threlfall, D. R., eds. (1976). "Biochemical Aspects of Plant-Parasite Relation-ships." Academic Press, New York.

Griffiths, D. A. (1973). An electron microscopic study of host reaction in roots following invasion by *Verticillium dahliae*. *Shokubutsu Byogai Kenkyu* **8**, 147–154.

Griffiths, D. A., and Lim, W. C. (1964). Mechanical resistance in root hairs to penetration by species of vascular wilt fungi. *Mycopathol. Mycol. Appl.* **24**, 103–112.

Grover, R. K. (1971). Participation of host exudate chemicals in appressorium formation by *Col-letotrichum piperatum*. *In* "Ecology of Leaf Surface Micro-Organisms" (T. F. Preece and C. H. Dickinson, eds.), pp. 509–518. Academic Press, New York.

Guggenheim, R., and Harr, J. (1978). Contributions to the biology of *Hemileia vastatrix*. II. SEM-investigations on sporulation of *Hemileia vastatrix* on leaf surfaces of *Coffea arabica*. *Phytopathol. Z.* **92**, 97–100.

Harvey, I. C. (1977). Studies of the infection of lupin leaves by *Pleiochaeta setosa*. *Can. J. Bot.* **55**, 1261–1275.

Hashioka, Y., and Ando, N. (1975). Fine structure of the rice blast. XIII. Ultrastructure of *Pyricularia oryzae*—Nonhost interfaces. *Res. Bull. Fac. Agric., Gifu Univ.* **38**, 39–47.

Hashioka, Y., and Kusadome, H. (1975). Fine structure of the rice blast. XII. The mode of pseudoin-fection of *Pyricularia oryzae*. Cav. to the non-host plants. *Res. Bull. Fac. Agric., Gifu Univ.* **38**, 29–37.

Hashioka, Y., and Nakai, Y. (1974). Fine structure of the rice blast. XI. Outthrust of conidiophore cells of *Pyricularia oryzae*, etc., through an epidermal outer wall of a host leaf. *Res. Bull Fac. Agric., Gifu Univ.* **36**, 9–18.

Hashioka, Y., Ikegami, H., Horino, O., and Kamei, T. (1967). Fine structure of the rice blast. II. Electronmicrographs of the initial infection. *Res. Bull. Fac. Agric., Gifu Univ.* **24**, 78–90.

Heath, M. C. (1974). Light and electron microscope studies of the interactions of host and non-host plants with cowpea rust—*Uromyces phaseoli* var. *vignae*. *Physiol. Plant Pathol.* **4**, 403–414.

Heath, M. C. (1977). A comparative study of non-host interactions with rust fungi. *Physiol. Plant Pathol.* **10**, 73–88.

Heath, M. C., and Wood, R. K. S. (1969). Leaf spots induced by *Ascochyta pisi* and *Mycos-phaerella pinodes*. *Ann. Bot. (London)* **33**, 657–670.

Hess, W. M. (1969). Ultrastructure of onion roots infected with *Pyrenochaeta terrestris*, a fungus parasite. *Am. J. Bot.* **56**, 832–845.

Higgins, V. J. (1966). Phytoalexin production by alfalfa in response to infection by *Stemphylium botryosum*, *S. loti*, *Helminthosporium turcicum*, and *Colletotrichum phomoides*. M.S. Thesis, Cornell University, Ithaca, New York.

Higgins, V. J., and Millar, R. L. (1968). Phytoalexin production by alfalfa in response to infection by *Colletotrichum phomoides*, *Helminthosporium turcicum*, *Stemphylium loti*, and *S. bot-ryosum*. *Phytopathology* **58**, 1377–1383.

Hirata, K. (1967). Notes on haustoria, hyphae, and conidia of the powdery mildew fungus of barley *Erysiphe graminis* f.sp. *hordei*. *Mem. Fac. Agric., Niigata Univ.* **6**, 207–259.

Hoch, H. C. (1977a). Mycoparasitic relationships; *Gonatobotrys simplex* parasitic on *Alternaria tenuis*. *Phytopathology* **67**, 309–314.

Hoch, H. C. (1977b). Mycoparasitic relationships. III. Parasitism of *Physalospora obtusa* by *Cal-carisporium parasiticum*. *Can. J. Bot.* **55**, 198–207.

Hoch, H. C. (1978). Mycoparasitic relationships. IV. *Stephanoma phaeospora* parasitic on a species of *Fusarium*. *Mycologia* **70**, 370–379.

Hooker, A. L. (1967). The genetics and expression of resistance in plants to rusts of the genus *Puccinia*. *Annu. Rev. Phytopathol.* **5**, 163–182.

Huang, H. C., and Tinline, R. D. (1976). Histology of *Cochliobolus sativus* infection in subcrown internodes of wheat and barley. *Can. J. Bot.* **54**, 1344-1354.

Hughes, S. J. (1953). Some foliicolous hyphomycetes. *Can. J. Bot.* **31**, 560-576.

Hunt, P. (1968). Cuticular penetration by germinating uredospores. *Trans. Br. Mycol. Soc.* **51**, 103-112.

Ingram, D. S., Sargent, J. A., and Tommerup, I. C. (1976). Structural aspects of infection by biotropic fungi. *In* "Biochemical Aspects of Plant-Parasite Relationships" (J. Friend and D. R. Threlfall, eds.), pp. 43-78. Academic Press, New York.

Johnson, L. E. B., Zeyen, R. J., and Bushnell, W. R. (1978). Defense patterns in inappropriate higher plant species in two powdery mildew fungi. *Phytopathol. News* **12**, 89 (abstr.).

Johnson, R., and Allen, D. J. (1975). Induced resistance to rust diseases and its possible role in the resistance of multiline varieties. *Ann. Appl. Biol.* **80**, 359-363.

Jones, P., and Ayers, P. G. (1974). *Rhynchosporium* leaf blotch of barley studied during the subcuticular phase by electron microscopy. *Physiol. Plant Pathol.* **4**, 229-233.

Knox-Davies, P. S. (1974). Penetration of maize leaves by *Helminthosporium turcicum*. *Phytopathology* **64**, 1468-1470.

Kosuge, T. (1978). The capture and use of energy of diseased plants. *In* "Plant Disease: An Advanced Treatise" (J. G. Horsfall and E. B. Cowling, eds.), Vol. 3, pp. 85-116. Academic Press, New York.

Kuć, J. (1976). Phytoalexins. *In* "Physiological Plant Pathology" (R. Heitefuss and P. H. Williams, eds.), pp. 632-652. Springer-Verlag, Berlin and New York.

Kuć, J., Shockley, G., and Kearney, K. (1975). Protection of cucumber against *Colletotrichum lagenarium* by *Colletotrichum lagenarium*. *Physiol. Plant Pathol.* **7**, 195-199.

Lazarovits, G., and Higgins, V. J. (1976a). Histological comparison of *Cladosporium fulvum* race 1 on immune, resistant, and susceptible tomato varieties. *Can. J. Bot.* **54**, 224-234.

Lazarovits, G., and Higgins, V. J. (1976b). Ultrastructure of susceptible, resistant, and immune reactions of tomato to races of *Cladosporium fulvum*. *Can. J. Bot.* **54**, 235-249.

Lewis, B. G., and Day, J. R. (1972). Behavior of uredospore germ-tubes of *Puccinia graminis tritici* in relation to the fine structure of wheat leaf surfaces. *Trans. Br. Mycol. Soc.* **58**, 139-145.

Lewis, D. H. (1973). Concepts in fungal nutrition and the origin of biotrophy. *Biol. Rev. Cambridge Philos. Soc.* **48**, 261-278.

Littlefield, L. J., and Bracker, C. E. (1972). Ultrastructural specialization at the host-pathogen interface in rust-infected flax. *Protoplasma* **74**, 271-305.

Locci, R., and Bisiach, M. (1971). Scanning electron microscopy of the invasion of leaf tissues by the apple scab fungus. *Riv. Patol. Veg.* [4] **7**, 15-29.

Locci, R., and Quaroni, S. (1971). Scanning electron microscopy detected maize leaf modifications caused by *Helminthosporium maydis* and other microorganisms. *Riv. Patol. Veg.* [4] **7**, 109-125.

Locci, R., and Quaroni, S. (1974). Studies on powdery mildews. II. Investigations on grapevine leaf tissue infection by *Oidium tuckeri*. *Riv. Patol. Veg.* [4] **10**, 343-353.

Luttrell, E. S. (1963). Taxonomic criteria in *Helminthosporium*. *Mycologia* **55**, 643-674.

Luttrell, E. S. (1964). Morphology of *Trichometasphaeria turcica*. *Am. J. Bot.* **51**, 213-219.

Luttrell, E. S. (1976). Ovarian infection of *Sporobolus poiretii* by *Bipolaris ravenelii*. *Phytopathology* **66**, 260-268.

Luttrell, E. S. (1977). The disease cycle and fungus-host relationships in dallisgrass ergot. *Phytopathology* **67**, 1461-1468.

MacDonald, W. L., and McNabb, H. S., Jr. (1970). Fine-structural observations of the growth of *Ceratocystis ulmi* in elm xylem tissue. *BioScience* **20**, 1060-1061.

Mace, M. E. (1963). Histochemical localization of phenols in healthy and diseased banana roots. *Physiol. Plant.* **16**, 915-925.

Mace, M. E. (1978). Contributions of tyloses and terpenoid aldehyde phytoalexins to *Verticillium* wilt resistance in cotton. *Physiol. Plant Pathol.* **12**, 1–11.

Mace, M. E., Veech, J. A., and Beckman, C. H. (1972). *Fusarium* wilt of susceptible and resistant isolines: Histochemistry of vascular browning. *Phytopathology* **62**, 651–654.

McKeen, W. E. (1974). Mode of penetration of epidermal cell walls of *Vicia faba* by *Botrytis cinerea*. *Phytopathology* **64**, 461–467.

McKeen, W. E., and Rimmer, S. R. (1973). Initial penetration process in powdery mildew infection of susceptible barley leaves. *Phytopathology* **63**, 1049–1053.

Maeda, K. M. (1970). An ultrastructural study of *Venturia inaequalis* (Cke.) Wint. infection of *Malus* hosts. M. S. Thesis, Purdue University, Lafayette, Indiana.

Mangin, M. L. (1899). Sur le Piétin ou maladie du pied du lilé. *Bull. Soc. Mycol.* **15**, 210–239.

Martin, J. T. (1964). Role of cuticle in the defense against plant disease. *Annu. Rev. Phytopathol.* **2**, 81–100.

Mason, D. L. (1973). Host-parasite relations in the *Gibberidea* disease of *Helianthus strumosus*. *Mycologia* **65**, 1158–1170.

Mason, D. L., and Backus, M. P. (1969). Host-parasite relations in spot anthracnose of *Desmodium*. *Mycologia* **61**, 1124–1141.

Mason, D. L., and Wilson, C. L. (1978). Fine-structure analysis of host-parasite relations in the spot anthracnose of *Desmodium*. *Phytopathology* **68**, 65–73.

Matta, A. (1971). Microbial penetration and immunization of uncongenial host plants. *Annu. Rev. Phytopathol.* **9**, 387–410.

Mendgen, K. (1978). Der Infektionsverlauf von *Uromyces phaseoli* bei anfälligen und resistenten Bohnensorten. *Phytopathol. Z.* **93**, 295–313.

Mercer, P. C., Wood, R. K. S., and Greenwood, A. D. (1971). Initial infection of *Phaseolus vulgaris* by *Colletotrichum lindemuthianum*. *In* "Ecology of Leaf Surface Micro-Organisms" (T. F. Preece and C. H. Dickinson, eds.), pp. 381–389. Academic Press, New York.

Mercer, P. C., Wood, R. K. S., and Greenwood, A. D. (1974). Resistance to anthracnose of French bean. *Physiol. Plant Pathol.* **4**, 291–306.

Mercer, P. C., Wood, R. K. S., and Greenwood, A. D. (1975). Ultrastructure of the parasitism of *Phaseolus vulgaris* by *Colletotrichum lindemuthianum*. *Physiol. Plant Pathol.* **5**, 203–214.

Murray, G. M., and Maxwell, D. P. (1975). Penetration of *Zea mays* by *Helminthosporium carbonum*. *Can. J. Bot.* **53**, 2872–2883.

Myers, D. F., and Fry, W. E. (1978). The development of *Gloeocercospora sorghi* in *Sorghum*. *Phytopathology* **68**, 1147–1155.

Nicholson, R. L., Kuć, J., and Williams, E. B. (1972). Histochemical demonstration of transitory esterase activity in *Venturia inequalis*. *Phytopathology* **62**, 1242–1247.

Nusbaum, C. J., and Keitt, G. W. (1938). A cytological study of host-parasite relations of *Venturia inaequalis* on apple leaves. *J. Agric. Res.* **56**, 595–618.

Ouchi, S., Hibino, C., and Oku, H. (1976a). Effect of earlier inoculation on the establishment of a subsequent fungus as demonstrated in powdery mildew of barley by a triple inoculation procedure. *Physiol. Plant Pathol.* **9**, 25–32.

Ouchi, S., Oku, H., and Hibino, C. (1976b). Localization of induced resistance and susceptibility in barley leaves inoculated with the powdery mildew fungus. *Phytopathology* **66**, 901–905.

Paddock, W. C. (1953). Histological study of suscept-pathogen relationships between *Helminthosporium victoriae* and seedling oat leaves. "*Cornell Univ. Agric. Exp. Stn.*" Mem. **315**.

Patton, R. F., and Spear, R. N. (1978). Scanning electron microscopy of infection of Scotch pine needles by *Scirrhia acicola*. *Phytopathology* **68**, 1700–1704.

Paus, F., and Raa, J. (1973). An electron microscope study by infection and disease development in cucumber hypocotyls inoculated with *Cladosporium cucumerinum*. *Physiol. Plant Pathol.* **3**, 461–464.

Pearson, N. L. (1931). Parasitism of *Gibberella saubinetii* on corn seedlings. *J. Agric. Res.* **43,** 569–596.

Peterson, G. W., and Walla, J. A. (1978). Development of *Dothistroma pini* upon and within needles of Austrian and Ponderosa pines in eastern Nebraska. *Phytopathology* **68,** 1422–1430.

Phipps, P. M., and Stipes, R. J. (1976). Histopathology of *Mimosa* infected with *Fusarium oxysporum* f.sp. *perniciosum*. *Phytopathology* **66,** 839–843.

Pitt, D., and Galpin, M. (1973). Role of lysosomal enzymes in pathogenicity. *In* "Fungal Pathogenicity and the Plant's Response" (R. J. W. Byrde and C. V. Cutting, eds.), pp. 449–467. Academic Press, New York.

Plotnikova, Y. M., Littlefield, L. J., and Miller, J. D. (1979). Scanning electron microscopy of the haustorium-host interface regions in wheat infected with *Puccinia graminis* f.sp. *tritici*. *Physiol. Plant Pathol.* **14,** 37–39.

Politis, D. J. (1976). Ultrastructure of penetration by *Colletotrichum graminicola* of highly resistant oat leaves. *Physiol. Plant Pathol.* **8,** 117–122.

Politis, D. J., and Wheeler, H. (1973). Ultrastructural study of penetration of maize leaves by *Colletotrichum graminicola*. *Physiol. Plant Pathol.* **3,** 465–471.

Rathaiah, Y. (1976). Infection of sugarbeet by *Cercospora beticola* in relation to stomatal condition. *Phytopathology* **66,** 737–740.

Rathaiah, Y. (1977a). Spore germination and mode of cotton infection by *Ramularia areola*. *Phytopathology* **67,** 351–357.

Rathaiah, Y. (1977b). Stomatal tropism of *Cercospora beticola* in sugarbeet. *Phytopathology* **67,** 358–362.

Richmond, S., and Kuć, J. (1978). Penetration of the epidermis of cucumber by *Colletotrichum lagenarium* is inhibited in plants previously inoculated with the pathogen. *Phytopathol. News* **12,** 182 (abstr.).

Salmon, E. S. (1906). On *Oidiopsis taurica* (Lév.), an endophytic member of the Erysiphaceae. *Ann. Bot. (London)* **20,** 187–200.

Sargent, J. A., Tommerup, I. C., and Ingram, D. S. (1973). The penetration of a susceptible lettuce variety by the downy mildew fungus *Bremia lactucae* Regel. *Physiol. Plant Pathol.* **3,** 231–239.

Schönbeck, F., and Schlösser, E. (1976). Preformed substances as potential protectants. *In* "Physiological Plant Pathology" (R. Heitefuss and P. H. Williams, eds.), pp. 653–678. Springer-Verlag, Berlin and New York.

Sharman, S., and Heale, J. B. (1977). Penetration of carrot roots by the grey mould fungus *Botrytis cinerea* Pers. ex. Pers. *Physiol. Plant Pathol.* **10,** 63–71.

Simmonds, P. M. (1928). Studies in cereal diseases. III. Seedling blight and foot-rots of oats. *Can. Dep. Agric., Bull.* **105,** 3–43.

Sinha, A. K., and Das, N. C. (1972). Induced resistance in rice plants to *Helminthosporium oryzae*. *Physiol. Plant Pathol.* **2,** 401–410.

Skipp, R. A., and Deverall, B. J. (1972). Relationships between fungal growth and host changes visible by light microscopy during infection of bean hypocotyls (*Phaseolus vulgaris*) susceptible and resistant to physiological races of *Colletotrichum lindemuthianum*. *Physiol. Plant Pathol.* **2,** 357–374.

Skipp, R. A., and Deverall, B. J. (1973). Studies on cross-protection in the anthracnose disease of bean. *Physiol. Plant Pathol.* **3,** 299–313.

Smith, G. (1900). The haustoria of the Erysipheae. *Bot. Gaz. (Chicago)* **29,** 153–184.

Spencer, J. A. (1973). Leaf-streak of daylily: Infection, disease development, and pathological histology. *Phytopathology* **63,** 864–866.

Stanbridge, B., Gay, J. L., and Wood, R. K. S. (1971). Gross and fine structural changes in *Erysiphe graminis* and barley before and during infection. *In* "Ecology of Leaf Surface

Micro-Organisms'' (T. F. Preece and C. H. Dickinson, eds.), pp. 367–379. Academic Press, New York.

Staub, T., Dahmen, H., and Schwinn, F. J. (1974). Light- and scanning electron microscopy of cucumber and barley powdery mildew on host and nonhost plants. *Phytopathology* **64**, 364–372.

Swinburne, T. R. (1976). Stimulants of germination and appressoria formation by *Colletotrichum musae* (Berk. & Curt.) Arx. in banana leachate. *Phytopathol. Z.* **87**, 74–90.

Talboys, P. W. (1972). Resistance to vascular wilt fungi. *Proc. R. Soc. London, Ser. B* **181**, 319–332.

Tani, T., Ouchi, S., Onoe, T., and Naito, N. (1975). Irreversible recognition demonstrated in the hypersensitive response of oat leaves against the crown rust fungus. *Phytopathology* **65**, 1190–1193.

Tjamos, E. C., and Smith, I. M. (1975). The expression of resistance to *Verticillium albo-atrum* in monogenically resistant tomato varieties. *Physiol. Plant Pathol.* **6**, 215–225.

Tsuneda, A., and Skoropad, W. P. (1978). Behavior of *Alternaria brassicae* and its mycoparasite *Nectria inventa* on intact and on excised leaves of rapeseed. *Can J. Bot.* **56**, 1333–1340.

Vance, C. P., and Sherwood, R. T. (1977). Lignified papilla formation as a mechanism for protection in reed canarygrass. *Physiol. Plant Pathol.* **10**, 247–256.

Vandermolen, G. E., Beckman, C. H., and Rodehorst, E. (1977). Vascular gelation: A general response phenomenon following infection. *Physiol. Plant Pathol.* **11**, 95–100.

Waterman, M. A., Aist, J. R., and Israel, H. W. (1978). Centrifugation studies help to clarify the role of papilla formation in compatible barley powdery mildew interactions. *Phytopathology* **68**, 797–802.

Wheeler, H. (1977). Ultrastructure of penetration by *Helminthosporium maydis*. *Physiol. Plant Pathol.* **11**, 171–178.

Williams, P. H., Aist, J. R., and Bhattacharya, P. K. (1973). Host-parasite relations in cabbage clubroot. *In* "Fungal Pathogenicity and the Plant's Response" (R. J. W. Byrde and C. V. Cutting, eds.), pp. 141–158. Academic Press, New York.

Wilson, C. L. (1973). A lysosomal concept for plant pathology. *Annu. Rev. Phytopathol.* **11**, 247–272.

Wood, R. K. S. (1960). Chemical ability to breach the host barriers. *In* "Plant Pathology: An Advanced Treatise" (J. G. Horsfall and A. E. Dimond, eds.), Vol. 2, pp. 233–272. Academic Press, New York.

Wood, R. K. S. (1967). "Physiological Plant Pathology." Blackwell, Oxford.

Wood, R. K. S. (1972). Introduction: Disease resistance in plants. *Proc. R. Soc. London, Ser. B* **181**, 213–232.

Wynn, W. K. (1976). Appressorium formation over stomates by the bean rust fungus: Response to a surface contact stimulus. *Phytopathology* **66**, 136–146.

Yarwood, C. E. (1959). Predisposition. *In* "Plant Pathology: An Advanced Treatise" (J. G. Horsfall and A. E. Dimond, eds.), Vol. 1, pp. 521–562. Academic Press, New York.

Yarwood, C. E. (1976). Modification of the host response—Predisposition. *In* "Physiological Plant Pathology" (R. Heitefuss and P. H. Williams, eds.), pp. 703–718. Springer-Verlag, Berlin and New York.

18

Food Spoilage and Biodeterioration

John I. Pitt

Biology of Conidial Fungi, Vol. 2
Copyright © 1981 by Academic Press, Inc.
All rights of reproduction in any form reserved.
ISBN 0-12-179502-0

I. INTRODUCTION

The world ecosystem relies on fungi and bacteria to recycle used biological materials. In this broad sense, the activities of nearly all fungi are biodeteriorative, because their prime sources of carbon, nitrogen, and other nutrients are living organisms or materials derived from them. However, the term "biodeterioration" has come to have a more limited application: the degradation of materials utilized or manufactured by man.

From the human viewpoint, food spoilage is a problem distinct from biodeterioration, but both phenomena are governed by the same underlying microbiological principles. Just as biodeterioration is a special form of normal fungal degradation, so food spoilage is a special form of biodeterioration.

Generally speaking, both food spoilage and biodeterioration are concerned with the action of fungi on nonliving substrates, so the major factors influencing the growth of particular fungi are environmental in nature and of broad applicability. Thus an understanding of the influence of these environmental factors provides a basis for understanding the varied capabilities of different fungi to cause food spoilage or biodeterioration. It is appropriate therefore that these factors be discussed first.

Section III will deal in some detail with the more important conidial fungi which cause food spoilage and biodeterioration, and Section IV will discuss the role of conidial fungi in the spoilage of particular classes of foods.

II. FACTORS INFLUENCING SAPROPHYTIC FUNGAL GROWTH

Seven principal factors influence—indeed govern—the germination and growth of fungi in a saprophytic environment. For a particular substrate, a knowledge of the state of each of these factors will, at least in theory, enable prediction of the particular fungi responsible for spoilage or degradation.

It must be pointed out, however, that our knowledge is incomplete concerning the effect of each of these variables on even commonly occurring fungi. Further, interactions among these parameters are to be expected, and only in rare instances are such interactions understood in other than qualitative terms. Nevertheless current knowledge is sufficient for the approach used here to be of value.

The seven principal factors governing germination and growth of saprophytic fungi are, (1) water activity; (2) temperature; (3) oxygen tension; (4) hydrogen ion concentration; (5) nutritional status; (6) specific solute effects; and (7) preservatives. As these factors are not specifically discussed elsewhere in this book, each is considered in turn below, the more significant ones in some detail.

A. Water Activity

The concept of water activity (a_w) was introduced to microbiologists by Scott (1957), who showed that a_w effectively quantitated the relationship between moisture in foods and other commodities and the ability of microorganisms to grow on them.

Water activity is defined as the ratio

$$a_w = p/p_0$$

where p is the partial pressure of water vapor under test conditions and p_0 the saturation vapor pressure of water under the same conditions.

Water activity is numerically equal to the equilibrium relative humidity (ERH) expressed as a decimal. In simple terms, if a piece of food or other substrate is held at constant temperature in a moisture-proof enclosure until equilibration occurs between the water in the sample and in the enclosed air space, then

$$a_w \text{ (food)} = \text{ERH}/100 \text{ (air)}$$

where ERH refers to the air.

This equality provides a means of measuring a_w and of controlling it during experiments. However, practical aspects of the measurement and control of a_w are outside the scope of this chapter (for further information, see Pitt, 1975; Scott, 1957).

In food spoilage and biodeterioration we are most often concerned with mold growth at temperate ambient temperatures with an abundant oxygen supply, with a pH near neutral, with adequate nutrients, and in the absence of preservatives. Under these conditions, clearly, a_w is the predominant factor affecting fungal growth. Information about the water relations of particular fungi will then enable prediction of their ability to grow on substrates of any a_w.

Fungal Water Relations

Like all other living systems, fungi are profoundly affected by the availability of water. On the a_w scale, life as we know it exists over the range 0.9999+ to 0.60. Growth of animals is confined to 1.0–0.99 a_w; the permanent wilt point of mesophytic plants is near 0.98 a_w; and most microorganisms are limited to 0.95 a_w and above (Pitt, quoted by Brown, 1976), although a few halophilic algae and bacteria can grow in saturated sodium chloride (0.75 a_w). Of all the organisms able to tolerate water activities below 0.9, ascomycetous fungi and conidial fungi of ascomycetous origin are by far the most numerous. Such organisms, able to thrive in the presence of extraordinarily high solute concentrations both inside and out, must be ranked among the most highly evolved organisms on earth.

The degree of tolerance of low a_w is most simply, perhaps simplistically, expressed in terms of the minimum a_w at which germination and growth can occur (Table I). Fungi able to grow at low a_w are termed xerophiles, one definition being those able to grow below 0.85 a_w under at least one set of environmental conditions (Pitt, 1975). It is apparent from Table I that few genera even of conidial fungi meet this definition.

The concept of xerophily is of much more than theoretical interest. An a_w of 0.85 or thereabouts provides a natural boundary line between fungi which can

TABLE I

Minimum Water Activities Permitting Germination of Conidial Food Spoilage and Biodeteriorative Fungi

Fungus	Minimum reported a_w	Time (days)[a]	Temperature (°C)	Reference
Aspergillus spp.	0.62–0.85	Various	Various	See Table III
Beauveria bassiana (Bals.) Vuill.	0.92	56	20	Schneider (1954)
Botrytis cinerea Pers. ex Fr.	0.93	2	20	Snow (1949)
Chrysosporium fastidium Pitt	0.69	49	25	Pitt and Christian (1968)
C. xerophilum Pitt	0.71	37	25	Pitt and Christian (1968)
Cladosporium herbarum (Pers.) Link	0.88	7	20	Snow (1949)
Fusarium spp.	0.88–0.91	Various	Various	See Table IV
Geotrichum candidum Link ex Pers.	0.90	15	20	Heintzeler (1939)
Paecilomyces variotii Bain.	0.84	9	25	Pitt and Christian (1968)
Penicillium spp.	0.78–0.90	Various	Various	See Table V
Scopulariopsis brevicaulis (Sacc.) Bain.	0.90	14	25	Galloway (1935)
Stachybotrys atra Corda	0.94	2	23	Ayerst (1969)
Stachybotrys sp.	0.90	14	25	Galloway (1935)
Stemphylium sp.	0.90	14	25	Galloway (1935)
Thielaviopsis sp.	0.95	14	25	Galloway (1935)
Trichoderma viride (Pers.) Link ex Gray	0.91	7	25	Tomkins (1929)
Trichothecium roseum Link	0.90	4	20	Snow (1949)
Wallemia sebi (Fr.) von Arx	0.75	NS	22	Pelhate (1968)

[a] NS, Not stated.

spoil dried foods or commodities in equilibrium with an atmosphere of normal humidity and those restricted to moist habitats—fresh foods, soils, water, and so on, including of course abnormally moist atmospheres such as those in mines, canneries, and some tropical regions.

A knowledge of the a_w of a substrate, or of the humidity of the air with which it is in equilibrium, immediately provides a basis for assessing the kind of fungus likely to cause spoilage.

Although *Aspergillus echinulatus* Thom & Church, the ascomycete *Xeromyces bisporus* Fraser, and the yeast *Saccharomyces rouxii* Boutroux have lower limits for growth below 0.65 a_w, it is generally considered that 0.70 a_w is a safe limit for the storage of foods and other commodities. Conversely, storage above 0.70 a_w—or in an atmosphere consistently above 70% relative humidity—will eventually lead to the appearance of fungal growth. Artifacts in museums, paintings in galleries, and books in libraries are normally held in air-conditioned surroundings at less than 70% humidity: if fungal growth occurs, then a local source of moisture (e.g. a damp wall or new and improperly dried accessions) or air-conditioning failure is usually responsible.

In agricultural practice, experience has enabled the establishment of safe water contents, usually expressed as percent moisture, for the storage of particular commodities. Because the a_w is greatly influenced by the type and quantity of soluble solids present, safe water contents vary from food to food and even from region to region. For example, the safe water content of peanuts is about 8%, of soybeans 12.5%, of cereal grains 13.5–14.5%, of polished rice 15%, and of Australian prunes as high as 23% (Pitt, 1965; Christensen, 1978). In each case, however, the safe water content corresponds to 0.70 a_w.

The fungi involved in spoilage or deterioration of the types outlined above must by definition be xerophiles. A key to xerophile genera is given by Pitt (1975).

The types of fungi involved in spoilage and deterioration above the boundary line of 0.85 a_w are much less clearly defined. Fast-growing xerophiles such as *Aspergillus* species compete with other diverse conidial fungi over almost the entire range of water availability. Other factors then determine the course of spoilage or deterioration.

B. Temperature

Perhaps because the earth's surface temperature mostly lies in the range permitting growth of mesophilic fungi, relatively little attention has been paid to the precise effects of temperature on their growth.

Available data indicate that temperature influences fungal growth in a manner similar to its effect on other microorganisms. Growth at low temperatures is slow, with a relatively indefinite cutoff point; i.e., the longer the time of incuba-

John I. Pitt

TABLE II

Cardinal Temperatures of Some Conidial Fungi

Fungus	Minimum (°C)	Optimum (°C)	Maximum (°C)[a]	Reference
Psychrotolerant				
Alternaria solani Sorauer	2	7	45	Hunter and Barnett (1974)
Botrytis cinerea Pers. ex Fr.	−2	22–25	30–33	Panasenko (1967)
Cladosporium herbarum (Pers.) Link ex Fr.	−5	24–25	30–32	Panasenko (1967)
Fusarium poae (Peck) Wollenweber	−7	ca. 25	NS	Joffe (1962)
F. sporotrichioides Sherb.	−2	ca. 25	NS	Joffe (1962)
Penicillium brevi-compactum Dierckx	−2	23	30	Mislivec and Tuite (1970)
P. expansum Link ex Gray	−2	23	30	Mislivec and Tuite (1970)
Mesophilic				
Aspergillus flavus Link ex Fr.	6–8	35–37	42–45	Panasenko (1967)
A. niger van Tieghem	6–8	35–37	45–47	Pansenko (1967)
Beauveria bassiana (Bals.) Vuill.	3–5	25–27	36–38	Panasenko (1967)
Diplodia zeae (Schw.) Lév.	10	30	35	Hunter and Barnett (1974)
Paecilomyces variotii Bain.	−2	35–40	45–48	Panasenko (1967)
Penicillium chrysogenum Thom	4	23	36	Mislivec and Tuite (1970)
P. citrinum Thom	12	30	37–40[b]	Mislivec and Tuite (1970)
P. piceum Raper & Fennell	12	37–40	48	Evans (1971)
Stachybotrys atra Corda	2–3	25–27	37–40	Panasenko (1967)
Verticillium albo-atrum Reinke & Berthold	5	25	35	Hunter and Barnett (1974)

(continued)

tion, the lower the temperature at which growth will eventually occur. Increases in temperature cause increases in growth rate, with an optimum usually only a little below the maximum temperature for growth. The latter temperature is usually quite sharp, because if germination does not occur within a short time, conidia held at these temperatures rapidly become nonviable (J. I. Pitt, unpublished).

TABLE II

Fungus	Minimum (°C)	Optimum (°C)	Maximum (°C)[a]	Reference
Thermotolerant				
Aspergillus candidus Link ex Fr.	10–15	45–50	50–55	Tansey and Brock (1978)
A. fumigatus Fres.	12	40–42	55	Evans (1971)
Chrysosporium pruinosum (Gilman & Abbott) Carmichael	20	37.5	50	Rosenberg (1975)
C. thermophile (Apinis) Klotopek	18–24	40–50	55	Tansey and Brock (1978)
Geosmithia emersonii (Stolk) Pitt[c]	30	47.5	60	Rosenberg (1975)
Humicola insolens Cooney & Emerson	25	45	60	Rosenberg (1975)
H. lanuginosa (Griffin & Maubl.) Bunce	28–30	45–55	60	Crisan (1973)
Malbranchea pulchella var. *sulfurea* (Miehe) Cooney & Emerson	25–30	45–46	53–57	Crisan (1973)
Paecilomyces byssochlamydoides Stolk & Samson[d]	25	40	55	Awao and Otsuka (1974)
P. crustaceus Apinis & Chesters[e]	20	35–40	55	Awao and Otsuka (1974)
Penicillium dupontii Griffin & Maubl.[f]	25–30	45–50	57–60	Tansey and Brock (1978)
Torula thermophila Cooney & Emerson	20	40–45	55–60	Awao and Otsuka (1973)

[a] NS, Not stated.
[b] J. I. Pitt, unpublished observations.
[c] Teleomorph *Talaromyces emersonii* Stolk.
[d] Teleomorph *Talaromyces byssochlamydoides* Stolk & Samson.
[e] Teleomorph *Dactylomyces crustaceus* Apinis & Chesters.
[f] Teleomorph *Talaromyces thermophilus* Stolk.

Quite a variety of conidial fungi are capable of growth at low temperatures, such as are used in refrigerated storage, or even below the freezing point. Available data suggest that *Cladosporium, Fusarium,* and *Penicillium* species are in many cases capable of growth near the freezing point (Mislivec and Tuite, 1970; Pitt, 1973; Joffe, 1962). Further information is given in Section III, D, E, and J, and in Table II.

Many more data have accumulated on the growth of conidial fungi at high temperatures. According to Tansey and Brock (1972, 1978), the upper limit for growth of fungi is very close to 60°C, much lower than that for prokaryotes. Thermophiles as such are of relatively little importance in food spoilage and biodeterioration but, as many of them are very active agents of decay, they are frequently significant at the lower temperatures (below 50°C) which usually prevail in the situations of interest here. Cardinal temperatures for some thermophiles are listed in Table II. For further information the reviews by Crisan (1973) and Tansey and Brock (1978) should be consulted.

By far the majority of conidial fungi are mesophiles, with minimum temperatures for growth between 0° and 15°C, optima between 25° and 35°C, and maxima below 50°C. A few for which data on temperature relations are available are listed in Table II. The great majority of species of *Aspergillus, Fusarium, Paecilomyces,* and *Penicillium* can be safely assumed to be mesophilic, although their actual temperature relations will vary widely. *Aspergillus* and *Paecilomyces* species generally prefer warmer conditions than the other two genera mentioned above.

C. Oxygen Tension

Without exception, conidial fungi require oxygen for growth. However, many species appear to be efficient oxygen scavengers, so that the oxygen *tension* which will permit growth is often quite low.

Oxygen tension is of little importance in most food spoilage and biodeterioration problems, where oxygen is abundantly available. However in soil, fungal growth and hence biodeteriorative attack is controlled to some extent by oxygen availability, while the absence of oxygen is an important factor in controlling spoilage in some kinds of foods, e.g., grain in silos, packaged meat, canned fruit, and bottled beverages.

Golding (1940a) showed that *Penicillium roquefortii* grew normally at reduced pressures down to 75 mm Hg, i.e., only 10% of standard atmospheric pressure. In atmospheres containing reduced oxygen concentrations, Golding (1940a, 1945) showed that growth of *P. roquefortii, P. expansum, Aspergillus flavus, A. niger,* and *Geotrichum candidum* was little affected until the oxygen content was below 4.2%. Growth of *P. expansum* was virtually unaffected by even 2.1% oxygen over the entire temperature range supporting growth. Where reduction in rates did occur, it was greater at higher temperatures.

In similar experiments, Golding (1940b, 1945) also studied the influence of carbon dioxide on fungal growth. He showed that growth of the species mentioned above was stimulated by increases in carbon dioxide concentration up to 15% in air, but that further increases caused a decline in growth rates. In gas of composition 80% carbon dioxide, 4.2% oxygen, and 15.8% nitrogen, growth rates of *Penicillium roquefortii* were still 30% of those in air, at 20°C and above, but much less at lower temperatures. Similar results with the other molds led to the conclusion that the level of carbon dioxide dissolved in the medium, not the atmospheric partial pressure, controlled the effect on growth.

The effect of oxygen is similarly dependent on the concentration dissolved in the substrate, not on that in the atmosphere (Miller and Golding, 1949).

Such tolerance of wide changes in atmospheric composition appears to be widespread in conidial fungi. For example, Griffin (1966) reported that an atmosphere containing 0.93% oxygen and 17.6% carbon dioxide had little effect, in comparison with air, on the growth on plant stems of *Gliocladium roseum* (Link ex Pers.) Bain., *Trichoderma koningii* Oudem., or a *Fusarium* species. Macauley and Griffin (1969) showed that a reduction in oxygen tension from 20 to 0.2% was necessary to reduce significantly the growth of *F. oxysporum* Schlecht. However, Burges and Fenton (1953) reported that carbon dioxide levels above 5% greatly inhibited growth by *Penicillium nigricans* Bain.

D. Hydrogen Ion Concentration

In general, conidial fungi and yeasts share the property of being less affected by pH than other organisms. Most show optimal growth under slightly acid conditions, about pH 5, and are relatively unaffected by pH in the range 3–8. Some species are able to grow down to pH 2: *Aspergillus niger* (J. I. Pitt, unpublished); *A. oryzae* Cohn, *Penicillium italicum* Wehmer, and *Fusarium oxysporum* (Hunter and Barnett, 1974). This list is undoubtedly incomplete.

The effect of pH on 21 thermophilic fungi was studied by Rosenberg (1975). Optimum pH values proved to be surprisingly diverse: that for *Penicillium emersonii* Stolk was 3.4–5.4, while that for *Penicillium dupontii* Griffin & Maubl. was 7.2–8.1. Growth at pH 9 was not uncommon in this specialized group of fungi.

The most important influence of pH on food spoilage and biodeterioration is to determine whether spoilage will be bacterial or fungal. Under conditions of high a_w and high pH, bacterial spoilage is more likely. From the food technologist's viewpoint, pH 4.5 is critical: above this pH, spores of the highly toxigenic bacterium *Clostridium botulinum* can germinate, so all so-called low-acid, high-a_w foods must be processed by retorting (autoclaving) or preserved in such a way as positively to prevent growth of this microorganism. Below pH 4.5, lactobacilli and yeasts compete for food and space with conidial fungi.

E. Nutritional Factors

The wide variation in carbon and nitrogen sources existing in natural substrates demands that, the more successful (in this sense, ubiquitous) a conidial fungus, the less nutritionally demanding it must be and the broader its ability to assimilate sources of carbon.

1. Carbon Sources

Perlman (1965) has listed more than 150 carbon sources known to be assimilated by fungi, many as a sole source of carbon. He concludes that "fungi may be found which will metabolise any carbon-containing compound ... (perhaps the only exceptions are certain plastic, fluorine-containing compounds and 'non-biodegradable' detergents)".

While Perlman (1965) is fundamentally right, two points are of importance here. First, his list does not include cellulose, lignin, rubbers, or hydrocarbons, all compounds of great importance in biodeterioration; and second, fungi vary widely in their ability to grow at the expense of different chemicals.

In food spoilage, nutrient status is rarely of importance, because food by definition should contain readily assimilated carbon compounds. In biodeterioration, however, this is often not the case. The only available carbon may be lignin, cellulose, or chitin, and many common conidial fungi utilize these materials poorly if at all.

From the practical viewpoint, materials not containing effective preservatives and subjected to high humidities will almost all eventually support mold growth, and it may be of little consequence what the mold is. But from the viewpoint of the mycologist attempting to understand the parameters governing mold growth, the carbon source can be of great significance.

2. Nitrogen Sources

Conidial fungi of interest here are almost all indifferent to nitrogen source, using nitrate, ammonium ions, or organic nitrogen sources with equal ease. Some species achieve only limited growth if amino acids or proteins must be utilized as a source of carbon as well as nitrogen.

A few species of xerophilic fungi are known to be more fastidious. Ormerod (1967) showed that growth of *Wallemia sebi* was strongly stimulated by yeast autolysate, an amino acid mixture, or tryptone. Proline was found to be the factor responsible for this effect. Xerophilic *Chrysosporium* species are also nutritionally fastidious, but the factors involved have not been defined (Pitt, 1975).

3. Vitamins

The literature on the requirements of fungi for vitamins and other organic growth factors was reviewed by Fries (1965). Although a great deal of information has accumulated since then, the fundamental fact has not changed: very few biodeteriorative and food spoilage fungi have any specific growth factor requirements.

F. Specific Solute Effects

Although the overriding effect of growth under conditions of low water availability can be accounted for in terms of a_w, some solutes present in foods can have an additional effect on growth rates. Scott (1957) reported that *Aspergillus amstelodami* grew 50% faster at its optimal a_w (0.96) when sucrose or glucose was used to control the a_w rather than magnesium chloride, sodium chloride, or glycerol. Pitt and Hocking (1977) showed similar effects for *Aspergillus chevalieri* and *Chrysosporium fastidium*. The latter mold in fact failed to grow beyond malformed germ tubes in sodium chloride at pH 6.5. On the other hand, Pitt and Hocking (1977) and Hocking and Pitt (1979) showed that germination and growth of a number of xerophilic *Aspergillus* and *Penicillium* species were little different whether glucose–fructose, glycerol, or sodium chloride was the major solute present in the medium.

G. Preservatives

Preservatives fall into two broad classes: those for use in foods to prevent spoilage, and those for use in industrial situations to prevent degradation. The former must of course be suitable for human consumption and approved as such. In practice the food technologist is limited to weak acids such as sorbic, benzoic, acetic, and propionic and, in some countries, some esters of these, and sulfur dioxide. With the exception of the latter, these are relatively ineffective against conidial fungi and, in any case, many countries have food laws which greatly restrict their use except in specific situations.

In industrial practice, in contrast, the type of preservative used is limited only by human ingenuity and fungal adaptability. Most favored compounds contain copper or other heavy metals as the active ingredients, copper naphthenate, copper 8-quinolinate, and more recently copper–chrome–arsenic complexes being most commonly used. Phenolic compounds—creosote, *m*-cresol, *o*-phenylphenate, pentachlorophenol—have also been widely used.

Detailed discussion of the properties and uses of preservatives is beyond the scope of this chapter. For a general discussion of the philosophy of preservatives and their uses see Kaplan (1968), and for a study of the mechanism of action of weak-acid preservatives see Warth (1977).

III. IMPORTANT GENERA OF SPOILAGE AND BIODETERIORATIVE FUNGI

Although from time to time a very wide variety of conidial fungi are isolated from spoiled or deteriorating materials, the species of importance are confined to a few genera. These are briefly discussed below, with emphasis on the parameters governing their growth, and their common habitats.

A. *Aspergillus* Mich. ex Fr.

Aspergillus shares with *Penicillium* the role of being the most widespread and destructive agent of decay on earth. *Aspergillus* species have in general a requirement for higher growth temperatures than *Penicillium* and are therefore more commonly found (if that is possible) in tropical rather than frigid or temperate zones. The common *Aspergillus* species encountered in the fields of interest here are mostly xerophilic (Table III), with somewhat lower requirements for moisture than many *Penicillium* species. The dominance of *Aspergillus* in the world of spoilage is undoubtedly due to many factors, but the cardinal ones are

TABLE III

Minimum Water Activities Permitting Germination of *Aspergillus* Species

Species	Minimum reported a_w	Time (days)[a]	Temperature (°C)	Reference
Aspergillus amstelodami (Mangin) Thom & Church[b]	0.70	120	25	Armolik and Dickson (1956)
A. candidus Link ex Fr.	0.75	14	25	Galloway (1935)
A. carnoyi (Biourge) Thom & Raper[b]	0.74	19	25	Pitt and Christian (1968)
A. chevalieri (Mangin) Thom & Church[b]	0.71	NS	33	Ayerst (1969)
A. conicus Blochwitz	0.70[c]	NS	22	Pelhate (1968)
A. echinulatus (Delacr.) Thom & Church[b]	0.62	730	25	Snow (1949)
A. flavus Link ex Fr.	0.78	NS	33	Ayerst (1969)
A. fumigatus Fres.	0.82	NS	40	Ayerst (1969)
A. halophilicus Christ., Papav. & C. R. Benj.	0.68	NS	26–30	Christensen (1978)
A. manginii Thom & Raper[b]	0.74	14	25	Pitt and Christian (1968)
A. nidulans (Eidam) Wint.	0.78	NS	37	Ayerst (1969)
A. niger van Tiegh.	0.77	NS	35	Ayerst (1966)

(*continued*)

probably small conidia prolifically produced, biochemical diversity, and ability to grow at low water activities.

Although *Aspergillus* is an ubiquitous genus, differing properties of individual species mean that some are dominant in particular habitats—often for reasons not well understood. Some species are discussed below.

The standard taxonomic treatment of the genus is that by Raper and Fennell (1965); Smith (1969) has provided an abridged version. Teleomorphs of *Aspergillus* are discussed by Subramanian (1972) and Malloch and Cain (1972a,b, 1973) and Kendrick and DiCosmo (1979).

Perhaps no other microorganism has had so much effect on the food-processing industry in the past 20 yr as *Aspergillus flavus* Link ex Fr. The discovery of aflatoxins produced by this species and its sibling *A. parasiticus* Speare has led to the realization that many common food-borne molds are poten-

TABLE III

Species	Minimum reported a_w	Time (days)[a]	Temperature (°C)	Reference
A. ochraceus Wilhelm	0.77	57	25	Pitt and Christian (1968)
A. oryzae (Ahlburg) Cohn	0.85	14	25	Galloway (1935)
A. repens de Bary[b]	0.71	NS	21	Ayerst (1969)
A. restrictus G. Smith	0.75	15	25	Snow (1949)
A. ruber (Kon., Spieck., & Brem.) Thom & Church[b]	0.70	120	25	Snow (1949)
A. sydowi (Bain. & Sart.) Thom & Church	0.78	24	20	Snow (1949)
A. tamarii Kita	0.78	NS	33	Ayerst (1969)
A. terreus Thom	0.78	NS	37	Ayerst (1969)
A. versicolor (Vuill.) Tiraboschi	0.78	24	25	Snow (1949)
A. wentii Wehmer	0.84	8	25	Pitt and Christian (1968)

[a] NS, Not stated.

[b] Teleomorph in *Eurotium* Link ex Fr. Nomenclature of Raper and Fennell (1965). The names given are in common usage but are not necessarily validly published.

[c] Germination without colony development.

tially or actually toxigenic. As a result, a whole new field of work, mycotoxicology, has been born, a field notable for its interdisciplinary character, as it involves cooperation among mycologists, chemists, biochemists, veterinarians, medical practitioners, and pathologists.

Aspergillus flavus and *A. parasiticus* form rapidly growing colonies colored a distinctive, persistent yellow-green. Heads of *A. flavus* commonly consist of closely packed metulae and phialides, while those of *A. parasiticus* are formed from phialides alone.

As noted in Tables II and III, *Aspergillus flavus* is a mesophilic fungus, preferring higher rather than lower temperatures, and is quite xerophilic. Its principal habitats appear to be oilseeds and maize, and certainly these foods induce the highest levels of aflatoxin production. It also occurs commonly as a biodeteriorative agent on textiles and other commodities.

Unlike most other Aspergilli, *Aspergillus fumigatus* Fresenius is a true thermophile (Table II), hence is found most frequently in tropical areas or in warm, decaying vegetation. It is not confined to high temperatures, however, and may sometimes be found as a cause of spoilage in foods or other commodities in temperate zones.

Colonies of *Aspergillus fumigatus* are blue-green and grow very rapidly at 37°C. Conidia are born in long columns, while heads bear phialides only.

This species has sometimes been used in durability testing of textiles, paper and cardboard. Such a practice is not recommended because the conidia of *A. fumigatus* are both allergenic and pathogenic.

Raper and Fennell (1965) have placed all the *Aspergillus* species with teleomorphs in *Eurotium* in a single taxonomic group. All the species included are strongly xerophilic and, taken together, are ubiquitous in situations where commodities are stored just above the safe water content or where the relative humidity is maintained above 70%. Among their common habitats are cereals (Christensen and Kaufmann, 1965), dried fruit (Pitt and Christian, 1968), glass (Ohtsuki, 1962), tobacco (Mitchell and Stauber, 1975), and textiles (Galloway, 1935), and from personal observation, jams, cakes, puddings, spices, nuts, bread, confectionary, and cardboard.

The taxonomy of these species is not difficult once it is recognized that they do not grow typically on media containing less than 20% sucrose or other sugar. Classification is based principally on ascospore ornamentation (Raper and Fennell, 1965), as the *Aspergillus* anamorphs are all quite similar. Nomenclature has been discussed by Malloch and Cain (1972a), while Blaser (1975) has described some new species.

Aspergillus niger van Tieghem has temperature and water relations similar to those of *A. flavus* but is more commonly isolated from commodities dried in the sun, such as vine fruits, than is the latter species. Probably the black pigmentation of the conidia of *A. niger* renders them especially resistant to sunlight.

Aspergillus niger is but one of a closely related series of species which are not easy to distinguish (see Raper and Fennell, 1965).

Although the temperature and water relations of *Aspergillus ochraceus* Wilhelm are similar to those of *A. flavus* and *A. niger*, *A. ochraceus* is more commonly found in cereals (Christensen and Kaufmann, 1965) than in oilseeds or commodities exposed to sunlight. It appears to be encountered in biodeterioration studies only rarely.

As the name implies, *Aspergillus ochraceus* colonies are golden brown or ochre. Stipes are long, heads are at first globose but then split in age, and vesicles bear closely packed metulae and phialides. There are several related species (Raper and Fennell, 1965; Smith, 1969).

Aspergillus restrictus G. Smith and the closely related species *A. penicilloides* Spegazzini are slow-growing xerophiles, and as such are often overlooked in the routine examination of foods. *Aspergillus restrictus* will grow, although tardily, on high-a_w isolation media such as DRBC (King *et al.*, 1979) but, for the isolation of *A. penicilloides,* a medium of reduced a_w such as malt salt or DG18 (Hocking and Pitt, 1980) is essential.

Aspergillus restrictus is an important spoilage agent in cereals (Christensen, 1955), and both species have recently been isolated from a wide variety of commodities in our laboratory (A. D. Hocking, unpublished).

Both species grow slowly and tardily on ordinary media and only a little faster on media containing 20% sucrose. Colonies are gray-green, with small conidiophores bearing phialides only. They have sometimes been mistaken for *Penicillium* species, but are recognizable unequivocally as Aspergilli by the simultaneous production of phialides from their vesicles.

Aspergillus versicolor (Vuill.) Tiraboschi and *A. sydowi* (Bain. & Sartory) Thom & Church grow more slowly than other Aspergilli (with the exception of *A. restrictus* and related species) and produce green and blue conidia, respectively. The heads of both species have ellipsoidal vesicles which bear both metulae and phialides.

Aspergillus versicolor and *A. sydowi* are both strongly xerophilic and are found in a wide variety of spoilage situations. In our experience *A. versicolor* has a preference for cereals, while *A. sydowi* is remarkably well adapted to chemical environments. *Aspergillus sydowi* regularly grows in saturated analytic grade potassium chloride solutions (0.85 a_w) in our laboratory.

B. *Aureobasidium pullulans* (De Bary) Arnaud

Despite a very high moisture requirement, *Aureobasidium pullulans* is a very common spoilage agent, although spoilage is more often unsightly than biodeteriorative. Colonies are at first cream-colored and yeastlike but soon darken. Conidia are borne as blastospores from denticles. Black areas on shower curtains, refrigerator gaskets, and bathroom ceilings are a universal problem and are almost always due to this mold. It is an important cause of slime in paper mills and other factories using water in continuous processing, and in damp areas where cleaning is intermittent. Judged from its ability to grow on almost any surface where free moisture is available, its nutrient requirements must be very low and nonexacting.

Taxonomic treatments of *Aureobasidium pullulans* are given by Cooke (1959) and Hermanides-Nijhof (1977). In earlier literature this species was often referred to as *Pullularia pullulans* or *Dematium pullulans*.

C. *Chrysosporium* Corda

This genus was revived by Carmichael (1962) for fungi producing solitary aleuriospores. In the present context, the genus is of interest because *C. fastidium* Pitt is the most xerophilic purely conidial fungus known. It is an important cause of spoilage of Australian prunes (Pitt and Christian, 1968) but has not been recorded elsewhere.

Recently *C. inops* Carmichael has been reported from corn samples stored for at least a year at 0.75 a_w (Christensen, 1978). *Chrysosporium inops* is not regarded as a xerophile but is similar morphologically to *C. xerophilum* Pitt (Pitt, 1966), and it is very likely that Christensen has isolated the latter species.

D. *Cladosporium* Link ex Fr.

Cladosporium herbarum Link ex Fr. is one of the most common molds on earth and is efficiently distributed in air currents (Upsher, 1968). It is a conspicuous biodeteriogen, both because of its wide metabolic capability and its dark coloration. *Cladosporium herbarum* is capable of growth at very low temperatures (Table II) and was for a long time a problem in chilled beef.

Cladosporium resinae (Lindau) de Vries is the cause of a serious problem in kerosene fuel for jet aircraft. It is a soil fungus with the ability to utilize a wide variety of carbon sources including vegetable oils and waxes, creosote, and kerosene (Parbery, 1968). Jet fuel inevitably contains water, some of which adheres to the sides of aircraft fuel tanks. *Cladosporium resinae* grows in this water and causes corrosion of the tank walls which form the structure of the aircraft (Park, 1978). The problem is kept under control by biocides and inspection programs.

Cladosporium cladosporioides (Fres.) de Vries and *C. sphaerospermum* Penzig are other commonly occurring species. For a taxonomic treatment see de Vries (1952).

E. *Fusarium* Link ex Fr.

Fusarium is characterized by the production of crescent-shaped, fusiform conidia with two to several septa and sparse, spreading colonies which are generally pink, red, or mauve. Species for which water relations data are available are listed in Table IV. Data on temperature relations are limited, but Joffe (1962) showed that some species were capable of growth at temperatures below the freezing point. He also claimed that toxigenic isolates of these and other fungi grew at lower temperatures than similar nontoxic isolates, but this has not been confirmed elsewhere.

Fusarium species are best known as plant pathogens. As a result of this activity they are quite frequently found as causes of spoilage in fresh fruit and vegetables or moist grain. In the latter context, it has now been unequivocally established that *F. poae* and *F. sporotrichioides* caused alimentary toxic aleukia, the notorious mycotoxicosis which killed thousands of Russians between 1942 and 1947 (Joffe, 1978). It has also been established (Yagen and Joffe, 1976; Joffe, 1978) that the toxin responsible for this disease was T-2 toxin.

Fusarium species are also, rather infrequently, isolated from wood pulp, paints, and textiles. Most references in the biodeterioration literature do not provide species identifications, and the role of any particular species is therefore uncertain. Booth (1971) has provided the most workable recent taxonomic treatment.

F. *Geotrichum* Link ex Pers.

Geotrichum candidum Link ex Pers. is a very common species in soil (Barron, 1968) and thus readily contaminates dairy products, in which it is a frequent

TABLE IV

Minimum Water Activities Permitting Germination of *Fusarium* Species[a]

Species	Minimum reported a_w	Time (days)	Temperature (°C)	Reference
Fusarium avenaceum (Fr.) Sacc.	91	56	20	Schneider (1954)
F. culmorum (W. G. Smith) Sacc.	91	56	20	Schneider (1954)
F. decemcellulare Brick	90	56	20	Schneider (1954)
F. lateritium Nees	89	56	20	Schneider (1954)
F. moniliforme Sheldon	87	120	25	Armolik and Dickson (1956)
F. oxysporum Schlecht.	89	56	20	Schneider (1954)
F. solani (Mart.) Sacc.	90	56	20	Schneider (1954)
F. solani var. *coeruleum* (Sacc.) Booth	89	56	20	Schneider (1954)
F. sporotrichioides Sherb.	88	56	20	Schneider (1954)

[a] Nomenclature of Booth (1971).

cause of spoilage. Sour rot of citrus fruit is also caused by this mold (Butler *et al.*, 1965). Carmichael (1957) has studied the taxonomy of *G. candidum* and provided a synonymy.

G. *Myrothecium* Tode ex Fr.

This genus is characterized by the production of viscous masses of dark-green phialoconidia in sporodochia. One species, *M. verrucaria* Ditmar ex Fr., is the cause of loss of strength in cotton (Selby, 1961, Selby *et al.*, 1963). For a taxonomic treatment see Tulloch (1972).

H. *Oidiodendron* Robak

Robak (1932) described this genus, with four species, from Norwegian wood pulp. All species grow slowly, and from long, dark conidiophores produce elaborate sets of branching hyphae which, from the apices back, differentiate into arthroconidia. For a taxonomic treatment see Barron (1962).

I. *Paecilomyces* Bain.

The majority of *Paecilomyces* species, as circumscribed by Samson (1974), are insect pathogens, while most others are strictly soil fungi. However, *P. variotii* Bain. is a very common food spoilage and biodeteriorative organism. It is weakly xerophilic (Pitt and Christian, 1968) and is capable of growth from $-2°$ to $45°$-$48°C$ (Panasenko, 1967). In culture *P. varioti* is readily recognized by rapidly growing brown colonies which produce irregularly verticillate penicilli bearing ellipsoidal conidia.

J. *Penicillium* Link ex Gray

Penicillium species dominate spoilage at low temperature as *Aspergillus* species do in warmer climates. Refrigerated storage rooms and domestic refrigerators operating at $-2°$ to $5°C$ are invariably contaminated with Penicillia which will grow on product, container, structural woodwork, and walls alike. Remedial action is either maintenance of reduced humidity, not practical with high-a_w products such as meat and cheese, or regular fumigation with chlorine and/or formaldehyde. Ultraviolet lamps are frequently employed in such rooms, but their value is marginal.

The most troublesome *Penicillium* species in cold stores are those which Pitt (1979) classifies in subgenus *Penicillium* section *Penicillium* (section *Asymmetrica* subsections *Fasciculata* and *Velutina* of Raper and Thom, 1949). The most commonly occurring species are *P. aurantiogriseum* Dierckx (*P. cyclopium* Westling),

P. chrysogenum Thom, *P. brevicompactum* Dierckx, *P. crustosum* Thom, *P. viridicatum* Westling and *P. griseofulvum* Dierckx (*P. urticae* Bain.). All of these for which data are available are both xerophilic (Table V) and psychrotolerant (Mislivec and Tuite, 1970; Pitt, 1973), and it is likely that the others listed are also.

Also in section *Penicillium* is classified *P. expansum* Link ex Gray, responsible for a destructive rot of apples, pears and occasionally other fruits and vegetables. The cost to world food production annually, in actual losses and in preventative measures, is high indeed.

Furthermore, each of the species mentioned so far produces at least one known mycotoxin, the full significance of which has yet to be established.

Two other universally occurring fruit-rotting molds are classified by Pitt (1979) in subgenus *Penicillium* section *Cylindrosporum: P. digitatum* (Pers. ex Fr.) Sacc. and *P. italicum* Wehmer, the cause of rots in orange and lemon fruits, respectively. Again the annual cost to world agriculture must be measured in millions of dollars.

Among species classified by Pitt (1979) in subgenus *Aspergilloides* (section *Monoverticillata* of Raper and Thom, 1949), the most important spoilage organisms are probably *P. spinulosum* Thom, *P. glabrum* (Wehmer) Westling (*P. frequentans* Westling), *P. implicatum* Biourge, and *P. phoeniceum* van Beyma. *Penicillium spinulosum* is a biodeteriogen with a high resistance to preservatives and other chemicals. It is xerophilic (Pelhate, 1968) and capable of growth below 5°C (Pitt, 1973). *Penicillium implicatum* and *P. phoeniceum* are the most xerophilic *Penicillium* species known to date, being able to grow down to 0.78 a_w at least (Hocking and Pitt, 1979).

Penicillium species producing biverticillate penicilli are divided into those belonging in subgenus *Furcatum* by Pitt (1979) or subsections *Divaricata* and *Velutina* by Raper and Thom (1949), and those in subgenus *Biverticillium* (section *Biverticillata-Symmetrica*). Among the former *P. citrinum* Thom, *P. oxalicum* Currie & Thom, *P. simplicissimum* (Oudem.) Thom and *P. corylophilum* Dierckx are important in the present context. *Penicillium citrinum* is an ubiquitous xerophilic mesophile commonly found in paper, cardboard, textiles, corks, and many kinds of foods. *Penicillium corylophilum* has similar temperature and water relations but rather surprisingly has been found in my laboratory most often in jams of high a_w. *Penicillium oxalicum* and *P. simplicissimum* prefer higher temperatures and are associated with spoilage in warm and damp situations.

Species classified in subgenus *Biverticillium* require more moisture for growth than most other *Penicillium* species mentioned so far. Only *P. islandicum* Sopp and *P. purpurogenum* Stoll are marginal xerophiles. *Penicillium islandicum* is infamous as the main causal organism of "yellow rice," a product with which it is associated in the same way as *Aspergillus flavus* is with peanuts (Saito *et al.*,

TABLE V

Minimum Water Activities Permitting Germination of *Penicillium* Species[a]

Species	Minimum reported a_w	Time (days)[c]	Temperature (°C)	Reference
Penicillium aurantiogriseum Dierckx[b]	0.81	120	25	Armolik and Dickson (1956)
P. brevicompactum Dierckx	0.78	24	25	Hocking and Pitt (1979)
P. chrysogenum Thom	0.78	82	25	Hocking and Pitt (1979)
P. citrinum Thom	0.80	14	25	Galloway (1935)
P. corylophilum Dierckx	0.80	38	25	Hocking and Pitt (1979)
P. digitatum (Pers. ex Fr.) Sacc.	0.90	1	25	Hocking and Pitt (1979)
P. duclauxii Delacr.	0.85	14	25	Galloway (1935)
P. expansum Link ex Gray	0.82	22	25	Hocking and Pitt (1979)
P. fellutanum Biourge	0.78	89	25	Hocking and Pitt (1979)
P. funiculosum Thom	0.90	NS	23	Mislivec and Tuite (1970)
P. glabrum (Wehmer) Westling[d]	0.81	20	23	Mislivec and Tuite (1970)
P. granulatum Bain.	0.86	18	25	Hocking and Pitt (1979)
P. griseofulvum Dierckx[e]	0.81	60	23	Mislivec and Tuite (1970)
P. herquei Bain.	0.88	6	25	Hocking and Pitt (1979)
P. implicatum Biourge	0.78	10	25	Hocking and Pitt (1979)
P. islandicum Sopp	0.83	NS	31	Ayerst (1969)
P. janczewskii Zaleski[f]	0.78	23	25	Hocking and Pitt (1979)
P. kloeckeri Pitt[g]	0.88	8	25	Hocking and Pitt (1979)

(continued)

1971). *Penicillium purpurogenum* and other nonxerophiles, e.g., *P. variable* Sopp, *P. funiculosum* Thom, *P. minioluteum* Dierckx, *P. pinophilum* Hedgcock, and *P. verruculosum* Peyronel, are ubiquitous biodeteriogens with very diverse degradative enzymes and the capacity to grow in almost any moist environment. Textiles and canvas in tropical regions, mine equipment, cooling towers, corks, and rubber are all susceptible to invasion by these species.

The taxonomy of *Penicillium* is relatively difficult and has been only touched upon here. For a detailed account see Pitt (1979).

K. *Scopulariopsis* Bain.

Most *Scopulariopsis* species are confined to saprophytic habitats in soil, but *S. brevicaulis* (Sacc.) Bain. is a common biodeteriogen. According to Smith (1969) it flourishes on high-protein substrates such as meat and cheese.

For a taxonomic treatment of *Scopulariopsis* see Morton and Smith (1963).

TABLE V

Species	Minimum reported a_w	Time (days)[c]	Temperature (°C)	Reference
P. ochrosalmoneum Udagawa	0.88	6	25	Hocking and Pitt (1979)
P. oxalicum Currie & Thom	0.86	5	23	Mislivec and Tuite (1970)
P. phoeniceum van Beyma[h]	0.78	13	25	Hocking and Pitt (1979)
P. puberulum Bain.	0.81	28	23	Mislivec and Tuite (1970)
P. purpurogenum Stoll	0.84	14	25	Hocking and Pitt (1979)
P. restrictum Gilman & Abbott	0.82	14	25	Hocking and Pitt (1979)
P. rugulosum Thom	0.86	9	20	Snow (1949)
P. simplicissimum (Oudem.) Thom	0.86	11	25	Hocking and Pitt (1979)
P. spinulosum Thom	0.80	NS	22	Pelhate (1968)
P. vanbeymae Pitt[i]	0.82	12	25	Hocking and Pitt (1979)
P. variabile Sopp	0.86	7	25	Hocking and Pitt (1979)
P. viridicatum Westling	0.80	21	25	Hocking and Pitt (1979)

[a] Nomenclature of Pitt (1979).
[b] NS, Not stated.
[c] Synonym *P. cyclopium* Westling.
[d] Synonym *P. frequentans* Westling.
[e] Synonyms *P. patulum* Bain. and *P. urticae* Bain.
[f] Synonym *P. nigricans* Bain.
[g] Stat. anam. of *Eupenicillium wortmannii* Klöcker.
[h] Stat. anam. of *Eupenicillium cinnamopurpureum* Abe ex Udagawa.
[i] Stat. anam. of *Eupenicillium baarnense* van Beyma.

L. *Trichoderma* Pers.

The only industrially important species in *Trichoderma* is *T. viride* Pers., which produces rapidly growing, floccose colonies which become green as conidiogenesis proceeds. *Trichoderma viride* produces powerful cellulolytic enzymes, and it is a common cause of spoilage in textiles, paper, and timber.

M. *Wallemia* Johan-Olsen

The single species in *Wallemia, W. sebi* (Fr.) von Arx is quite distinctive, forming small, brown, powdery colonies. Conidiogenesis is equally distinctive: phialides produce elements which disarticulate into arthrospores and then round up into rough-walled, brown conidia. *Wallemia sebi* is a salt-tolerant xerophile found on salt fish (Frank and Hess, 1941) and paper (Wang, 1965). It is also very common in cereals (Christensen 1978; A. D. Hocking, unpublished) and, at least in Australia, in jams, bread, and many kinds of dried foods.

IV. SPOILAGE OF SPECIFIC FOODS BY CONIDIAL
FUNGI

In this section, a brief account will be given of some of the problems inherent in preserving foods and of the role played by conidial fungi in their spoilage.

A. Fresh and High Water Activity Foods

1. Milk and Milk Products

Milk is a food of high pH and a_w, so spoilage is overwhelmingly bacterial in nature. From time to time, however, *Geotrichum candidum* has been implicated in the spoilage of milk and cream.

Yogurts, naturally fermented milk products, are high in acid and are thus protected from bacterial spoilage. Yeasts, however, are little affected by the pH change and cause most yogurt spoilage. *Geotrichum candidum* may also be involved.

2. Cheeses

As low-acid proteinaceous products of 0.95–0.90 a_w, cheeses are susceptible to spoilage by bacteria, yeasts, and molds. Preservative techniques include encasing in dry rinds, adding preservatives such as nisin to wrappers, refrigeration, pasteurizing in cans or aluminium foil, and maturing in the presence of desirable bacteria or molds. Stilton, Gloucester, Camembert, Roquefort and other cheeses are matured in the latter manner.

Despite all these precautions, cheeses are readily spoiled by psychrotolerant molds, usually *Penicillium* species from the subgenus *Penicillium*. As most of these species are mycotoxigenic, moldy cheese should be treated with caution.

3. Fresh Fruit

The acid nature of most fruits makes them resistant to bacterial rots. As they are of living tissue and highly adapted by evolutionary processes, they are also quite resistant to attack by the overwhelming majority of conidial fungi. Quite specific host–parasite interactions, mostly poorly understood, determine the species which can cause fruit spoilage. Unfortunately fungi which are able to overcome the natural defenses of the fruit usually do so totally. If unchecked, vast losses can result from attack on citrus fruits by *Penicillium digitatum* and *P. italicum*, pome fruits by *P. expansum*, and stone fruits by *Monilinia fructicola*, *M. fructigena*, and *M. laxa*. To a lesser extent, tomatoes are damaged by *Fusarium* and *Cladosporium* species, and tropical fruits by *Colletotrichum* species. For further information see Splittstoesser (1978).

Control measures in the field involve controlled spraying programs and, in marketing, physical separation such as wrapping or interleaving is often used.

4. Meat

Fungal spoilage of fresh meat was a serious problem until recently. The advent of careful control over slaughter hygiene and cool room temperatures and humidities has greatly reduced the problem, especially in developed countries.

Psychrotolerant molds, especially *Penicillium* species and *Cladosporium herbarum* were the principal hyphomycetous spoilage organisms involved (Ayres, 1960).

Semipreserved European sausages, such as wursts and salamis, are expected to have a much longer shelf life than fresh meats and are a separate problem. Many such products are sold unrefrigerated, so surface mold growth is to be expected and on some products is even encouraged. A survey of *Penicillium* species growing on European salamis (Ciegler *et al.*, 1972; Leistner and Pitt, 1977) showed that a high proportion (73%) of molds contaminating such products were potentially toxigenic. Of these, 25% produced identifiable mycotoxins in the laboratory. As a result the Bundesanstalt für Fleischforschung in Kulmbach, West Germany, is now marketing a mold strain with desirable properties for use as a salami coating (Mintzlaff and Leistner, 1972).

B. Canned Foods

Almost all canned foods are processed at temperatures sufficiently high to destroy conidia, which are not highly heat-resistant (Pitt and Christian, 1970). Spoilage of canned foods by fungi is rare and is due to ascomycetes of high heat resistance: *Byssochlamys fulva* Olliver & Smith (Olliver and Rendle, 1934), *B. nivea* Westling (Put and Kruiswijk, 1964) and *Talaromyces flavus* (Klöcker) Stolk & Samson [synonym *T. vermiculatus* (Dang.) C. R. Benjamin; van der Spuy, 1972]. Isolations from such spoiled canned foods may sometimes yield only the anamorphs: *Paecilomyces fulvus* Samson (very similar to *P. varioti*), *P. niveus* Samson and *Penicillium dangeardii* Pitt (synonym *P. vermiculatus* Dangeard), respectively.

Condensed milk, a product of low a_w and so sometimes canned without a heat process, is prone to spoilage by *Wallemia sebi*, which forms brown spots in it.

C. Dried or Concentrated Foods

1. Dried Fruit

Before drying, most peaches, pears, and apricots are impregnated with high levels of sulfur dioxide as a color preservative, and this practice effectively eliminates microbial spoilage. Unsulfured fruit, i.e., prunes, many vine fruits, and "health food" packs of apricots and peaches, however, are susceptible to spoilage from molds. If drying is carried out in the sun, the fruit is vulnerable to rain. California vine fruits, normally field-dried, are damaged by abnormal rains once every few years. The practice in Australia, where storms during the drying season are to be expected, is to dry vine fruits on racks under shelter, and losses are much reduced by this technique. Australian dried vine fruits normally carry a high count of *Aspergillus niger* (A. D. Hocking, unpublished) but without apparent damage.

Dried vine fruits are commonly sold at moisture contents low enough to prevent spoilage, i.e., at a_w values below 0.70. Prunes, however, are usually marketed as moist packs of 0.80–0.85 a_w, and are highly susceptible to spoilage by xerophilic fungi. Until recently, Australian prunes were by law marketed as sterile packs, without preservatives. In the course of a 3-yr project, the author recovered nearly every known xerophilic fungus from spoiled packs of such fruit (Pitt, 1965; Pitt and Christian, 1968). As is the practice elsewhere, Australian food laws now permit the use of sorbic or benzoic acids as preservatives in prunes.

2. Nuts

Most nuts are borne on trees and, protected by stout shells, are relatively free from fungal invasion during harvest and drying. From time to time, however, conidial fungi cause spoilage in tree nuts, especially pecans (Wells and Payne, 1976) and Brazil nuts. More importantly, the growth of *Aspergillus flavus* causes aflatoxin production (Stoloff, 1976).

Peanuts, because of their hypogeous nature, are a special problem. Drought during growth of the nuts can cause minute damage to the shells (''checking''), which will permit the entry of soil fungi (McDonald, 1970). Little harm is done unless drying in the field after harvest is accompanied by excessively humid weather or rain. As the nut begins to dry and becomes dormant, weakening of its defense mechanisms will allow rapid growth of spoilage fungi if drying is interrupted. *Aspergillus flavus* and, in tropical regions, *A. parasiticus* appear to be especially favored by these conditions, and as a result aflatoxins may be produced. A single moldy nut may contain as much as 0.1% aflatoxin (Lee *et al.*, 1967).

The challenge to the peanut industry to ensure removal of such a nut from commerce, and indeed of virtually all aflatoxin-containing nuts, has been met by the use of electronic sorting equipment which detects and rejects nuts discolored by mold growth. This procedure, combined with hand sorting and aflatoxin assays on all batches of nuts at receipt and before marketing, has been remarkably successful in the United States and other countries.

3. Cereals and Other Grains

The extent of problems with conidial fungi in cereals and other grains depends on the kind of crop, the locality, and the drying conditions after harvest.

a. Wheat. Wheat crops in Australia, the United States, and Canada at least appear to suffer little damage from molds in normal years. In seasons with an abnormally wet harvest mold damage is to be expected, resulting in grain of poor quality. In Australia such grain is not salable but is used as animal feed on the farm of origin, so that possible toxicity is readily observed and controlled. A lack of reports to the contrary suggests that wheat becomes toxic relatively rarely, although it is quite a suitable substrate for mycotoxin production in the laboratory (Newberne *et al.*, 1964).

b. Rice. Unlike wheat, rice is usually grown in humid tropical climates and, if not mechanically dried, is susceptible to mold damage. Moldy rice is frequently toxic (Saito *et al.*, 1971) because of the growth of *Penicillium citrinum, P. islandicum,* and *P. citreoingrum* Dierckx, but fortunately these species all produce yellow pigments which make their detection possible. Japan banned the sale of yellow rice as long ago as 1910, and in so doing wiped out the disease "acute cardiac beriberi" of which the probable cause was citreoviridin produced by *P. citreonigrum* (syn. *P. citreoviride* Biourge) (Uraguchi, 1971).

c. Barley. The mycoflora of barley before and after harvest and the influence of harvesting and storage factors have been extensively studied by Flannigan (1969, 1974). Conidial fungi with relatively high water requirements, *Alternaria, Cladosporium,* and *Fusarium* species, are slowly replaced by xerophiles, especially the *Aspergillus glaucus* series, other Aspergilli and *Penicillium* species. The initial contamination by xerophiles takes place at harvest (Flannigan, 1978).

Barley has been shown to be an important source of the mycotoxins responsible for porcine nephropathy in Scandinavia (Krogh *et al.*, 1973). Although ochratoxin A was first found as a metabolite of *Aspergillus ochraceus,* it is also produced by *Penicillium viridicatum* (Van Walbeek *et al.*, 1969), and in Scandinavia at least the latter fungus is probably the dominant producer of ochratoxin A. As ochratoxin A is fat-soluble, it can occur in pork and bacon (Krogh, 1976). The ramifications of the occurrence of ochratoxin A in human foods are still being explored (Krogh, 1978).

d. Maize. Much of the world's maize is grown in Africa and the United States under conditions of high humidity. Drying to a safe moisture content often occurs slowly, allowing time for a wide variety of xerophiles and other fungi to grow.

The most serious problem in maize, as in peanuts, is the growth of *Aspergillus flavus,* and the consequent production of aflatoxin. Recent research (Lillehoj *et al.*, 1976) has shown that the primary time of *A. flavus* invasion is not after harvest as might be expected, but during growth of the ear, as a result of insect damage. A reduction in insect damage can have a dramatic effect on aflatoxin levels in maize.

e. Storage of Cereals. Demographic, economic, agronomic, and climatological factors demand that vast quantities of cereals and other grains be stored for periods of a year or more and shipped long distances in huge tonnages. Storage conditions are of critical importance in preventing spoilage by conidial fungi. Clearly, the most satisfactory way to ensure this is to store these commodities below their safe moisture content (see Section II,A). Many factors may be responsible for mold growth; among them are adverse weather conditions, the high cost of mechanical drying, insect and rodent damage, and temperature differentials which cause moisture movement during storage and transport. Mold growth in turn will cause loss of germinability, discoloration, loss of milling properties, rancidity from free fatty acid production, spoilage, and mycotoxin production.

C. M. Christensen of the University of Minnesota, who has devoted a lifetime to the study of wheat storage, considers that the most destructive fungi are *Aspergillus restrictus, A. halophilicus* and other members of the *A. glaucus* series, *A. candidus, A. flavus, A. ochraceus,* and *Penicillium* species (Christensen, 1978).

Various methods have been proposed for controlling fungi in moist stored grain. Low temperature is effective if moisture contents are only a little too high. The maintenance of 5°–10°C in grain bulks by means of refrigerated aeration systems can produce safe long-term storage for wheat of 15–18% moisture (Papavizas and Christensen, 1958). However, Ciegler and Kurtzman (1970) reported that maize of high but unspecified moisture was spoiled and made toxic by *Penicillium aurantiogriseum* (syn. *P. martensii* Biourge) when stored at 1°–10°C.

Milner *et al.* (1947) tested more than 100 fungicidal chemicals for their ability to control xerophilic fungi in wheat, with limited success. As Christensen (1978) points out, most fungicides are effective only when in solution, and at 0.75 a_w their solubility may be very low. More recently it has been found that high-moisture grain intended for feed can be effectively preserved by the application of propionic acid in combination with acetic acid or formaldehyde (Christensen, 1978).

High carbon dioxide levels produced by respiring grain in airtight storage can prevent mold growth, but such levels are not readily attainable in commercial practice. As indicated in Section II,C, carbon dioxide levels must be very high to prevent positively the growth of some conidial fungi. Nevertheless, Peterson *et al.* (1956) reported that even in the presence of 21% oxygen, 13–18% carbon dioxide markedly improved the quality of wheat stored at 18% moisture (0.85 a_w). Clearly the effect of carbon dioxide is enhanced by the reduced a_w of storage. On the other hand, Peterson *et al.* (1956) found that a reduction in oxygen levels down to 0.2% (in nitrogen) did not completely suppress mold growth.

4. Fish

Fungi are rarely implicated in the spoilage of fresh fish, either living or dead. However, *Fusarium solani* (Mart.) Sacc. has recently been reported as a disfiguring pathogen on lobsters (Fisher *et al.*, 1978), and this disease is a cause for commercial concern.

Smoked and dried fish may be invaded by xerophilic fungi. Salted and dried fish frequently become brown from the growth of *Wallemia sebi* (Frank and Hess, 1941).

5. Jams and Jellies

Traditional jams and jellies are shelf-stable, having a_w values below 0.7. In modern manufacturing practice, to reduce sweetness, calories, or cost, jams and jellies are often finished to a_w values nearer 0.8, and reliance is then placed on sterile filling, refrigeration or, in some countries, preservatives, for microbial stability. In the author's laboratory, species from the *Aspergillus glaucus* series, *Penicillium corylophilum,* and *Wallemia sebi* have been the molds most commonly isolated from these sources.

6. Other Dried Foods

A variety of other types of foods, for example, confectionery, pepper and spices, dried meats such as jerky and bilthong, traditional puddings, cakes and biscuits, depend entirely on a low a_w for their stability. All are capable of supporting the growth of xerophilic molds. Some, because of place of origin or composition, possess quite distinctive floras, but space does not permit further discussion here.

REFERENCES

Armilok, N., and Dickson, J. G. (1956). Minimum humidity requirements for germination of conidia associated with storage of grain. *Phytopathology* **46,** 462–465.

Awao, T., and Otsuka, S. (1973). Notes on thermophilic fungi in Japan. (2). *Trans. Mycol. Soc. Jpn.* **14,** 221–236.

Awao, T., and Otsuka, S. (1974). Notes on thermophilic fungi in Japan. (3). *Trans. Mycol. Soc. Jpn.* **15,** 7–22.

Ayerst, G. (1966). The influence of physical factors on deterioration by moulds. SCI *Monogr.* **23,** 14–20.

Ayerst, G. (1969). The effects of moisture and temperature on growth and spore germination in some fungi. *J. Stored Prod. Res.* **5,** 127–141.

Ayres, J. C. (1960). Temperature relationships and some other characteristics of the microbial flora developing on refrigerated beef. *Food Res.* **25,** 1–18.

Barron, G. L. (1962). New species and records of *Oidiodendron*. *Can. J. Bot.* **40,** 589–607.

Barron, G. L. (1968). "The Genera of Hyphomycetes from Soil." Williams & Wilkins, Baltimore, Maryland.

Blaser, P. (1975). Taxonomische und physiologische Untersuchungen über die Gattung *Eurotium* Link ex Fries. *Sydowia* **28**, 1–49.

Booth, C. (1971). "The genus *Fusarium.*" Commonwealth Mycological Institute, Kew, Surrey.

Brown, A. D. (1976). Microbial water stress. *Bacteriol. Rev.* **40**, 803–846.

Burges, A., and Fenton, E. (1953). The effect of carbon dioxide on the growth of certain soil fungi. *Trans. Br. Mycol. Soc.* **36**, 104–108.

Butler, E. E., Webster, R. K., and Eckert, J. W. (1965). Taxonomy, pathogenicity and physiological properties of the fungus causing sour rot of citrus. *Phytopathology* **55**, 1262–1268.

Carmichael, J. W. (1957). *Geotrichum candidum. Mycologia* **49**, 820–830.

Carmichael, J. W. (1962). *Chrysosporium* and some other aleuriosporic hyphomycetes. *Can. J. Bot.* **40**, 1137–1173.

Christensen, C. M. (1955). Grain storage studies. XVIII. Mold invasion of wheat stored for sixteen months at moisture contents below 15 per cent. *Cereal Chem.* **32**, 107–116.

Christensen, C. M. (1978). Storage fungi. *In* "Food and Beverage Mycology" (L. R. Beuchat, ed.), pp. 173–190. Avi Publ. Co., Westport, Connecticut.

Christensen, C. M., and Kaufmann, H. H. (1965). Deterioration of stored grains by fungi. *Annu. Rev. Phytopathol.* **3**, 69–84.

Ciegler, A., and Kurtzman, C. P. (1970). Penicillic acid production by blue-eye fungi on various agricultural commodities. *Appl. Microbiol.* **20**, 761–764.

Ciegler, A., Mintzlaff, H.-J., Machnik, W., and Leistner, L. (1972). Untersuchungen über das Toxinbildungsvermögen von Rohwürsten isolierter Schimmelpilze der Gattung *Penicillium*. *Fleischwirtschaft* **52**, 1311–1314, 1317–1318.

Cooke, W. B. (1959). An ecological life history of *Aureobasidium pullulans* (de Bary) Arnaud. *Myopathologia* **12**, 1–45.

Crisan, E. V. (1973). Current concepts of thermophilism and the thermophilic fungi. *Mycologia* **65**, 1171–1198.

de Vries, G. A. (1952). "Contribution to the Knowledge of the Genus *Cladosporium* Link ex Fr." Hollandia Press, Baarn.

Evans, H. C. (1971). Thermophilous fungi of coal spoil tips. II. Occurrence, distribution and temperature relationships. *Trans. Br. Mycol. Soc.* **57**, 255–266.

Fisher, W. S., Nilson, E. H., Steenbergen, J. F., and Lightner, P. V. (1978). Microbial diseases of cultured lobsters: A review. *Aquaculture* **14**, 115–140.

Flannigan, B. (1969). Microflora of dried barley grain. *Trans. Br. Mycol. Soc.* **53**, 371–379.

Flannigan, B. (1974). Distribution of seed-borne micro-organisms in naked barley and wheat before harvest. *Trans. Br. Mycol. Soc.* **62**, 51–58.

Flannigan, B. (1978). Primary contamination of barley and wheat grain by storage fungi. *Trans. Br. Mycol. Soc.* **71**, 37–42.

Frank, M., and Hess, E. (1941). Studies on salt fish. V. Studies on *Sporendonema epizoum* from "dun" salt fish. *J. Fish. Res. Board Can.* **5**, 276–286.

Fries, N. (1965). The chemical environment for fungal growth. 3. Vitamins and other organic growth factors. *In* "The Fungi" (G. C. Ainsworth and A. S. Sussman, eds.), Vol. 1, pp. 491–523. Academic Press, New York.

Galloway, L. D. (1935). The moisture requirements of mold fungi with special reference to mildew of textiles. *J. Text. Inst.* **26**, T123–129.

Golding, N. S. (1940a). The gas requirements of molds. II. The oxygen requirements of *Penicillium roquefortii* (three strains originally isolated from blue veined cheese) in the presence of nitrogen as diluent and the absence of carbon dioxide. *J. Dairy Sci.* **23**, 879–889.

Golding, N. S. (1940b). The gas requirements of molds. III. The effect of various concentrations of carbon dioxide on the growth of *Penicillium roquefortii* (three strains originally isolated from blue veined cheese) in air. *J. Dairy Sci.* **23**, 891–898.

Golding, N. S. (1945). The gas requirements of molds. IV. A preliminary interpretation of the growth rates of four common mold cultures on the basis of absorbed gases. *J. Dairy Sci.* **28**, 737–750.

Griffin, D. M. (1966). Soil physical factors and the ecology of fungi. IV. Influence of the soil atmosphere. *Trans. Br. Mycol. Soc.* **49**, 115–119.

Heintzeler, I. (1939). Das Wachstum der Schimmelpilze in Abhängigkeit von der Hydratnurverhältnissen unter verschiedenen Aussenbedingungen. *Arch. Mikrobiol.* **10**, 92–132.

Hermanides-Nijhof, E. J. (1977). *Aureobasidium* and allied genera. *Stud. Mycol., Baarn* **15**, 141–177.

Hocking, A. D., and Pitt, J. I. (1979). Water relations of some *Penicillium* species at 25°C. *Trans. Br. Mycol. Soc.* **73**, 141–145.

Hocking, A. D., and Pitt, J. I. (1980). Dichloran-glycerol medium for enumeration of xerophilic fungi from low-moisture foods. *Appl. Environ. Microbiol.* **39**, 488–492.

Hunter, B. B., and Barnett, H. L. (1974). Deuteromycetes (Fungi Imperfecti). *In* "Handbook of Microbiology" (A. I. Laskin and H. A. Lechevalier, eds.), pp. 448–482. CRC Press, Cleveland, Ohio.

Joffe, A. Z. (1962). Biological properties of some toxic fungi isolated from overwintered cereals. *Mycopathol. Mycol. Appl.* **16**, 201–221.

Joffe, A. Z. (1978). *Fusarium poae* and *F. sporotrichioides* as principal causal agents of alimentary toxic aleukia. *In* "Mycotoxic Fungi, Mycotoxins, Mycotoxicoses" (T. D. Wyllie and L. G. Morehouse, eds.), Vol. III, pp. 21–86. Dekker, New York.

Kaplan, A. M. (1968). The control of biodeterioration by fungicides—philosophy. *In* "Biodeterioration of Materials. Microbiological and Allied Aspects" (A. H. Walters and J. J. Elphick, eds.), pp. 196–204. Elsevier, Amsterdam.

Kendrick, W. B., and DiCosmo, F. (1979). Teleomorph-Anamorph connections in Ascomycetes. *In* "The Whole Fungus" (W. B. Kendrick, ed.), Vol. 1, pp. 283–410. National Museums of Canada, Ottawa.

King, A. D., Hocking, A. D., and Pitt, J. I. (1979). Dichloran-rose bengal medium for enumeration and isolation of molds from foods. *Appl. Environ. Microbiol.* **37**, 959–964.

Krogh, P. (1976). Mycotoxic nephropathy. *Adv. Vet. Sci. Comp. Med.* **20**, 147–170.

Krogh, P. (1978). Ochratoxins. *In* "Mycotoxins in Human and Animal Health" (J. V. Rodricks, C. W. Hesseltine, and M. A. Mehlman, eds.), pp. 489–498. Pathotox Publ., Park Forest South, Illinois.

Krogh, P., Hald, B., and Pedersen, E. J. (1973). Occurrence of ochratoxin A and citrinin in cereals associated with mycotoxic porcine nephropathy. *Acta Pathol. Microbiol. Scand., Sect. B* **81**, 689–695.

Lee, L. S., Yatsu, L. Y., and Goldblatt, L. A. (1967). Aflatoxin contamination. Electron microscopic evidence of mold penetration. *J. Am. Oil Chem. Soc.* **44**, 331–332.

Leistner, L., and Pitt, J. I. (1977). Miscellaneous *Penicillium* toxins. *In* "Mycotoxins and Human and Animal Health" (J. V. Rodricks, C. W. Hesseltine, and M. A. Mehlman, eds.), pp. 639–653. Pathotox Publ., Park Forest South, Illinois.

Lillehoj, E. B., Fennell, D. I., and Kwolek, W. F. (1976). *Aspergillus flavus* and aflatoxin in Iowa corn before harvest. *Science* **193**, 495–496.

Macauley, B. J., and Griffin, D. M. (1969). Effects of carbon dioxide and oxygen on the activity of some soil fungi. *Trans. Br. Mycol. Soc.* **53**, 53–62.

McDonald, D. (1970). Fungal infection of groundnut fruit before harvest. *Trans. Br. Mycol. Soc.* **54**, 453–460.

Malloch, D., and Cain, R. F. (1972a). New species and combinations of cleistothecial Ascomycetes. *Can. J. Bot.* **50**, 61–72.

Malloch, D., and Cain, R. F. (1972b). The Trichocomataceae: Ascomycetes with *Aspergillus, Paecilomyces* and *Penicillium* imperfect states. *Can. J. Bot.* **50**, 2613–2628.

Malloch, D., and Cain, R. F. (1973). The Trichocomaceae (Ascomycetes): Synonyms in recent publications. *Can. J. Bot.* **51,** 1647–1648.

Miller, D. D., and Golding, N. S. (1949). The gas requirements of molds. V. The minimum oxygen requirements for normal growth and for germination of six mold cultures. *J. Dairy Sci.* **32,** 101–110.

Milner, M., Christensen, C. M., and Geddes, W. F. (1947). Grain storage studies. VII. Influence of certain mold inhibitors on respiration of moist wheat. *Cereal Chem.* **24,** 507–517.

Mintzlaff, H.-J., and Leistner, L. (1972). Untersuchungen zur Selektion eines technologisch geeigneten und toxikologisch unbedenklichen Schlimmerpilz-Stammes für die Rohwurst-Herstellung. Zentbl. VetMed. **19,** 291–300.

Mislivec, P. B., and Tuite, J. (1970). Temperature and relative humidity requirements of species of *Penicillium* isolated from yellow dent corn kernels. *Mycologia* **62,** 75–88.

Mitchell, T. G., and Stauber, P. C. (1975). Biodeterioration of tobacco. *In* "Microbial Aspects of Deterioration of Materials" (R. J. Gilbert and D. W. Lovelock, eds.), pp. 203–211. Academic Press, New York.

Morton, F. J., and Smith, G. (1963). The genera *Scopulariopsis* Bainier, *Microascus* Zukal and *Doratomyces* Corda. *Mycol. Pap.* **86,** 1–96.

Newberne, P. M., Wogan, G. N., Carlton, W. W., and Abdel Kader, M. M. (1964). Histopathological lesions in ducklings caused by *Aspergillus flavus* cultures, culture extracts, and crystalline aflatoxins. *Appl. Pharmacol.* **6,** 542–556.

Ohtsuki, T. (1962). Studies on the glass mould. V. On two species of *Aspergillus* isolated from glass. *Bot. Mag.* **75,** 436–442.

Olliver, M., and Rendle, T. (1934). A new problem in fruit preservation. Studies on *Byssochlamys fulva* and its effect on the tissues of processed fruit. *J. Soc. Chem. Ind., London* **53,** 166–172.

Ormerod, J. G. (1967). The nutrition of the halophilic mold *Sporendonema epizoum. Arch. Mikrobiol.* **56,** 31–39.

Panasenko, V. T. (1967). Ecology of microfungi. *Bot. Rev.* **33,** 189–215.

Papavizas, G. C., and Christensen, C. M. (1958). Grain storage studies. 26. Fungus invasion and deterioration of wheats stored at low temperatures and moisture contents of 15 to 18 per cent. *Cereal Chem.* **35,** 27–34.

Parbery, D. G. (1968). The soil as a natural source of *Cladosporium resinae. In* "Biodeterioration of Materials: Microbiological and Allied Aspects" (A. H. Walters and J. J. Elphick, eds.), pp. 371–380. Elsevier, Amsterdam.

Park, P. B. (1978). Biodeterioration in aircraft fuel systems. *In* "Microbial Aspects of Deterioration of Materials" (R. J. Gilbert and D. W. Lovelock, eds.), pp. 105–126. Academic Press, New York.

Pelhate, J. (1968). Recherche des besoins en eau chez quelques moisissures des grains. *Mycopathol. Mycol. Appl.* **36,** 117–128.

Perlman, D. (1965). The chemical environment for fungal growth. 2. Carbon sources. *In* "The Fungi" (G. C. Ainsworth and A. S. Sussman, eds.), Vol. 1, pp. 479–489. Academic Press, New York.

Peterson, A., Schlegel, V., Hummel, B., Cuendet, L. S., Geddes, W. F., and Christensen, C. M. (1956). Grain storage studies. XXII. Influence of oxygen and carbon dioxide concentrations on mold growth and grain deterioration. *Cereal Chem.* **33,** 53–63.

Pitt, J. I. (1965). Microbiological problems in prune preservation. M.Sc. Thesis, University of New South Wales, Kensington.

Pitt, J. I. (1966). Two new species of *Chrysosporium. Trans. Br. Mycol. Soc.* **49,** 467–470.

Pitt, J. I. (1973). An appraisal of identification methods for *Penicillium* species: novel taxonomic criteria based on temperature and water relations. *Mycologia* **65,** 1135–1157.

Pitt, J. I. (1975). Xerophilic fungi and the spoilage of foods of plant origin. *In* "Water Relations of Foods" (R. B. Duckworth, ed.), pp. 273–307. Academic Press, New York.

Pitt, J. I. (1979). "The Genus *Penicillium* and its Teleomorphic States *Eupenicillium* and *Talaromyces*. Academic Press, New York.

Pitt, J. I., and Christian, J. H. B. (1968). Water relations of xerophilic fungi isolated from prunes. *Appl. Microbiol.* **16**, 1853–1858.

Pitt, J. I., and Christian, J. H. B. (1970). Heat resistance of xerophilic fungi based on microscopical assessment of spore survival. *Appl. Microbiol.* **20**, 682–686.

Pitt, J. I., and Hocking, A. D. (1977). Influence of solute and hydrogen ion concentration on the water relations of some xerophilic fungi. *J. Gen. Microbiol.* **101**, 35–40.

Put, H. M. C., and Kruiswijk, J. T. (1964). Disintegration and organoleptic deterioration of processed strawberries caused by the mould *Byssochlamys nivea*. *J. Appl. Bacteriol.* **27**, 53–58.

Raper, K. B., and Fennell, D. I. (1965). "The Genus *Aspergillus*." Williams & Wilkins, Baltimore, Maryland.

Raper, K. B., and Thom, C. (1949). "A Manual of the Penicillia." Williams & Wilkins, Baltimore, Maryland.

Robak, H. (1932). Investigations regarding fungi on Norwegian ground woodpulp and fungal infections at woodpulp mills. *Saertr. Nyt Mag. Naturv.* **71**, 185–330.

Rosenberg, S. L. (1975). Temperature and pH optima for 21 species of thermophilic and thermotolerant fungi. *Can. J. Microbiol.* **21**, 1535–1540.

Saito, M., Enomoto, M., and Tatsuno, T. (1971). Yellowed rice toxins. Luteoskyrin and related compounds, and citrinin. *In* "Microbial Toxins" (A. Ciegler, S. Kadis, and S. J. Ajl, eds.), Vol. 6, pp. 299–380. Academic Press, New York.

Samson, R. A. (1974). *Paecilomyces* and some allied hyphomycetes. *Stud. Mycol., Baarn* **6**, 1–119.

Schneider, R. (1954). Untersuchungen über Feuchtigkeitsansprüche parasitischer Pilze. *Phytopathol. Z.* **21**, 63–78.

Scott, W. J. (1957). Water relations of food spoilage microorganisms. *Adv. Food Res.* **7**, 83–127.

Selby, K. (1961). The degradation of cotton cellulose by the extracellular cellulase of *Myrothecium verrucaria*. *Biochem. J.* **79**, 562–566.

Selby, K., Maitland, C. C., and Thompson, K. V. A. (1963). The degradation of cotton cellulose by the extracellular cellulase of *Myrothecium verrucaria*. 2. The existence of an "exhaustible" cellulase. *Biochem. J.* **88**, 288–296.

Smith, G. (1969). "An Introduction to Industrial Mycology," 6th ed. Arnold, London.

Snow, D. (1949). The germination of mould spores at controlled humidities. *Ann. Appl. Biol.* **36**, 1–13.

Splittstoesser, D. F. (1978). Fruits and fruit products. *In* "Food and Beverage Mycology" (L. R. Beuchat, ed.), pp. 83–109. Avi Publ. Co., Westport, Connecticut.

Stoloff, L. (1976). Occurrence of mycotoxins in foods and feeds. *Adv. Chem. Ser.* **149**, 23–50.

Subramanian, C. V. (1972). The perfect states of *Aspergillus*. *Curr. Sci.* **41**, 755–761.

Tansey, M. R., and Brock, T. D. (1972). The upper temperature limit for eukaryotic organisms. *Proc. Natl Acad. Sci. U.S.A.* **69**, 2426–2428.

Tansey, M. R., and Brock, T. D. (1978). Microbial life at high temperatures: Ecological aspects. *In* "Microbial Life in Extreme Environments" (D. J. Kushner, ed.), pp. 159–216. Academic Press, New York.

Tomkins, R. G. (1929). Studies of the growth of moulds. I. *Proc. R. Soc. London, Ser. B* **105**, 375–401.

Tulloch, M. (1972). The genus *Myrothecium* Tode ex Fr. *Mycol. Pap.* **130**, 1–42.

Upsher, F. J. (1968). Fungal spora of the air at the joint Tropical Research Unit, Innisfail, Queensland. *In* "Biodeterioration of Materials. Microbiological and Allied Aspects" (A. H. Walters and J. J. Elphick, eds.), pp. 131–142. Elsevier, Amsterdam.

Uraguchi, K. (1971). Yellowed rice toxins. Citreoviridin. *In* "Microbial Toxins" (A. Ciegler, S. Kadis, and S. J. Ajl, eds.), Vol. 6, pp. 367–380. Academic Press, New York.

Van Der Spuy, J. E. (1972). Apple beverages: heat-resistant moulds in apple juice. 1966–1971. Final report. *Agric. Res., Pretoria* p. 82.

Van Walbeek, W., Scott, P. M., Harwig, J., and Lawrence, J. W. (1969). *Penicillium viridicatum:* a new source of ochratoxin A. *Can. J. Microbiol.* **15,** 1281–1285.

Wang, C. J. K. (1965). Fungi of pulp and paper in New York. *State Univ. Coll. For. Syracuse Univ., Tech. Publ.* **87.**

Warth, A. D. (1977). Mechanism of resistance of *Saccharomyces bailii* to benzoic, sorbic and other weak acids used as food preservatives. *J. Appl. Bacteriol.* **43,** 215–230.

Wells, J. M., and Payne, J. A. (1976). Toxigenic species of *Penicillium, Fusarium* and *Aspergillus* from weevil-damaged pecans. *Can. J. Microbiol.* **22,** 281–285.

Yagen, B., and Joffe, A. Z. (1976). Screening of toxic isolates of *Fusarium poae* and *F. sporotrichioides* involved in causing alimentary toxic aleukia. *Appl. Environ. Microbiol.* **32,** 432–427.

19

Use of Conidial Fungi in Biological Control

T.E. Freeman

I. INTRODUCTION

Since emergence of the lowest form of life, a delicate balance has existed between organisms and their external environment. Physical and biotic factors of the environment have an impact upon the well-being and survival of each organism. The impact can be either beneficial or harmful and can affect a species over a large geographical range and a long time span. Harmful impacts tend to limit populations, whereas beneficial ones favor population growth. In either case, the extent to which species of the population are affected is related to the severity of the impact on the target species.

When one biotic agent acts upon another in such a manner as to limit its population, a state of biological control is operating. In other words, a biological control agent, or more simply a biocontrol agent, is active in limiting the popula-

Biology of Conidial Fungi, Vol. 2

tion of a target species. H. S. Smith (1919) first defined biological control in its present-day context as ''regulation of pest populations by natural enemies.'' Biological control in its broadest sense has existed as a natural phenomenon since the beginning of life. Man has recognized this process and attempted to use it to their advantage. Admittedly, until recently human attempts have been disorganized. An effective biological control research program was undertaken in California in the late 1880s, which achieved control of the cottony cushion scale on citrus. The success of this effort, which pitted insect against insect, provided a stimulus for further research. Almost without exception, ensuing programs were directed by entomologists. Accordingly, as Huffaker (1976) noted, insects were used as the principal biological control agent against target species of other insect and weed pests. Only recently have additional biocontrol agents and target species, including fungi, begun to receive concentrated attention.

General aspects of the biological control of plant pathogens have been treated by Baker and Cook (1974) and Snyder *et al.* (1976). These reviews deal primarily with interactions affecting soil-borne pathogens; less detailed attention has been paid to airborne pathogens. The use of plant pathogens as biocontrol agents for weeds has been reviewed by Wilson (1969), Zettler and Freeman (1972), and more recently by Freeman *et al.* (1976a). Conidial fungi as specific biological control agents will be examined in this chapter.

II. CONIDIAL FUNGI AS BIOLOGICAL CONTROL AGENTS

A. Advantages

Theoretically, conidial fungi have many advantages that make them desirable candidates for biocontrol agents. They are numerous and diverse in their habitat and environmental requirements. Thus it should be possible to find a conidial fungus that fits into a wide range of biocontrol situations. Most species can be easily cultured en masse, and their reproductive capacity either in culture or in nature is usually quite high. These attributes are extremely important in establishing and maintaining populations of biocontrol agents. In addition, most conidial fungi are capable of surviving in a saprophytic state for extended periods. Therefore, as the conidial fungus reduces the population of the target species, it does not necessarily lose its inoculum potential. Even if the agent's population should be drastically reduced, its reproductive capacity would quickly enable it to reestablish a threshold level of the propagules needed to maintain an acceptable level of control. In this regard it should be pointed out that the use of a conidial fungus as a biocontrol agent would probably not result in the complete elimination of a target species, even in restricted areas. This is especially true when biocontrol ability is dependent upon parasitic activity of the agent. An efficient parasite does not eliminate its host (Baker and Cook, 1974). Some argue, however, that in general conidial fungi are not efficient parasites. Conidial fungi also have distinct advantages as biocontrol agents in being diverse in their mode of action. They may act simply as competitors for space, food, and infection sites. However, in competition for food and space their action may affect nontarget as well as target species. Another mode of action occurs through the production of toxic substances such as antibiotics and phytotoxins (Wheeler and Luke, 1963). The former are usually encountered in the suppression of soil-borne target species by soil-inhabiting biocontrol agents, a condition called antibiosis. This may be a general mode of action in cases where competition occurs. Indeed, the function of antibiosis may well be to give the biocontrol agent a competitive edge. Phytotoxin production is encountered in pathological relationships of conidial fungi and higher plants (Wheeler and Luke, 1963), and such compounds may be highly specific for a target species. Conidial fungi can be pathogenic, parasitic, or both on higher plants and other fungi. The parasitic types may be either mycoparasites or hyperparasites. Despite the fact that the two terms are often used synonymously, they are not necessarily one and the same, being entirely dependent upon whether or not the target species is itself a parasite. Predation has even been attributed to conidial fungi, as in the case of species that entrap nematodes.

One final extremely important advantage of conidial fungi in biocontrol programs is that so far conidial fungi that exhibit biocontrol potential for pest species have not been shown to have detrimental effects on humans, domestic animals, or wildlife. Admittedly, however, few species have been thoroughly tested.

B. Disadvantages

Despite their many advantages, conidial fungi also have disadvantages. They are subject to the rigors of competition, predation, and parasitism. Any of these conditions could limit their usefulness in biocontrol systems. Baker and Cook (1974) are pessimistic about the effectiveness of fungi as biocontrol agents of plant pathogens. Snyder et al. (1976) consider that mycoparasites have some promise as biocontrol agents, though successes have been very limited. They consider their use in the control of pathogens on aerial plant parts, however, to be highly speculative. These authors are more hopeful about control in the soil environment but note that, here too, successes have been limited. Baker and Cook (1974) consider fungi third in importance behind bacteria and actinomycetes as biocontrol agents. Both Baker and Cook (1974) and Snyder et al. (1976) consider low inoculum density an important factor in reducing biocontrol potential of fungi. However, it appears possible to overcome this limitation, especially when using conidial fungi as mycopesticides where massive inoculum dosages could be used. Only further efforts will ultimately determine if laboratory successes in the control of plant pathogens can be projected to practical usage in pest management systems. The disadvantages of using conidial fungi in the biological control of weed pests are apparently fewer. At least, successes and the potential for success are more heartening. However, conidial fungi are often not considered as host-specific as more specialized groups of fungal pathogens of plants. Thus extensive host range tests must be conducted at considerable expense of time and money.

Other disadvantages, which may not be definable in the light of present knowledge, are suggested by questions which haunt progenitors of biocontrol programs. What are the long-term effects on nontarget species and what will be the eventual influence on the total system, whether an agro or a natural ecosystem? Is the reward worth the risk? These questions must be answered before any biological control program is put into effect.

III. RECENT ADVANCES

A. Control of Plant Pathogens

Despite the fact that successes have been few and often demonstrated only under experimental conditions, the control of plant pathogens with fungi holds considerable promise. During the last decade a more direct approach to biological control of plant pathogens began to evolve. Plant pathologists are taking the classic approach entomologists have used for many years, i.e., pitting a specific organism against a pest. It is hoped that this approach will be more successful than earlier efforts. Biological control of plant pathogens had earlier revolved around the manipulation of factors that tip the balance in favor of organisms antagonistic to plant pathogens. This aspect has been adequately covered by Baker and Cook (1974).

Since the pioneering work of Weindling (1932), *Trichoderma* spp. have probably been more extensively studied as biocontrol agents for plant pathogens than any other group. In these studies soil-borne pathogens have been the primary target species. Weindling (1932, 1934) first demonstrated that *Trichoderma viride* was both parasitic on and antagonistic to a pathogen such as *Rhizoctonia solani*. *Trichoderma viride* readily parasitizes and kills hyphae of *R. solani*. In addition, the fungus produces the antibiotic gliotoxin, which has proved toxic to several species. This fungus was thus shown to be capable of exerting dual effects, both with biocontrol potential. Since this early work, *Trichoderma* spp. have often been implicated in reducing populations of soil-inhabiting pathogens. Bliss (1951) showed that control of *Armillaria mellea* in citrus attributed to carbon disulfide actually resulted from the action of *Trichoderma*. This antagonist rapidly colonized and reproduced in the fumigated soil and was actually the factor controlling *A. mellea* rather than the fumigation per se. Ohr *et al.* (1973) showed that methyl bromide fumigation had a similar sequel. Apparently *Trichoderma* is more resistant to methyl bromide than *A. mellea* and is effective in reducing the activity of this causative agent of citrus root disease. The studies of Ohr *et al.* (1973) suggested the concept of an integrated chemical-biological control which was more effective than either one alone. For this system to be successful, the biocontrol agent must be less sensitive to the chemical than the target species.

Prior to 1970, most of the work with *Trichoderma* involved indirect enhancement of natural populations of the fungus by various methods of soil treatment. However, the present trend is to apply *Trichoderma* in a manner reminiscent of that used for other pesticides. Wells *et al.* (1972) reported the isolation of *T. harzianum* from diseased sclerotia of *Sclerotium rolfsii* collected from a blue-lupine field in Gainesville, Florida. The isolated fungus proved pathogenic *in vitro* to *S. rolfsii*, *Sclerotinia trifoliorum*, and *Botrytis cinerea*, but innocuous to *R. solani*, *Pythium aphanidermatum*, and *P. myriotylum*. In further studies, these authors found that *T. harzianum* grown in culture and applied to the soil controlled *S. rolfsii* on blue lupine and tomatoes in the greenhouse. Under natural field conditions, one to three applications of *T. harzianum* effectively controlled *S. rolfsii* on peanuts. These findings prompted intensification of research into inoculum production, application, and efficiency of *T. harzianum* as a biocontrol for *S. rolfsii* and other soil-borne pathogens. Backman and Rodriguez-Kabana (1975) reported studies in which they successfully grew *T. harzianum* on diatomaceous earth granules impregnated with a 10% molasses solution. They obtained significant control of *S. rolfsii* on peanuts when 140 kg/ha of the granules was applied to the soil surface. The control obtained equaled that resulting from the use of 10% pentachloronitrobenzene (PCNB) at a dose of 112 kg/ha. The work with *T. harzianum* in biological control is very promising; at least one commercial pharmaceutical company is vigorously exploring the production of a marketable form of *T. harzianum* for biological control of *S. rolfsii*. Its potential usefulness for other pathogens remains to be determined. Kelly (1976) reported that *T. harzianum*-impregnated clay granules failed to provide control of *Phytophthora cinnamomi*, which causes damping-off of pine seedlings. There are also other factors to be considered in the use of *T. harzianum* as a biocontrol agent of *S. rolfsii*. Backman *et al.* (1975) have shown that fungicidal sprays used for leaf spot (*Cercospora*) control on peanuts reduce natural populations of *T. viride*, resulting in an increase in stem blight caused by *S. rolfsii*. According to Peeples *et al.* (1976), use of the herbicide EPTC also reduces the biocontrol potential of *T. viride* on *S. rolfsii*.

Several other specialized usages of *Trichoderma* in biological control programs have been reported. De Trogoff and Ricard (1976) noted that *Trichoderma* spray concentrate was available in France at a price competitive with that of benomyl. They proposed that it be employed in spraying casing soil used in commercial mushroom production for control of *Verticillium malthousei* on *Agaricus bisporus*. The fungus effectively controlled the pathogen and did not invade *A. bisporus*. In addition, *Trichoderma* is nontoxic up to 4000 mg/kg of body weight of laboratory animals, making it an extremely safe pesticide. Also in France, Grosclaude *et al*. (1975) showed that application of *Trichoderma* spores to wounds of plum trees caused by pruning controlled invasion by *Stereum purpureum*. The conidial fungus was effective either when applied as spray or when injected into the tree with special pruning shears. A somewhat similar technique was researched by Hunt *et al*. (1971) in attempts to control *Fomes (Heterobasidion) annosus* on stumps using *Trichoderma*. These authors placed spores of *Trichoderma* in chain saw oil so that the fungal propagules would be deposited on the stump during the cutting operation. The system worked on wood disks enclosed in plastic bags but was not effective on stumps. They postulated that the humidity was too low on the exposed stumps. It seems feasible to cover cut stumps with plastic bags until colonization by *Trichoderma* is established, but they did not explore this possibility. In additional work on biological control of wound parasites, Pottle and Shigo (1975) and Pottle *et al*. (1977) demonstrated the effectiveness of *T. viride* and *T. harzianum* in controlling invasion of wounds by Hymenomycetes. No Hymenomycetes could be isolated from wounds of *Acer rubrum* 17 months after treatment with *T. viride*. Hymenomycetes were isolated from 6% of untreated wounds. In another somewhat specialized area, Harder and Troll (1973) proposed the use of *Trichoderma* species to parasitize sclerotia of *Typhula incarnata;* they wanted to reduce the inoculum potential of this pathogen on golf course greens. These authors based their proposal on studies showing that *Trichoderma* species were antagonistic toward sclerotia of *T. incarnata*. The hypothesis has apparently not been tested.

The common soil-inhabiting fungi which belong to the genus *Gliocladium* were shown by Barnett and Lilly (1963) to be destructive mycoparasites of other species of fungi. They reported that species of *Gliocladium* were especially parasitic on *Ceratocystis fagacearum* but also attacked several other fungi. However, *Gliocladium* spp., which are in many ways similar to *Trichoderma* spp. in their ability to parasitize other fungi and produce gliotoxin, have not been studied extensively. Overmier and Roncadori (1977) have reported that *Gliocladium virens* and *T. harzianum* produce substances which inhibit germination of *Diplodia gossypina*. The latter is the causal agent of boll rot in cotton. The fungus overwinters in soil where it is especially vulnerable to the action of antagonistic organisms. These authors have demonstrated that the substances produced by the two antagonistic fungi also inhibit germination of *Verticillium* and *Fusarium* conidia. Ricard *et al*. (1974) have shown antagonism between *Gliocladium roseum* and *Eutypa armeniacae*. The latter is the causal agent of a canker in apricots. Since *E. armeniacae* enters through pruning wounds, they feel that *G. roseum* may be effective in controlling the disease. The antagonist could be applied either as a spray or with specially adapted pruning shears.

Some work has also been conducted on seed treatment with conidial fungi to control seedling disorders. Chang and Kommedahl (1968) reported that applying *Chaetomium globosum* to corn seed gave protection from *Fusarium roseum* f.sp. *cerealis* 'Graminearum.' Grinding cultured *C. globosum* in a blender and then applying it to corn seed gave protection equal to that obtained by treating seeds with either thiram or captan. In addition, there was increased emergence, root vigor, stand, fresh weight, and dry weight of roots. *Chaetomium globosum* secreted an unidentified substance that adversely affected *F. roseum*. Windels and Kommedahl (1977) reported control of seedling blight of peas by treating each seed with 6×10^6 spores of *Penicillium*. The control obtained was virtually equal to that obtained by treatment with captan. H. H. Luke (personal communication, 1978) has shown that the natural occurrence of antagonists (primarily species of *Aspergillus* and *Penicillium*) apparently provides natural protection to oat seed from infestation by pathogenic species of *Helminthosporium*. This fact, coupled with the successes reported above, indicates that the protection of seeds with biological control agents is both feasible and practical.

Cross-protection with conidial fungi as a type of biological control is beginning to receive close attention. Among the soil-borne pathogens those belonging to the genera *Fusarium* and *Verticillium* have attracted researchers in biological control. Cross-protection has been demonstrated both within and between species of these genera. Schnathorst and Mathre (1966) showed that weakly pathogenic strains of *Verticillium albo-atrum* protected cotton plants from more virulent strains of the same pathogen. Several years later, Zaki *et al.* (1972) reported studies which indicated that cross-protection was due to the accumulation in the xylem of antifungal compounds (phytoalexins) produced by the cotton plant in response to inoculation with the mildly pathogenic strain of *V. albo-atrum*. Apparently, a similar mechanism is also responsible for protection between species of this genus. Melonk and Horner (1975) reported that peppermint and spearmint plants were cross-protected against disease induced by a virulent isolate of *Verticillium dahliae* when inoculated first with the weak pathogen *Verticillium nigrescens*. Their results indicated that cross-protection was not due to competition but resulted from the presence of either inhibitory compounds or an alteration in host metabolism. The latter mechanism was also considered by Davis (1967) to be the most likely way that *Fusarium oxysporum* provided protection against *Fusarium* wilt of tomato, flax, carnation, cabbage, and watermelon. This author postulated that there were at least five ways in which this protection could be mediated: (1) formation and accumulation of gums and tyloses; (2) phytoalexin production; (3) altered host metabolism; (4) antibiotic production; and (5) formation of antibodies. He considered altered host metabolism the most likely mechanism of action. However, in cases of cross-protection between many unrelated organisms, competition (perhaps for infection sites) may well be operative. Phillips *et al.* (1967) reported that reduction of *Fusarium* wilt by a *Cephalosporium (Acremonium)* sp. was due to the latter blocking the entry of *F. oxysporum* f. sp. *lycopersici* into tomato tissue.

Cross-protection has also been reported between conidial fungi that inhabit and invade aerial portions of plants. Kuć *et al.* (1975) reported that cucumber was protected from *Colletotrichum lagenarium* by prior inoculation with *C. lagenarium.* They suggested that the protection resembled the immunization process in animals, but they did not investigate this process further at the molecular level. Kuć and Richmond (1977) later showed that inoculation of the first two leaves of cucumber with *C. lagenarium* systematically protected plants against disease caused by a subsequent challenge inoculation with the same fungus. The same year, Caruso and Kuć (1977) showed that similar protection provided by *C. lagenarium* against subsequent *C. lagenarium* infection also occurred in muskmelon and watermelon. Cross-protection has also been demonstrated between species. In this latter case a different mechanism is perhaps operative; antagonism may be responsible for the protection rather than a host-pathogen interaction. Van den Heuvel (1971) reported that the saprophytic fungus *Alternaria tenuissima* was antagonistic to the pathogen *Alternaria zinniae* on bean leaves and thus provided protection. He suggested that a substance produced by *A. tenuissima* inhibited germination of spores of *A. zinniae*. Spurr (1972, 1977) reported a similar situation in which a nonpathogenic *Alternaria* sp. applied to leaves of tobacco before inoculation with the brown-spot pathogen *Alternaria alternata* reduced infection. Brown spot was reduced by 60% in the laboratory and by 65% in the field. Spurr (1977) was of the opinion that other mycoflora in addition to *Alternaria* may have been involved in reducing *Alternaria* brown spot on tobacco leaves. Silverthron-Staroba and McCain (1975) reported that use of the fungicide benomyl resulted in an increase in *Pleospora* calyx rot of carnation. They postulated that benomyl reduced the population of *Penicillium,* which is a natural inhibitory resident of calyx tissue, allowing *Pleospora herbarum* to invade. Cross-protection of the type involving non-pathogenic strains or species appears to hold considerable promise as a method of biological control of foliar pathogens. In fact, manipulation of leaf mycoflora may well prove to be at least as viable a procedure as manipulation of soil mycoflora. Certainly the mycoflora on leaf surfaces is diverse. No doubt similar competitive forces exist in the phyllosphere and in the soil.

Hyperparasites among conidial fungi have also been investigated as potential biological controls for pathogens affecting aerial portions of plants. Such hyperparasites are relatively host-specific and thus, according to Huffaker *et al.* (1976), more likely to be successful as biological control agents than more general parasites and predators. Nevertheless, few examples of conidial fungal hyperparasites have been shown to have potential in biocontrol. This situation may merely indicate a lack of concerted effort in their study. For example, I have noted the constant parasitism by *Darluca (Sphaerellopsis) filum* of *Puccinia stenotaphri* on St. Augustine grass in Florida and feel that this hyperparasite is a factor in maintaining *P. stenotaphri* at a low level of destructiveness on this important turf grass of the coastal areas of the southern United States. *Darluca (Sphaerellopsis) filum* and other hyperparasites of obligate parasites have been the object of recent research on their biocontrol potential. Kuhlmann and Matthews (1976) reported that *D. filum* had potential for control of the rust fungi *Cronartium strobilinum* and *Cronartium fusiforme* on oak trees. They obtained positive infection of *C. fusiforme* in the field, but results were inconclusive. They concluded that, because of the nature of the life cycle of *C. strobilinum,* it was probably a more likely target species for biocontrol. Later, Kuhlmann *et al.* (1978) reported that *D. filum* observed in Florida seemed to exert natural control of *C. strobilinum* in pure stands of dwarf live oaks, *Quercus minima.* They found that 93% of the sori were infected and only 0.8% had telia, whereas in less dense stands only 36% of the rust sori were infected and 23% had telia. In further studies with *D. filum,* these authors confirmed that the fungus did not hold as much promise for control of *C. fusiforme* on water oak and red oak. *Darluca filum* infected *C. fusiforme* when the rust was inoculated on trees 4–21 days prior to inoculation with the hyperparasite. However, despite the infection, the number of teliospores produced was not reduced. There was, however, some reduction in basidiospore production due to inoculation of 8-to-14-day-old rust infections with *D. filum.* Nevertheless, Kuhlmann *et al.* (1978) concluded that, because of the short, irregular cycle of *C. fusiforme* on the water oak and red oak, there was not enough time for *D. filum* to colonize the rust and effect biological control. *Cronartium fusiforme* was active on oaks for only a few weeks, whereas *C. strobilinum* was active all summer. The latter was therefore adequately colonized by the hyperparasite. *Darluca filum* infection of *Puccinia recondita* on wheat has been examined by Swendsrud and Calpouzos (1972). They found that the best infection of the rust by *D. filum* was obtained by either simultaneous inoculation of the two organisms or inoculation of the hyperparasite subsequent to rust infection. These authors concluded that spraying *D. filum* onto wheat prior to rust infection was useless and in fact may even increase rust infection.

Tuberculina maxima is another hyperparasite of rust fungi that has been credited with biocontrol potential. Kimmey (1969) considered *T. maxima* responsible for the reduced number of lethal-type trunk cankers caused by *Cronartium ribicola* on white pine plantings surveyed in 1941 and again in 1965. Wicker and Wells (1968) noted that *T. maxima* could overwinter in *C. ribicola* cankers as conidia, as sporodochia, or in the mycelial stage within the pine cortex. Therefore a viable inoculum would be available for the invasion of cankers during favorable periods, and this would enhance the biological control potential of *T.*

maxima for white pine blister rust. Another hyperparasite that has been studied as a biological control agent is *Ampelomyces quisqualis,* which parasitizes powdery mildew. In fact, one of the earliest reports of biological control of a plant pathogen was that of Yarwood (1932) concerning the effects of this fungus on powdery mildew of clover caused by *Erysiphe polygoni.* This author reported that 8 days after inoculation of mildew-infected clover plants at 25°C with *A. quisqualis,* conidium production by *E. polygoni* ceased. Forty-five years later, Jarvis and Singsby (1977) reported that powdery mildew, caused by *Sphaerotheca fuliginea,* could be successfully controlled on cucumbers in greenhouses by using *A. quisqualis* and water sprays. Unfortunately, this hyperparasite is also pathogenic on foliage and fruit of the cucumber. However, damage is not too objectionable on the fruit, and the method is considered to be of practical value for the control of powdery mildew on greenhouse-grown cucumber.

Mower *et al.* (1975) reported intriguing biological control of a pathogen, which was accompanied by biodegradation of toxic alkaloids produced by the pathogenic fungus. They reported, after testing several hyperparasites, that *Fusarium roseum* 'Sambucinum' showed promise in the greenhouse and field for biological control of *Claviceps purpurea* on wheat. The authors further noted that the hyperparasite broke down ergotamine to psychotropically inert substances. In fact, they concluded that their strain of *F. roseum* appeared capable of detoxifying many, if not all, ergot alkaloids. In addition, the hyperparasite did not affect either rats or rabbits in limited toxicity tests.

B. Control of Weeds

Efforts by plant pathologists to control weeds biologically with plant pathogens have lagged behind those of entomologists in their efforts to control weeds with insects. However, Freeman *et al.* (1976a) noted that there has been a rapid increase since 1970 in research on plant pathogens for weed control. Prior to this time, the limited efforts (Wilson, 1969; Zettler and Freeman, 1972) had resulted more in hopeful indications than in definite success. A wide variety of pathogens, ranging from viruses (Charudattan *et al.,* 1976b) to nematodes (Watson, 1976), have been or are being studied. Numerous conidial fungi are among those being investigated. The array of plants selected as target species is equally diverse, ranging from oak trees in conifer forests (French and Schroeder, 1969) to blue-green algae (Cannon, 1975) in aquatic environments.

Efforts by plant pathologists have developed along two lines: the use of endemic pathogens as types of biological herbicides, and the search for and introduction of exotic pathogens. The latter is a classic approach used successfully by entomologists, and the rationale is based simply on the fact that many of the more troublesome weeds are imported into an area without their natural enemies. In many instances, this importation is accidental. In others, introduction is intentional, but the plant escapes from its original confines. Examples of both types of introduction can be found among aquatic weeds in North America. Alligator weed *(Alternanthera philoxeroides)* was flushed from the ballast of ships arriving from South America, whereas the water hyacinth *(Eichhornia crassipes)* was brought from South American because of its showy flowers. At any rate, the fact that these weeds were imported dictates that a search be conducted for natural enemies from the plant's native habitat. However, this aspect has not been exploited with plant pathogens to the extent that it has with insects.

Control of terrestrial weeds with plant pathogens seems feasible, since the fortuitous introduction of soft-rot bacteria along with the insect *Cactoblastis* for control of prickly pear *(Opuntia)* in Australia and Hawaii (Dodd, 1927; Fullaway, 1954). The bacterium is now considered to have played an important part in the successful control of this weed and no doubt prompted some of the early studies reviewed by Wilson (1969). These investigations in turn led to the more recent ones in which the evaluation of several species of conidial fungi has been prominent (Freeman *et al.*, 1976a).

The genus *Colletotrichum* has been of special interest to plant pathologists working on aspects of biocontrol. Butler (1941) reported that *Colletotrichum xanthii,* the causal agent of anthracnose and blight of Bathurst burr *(Xanthium spinosum)* in Australia, was artificially disseminated as a biocontrol agent against this weed in New South Wales with some success. More recently, other species of *Colletotrichum* have been studied for the control of terrestrial weeds. Daniel *et al.* (1973) reported control of northern joint vetch, *Aeschynomene virginica,* in Arkansas rice fields by *Colletotrichum gloeosporioides* f.sp. *aeschynomene,* an endemic fungal pathogen. These authors had originally noted the disease in 1969 at the University of Arkansas Rice Experiment Station at Stuttgart, Arkansas, where it virtually eradicated the weed. The causal fungus was isolated from diseased joint vetch, and intensive work was begun on its use as a biological control agent. Extensive tests were conducted with the pathogen in the greenhouse to determine its virulence and host range. When a spore suspension containing 2×10^6 spores/ml was sprayed onto plants to the point of runoff, northern joint vetch plants from 5 to 30 cm tall were infected and killed by the organism. The fungus also infected Indian joint vetch, *Aeschynomene indica*, but did not kill this plant. No infection was noted on the other 165 crops and native plant species tested. Replicated field experiments were begun in 1970 and continued through 1973. In field plots, control was achieved by spraying plants with spore concentrations of $2-6 \times 10^6$ spores/ml at the rate of 374 liter/ha. Control was most rapid on young plants but was 99% effective on plants as tall as 66 cm.

In further studies on biological control of northern joint vetch, Smith *et al.* (1973) reported the occurrence of *C. gloeosporiodes* f.sp. *aeschynomene* in 31 rice-growing counties of Arkansas. However, the fungus did not cause as much damage to the weed under natural conditions as it did in the biological control experiments (Daniel *et al.*, 1973). Smith *et al.* (1973) concluded that a low inoculum level in the field prevented establishment of the disease. This obstacle was apparently overcome in the biocontrol experiments by applying massive numbers of spores. Thus the Arkansas investigators developed the concept of using the fungus as a "mycoherbicide" (Templeton *et al.*, 1976). The strategy revolved around the application of massive doses of the fungus to infested rice fields by conventional methods used with other pesticides. Freeman and Charudattan (1974) proposed a similar approach when using *Cercospora piaropi* as a biological herbicide to control water hyacinth. The term "biological herbicide" is more inclusive than "mycoherbicide," which could be limited to the specific use of fungi as the biological control agent. Whatever the terminology, the approach is the same and holds considerable promise. Templeton *et al.* (1976) reported that they obtained a 95–100% kill of northern joint vetch in 17 rice fields totaling 600 acres (240 ha). Spores of *C. gloeosporioides* f.sp. *aeschynomene* were dispersed on the fields by airplane at the rate of 1.5×10^6/ml in 96 liters of water per hectare. At the time of application, the northern joint vetch was 41–68 cm tall. TeBeest *et al.* (1976) reported that temperature, dew period, and inoculum level affected the infection and severity of *C. gloeosporioides* f.sp. *aeschynomene* on northern joint vetch. One hundred percent infection occurred with concentrations above 1×10^5 spores/ml. In addition, 12 h of free moisture was needed at 28°C for 100% infections. Disease development was rapid between 28° and 32°C. Following these results, the Arkansas group patented the fungus as a mycoherbicide, obtained an experimental use permit from the Environmental Protection Agency, and entered into an agreement with a pharmaceutical company to produce the fungus in large quantities (Templeton *et al.*, 1976). They foresee no biological or technological reasons that would preclude commercialization of the fungus as a mycoherbicide.

In additional work in Arkansas, Templeton (1974) noted the potential of *Colletotrichum malvarum* as a control agent of prickly sida, *Sida spinosa*, a pest of cotton and soybean. Gudauskas *et al.* (1977) reported that yet another species of *Colletotrichum* may have potential as a mycoherbicide. In their study, *Colletotrichum dematium* f.sp. *truncata* was identified as the causal agent of anthracnose of coffee senna, *Cassia occidentalis*. However, there was some concern that the weed may serve as a source of inoculum for the pathogen on soybean.

Many species of ornamental plants introduced into the island state of Hawaii have escaped and become weeds of economic importance. Several have been target species for biological control using conidial fungi. Trujillo and Obrebo (1972, 1976) reported that severe dieback of kolomana (*Cassia surattensis*) was recognized by a rancher in 1968. They isolated a species of *Cephalosporium*

(Acremonium) from the diseased plants and determined that the fungus was pathogenic when inocula obtained from wounds of diseased plants were transferred to healthy seedlings of *Cassia surattensis*. Inoculated seedlings showed typical disease symptoms—wilting of twigs, followed by chlorosis of leaflets and final defoliation as the twigs became necrotic and died. There was a systemic invasion of the xylem characterized by a reddish-brown discoloration of the vascular tissue. The fungus was tested for its host range on eight other species of *Cassia* and *Leucaena glauca*. From these studies, the fungus appeared to be host-specific; it did not attack any of the test plants. This fact, coupled with the ability of the fungus to kill seedlings of *Cassia surattensis*, indicated biocontrol potential. In addition, Wilson (1965) reported the use of *Cephalosporium diospyri* to control persimmon *(Diospyros virginiana)* in Arkansas, thus indicating the biocontrol potential of other species of the genus. Accordingly, field experiments were begun in Hawaii on the island of Kauai in three isolated areas that differed in rainfall (100, 130, and 190 cm annually). At the outset, *Cassia surattensis* in the three areas ranged in age from 3 to 5 yr and showed no disease symptoms. Plants were inoculated by cutting the cortex of the stems and spraying the exposed xylem with a spore suspension of a *Cephalosporium* sp. Less than 5% of the plants in each area were randomly inoculated. Within 4 months, all inoculated trees died in the area with less than 100 cm of annual rainfall. In the case of sites with 130 and 190 cm of precipitation, some inoculated plants were still alive after 4 months. After 1 yr, there was 100% mortality of all host plants in the low-rainfall area and 100% infection in the other two. Three years later, the *Cassia surattensis* population in all areas had been reduced from an 80% infestation to less than 1%. Unfortunately, lantana *(Lantana camara)* and other weed species were successors.

Trujillo (1976) has also reported successful biological control of hamakua pomkami *(Ageratina riparia),* using an exotic fungal pathogen imported from Jamaica. The fungus, *Cercosporella riparia,* was probably introduced into Hawaii in 1974. Since that time, it has drastically reduced populations of *A. riparia* at elevations ranging from 1500 to 6500 ft, where ideal temperatures for disease development coincide with increased spring and summer rainfall. Proliferation of the weed had been most serious in rangelands at elevations between 800 and 6500 ft, and Trujillo (1976) considers this success a classic example of biocontrol with a plant pathogen. E. E. Trujillo (personal communication, 1978) has also suggested using *Cercospora lantanae* for control of *Lantana camara*. The potential of *C. lantanae* and other pathogens for biocontrol of *L. camara* is now being studied in a cooperative effort between plant pathologists in Florida and scientists from the Australian CSIRO (T. E. Freeman and K. L. H. Harley, unpublished data, 1978). Interest in further study of *Cercospora* spp. for biological weed control in Australia was generated by Dodd's (1961) report of successful control of the crofton weed, *Eupatorium adenophorum,* by *Cercospora eupatorii* in Queensland.

In the mid-1970s, the U.S. Department of Agriculture established a research program for biological control of terrestrial weeds in Mississippi. At least some of this work has involved conidial fungi. Ohr *et al.* (1977) reported the occurrence of *Alternaria macrospora* on the weed *Anoda cristata.* The fungus ordinarily causes a disease of cotton, and *A. cristata* is a weed pest in the cotton and soybean fields of Mississippi. However, the fungal isolate was apparently a strain that does not attack cotton, since *A. cristata* was severely affected while the surrounding cotton plants were not. As a result of these studies, Ohr *et al.* (1977) suggested that the fungus may have biocontrol potential.

Studies on the control of aquatic weeds with plant pathogens have paralleled those on terrestrial weeds. As early as 1932, Agharkar and Banerjee reported a species of *Fusarium* that attacked water hyacinth *(E. crassipes).* They considered the potential of the fungus in the control of water hyacinth but concluded that the disease progressed too slowly (because of the high resisting power of the plant) to curtail spread of the weed effectively. The fungus was later identified by Banerjee (1942) as *Fusarium equiseti,* and Snyder and Hanson (1945) later considered this species synonymous with *F. roseum.* Even prior to this report, Webber (1897) had noted the occurrence of a zonate leaf spot on water hyacinths in the St. John's River in Florida. Noting that the disease damaged the plant severely, he thought that this might prove important in combating the plant. Although no causal organism was identified, Freeman (1977) felt that the disease reported by Webber was caused by *Acremonium zonatum.* These early investigations prompted the initiation of two intensive studies on pathogens of aquatic weeds and their usefulness as biological control agents for these plants. In India, work supported primarily by PL480 funds provided by the United States was concentrated at the Commonwealth Institute of Biological Control, Indian Station, at Bangalore. In the United States research has been conducted primarily at the University of Florida. The floating water hyacinth, probably the world's most prolific aquatic weed, and hydrilla *(Hydrilla verticillata),* a submerged plant, have been the principal weeds studied.

Two species of *Cercospora* have been found that affect water hyacinths. *Cercospora piaropi* has been reported to attack water hyacinths in both India and the United States. According to Nag Raj and Ponappa (1967), *C. piaropi* causes negligible damage and appears to be of little value in reducing the vigor of water hyacinth populations. Freeman and Charudattan (1974) were of somewhat the same opinion when they reported the occurrence of *C. piaropi* in Florida. However, they noted that the fungus may have potential if used as a biological herbicide. This aspect has been more thoroughly researched with another species of *Cercospora* described from water hyacinths in Florida by Conway (1976a).

This fungus was isolated from declining water hyacinths in the Rodman Reservoir in north central Florida. From conidial size and morphology, as well as the symptoms incited, Conway (1976a) determined the fungus to be different from *C. piaropi* and named it *Cercospora rodmanii*. In later tests, this fungus was found to hold considerable promise as a biocontrol agent for water hyacinths (Conway, 1976b; Conway and Freeman, 1976a).

Cercospora rodmanii initially causes small, punctate spots on the leaves and petioles of infected plants. Severely diseased leaves become chlorotic, and the bulbous petioles become spindly. Secondary root rot (cause undetermined) is frequently encountered on weakened plants. Severely affected water hyacinths eventually die and sink below the surface (Conway, 1976a). *Cercospora rodmanii* was considered by Conway (1976a) to be the causal agent of the widespread destruction of water hyacinths in the Rodman Reservoir in 1971 and to a lesser extent in 1972 and 1973 (Freeman *et al.,* 1975). In fact, Conway (1976b) has duplicated the symptoms noted on plants in the Rodman Reservoir by spraying *C. rodmanii* on water hyacinths under field conditions. He demonstrated that the fungus could be grown in culture on potato dextrose broth fortified with 0.5% yeast extract. The cultures were then macerated in a blender and sprayed on the plants. A high degree of infection and subsequent damage resulted. The fungus spread from a small inoculated area of less than 65 m^2 to damage severely water hyacinths in a 1.7-ha pool. Approximately 1 kg wet weight of fungal mycelium was used to inoculate the small area. Conway and Freeman (1976a) have since shown that similar results can be obtained in larger bodies of water (over 100 ha). The damaging effects are augmented when *C. rodmanii* is combined with insects *(Arzama densa* and *Neochetina eichhorniae)* and other fungi (e.g., *Acremonium zonatum*). Even larger-scale tests of *C. rodmanii* are planned by the Florida group in conjunction with the U.S. Army Corps of Engineers.

The fungus *C. rodmanii* has a limited host range (Conway and Freeman, 1976b; Freeman *et al.,* 1976b). In fact, of over 80 different plants of ecological and economic importance tested in the greenhouse and field, only healthy water hyacinths were vigorously attacked. The fungus sporulated on senescent leaves of lettuce and certain cucurbits but did not attack healthy foliage. Thus its use as a biocontrol agent for water hyacinth appears to hold little danger for other plants. Toxicological properties toward humans and other warm-blooded animals have not been determined. However, Conway and Cullen (1978) have recently shown that the fungus is not toxic to mosquito fish (a *Gambusia* sp.) in aquaria tests. Nevertheless, in the United States extensive toxicological tests must be conducted before the Environmental Protection Agency will allow the fungus to be used on a wide scale. Work is now underway to reach the goal of utilizing *C. rodmanii* in a biological control program for water hyacinth. Its potential is too great not to continue this research with maximum effort. The University of

Florida has recently patented the fungus and entered into an agreement with a pharmaceutical firm to produce and distribute it as a biological herbicide on a commercial basis.

Cephalosporium eichhorniae, first described in India by Padwick (1946), was considered by Nag Raj and Ponappa (1967) to have some biocontrol potential. Rintz (1973) considered *C. eichhorniae* synonymous with *C. zonatum,* described from fig by Tims and Olive (1948). In the meantime, Gams (1971) placed the name *Cephalosporium* in synonymy with *Acremonium,* and the fungus is now referred to as *Acremonium zonatum.* Rintz (1973) concluded that this fungal pathogen did not seem capable of killing water hyacinths or seriously hindering their prolific growth. In addition, he was of the opinion that in areas where the disease was first noted (i.e., Louisiana) it did not appear to be causing significant damage or even infection. Rintz appears to be correct in this latter assessment, at least for certain latitudes. As noted previously, Webber (1897) had observed the same apparent zonal leaf spot in the St. John's River in northern Florida and suggested that the disease-causing microorganism might prove useful in combating the water hyacinth in its native home. Eighty years later, the disease noted by Webber certainly has not controlled proliferation of the water hyacinth in northern Florida. However, in southern Florida, *A. zonatum* is omnipresent on water hyacinths and no doubt exerts pressure on plant populations, especially in the presence of insects (Charudattan *et al.,* 1978). In addition, Freeman *et al.* (1976b) have noted severe damage due to *A. zonatum* in more tropical areas such as Puerto Rico, Trinidad, and El Salvador.

Considerably more work is needed before the true potential of *A. zonatum* can finally be assessed. The fungus is easily cultured in large volumes on a simple medium such as potato dextrose broth fortified with 0.5% yeast extract. Water hyacinths in the field can be inoculated by spraying the plants with ground-up fungal mats. With such a procedure, small-scale tests have been performed in Florida using the fungus alone (Freeman *et al.,* 1976b) and also with insects (Charudattan *et al.,* 1978). In Louisiana the fungal pathogen is used in combination with insects and other fungi (Conway and Freeman 1976a). Results indicate that *A. zonatum* alone does not generally cause extensive damage, but that its effects are enhanced when it is used in combination with other microorganisms and insects. In addition, there is variation in the disease reaction among water hyacinth populations. Some plants are damaged severely as a result of inoculation with the fungus, while others are not. Martyn (1977) has correlated this with variations in phenolic acid and polyphenol oxidase activity. These factors will have to be taken into consideration in a biocontrol program. Also, Rintz (1973) was concerned about the apparently wide host range of *A. zonatum.* Out of the 17 plant species he tested 16 were susceptible to some degree. However, his tests were conducted with inoculated plants covered with plastic bags for 7 days. Needless to say, these conditions were abnormal. Freeman *et al.* (1976a) tested

52 different crop species under field conditions where the plants were sprayed with the fungus in the same manner as water hyacinths. None of the crop species was susceptible under these conditions. However, the fungus has been reported occurring naturally on two important crops: fig (*Ficus carica* L.) in Louisiana (Tims and Olive, 1948), and coffee (*Coffea arabica* L.) in Africa, India, and Central America (Deighton, 1954; Nag Raj and George, 1962; Salas and Echandi, 1967).

Bipolaris (Helminthosporium) stenospila was reported by Charudattan *et al.* (1976a) to be a virulent pathogen of water hyacinth. The fungus was isolated from blighted plants in the Dominican Republic. However, its biocontrol potential has not been tested beyond the quarantine greenhouse stage. Bermuda grass *(Cynodon dactylon)* and sugarcane *(Saccharum officinarum)* are known hosts of this fungus (Freeman, 1957; Faris, 1928). Its pathogenicity to important crop plants may preclude its use as a biocontrol agent, especially in view of the fact that the water hyacinth isolates of the fungus also infect certain crop plants tested under greenhouse conditions (Charudattan *et al.*, 1976a; Freeman *et al.*, 1976b).

Nag Raj and Ponappa (1970) reported a blight of water hyacinth caused by *Alternaria eichhorniae*. The fungus produced a toxin and had a narrow host range. These authors considered the fungus to have biocontrol potential, but no further studies have been reported. Likewise, Ponappa (1970) reported the occurrence of *Myrothecium roridum* var. *eichhorniae* pathogenic to water hyacinth. Although it appeared to have biocontrol potential, it also had a wide host range which he felt would restrict its use.

Work on pathogens of hydrilla, a submerged aquatic weed, has been concentrated in Florida. Charudattan and Lin (1974) reported that species of *Penicillium, Aspergillus,* and *Trichoderma* produced toxins which caused lysis of the plant cells in test tubes. From these results they concluded that the fungi might have biocontrol potential. Charudattan and McKinney (1977) also reported the isolation of *Fusarium roseum* from a *Stratioites* sp. in Holland. The fungus proved to be highly pathogenic to hydrilla in test tubes and aquaria. Aside from this work, the search for pathogens for biocontrol of hydrilla has not been fruitful.

Freeman (1977) has noted the occurrence of other conidial fungi on miscellaneous aquatic weeds. However, none of them appear to have biocontrol potential.

C. Control of Insects and Other Pests

Several genera of conidial fungi which parasitize insects may be useful biocontrol agents for these pests. (See Roberts and Humber, this volume, Chapter 21.) In addition, there are conidial fungi which infect other pests. For example, species of *Arthrobotrys, Candelabrella, Dactylella,* and *Monacrosporium* are parasites or predators of nematodes (see Barron, this volume, Chapter 20).

However, they have not been studied extensively as potential biological control agents for these pests. If the imagination is allowed latitude, there are no doubt other cases in which conidial fungi could play a role in the biological control of invertebrate and perhaps even vertebrate pests.

IV. CONSIDERATIONS

In utilizing conidial fungi, as with other biological control agents, caution must be the rule. Their effect on humans, domestic animals, and wildlife must be determined. Their relationship with other nontarget species must be continually assessed and reevaluated. Genetic stability is another important factor; some idea of genetic variation should be obtained before their wide-scale use is undertaken. There should be a system of checks and balances in the form of effective control measures for the proposed agent before its use is recommended to the public. Certainly there will be many problems in the successful implementation and completion of such programs. The Environmental Protection Agency has taken a keen interest in the use of pathogens in the environment, especially the aquatic one. Yet a large segment of the general public is affected with "pathophobia." Therefore researchers will have to prove their case well and document beyond reasonable doubt the effectiveness and safety of a pathogenic biocontrol agent.

V. SUMMARY AND CONCLUSIONS

Biological control is recognized as a natural phenomenon in ecosystems which has probably existed since the beginning of life. However, only recently have humans attempted to use this phenomenon to their advantage in the control of pests. Entomologists were pioneers in this area. Plant pathologists have now begun to look at pathogens as biocontrol agents for other pathogens and weeds. Conidial fungi are prominent among those being studied and have a number of advantages as biocontrol agents. For example, they are numerous and diverse in habitat and mode of action; most species are easily cultivated and disseminated; they are not likely to eradicate their host; and they are not normally considered parasitic or toxigenic to animals. However, there are also disadvantages to their use as biocontrol agents. Fewer disadvantages are recognized in their use against weed pests than against other pathogens and pests. At least successes against weeds as target species have in many instances progressed beyond the laboratory stage. Researchers are now using the more classic approach that entomologists have used successfully in pitting organism against organism. This is especially evident in studies on cross-protection and the use of antagonists such as *Trichoderma* spp. for the control of pathogens. The same approach has been taken in the control of both terrestrial and aquatic weeds with endemic and exotic pathogens.

On the basis of these early studies, we can conclude that the use of conidial fungi for biocontrol of pest species is in its infancy. However, the successes thus far achieved should serve as building blocks upon which to base future studies. The opportunities are too great to ignore.

REFERENCES

Agharkar, S. P., and Banerjee, S. N. (1932). *Fusarium* sp. causing disease of *Eichhorniae crassipes* Solms. *Proc. Indian Sci. Congr., 19th* p. 298.

Backman, P. A., and Rodriguez-Kabana, R. (1975). A system for the growth and delivery of biological control agents to the soil. *Phytopathology* **65,** 819-821.

Backman, P. A., Rodriguez-Kabana, R., and Williams, J. C. (1975). The effect of peanut leafspot fungicides on the nontarget pathogen, *Sclerotium rolfsii. Phytopathology* **65,** 773-776.

Baker, K. F., and Cook, R. J. (1974). "Biological Control of Plant Pathogens." Freeman, San Francisco, California.

Banerjee, S. N. (1942). *Fusarium equiseti* (Cda.) Sacc. (*Fusarium falcatum* App. et Wr.) causing a leafspot of *Eichhorniae crassipes. J. Dep. Sci. Calcutta Univ.* **1,** 29-37.

Barnett, H. L., and Lilly, V. G. (1963). A destructive mycoparasite *Gliocladium roseum. Mycologia* **54,** 72-77.

Bliss, D. E. (1951). The destruction of *Armillaria mellea* in citrus soils. *Phytopathology* **41,** 665-683.

Butler, F. C. (1941). Anthracnose and seedling blight of Bathurst burr caused by *Colletotrichum xanthii. Aust. J. Agric. Res.* **2,** 401-410.

Cannon, R. (1975). Field and ecological studies on blue-green algal viruses. *Proc. Symp. Water Qual. Manag. Through Biol. Control* pp. 112-117.

Caruso, F. L., and Kuć, J. (1977). Protection of watermelon and muskmelon against *Colletotrichum lagenarium* by *Colletotrichum lagenarium. Phytopathology* **67,** 1285-1289.

Chang, I. P., and Kommedahl, T. (1968). Biological control of seedling blight of corn by coating kernels with antagonist organisms. *Phytopathology* **58,** 1395-1401.

Charudattan, R., and Lin, C. (1974). *Penicillium, Aspergillus* and *Trichoderma* isolates toxic to hydrilla and other aquatic plants. *Hyacinth Control J.* **12,** 70-73.

Charudattan, R., and McKinney, D. E. (1977). A *Fusarium* disease of the submersed aquatic weed *Hydrilla verticillata. Proc. Am. Phytopathol. Soc.* **4,** 222 (abstr.).

Charudattan, R., Conway, K. E., and Freeman, T. E. (1976a). A blight of waterhyacinth, *Eichhornia crassipes,* caused by *Bipolaris stenospili (Helminthosporium stenospilum). Proc. Am. Phytopathol. Soc.* **2,** 65 (abstr.).

Charudattan, R., Cordo, H. A., Silveira-Guido, A., and Zetter, F. W. (1976b). Obligate pathogens of the milkweed vine *Morrenia odorata,* as biocontrol agents. *Proc. Int. Symp. Biol. Control Weeds, 4th, 1976* p. 241 (abstr.).

Charudattan, R., Perkins, B. D., and Littrell, R. C. (1978). The effects of fungi and bacteria on the decline of anthropod-damaged waterhyacinths in Florida. *Weed Sci.* **26,** 101-107.

Conway, K. E. (1976a). *Cercospora rodmanii,* a new pathogen of waterhyacinth with biocontrol potential. *Can. J. Bot.* **54,** 1979-1083.

Conway, K. E. (1976b). Evaluation of *Cercospora rodmanii* as a biological control of waterhyacinth. *Phytopathology* **66,** 914-917.

Conway, K. E., and Cullen, R. . (1978). The effect of *Cercospora rodmanii,* a biological control of waterhyacinth, on the fish *Gambusia affinis. Mycopathologia* **66,** 113–116.

Conway, K. E., and Freeman, T. E. (1976a). Field evaluation of *Cercospora rodmanii* as a biological control for waterhyacinth. *Proc. Am. Phytopathol Soc.* **3,** 272 (abstr.).

Conway, K. E., and Freeman, T. E. (1976b). The potential of *Cercospora rodmanii* as a biological control for waterhyacinth. *Proc. Int. Symp. Biol. Control Weeds, 4th, 1976* pp. 207–209.

Daniel, J. T., Templeton, G. E., Smith, R. J., and Fox, W. T. (1973). Biological control of northern jointvetch in rice with an endemic fungal disease. *Weed Sci.* **21,** 303–307.

Davis, O. (1967). Cross protection in *Fusarium* wilt disease. *Phytopathology* **57,** 311–314.

Deighton, F. C. (1954). Plant pathology report. *Rep. Dep. Agric. Sierra Leone, 1952* pp. 29–30.

de Trogoff, H., and Ricard, J. L. (1976). Biological control of *Verticillium malthousei* by *Trichoderma viride. Plant Dis. Rep.* **60,** 677–680.

Dodd, A. P. (1927). The biological control of prickly pear in Australia. *Aust., Commw., Counc. Sci. Ind. Res., Bull.* **34,** 1–44.

Dodd, A. P. (1961). Biological control of *Eupatorium andenophorum* in Queensland, Australia. *Aust. J. Sci.* **23,** 356–365.

Faris, J. A. (1928). Three helminthosporium diseases of sugar cane. *Phytopathology* **18,** 753–774.

Freeman, T. E. (1957). A new helminthosporium disease of bermuda grass. *Plant Dis. Rep.* **58,** 277–278.

Freeman, T. E. (1977). Biological control of aquatic weeds with plant pathogens. *Aquat. Bot.* **3,** 174–184.

Freeman, T. E., and Charudattan, R. (1974). Occurrence of *Cercospora piaropi* in waterhyacinth in Florida. *Plant Dis. Rep.* **58,** 277–278.

Freeman, T. E., Charudattan, R., and Conway, K. E. (1975). Use of plant pathogens for bioregulation of aquatic macrophytes. *Proc. Water Qual. Manage, Through Biol. Control, 1975* pp. 20–23.

Freeman, T. E., Charudattan, R., and Conway K. E. (1976a). Status of the use of plant pathogens in the biological control of weeds. *Proc. Int. Symp. Biol. Control Weed, 4th, 1976* pp. 201–206.

Freeman, T. E., Charudattan, R., Conway, K. E., and Zettler, F. W. (1976b). "Biological Control of Aquatic Weeds with Plant Pathogens," Contract Rep. A–76–2. U.S. Army Corps Engineers, Vicksburg, Mississippi.

French, D. W., and Schroeder, D. B. (1969). Oak wilt fungus, *Ceratocystis fagacearum,* as a selective herbicide. *For. Sci.* **15,** 198–203.

Fullaway, D. T. (1954). Biological control of cactus in Hawaii. *J. Econ. Entomol.* **47,** 696–670.

Gams, W. (1971). "Cephalosporiumartige Schimmelpilze (Hyphomycetes)." Fischer, Stuttgart.

Grosclaude, D., Ricard, J., and Dubos, B. (1975). Inoculation of *Trichoderma viride* spores via pruning shears for biological control of *Stereum purpureum* on plum tree wounds. *Plant Dis. Rep.* **57,** 25–28.

Gudauskas, R. T., Teem, D. H., and Morgan-Jones, G. (1977). Anthracnose of *Cassia occidentalis* caused by *Colletotrichum dematium* f.sp. *truncata. Plant Dis. Rep.* **61,** 468–470.

Harder, P. R., and Troll, J. (1973). Antagonism of *Trichoderma* spp. to sclerotia of *Typhula incarnata. Plant Dis. Rep.* **57,** 924–926.

Huffaker, C. B. (1976). An overview of biological control, with particular commentary on biological weed control. *Proc. Int. Symp. Biol. Control Weeds, 4th, 1976* pp. 3–12.

Huffaker, C. B., Simmonds, F. J., and Laing, J. E. (1976). The theoretical and empirical basis of biological control. *In* "Theory and Practice of Biological Control" (C. B. Huffaker and P. S. Messenger, eds.), pp. 41–73. Academic Press, New York.

Hunt, R. S., Parmeter, F. R., and Cobb, F. W., Jr. (1971). A stump treatment technique for biological control of forest root pathogens. *Plant Dis. Rep.* **61,** 728–730.

Jarvis, W. R., and Singsby, K. (1977). The control of powdery mildew on greenhouse cucumber by water sprays and *Ampelomyces quisqualis*. *Plant Dis. Rep.* **61**, 728-730.

Kelly, W. D. (1976). Evaluation of *Trichoderma harzianum* impregnated clay granules as a biocontrol for *Phytophthora cinnamomi* causing damping-off of pine seedlings. *Phytopathology* **66**, 1023-1027.

Kimmey, J. W. (1969). Inactivation of lethal-type blister rust canker on western white pine. *J. For.* **67**, 296-299.

Kuć, J., and Richmond, S. (1977). Aspects of the protection of cucumber against *Colletotrichum lagenarium* by *Colletotrichum lagenarium*. *Phytopathology* **67**, 533-536.

Kuć, J., Shockley, G., and Kearney, K. (1975). Protection of cucumber against *Colletotrichum lagenarium* by *Colletotrichum lagenarium*. *Physiol. Plant Pathol.* **7**, 195-199.

Kuhlmann, E. G., and Matthews, F. R. (1976). Occurrence of *Darluca filum* on *Cronartium strobilinum* and *Cronartium fusiforme* infecting oaks. *Phytopathology* **66**, 1195-1197.

Kuhlmann, E. G., Matthews, F. H., and Tillerson, H. P. (1978). Efficacy of *Darluca filum* for biological control of *Cronartium fusiforme* and *Cronartium strobilinum*. *Phytopathology* **68**, 507-511.

Leach, C. L. (1946). A disease of dodder caused by the fungus *Colletotrichum destructivum*. *Plant Dis. Rep.* **42**, 827-829.

Martyn, R. D. (1977). Disease resistance mechanisms in waterhyacinths and their significance in biocontrol programs with phytopathogens. Ph.D. Dissertation, University of Florida, Gainesville.

Melonk, H. A., and Horner, C. E. (1975). Cross protection in mints by *Verticillium nigrescens* against *Verticillium dahliae*. *Phytopathology* **65**, 767-769.

Mower, R. L., Snyder, W. C., and Hancock, J. G. (1975). Biological control of ergot by *Fusarium*. *Phytopathology* **65**, 6-10.

Nag Raj, T. R., and George, K. V. (1962). Occurrence of zonal leaf spot of coffee caused by *Cephalosporium zonatum*. *Curr. Sci.* **31**, 104-105.

Nag Raj, T. R., and Ponappa, K. M. (1967). Some interesting fungi of India. *Commonw. Inst. Biol. Control, Tech. Bull.* **9**, 31-43.

Nag Raj, T. R., and Ponappa, K. M. (1970). Blight of waterhyacinth caused by *Alternaria eichhorniae*. *Trans. Br. Mycol. Soc.* **55**, 123-150.

Ohr, H. D., Munnecke, D. E., and Bricker, J. L. (1973). The interaction of *Armillaria mellea* and *Trichoderma* spp. as modified by methyl bromide. *Phytopathology* **63**, 965-973.

Ohr, H. D., Pollack, F. G., and Ingler, B. F. (1977). The occurrence of *Alternaria macrospora* on *Anoda cristata* in Mississippi. *Plant Dis. Rep.* **61**, 208-209.

Overmier, K., and Roncadori, R. W. (1977). Factors influencing antagonism of *Gliocladium virens* and *Trichoderma harzianum* to *Diplodia gossypina*. *Proc. Am. Photopathol. Soc.* **4**, 128 (abstr.).

Padwick, G. W. (1946). Notes on Indian fungi. *Mycol. Pap.* **17**, 1-12.

Peeples, J. L., Curl, E. A., and Rodriguez-Kabana, R. (1976). Effect of the herbicide EPTC on the biocontrol activity of *Trichoderma viride* against *Sclerotium rolfsii*. *Plant Dis. Rep.* **60**, 1050-1054.

Phillips, D. V., Leben, C., and Allison, D. C. (1967). Mechanism for the reduction of *Fusarium* wilt by *Cephalosporium* sp. *Phytopathology* **57**.

Ponappa, K. M. (1970). On the pathogenicity of *Myriothecium roridum-Eichhornia crassipes* isolate. *Hyacinth Control J.* **8**, 18-20.

Pottle, H. W., and Shigo, A. L. (1975). Biological control of hymenomycetes in *Acer rubrum* wounds by *Trichoderma viride*. *Proc. Am. Phytopathol. Soc.* **2**, 104 (abstr.).

Pottle, H. W., Shigo, A. L., and Blanchard, R. O. (1977). Biological control of hymenomycetes by *Trichoderma harzianum*. *Plant Dis. Rep.* **61**, 687–690.

Ricard, J. L., Grosclaude, C., and Ale-Agha, N. (1974). Antagonism between *Eutypa armeniacae* and *Gliocladium roseum*. *Plant Dis. Rep.* **58**, 983–984.

Rintz, R. E. (1973). A zonal leaf spot of waterhyacinth caused by *Cephalosporium zonatum*. *Hyacinth Control J.* **11**, 41–44.

Rudakov, O. L. (1960). Biological control of *Cuscuta* with *Alternaria cuscutacidae* in USSR. Cited in Simmonds *et al.* (1976).

Salas, A., and Echandi, E. (1967). Mancha zonal de la hoja de los cafetos (*Coffea arabica* L.) causada por el hongo *Cephalosporium zonatum*. *Turrialba* **17**, 292–295.

Schnathorst, W. C., and Mathre, O. E. (1966). Cross protection in cotton by strains of *Verticillium albo-atrum*. *Phytopathology* **56**, 1204–1209.

Silverthron-Staroba, M., and McCain, A. H. (1975). Effect of *Penicillium* and *Bacillus* spp. and benomyl on development of *Pleospora* calyx rot of carnation. *Proc. Am. Phytopathol. Soc.* **2**, 118 (abstr.).

Simmonds, F. J., Franz, J. M., and Sailer, R. (1976). *In* "Theory and Practice of Biological Control" (C. B. Huffaker and P. S. Messenger, eds.), pp. 17–39. Academic Press, New York.

Smith, H. S. (1919). Some phases of insect control by the biological methods. *J. Econ. Entomol.* **12**, 288–292.

Smith, R. J., Jr., Daniel, J. T., Fox, W. T., and Templeton, G. E. (1973). Distribution in Arkansas of a fungus disease used for biocontrol in northern jointvetch. *Plant Dis. Rep.* **57**, 695–697.

Snyder, W. C., and Hansen, H. N. (1945). The species concept in *Fusarium* with reference to discolor and other sections. *Am. J. Bot.* **32**, 657–666.

Snyder, W. C., Wallis, G. W., and Smith, S. N. (1976). Biological control of plant pathogens. *In* "Theory and Practice of Biological Control" (C. B. Huffaker and P. S. Messenger, eds.), pp. 521–539. Academic Press, New York.

Spurr, H. W., Jr. (1972). Biological control of tobacco brown spot. *Phytopathology* **62**, 807 (abstr.).

Spurr, H. W., Jr. (1977). Protective applications of conidia of nonpathogenic *Alternaria* sp. isolates for control of tobacco brown spot disease. *Phytopathology* **67**, 128–132.

Swendsrud, D. P., and Calpouzos, L. (1972). Effect of inoculation sequence and humidity on infection of *Puccinia recondita* by the mycoparasite *Darluca filum*. *Phytopathology* **62**, 931–932.

TeBeest, D. O., Templeton, G. E., and Smith, R. J. (1976). Epidemiology of northern jointvetch anthracnose—A proposed mycoherbicide. *Proc. Am. Phytopathol. Soc.* **3**, 271 (abstr.).

Templeton, G. E. (1974). Endemic fungus disease for control of prickly sida in cotton and soybean. *Ark. Farm Res.* **23**, (4), 12.

Templeton, G. E., TeBeest, D. O., and Smith, R. J. (1976). Development of an endemic fungal pathogen as a mycoherbicide for biocontrol of northern jointvetch in rice. *Proc. Int. Symp. Biol. Control Weeds, 4th, 1976* pp. 214–216.

Tims, E. C., and Olive, L. S. (1948). Two interesting leaf spots of fig. *Phytopathology* **38**, 706–715.

Trujillo, E. E. (1976). Biological control of hamakua pamakaui with plant pathogens. *Proc. Am. Phytopathol. Soc.* **3**, 298 (abstr.).

Trujillo, E. E., and Obrero, F. P. (1972). The biological control of *Cassia surrattensis* burn brush weed of pastures in Hawaii with *Cephalosporium* sp. *Phytopathology* **62**, 793 (abstr.).

Trujillo, E. E., and Obrebo, F. P. (1976). *Cephalosporium* wilt of *Cassia surrattensis* in Hawaii. *Proc. Int. Symp. Biol. Control Weeds, 4th, 1976* pp. 217–220.

van den Heuvel, J. (1971). Antagonism between pathogenic and saprophytic *Alternaria* species on bean leaves. *In* "Ecology of Leaf Surface Micro-Organisms" (T. E. Preece and C. H. Dickinson, eds.). Academic Press, New York.

Watson, A. K. (1976). The biological control of Russian knapweed with a nematode. *Proc. Int. Symp. Biol. Control Weeds, 4th, 1976* pp. 221–223.

Webber, H. J. (1897). The water hyacinth and its relation to navigation in Florida. *U.S. Dep. Agric., Div. Bot. Bull.* **18**, 1–20.

Weindling, R. (1932). *Trichoderma lignorum* as a parasite on other soil fungi. *Phytopathology* **22**, 837–845.

Weindling, R. (1934). Studies on a lethal principle effective in the parasite action of *Trichoderma lignorum* on *Rhizoctonia solani* and other soil fungi. *Phytopathology* **24**, 1153–1179.

Wells, H. D., Bell, D. K., and Jaworski, C. A. (1972). Efficacy of *Trichoderma harzianum* as a biocontrol for *Sclerotium rolfsii*. *Phytopathology* **62**, 442–447.

Wheeler, H., and Luke, H. H. (1963). Microbial toxins in plana disease. *Annu. Rev. Microbiol.* **17**, 223–242.

Wicker, E. F., and Wells, J. M. (1968). Overwintering of *Tuberculina maxima* on white pine blister rust canker. *Phytopathology* **58**, 391.

Wilson, C. L. (1965). Consideration of the use of persimmon wilt as a *silvercide* for weed persimmon. *Plant Dis. Rep.* **49**, 780–791.

Wilson, C. L. (1969). Use of plant pathogens in weed control. *Annu. Rev. Phytopathol.* **7**, 411–434.

Windels, C. E., and Kommedahl, T. (1977). Factors affecting biological seed treatment in controlling seedling blight of peas. *Proc. Am. Phytopathol. Soc.* **4**, 157 (abstr.).

Yarwood, C. E. (1932). *Ampelomyces quisqualis* on clover mildew. *Phytopathology* **22**, 31.

Zaki, A. I., Keen, N. T., and Erwin, D. C. (1972). Implications of vergosin and hemigossypol in the resistance of cotton to *Verticillium albo-atrum*. *Phytopathology* **62**, 1402–1406.

Zettler, F. W., and Freeman, T. E. (1972). Plant pathogens as biocontrols of aquatic weeds. *Annu. Rev. Phytopathol.* **10**, 455–470.

20

Predators and Parasites of Microscopic Animals

G.L. Barron

I. INTRODUCTION

Soil and organic debris are the habitats of a wealth of microscopic life forms. Coexisting in these terrestrial microhabitats is a remarkable group of fungi which specialize in capturing the consuming minute animals. These predatory fungi are scattered among several classes but are particularly prevalent in the Hyphomycetes (Drechsler, 1941a). Predators produce a sparse but extensive mycelial

Biology of Conidial Fungi, Vol. 2

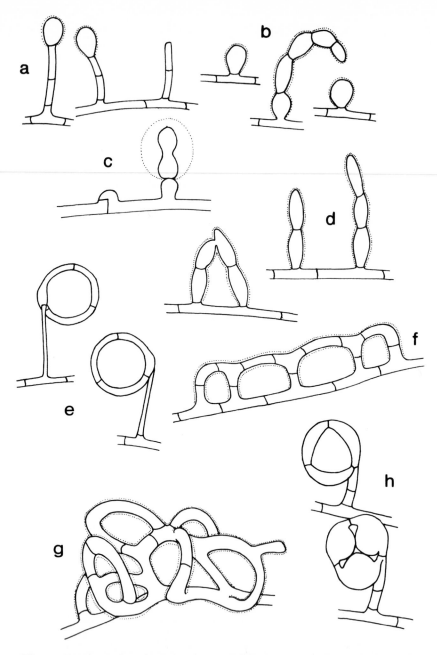

Fig. 1. Trapping devices of predatory fungi. (a) Stalked adhesive knobs *(Dactylaria candida);* (b) sessile adhesive knobs *(Dactylella parvicollis);* (c) hourglass adhesive knob *(Nematoctonus concurrens);* (d) adhesive branches *(Dactylella cionopaga);* (e) nonconstricting rings *(Dactylella leptospora);* (f) scalariform adhesive trap *(Dactylella gephyropaga);* (g) reticulate adhesive trap *(Arthrobotrys oligospora);* (h) constricting rings *(Arthrobotrys dactyloides).* (From Barron, 1977a.)

system in the environment. Along the length of the hyphae, specialized organs of capture are produced from modified branches or branch systems.

A second group of Hyphomycetes also attack microanimals, but no mycelium is produced in the environment. The entire hyphal development occurs inside the body of the animal host. Once the host contents have been assimilated, the fungus then breaks through the cuticle at a number of points and conidiophores and conidia are produced outside the body of the host. This second group is referred to as endoparasitic and exists in soil only as conidia or resting spores.

Animals captured or parasitized include rhizopods, testaceous rhizopods, rotifers, tardigrades, and occasionally springtails, but it is as predators or parasites of nematodes that these fungi are best known and most intensively studied (Duddington, 1962; Barron, 1977a). The scarcity of reports on animals other than nematodes reflects in part the limitations of our techniques for studying the more difficult groups, and it is probable that, with critical and imaginative approaches, many more predators and parasites of microscopic animals will be discovered.

II. PREDATORS OF NEMATODES

Under the low-nutrient conditions of water agar plates, and probably in their natural habitats, the predatory Hyphomycetes produce a sparse but extensive system of anastomosing hyphae. These hyphae radiate out from captured nematodes, producing trapping devices at intervals along their length. Organs of capture take a variety of forms (Fig. 1) but are usually categorized as adhesive nets or branches, adhesive knobs, nonconstricting rings, and constricting rings. In the predatory Zoopagales (Drechsler, 1941a) the hyphae themselves function as organs of capture by secreting an adhesive material. In the Hyphomycetes, however, only *Arthrobotrys botryospora* (Barron, 1979c) is known to capture nematodes consistently by means of adhesive hyphae, and even in this species adhesive nets are the primary trapping device.

A. Generalized Life Cycle

Irrespective of the type of trapping device used to capture the prey, subsequent events resulting in penetration, colonization, and exploitation of the victim are very similar for most predatory fungi. The sequence of events after capture is outlined here for *Arthrobotrys oligospora* (Drechsler, 1937a).

The cuticle is penetrated by a narrow infection peg (Shepherd, 1955) which swells inside the body and forms a spherical postpenetration bulb. This bulb varies considerably in size within and between species but may be large enough to fill the body diameter of the nematode completely. In some predatory fungi, particularly the clamp-forming genus *Nematoctonus,* postpenetration bulbs are not produced (Drechsler, 1941b). Assimilative hyphae develop rapidly from the bulb and digest the contents of the nematode. Protoplasm in the assimilative hyphae is then translocated outside the corpse. All that now remains is the cuticular shell of the nematode filled with the empty hyphae of the predator. Within a few days these remains are degraded by other microbial life forms. The material translocated from the victim is used for further hyphal growth and trap formation or may be channeled into the production of dispersal spores (conidia) or persistent spores (chlamydospores).

In *A. oligospora* the conidia are produced successively to form a cluster at the apex of a tall, slender conidiophore (Fig. 2). As the conidiophore matures, a succession of clusters is produced from a sympodially elongating apex. The conidia are large (22–32 × 12–20 μm), hyaline, uniseptate, and dry. They are apparently thin-walled and shrivel quickly when exposed to dry air or collapse in age while still attached to the conidiophore. The method of spore dispersal is not known with certainty. Conidia may be wind-borne; it is probable, however, that they are dispersed on the body hairs of arachnids and insects.

The large spore carries a considerable food reserve. After dispersal, this allows the germinating spore to produce a hyphal system complete with one or more trapping devices independent of any external nutrient source.

B. Adhesive Knobs

Nearly 20 species of Hyphomycetes trap nematodes by means of adhesive knobs. A knob is a morphologically distinct cell which is either sessile, as in *Dactylella phymatopaga* (Drechsler, 1954a), or more often borne aloft on a two- or three-celled stalk, as in *Dactylella candida* (Drechsler, 1937a). A thin layer of adhesive covers each knob, but the support stalks are not adhesive. Four species that produce knobs also produce nonconstricting rings as an additional trapping device (Cooke and Godfrey, 1964). *Dactylella parvicollis* (Drechsler, 1961) is exceptional in that the adhesive knobs often continue growth from their apex (Jarowaja, 1968) to produce a curved adhesive branch (Fig. 1b) which may eventually form an adhesive ring.

Ultrastructural studies by Heintz and Pramer (1972) on *Dactylella drechsleri* and by Dowsett and Reid (1977) on *Dactylaria candida* showed that the knobs were usually multinucleate and contained numerous osmiophilic inclusions, especially around the periphery of the cells. These inclusions were considered to have a secretory function related to the production of additional adhesive material following capture of the prey. Dowsett and Reid have shown that, after capture, there is a flattened mass of adhesive material at the point of capture. In scanning

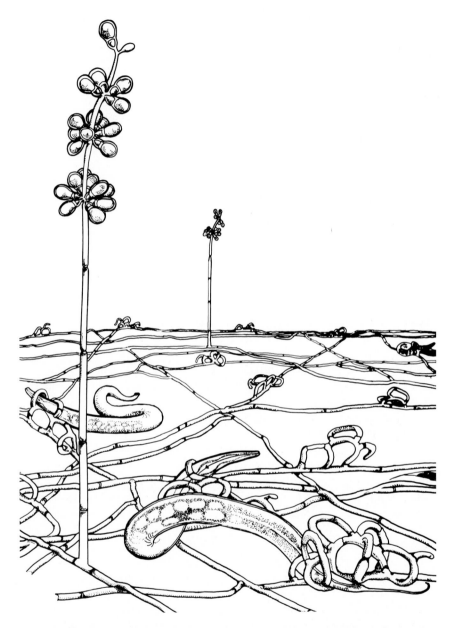

Fig. 2. *Arthrobotrys oligospora*. Diagrammatic representation of the hyphae and adhesive traps with captured nematodes. Conidiophore with clusters of conidia in left foreground. (From Barron, 1977a. Drawing by Rick Hurst.)

electron microscopy (SEM) studies of an Ontario isolate of *Dactylaria dasguptae* this appressorium-like pad (Fig. 3b) is clearly visible.

Drechsler (1937a) thought that stalked knobs in *Dactylaria candida* were rarely operative and then only in the capture of the smallest or feeblest of nematodes. Duddington (1962) supported Drechsler's findings and observed that, in fungi possessing both rings and knobs, knobs were usually secondary in action to rings and in some cases appeared to be functionless. It was shown by Barron (1975), however, that nematodes were captured by knobs, but that in the ensuing struggle the latter often broke off at the point of attachment to the stalk. The nematode then swam off with the detached knob anchored firmly in its cuticle. The presence of a knob or several knobs attached to its cuticle did not seem to interfere with the locomotion of the animal. The presence of detachable knobs has been noted in another strain of *D. candida* (Dowsett and Reid, 1977).

Detachable knobs are an advantage to predatory fungi. A nematode can travel a considerable distance before penetration and subsequent growth incapacitate it. The predator arrives at a new site for further predation with an immediately available food source and can use this as the energy base to produce many more trapping devices. Detachable knobs are particularly important for species which depend entirely on knobs for predation. Penetration and subsequent assimilation for adhesive knobs is the same as for *A. oligospora*.

The conidia of species that produce adhesive nets or knobs often germinate directly from the spore apex and produce a short adhesive structure which allows the spore to attach directly to the cuticle of the nematode. In some knob-forming species, the conidia produce adhesive knobs while still attached to the co-nidiophore (Drechsler, 1937a). Thus prepared, the conidia can attach directly to the nematode and circumvent any fungistatic effects in the environment which might prevent spore germination.

Two species of nematode-destroying fungi whose hyphae possess clamp connections were described by Drechsler (1941a) in a new genus, *Nematoctonus*. Since these original descriptions an additional seven species have been described in this genus (Giuma and Cooke, 1972). Predatory *Nematoctonus* species capture nematodes in the same manner as *Dactylaria candida*, except that the knobs are strongly anchored to the parent hyphae and rarely detach. In *Nematoctonus* the adhesive cell is very distinctive; it consists of a secretory cell shaped like an hourglass (Fig. 1c). At maturity this cell is engulfed in a large, spherical ball of viscous material. The hyphae are usually prostrate and extremely fine (Drechsler, 1941b), the knobs being produced either directly on the hyphae or on lateral branches. Conidia in *Nematoctonus* are produced individually on short denticles from the vegetative hyphae. Following dispersal, and in the presence of nematodes, each conidium produces an adhesive knob from its tip which can become attached to a nematode and initiate the predatory cycle.

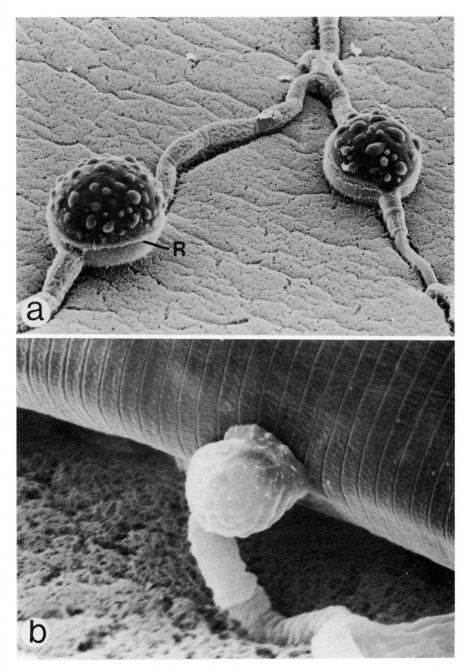

Fig. 3. (a) Chlamydospores of *Arthrobotrys flagrans* showing ruptured primary wall (R) and warty encrustations on expanded inner wall. (b) Adhesive knob of *Dactylaria dasguptae* with appressorium-like pad of adhesive at the point of attachment.

C. Adhesive Nets and Branches

Nearly 40 species of fungi capture nematodes by means of adhesive nets or adhesive branches. In its simplest form, as in *Dactylella cionopaga* (Drechsler, 1950a), the trap consists of an erect branch composed of one to three cells (Fig. 1d). Branches in close proximity often anastomose and produce adhesive hoops. In *Dactylella gephyropaga* (Drechsler, 1937a) the adhesive branches arise in a close, regular arrangement. Individual branches may capture nematodes directly, but more often lateral anastomoses produce scalariform nets (Fig. 1f).

In most predatory Hyphomycetes, however, the trapping device is a complex three-dimensional net, as in *Arthrobotrys oligospora* (Fig. 1g). In this species an erect lateral branch grows from a prostrate vegetative hypha, curves around, and grows down to fuse with the parent hypha. A branch from this primary loop, or another branch from a prostrate hypha, repeats this process and through hyphal anastomoses a network of adhesive hyphae is constructed. The hyphae composing the net are often more robust than the vegetative hyphae.

Nets are apparently adhesive almost from inception, and short branches destined to be primary loops or adhesive networks are capable of capturing nematodes. Nordbring-Hertz and Stahlhammar-Carlemalm (1978) induced net formation in cultures of *A. oligospora* by adding morphogenic peptides. Transmission electron microscopy (TEM) of these nets showed that a thin film of adhesive was present over their entire surface even before nematodes were added.

A moving nematode can brush against an adhesive trap without being captured. If the nematode stops briefly in contact with the trap, however, it is held firmly. In its struggle to escape, the victim frequently contacts or becomes entangled in another net system. Attachment at several points considerably limits the ability of the nematode to exert escape leverage. During the struggle there is considerable movement of the trap and associated hyphae. Barron (1977a) has suggested that this hyphal drag plays a significant part in exhausting the nematode and making its escape more difficult.

In their pioneering TEM studies on several species of predatory Hyphomycetes, Heintz and Pramer (1972) found that the vegetative hyphae were similar to those described for other fungi. The trapping devices, however, possessed distinctive ultrastructural features. They found that the nets of *Monacrosporium rutgeriensis* were multinucleate and contained dense inclusions, which they suggested were for storage of adhesive prior to secretion. They also found an accumulation of membrane-bound inclusions possibly functional in the formation and transport of adhesive. In an excellent study on *Arthrobotrys oligospora*, Nordbring-Hertz and Stahlhammar-Carlemalm (1978) induced nets by the addition of peptides to pure cultures of the fungus and subsequently added bacteria-free cultures of nematodes. They were thus able to study a time sequence of events after capture, while eliminating possible influences of contaminating organisms. Ultrastructural studies showed that walls of traps were four to five times as thick as those of vegetative hyphae (Fig. 4a). They also found numerous electron-dense vesicles which seemed to empty their contents at the cytoplasmic membrane and gradually disappeared after invasion of the nematode. These inclusions were not found in either the vegetative hyphae or the infection hyphae. They were therefore considered storage sites for the adhesive prior to capture of the nematode. It is possible that the thick wall of the trap has a special structure permitting ready passage of the reserve adhesive to the outside. After capture there is an increased secretion of adhesive (Nordbring-Hertz and Stahlhammar-Carlemalm, 1978), and a thick deposit is laid down at the point of contact within 15 min of capture and before the cuticle is penetrated (Fig. 4b). In their SEM studies on *Arthrobotrys robusta,* Estey and Tzean (1976) illustrated a thick appressorium-like pad at the capture point.

Using a light microscope, Shepherd (1955) studied the penetration of nematodes captured by *A. oligospora.* She found that penetration and production of the postpenetration bulb took about an hour, by which time movement of the nematode had virtually ceased. In their TEM study on early infection stages, Nordbring-Hertz and Stahlhammar-Carlemalm (1978) confirmed these findings and noted that the body cavity of the nematode was filled with assimilative hyphae within 6 h of capture.

In their discussion, Nordbring-Hertz and Stahlhammar-Carlemalm noted that a nematode caught by a net could be penetrated at any point, whereas vegetative hyphae could penetrate only through a body orifice of the nematode. They concluded from this that nets produced extracellular enzymes (functional in penetration) absent from ordinary hyphae. Whether penetration of the cuticle is enzymatic or mechanical is not clear. The evidence indicates that penetration is very rapid and follows the formation of a thick adhesive pad. It is probable that this adhesive pad functions as an appressorium, anchoring the trap to the cuticle. Thus anchored, the fungus can exert the intrusive force necessary to penetrate the cuticle by growth pressure. It is possible that penetration is in part mechanical and in part enzymatic (see Fig. 4a).

D. Nonconstricting Rings

Only four species of predatory fungi capture their prey by means of nonconstricting rings. The ring in *Dactylella leptospora* (Drechsler, 1937a) is initiated by a very slender, erect hypha which widens in its upper part and then curves in a circular pathway and fuses with the stalk at the point where it widens (Fig. 1e). The ring is usually three-celled, but four-celled rings are not uncommon. There is a marked swelling of the ring cell just above the point of contact

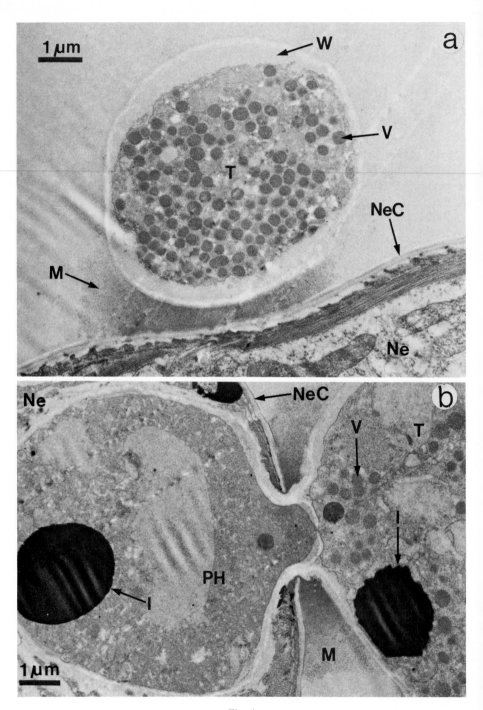

Fig. 4.

with the stalk. As the ring ages, the stalk often shows signs of collapse just below the point of attachment to the ring (Barron, 1977a).

Nonconstricting rings are passive in their action. When a moving nematode enters the ring, the forward motion causes the ring to wedge around the body, sometimes with sufficient force to cause a marked constriction of the cuticle. In the nematode's ensuing struggle to escape, the ring often breaks off at the point of weakness near the stalk apex. The host then moves off with the ring wedged firmly around its body. This process may be repeated until a succession of rings encircle the victim. Detached rings remain viable, and the nematode is penetrated and colonized as described for other predators (Drechsler, 1937a).

All fungi that produce nonconstricting rings also produce adhesive knobs. Sometimes, in soil plates, the knobs will be common and the rings rare. In one strain of *Dactylaria candida* only a single ring was found during several weeks' observation (G. L. Barron, unpublished data). The factors that favor the development of knobs rather than rings, or vice versa, are not known. It is possible that predators known to produce only knobs will also produce nonconstricting rings when suitably stimulated.

E. Constricting Rings

Twelve species of predatory fungi capture nematodes by means of constricting rings. Rings may vary considerably in size both within and among species. In *Dactylella stenobrocha* (Drechsler, 1950a) the rings are small, ranging from 20 to 31 μm in diameter. In *Arthrobotrys anchonia* (Drechsler, 1954a) rings range from 20 to 42 μm in diameter, and rings up to 50 μm in diameter have been recorded in some strains (G. L. Barron, unpublished observation). Drechsler (1937a) gave the ring dimensions for *Dactylaria brochopaga* as 20–35 μm in diameter. This species, however, has proved to be most variable, and recent reports show that *D. brochopaga* can produce giant rings up to 130 μm in diameter (Tsai *et al.*, 1975; Insell and Zachariah, 1977, 1978a), and at the other extreme tiny rings no more than 10–15 μm in diameter (Barron, 1979a).

1. Ring Formation

In critical studies on *Dactylella aphrobrocha*, Drechsler (1950a) described precisely the sequence of events leading to formation of the constricting ring. He

---

Fig. 4.   Ultrastructure of *Arthrobotrys oligospora*. (a) Transverse section through trap (T) attached to the nematode cuticle (NeC). Note thickened wall (W) and numerous electron-dense vesicles. (V). The trap is attached to the nematode by a thick adhesive secretion (M). The nematode's cuticle appears to be thinner in the mucilaginous zone. (b) The nematode's cuticle has been penetrated, and the assimilative hyphae (PH) are inside the body of the nematode (Ne). Electron-dense vesicles are still present inside the trap cell but not in the penetration hypha. (From Nordbring-Hertz and Stahlhammar-Carlemalm, 1978.)

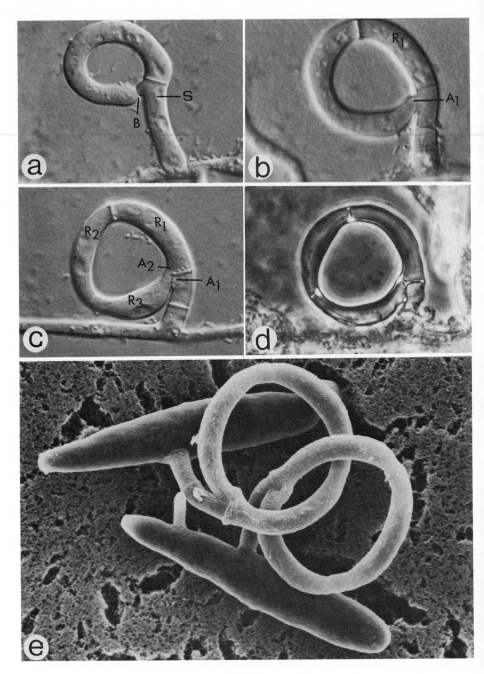

Fig. 5.

showed that its formation was quite complex and involved two separate and sequential anastomoses. I have found that Drechsler's interpretation applies also to *Arthrobotrys anchonia, Arthrobotrys dactyloides,* and *Dactylaria brochopaga,* and it is probably common to the other ring-forming species. The series of steps described by Drechsler is illustrated here for *Dactylaria brochopaga* (Fig. 5) and described in detail below.

The ring is first recognized as a circinately curved lateral branch arising from a prostrate vegetative hypha. Cross-walls soon appear in the proximal part, delimiting the stalk cells (S in Fig. 5a). As the tip of the recurving branch approaches the stalk, a lateral bud (B in Fig. 5a) grows out from the distal portion of the upper stalk cell, and the tip of the recurving branch anastomoses (A_1 in Fig. 5c) with this bud. This event forms a rudimentary ring that incorporates a portion of the stalk cell. About the same time the first and second ring cells (R_1 and R_2 in Fig. 5c) are delimited. The branch tip continues to grow, bridges the septal barrier between the stalk cell and the first ring cell, and anastomoses with the base of the first ring cell. This second anastomosis (A_2 in Fig. 5c) unites the first and third ring cells. Septum formation across this anastomosis completes the ring. An additional septum forms at the base of the bud, separating the third ring cell from the stalk cell. The results of these two anastomoses are clearly seen in the scanning electron micrograph of *A. dactyloides* in Fig. 5e.

The tip of the curving ring homes in on the bud of the stalk cell with unerring accuracy. I observed one exceptional case in which the ring curved in one direction but the bud developed on the opposite side of the stalk. The tip grew past the stalk and then reversed its direction, turning back on itself to fuse with the bud. It is tempting to hypothesize that gaseous emanations guide the tip to fuse with the bud.

Using time-lapse photomicrography, Higgins and Pramer (1967) published an excellent sequence confirming the formation of the stalk bud in *Arthrobotrys dactyloides.* They noted that the bud was invariably formed by the upper stalk cell in opposition to the advancing branch tip with which it fused, producing a closed ring.

2. Ring Mechanism

When stimulated, the cells of the constricting ring expand rapidly inward (Drechsler, 1933). Closure is irreversible, takes place in about $\frac{1}{10}$ s, and is accompanied by a threefold increase in cell volume (Commandon and de Fonbrune, 1939; Müller, 1958). Rings can be activated by touching or rubbing the

Fig. 5. Formation of the constricting ring in *Dactylaria brochopaga*. (a) Tip of curving ring branch is attracted to bud (B) at apex of stalk cell (S). (b) Ring cell tip anastomoses (A^1) with bud, and first ring cell (R^1) is delimited. (c) Advancing tip anastomoses with base of first ring cell (A^2), and second (R^2) and third (R^3) ring cells are delimited. (d) Mature ring. (e) Scanning electron micrograph of rings of *Arthrobotrys dactyloides* showing anastomoses with both stalk cell and ring cells.

inside (luminal) surface of even a single ring cell with a glass microneedle (Commandon and de Fonbrune, 1939). Closure can also be effected by the addition of hot water (Couch, 1937) or a hot scalpel in the vicinity (Müller, 1958). The osmotic potential of the ring cells is the same before and after closure (Müller, 1958), indicating a rapid threefold increase in osmotically active materials in the ring cells either during or very shortly after closure.

Couch (1937) suggested that the water necessary for cell expansion was imbibed through the stalk and hyphal cells. Müller (1958), however, calculated that the volume of water required could not pass through several septa of the stalk cells in the time available. According to Müller, stimulation of the inner wall of the ring cell induces an instantaneous decrease in wall pressure and an increase in membrane permeability. An increase in the suction potential draws in water from the surrounding medium. As this uptake proceeds, the level of osmotically active substances in the cell rises as a result of the hydrolysis of large molecules present, which maintains the inflow of water. The osmotic potential is maintained at about $0.6\ M,$ never rising much above a level necessary to maintain the plastic extension of the wall. This process continues until the inflation is complete.

As an alternative hypothesis, Müller has suggested that the osmotic recovery is rather more slow. The cell would be able to expand by virtue of its initial osmotic potential, which would fall during expansion from a level equivalent to $0.6\ M$ sucrose to one finally equivalent to $0.2\ M$ sucrose, as a result of water uptake. The cell would then recover its initial osmotic potential after inflation by relatively slow hydrolysis of polymers present within it, and/or by relatively slow transport of solutes from the stalk cells. Müller noted that the time interval between cell inflation and the plasmametric estimation of its osmotic pressure would allow this recovery of osmotic potential to take place undetected.

Müller noted that the second version of his hypothesis was the less physiologically exacting. It envisaged stimulation of the cell as inducing a change in wall structure and a change in permeability but did not require in addition any exceptionally rapid increase in the amount of osmotically active solutes in the cell.

Rudek (1975) believed that closure could be explained in part by imbibition through the stalk and in part by a change in membrane permeability. It was demonstrated by Insell and Zachariah (1978b) and Barron (1979a), however, that detached rings were fully functional and would close normally when stimulated by heat. Barron showed further that particulate inclusions in the stalk cells and adjacent hyphae did not shift position during trap closure. Movement of water through the stalk from the hyphal system is apparently not necessary for ring closure.

A major contribution to our understanding of the ring mechanism was made by Heintz and Pramer (1972), who studied the ultrastructure of the constricting rings of *Arthrobotrys dactyloides* before and after closure. They found that the outer wall of the ring cell did not expand into the lumen, but rather the wall ripped open along a median line and it was the inner wall which ballooned into the lumen to effect closure. The irreversibility of the process was thus explained.

In the uninflated ring cells Heintz and Pramer noted numerous membrane-bound, electron-dense inclusions localized beneath the plasma membrane on the luminal side of the ring only. These inclusions appeared to be able to fuse with each other and to form a labyrinthine matrix enclosed by involutions of the expanded and invaginated plasma membrane. Open ring cells also possessed an electron-lucent layer containing flocculent material, between the plasma membrane and the outer wall on the luminal side. After closure, the membrane-bound inclusions and labyrinthine networks had disappeared and the electron-lucent layer was no longer evident.

As emphasized by Heintz and Pramer, the threefold increase in cell volume needs additional wall and membrane material to accommodate the enlarged cell surface which cannot be explained by *de novo* synthesis. They have suggested that labyrinthine structures produced by fusion of the membrane-bound inclusions with each other and with the plasma membrane represent a preformed reserve surface able to accommodate the expanding protoplast on demand. They regard the electron-lucent layer as a suitable site for membrane fusion and as the source of reserve wall material.

In ultrastructural studies of constricting rings in *Dactylaria brochopaga*, Dowsett *et al.* (1977) found that the wall of the luminal side was composed of four or five layers (Fig. 6), whereass the nonluminal wall was composed of only two layers. They also found that the membrane-bound inclusions could persist in inflated rings which had not trapped nematodes. They proposed that the vesicles functioned as a binding surface to maintain the integrity of the inner wall during expansion and suggested that they might also contain lytic or hydrolytic enzymes which aid the fungus in digesting the prey.

3. An Explanation of the Ring Mechanism

Current theories on the ring mechanism by Heintz and Pramer (1972), Rudek (1975), Dowsett *et al.* (1977), Insell and Zachariah (1978a), and others support Müller's explanation or a modification thereof. These theories propose an instantaneous increase in membrane permeability as the primary force initiating ring closure. I propose a simpler theory to explain the ring mechanism, which does not require any unique membrane properties and at the same time accommodates all the known information.

It is well established that many fungi produce lines of weakness at predetermined positions in their cell walls, which will rupture at a critical moment. In Ascomycetes this results in the explosive discharge of ascospores; in other fungi sporangia or spores may be violently shot off. I consider closure of the constricting ring analogous to a controlled explosive discharge.

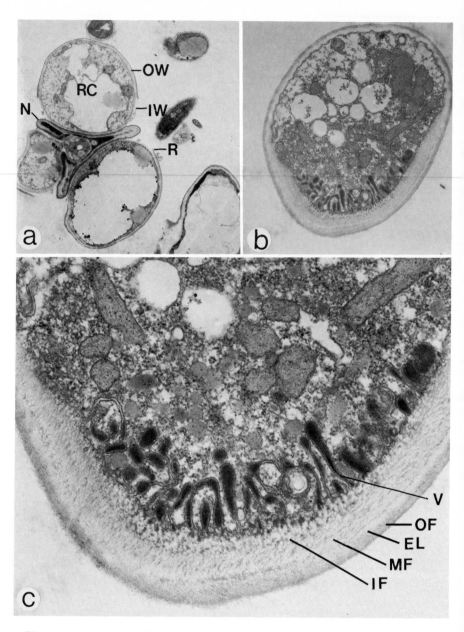

Fig. 6. Ultrastructure of *Dactylaria brochopaga* rings. (a) Captured nematode (N) showing triangulate configuration. Outer wall (OW) of the expanded ring cell (RC) has ruptured (R), and the inner wall (IW) has swollen inward to constrict the nematode. (b) Section through unexpanded ring cell. (c) Higher magnification of (b). Note complex wall structure consisting of outer fibrillar (OF), electron-luscent (EL), middle fibrillar (MF), and inner fibrillar (IF) layers. Labyrinthine formation of electron-dense vesicles (V) can be seen just inside the inner wall. (From Dowsett *et al.* 1977.)

A line of weakness exists along the middle of the wall on the luminal face. This is at a critical point such that a physical disturbance will cause the line of weakness to break. The membrane does not change its permeability; it is already freely permeable to water. The release of wall pressure due to rupture, however, results in a rapid uptake of water. This causes the elastic inner wall to balloon into the lumen, making use of preformed membrane and wall materials as described by Heintz and Pramer and Dowsett et al.

Swelling of one cell alone will exert sufficient mechanical stress on the remaining ring cells that they too will rupture and expand, either concurrently or in rapid succession. Presumably, expansion forces resulting from sudden heat will also cause enough mechanical stress to effect closure.

The sudden threefold increase in cell volume results in an equivalent reduction in osmotic pressure. The ring therefore is not always capable of constricting the nematode immediately. As indicated by Barron (1979a), closure is often biphasic. In the first phase the ring cells cushion over the nematode's cuticle with little constriction. As the osmotic potential of the ring cells increases, the resistance of the host cuticle is eventually overcome, at which time the second phase brings about sudden and severe constriction of the victim.

4. Giant Rings

In an isolate of *Dactylaria brochopaga* recovered from Taiwan soils, Tsai *et al.* (1975) reported the occurrence of occasional giant rings and illustrated a ring measuring about 90 μm in diameter. Later, and apparently independently, Insell and Zachariah (1977, 1978a,b) found giant rings (megatraps) in Drechsler's original strain (ATCC 13897) after treatment with a chemical mutagen. Their mutant strain (MG20) regularly produced giant rings with diameters recorded up to 130 μm (Insell and Zachariah, 1978a). The yield of giant traps was estimated to be 12–15% of total trap production.

Megatraps were found to be functional in that they would inflate and constrict on nematodes. It was not confirmed, however, that the captured prey could be invaded and assimilated. While large numbers of giant rings were produced by the mutant strain, it was shown later (Turnbull and Zachariah, 1978) that giant rings could be produced, less abundantly, by the wild-type strain under conditions of poor ventilation.

In an Ontario strain of *Dactylaria brochopaga* we found occasional giant rings (Fig. 7) in a bench-incubated culture on potato dextrose agar 7 days after adding nematodes (G. L. Barron, unpublished data). Thus stagnant air is not obligatory for giant ring formation, although it apparently encourages it. It should be pointed out that all occurrences of giant rings have been reported using nutrient agar. We have never found giant rings on water agar with nematodes as the only nutrient source. On nutrient agar there is prolific hyphal growth which gives the fungus a

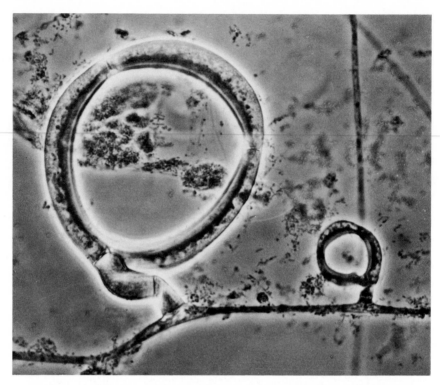

Fig. 7. *Dactylaria brochopaga.* Giant and standard rings arising from the same hypha.

much greater and more compact nutrient base to supply the increased energy required for giant ring formation.

5. Microconidia

Seven species of predatory Hyphomycetes possess microconidial states (Barron, 1979a). In *A. anchonia* standard conidia measure 29–43 × 15–19 μm, whereas microconidia measure only 18–22 × 4–5 μm. Barron showed that, in the presence of nematodes, microconidia of *A. anchonia* and *D. brochopaga* germinated to form tiny rings ranging from 8–15 μm in diameter, and that these rings were fully functional in attacking nematodes. The importance of microconidia in the biology of predatory fungi, and of factors which control their occurrence, has not been established.

F. Morphogenesis of Traps

Early studies on predatory fungi showed that trapping devices were seldom developed in pure cultures but were produced abundantly in the presence of nematodes (Couch, 1937) or nematode filtrates (Commandon and de Fonbrune,

1938). It was subsequently found that many compounds of animal origin, particularly blood serum and earthworm extract, stimulated trap morphogenesis (Roubaud and Deschiens, 1939). The term "nemin" was coined by Pramer and Stohl (1959) to categorize substances inducing trap formation. Wootton and Pramer (1966) found that leucine, isoleucine, and valine could induce net formation in *Arthrobotrys conoides*. In *Arthrobotrys oligospora*, however, Nordbring-Hertz (1973, 1974) found valyl peptides were more effective in net induction, especially when their low-nutrient medium (LNM) was supplemented with mineral salts. This is in contrast to the findings of Balan and Lechevalier (1972) who suggested that abundant trap formation in *Arthrobotrys dactyloides* (constricting rings) was induced by starvation conditions in the presence of nematodes. In *A. conoides* (adhesive nets) trap formation was found to be carbon dioxide-dependent, and in the presence of nemin different strains of this species responded differently to varying concentrations of carbon dioxide (Bartnicki-Garcia *et al.*, 1964).

Nematodes themselves are much more effective in trap induction than the suggested trap-inducing agents. Feder *et al.* (1963) showed that a single nematode could initiate trap formation (constricting rings) over a 1-cm-diameter column containing *Dactylella doedycoides*. This point was emphasized by Nordbring-Hertz (1977) who found that living nematodes induced trap formation in *A. oligospora* much more rapidly than either peptides (Phe-Val) or crude extracts from nematodes (Table I). The response to nematodes was independent of the number of nematodes used. In the presence of nematodes, trap induction began within 2 h, and the maximum number of traps was initiated within 3-4 h. Pure peptides and crude extracts of nematodes did not initiate trap production

TABLE I

Trap Formation in Response to Number of Living Nematodes,
A Nematode Extract, and a Pure Peptide[a]

Additions[b]	Trap formation on LNM[c]				
	1 h	2 h	3 h	5 h	24 h
4–6 nematodes	0	+	+	+++	n.d.
22–26 nematodes	0	+	+	+++	n.d.
50–60 nematodes	0	+	+	>500	n.d.
400 μg Phe-Val	0	0	0	+	>500
Crude nematode extract	0	0	0	0	+

[a] From Nordbring-Hertz, 1977.

[b] Additions were made to 3-day-old cultures on LNM. Phe-Val, Phenylalanine-valine.

[c] 0, No traps; +, ++, +++ indicate increasing number of traps; n.d., not determined.

until 6–24 h after addition. Nordbring-Hertz concluded that peptides and amino acids were only partly responsible for trap induction, and that additional effects were due to another mechanism. She proposed that a combination of volatiles, environmental factors, and permeability factors might result in a facilitated intake of ions and/or morphogenic substances. The mechanism might also involve organisms other than nematodes.

G. Persistence of Predatory Hyphomycetes

In the absence of a sexual spore, persistence in the Hyphomycetes is often accomplished by a thick-walled chlamydospore. Chlamydospores have been described for a number of predatory Hyphomycetes by Drechsler (1937a). These resting spores occur in *Arthrobotrys oligospora, A. musiformis,* and *A. conoides* and are abundant in *A. amerospora* (Schenck *et al.,* 1977). In *A. flagrans,* chlamydospores are so prolific that they may outnumber the conidia. In this species they have a striking appearance under SEM (Barron, 1979a). As the chlamydospore enlarges, the primary hyphal wall containing it eventually ruptures and the developing chlamydospore grows out of this casing (Fig. 3a). At maturity the primary hyphal wall appears as a frill around the base of the chlamydospore (arrow), and warty protuberances arise only from the "new" outside wall of the spore.

Chlamydospores have not been found in most species of predatory fungi. Under the peculiar conditions of petri dishes, where most species have been discovered, chlamydospores may not develop. The conidia of the predatory fungi appear thin-walled, collapse readily, and lyse quickly on the biosurface. These conidia do not seem capable of functioning as persistent spores. It is not clear therefore how most of the predatory fungi persist through adverse conditions.

It is possible that teleomorphs exist but have yet to be discovered. A good clue was provided by Drechsler (1937a) who found a small discomycete with apothecia 0.5–0.8 mm in diameter in a culture of *Arthrobotrys superba* infested with nematodes. Noteworthy is that *A. superba* is one of the species in which chlamydospores have never been found.

The ascospores in Drechsler's tiny apothecia were colorless, tear-shaped, and about 5 μm long and 1.3 μm wide. Drechsler suggested the possibility that infection might be initiated by adhesion of the spore to the nematode's cuticle, a method of infection not unusual in other predatory fungi (Barron, 1977a).

Barron and Dierkes (1977) discovered a predatory species of *Nematoctonus* which produced small basidiomata in pure cultures on nutrient agar. The basidiomata were typical of the genus *Hohenbuehelia,* a gill-forming Basidiomycete. At maturity, viable basidiospores were discharged, and these germinated in the presence of nematodes, producing hourglass-shaped adhesive cells. Adhesive basidiospores could initiate haploid infections and dikaryotized in the nematode's body when the victim was attacked simultaneously by both mating types.

Recently, Tzean and Estey (1978) showed that *Arthrobotrys oligospora, A. robusta,* and *A. superba* could behave as destructive mycoparasites. They were found to attack *Matruchotia varians, Rhizoctonia solani,* and a *Geotrichum* species. The ability of predatory Hyphomycetes to persist as mycoparasites in soil or organic debris in nature has not yet been established.

H. Attractants

It is interesting to consider that predatory fungi produce chemicals that lure nematodes to the site of capture. Such chemicals might be produced by the hyphae and attract nematodes to the general area of the predator, or chemicals might be secreted by the trapping devices themselves and attract nematodes directly to the organs of capture.

It was shown by Townshend (1964) that the fungus-feeding nematode *Aphelenchus avenae* was attracted to 57 out of 59 fungi tested. Although no predatory fungi were tested, there is no reason to suppose they might be exceptions. The host-finding mechanism of the plant parasitic nematode *Neotylenchus linfordi* was studied by Klink *et al.* (1970), who found that the nematode was attracted to filtrates from several species of fungi, particularly to *Gliocladium roseum,* a common soil fungus. They did not characterize the attractants precisely but found them to be small, thermostable molecules.

It is possible that volatile compounds serve as attractants for predatory fungi, and it is well established that carbon dioxide plays an important role in attracting free-living nematodes (Nicholas, 1975). Katznelson and Henderson (1963) showed that *Rhabditis* nematodes were attracted to certain fungi and Actinomycetes, and suspected that ammonia production by the organisms formed the basis of this attraction. Their tests demonstrated that ammonium chloride (50 μg/liter) produced an effect equal to that of the culture filtrate. It was found by Balan and Gerber (1972) that predatory fungi produced significant quantities of ammonia. These workers, however, believed that the ammonia played a significant role as a nematicide rather than as an attractant. We might expect in the natural scheme of things that organisms such as nematodes would be attracted to sites of protein degradation (ammonia released) or high metabolic activity (carbon dioxide released) as part of their food-finding mechanisms. Predatory fungi could take advantage of this.

Monoson and Ranieri (1972) studied the nematode *Aphelenchus avenae* in association with the net-forming predator *Arthrobotrys musiformis* and observed that movement of nematodes did not appear to be random. They suggested that nematodes were attracted to the trapping organs of the predator because of the presence of a nematode-attracting substance (NAS) produced by the trapping devices. Monoson *et al.* (1973) claimed that extracts of *Arthrobotrys musiformis* and *Monacrosporium doedycoides,* from hyphae in which traps had been induced, contained an NAS, while extracts from hyphae without such traps did not display any attraction potential. The free-living nematode *Panagrellus redivivus* and the predatory *Arthrobotrys dactyloides* (constricting rings) were studied by Balan and Gerber (1972). They showed that *A. dactyloides* produced attractants and suggested that this attraction was explained in part by carbon dioxide. Balan *et al.* (1974) studied attraction by the net-forming species *Arthrobotrys conoides, A. oligospora,* and *Monacrosporium rutgeriensis*. They found that cultures of these species produced both nematode-attracting substances and nematicidal substances. Attraction to a culture growing on cornmeal agar was no greater than to cornmeal agar alone. If, however, trapping devices were induced, then there was a marked enhancement in attracting nematodes, and this effect was measurable within 5 min of adding a nematode suspension. In their experiments, nets were induced by adding a 1-ml suspension containing approximately 10,000 nematodes. The methods section of their paper, however, did not make clear how these nematodes were removed prior to assaying for NAS.

In a subsequent paper, Balan *et al.* (1976) outlined a rapid method which allowed them to establish the presence of NAS within 2 h. Using thin-layer chromatographic techniques they were able to show that *Monacrosporium rutgeriensis* produced three different compounds attractive to *Panagrellus redivivus*. Their primary extract was prepared from 15 petri dishes containing 21-day-old cultures of the fungus. While it is not stated in the methods section, it is presumed that nets were induced in the cultures prior to extraction.

The production of attractants by the trapping devices of predatory fungi was shown simply and clearly by Field and Webster (1977). They tested five species of predatory fungi (Table II) against either *Rhabditis* or *Aphelenchus* and analyzed the results statistically. It was found that fungal hyphae bearing traps,

TABLE II

Mean Numbers of Nematodes Moving Toward Stimulated and Nonstimulated Disks of Predaceous Fungi after 12 h[a]

Fungus	Trap stimulant	Nematode	Mean number of nematodes moving toward stimulated disk	Mean number of nematodes moving toward nonstimulated disk
Arthrobotrys anchonia	Horse serum	*Rhabditis*	36.5 ± 1.5	6.2 ± 1.0
A. dactyloides	Horse serum	*Rhabditis*	45.8 ± 3.6	7.0 ± 0.6
A. oligospora	Horse serum	*Rhabditis*	40.2 ± 1.1	8.7 ± 1.4
A. conoides	Horse serum	*Rhabditis*	40.8 ± 2.0	5.8 ± 1.2
A. anchonia	Nematode extract	*Rhabditis*	36.8 ± 1.0	6.7 ± 0.7
A. anchonia	Horse serum	*Aphelenchus*	39.8 ± 2.2	8.5 ± 1.3
Monacrosporium eudermatum	Horse serum	*Rhabditis*	38.5 ± 0.9	6.7 ± 1.2

induced by either horse serum or nematode extracts, were attractive to nematodes. Their results were significant at the $P = 0.01$ level. They showed further that neither horse serum, unstimulated cultures (no trapping devices present), nematode extracts, nor the nutrient agar used had any attraction capability.

The weight of evidence now supports the original contention of Monoson and Ranieri that nematodes are indeed attracted to the site of their destruction. It is unfortunate that in the mycological literature the word "sirenin" has already been preempted for a sex hormone. Its application to the attractants used by the predatory fungi would surely have been much more appropriate.

III. PREDATORS OF OTHER ANIMALS

A. Arthropods

Larger soil animals such as mites, springtails, and other arthropods are often parasitized by fungi but are seldom captured by predatory fungi. This is undoubtedly related to the physical strength of these animals, associated with the traction afforded by their leg action which allows them to exert considerable force against restraint. There is a notable exception to this generalization: *Arthrobotrys entomopaga* was discovered by Drechsler (1944) capturing a small species of springtail belonging to the genus *Sminthurides* (Collembola). The captured springtails ranged from 125 to 130 μm in length.

The fungus produces, through numerous anastomoses, a prostrate hyphal network on the surface of the substrate. Scattered over each network a cluster of two-celled trapping devices is formed. Each trap consists of a stout basal stalk cell and an ellipsoid, glandular, adhesive cell. Unlike the adhesive knobs of the nematode-trapping Hyphomycetes, these cells produce copious quantities of adhesive which form glistening droplets 15–20 μm in diameter. The adhesive cells attach to the ventral surface or legs of the springtails. The large amount of exudate ensures extensive adhesion, and the clustered arrangement of the adhesive cells results in simultaneous attachment at several points. The low body position and the weak leg action of the animals prevent them from developing sufficient traction to escape.

After capture, the springtails are penetrated, and a series of swollen cells is produced from which the assimilative hyphae develop and colonize the body. Drechsler pointed out that minute springtails thoroughly infest decaying porous material and roam the minute, deeply ramifying passages in the organic debris. The predatory fungus can line such channels with its adhesive capture organs, and the high humidity prevents desiccation of the fluid adhesive.

Reproduction of *A. entomopaga* is reminiscent of that of *A. cladodes* (Drechsler, 1937a), with successive clusters of conidia (15–28 × 4.5–5.5 μm) arising from an elongating conidiophore.

Drechsler found that *A. entomopaga* could also capture nematodes. Assimilative hyphae in this prey were unlike those of other nematode predators and had an atypical, knotted appearance. Drechsler drew attention to the fact that several fungi have been described which, although similar in appearance to known predators, fail to capture nematodes. He suggested that they might represent other predators of springtails.

In general, adhesive knobs are not produced by didymosporous *Arthrobotrys* species. Adhesive knobs are more typically found in forms with phragmosporous conidia, such as *Dactylella* or *Dactylaria*. [Schenck *et al.* (1977) considered spore septation an inadequate basis for generic segregation of many nematode-trapping fungi and transferred most of them into *Arthrobotrys*. *Dactylaria* was reserved for lignicolous, nonnematophagous species.] The only other known didymosporous species which produces adhesive knobs was described by McCulloch (1977) from Australia as *Arthrobotrys pauca,* which captures nematodes.

B. Rhizopods

Many soil and litter fungi were found by Drechsler (Duddington, 1955) to capture and consume rhizopods or testaceous rhizopods. Only six of the species that trap amebas, however, belong to the Hyphomycetes. Of these, *Dactylella tylopaga* (Drechsler, 1935) subsists on nontestaceous amebas. *Tridentaria carnivora, T. glossopaga, Dactylella passalopaga, Pedilospora dactylopaga,* and *Triposporina quadridens* capture testaceous amebas (Drechsler, 1936, 1937b, 1961). It is worth noting that Zoopagales capture ordinary amebas for the most part, testaceous amebas being the rare exception. Hyphomycetes, on the other hand, capture mostly testaceous amebas, nontestaceous amebas being the exception. This may be related to the feeding habits of the testaceous rhizopods.

In all cases capture of the prey is by means of adhesion, either directly to the hyphae or on specialized organs of capture arising from the hyphae. In *D. passalopaga* (Fig. 8f–h) there are no obvious traps arising from the hyphae, although slight protuberances, on hyphae or on germinating conidia, may represent rudimentary trapping devices. This fungus captures *Geococcus vulgaris* which, as pointed out by Drechsler (1936), feeds on resting spores and hyphae of fungi by fastening to the wall and exerting a sucking action to withdraw the protoplasm. In the case of *D. passalopaga,* however, the roles become reversed and the fungus responds by producing a bulbous outgrowth slightly larger than the oral aperture of the rhizopod which is thus held securely. From this outgrowth one or several assimilative hyphae develop into the rhizopod and digest its contents. *Triposporina quadridens* (Fig. 8d and e) is similar to *D. passalopaga* in that no specialized trapping devices are produced and the rhizopods are held by a compact mass of swollen lobes formed just inside the oral aperture.

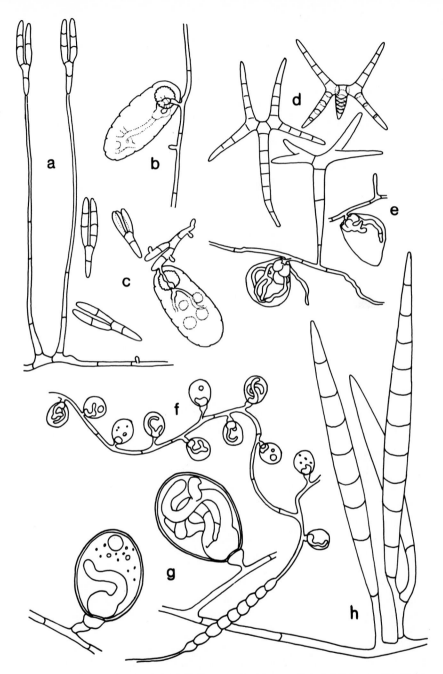

Fig. 8. Ameba-trapping fungi. (a–c) *Pedilospora dactylopaga;* (d and e) *Triposporina quadridens;* (f–h) *Dactylella passalopaga.* (Figs. 8a–e from Drechsler, 1934, 1961.)

In *Pedilospora dactylopaga* (Fig. 8a–c), fingerlike or ellipsoidal protuberances with adhesive tips are produced on the vegetative hyphae. Adhesive knobs also develop on the detached conidia (Fig. 8c). The animals become attached to the traps and then are invaded by the fungus.

Drechsler pointed out that, as in nematode-trapping fungi, the organs of capture are not produced in pure cultures of the fungus. This suggests that substances produced by the rhizopods are responsible for morphogenesis of the trapping devices, as has been shown for nematode trappers.

Dactylella tylopaga and *Tridentaria glossopaga* also capture their prey by means of sessile adhesive knobs. In *D. tylopaga* the uppermost cell of the conidium becomes empty of protoplasm, producing a long, empty, apical appendage. Such appendages are commonly found in the ameba-trapping Zoopagales and probably function as aids in flotation dispersal of the conidia.

IV. ENDOPARASITES OF NEMATODES

More than 30 species of Hyphomycetes are known to parasitize nematodes or their eggs. Compared with predators, endoparasites are relatively small and inconspicuous. Although a technique for their isolation has been described by Aschner and Kohn (1958), nevertheless, endoparasites are slow growing and often difficult to obtain in pure culture. Maintaining endoparasites for extended periods in nonaxenic cultures along with their nematode host is not always successful, as several are host specific (Barron, 1977a). As a result, endoparasites are less well understood than predators, and there is little hard information on their biology.

Endoparasites are categorized as either adhesive or ingested. In adhesive species conidia attach directly to the host cuticle. In ingested species, a nematode attempts to swallow the conidia as food and they lodge in the mouth or esophagus or germinate in the gut.

A. Adhesive Endoparasites

In this group the conidium produces an adhesive substance which aids it in attaching to the host cuticle. The nematode frequently contacts conidia during its forward motion or as it attempts to consume them as food. Therefore, although conidia may adhere to any part of the cuticle, most are found in the head region (Fig. 9d). In general, conidia of endoparasites are small and are globose, subglobose, cuneiform, or clavate in shape. Because of their small size and/or streamlined shape, they are difficult to dislodge once attached. When the shape of the conidium has permitted critical observations, it has been found (Barron, 1970) that only the distal end of the conidium produces an adhesive (Fig. 9c). It

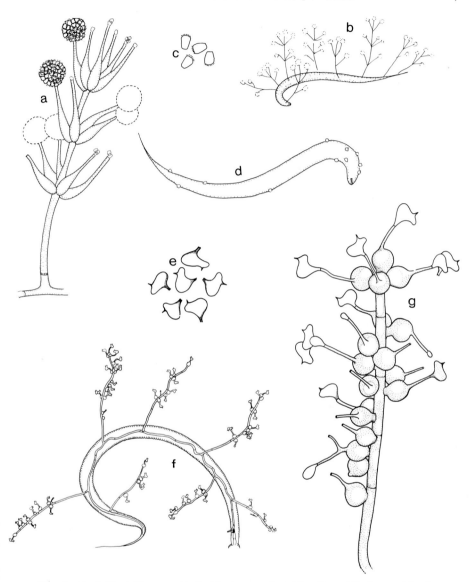

Fig. 9. (a–d) *Verticillium* sp. (a) Conidiophores and conidia; (b) conidiophores arising from body of parasitized host; (c) kernel-shaped conidia with apical adhesive; (d) conidia attached to cuticle of host. (e–g) *Harposporium rhynchosporum*. (e) Odd-shaped conidia which lodge in mouth; (f) solitary assimilative hypha running the length of the host's body and producing external conidiophores; (g) conidiophore with scattered conidiogenous cells. (Figs. 9e–g from Barron, 1977b.)

is also probable that contact with the cuticle results in increased secretion of the adhesive material to strengthen the attachment.

Several species of *Verticillium* (Fig. 9a–d) have been described by Drechsler (1941b, 1942, 1946) as adhesive endoparasites. Once the conidium is firmly attached, a narrow infection peg penetrates the cuticle and assimilative hyphae proliferate rapidly throughout the body cavity of the host. The host is completely digested in a day or two, by which time the body is packed with hyphae of the parasite.

A number of hyphae break through the cuticle and produce erect or suberect conidiophores (Fig. 9a and b) consisting of a simple main axis bearing verticils of phialides at intervals. The conidia are produced in gloeoid balls at the tips of the phialides and are readily dispersed in water. Small soil animals may also aid in the dispersal of conidia.

With the exception of *Haptocara latirostrum* (Drechsler, 1975) all adhesive-spored endoparasitic Hyphomycetes seem taxonomically related to *Verticillium*. Species have been described under *Cephalosporium* (Drechsler, 1941b), *Harposporium* (Drechsler, 1950b), *Meria* (Drechsler, 1941b), *Plesiospora* (Drechsler, 1970), *Cephalosporiopsis* (Drechsler, 1968), and *Paecilomyces* (Drechsler, 1941b). Despite the scattering of adhesive endoparasites among many anamorph-genera, they are probably all derived from a verticillate ancestral form and are more related to each other than to other species of the genera in which they now reside.

B. Ingested Endoparasites

All ingested endoparasitic Hyphomycetes belong to the genus *Harposporium* (Fig. 9e–g). *Harposporium* is taxonomically closely related to *Verticillium*. The typical vase-shaped phialide of *Verticillium* is no longer found. The conidiogenous cell has become modified and produces a spherical inflated base bearing a narrow, cylindrical mouth (Fig. 9g). In *Harposporium* the verticillate arrangement of phialides is inconsistent, and they arise not only in nodal groups but more often scattered irregularly along the length of the conidiophore (Fig. 9f and g). Conidia of *Harposporium* are usually of an unusual shape, which aids them in lodging in the mouth, esophagus, or gut of the host.

The conidia of *Harposporium* species that lodge in the esophagus are either helicoid (Drechsler, 1941b) or crescent-shaped, with the ends out of alignment such that they form part of a wide helix. Conidia have a very sharp point at the distal end and often a droplet of mucus at the proximal end. Observations show that the conidial tip is reinforced with a thick deposit of wall material producing a tough, rigid point. The function of the mucus at the proximal end of the conidium is not known. It has no apparent adhesive properties and may make conidia more attractive to the victim. After ingestion, pumping action of the esophageal muscle may cause rotation of the helicoid conidium, driving the sharp point between the striated muscle fibers of the esophagus.

Conidia that lodge in the buccal cavity (mouth) are also odd in appearance. They have been variously described as resembling a humerus bone, a high-heeled shoe, and a legless chick (Fig. 9e) and in general defy common descriptive terms. These spores usually have one or more pointed and reinforced, often reflexed, tips which aid them in lodging.

Once lodged in the mouth or esophagus, conidia germinate by a narrow germ tube which passes into the body cavity of the victim and produces assimilative hyphae. In the case of *H. rhynchosporum* (Barron, 1977b) a solitary assimilative hypha runs the length of the body from the initial infection site in the mouth to the tip of the tail (Fig. 9f).

The conidia of both *H. helicoides* and *H. arcuatum* are sharply pointed at one end and curved to form part of a wide helix. Their appearance suggests that they should lodge in the esophagus. Barron (1970, 1980), however, has noted that large numbers of conidia are swallowed by both species within a hour of exposure and that the conidia eventually form a tangled mass in the lower gut. Conidia are not digested by the host enzymes and apparently cannot be defecated because of their shape. Conidia of *H. helicoides* and *H. arcuatum* then germinate and digest the body contents. *Harposporium arcuatum* sometimes lodges in the esophagus, but this has not been observed in *H. helicoides*.

For many fungi, sexual spores function as the persistent stage. In the endoparasitic Hyphomycetes no teleomorphs are known, and their function is taken over by other spore forms. In *Harposporium anguillulae, H. latirostrum, H. crassum, H. cycloides,* and some other species, chlamydospores act as the persistent spore form.

In *H. anguillulae,* after the conidiophores and conidia have developed and the contents of the nematode are virtually exhausted, some hyphal cells inside the host swell up, become densely protoplasmic, secrete a thick wall, and become brown-pigmented. These chlamydospores are eventually released into the environment by disintegration of the host cuticle. Under suitable conditions chlamydospores may germinate, producing conidiophores and conidia (Karling, 1938) which constitute the primary inoculum of a new infection cycle. *Harposporium bysmatosporum* was found by Drechsler (1954b) to produce aerial chains of arthrospores. This species does not possess chlamydospores.

Another species that produces aerial arthrospores and lacks chlamydospores was described by Barron (1979b) as *H. arthrosporum*. Barron demonstrated that the arthrospores could germinate and produce conidia and thus might represent the persistent spore in this species. For many species the persistent spore state is now known, and the possibility of an undiscovered teleomorph cannot be discounted.

The value of nematode-destroying fungi in biological control was reviewed by Duddington (1962) and summarized by Barron (1977a). The consensus was that biological control of plant parasitic nematodes using nematode-destroying fungi did not show much promise. The recent discovery by Stirling and Mankau (1978) of *Dactylella oviparasitica*, however, must cause us to reflect on this conclusion. Stirling *et al.* (1978) found that several old peach orchards in the San Joaquin Valley of California had unexpectedly low counts of root-knot nematodes (*Meloidogyne* spp.). Investigations revealed that natural biological control had been effected by *D. oviparasitica*. The fungus attacked and destroyed the egg masses of the nematodes. In some cases the root-knot nematode females were also attacked. This is an exciting and significant discovery, and an understanding of this relationship will give us renewed hope for biological control of nematodes in special situations.

V. ENDOPARASITES OF OTHER MICROFAUNA

No Hyphomycetes are known to be endoparasitic on amebas or testaceous amebas. Individual amebas are probably too small an energy base to support the production of conidiophores and conidia of the relatively large Hyphomycetes. Endoparasites of larger protozoa may eventually be discovered.

Rotifers and tardigrades are much larger animals, similar in size to nematodes, and thus Hyphomycete endoparasites might be expected to attack them. I have found a *Harposporium* species similar to *H. anguillulae* endoparasitic on young tardigrades on two separate occasions. To my knowledge this is the only record of a Hyphomycete attacking tardigrades.

Records of Hyphomycetes parasitic on rotifers are not common. Drechsler (1942) described *Acrostalagmus tagenophorus* and later *Harposporium cocleatum* (Drechsler, 1964) as endoparasitic on rotifer hosts. The only other record is by Barron (1973) who described *Verticillium reniformis* as a parasite of an unidentified rotifer species. Virtually nothing is known regarding the details of the method of initial infection or other factors concerning the above relationships. It is clear from the paucity of information that studies on endoparasites of meio- and microfauna other than nematodes would prove a fruitful field for study.

REFERENCES

Aschner, M., and Kohn, S. (1958). The biology of *Harposporium anguillulae*. *J. Gen. Microbiol.* **19**, 182–189.
Balan, J., and Gerber, N. N. (1972). Attraction and killing of the nematode *Panagrellus redivivus* by the predaceous fungus *Arthrobotrys dactyloides*. *Nematologica* **18**, 163–173.

Balan, J., and Lechevalier, H. A. (1972). The predaceous fungus *Arthrobotrys dactyloides:* Induction of trap formation. *Mycologia* **64,** 919–922.

Balan, J., Križková, L., Nemec, P., and Vollek, V. (1974). Production of nematode-attracting and nematicidal substances by predaceous fungi. *Folia Microbiol. (Prague)* **19,** 512–519.

Balan, J., Križková, L., Nemec, P., and Rolozsvary, A. (1976). A qualitative method for detection of nematode attracting substances and proof of production of three different attractants by the fungus *Monacrosporium rutgeriensis. Nematologica* **22,** 306–311.

Barron, G. L. (1970). Nematophagous Hyphomycetes: Observations on *Harposporium helicoides. Can. J. Bot.* **48,** 329–331.

Barron, G. L. (1973). A new species of *Verticillium* parasitic on rotifers. *Mycopathologia* **50,** 271–274.

Barron, G. L. (1975). Detachable adhesive knobs in *Dactylaria. Trans. Br. Mycol. Soc.* **65,** 311–312.

Barron, G. L. (1977a). ''The Nematode-Destroying Fungi.'' Canadian Biological Publ., Guelph, Ontario.

Barron, G. L. (1977b). Nematophagous fungi: A new *Harposporium* parasitic on *Prismatolaimus. Can. J. Bot.* **55,** 892–895.

Barron, G. L. (1979a). Observations on predatory fungi. *Can. J. Bot.* **57,** 187–193.

Barron, G. L. (1979b). Nematophagous fungi: A new *Harposporium* producing aerial arthroconidia. *Can. J. Bot.* **57,** 892–895.

Barron, G. L. (1979c). Nematophagous fungi: A new *Arthrobotrys* with non-septate conidia. *Can. J. Bot.* **57,** 1371–1373.

Barron, G. L. (1980). Nematophagous fungi: A new *Harposporium* from soil. *Can. J. Bot.* **58,** 447–450.

Barron, G. L., and Dierkes, Y. (1977). Nematophagous fungi: *Hohenbuehelia,* the perfect state of *Nematoctonus. Can. J. Bot.* **55,** 3054–3062.

Bartnicki-Garcia, S., Eren, J., and Pramer, D. (1964). Carbon dioxide dependent morphogenesis in *Arthrobotrys conoides* (nematode trapping fungus). *Nature (London)* **204,** 804.

Commandon, J., and de Fonbrune, P. (1938). Recherches expérimentales sur les champignons prédateurs des nématodes du sol: Les pièges garrotteurs. *C.R. Seances Soc. Biol. Ses Fil.* **129,** 620–625.

Commandon, J., and de Fonbrune, P. (1939). De la formation et du fonctionnement des pièges des champignons prédateurs des nématodes: Recherches effectuées à l'aide de la micromanipulation et de la cinénatographie. *C.R. Hebd. Seances Acad. Sci.* **207,** 304–305.

Cooke, R. C., and Godfrey, B. E. S. (1964). A Key to the nematode-destroying fungi. *Trans. Br. Mycol. Soc.* **47,** 61–74.

Couch, J. N. (1937). The formation and operation of the traps in the nematode-catching fungus, *Dactylella bembicodes* Drechsler. *J. Elisha Mitchell Sci. Soc.* **53,** 301–309.

Dowsett, J. A., and Reid, J. (1977). Transmission and scanning electron microsocpe observations on the trapping of nematodes by *Dactylaria candida. Can. J. Bot.* **55,** 2963–2970.

Dowsett, J. A., Reid, J., and Van Caeseele, L. (1977). Transmission and scanning electron microscope observations on the trapping of nematodes by *Dactylaria brochopaga. Can. J. Bot.* **55,** 2945–2955.

Drechsler, C. (1933). Morphological diversity among fungi capturing and destroying nematodes. *J. Wash. Acad. Sci.* **23,** 138–141.

Drechsler, C. (1934). *Pedilospora dactylopaga* n.sp.—A fungus capturing and consuming testaceous rhizopods. *J. Wash. Acad. Sci.* **24,** 395–402.

Drechsler, C. (1935). A new mucedinaceous fungus capturing and consuming *Amoeba verrucosa. Mycologia* **27,** 216–223.

Drechsler, C. (1936). A *Fusarium*-like species of *Dactylella* capturing and consuming testaceous rhizopods. *J. Wash. Acad. Sci.* **26,** 397–404.

Drechsler, C. (1937a). Some Hyphomycetes that prey on free-living terricolous nematodes. *Mycologia* **29**, 447–552.

Drechsler, C. (1937b). A species of *Tridentaria* preying on *Difflugia constricta*. *J. Wash. Acad. Sci.* **27**, 391–398.

Drechsler, C. (1941a). Predaceous fungi. *Biol. Rev. Cambridge Philos. Soc.* **16**, 265–290.

Drechsler, C. (1941b). Some Hyphomycetes parasitic on free-living terricolous nematodes. *Phytopathology* **31**, 773–802.

Drechsler, C. (1942). Two zoophagous species of *Acrostalagmus* with multicellular *Desmidospora*-like chlamydospores. *J. Wash. Acad. Sci.* **32**, 343–350.

Drechsler, C. (1944). A species of *Arthrobotrys* that captures springtails. *Mycologia* **36**, 382–399.

Drechsler, C. (1946). A new hyphomycete parasitic on a species of nematodes. *Phytopathology* **36**, 212–217.

Drechsler, C. (1950a). Several species of *Dactylella* and *Dactylaria* that capture free-living nematodes. *Mycologia* **42**, 1–79.

Drechsler, C. (1950b). A *Harposporium* infecting eelworms by means of externally adhering awl-shaped conidia. *J. Wash. Acad. Sci.* **40**, 405–409.

Drechsler, C. (1954a). Some Hyphomycetes that capture eelworms in southern states. *Mycologia* **46**, 762–782.

Drechsler, C. (1954b). Production of aerial arthrospores by *Harposporium bysmatosporum*. *Bull. Torrey Bot. Club* **81**, 411–413.

Drechsler, C. (1961). Some clampless Hyphomycetes predacious on nematodes and rhizopods. *Sydowia* **15**, 9–25.

Drechsler, C. (1964). A *Harposporium* parasitic on rotifers. *Mycopathologia* **27**, 285–288.

Drechsler, C. (1968). A nematode-destroying species of *Cephalosporiopsis*. *Sydowia* **22**, 194–198.

Drechsler, C. (1970). A nematode-destroying parasite bearing lageniform conidiiferous branches on endozoic hyphae. *Sydowia* **24**, 173–176.

Drechsler, C. (1975). A nematode-destroying Hyphomycete forming parallel multiseptate hyaline conidia in circular arrangements *Am. J. Bot.* **62**, 1073–1077.

Duddington, C. L. (1955). Fungi that attack microscopic animals. *Bot. Rev.* **21**, 377–439.

Duddington, C. L. (1962). Predacious fungi and the control of eelworms. *Viewpoints Biol.* **1**, 151–200.

Estey, R. H., and Tzean, S. S. (1976). Scanning electron microscopy of fungal nematode-trapping devices. *Trans. Br. Mycol. Soc.* **66**, 520–522.

Feder, W. A., Everard, C. O. R., and Wootton, L. M. O. (1963). Sensitivity of several species of the nematophagous fungus *Dactylella* to a morphogenic substance derived from free-living nematodes. *Nematologica* **9**, 49–54.

Field, J. I., and Webster, J. (1977). Traps of predaceous fungi attract nematodes. *Trans. Br. Mycol. Soc.* **68**, 467–469.

Giuma, A. Y., and Cooke, R. C. (1972). Some endozooic parasites on soil nematodes. *Trans. Br. Mycol. Soc.* **59**, 213–218.

Heintz, C. E., and Pramer, D. (1972). Ultrastructure of nematode-trapping fungi. *J. Bacteriol.* **110**, 1163–1170.

Higgins, M. L., and Pramer, D. (1967). Fungal morphogenesis: Ring formation and closure by *Arthrobotrys dactyloides*. *Science* **155**, 345–346.

Insell, J. P., and Zachariah, K. (1977). A biometrical analysis of the giant constricting ring mutant of the predacious fungus *Dactylella brochopaga*. *Protoplasma* **93**, 305–310.

Insell, J. P., and Zachariah, K. (1978a). Some ring-trap mutants of the fungus *Dactylella brochopaga* Drechsler. *Arch. Microbiol.* **117**, 221–226.

Insell, J. P., and Zachariah, K. (1978b). The mechanism of the ring trap of the predacious hyphomycete *Dactylella brochopaga* Drechsler. *Protoplasma* **95**, 175–191.

Jarowaja, N. (1968). *Dactylella parvicollis* Drechsler: An interesting nematode-killing fungus new for the Polish mycoflora. *Bull. Acad. Pol. Sci., Ser. Sci. Biol.* **16**, 253–256.

Karling, J. S. (1938). *Harposporium anguillulae. Mycologia* **30**, 512–519.

Katznelson, H., and Henderson, V. E. (1963). Ammonium as an attractant for a soil nematode. *Nature (London)* **198**, 907–908.

Klink, J. W., Dropkin, V. H., and Mitchell, J. E. (1970). Studies on the host-finding mechanism of *Neotylenchus linfordi. J. Nematol.* **2**, 106–117.

McCulloch, J. S. (1977). New species of nematophagous fungi from Queensland. *Trans. Br. Mycol. Soc.* **68**, 173–179.

Monoson, H. L., and Ranieri, G. M. (1972). Nematode attraction by an extract of a predaceous fungus. *Mycologia* **64**, 628–631.

Monoson, H. L., Galsky, A. G., Griffin, J. A., and McGrath, J. J. (1973). Evidence for and partial characterization of a nematode-attraction substance. *Mycologia* **65**, 78–86.

Müller, H. G. (1958). The constricting ring mechanism of two predacious Hyphomycetes. *Trans. Br. Mycol. Soc.* **41**, 341–364.

Nicholas, W. L. (1975). "The Biology of Free-Living Nematodes" Oxford Univ. Press (Clarendon), London and New York.

Nordbring-Hertz, B. (1972). Scanning electron microscopy of the nematode-trapping organs in *Arthrobotrys oligospora. Physiol. Plant.* **26**, 279–284.

Nordbring-Hertz, B. (1973). Peptide-induced morphogenesis in the nematode-trapping fungus *Arthrobotrys oligospora. Physiol. Plant.* **29**, 223–233.

Nordbring-Hertz, B. (1974). Qualitative characterization of some peptides inducing morphogenesis in the nematode-trapping fungus *Arthrobotrys oligospora. Physiol. Plant.* **31**, 59–63.

Nordbring-Hertz, B. (1977). Nematode-induced morphogenesis in the predacious fungus *Arthrobotrys oligospora. Nematologica* **23**, 443–451.

Nordbring-Hertz, B., and Stalhammar-Carlemalm, M. (1978). Capture of nematodes by *Arthrobotrys oligospora,* an electron microscope study. *Can. J. Bot.* **56**, 1297–1307.

Nordbring-Hertz, B., Jansson, H. B., and Stalhammar-Carlemalm, M. (1977). Interactions between nematophagous fungi and nematodes. *Ecol. Bull. (Stockholm)* **25**, 483–484.

Roubaud, M. E., and Deschiens, R. (1939). Sur les agents de formation des dispositifs de capture chez les Hyphomycètes prédateurs de nématodes. *C.R. Seances Soc. Biol. Ses Fil.* **209**, 77–79.

Rudek, W. T. (1975). The constriction of the trapping rings in *Dactylaria brochopaga. Mycopathologia* **53**, 193–197.

Pramer, D., and Stohl, N. R. (1959). Nemin: A morphogenic substance causing trap formation by predaceous fungi. *Science* **129**, 966–967.

Schenck, S., Kendrick, W. B., and Pramer, D. (1977). A new nematode-trapping hyphomycete and a reevaluation of *Dactylaria* and *Arthrobotrys. Can. J. Bot.* **55**, 977–985.

Shepherd, A. M. (1955). Formation of the infection bulb in *Arthrobotrys oligospora* Fresenius. *Nature (London)* **175**, 475.

Stirling, G. R., and Mankau, R. (1978). *Dactylella oviparasitica,* a new fungal parasite of *Meloidogyne* eggs. *Mycologia* **70**, 774–783.

Stirling, G. R., McKenry, M. V., and Mankau, R. (1978). Biological control of root-knot nematode on peach. *Calif. Agric.* **32**, 6–7.

Townshend, J. L. (1964). Fungus hosts of *Aphelenchus avenae* Bastian, 1865 and *Bursaphelenchus fungivorus* Franklin and Hooper 1962 and their attractiveness to these nematode species. *Can. J. Microbiol.* **10**, 727–737.

Tsai, B. Y., Kao, J., Huang, C. S., and Chang, H. S. (1975). Morphological and ecological characteristics of some nematode-trapping fungi found in Taiwan. *Plant Prot. Bull. (Taiwan)* **17**, 272–284.

Turnbull, J. R., and Zachariah, K. (1978). The induction of giant ring traps and regulation of conidio-genesis in the predacious fungus *Dactylella brochopaga*. *Can. J. Microbiol.* **24,** 1182–1189.

Tzean, S. S., and Estey, R. H. (1978). Nematode-trapping fungi as mycopathogens. *Phytopathology* **68,** 1266–1270.

Wootton, L. M. O., and Pramer, D. (1966). Valine-induced morphogenesis in *Arthrobotrys conoides*. *Bacteriol. Proc.* p. 75.

21

Entomogenous Fungi

Donald W. Roberts and Richard A. Humber

I. INTRODUCTION

Human appreciation of the fungi attacking insects is by no means limited to the modern concern in using them for the biological control of insect pests. Two millennia ago, the Chinese were aware of the mummification of silkworms and cicadas by species of *Cordyceps* and *Isaria,* and placed semiprecious and precious stone effigies of these insects in the mouths of their dead in an attempt to confer a similar degree of immortality (Kobayasi, 1977). *Cordyceps*-infected caterpillars have been mentioned in the Chinese folk pharmacopoeia for at least 1000 yr for the treatment of a large number of ailments ranging from opium addic-

Biology of Conidial Fungi, Vol. 2

TABLE I

Genera of Entomopathogenic and Entomoparasitic Fungi Forming Conidia
or Having Conidial States[a]

Entomogenous genera	Known or suspected entomogenous anamorphs, teleomorphs, or associated alternate states
Zygomycotina	
Zygomycetes–Entomophthorales[b]	
Conidiobolus	
Entomophthora[c]	
Massospora	
Strongwellsea	
Zygaenobia (not validly published)	
Ascomycotina	
Plectomycetes–Eurotiales	
Pseudeurotium[d]	*Beauveria*
Sartorya[d]	*Aspergillus*
Pyrenomycetes–Clavicipitales	
Cordyceps	*Acremonium, Akanthomyces, Hirsutella, Hymeno-stilbe, Paecilomyces, Pseudogibellula, Sporothrix, Stilbella*
Hypocrella	*Aschersonia*
Torrubiella	*Acremonium, Akanthomyces, Cylindrophora,[c] Gibellula, Hirsutella, Paecilomyces, Stilbella*
Pyrenomycetes–Hypocreales	
Hypomyces[d]	*Trichothecium*
Nectria	*Fusarium*
Pseudonectria[d]	*Tilachlidium*
Loculoascomycetes–Pleosporales	
Podonectria	*Tetracrium, Tetranacrium*
Basidiomycotina	
Phragmobasidiomycetes–Septobasidiales	
Septobasidium	*Harpographium* and other unnamed genera
Uredinella (=*Septobasidium*?)	
Deuteromycotina	
Hyphomycetes	
Acariniola	
Acremonium	*Cordyceps, Torrubiella*
Akanthomyces	*Cordyceps, Torrubiella*
Amphoromorpha	
Antennopsis	
Aphanocladium	
Arthrobotrys	
Aspergillus	(*Sartorya* and other genera)[d]
Beauveria	(*Pseudeurotium*)[d]
Chantransiopsis	
Coreomycetopsis	
Culicinomyces	
Endosporella	
Fusarium	*Nectria*

(*continued*)

TABLE I

Entomogenous genera	Known or suspected entomogenous anamorphs, teleomorphs, or associated alternate states
Gibellula	*Torrubiella, Cylindrophora*[e]
Granulomanus[e]	*Torrubiella, Gibellula*
Harpographium	*Septobasidium*
Hirsutella	*Cordyceps, Torrubiella*
Hymenostilbe	*Cordyceps*
Mattirolella	
Metarrhizium	
Muiaria	
Muiogone	
Nomuraea	
Paecilomyces	*Cordyceps, Torrubiella*
Peziotrichum	
Pseudogibellula	*Cordyceps*
Rhinotrichum[f]	
Sorosporella	*Synpliocladium*
Sporothrix	*Cordyceps?*
Stilbella	*Cordyceps, Torrubiella*
Syngliocladium	*Sorosporella*
Synnematium	
Termitaria	
Tetracrium	*Podonectria*
Thaxteriola	
Tilachlidiopsis	
Tilachlidium	*Pseudonectria*
Tolypocladium	
Trichothecium	
Verticillium	
Coelomycetes	
Aschersonia	*Hypocrella*
Tetranacrium	*Podonectria*
Mycelia Sterilia	
Aposporella	
Aegerita	
Endosclerotium (not validly published)	
Hormiscioideus	
Hormiscium	

[a] Classification follows Ainsworth *et al.*, 1973a,b. For characterizations and further information on all genera, see Humber and Soper, 1980.

[b] See note, p. 204, regarding entomophthoralean conidia.

[c] Including *Culicicola, Entomophaga, Triplosporium,* and *Zoophthora*; Batko's (1964a–d) segregate genera are discussed by King and Humber (1980) and Humber and Soper (1980). A substantially different opinion regarding the genera of the Entomophthorales is offered by Remaudière and Hennebert (1980) and Remaudière and Keller (1980).

[d] Not entomopathogenic.

[e] *Cylindrophora aranearum* Petch = *Granulomanus aranearum* (Petch) de Hoog & Samson (see de Hoog, 1978).

[f] *Engyodontium* (de Hoog, 1978) = *Rhinotrichum* (pro parte) and *Beauveria* (pro parte).

tion to consumption, as a general tonic, and even as an aphrodisiac (Lloyd, 1918; Hoffman, 1947).

Insect fungi never assumed roles in occidental folk traditions, but they have long aroused the curiosity and speculations of their observers: In 1779, de Geer (in Ramsbottom, 1914) characterized an *Entomophthora* infection of flies and supposed sympathetically that they might have eaten something harmful. By the middle of the next century, the feasibility of using fungal diseases to control insect pests was being seriously considered.

The fungi which attack living insects comprise a large, diverse group. Every major fungal subdivision and more than 100 genera (not counting the even more numerous genera in the Laboulbeniales) are represented. The most important entomopathogenic and entomoparasitic fungi produce conidia at some point in their life cycle (see Table I). In addition to the many deuteromycetes from insects, entomogenous teleomorphs (sexual states in the Ascomycetes and Basidiomycetes) having conidial anamorphs (imperfect states), the few aposporic entomoparasites of insects, and the entomogenous Entomophthorales (whose forcibly discharged spores have long been regarded as conidial) will be treated here. [It is unknown if the primary spores of the Entomophthorales meet Dykstra's (1974a) criteria for zygomycetous conidia. Electron microscope studies of sporogenesis in *Strongwellsea magna* indicate that the uninucleate spore type with a separable outer wall layer (the most controversial entomophthoralean spore type) cannot be a monosporic sporangiole but might be a true conidium (Humber, 1975).]

Entomogenous fungi, which will not be treated here (because they form no conidia), are the limited numbers of water molds in *Coelomomyces, Coelomycidium, Myiophagus, Lagenidium, Leptolegnia,* and other aquatic genera with species attacking insect eggs (or whose pathogenicity is doubtful), the endocommensal Trichomycetes, the mucoralean genus *Sporodiniella* (and allied genera with species of doubtful pathogenicity), yeasts in the Hemiascomycetes (Steinhaus, 1949), the Loculoascomycetes in the Myriangiales (*Myriangium, Angatia,* and *Uleomyces*), and the Pyrenomycetes in the Laboulbeniales and its related genus *Laboulbeniopsis;* all these fungi are discussed by Humber and Soper (1980).

Significant reviews of the biology of fungal pathogens and parasites of insects are given by MacLeod (1963), Madelin (1963, 1966b), McEwen (1963), Müller-Kögler (1965), Roberts and Yendol (1971), and Ferron (1978) and are included in Roberts (1980a) and Burges (1980).

A. Types of Fungus–Insect Associations

The term ''entomogenous fungus'' was originally intended to designate pathogens and parasites of insects (e.g., Steinhaus, 1949; Madelin, 1966b; Ferron,

1978). However, the expansion of insect mycology has extended this term to every possible type of fungus–insect association and it necessarily includes the fungal associates of all other terrestrial arthropods—mites, spiders, centipedes, millipedes, symphylans, and so on—as well as insects. Unfortunately, no good alternative term has emerged to solve the resulting semantic problems.

The taxonomic and biological diversity of entomogenous fungi, particularly among the pathogens and parasites of insects, suggests that entomogenous habits have been adopted several times during the course of fungal evolution. Most fungi associated with arthropods can be placed in one of the following categories:

1. Pathogens cause the early death of the host by penetrating and proliferating inside the host, which is killed by being deprived of soluble nutrients in its hemolymph, by the invasion or digestion of its tissues, and/or by the release of toxins from the fungus. Examples are *Beauveria, Metarrhizium,* and *Entomophthora.*

2. Parasites may impair host activities and/or cause severe debilitation, but do not cause early death. These fungi are ectoparasites such as the Laboulbeniales, *Termitaria, Antennopsis,* and *Coreomycetopsis.* Very few of these ectoparasites have haustoria which completely penetrate the cuticle, and their growth may be nutrient-limited. The coiled haustoria of *Septobasidium* are the only fungal structures that penetrate host scale insects. Endoparasitic fungi are notably absent, although several entomophthoralean fungi (all species of *Massospora* and *Strongwellsea* and a few of *Entomophthora*) release their conidia before the host's death from the mycosis. *Strongwellsea,* however, appears to be evolving away from its pathogenic relatives toward a nonlethal relationship with its host flies (Humber, 1975, 1976).

3. Facultative pathogens are weak pathogens which usually attack only old, weakened, diseased, or wounded hosts. The fungus may be capable of penetrating the cuticle but does so only occasionally. Examples are *Conidiobolus* species (*C. coronatus,* and so on) and conidioboloid species of *Entomophthora;* these fungi often occur together with (and frequently mask the presence of) a more virulent fungus or other microbe.

4. Wound pathogens are incapable of penetrating an intact host cuticle but can invade the hemocoel through wound repair sites on the cuticle. They are lethal to the insect if entry is accomplished. Examples are *Pythium* (Clark *et al.,* 1966), *Mucor* (Heitor, 1962), and *Trichothecium* (Madelin, 1966a).

5. Saprobes colonizing insect cadavers include a wide variety of common molds which may occur on dead insects but for which little or no evidence of pathogenicity exists. Examples are species of *Cladosporium, Penicillium,* and *Scopulariopsis,* and many members of the Mucorales.

6. Commensals and symbiotes include a broad variety of fungi which have adopted this sort of association. Trichomycetes are found attached in the gut of many insects and other arthropods (Lichtwardt, 1973), but no evidence suggests

any parasitic, pathogenic, or even mutualistic role for these fungi despite the close coupling of their sporulation with the host's molting cycle. A rich flora of sooty molds grows on or over scale insects without parasitizing them (Hughes, 1976). Ambrosia fungi (a diverse group of yeasts and molds) play important mutualistic roles in the biology of some wasps and wood-inhabiting beetles (Batra, 1967, 1979; Graham, 1967). Also, symbiotic yeasts may grow in the hemolymph or intracellularly in the mycetomes or mycetocytes of Homoptera and Coleoptera (Steinhaus, 1947, 1949; Buchner, 1965; Graham, 1967).

7. Insect dispersal of conidia is as important for some fungi as the role of insects is in the pollination of some flowering plants. The spread of saprobic and phytopathogenic fungi by insects is discussed by Ingold (1971). The long-range disperal of entomopathogenic fungi often depends on mobile hosts bearing still ungerminated conidia or the early stages of a developing mycosis. Some entomophthoralean fungi, as mentioned above, disperse their conidia from living (but doomed) hosts. Other adaptations for dispersal of entomopathogenic fungi are mentioned in Section III,B.

8. Insect mycophagy is both an important source of nutrition for some insects and an important means of dispersing fungal spores which can withstand passage through the insect gut. Fogel (1975) compiled an extensive bibliography on this underappreciated interaction between fungi and insects.

B. Taxonomic Distribution of Entomopathogenic and Entomoparasitic Fungi

The distribution of fungal genera in Table I indicates that the entomopathogenic and entomoparasitic habits are neither so widely nor so uniformly distributed among the fungi as is widely believed. This distribution can be used to extract a considerable amount of information.

Excluding the Laboulbeniales, more genera of entomogenous fungi occur in the "Deuteromycotina" than in any other fungal subdivision, but many of these deuteromycete genera contain only one to a few species attacking living insects. The distribution of numbers of entomogenous species, however, is much more nearly balanced among all the fungal subdivisions—so long as the ca. 1300 species and varieties of Laboulbeniales are again excluded.

Nearly all deuteromycetous fungi attacking insects are Hyphomycetes. In the Saccardoan classification, most of these fungi fall into the Moniliaceae, Stilbellaceae, and Tuberculariaceae of the Moniliales. Parasitic or pathogenic species with dark hyphae and/or conidia (in the Saccardoan families Dematiaceae and Stilbellaceae) form a conspicuously small proportion of entomogenous species and are only infrequently reported: *Antennopsis* (a genus of termite ectoparasites) may be the most widespread and best studied of these fungi (Gouger and Kimbrough, 1969). Thaxter (1914, 1920) is apparently the only one to have reported on *Muiaria* or *Muiogone*. *Peziotrichum lachnella* is a relatively rare synnematous pathogen of scale insects in India, Sri Lanka, and Southeast Asia (Petch, 1927). The other dematiaceous fungi reported from insects—e.g., species of *Acremoniella, Cladosporium, Desmidiospora* (q.v., Humber and Soper, 1980), and *Scopulariopsis*—are probably saprobes colonizing insect cadavers. *Aegerita webberi* forms dark-brown "sporodochia" on which no true conidia are produced, but these sporodochia are easily dislodged and act as propagules.

Only two genera of Coelomycetes have entomogenous species: An undescribed *Tetranacrium* sp. is associated with *Podonectria gahnia* (Rossman, 1978), and *Aschersonia* species are anamorphs of *Hypocrella* species (Mains, 1959). However, the "pycnidia" of *Aschersonia* are little more than irregular convolutions of the stromatic surface and may be absent altogether from sporulating cultures; *Aschersonia* might be placed equally well among the tubercularioid genera of the Hyphomycetes.

A small number of rarely encountered deuteromycetes must be placed among the Mycelia Sterilia. One invalidly published genus, *Endosclerotium,* appears to be the sclerotial form of a synnematous *Cordyceps* anamorph (see Humber and Soper, 1980). The assignment to *Aegerita* of two species from larval whiteflies and scales should be reexamined; *A. insectorum* may form conidia in addition to the sterile sporodochia which are the usual dispersal and infective units for *A. webberi* (Petch, 1926, 1937). Most aposporic fungi, however, are ectoparasites with very limited hyphal or thallic growth (e.g., *Aposporella, Hormiscioideus, Hormiscium*).

The sporocarps of entomogenous ascomycetes are either perithecia or pseudothecia; no apothecial ascomycetes attack insects. Nearly all perithecial entomogenous fungi belong in the Hypocreales or Clavicipitales; with a few little understood exceptions—*Cordycepioideus bisporus, Xylaria apum* (both in the Xylariales), and *Cantharosphaeria chilensis* (Sphaeriales)—the remaining orders of Pyrenomycetes contain no entomogenous species.

Very few Loculoascomycetes are entomogenous. The anamorphs of *Podonectria* species are unique among entomogenous fungi for their tetraradiate aleurioconidia produced on a sporodochium (*Tetracrium* spp.) or in a pycnidium (*Tetranacrium* sp.) (Rossman, 1978). The only other bitunicate ascomycetes affecting insects are species of *Myriangium* and allied genera; these fungi produce no conidia.

No cleistothecial teleomorphs are entomopathogenic, even though the conidial states of several fungi in the Eurotiales are entomopathogens. The anamorph of *Pseudeurotium bakeri* is a *Beauveria,* possibly *B. bassiana* (Booth, 1961). A few entomopathogenic aspergilli are associated with cleistothecial teleomorphs (Raper and Fennell, 1965), but cleistothecia almost never appear on affected arthropods. Species of *Paecilomyces* section *Paecilomyces* have teleomorphs in the Eurotiales, but none of these fungi are entomogenous; all the entomopathogenic species belong in section *Isarioidea* and have teleomorphs (where they are known) which belong in the Clavicipitales (Samson, 1974).

No conidial form is known for any ascomycete in the Laboulbeniales. Furthermore, no evidence suggests any linkage between the Laboulbeniales and the species of *Acariniola*, *Amphoromorpha*, or *Thaxteriola*—a group of ectoparasitic Hyphomycetes with reduced thalli strongly resembling those of the Laboulbeniales. *Coreomycetopsis*, however, might be an anamorph of *Laboulbeniopsis*, an ascomycete which is clearly allied to the Laboulbeniales but which cannot be placed in that order (see Blackwell and Kimbrough, 1976a,b).

Septobasidium and the allied (or synonymous) genus *Uredinella* are the only entomoparasitic basidiomycetes. Many of these remarkable fungi produce conidia (Couch, 1938), but almost no anamorphic states have been named except for *Harpographium corynelioides* (Coles and Talbot, 1977). The role of conidia in the life cycle of these fungi remains uninvestigated.

C. Types of Conidium Production in the Entomogenous Deuteromycotina

Conidia are released from phialides in the great majority of entomogenous deuteromycetes. Aleurioconidia are found in the dictyoconidial genera *Muiaria* and *Muiogone* and in the stauroconidial genera *Tetracrium* and *Tetranacrium*. It seems likely that botryoblastoconidia are produced by *Granulomanus*, *Hymenostilbe*, *Rhinotrichum*, and *Sporotrichum*. Conidia form sympodially on a denticulate rachis in *Beauveria* and as raduloconidia in *Arthobotrys* and *Pseudogibellula*. *Antennopsis* seems to be the only entomogenous genus whose conidia are produced on (modified) annellophores.

Several types of conidiogenesis are absent among the ranks of the entomogenous deuteromycetes. These include blastoconidia, meristem blastoconidia, arthroconidia, and meristem arthroconidia (although the conidia of *Trichothecium* might be some modified form of arthroconidia according to Barron, 1968). The absence of poroconidial genera from the fungi attacking insects seems to reflect the rarity of dematiaceous fungi on insects.

The teleomorphic forms of many entomogenous Hyphomycetes remain unknown, and the teleomorph–anamorph connection is often difficult to prove (see Humber and Soper, 1980). While recognizing these difficulties, the correlation between conidial types and their teleomorphs seems to be instructive: The known or suspected teleomorphs for anamorphic fungi producing ameroconidia belong to the Clavicipitales, Eurotiales, Pleosporales, and Septobasidiales; with phragmoconidia, to the Hypocreales; and with stauroconidia, to the Pleosporales. The teleomorphs are unknown for the only entomogenous genera with dictyoconidia—

Muiaria and *Muiogone*. No entomogenous deuteromycetes produce scolecoconidia. Except for the synnematoid habit of *Peziotrichum lachnella* (which may be the anamorph of *Podonectria coccorum*), the teleomorphic forms of synnematous fungi affecting insects belong almost exclusively to *Cordyceps* or *Torrubiella* in the Clavicipitales.

D. Host Specificity

1. Fungal Specificities for Insects

The problem of host specificity can be regarded in several ways. Fargues and Remaudière (1977, 1980) have reviewed the mechanical and biological aspects of specificity which revolve around the success or failure of a fungus to complete the infection process (see Section II). A broader aspect of this topic is the cataloging of the pathogens and parasites to which a host group is susceptible.

When adult insects occupy ecological niches different from those of their larval or nymphal stadia, it is common to find correspondingly different spectra of fungal pathogens for adult hosts versus their immature forms. For example, nymphal and adult aphids occur together on their host plants, and the numerous entomophthoralean and hyphomycete species which affect these insects do not discriminate between immature or adult hosts; on the other hand, the larvae and adults of most flies usually occupy completely different habitats, and the many species of *Entomophthora* and *Strongwellsea* affecting dipterans are known only from adult hosts. Insect eggs have remarkably few fungal pathogens whether they remain in the body of a mycotized female or have been deposited (see Humber and Soper, 1980).

Relatively few conidial fungi attack insects living beneath the surface of water. Major exceptions to this statement, however, are fungi such as *Metarrhizium anisopliae* (see Section IV,C), *Culicinomyces clavosporus,* and *Entomophthora aquatica*. These fungi attack aquatic larvae of mosquitoes and other medically important insects. Some *Entomophthora* spores may assume tetraradiate forms in water (Webster *et al.,* 1978), but the hosts for these fungi seem to be adult dipterans.

Social insects also appear to be relatively free of serious fungal diseases, possibly because they routinely clean their nests and eject sick or weakened members. Species of *Ascosphaera* and *Aspergillus* cause the major fungal diseases of bees; the most common fungal pathogens of ants are *Cordyceps* species and their diverse anamorphs. Despite relative freedom from pathogens, termites can be infested with a rich variety of minute ectoparasitic deuteromycetes and ascomycetes (see Blackwell and Kimbrough, 1978).

Larvae of coleopterans and lepidopterans may be attacked by a wide variety of hyphomycete and entomophthoralean fungi, but the adults in both these orders are susceptible to a much narrower selection of the same fungi. Beetle larvae and other insects feeding and remaining in the soil seem to be particularly subject to infection by *Beauveria, Metarrhizium,* and *Cordyceps* species.

Among all insect hosts, larval scale insects and whiteflies (Homoptera: Coccidae and Aleyrodidae) are susceptible to the widest spectrum of fungi: These hosts may be attacked by Chytridiomycetes (*Myiophagus*), by Zygomycetes (*Entomophthora, Conidiobolus*), by Pyrenomycetes in the Hypocreales (*Nectria, Lisea, Stereocrea*) and Clavicipitales (*Hypocrella, Cordyceps*), by Loculoascomycetes in the Pleosporales (*Podonectria*) and Myriangiales (*Myriangium, Angatia, Uleomyces*), by all basidiomycetes in the Septobasidiales, and by a wide variety of deuteromycetous genera—species of *Aegerita, Aschersonia, Fusarium, Peziotrichum, Tetracrium, Tetranacrium,* and *Verticillium*. Because most of these fungi are stromatic and relatively slow-growing, it is tempting to assume that the immobile habit of these hosts leaves them particularly vulnerable to fungal attack. After considerable attention in the early part of this century (see Section IV,A), the diverse fungal flora of these insects has been inexplicably ignored for use in their biological control. However, serious efforts are now under way in several countries to use *Verticillium lecanii* against aphids and whiteflies (particularly in greenhouses) and *Aschersonia* species against scales and whiteflies (on citrus crops).

2. Potential Health Hazards of Entomopathogenic Fungi in Man

Entomophthora and its allied fungi attack a broad spectrum of insects, mites, spiders, nematodes, and tardigrades. Considerable attention has been given to the fact that *Conidiobolus coronatus,* a ubiquitous saprobe, can attack mammals (including humans) as well as insects. This species does not seem to be strongly pathogenic, even though the mycosis, once contracted, can be fatal for some affected vertebrates and is always fatal for arthropods. Insects attacked by *C. coronatus* are usually previously weakened or diseased and often already are infected by some more actively pathogenic species of *Entomophthora*. While little is known of the etiology of mammalian entomophthoroses, it seems likely that the rhinoentomophthoroses caused by *C. coronatus* result from chronic challenges to the nasal mucosa such as might occur when sleeping on decaying vegetable material. Two other entomophthoralean species—*Basidiobolus haptosporus* (=*meristosporus*) and *C. incongruus*—also cause human mycoses (Emmons *et al.,* 1977; King and Jong, 1976). At the present time, all evidence suggests that the host ranges of the truly entomogenous species of the Entomophthorales (cf. King and Humber, 1980) are restricted entirely to the Insecta and Arachnida, and that these fungi present no health hazard for vertebrates.

No other entomogenous fungi are known with certainty to cause mycoses of humans or other homeotherms. It is possible, however, that chronic exposure to and inhalation of high densities of conidia of fungi such as *Beauveria bassiana* or *Metarrhizium anisopliae* (fungi being mass-produced for biological control) could lead to allergic sensitization. Recent mammalial tests with *Nomuraea rileyi* indicated no health hazards from this fungus (Ignoffo and Garcia, 1978; Ignoffo *et al.*, 1979).

II. THE INFECTION PROCESS

A. Fungal Invasion of the Host

Unlike the insect-pathogenic viruses, bacteria, and protozoa which must be ingested to initiate disease, entomopathogenic fungi normally invade through the host's cuticle. Invasion to the hemocoel by germ tubes from ingested conidia has been noted (e.g., Yendol and Paschke, 1965; Roberts, 1970); but, except for *Culicinomyces* (Sweeney, 1975a), this is an unusual route of infection. It should be noted that in the case of *Culicinomyces* invasion occurs only through the foregut and hindgut (which, in insects, are ectodermal in origin).

The disease development cycle of most conidial entomopathogenic fungi can be divided into 10 steps:

1. Attachment of the conidium to the insect cuticle. The conidia of entomopathogenic fungi apparently are adapted for attachment to insects, but the physical and chemical characteristics of conidial and cuticular surfaces responsible for attachment are unknown. The importance of this step for infection was demonstrated with a mutant of *Metarhizium anisopliae* which was hypovirulent for mosquito larvae (Al-Aidroos and Roberts, 1978): In comparison to the virulent wild type, very few conidia of the mutant attached to the perispiracular valves of *Culex pipiens* larvae.

2. Germination of the conidium on the insect cuticle. In general, high relative humidity (>90%) is needed for germination (Roberts and Campbell, 1977), but the microenvironment on foliage, particularly during periods of dew, frequently affords the proper conditions even though the macroclimate is too dry. Both germination stimulators and inhibitors have been reported to be on the cuticle. Bacteria on the cuticle can have a strongly inhibiting effect on the germination of *M. anisopliae* (Schabel, 1976).

3. Penetration of the cuticle. The germ tube may penetrate directly into the cuticle, or an appressorium may be formed which attaches firmly to the cuticle, and a narrow infection peg sent into the cuticle (Zacharuk, 1973). Cuticular invasion involves both enzymatic and physical activities.

The enzymes elaborated by germinating conidia have not been identified, but it is known that colonies of most entomopathogenic conidial fungi produce proteases, lipases, and chitinases in liquid and agar media (see Roberts, 1980b).

With *B. bassiana*, these enzymatic activities did not dissolve excised insect cuticle when applied individually, but the cuticle was dissolved by a mixture. If the partially purified enzyme activities were applied sequentially, the chitinase had to be introduced first to obtain dissolution (Samsinakova *et al.*, 1971). In the heavily sclerotized cuticle of elaterid (Coleoptera) larvae, the hyphae tend to grow between the cuticular lamellae. In these cases, penetration to the hemocoel may take several days (Zacharuk, 1973). Insects which molt before the hemocoel is invaded may discard the fungus totally in the molting process (Fargues and Vey, 1974).

4. Growth of the fungus in the hemocoel. The fungus usually grows in the hemocoel as yeastlike hyphal bodies, essentially blastospores, which multiply by budding. In some instances, hyphae rather than hyphal bodies occur (e.g., Prasertphon and Tanada, 1968), and several Entomophthorales produce protoplasts (cells without cell walls) in the hemocoel (Tyrrell, 1977).

5. The production of toxins. Many entomopathogenic fungi overcome their hosts before extensive invasion of organs takes place, and toxins are presumed to be responsible for host mortality (Roberts, 1980b). Although compounds toxic to insects have been reported from culture filtrates and/or mycelia of several entomopathogenic fungi and from fungi not known to affect insects naturally, virtually no searches for toxins in fungus-infected insects have been conducted. However, the depsipeptides destruxin B and desmethyldestruxin B were detected in *Metarrhizium anisopliae*-infected silkworm larvae at levels known to be lethal to this species (Suzuki *et al.*, 1971). All presently known toxins were isolated and identified from mycelia or culture filtrates of fungi grown *in vitro*. The majority are small molecules, primarily depsipeptides. Enzymes, particularly proteases, are present in culture filtrates. When concentrated, these are detrimental to certain insects on intrahemocoelic injection, but their role in disease development is unknown. Toxins are usually assumed to be produced by the fungus alone, but it is possible as with some plant pathogens, that toxic compounds are produced by the diseased host as well. The extensive fungal development prior to death of some fungus–insect combinations indicates that toxins are not produced. Death is very rapid (48 h or less) with some fungi with high inoculum (e.g., Baird, 1954); the cause of death in these instances may be toxins.

6. Death of the host. This may be preceded by behavior changes such as tremors, loss of coordination, or climbing to an elevated position.

7. Growth in the mycelial phase with invasion of virtually all organs of the host. Because of the replacement of internal organs with mycelia, the insect immediately after death is very nearly normal in appearance. Small, melanized spots at the sites of infection may be apparent, however, and in some cases a reddish cast (*Beauveria bassiana*) or blackening (*Entomophthora*) of the host is

detected. In addition to hyphal filaments, chlamydospores may be produced. These fungus-filled insects ("mummies") can serve as reservoirs of the fungus through periods of adverse conditions, such as dry or cold weather.

8. Penetration of hyphae from the interior through the cuticle to the exterior of the insect. If the "mummy" is held under conditions of moderate or low relative humidity, the fungus will remain within the cadaver, but the fungus will grow through the cuticle when the latter is placed in a humid environment. Intersegmental membranes afford less resistance to penetration, and this is the area of most reemergence of fungi, particularly the Entomophthorales.

9. Production of infective units (usually conidia) on the exterior of the host.

10. The final step is the dispersing of infective units to locations where they are likely to encounter susceptible insects for the initiation of new cases of disease. As with plant pathogens, this is usually done by wind or water. The spores of the Entomophthorales, as mentioned elsewhere, are forcibly discharged, while the conidia of other entomopathogenic fungi are passively dispersed.

B. Host Responses to Fungal Invasion

Because nearly all entomopathogenic fungi enter the host through the cuticle, this noncellular organ forms the primary defense against microbial challenges to the host. The most common cuticular reaction to fungal penetration is a localized melanization around and in front of the organ of penetration (Brobyn and Wilding, 1977; Travland, 1979); this melanization is enough to stop some fungi, but successful pathogens overcome any such reaction and penetrate to the hemocoel. Entomopathogens in the Mastigomycotina (e.g., *Coelomomyces, Lagenidium*) and Zygomycotina (*Entomophthora*) elicit no further host response after reaching the hemocoel, but Hyphomycetes such as *Beauveria* and *Metarrhizium* may trigger more-or-less extensive granulomatous reactions which apparently are important to host specificity (Fargues *et al.,* 1976). The reasons why some fungi elicit such strong host responses while other pathogenic fungi may colonize the host with impunity remain a matter for much investigation.

Wall-less protoplasts occur naturally during early developmental stages of mycoses caused by *Coelomomyces* (Powell, 1976), some species of *Entomophthora* (Tyrrell, 1977; MacLeod *et al.,* 1980), and *Massospora* (R. S. Soper, unpublished). Electron microscopy of vegetative hyphae of *Strongwellsea magna* and *Entomophthora sphaerosperma* reveals that the cell wall may be indistinct or absent altogether (Humber, 1975; R. S. Soper and R. J. Milner, unpublished). The role of these protoplasts and "pseudoprotoplasts" in the host remains unclear but might somehow aid in the fungus' escape from the host's immune system.

III. SPECIAL PROBLEMS: DISPERSAL, SURVIVAL, AND ADAPTATION TO THE HOST

A. Adaptations for Conidial Dispersal

The synnematous habit is particularly common among the entomopathogenic Hyphomycetes and for many of their *Cordyceps* teleomorphs. Twice as many synnematous Hyphomycetes produce dry conidia as those with conidia in a quantity of slime. The positive phototropic orientation of most synnemata ensures that conidia will be produced in free air where passing insects can pick up dry or slimed conidia or where air current can disperse dry conidia.

The release of conidia in slime drops (e.g., *Acremonium, Synnematium, Verticillium*) or with slime coats (e.g., *Hirsutella*) seems to be an important adaptation for conidial attachment to insects as well as for the dispersal of conidia (Ingold, 1961, 1978; Fletcher, 1977).

The spores of numerous entomopathogens are forcibly discharged. With the exception of the Myriangiales, the ascospores of entomopathogenic ascomycetes are discharged from dehiscent asci and become part of the air spora. Ascospores of *Cordyceps* species have the dual advantage of being forcibly discharged from their perithecia atop the columnlike stromata which characterize this genus. In some *Cordyceps* species, the stroma serves the special purpose of allowing the aerial dispersal of ascospores from infected hosts which may be buried many centimeters into the soil.

The conidia of *Beauveria bassiana,* which frequently infect insects on foliage or tree trunks, are single, dry, and easily carried by wind currents, whereas the conidia of *Metarrhizium anisopliae,* a frequent pathogen of subterranean insects, are produced in large aggregates which stay on or in the soil rather than being airborne. Spore trap studies indicate that Entomophthorales spores are dispersed by wind (Wilding, 1970a).

Three distinct mechanisms of spore discharge occur in the Entomophthorales (King and Humber, 1980). Once discharged, the primary spores of these Zygomycetes adhere where they land because of a sticky spore surface (on spores discharged by the rounding off of turgid cells) or a droplet of cytoplasm ejected along with the conidium (as in *Entomophthora muscae*) (Ingold, 1971). Spores of some *Entomophthora* species occurring on small dipterans in riparian habitats may become tetraradiate when trapped below the surface film of the water (Webster *et al.,* 1978); this adaptation may aid dispersal but has no obvious benefit for the infection of new hosts. A small number of entomophthoralean fungi produce and discharge their spores while the host is still alive and motile.

Entomophthoralean spores are exceptional in escaping from unsuitable substrates by forming forcibly discharged secondary spores (Page and Humber, 1973; King and Humber, 1980). These secondary spores in turn may also form further forcibly discharged spores until their nutrient reserves are exhausted. A completely different form of spore can be formed by many entomophthoralean fungi; these spores (which may also form upon germination of thick-walled resting spores) have a more-or-less sticky region (haptor) on the distal tip and are produced atop a very thin, upright capillary sporophore which rises up to 250 μm from the substrate (King and Humber, 1980). "Anadhesive" spores atop the capillaries are not dislodged by strong air currents, but the spores are placed where the haptors can contact and adhere to a passing insect. Such spores are known to be infective (Wilding, 1970b; Nemoto and Aoki, 1975; Brobyn and Wilding, 1977).

Fungus-mediated modifications of host behavior which favor efficient conidial dispersal occur in moribund hosts with several entomophthoralean fungi. Dying insects frequently are observed to seek out abnormally high positions, often climbing during periods when uninfected hosts remain close to the ground. The host insects become affixed at these elevated positions, die, and rain down spores on potential new hosts (Thaxter, 1888; Batko, 1974; Loos-Frank and Zimmermann, 1976). The apparent lack of fungal effect upon host behavior, however, may be critical for dispersal: The attempted copulations of healthy cicadas with those bearing sporulating *Massospora* infections is a direct, efficient means of transmission (Soper *et al.,* 1976a). Similarly, the behavior of flies bearing a sporulating *Strongwellsea* infection in their abdomen is not notably altered despite the presence of hyphae throughout the nervous system and the parasitic castration of females (Humber, 1976). *Entomophthora*-infected grasshoppers tend to climb shrubs before death and thereby permit primary spores produced on the cadaver to be rained down on healthy individuals on the ground.

B. Fungal Responses to Adverse Conditions

There may be substantial periods in a given year when the natural hosts of entomopathogenic fungi are more or less unavailable. In temperate zone regions, such times include winter with its usual dearth of active hosts, the early spring when host populations are resuming their activities but are still sparse, late in the collapse of a population due to predation and/or disease, and late summer or autumn when the annual die-off of the host population occurs.

Nearly all entomogenous fungi in the temperate zones must have some means of surviving the winter while a susceptible host population is generally unavailable. The usual overwintering forms for the insects themselves are eggs or diapausing larvae or nymphs. Insect eggs are affected by very few fungi (see Humber and Soper, 1980) and usually remain unpenetrated even when they are the only recognizable host structure in a heavily mycotized host. Little evidence suggests that diapausing insect populations regularly suffer much mortality from mycoses.

Many entomopathogens grow well in pure culture on nutritionally simple media but fare poorly in the presence of bacterial or fungal contaminants. Despite the seeming possibility that entomopathogenic species might survive as saprobes in soil or detritus (e.g., Gustafsson, 1965), vegetative forms of these fungi generally seem unable to compete against antagonistic microbes (Fargues and Remaudière, 1977). Furthermore, species known to be entomopathogenic are almost never included among lists of saprobic fungi isolated from various diverse habitats.

No evidence suggests that entomopathogenic fungi adopt alternate hosts during the winter or during periods when susceptible hosts are available but remain uninfected (especially early in the season).

A common strategy for surviving unfavorable climatic conditions or the absence of hosts is formation of a resistant or resting state: Among the higher fungi (ascomycetes, basidiomycetes, and deuteromycetes), the most common specialized resistant structures are stromata, sclerotia, and chlamydospores. The hyphomycete genus *Sorosporella* is based on red chlamydospores which germinate to produce an alternate *Synsliocladium* state (Petch, 1942). Among the phycomycetous fungi, the predominant resistant stages are thick-walled structures such as resistant sporangia, oospores, zygospores, and azygospores.

Most species of the Entomophthorales produce resting spores (zygospores or azygospores) (MacLeod, 1963). In some of these fungi, the conidia and resting spores may be formed together in a single host, and both forms may be found throughout the season. More typically, however, the resting spores occur only late in the course of an epizootic or during the annual decline of the host population, whereas the conidia formed earlier augment the inoculum levels (MacLeod, 1963; Soper *et al.*, 1976b). In *Massospora, Strongwellsea,* and most species of *Entomophthora,* forcibly discharged and resting spores are never found together in an individual host; the reasons for such mutual exclusion of these two reproductive modes remain unknown.

Conidia or ascospores may be able to survive extreme climatic conditions of temperature and/or dryness but do not tend to be so durable as the more specialized forms listed above. The function of the conidium as a resting state for prolonged fungal survival is emphasized in fungi such as *Metarrhizium* where contact with the host cuticle stimulates conidial germination (Walstad *et al.,* 1970) and in which no other specialized resistant structures are formed.

Relatively little is known of the means by which fungal infections or epizootics begin in any new season. Viable conidia produced in the previous season are presumed to be the usual inoculum for many Hyphomycetes. However, when a fungus overwinters as a chlamydospore or some other resting spore form, environmental conditions are assumed to be responsible for breaking spore dormancy. The resting spores of the Entomophthorales are those which are best understood at this time: MacLeod (1963) discussed some of the factors believed to stimulate the germination of resting spores of *Entomophthora*, but little rigorous information is available for many species. The germination of resting spores of an *Entomophthora* from the woolly pine-needle aphid is governed both by day length (Wallace *et al.*, 1976) and temperature (Payandeh *et al.*, 1978), so that the fungus does not become active until well after the host population is established. Soper *et al.* (1976a) has indicated that the resting spores of *Massospora* germinate only in the presence of cicada nymphs just at the time when they will emerge for their final molt to the imago. Latteur (1977) and Latgé *et al.* (1978) have demonstrated that *Entomophthora* resting spores in the soil during the winter can germinate in the spring to infect susceptible aphids but have not indicated how germination is stimulated. Artificial means for inducing germination of *E. virulenta* resting spores were reported by Soper *et al.* (1975) and by Matanmi and Libby (1976); these methods could help the early initiation of epizootics before any naturally occurring inoculum might become activated but give little indication of the factors which might control germination in the field.

Relatively few Hyphomycetes are capable of shifting from the production of conidia to that of a more resistant form as the host populations decrease from disease or with the encroaching end of the field season. *Syngliocladium–Sorosporella* alternate between conidia and chlamydospore production (Petch, 1942), but the mechanism(s) controlling the switch between spore forms remains unknown. A few Hyphomycetes produce sclerotia or stromata (e.g., *Synnematium, Fusarium*). Deuteromycetes which have pyrenomycetous teleomorphs presumably produce conidia first and perithecia later; *Nectria* species in which synnematous anamorphs and perithecia occur together on the same stroma have been classified superfluously as *Sphaerostilbe* (see Humber and Soper, 1980).

The problems of survival may be very different for entomopathogenic fungi in tropical regions, but many of the same mechanisms still occur. Unfortunately, the biologies and significance of many of these tropical entomopathogens are less well understood than those of their temperate zone counterparts.

C. Some Exceptional Examples of Fungal Adaptation to the Host

Not surprisingly, the long association of fungi with insects has produced some extraordinary adaptations allowing fungi to exploit these particular ecological niches. Despite the fact that very few people actively study them, the Laboulbeniales continually fascinate most mycologists with their extraordinary taxonomic diversity, morphological complexity, and biological specializations (see Benjamin, 1971). Some equally notable and occasionally bizarre adaptations of conidial fungi to their insect hosts are noted below:

Culicinomyces clavosporus, a pathogen of mosquito larvae, is unique among the entomogenous Hyphomycetes in being a fungus for which (1) conidia are produced and dispersed under water and (2) infection occurs through the fore- or hindgut rather than through the exoskeleton (Couch *et al.,* 1974; Sweeney, 1975a). This fungus also infects chironomid midges (Sweeney, 1975b) and blackflies (Simuliidae) (A. W. Sweeney, unpublished). Even though infection normally occurs through the gut with *Culicinomyces,* it can occur through the anal papillae also (Sweeney, 1979).

The species of *Massospora* are obligate entomophthoralean pathogens of cicadas. These fungi are restricted to the terminal abdominal segments of the hosts; primary spores or resting spores are passively dispersed when the fungal mass is exposed by disarticulation of the living host's exoskeleton. There is no evidence of any alternate host or saprobic phase for this genus, so its resting spores apparently must remain dormant for up to 17 yr to germinate on cue and to infect the next emergence of the periodical cicada (Soper *et al.,* 1976a). Once the fungus is established at a site, it is always present with each successive emergence and may constitute the single largest mortality factor for emergent cicadas (White *et al.,* 1979). This fungus-host relationship has forced *M. cicadina* (from 17-yr cicada) to adopt what must be the longest life cycle of any fungus.

Septobasidium has the most complex relationship with its host of any entomogenous fungus. The numerous species of *Septobasidium* shelter and husband their scale insect hosts while parasitizing only a portion of the included host population. This scale–fungus association is distinctly harmful to the trees on which it occurs (Couch, 1938). The septal and basidial morphology of *Septobasidium* are identical to those of rust fungi (Dykstra, 1974b; Couch, 1938), despite the traditional placement of these fungi among the tremelloid Hymenomycetes.

Because most entomopathogenic fungi sporulate after the death of the host, it is possible to divide the development of most insect mycoses into parasitic and saprobic stages (Ferron, 1978). The length of the saprobic stage varies among species from fungi in which extensive fungal development occurs before death and the immediate onset of sporulation (e.g., most of the Entomophthorales) to those in which the insect dies very soon after infection and before any extensive vegetative growth by the fungus. In the latter instance, death usually results from the action of a mycotoxin and is followed by extensive fungal development in the cadaver in a manner similar to the toxic effects of some phytopathogenic fungi which proliferate in freshly killed plant tissues. A variety of mycotoxins are now known from diverse entomopathogenic fungi such as *Beauveria, Metarrhizium, Paecilomyces,* and even some *Conidiobolus*-like species of *Entomophthora* (Roberts, 1980a). The antibiotic activity of these fungi is capable of suppressing competition from the host's gut flora and from other saprobes and allows mummification of the host body as was so well appreciated by the ancient Chinese (see Section I).

In addition to their more obvious effects, a few insect mycoses cause parasitic castration of the living hosts without directly destroying the gonads. *Strongwellsea* species (Entomophthorales) are the most notable of these fungi, and castration appears to occur only by the prevention of oviposition. The gonads in either sex appear to be free from any fungus-mediated histopathology (Humber, 1976); however, it is probable that the physical pressure of the highly organized fungal mass against the dorsal abdominal wall collapses the oviduct and thereby prevents the passage of the fully formed eggs. Affected female flies are so severely weakened by the nutrient drain to both the fungus and oogenesis that they usually die while attempting to deposit eggs (Humber, 1975). *Trichothecium acridiorum* grows as a cuticular parasite of grasshoppers and locusts, but it also can suffocate the host when it invades the tracheae (Madelin, 1966a); this fungus may cause enough cuticular hypertrophy to distort the abdominal segments and thereby prevent oviposition. In contrast, *Entomophthora erupta* also strongly distorts the abdominal shape of affected mirid bugs (Wheeler, 1972), but no parasitic castration results. Instead, the ovaries and all other abdominal organs are destroyed indiscriminately in a host which remains active until this fungal evisceration is more or less complete.

IV. MICROBIAL CONTROL

A. Historical Perspectives

The eminent Harvard University mycologist Roland Thaxter began his life-long studies in insect mycology because he was impressed with the devastation the Entomophthorales wrought on his fly colonies in the laboratory (Thaxter, 1888). Epizootics of fungal diseases are frequently noted in the field in a wide variety of insect pests, and they have inspired entomologists, mycologists, and plant pathologists to try to initiate epizootics artifically (Roberts, 1978b). The first such attempts were conducted by Metchnikoff (1879) and Krassilstchik (1888) using mass-produced *M. anisopliae* conidia against larvae of the wheat cockchafer (*Anisoplia austriaca*) and the sugar beet curculio (*Cleonus punctiventris*). At least partial control seems to have been achieved; but, as in many of the early fungal studies, critical evaluation was not reported.

At approximately the same time, ca. 1890, a project was initiated in the central United States to control a serious pest of cereals, the chinch bug (*Blissus leucopterus*), with *Beauveria bassiana* (summarized by Steinhaus, 1956). Unfortunately, this project was not based on sound experimental work and did not afford the expected level of pest control. The effort covered a 10-yr period, and at its termination the prospect for fungal control of insects in general, and for chinch bugs in particular, was estimated to be very low. Several scientists stated that introductions of fungal spores did not increase disease incidence because *Beauveria* was present naturally at levels sufficiently high to induce epizootics. The limiting factor was assumed to be weather. Warm temperatures coupled with high humidity appeared to encourage epizootics. This project was abandoned totally, and there have been no publications on *B. bassiana* on chinch bugs for more than 60 yr, even though the basic research on this host–pathogen relationship was not completed in the pre-1900 period.

The second wave of microbial control activity with fungi in the United States took place in the citrus fields of Florida. Beginning in 1906 through the mid-1940s several entomologists encouraged the use of "friendly fungi"—a group of ascomycetes and deuteromycetes—for scale (coccid) and whitefly (aleyrodid) control (summarized in Steinhaus, 1975). In addition to advising citrus growers to transfer diseased insects into orchards to increase fungal inoculum, these entomologists encouraged the restriction of fungicides used for control of plant pathogens to reduce damage to the friendly fungi. The situation was thoroughly confused in the late 1940s by the claim that the friendly fungi were basically saprophytes growing on insects killed by some other cause, primarily infection of the chytridiomycete *Myiophagus ucrainicus* (Fisher, 1947). Because *Myiophagus* requires free water for dissemination of its zoospores, its potential for insect control was considered less than that claimed for the friendly fungi. Except for *Aschersonia*, the friendly fungi are now ignored in Florida (C. M. McCoy, personal communication). Unfortunately, the importance of *Myiophagus* in scale insect natural mortality has not been evaluated quantitatively with modern methodology, and interpretation of the early claims concerning this fungus is difficult. It is apparent that a critical study of the friendly fungi is needed. Several genera (*Aschersonia, Aegerita, Verticillium, Nectria, Podonectria, Myriangium,* and others) are included in the group, and it seems unreasonable to declare all to be saprophytes when only a few have been closely observed. Except for *Aschersonia* and possibly *Verticillium*, there are at present no such studies underway.

Although there were some notable later successes, microbial control of insects with fungi obviously had difficult beginnings; and it is only within the past two decades that mycoinsecticides have again assumed prominence in insect control projects. Conceptual approaches to microbial control and some examples of recent and/or current research with fungi follow.

B. Conceptual Approaches

There are three basic approaches to the use of fungi for insect control: colonization, mycoinsecticide, and integration with other control measures (Yendol and

Roberts, 1970). The first of these, colonization, is the most appealing because it is presumably the easiest, least expensive, and longest lasting. It simply entails release of the proper fungal isolate in one or a few locations in the pest population. To be rated successful the fungus must establish itself permanently in the pest population and hold the population below its economic threshold. Examples of this are very rare (Roberts, 1978a). The second approach, use as a mycoinsecticide, calls for large-scale application of the fungus each time the pest population exceeds its economic threshold. This of course is appealing to industrial concerns producing the fungi, because repeated sales are probable. As will be discussed later, a single mass introduction each season to hasten the initiation of epizootics which normally occur, but occur too late to protect the crop, has been proposed for *Hirsutella thompsonii* and *Nomuraea rileyi*. This approach is a combination of colonization with a mycoinsecticide. The third approach, integrated pest management, is the most difficult because it requires a great deal of information on the pest, the host, other natural enemies, chemical pesticides safe to the fungus, and the fungus itself. Ideally, all these factors will be sufficiently well known to predict disease significance, whether the fungus is naturally occurring or artificially applied.

The failure of early researchers to recognize the importance of physiological races of entomopathogenic fungi probably impeded development of these organisms for insect control. Wide ranges in virulence, host range, and conidial productivity have been demonstrated in recent years for all entomopathogenic species examined. The method most commonly used, that of isolating a fungus from the host in the locality of interest, mass-producing the new isolate, and reintroducing it into a pest population, neglects one of the advantages of utilizing microorganisms, namely, their genetic diversity. Fungal isolates from other insects or from other populations of the target pest are likely to be more effective than an isolate which has come to an accommodation with the pest population (Lappa and Goral, 1980). Genetic manipulation, e.g., parasexuality of imperfect fungi and induction of mutations, has not been utilized to any significant degree in improving insect pathogenic fungal isolates. An apparently very efficient selection technique involves culturing fungi from the hemocoel of insects surface-sterilized at or before death (Al-Aidroos and Roberts, 1978).

C. Progress and Prospects for Some Major Entomopathogens

1. *Entomophthora*

The Entomophthorales frequently occur in spectacular epizootics which decimate very large pest populations. The most common hosts are aphids [*Entomophthora aphidis*, *E. virulenta*, *E. obscura* (= *ignobilis* = *thaxteriana*), and

E. sphaerosperma], muscoid flies (*E. muscae*), lepidopterous larvae [*E. sphaerosperma, E. aulicae* (= *egressa*) and *E. gammae*], and grasshoppers (*E. grylli*). In addition to adequate fungal inoculum, high host density is critical to the development of epizootics (Soper, 1980). Weather, particularly temperature and humidity, also vitally affects the development of epizootics. However, epizootics of *E. sphaerosperma* in aphids have been noted in cultivated crops in one of the world's driest deserts (Kenneth and Olmert, 1975), and *E. grylli* on grasshoppers at Alice Springs in the middle of the Australian outback (R. S. Soper, personal communication) presumably because the microclimate on the plant is appropriate for disease induction; in general, humid climates are most conducive to epizootics of *Entomophthora* (Roberts and Campbell, 1977). One of the best documented cases is that of pea aphids on various legumes in the USSR (Voronina, 1971). Surveys conducted over much of the western USSR from 1962 to 1969 revealed a close correlation between rainfall and incidence of entomophthoroses in pea aphid populations. By dividing the nation into regions based on hydrothermal coefficient, the incidence of disease and whether or not chemical insecticide application will be needed, can be predicted. Very large areas of the alfalfa-producing portion of the USSR are not treated with pesticides at present as a result of this study. Some large marginal areas are pesticide-treated as needed, and a third large area is routinely treated, since the index indicates that *Entomophthora* will seldom be present in sufficient amounts to reduce the pest populations.

Entomophthora sphaerosperma was introduced into Australian populations of the spotted alfalfa aphid [*Therioaphis trifolii* (=*maculata*)] in 1979 (R. S. Soper and R. Milner, personal communcation). This insect was first detected in Australia in March 1977. *Entomophthora* spp. are considered major mortality factors for *T. trifolii* in the United States (Hall and Dunn, 1958), but the aphid arrived in Australia free of virulent *Entomophthora* pathogens. It will be very interesting to follow the fate of the introduced fungus, which was collected originally in Israel. In one experiment diseased aphids were collected 60 m from a small focus where conidia were released 1 week earlier. In other experiments on $\frac{1}{2}$-ha plots epizootics were initiated within 3 weeks following multipoint releases.

The use of Entomophthorales as mycoinsecticides has been hindered by the fragility of their forcibly discharged spores, most of which live only a few hours after discharge. A recent finding, however, has provided promise for exploiting the very long-lived resting spores which many Entomophthorales species produce (Soper *et al.*, 1975). Germination of the spores in the laboratory was not possible previously, and at present only one species readily germinates. Nevertheless, the devising of a method (involving vigorous agitation in a blender for 30 min with 0.5% ethanol) which stimulates germination of one species indicates that methods can be devised for other species. When such methods are available, mass-produced resting spores can be applied to fields with conventional equipment.

2. *Beauveria*

Agostino Bassi established the germ theory of disease in animals in the mid-1830s with his studies on *Beauveria bassiana* infections of silkworm larvae (Steinhaus, 1956). The fungus most frequently isolated from dead (particularly from overwintering) insects collected in the field is *B. bassiana*. The host range of the species is extensive and includes most orders of insects. A majority of the microbial control tests with fungi have utilized either *B. bassiana* or *B. brongniartii* ($=tenella$). As mentioned previously, the first major efforts with *B. bassiana* were its promising but ultimately ineffective introductions against chinch bugs in the United States.

The production of large amounts of *Beauveria* spp. for field application has been accomplished with a wide variety of technological approaches. It is produced in mainland China on vegetable materials in bottles or in troughs in the soil (Anonymous, 1977). The fungal inoculum presumably must be very large, since contamination with other fungi apparently is not a serious problem. Twenty-two metric tons of *B. bassiana* were produced in the USSR in 1977 at a permanent production facility and used commercially for insect control, primarily of Colorado potato beetle (A. A. Yevlakhova, 1978, and personal communication). In this instance, the fungus is grown in an unagitated liquid medium. The isolate used apparently is a remarkable producer of antibiotics, since the medium is not sterilized prior to inoculation.

The most widely used method for mass production of fungi at present in the USSR is the submerged-surface method. This consists of initiating growth in fermenters or shake flasks and then placing the cultures in thin layers in shallow pans for sporulation (Lappa and Goral, 1980). Conidia produced in this way have a high virulence and good storage characteristics. The method, however, requires considerable space for the surface phase. Semisolid fermentation technology has been utilized in some cases to alleviate the space problems associated with surface methods. The most elementary form is the use of an unagitated particulate substrate, such as rice kernels (as described in Section IV,C,3) to provide a large surface area in a small volume. The more advanced forms include high levels of aeration and gentle agitation of the substrate. This, at present, is the method usually selected by industry in the West.

In the USSR, although most mass production currently utilizes surface or submerged-surface techniques, the major research emphasis is on submerged (fermenter) production. Media and conditions which encourage formation of blastospores in submerged culture are readily found for most isolates of *B. bassiana* (Samsinakova, 1966). As mentioned in Section II, blastospore production in the host hemolymph is a normal stage in the development of most entomopathogenic conidial fungi. Unfortunately, blastospores are essentially short hyphae, and they lack the heavy wall which makes conidia somewhat resistant to adverse conditions such as drying. Because of the ease of production, blastospores of *B. bassiana* and *B. brongniartii* have been utilized in a number of laboratory and experimental-scale field studies. In general, however, these spores are considered too short-lived for insect control. The recent observation that blastospores in silicate clays (clarsol and montmorillonite) remain active in soil much longer than non-silicate-coated blastospores may rekindle interest in them as microbial control agents (Reisinger *et al.*, 1977; Fargues *et al.*, 1978). In the USSR research emphasis at present is on submerged (fermenter) production of conidia rather than blastospores. In the early 1970s, V. Goral was granted USSR patents on methods for producing *B. bassiana* and *M. anisopliae* conidia in submerged culture. *Beauveria bassiana* conidia were produced in submerged culture using media with a neutral initial pH, approximately 11% amino nitrogen, low aeration, and slightly elevated temperatures. These conidia usually had reduced virulence and longevity in comparison to surface-produced conidia (Goral, 1978; Lappa and Goral, 1980), although preparations with unimpaired virulence have been reported (Kononova, 1978).

Although utilized for immediate kill of pest insects, one of the major advantages of *B. bassiana* over conventional pesticides is its long-term effect on the host population. Reduced adult longevity and high mortality rates in larvae from these adults has been noted in insect populations exposed to *B. bassiana* or *B. brongniartii* as larvae or adults (Bajan *et al.*, 1976, 1977; Müller-Kögler and Stein, 1970; Keller, 1978; Lappa and Goral, 1980). The duration of activity of *B. bassiana* conidia introduced into the soil against Colorado potato beetle was estimated to be 2 yr (Wojciechowska *et al.*, 1977).

As summarized by Ferron (1978), the numbers of fungal spores (or colony-forming units, if determined by dilution plate counts instead of hemocytometer counts) currently utilized in experimental plots for insect control usually fall in the range of 10^{13}–10^{17}/ha. Such high levels are difficult to attain in practice, but it is presumed that formulation, genetic alteration, and improved application technology will permit drastic reductions in dosage. It should be noted that Ignoffo *et al.* (1976a) induced epizootics of *Nomurae rileyi* in soybean caterpillars by introducing 2.7×10^{10} spores/ha.

Sublethal doses of chemical insecticides in combination with a fungus have been advocated as a means of permitting reduced doses of both fungus and insecticide (usually one-fifth of the amount of each needed for control if used alone). The insecticide is presumed to debilitate the host sufficiently to facilitate infection by the fungus. The host species is critical to the efficiency of this approach. After considerable effort with *B. bassiana* plus insecticides against the Colorado potato beetle in the USSR, it has been concluded that the action of the two agents is additive, not synergistic (Lappa and Goral, 1980). Similar results were obtained in France with the same host–pathogen combination (Fargues, 1975). Synergism was noted, on the other hand, when low doses of insecticides and *B. brongniartii* were tested against larvae of the beetle *Melolontha melolontha* (Ferron, 1970, 1971).

Beauveria bassiana is being developed at present as a control agent of the Colorado potato beetle (*Leptinotarsa decemlineata*) in the USSR, Poland, France, and United States. The insect entered Eastern Europe after World War II, where it has been an exceptionally devastating pest of potatoes. The populations are resistant to virtually all commercial insecticides in Rhode Island and Long Island, two potato-producing areas of the United States. Soviet Scientists have proposed that the pest population be carefully monitored and the fungus be utilized when the larvae are predominantly small (first and second instar) and the adult population small (Lappa and Goral, 1980). Insecticide-fungus combinations or pesticides alone are recommended for large populations. Inclusion of the fungus is encouraged because of its long-term effects. Other target species for this fungus are the codling moth (*Laspeyresia pomonella*) in the USSR and the European corn borer (*Ostrinia nubilalis*) in the Peoples Republic of China.

The other widespread *Beauveria* species, *B. brongniartii* is utilized almost exclusively against soil-inhabiting scarabeid beetles, primarily *Melolontha melolontha* in Western Europe (Ferron, 1974; Keller *et al.*, 1979).

3. *Metarrhizium*

Metarrhizium anisopliae occurs in two forms based on conidial size: *M. anisopliae* var. *anisopliae* (3.5–9.0 μm) and *M. anisopliae* var. *major* (9.0–18.0 μm) (Tulloch, 1976). The latter tend to be rather host-specific and infect primarily scarabeid beetles. One of the most serious pests in coconut-growing regions of the South Pacific, the rhinoceros beetle (*Oryctes rhinoceros*), is a scarabeid susceptible to *M. anisopliae* var. *major*. The Rhinoceros Beetle Project in Apia, Western Samoa, has demonstrated that this fungus can be effectively used to control the pest (K. J. Marschall, 1978, and personal communication). Since the insect entered Western Samoa about 50 yr ago, no new plantings of coconut trees had survived, and virtually no fruit had been produced on older surviving trees. In 1968, the fungus was cultured on grain and distributed by hand (with a spoon as whole kernels of infested grain or with a watering can as a spore suspension) in approximately one-half of the larval breeding sites on Fagaloa Bay. Within 1 yr the standing trees were producing good yields of coconuts, and newly planted trees were surviving. Preliminary experiments with an isolate of this fungus from Sri Lanka tested against larval *Oryctes* beetles in Coimbatore, India, also gave positive results (P. C. S. Babu, personal communication).

The small-spored forms usually have wide host ranges, and they are currently being considered for control of several insect species. Mosquito larvae (*Anopheles, Aedes,* and *Culex*) are highly susceptible to some isolates (Al-Aidroos and Roberts, 1978; R. A. Daoust and D. W. Roberts, unpublished), and outdoor trials with one of the less virulent isolates indicated that *M. anisopliae* may be an effective mosquito larvicide (Roberts, 1977). The largest efforts with this group of fungi are the sugarcane–spittlebug (Homoptera: Cercopidae) control projects in Brazil. Several grower associations maintain staff with expertise on *Metarrhizium*. Their function is primarily to mass-produce the fungus for large-scale field application. At least 146,000 ha of sugarcane was treated with *M. anisopliae* during the 1979–1980 summer in the states of Pernambuco and Alagoas. The methodology for mass production is very simple and could be utilized virtually any place in the world. Rice is autoclaved with a small amount of water in polypropylene bags closed with metal clips or ties. After cooling, the bags are inoculated with spores and incubated at room temperature for approximately 2 weeks. The product is harvested by either milling the rice kernels or by floating the spores off them in a large water bath. Introduction of whole, fungus-infested kernels into the field as "granules" also has been proposed (Villacorta, 1976). The doses used are 600 g/ha of milled fungus-infested rice or 50 g/ha of pure conidia. The appropriateness of these doses has not been tested in recent years. Although the results of the field applications have not been reported since 1974 (Guagliumi *et al.,* 1974), growers prefer the fungus because it kills nymphs as well as adult spittlebugs whereas carbaryl, the most commonly used insecticide, kills only adults. At present, the demand for fungal preparations outstrips production capability. Sugarcane is a very important crop in Brazil, because it is not only a food crop but it is also used for alcohol production (gasoline now contains 20% ethanol in many parts of the country). The most important pests of sugarcane in South America and the Caribbean are spittlebugs and stem-boring Lepidoptera. Spittlebugs, different species from those on sugarcane, are the most important pests on the ca. 150 million ha of Brazilian pastureland. These insects are a principal reason for Brazil's importation rather than exportation of meat. The grasses planted on cleared land are all from Africa or Australia, whereas the pests are native to Brazil. *Metarrhizium* occurs in natural epizootics in some pastures and sugarcane fields, particularly late in the growing season. Nevertheless, it is thought that introductions of large amounts of fungus early in the season may keep the spittlebugs below damaging levels, and several laboratories are now studying the fungus in both the field and the laboratory. As of early 1979, five commercial concerns expressed interest in obtaining governmental permission to market *M. anisopliae* preparations for pasture spittlebug control, and at least temporary approval was granted to two of the firms.

  Metarhizium anisopliae is most frequently isolated from soil-inhabiting insects, and this group of pests is a promising target of the species for microbial control. Soil frequently is physically appropriate for the fungus (offering high temperature, high humidity, and protection from sunlight). Nevertheless, soil may be inappropriate chemically, primarily because of fungistatic compounds produced by other microorganisms.

4. *Hirsutella*

Epizootics of *Hirsutella thompsonii* occur each year in citrus rust mite (*Phyllocoptruta oleivora*) populations in Florida (McCoy, 1980). Unfortunately, the epizootics occur after the fruit is damaged. Damage from this pest is only cosmetic, since it does not injure the interior of the fruit; it is very costly, however, because it causes the fruit to be downgraded to sale for juice rather than as fresh fruit. Field applications of the fungus to initiate epizootics earlier in the season have given variable results, but good control has been obtained in some years. The method of spore production and formulation of the product appears to be critical. For example, freeze-dried conidia remain viable but are no longer pathogenic to mites (T. Couch and C. McCoy, personal communication); fortunately, this response to freeze-drying has not been noted with other entomopathogenic fungi. High-quality conidia are obtained by spraying mycelial fragments and nutrients onto the trees where the fungus then conidiates. This approach requires that the environmental conditions—particularly humidity—remain satisfactory for many hours to allow for conidiogenesis and mite infection.

The potential of this fungus is high, and development of it as a commercial product is being pursued by at least one U.S. firm. This is the only arthropod-killing fungus for which an Experimental Use Permit (EUP) has been granted by the U.S. Environmental Protection Agency (EPA). This an an intermediate status between ''experimental'' and ''registered''. It should be added that EUPs have not been sought by U.S. scientists for any other entomopathogenic fungi. Guidelines for full registration of mycoinsecticides are currently being drawn up by the EPA, and, when they are completed, *H. thompsonii* probably will be the first arthropod-pathogenic fungus for which U.S. registration is sought.

5. *Nomuraea*

Epizootics of *Nomuraea* (=*Spicaria*) *rileyi* in noctuid caterpillars occur annually in all major soybean-growing areas of the United States and Brazil. This fungus has been studied extensively both in the laboratory and the field during the past 5 yr in the central and southern United States. Soybeans are most sensitive to defoliation by caterpillars during pod formation and development of beans to full size. Defoliaters such as the green cloverworm (*Plathypena scabra*) usually reach their peak population at this time. The fungus is normally detected 2–3 weeks after green cloverworms are found. Although the incidence of disease increases with time, the epizootic frequently develops too slowly to suppress pest damage during the critical pod-forming and bean development stages. Heavy application of conidia produced on artificial media or introduction of fragmented fungus-killed larvae advanced the peak incidence of *N. rileyi* by at least 2 weeks in comparison to the naturally occurring local epizootic, and this afforded crop protection (Sprenkel and Brooks, 1975; Ignoffo *et al.*, 1976a,b; Hostetter and Ignoffo, 1978). The fungus apparently overwinters, and the first trifoliates are contaminated while passing through conidium-bearing soil or by wind-borne conidia; insects feeding on these leaves become infected (Ignoffo *et al.*, 1977). The inoculum then accumulates slowly over the season to high levels. Artificial introduction of the fungus affords high inoculum levels when the insects are too small to cause economic damage. Cultural practices—particularly early, close planting—also enhance disease development in soybeans (Sprenkel *et al.*, 1979); the same was found for lepidopterous pests of cotton (Burleigh, 1975).

A model has been proposed for predicting the level in Florida soybean fields of *N. rileyi* infection in velvetbean caterpillar (*Anticarsia gemmatalis*) populations on a given day (Kish and Allen, 1978). The model is still under development, but at present the basic information required includes (1) inoculum density, (2) weather conditions and their effects on the inoculum, and (3) the relationship between inoculum density and infection levels (Allen and Kish, 1978). These factors are not easily estimated. Predicted values for 1976 (the last year for which data are available) were significant at the 10% level for 16 of 24 trials (66%) and at the 5% level for 12 of 24 trials (50%).

Ignoffo *et al.* (1978) have proposed that all lepidopterous pests of soybean could be controlled with microbial agents. *Nomuraea rileyi* would be integrated into the system by making one prophylactic heavy application of conidia 2–4 weeks before economic levels of insect defoliators were anticipated. If the environmental conditions were appropriate, this initial inoculum would increase to control defoliators during stages when soybeans were most sensitive to defoliation and also provide control of depodding damage by *Heliothis zea* later in the season. If the expected epizootic failed to develop, *Bacillus thuringiensis* could be applied in time to control the defoliators, while the nuclear polyhedrosis virus of *H. zea* could be used later to protect the pods. Chemical insecticides would be used as a last resort only.

V. CONCLUSION

The conidial fungi associated with insects are represented by members from virtually all taxonomic groups except the higher basidiomycetes and dematiaceous Hyphomycetes. More than 60 fungal genera are involved. The majority of published accounts deal with Entomophthorales and Hyphomycetes. Also, virtually all groups of insects have known associations with fungi. Although conidial fungi were studied initially in attempts to protect silkworm cultures and honeybee colonies, the emphasis at present, aside from basic studies, is on exploitation of their potential for control of agriculturally and medically important arthropods.

ACKNOWLEDGMENTS

R. A. H. gratefully acknowledges the support of the National Institutes of Health Research Service Award A105348 from the National Institutes of Allergy and Infectious Diseases, and of the U.S. Department of Agriculture Science and Education Administration—Agricultural Research.

REFERENCES

Ainsworth, G. C., Sussman, A. S., and Sparrow, F. K., eds. (1973a). "The Fungi: An Advanced Treatise," Vol. 4A. Academic Press, New York.

Ainsworth, G. C., Sussman, A. S., and Sparrow, F. K., eds. (1973b). "The Fungi: An Advanced Treatise," Vol. 4B. Academic Press, New York.

Al-Aidroos, K., and Roberts, D. W. (1978). Mutants of *Metarhizium anisopliae* with increased virulence toward mosquito larvae. *Can. J. Genet. Cytol.* **20,** 211–219.

Allen, G. E., and Kish, L. P. (1978). The role of entomopathogens in an integrated pest management system in soybean. *In* "Microbial Control of Insect Pests" (G. E. Allen, C. M. Ignoffo, and R. P. Jaques, eds.), pp. 164–185. University of Florida, Gainesville.

Anonymous (1977). "Insect Control in the People's Republic of China: A Trip Report of the American Insect Control Delegation," Committee on Scholarly Communication with the People's Republic of China, Rep. No. 2. Natl. Acad. Sci., Washington, D.C.

Baird, R. B. (1954). A species of *Cephalosporium* (Moniliaceae) causing a fungus disease in larvae of the European corn borer, *Pyrausta nubilalis* (Hbn.) (Lepidoptera: Pyraustidae). *Can. Entomol.* **86,** 237–240.

Bajan, C., Fedorko, A., Kmitowa, K., and Wojciechowska, M. (1976). Influence of pathogenic microorganisms on the Colorado beetle. *Bull. Acad. Pol. Sci., Ser. Sci. Biol.* **24,** 171–173.

Bajan, C., Kmitowa, K., Wojciechowska, M., and Fedorko, A. (1977). 17. The effect of entomopathogenic microorganisms introduced into the soil on the development of successive generations of the Colorado beetle. *Pol. Ecol. Stud.* **3,** 157–165.

Barron, G. L. (1968). "The Genera of Hyphomycetes from Soil." Williams & Wilkins, Baltimore, Maryland.

Batko, A. (1964a). Remarks on the genus *Entomophthora* Fresenius 1856 non Nowakowski 1883 (Phycomycetes: Entomophthoraceae). *Bull. Acad. Pol. Sci., Ser. Sci. Biol.* **12,** 319–321.

Batko, A. (1964b). On the new genera: *Zoophthora* gen. nov., *Triplosporium* (Thaxter) gen. nov., and *Entomophaga* gen. nov. (Phycomycetes: Entomophthoraceae). *Bull. Acad. Pol. Sci., Ser. Sci. Biol.* **12,** 323–326.

Batko, A. (1964c). Remarks on the genus *Lamia* Nowakowski 1883 vs. *Culicicola* Nieuwland 1916 (Phycomycetes: Entomophthoraceae). *Bull. Acad. Pol. Sci., Ser. Sci. Biol.* **12,** 399–402.

Batko, A. (1964d). Some new combinations in the fungus family Entomophthoraceae (Phycomycetes). *Bull. Acad. Pol. Sci., Ser. Sci. Biol.* **12,** 403–406.

Batko, A. (1974). Filogeneza a struktury taksonomiczne Entomophthoraceae. *In* "Ewolucja biologiczna: Szkice Teoretyczne i metodologiczne" (C. Nowinskiego, ed.), pp. 209–304. Polska Akad. Nauk, Institut Filozofii i Socjologii, Wroclaw.

Batra, L. R. (1967). Ambrosia fungi: A taxonomic revision and nutritional studies of some species. *Mycologia* **59,** 976–1017.

Batra, L. R., ed. (1979). "Insect-Fungus Symbiosis." Allanheld, Osmun & Co., Montclair, New Jersey.

Benjamin, R. K. (1971). Introduction and supplement to Roland Thaxter's "Contributions Toward a Monograph of the Laboulbeniaceae," Bibl. Mycol. Vol. 30. Cramer, Lehre, West Germany.

Blackwell, M., and Kimbrough, J. W. (1976a). Ultrastructure of the termite-associated fungus *Laboulbeniopsis termitarius. Mycologia* **68**, 541–550.

Blackwell, M., and Kimbrough, J. W. (1976b). A developmental study of the termite-associated fungus *Coreomycetopsis oedipus. Mycologia* **68**, 551–558.

Blackwell, M., and Kimbrough, J. W. (1978). *Hormiscioideus filamentosus* gen. and sp. nov., a termite-infesting fungus from Brazil. *Mycologia* **70**, 1274–1280.

Booth, C. (1961). Studies of Pyrenomycetes. VI. *Thielavia,* with notes on some allied genera. *Mycol. Pap.* **83**, 1–15.

Brobyn, P. J., and Wilding, N. (1977). Invasive and developmental processes of *Entomophthora* species infecting aphids. *Trans. Br. Mycol. Soc.* **69**, 349–366.

Buchner, P. (1965). "Endosymbiosis of Animals with Plant Microorganisms." Wiley (InterScience), New York.

Burges, H. D., ed. (1980). "Microbial Control of Insects, Mites, and Plant Diseases," Vol. 2. Academic Press, New York.

Burleigh, J. G. (1975). Comparison of *Heliothis* spp. larval parasitism and *Spicaria* infection in closed and open canopy cotton varieties. *Environ. Entomol.* **4**, 574–576.

Clark, T. S., Kellen, W. R., Lingren, J. E., and Sanders, R. D. (1966). *Pythium* sp. (Phycomycetes: Pythiales) pathogenic to mosquito larvae. *J. Invertebr. Pathol.* **8**, 351–354.

Coles, R. B., and Talbot, R. H. B. (1977). *Septobasidium clelandii* and its conidial state, *Harpographium corynelioides,* parasitizing female coccids. *Kew Bull.* **31**, 481–488.

Couch, J. N. (1938). "The Genus *Septobasidium.*" Univ. of North Carolina Press, Chapel Hill.

Couch, J. N., Romney, S. V., and Rao, B. (1974). A new fungus which attacks mosquitoes and related Diptera. *Mycologia* **66**, 374–379.

de Hoog, G. S. (1978). Notes on some fungicolous Hyphomycetes and their relatives. *Persoonia* **10**, 33–81.

Dykstra, M. J. (1974a). An ultrastructural examination of the structure of asexual propagules of four mucoralean fungi. *Mycologia* **66**, 477–489.

Dykstra, M. J. (1974b). Some ultrastructural features in the genus *Septobasidium. Can. J. Bot.* **52**, 971–971.

Emmons, C. W., Binford, C. H., and Utz, J. P. (1977). "Medical Mycology, "3rd ed. Lea & Febiger, Philadelphia, Pennsylvania.

Fargues, J. (1975). Etude expérimentale dans la nature de l'utilisation combinée de *Beauveria bassiana* et d'insecticides à dose reduite contre *Leptinotarsa decemlineata. Ann. Zool.—Ecol. Anim.* **7**, 247–64.

Fargues, J., and Remaudière, G. (1977). Considerations on the specificity of entomopathogenic fungi. *Mycopathologia* **62**, 31–37.

Fargues, J., and Remaudière, G. (1980). Host specificity and virulence of entomopathogenic fungi. *In* "Entomopathogenic Fungi" (D. W. Roberts, ed.). Allanheld, Osmun & Co., Montclair, New Jersey (in press).

Fargues, J., and Vey, A. (1974). Modalités d'infection des larves de *Leptinotarsa decemlineata* par *Beauveria bassiana* au cours de la mue. *Entomophaga* **19**, 311–323.

Fargues, J., Robert, P.-H., and Vey, A. (1976). Rôle du tégument et de la défense cellulaire des Coléoptères hôtes dans la spécificité des souches entomopathogènes de *Metarrhizium anisopliae* (Fungi Imperfecti). *C. R. Hebd. Seances Acad. Sci., Ser. D* **282**, 2223–2226.

Fargues, J., Reisinger, O., and Robert, P. H. (1978). Use of argil-coating to increase the resistance of spores of entomopathogenic fungi to biodegradation. *Abstr., Int. Colloq. Invertebr. Pathol., 1978* p. 35.

Ferron, P. (1970). Augmentation de la sensibilité des larves de *Melolontha melolontha* L. (Coléoptère: Scarabaeidae) *Beauveria tenella* (Delacr.) Siemaszko au moyen de quantités réduites de HCH. *Proc. Int. Colloq. Insect Pathol., 1970* pp. 66–79.

Ferron, P. (1971). Problèmes posés par la mise au point d'un procédé la lutte microbiologique contre *Melolontha melolontha* au moyen de le mycose à *Beauveria tenella* (Delacr.) Siemaszko. *Phytiatr.—Phytopharm.* **20,** 159–168.

Ferron, P. (1974). Essai de lutte microbiologique contre *Melolontha melolontha* par contamination du sol à l'aide de blastospores de *Beauveria tenella. Entomophaga* **19,** 103–114.

Ferron, P. (1978). Biological control of insect pests by entomogenous fungi. *Annu. Rev. Entomol.* **23,** 409–442.

Fisher, F. E. (1947). Insect disease studies. *Annu. Rep., Fla. Agric. Exp. Stn.* p.162.

Fletcher, H. (1977). Parallel evolution in insect-dispersed fungi and insectivorous plants? *Bull. Br. Mycol. Soc.* **11,** 50–51.

Fogel, R. (1975). Insect mycophagy: A preliminary bibliography. *U.S. Dep. Agric., For. Serv.* [*Gen. Tech. Rep.*] *PNW* **36,** 1–21.

Goral, V. M. (1978). Effect of cultivation conditions on the entomopathogenic properties of muscardine fungi. *In* "Proceedings of Project V: Microbial Control of Insect Pests" (C. M. Ignoffo, ed.), pp. 217–228. Am. Soc. Microbiol., Washington, D.C.

Gougher, R. J., and Kimbrough, J. W. (1969). *Antennopsis gallica* Heim and Buchli (Hyphomycetes: Gloeohaustoriales), an entomogenous fungus on subterranean termites in Florida. *J. Invertebr. Pathol.* **13,** 223–228.

Graham, K. (1967). Fungal-insect mutualism in trees and timber. *Annu. Rev. Entomol.* **12,** 105–126.

Guagliumi, P., Marques, E. J., and Vilas Boas, A. M. (1974). Contribuicão as estudo da cultura e aplicacao de *Metarrhizium anisopliae* (Metschn.) Sorokin no controle da "cigarrinha-da-folha" *Manhanarva posticata* (Stal) no nordeste do Brasil. *Bol. Tec. CODECAP (Commissao Executiva de Defesa Fitossanitaria da Lavoura Canavieira de Pernambuco)* No. 3, pp.11–54.

Gustafsson, M. (1965). On the species of the genus *Entomophthora* in Sweden. II. Cultivation and physiology. *Lantbrukhöegsk. Ann.* **31,** 405–457.

Hall, I. M., and Dunn, P. H. (1958). Artificial dissemination of entomophthorous fungi pathogenic to the spotted alfalfa aphid in California. *J. Econ. Entomol.* **51,** 341–344.

Heitor, F. (1962). Parasitisme de blessure par la champignon *Mucor hiemalis* Wehmer chez les insectes. *Ann. Epiphyt.* **13,** 179–205.

Hoffman, W. E. (1947). Insects as human food. *Proc. Entomol. Soc. Wash.* **49,** 233–237.

Hostetter, D. L., and Ignoffo, C. M. (1978). Induced epizootics: Fungi. *In* "Microbial Control of Insect Pests" (G. E. Allen, C. M. Ignoffo, and N. P. Jaques, eds.), pp.22–31. University of Florida, Gainesville.

Hughes, S. J. (1976). Sooty molds. *Mycologia* **68,** 693–820.

Humber, R. A. (1975). Aspects of the biology of an insect-parasitic fungus, *Strongwellsea magna* (Zygomycetes: Entomophthorales). Ph.D. Dissertation, University of Washington, Seattle.

Humber, R. A. (1976). The systematics of the genus *Strongwellsea* (Zygomycetes: Entomophthorales). *Mycologia* **68,** 1042–1060.

Humber, R. A., and Soper, R. S. (1980). Preparation, preservation, and identification of insect pathogenic fungi. *In* "Entomopathogenic Fungi" (D. W. Roberts, ed.). Allanheld, Osmun & Co., Montclair, New Jersey (in press).

Ignoffo, C. M., and Garcia, C. (1978). *In vitro* inactivation of conidia of the entomopathogenic fungus *Nomuraea rileyi* by human gastric juice. *Environ. Entomol.* 7(2), 217–218.

Ignoffo, C. M., Marston, N. L., Hostetter, D. L., and Puttler, B. (1976a). Natural and induced epizootics of *Nomuraea rileyi* in soybean caterpillars. *J. Invertebr. Pathol.* **27,** 191–198.

Ignoffo, C. M., Marston, N. L., Puttler, B., Hostetter, D. L., Thomas, G. D., Biever, K. D., and Dickerson, W. A. (1976b). Natural biotic agents controlling insect pests of Missouri soybeans. *World Soybean Res., Proc. World Soybean Res. Conf., 1975* pp.561–578.

Ignoffo, C. M., Garcia, C., Hostetter, D. L., and Pinnell, R. E. (1977). Laboratory studies of the entomopathogenic fungus *Nomuraea rileyi:* Soil-borne contamination of soybean seedlings and dispersal of diseased larvae of *Trichoplusia ni. J. Invertebr. Pathol.* **29,** 147–152.

Ignoffo, C. M., Hostetter, D. L., Biever, K. D., Garcia, C., Thomas, G. D., Dickerson, W. A., and Pinnell, R. (1978). Evaluation of an entomopathogenic bacterium, fungus, and virus for control of *Heliothis zea* on soybeans. *J. Econ. Entomol.* **71**(2), 165–168.

Ignoffo, C. M., Garcia, C., Kapp, N. W., and Coate, W. B. (1979). An evaluation of the risks to mammals of the use of an entomopathogenic fungus, *Nomuraea rileyi,* as a microbial insecticide. *Environ. Entomol.* **8**(2), 354–359.

Ingold, C. T. (1961). The stalked spore-drop. *New Phytol.* **60,** 181–183.

Ingold, C. T. (1971). "Fungal spores: Their Liberation and Dispersal." Oxford Univ. Press (Clarendon), London and New York.

Ingold, C. T. (1978). Role of mucilage in dispersal of certain fungi. *Trans. Br. Mycol. Soc.* **70,** 137–140.

Keller, S. (1978). Infectionsversuche mit dem Pilz *Beauveria tenella* an adulten Maikäfern (*Melolontha melolontha* L.). *Mitt. Schweiz. Entomol. Ges.* **51,** 13–19.

Keller, S., Keller, E., and Ramser, E. (1979). Ergebnisse eines Versuches zur mikrobiologischen Bekämpfung des Maikäfers (*Melolontha melolontha* L.) mit dem Pilz *Beauveria tenella. Mitt. Schweiz. Entomol. Ges.* **52,** 35–44.

Kenneth, R., and Olmert, I. (1975). Entomopathogenic fungi and their insect hosts in Israel: Additions. *Isr. J. Entomol.* **10,** 105–112.

King, D. S., and Humber, R. A. (1980). Entomophthorales: Identification. *In* "Microbial Control of Insects" (H. D. Burges, ed.), Vol. 2, pp. 107–127. Academic Press, New York.

King, D. S., and Jong, S. C. (1976). Identity of the etiological agent of the first deep entomophthoraceous infection of man in the United States. *Mycologia* **68,** 181–183.

Kish, L. P., and Allen, G. E. (1978). The biology and ecology of *Nomuraea rileyi* and a program of predicting its incidence on *Anticarsia gemmatalis* in soybean. *Fla., Agric. Exp. Stn., Bull.* **795,** 1–47.

Kobayasi, Y. (1977). Miscellaneous notes on the genus *Cordyceps* and its allies. *J. Jpn. Bot.* **52,** 269–272.

Kononova, E. V. (1978). Selection of commercial strains of the fungus *Beauveria bassiana* (Bals.) Vuill. *In* "Proceedings of Project V: Microbial Control of Insect Pests" (C. M. Ignoffo, ed.), pp. 172–191. Am. Soc. Microbiol., Washington, D.C.

Krassilstschik, I. M. (1888). La production industrielle des parasites végétaux pour la destruction des insectes nuisibles. *Bull. Sci. Fr. Belg.* **19,** 461–472.

Lappa, N. V., and Goral, V. M. (1980). Insect control with fungi in Eastern Europe. *In* "Entomopathogenic Fungi" (D. W. Roberts, ed.). Allanheld, Osmun, Montclair, New Jersey (in press).

Latgé, J.-P., Perry, C., Papierok, B., Coremans-Pelseneer, J., Remaudière, G. and Reisinger, O. (1978). Germination des azygospores d'*Entomophthora obscura* Hall & Dunn, rôle de sol. *C. R. Hebd. Seances Acad. Sci., Ser. D* **287,** 943–946.

Latteur, G. (1977). Sur le possibilité d'infection directe d'aphides par *Entomophthora* à partir de sols hebergeant un inoculum naturel. *C. R. Hebd. Seances Acad. Sci., Ser. D* **284,** 2253–2256.

Lichtwardt, R. W. (1973). Trichomycetes. *In* "The Fungi: An Advanced Treatise" (G. C. Ainsworth, A. S. Sussman, and F. K. Sparrow, eds.), Vol. 4B, pp.237–243. Academic Press, New York.

Lloyd, C. G. (1918). *Cordyceps sinensis,* from N. Gist Gee, China. *Mycol. Notes* **54,** 766–768.

Loos-Frank, B., and Zimmermann, G. (1976). Über eine dem *Dicrocoelium*-Befall analoge Verhaltensanderung bei Ameisen der Gattung *Formica* durch einen Pilz der Gattung *Entomophthora. Z. Parasitenkd.* **49,** 281–289.

McCoy, C. W. (1980). Fungi: Pest control by *Hirsutella thompsonii*. *In* "Microbial Control of Insects, Mites and Plant Diseases" (H. D. Burges, ed.), Vol. 2. Academic Press, New York (in press).

McEwen, F. L. (1963). *Cordyceps* infections. *In* "Insect Pathology: An Advanced Treatise" (E. A. Steinhaus, ed.), pp.273-290. Academic Press, New York.

MacLeod, D. M. (1963). Entomophthorales infections. *In* "Insect Pathology: An Advanced Treatise" (E. A. Steinhaus, ed.), Vol. 2, pp.273-290. Academic Press, New York.

MacLeod, D. M., Tyrrell, D., and Welton, M. A. (1980). Isolation and growth of the grasshopper pathogen, *Entomophthora grylli*. *J. Invertebr. Pathol.* **36**, 85-89.

Madelin, M. F. (1963). Diseases caused by hyphomycetous fungi. *In* "Insect Pathology: An Advanced Treatise" (E. A. Steinhaus, ed.), Vol. 2, pp.233-271. Academic Press, New York.

Madelin, M. F. (1966a). *Trichothecium acridiorum* (Trabut) comb. nov. on red locusts. *Trans. Br. Mycol. Soc.* **49**, 275-288.

Madelin, M. F. (1966b). Fungal parasites of insects. *Annu. Rev. Entomol.* **11**, 423-448.

Mains, E. B. (1959). Species of *Aschersonia* (Sphaeropsidales). *Lloydia* **22**, 215-221.

Marschall, K. J. (1978). Biological control of rhinoceros beetles: Experiences from Samoa. *Proc. Int. Conf. Cocoa Coconuts, 1978*, pp. 1-5.

Matanmi, B. A., and Libby, J. L. (1976). The production and germination of resting spores of *Entomophthora virulenta* (Entomophthorales: Entomophthoraceae). *J. Invertebr. Pathol.* **27**, 279-285.

Metchnikoff, E. (1879). "Diseases of the Larva of the Grain Weevil. Insects Harmful to Agriculture," Issue III. The Grain Weevil. Commission attached to the Odessa Zemstro Office for the Investigation of the Problem of Insects Harmful to Agriculture, Odessa (in Russian).

Müller-Kögler, E. (1965). "Pilzkrankheiten bei Insekten." Parey, Berlin.

Müller-Kögler, E., and Stein, W. (1970). Gewächshausversuche mit *Beauveria bassiana* (Bals.) Vuill. zur Infektion von *Sitona lineatus* (L.) (Coleopt., Curcul.) im Boden. *Z. Angew. Entomol.* **65**, 59-76.

Nemoto, H., and Aoki, J. (1975). *Entomophthora floridana* (Entomophthorales: Entomophthoraceae) attacking the Sugi spider mite, *Oligonychus hondoensis* (Acarina: Tetranychidae), in Japan. *Appl. Entomol. Zool.* **10**, 90-95.

Page, R. M., and Humber, R. A. (1973). Phototropism in *Conidiobolus coronatus*. *Mycologia* **65**, 335-354.

Payandeh, B., MacLeod, D. M., and Wallace, D. R. (1978). Germination of *Entomophthora aphidis* resting spores under constant temperatures. *Can. J. Bot.* **56**, 2328-2333.

Petch, T. (1926). Studies on entomogenous fungi. IX. *Aegerita*. *Trans. Br. Mycol. Soc.* **11**, 50-66.

Petch, T. (1927). Studies in entomogenous fungi. XII. *Peziotrichum lachnella; Ophionectria coccorum; Volutella epicoccum*. *Trans. Br. Mycol. Soc.* **12**, 44-52.

Petch, T. (1937). Notes on entomogenous fungi. *Trans. Br. Mycol. Soc.* **21**, 34-67.

Petch, T. (1942). Notes on entomogenous fungi. *Trans. Br. Mycol. Soc.* **25**, 250-265.

Powell, M. J. (1976). Ultrastructural changes in the cell surface of *Coelomomyces punctatus* infecting mosquito larvae. *Can. J. Bot.* **54**, 1419-1437.

Prasertphon, S., and Tanada, Y. (1968). The formation and circulation, in *Galleria,* of hyphal bodies of entomophthorous fungi. *J. Invertebr. Pathol.* **11**, 260-280.

Ramsbottom, J. (1914). Some notes on the history of the classification of the Phycomycetes. *Trans. Br. Mycol. Soc.* **5**, 324-350.

Raper, K. B., and Fennell, D. I. (1965). "The Genus *Aspergillus*." Williams & Wilkins, Baltimore, Maryland.

Reisinger, O., Fargues, J., Robert, P., and Arnould, M.-F. (1977). Effet de l'argile sur la conservation des microorganismes. I. Etude ultrastructurale de la biodégradation dans le sol de l'hyphomycète entomopathogène *Beauveria bassiana* (Bals.) Vuill. *Ann. Microbiol. (Paris)* **128b**, 271-287.

Remaudière, G., and Hennebert, G. L. (1980). Révision systématique de *Entomophthora aphidis* Hoffm. in Fres. Description de deux nouveaux pathogènes d'aphides. *Mycotaxon* **11**, 269-321.

Remaudière, G., and Keller, S. (1980). Révision systématique des genres d'Entomophthoraceae à potentialité entomopathogène. *Mycotaxon* **11**, 323-338.

Roberts, D. W. (1970). *Coelomomyces, Entomophthora, Beauveria,* and *Metarrhizium* as parasites of mosquitos. *Misc. Publ. Entomol. Soc. Am.* **7**, 140-155.

Roberts, D. W. (1977). Isolation and development of fungus pathogens of vectors. *In* "Biological Regulation of Vectors" (J. D. Briggs, ed.), DHEW Publ. No. (NIH) 77-1180, pp.85-93. USDHEW, Washington, D.C.

Roberts, D. W. (1978a). Introduction and colonization: Fungi. *In* "Microbial Control of Insect Pests" (G. E. Allen, C. M. Ignoffo, and R. P. Jacques, eds), pp.14-18. University of Florida, Gainesville.

Roberts, D. W. (1978b). Past history and current status of the development of entomopathogenic fungi in the United States. *In* "Proceedings of Project V: Microbial Control of Insect Pests" (C. M. Ignoffo, ed.), pp. 6-19. Am. Soc. Microbiol., Washington, D.C.

Roberts, D. W. (1980a). Fungi: Toxins. *In* "Microbial Control of Insects, Mites, and Plant Diseases" (H. D. Burges, ed.), Vol. 2. Academic Press, New York (in press).

Roberts, D. W., ed. (1980b). "Entomopathogenic Fungi." Allenheld, Osmun & Co., Montclair, New Jersey (in press).

Roberts, D. W., and Campbell, A. S. (1977). Stability of entomopathogenic fungi. *Misc. Publ. Entomol. Soc. Am.* **10**(3), 19-76.

Roberts, D. W., and Yendol, W. G. (1971). Use of fungi for microbial control of insects. *In* "Microbial Control of Insects and Mites" (H. D. Burges and N. W. Hussey, eds.), pp. 125-149. Academic Press, New York.

Rossman, A. Y. (1978). *Podonectria,* a genus in the Pleosporales on scale insects. *Mycotaxon* **7**, 163-182.

Samsinakova, A. (1966). Growth and sporulation of submerged cultures of the fungus *Beauveria bassiana* in various media. *J. Invertebr. Pathol.* **8**, 395-400.

Samsinakova, A., Misikova, S., and Leopold, J. (1971). Action of enzymatic systems of *Beauveria bassiana* on the cuticle of the greater wax moth larvae (*Galleria mellonella*). *J. Invertebr. Pathol.* **18**, 322-330.

Samson, R. A. (1974). *Paecilomyces* and some allied Hyphomycetes. *Stud. Mycol.* **6**, 1-119.

Schabel, H. G. (1976). Green muscardine disease of *Hylobius pales* (Herbst) (Coleoptera: Curculionidae). *Z. Angew. Entomol.* **81**, 413-421.

Soper, R. S. (1980). Epizootiology of entomopathogenic fungi. *In* "Entomopathogenic Fungi" (D. W. Roberts, ed.). Allanheld, Osmun & Co., Montclair, New Jersey (in press).

Soper, R. S., Holbrook, F. R., Majchrowicz, I., and Gordon, C. C. (1975). Production of *Entomophthora* resting spores for biological control. *Maine Life Sci. Agric. Exp. Stn., Tech. Bull.* **76**, 1-15.

Soper, R. S., Delyzer, A. J., and Smith, L. F. R. (1976a). The genus *Massospora* entomopathogenic for cicadas. Part II. Biology of *Massospora levispora* and its host *Okanagana rimosa,* with notes on *Massospora cicadina* on the periodical cicadas. *Ann. Entomol. Soc. Am.* **69**, 89-95.

Soper, R. S., Smith, L. F. R., and Delyzer, A. J. (1976b). Epizootiology of *Massospora levispora* in an isolated population of *Okanagana rimosa. Ann. Entomol. Soc. Am.* **69**, 275-283.

Sprenkel, R. K., and Brooks, W. M. (1975). Artificial dissemination and epizootic initiation of *Nomuraea rileyi,* an entomogenous fungus of Lepidopterous pests of soybeans. *J. Econ. Entomol.* **68**, 847-851.

Sprenkel, R. K., Brooks, W. M., Van Duyn, J. W., and Deitz, L. L. (1979). The effects of three cultural variables on the incidence of *Nomuraea rileyi,* phytophagous Lepidoptera, and their predators on soybeans. *Environ. Entomol.* **8**, 334-339.

Steinhaus, E. A. (1947). "Insect Microbiology." Cornell Univ. Press (Comstock), Ithaca, New York.

Steinhaus, E. A. (1949). "Principles of Insect Pathology." McGraw-Hill, New York.

Steinhaus, E. A. (1956). Microbial control—The emergence of an idea: A brief history of insect pathology through the nineteenth century. *Hilgardia* **26**, 107–160.

Steinhaus, E. A. (1975). "Disease in a Minor Chord." Ohio State Univ. Press, Columbus.

Suzuki, A., Kawakami, K., and Tamura, S. (1971). Detection of destruxins in silkworm larvae infected with *Metarrhizium anisopliae. Agric. Biol. Chem.* **35**, 1641–1643.

Sweeney, A. W. (1975a). The mode of infection of the insect pathogenic fungus *Culicinomyces* in larvae of the mosquito *Culex fatigans. Aust. J. Zool.* **23**, 49–57.

Sweeney, A. W. (1975b). The insect pathogenic fungus *Culicinomyces* in mosquitoes and other hosts. *Aust. J. Zool.* **23**, 59–64.

Sweeney, A. W. (1979). Infection of mosquito larvae by *Culicinomyces* sp. through anal papillae. *J. Invertebr. Pathol.* **33**, 249–251.

Thaxter, R. (1888). The Entomophthoreae of the United States. *Mem. Boston Soc. Nat. Hist.* **4**, 133–201.

Thaxter, R. (1914). On certain peculiar fungus-parasites of living insects. *Bot Gaz. (Chicago)* **58**, 235–253.

Thaxter, R. (1920). Second note on certain peculiar fungus-parasites of living insects. *Bot. Gaz. (Chicago)* 49, 1–27.

Travland, L. B. (1979). Initiation of infection of mosquito larvae (*Culiseta inornata*) by *Coelomomyces psorophorae. J. Invertebr. Pathol.* **33**, 95–105.

Tulloch, M. (1976). The genus *Metarhizium. Trans. Br. Mycol. Soc.* **66**, 407–411.

Tyrrell, D. (1977). Occurrence of protoplasts in the natural life cycle of *Entomophthora egressa. Exp. Mycol.* **1**, 259–263.

Villacorta, A. (1976). Technique for the mass culture of the entomophagous fungus *Metarrhizium anisopliae* in granular form. *An. Soc. Entomol. Bras.* **5**, 102–104.

Voronina, E. G. (1971). Entomophthorosis epizootics of the pea aphid *Acrythosiphon pisum* Harris (Homoptera, Aphidoidae). *Entomol. Rev.* **50**, 444–453.

Wallace, D. R., MacLeod, D. M., Sullivan, C. R., Tyrrell, D., and Delyzer, A. J. (1976). Induction of resting spore germination in *Entomophthora aphidis* by long-day light conditions. *Can. J. Bot.* **54**, 1410–1418.

Walstad, J. E., Anderson, R. F., and Stambaugh, W. J. (1970). Effects of environmental conditions on two species of muscardine fungi (*Beauveria bassiana* and *Metarrhizium anisopliae*). *J. Invertebr. Pathol.* **16**, 221–226.

Webster, J., Sanders, P. F., and Descals, E. (1978). Tetraradiate aquatic propagules in 2 species of *Entomophthora. Trans. Br. Mycol. Soc.* **70**, 472–479.

Wheeler, A. G., Jr. (1972). Studies on the arthropod fauna of alfalfa. III. Infection of the alfalfa plant bug, *Adelphocoris lineolatus* (Hemiptera: Miridae), by the fungus *Entomophthora erupta. Can Entomol.* **104**, 1763–1766.

White, J., Lloyd, M., and Zar, J. H. (1979). Faulty eclosion in crowded suburban periodical cicadas: Populations out of control. *Ecology* **60**, 305–315.

Wilding, N. (1970a). *Entomophthora* conidia in the air-spora. *J. Gen. Microbiol.* **62**, 149–157.

Wilding, N. (1970b). Resting spore formation and germination in *E. fresenii. Annu. Rep. Rothamsted Exp. Stn.* Pt. 1, p. 207.

Wojciechowska, M., Kmitowa, K., Fedorko, A., and Bajan, C. (1977). 15. Duration of activity of entomopathogenic microorganisms introduced into the soil. *Pol. Ecol. Stud.* **3**, 141–148.

Yendol, W. G., and Paschke, J. D. (1965). Pathology of an *Entomophthora* infection in the eastern subterranean termite *Reticulitermes flavipes* (Kollar). *J. Invertebr. Pathol.* **7**, 414–422.

Yendol, W. G., and Roberts, D. W. (1970). Is microbial control with entomogenous fungi possible? *Proc. Int. Colloq. Insect Pathol., 1970* pp. 28–42.

Yevlakhova, A. A. (1978). Basic trends in the use of entomopathogenic fungi in the Soviet Union. *In* "Proceedings of Project V: Microbial Control of Insect Pests" (C. M. Ignoffo, ed.), pp. 35–50. Am. Soc. Microbiol., Washington, D.C.

Zacharuk, R. Y. (1973). Electron-microscope studies of the histopathology of fungal infections by *Metarrhizium anisopliae. Misc. Publ. Entomol. Soc. Am.* **9,** 112–119.

22

Food Technology and Industrial Mycology

William D. Gray

I. INTRODUCTION

When one considers the thousands of different anamorph-species of conidial fungi (and, of these, the number of different strains), the varied synthetic capabilities of fungi in general, and the myriad materials used by man, it is especially surprising to find that numerically there are very few conidial fungi that man intentionally employs. Hence there are relatively few compounds synthesized or processed by conidial fungi for specific human use. No single answer explains this apparent lack of interest, but a variety of partial explanations may

Biology of Conidial Fungi, Vol. 2

be advanced. The average individual (especially the Occidental) has long viewed most fungi with indifference, with distaste, or in extreme cases with horror. Conidial and other fungi are viewed as disagreeable things which spoil human food (and are possibly poisonous), mildew clothing, or bring about the decay of something useful. In the same vein, molds are associated with damp, dark, disagreeable places. As an example of this attitude, Dyson (1928), in discussing certain Oriental foods prepared through the use of molds, stated ". . . the growth of which [molds] upon edible matter would be regarded by the Occidental as ample reason for assignment to the waste bin."

As the general field of biology developed and the field of plant pathology came into existence, many conidial fungi were shown to be the causal agents of plant disease. This of course did nothing to dispel the idea that fungi were harmful organisms that had best be avoided. Curiously enough, the fact that no more than 10% of all fungal species are plant pathogens does not seem to have been considered.

The strong and rapid development of the field of chemistry also contributed to the lack of interest in the synthesizing potential of conidial fungi. Why bother to find a fungus to synthesize a compound when a chemist could probably do it more quickly and more economically by straight chemical means? Fleming's (1929) discovery of the synthesis of a valuable medicine by a common conidial fungus led some at least to ponder the answer such a question implied.

Perhaps the greatest factor that militated against a more widespread exploitation of conidial fungi by humans was the circumstances under which the human species has lived until comparatively recent years. With low population levels and seemingly inexhaustible resources of all types, there seemed to be little reason for investigating (with a view to exploiting) organisms mostly considered undesirable. These circumstances have now changed markedly and will change with ever-increasing tempo as long as the world's population continues to increase at the current rate of about 1.8% per year. At this rate, the world's population will exceed 6 billion in the year 2000 A.D. Man will no longer be in a position to let any potential resource remain unexploited, and the conidial fungi will then of necessity receive their share of attention. When food and industrial and domestic energy are in imminent short supply, some difficult decisions may have to be made: whether to (1) burn waste organic material in order to obtain energy, (2) utilize fungi and other microorganisms to convert such waste material into food, or (3) employ fungi under controlled conditions to destroy such waste as a partial solution to the problems of pollution. With a bit of imagination one can visualize the time when the effluents from kitchen waste disposal units will drain, not to the sewer, but to a common, central collecting center where they can be microbially processed to human or livestock food—provided of course that there is sufficient energy available to run the disposal unit and pump the waste to the collection center.

II. FOOD TECHNOLOGY

The general lack of exploitation of conidial fungi is reflected in a brief examination of the area of food technology. While a few of these fungi have been used for many years to process certain foods, their numbers are largely confined to a few species of two common genera. Thus, in the Occident two types of cheese have been processed with species of *Penicillium*. In the Orient, where fungi are used more commonly, several species of *Aspergillus* have been employed for centuries to process foods with a soybean base. The preparation of many different foods of this nature has undoubtedly contributed in large measure to the avoidance of diet monotony in many Oriental countries. These foods also provide needed protein in the diet and prevent the great losses of vegetable protein which always occur in the naturally inefficient conversion of vegetable protein to animal protein.

A. Cheese Processing with Conidial Fungi

The discovery of how to make cheese was without question one of the very important early advances in food technology; it provided the only means of preserving the nutrients of milk for future use. When cheese was first made in the Western world is not known, but Scott (1968) reported that Pliny praised a mold-ripened cheese in the first century A.D. Since Pliny's time a tremendous number of different cheese varieties have been developed. As early as 1925, Thom and Fisk listed over 500 varieties, while in 1968 Scott listed 26 varieties of blue-veined cheeses alone. Yet, of the hundreds of types of cheese now available in the market only two depend upon conidial fungi for their processing, and in both instances species of *Penicillium* are involved.

1. Blue-Veined Cheeses

Probably the best-known blue-veined cheese is Roquefort from France. Similar cheeses from other countries include Gorgonzola, Stilton, Danish Blue, Blue Cheshire, Wensleydale, Blue Vinney, and Dolce Verde. Thom and Currie (1913) reported the isolation of *Penicillium roquefortii* from Roquefort, Stilton, and Gorgonzola cheeses, and Dattilo-Rubbo (1938) reported this same species from most of the other blue-veined cheeses listed above. The only exception was Dolce Verde, from which he isolated a species "related to *P. expansum.*"

Once the curd for a blue-veined cheese has been prepared, bread crumbs on which *P. roquefortii* has sporulated are mixed in, and the compressed, inoculated curd is placed in a room with controlled temperature and humidity where the fungus is allowed to grow to the extent desired. There is some evidence that inoculation of the curd may not be necessary in a blue-veined cheese factory of long standing, where *P. roquefortii* forms a major part of the microbial flora. However, omitting the inoculum may sometimes prove risky.

During the ripening period, *P. roquefortii* grows in the interior of the curd—a rather remarkable occurrence since most filamentous fungi are strongly aerobic and the center of a pressed curd is not well aerated. However, Thom and Currie (1913) showed quite early that this species was unusual in that it could grow in an environment of high carbon dioxide and low oxygen concentration. Nonetheless, the organism requires some oxygen, so holes are punched in the curd to permit the access of some air.

Blue-veined cheeses have characteristic odors and flavors which are due to the action of the fungus during the ripening period. Jensen (1904) suggested (without presenting any analytic data) that the odor and flavor were due to ethyl butyrate. However, Currie (1914) has stated that capric, caprylic, and caproic acids and their salts are responsible for the burning effect of Roquefort cheese on the tongue and palate and that these acids are formed from milk fat by a hydrolysis catalyzed by a *P. roquefortii* lipase. I cannot resist pointing out that these are the principal acids of goat perspiration, although most blue-veined cheese fanciers would probably prefer not to have this information. Starkle (1924) reported that *P. roquefortii* could oxidize fatty acids to methyl ketones, and Hammer and Bryant (1937) noted that the odor and taste of blue-veined cheeses were similar to the odor and taste of 2-heptanone, this ketone probably being formed from caprylic acid by *P. roquefortii*. Patton (1950) recovered 2-pentanone, 2-heptanone, and 2-nonanone, which he suggested were formed through the beta-oxidation of fatty acids in blue-veined cheese, and Girolami and Knight (1955) found that preformed mycelia of *P. roquefortii* oxidized fatty acids to methyl ketones of one less carbon atom and that these ketones (acetone, 2-butanone, 2-pentanone, 2-hexanone, 2-heptanone, 2-octanone?) were not further metabolized. On the basis of these more recent findings it appears that methyl ketones, particularly 2-heptanone, are largely responsible for the characteristic odor and flavor of blue-veined cheeses.

2. Camembert and Similar Cheeses

Like the blue-veined cheeses, Camembert, Thenay, Brie, Troyes, and Vendôme are ripened through the action of a *Penicillium*. Usually the species is *P. camembertii* or its white variety (*P. camembertii* var. *rogeri*), but Miall (1975) also includes *P. caseiolum* as a ripener for this type of cheese. In the ripening of blue-veined cheeses *P. roquefortii* is responsible for the production of characteristic odors and flavors, but in the ripening of cheeses like Camembert, the principal action of *P. camembertii* is to alter the texture of the cheese. Although changes in odor and flavor occur during ripening, Thom (1909) has suggested that the flavor of Camembert cheese is due to the action of lactic acid bacteria.

In blue-veined cheeses the fungus grows throughout the curd, but in the ripening of Camembert the fungus grows externally on the rind. As it ripens, the cheese softens to a smooth, buttery consistency from the outside to the center as the extracellular proteases of *P. camembertii* alter the milk proteins. According to Thom (1909), the curd is not inoculated with *P. camembertii* except when a new factory is being established. Once a factory is in operation, this fungus species apparently becomes a dominant part of the microbial flora. Since the temperature and humidity are closely controlled, it soon starts to grow on the rinds of the curing cheeses. There is some danger that a contaminant such as *P. roquefortii* or *P. brevicaule* may gain entry and impart a undesirable flavor.

B. Oriental Fungus-Fermented Foods

Considerably more initiative has been exhibited in the Orient in utilizing fungi as food and in the processing of food. Thus a greater number of food products are prepared using mold-type fungi as fermenting agents. The principal conidial fungi employed in this fashion are species of *Aspergillus*. The production of most such foods probably began as cottage industries for home use but, because of their great popularity, a number are now produced in large installations on a commercial basis. (Other fermented foods besides the ones discussed below are prepared in various parts of the Orient but, because they are processed with fungi other than conidial fungi, they have been excluded from the present discussion.)

1. *Shoyu*

Of the various fungus-processed Oriental foods, *shoyu* (soy sauce) is probably most widely known to Occidentals and has been made for many centuries. Groff (1919) reported that it was prepared in the time of Confucius. Hesseltine (1965) has stated that in terms of volume of production (annual per capita consumption approximately 3 gal in Japan) it leads all other fermented foods except *tofu*. The production of *shoyu* probably started as an uncontrolled cottage process which required up to 2 yr for completion.

Traditionally, *Aspergillus oryzae* was used in the preparation of *shoyu*. In addition, Lockwood (1947) has mentioned *A. flavus,* and both Hesseltine (1965) and Wood and Yong (1975) have stated that *Aspergillus soyae* is also used. Early American production (Staley, 1935) involved no fungal agents at all; instead, an imitation *shoyu* was produced by acid hydrolysis of soybean proteins.

Shoyu, made from wheat and soybeans, depends for its flavor on the proportions of these two ingredients. Hesseltine (1965) has stated that most Japanese *shoyu* is of the *koikuchi* type (wheat and soybeans in about equal amounts), while Chinese *shoyu* is of the *tamari* type (more soybeans than wheat). Early production in Japan involved the use of whole soybeans, but according to Shibasaki and Hesseltine (1962) these have been largely replaced by defatted soybean meal. As an example of the voluminous production of this foodstuff, Smith (1963) noted that 1,061,000 bushels of soybeans and 7,778,000 bushels of soybean meal were used in 1962 in Japan alone for the production of *shoyu*.

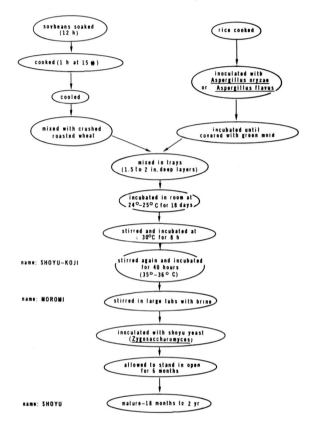

Fig. 1.   Processing steps in the manufacture of *shoyu* (from Gray, 1970).

Modern *shoyu* production is a fermentation process that involves the use of *Aspergillus oryzae*, a yeast, and a bacterium. The flow sheet in Fig. 1, based on the report of Dyson (1928), is probably representative of the process until the time of its modernization and refinement. It should be noted that there is no provision for inoculating with a bacterium, which probably entered accidentally as a contaminant. Lockwood (1947) applied pure culture techniques to *shoyu* production, using a conidial fungus (*A. oryzae*) followed by a bacterium (*Lactobacillus delbreuckii*) and a yeast (*Hansenula subpelliculosa*). This process required 37–100 days for completion. In the process described by Yong (1971) *A. oryzae* and *L. delbreuckii* were also employed, but *Saccharomyces rouxii* was used instead of *H. subpelliculosa;* furthermore, the yeast and bacterium were added simultaneously. Wood and Yong (1975) reported that by raising the temperature to 40°C for the souring and yeast fermentation stage they obtained *shoyu* of excellent quality in 1 month. For a more detailed discussion of *shoyu* preparation the papers of Yokotsuka (1960), Hesseltine and Wang (1967), and Wood and Yong (1975) should be consulted.

2. *Ketjap*

While *shoyu* is the best known and most widely used soy sauce in the Orient, another type is made in Indonesia. This soy sauce, known as *ketjap,* is made from black soybeans. The process by which it is prepared is much simpler and shorter than the *shoyu* process, since it is fermented with *Aspergillus oryzae* for only 2 or 3 days. This simpler process, outlined in Fig. 2, is based upon the description of Djien and Hesseltine (1961).

3. *Miso*

Lockwood and Smith (1950) consider *miso,* another conidial fungus-processed food, the most important soybean food product in Japan, but Hesseltine (1965) points out that in terms of volume production it is second to *shoyu* in that country. Like many other Oriental fermented foods, *miso* is produced in homes as well as in factories. Shibasaki and Hesseltine (1962) noted that, in the early 1960s, home production accounted for two-thirds as much as factory production. Apparently there are a variety of named types of *miso,* since Lockwood and Smith (1950) list white *miso,* red *miso,* and black *miso,* and Nakano (1959) classifies them as *mame, kome,* and *mugi.* However, Tamura *et al.* (1952) list *sendai, edo,* and *mame,* while Shibasaki and Hesseltine (1962) refer to *sendai*

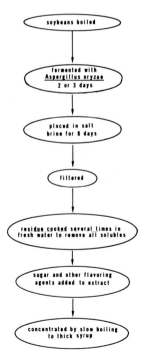

Fig. 2. Processing steps in the manufacture of *ketjap* (from Gray, 1970).

and *shinshu*. As Gray (1970) has noted, different names might be expected, since variations in ingredients, different proportions of ingredients, and different fermentation times and conditions result in slightly different products.

Miso, a paste which has the consistency of peanut butter and is used as a spread, can be produced from several combinations of ingredients. Lockwood and Smith (1950) reported its production from soybeans, wheat or wheat flour, and rice (2:1:1), but Hesseltine and Wang (1967) listed only soybeans and rice. Wood and Yong (1975) stated that *miso* could be made from beans and rice, beans and barley, or beans alone.

The production of *miso* involves two steps; first, a rice *koji* (starter) is prepared using *Aspergillus oryzae* or *A. soyae*. The *koji* is added to soybean grits and inoculated with *Saccharomyces rouxii* (Hesseltine and Shibasaki, 1961). A brief outline of the process based upon the reports of Hesseltine and Wang (1967), Shibasaki and Hesseltine (1961a,b, 1962), and Smith *et al.* (1961) is:

1. Washed, polished rice is soaked, drained of excess water, and then steamed.

2. After the steamed rice is cooled to 35°C it is inoculated with *Aspergillus oryzae* and incubated for 50 h at 27°–28°C. At the end of this time it is known as *koji*.

3. Before the *koji* is mature, soybeans are crushed to grits, washed, and soaked in water for $2\frac{1}{2}$ h. After draining the excess water the grits are steamed at 5 psi pressure for 1 h.

4. Upon cooling, the grits are mixed with *koji* and salt and inoculated with *Saccharomyces rouxii*.

5. The mixture is allowed to ferment 7 days at 28°C and then for 2 months at 35°C. It is then allowed to ripen for 2 weeks at room temperature.

6. The ripened mixture, mashed and blended, is now *miso*.

For a more detailed description and discussion of the *miso* process, the work of Shibasaki and Hesseltine (1962) is recommended.

4. *Hamanatto*

Stanton and Wallbridge (1969) refer briefly to the production of a soybean product known in Japan as *hamanatto*. Apparently *hamanatto* is produced by the fermentation of whole soybeans with *Aspergillus oryzae*. Hesseltine and Wang (1967) note that *hamanatto* is called *tu su* by the Chinese and *tao-si* by the Filipinos. Wood and Yong (1975) believe that *hamanatto* is especially suitable for development for the European market. The process (Smith, 1958) for preparing *hamanatto* is:

1. After soaking for 4 h, soybeans are steamed without pressure for 10 h.

2. They are then cooled to 30°C and inoculated with a *koji* made by culturing *Aspergillus oryzae* on roasted wheat or barley. After inoculation the beans are placed in trays.

3. When, after 20 h, the beans are covered with green mold, they are dried in the sun for 1 day.

4. The sun-dried beans are placed in baskets with strips of ginger. After soaking in salt brine they are aged under pressure for 6–12 months.

5. *Katsuobushi*

The preceding descriptions of fermented foods have all dealt with foods prepared from soybeans or soybeans mixed with cereal grains. Fish has always been an important part of the diet in the Orient, and so it was almost inevitable that some attempts at preparing foods by fermenting fish would be made. Thus several fermented foods using fish or shrimp as a starting material are also prepared. Hesseltine and Wang (1967) suggest that the oldest food fermentations may have been fish fermentations, a suggestion that may be correct if one excepts the view that alcoholic beverages are food. Thus *katsuobushi* is prepared in Japan by fermenting pieces of cooked bonito fish until they are dry. Thom and Church (1926) stated that members of the *Aspergillus glaucus* group were involved in the fermentation, a view with which Hesseltine and Wang concurred. Finished *katsuobushi* is dark, very hard, and very dry. It is shaved into ribbons and used for flavoring other foods.

6. *Ontjom*

In Indonesia two fungus-processed foods, *tempeh* and *ontjom,* are prepared. Fungal growth during their processing resembles the fungal curing of blue-veined and Camembert cheeses in the Occident. That is, *tempeh,* prepared from soybeans, is inoculated with the fungus throughout the mass, while *ontjom,* prepared from peanut press cake, is inoculated on the outer surface only. However, the similarity between these Oriental and Occidental foods exists only with respect to their modes of inoculation. *Tempeh* is produced through the action of a zygomycete (*Rhizopus oligosporus*) and will not be considered here, but *ontjom* is produced by employing the conidial phase of *Neurospora sitophila* and must be considered with the other foodstuffs that involve conidial fungi in their processing.

Both Ochse (1931) and Hesseltine and Wang (1967) have described the preparation of *ontjom*. Peanut oil is derived by pressing the seeds. The by-product, called press cake, is used as the basis for the production of *ontjom:*

1. Crumbled press cakes are soaked in water for 24 h. Any remaining oil that floats to the surface is discarded.

2. The press cake is washed, steamed, and placed in small molds (3 × 10 × 20 cm).

3. The molds are placed in a bamboo frame covered with banana leaves.

4. The molded cakes are inoculated by sprinkling with the orange-pink conidia of the *Monilia* anamorph of *Neurospora sitophila* from previously made *ontjom*. The inoculated masses are incubated in a shady place until the fungus has in-

vaded them. They are then cut into pieces, cooked, and consumed at home or taken to market.

7. Other Oriental Fungus-Fermented Foods

In addition to the various conidial fungus-fermented foods described above, Hesseltine (1965) has mentioned several others which will be briefly discussed here.

Fermented minchin is prepared by placing wheat gluten in a tightly covered container; in 2 or 3 weeks it is covered with a mixture of molds and bacteria. About 10% salt is added, and the mixture is allowed to age for 2 weeks. It is then cut into strips and used as a condiment. Although the mold population of *minchin* includes the zygomycete *Syncephalastrum,* it is largely made up of such conidial fungi as *Paecilomyces, Aspergillus, Cladosporium, Fusarium, Penicillium,* and *Trichothecium.* Apparently this fermented food has received little study, and the roles of the various fungi have not been determined.

Tao-cho is prepared from light-colored soybeans, which are soaked, dehulled, and boiled. After rice flour is added, the mixture is roasted until brown and inoculated with a species of *Aspergillus* from teak leaves. It is then fermented for 3 days and sun-dried. The dried cakes are dipped in brine, arenga sugar, and a paste of glutinous rice and exposed to the sun for at least a month.

Tao-si is also made from soybeans, but the beans are not dehulled. They are soaked and then boiled until soft. When cool, they are drained and the surfaces dried. Wheat flour is added until all surfaces are covered, and the mass is inoculated with *Aspergillus oryzae.* After 2 or 3 days' incubation the mass is covered with mycelia. It is then placed in salt brine in earthenware jars and is considered finished in about 2 months.

Taotjo, another soybean-based condiment, is prepared in the East Indies by mixing boiled soybeans with roasted wheat meal or glutinous rice. The mixture is wrapped in hibiscus leaves and inoculated with *Aspergillus oryzae.* Following 2 or 3 days' incubation the mass is placed in brine for several weeks, with palm sugar being added at intervals.

In addition to the above foods, all of which are processed with conidial fungi, many other fungus-fermented foods are prepared in the Orient. For example, *meitauza, sufu,* and *tempeh* are produced through the action of zygomycete species, and *ang-khak* is prepared with an ascomycete (*Monascus purpureus*). For those interested in this general area of food technology the paper of Hesseltine (1965) is strongly recommended. Attention is also called to the fact that the New-age Foods Study Center of Lafayette, California, has published a number of books and pamphlets in which specific instructions for preparing various fermented foods are presented.

The great emphasis on aflatoxin which began in the early 1960s raised some questions concerning the advisability of using molds for food processing. However, Yokotsuka *et al.* (1967) tested 73 strains of industrial molds used in Japan and found none that produced aflatoxin.

C. Fungal Aging and Flavoring of Meat

Apparently not many individuals today are familiar with the fact that under proper conditions conidial fungi may be used to tenderize meat and develop a better flavor. Even if such information were widespread, there is some question as to whether or not it would be utilized in the present era of instant living. In an earlier, more leisurely age it was not uncommon to allow steaks or roasts to hang in a cool (not cold) room until they became covered with green mold spores (frequently *Penicillium* spp.). The spores were wiped off with a clean cloth saturated with vinegar, and the meat was cooked as usual. Gray (1970) processed steaks in this manner and found them to be superior to steaks not so aged; he also noted that one midwestern establishment had earlier become rather famous for the quality of its steaks, which were aged by this process. It is doubtful that this uncontrolled method of aging meat would appeal to many people today in view of the widespread publicity that has been given to a few mycotoxins known to be produced by conidial fungi. However, the writer is unaware of any instance in which poisoning was due to mycotoxins produced in meats aged with fungi.

Williams (1957, 1962) was issued two patents for processes for the fungal aging of meats, but the organism he specified (*Thamnidium* sp.) was a zygomycete rather than conidial fungus. The author is not aware of any patents that have been issued involving the use of conidial fungi. While the older method of allowing fungi to grow on meat during the aging process has been largely relegated to the past, enzymes of conidial fungus origin are still used. Thus Blain (1975) notes that over 5% of United States beef is subjected to tenderization procedures, and Karmas (1970) cites the use of *Aspergillus niger* proteases in such procedures.

III. CONIDIAL FUNGI IN INDUSTRY

As in the area of food technology, only a few conidial fungi—namely, species of *Aspergillus* and *Penicillium*—have been exploited by industry. Whether the capability to synthesize materials that man needs and wants are confined to members of these two genera, or whether the potential of other conidial fungi has simply not been investigated, is not known. Van Tieghem's (1867a,b) discovery that aspergilli and penicillia were involved in the formation of gallic acid, soon followed by Wehmer's (1893) discovery that similar fungi were involved in the synthesis of citric acid, served to focus attention on these two genera. Hence early emphasis was placed upon investigations of the physiology and biochemistry of members of these two genera—an emphasis that has persisted. As a result, much more is known about these two genera than is known about all the other conidial fungi put together.

Many major contributions to the field of industrial mycology have emanated over a period of several years from the U.S. Department of Agriculture (USDA) Northern Regional Research Laboratory and its forerunner in the Bureau of Chemistry and Soils. For a brief history of the contributions of the USDA group the work of Ward (1970) should be read; however, for a more comprehensive picture the many individual USDA (Peoria) publications should be consulted.

A. Gallic Acid

Historically, the production of gallic acid (I) through the activities of a conidial fungus was the first such process investigated in the Occident. Although gallic acid had been made empirically from gallnuts for a long period of time, it was not until 1867 that Van Tieghem (1867a,b) showed that species of *Penicillium* and *Aspergillus* were the responsible agents. In connection with these investigations, Van Tieghem isolated and named a new species—*Aspergillus niger,* which has since become the most thoroughly investigated of all conidial fungi (excepting, perhaps, *Penicillium chrysogenum*).

Gallic acid is prepared from gallotannin by a hydrolysis involving a specific enzyme, tannase. The first crude method for obtaining this acid consisted of heaping up gallnuts (or other substances containing tannin) and keeping them moistened. Fungi developed throughout the mass, and after a natural fermentation of 1 month gallic acid was recovered by leaching. The fermentation was investigated by Knudson (1913a,b), and *Aspergillus niger* was found to be the fungus best suited to the job. Controlled fermentations are now conducted; for example, Calmette (cf. May and Herrick, 1930) has patented a process in which a clear tannin extract is fermented. The exact details of how this acid is made commercially are not available, but there is no question that it is made using closely controlled procedures, since Foster (1949) has stated that the fermentation is conducted with clear, aqueous, tannin-containing extracts. This fermentation is still being investigated, and Ikeda *et al.* (1972) has reported that *A. niger* is the best organism for use with Chinese gallotannin but that with another type of tannin *Penicillium chrysogenum* is more efficient.

Gallic acid is used in the printing and tanning industries and may also be employed in the synthesis of pyrogallol which is used as a photographic developer. During World War I it was used in the synthesis of a blue dye, gallo-cyanin. It may also be used as subgallate of bismuth in the treatment of certain skin disorders.

B. Citric Acid

Since Wehmer's (1893) first report of the synthesis of citric acid (II) by conidial fungi, this process has been the subject of a great many investigations. As a result, citric acid is produced by means of a fungus fermentation in greater

$$
\text{II} \quad
\begin{array}{l}
\text{H}_2\text{C}-\text{COOH} \\
\mid \\
\text{HOC}-\text{COOH} \\
\mid \\
\text{H}_2\text{C}-\text{COOH}
\end{array}
$$

amounts than any other chemical compound. Thus Cochrane (1948) reported that in 1948 about 35 million lb of this acid were produced by these means, and Lockwood (1975) has estimated that current annual production exceeds 90 million kg. Citric acid has wide usage in the manufacture of flavoring extracts, soft drinks, effervescent salts, medicines, and citrates and is also used as an ink ingredient, in dyeing, and in the silvering of mirrors.

Wehmer named a new genus (*Citromyces*) to include the responsible fungi, but today citric acid is made using a strain of the ubiquitous black mold *Aspergillus niger*. Foster (1949) noted that the capacity to form citric acid is widespread among fungi, especially the aspergilli and penicillia, and that the formation of this acid is probably one of the most widely distributed metabolic processes known in fungi.

Exact details of the processes employed in the large-scale production of citric acid by fermentation have always been well-kept trade secrets—so much so that for a number of years one company enjoyed a virtual monopoly in its manufacture. In all probability the greatest secret was the carefully selected strain of *A. niger* and knowledge of its sensitivity or nonsensitivity to various trace elements. Although there are still some carefully guarded trade secrets, enough information is available to state that there are three general methods in use.

In Japan a process has been developed in which layers of solid cooked vegetable residues are placed in trays or spread on floors and inoculated with *A. niger* (Yamada, 1965; Hisanga and Nakamura, 1966). After 5–8 days' incubation, citric acid is recovered by percolating water through the mass.

In Europe and America the earlier process involves the use of surface culture on liquid medium in shallow metal pans—a process which necessitates the use of acres of such pans to sustain a large-volume production. In fact, Lockwood (1975) notes that more than 30 acres are required in one large plant. The liquid medium is inoculated with the proper strain of *A. niger,* and the fermentation allowed to proceed for 8–10 days. For a description of a plant employing the shallow-pan process, the report of Mallea (1950) should be consulted.

In view of the disadvantages of a process involving the use of acres of cumbersome shallow pans, attention was directed toward the development of submerged culture methods using large tanks. Karow and Waksman (1947) stated that such attempts were largely unsuccessful. However, 2 yr later Perlman (1949) reviewed the relevant literature and pointed out that, in laboratory-scale submerged culture, yields were about the same as those obtained with the older surface culture method. Perlman suggested that a submerged culture process might become a reality if it were not already in operation. Lockwood (1975) has reported that submerged fermentation processes are used in America, Israel, and England. More detailed information concerning citric acid fermentation may be obtained from the works of Schweiger (1961), Batti (1966, 1967), and Lockwood (1975).

C. Gluconic Acid

One of the principal uses of gluconic acid (III) is as calcium gluconate which is

$$\text{III} \quad \begin{array}{c} \text{COOH} \\ | \\ \text{HCOH} \\ | \\ \text{HOCH} \\ | \\ \text{HCOH} \\ | \\ \text{HCOH} \\ | \\ \text{H}_2\text{COH} \end{array}$$

effective in cases of calcium deficiency. It can be injected into tissues without causing necrosis and has been used in the treatment of milk fever in cows and to increase the calcium content of the shells of eggs from calcium-deficient hens (Ramsbottom, 1936). According to Lockwood (1975), sodium gluconate also has wide use as a sequestering agent to prevent the precipitation of lime soap scum on cleaned products. Preparations of zinc gluconate and potassium gluconate are also marketed for use as diet supplements.

Although Boutroux (1880) reported the synthesis of gluconic acid by bacteria, it was not until over 20 yr later that Molliard (1922) found this product, along with citric and oxalic acids, to be a metabolite of *Aspergillus niger*. Butkewitsch (1923) found a strain of this species which, in the presence of calcium carbonate, formed gluconic acid almost to the exclusion of other acids. May *et al.* (1927) began the USDA group investigations, screened 172 different strains of fungi (many of which were conidial fungi), and selected *Penicillium purpurogenum* var. *rubisclerotium* for further study. *Penicillium chrysogenum* was later used (Moyer *et al.*, 1936), and ultimately a strain of *A. niger* was employed. Today *A. niger* is used (Ward, 1967), and the process is a submerged fermentation (Lockwood, 1975) in which calcium gluconate or sodium gluconate is produced as soon as gluconic acid is formed by the proper addition of calcium carbonate or sodium hydroxide.

The advantages of gluconic acid fermentation are threefold; while the principal product is gluconic acid, there are two valuable by-products. Glucono-δ-lactone, an intermediate in the biosynthesis of gluconic acid (used in baking powders), is recovered from the fermentation medium, and the enzyme, glucose aerodehydrogenase, is recovered from the mycelium.

D. Itaconic Acid

Itaconic acid (IV) has long been of interest as a substituted methacrylic acid

$$\text{IV} \quad \begin{array}{l} \text{COOH} \\ | \\ \text{C}{=}\text{CH}_2 \\ | \\ \text{CH}_2 \\ | \\ \text{COOH} \end{array}$$

and seemed to have potential use in the plastics industry. In fact, over 50 yr ago Hope (1927) patented a nonshatterable glass made by using polymerized itaconate esters. However, at that time itaconic acid was produced only by expensive chemical means and could not compete with cheaply prepared methacrylic acid.

Kinoshita (1929) first described the synthesis of itaconic acid by a fungus he called *Aspergillus itaconicus,* a species which Thom and Raper (1945) regarded as identical with *A. varians.* Calam *et al.* (1939) found that another species, *A. terreus,* also synthesized itaconic acid. Most subsequent investigations have been conducted with this species, since it was found that Kinoshita's strain of *A. itaconicus* produced only trace amounts of this acid (Moyer and Coghill, 1945). Early studies of this fermentation were made at the USDA laboratory in Peoria, Illinois (e.g., Lockwood and Reeves, 1945), as well as at other laboratories. Kane *et al.* (1945) received a broad patent covering an itaconic acid fermentation process. In this patent it was claimed that itaconic acid could be produced commercially in submerged, aerated cultures using selected fungus strains. Hollaender *et al.* (1945) and Raper *et al.* (1945) reported the production of better-yielding mutants derived from irradiated conidia of *A. terreus* (NRRL 265); in all probability this has been repeated in other laboratories.

One feature of the itaconic acid fermentation is that it must be conducted in a medium of low pH. Thus Lockwood and Reeves (1945) found that a pH below 2.3 was optimum for yield in surface cultures, and Lockwood and Nelson (1946) reported that pH 1.8 was optimum for submerged cultures. However, in the process described by Nubel and Ratajak (1962), when after 24 h the pH drops to 3.1, it is adjusted to pH 3.8 by the addition of lime or ammonia. Lockwood (1975) has reported that the commercial process now in use is based upon the Nubel and Ratajak (1962) patent; he has also stated that the principal use of itaconic acid is as a copolymer with acrylic resins.

E. Kojic Acid

Saito (1907) recovered a compound from the mycelium of *Aspergillus oryzae,* and Yabuta (1912), who studied it, named it kojic acid (V). In a later investigation Yabuta (1924) established its chemical structure, showing that it was 5-hydroxy-2-hydroxymethyl-γ-pyrone.

Tamiya and Hida (1929) showed that six other species of *Aspergillus* (in addition to *A. oryzae* and *A. flavus*) produced kojic acid, and Prescott and Dunn (1940) added five more species of *Aspergillus* and *Penicillium daleae* to this list. Birkinshaw *et al.* (1931) considered kojic acid production to be diagnostic for the *flavus-oryzae-tamarii* group of aspergilli.

The production of kojic acid was studied by various workers, mainly with the aim of yield improvement: e.g., Katagiri and Kitahara (1933), May *et al.* (1931, 1932), Kluyver and Perquin (1933), as well as others. Foster (1949) was of the opinion that the importance of kojic acid (as a starting compound for the synthesis of other compounds) had been exaggerated, but in 1955 Charles Pfizer and Company announced the large-scale production of kojic acid (Beesch and Shull, 1956). Whether or not it is produced in quantity today is not known. However, both Miall (1975) and Lockwood (1975) make only brief reference to kojic acid in their recent papers.

F. Penicillin

While in terms of actual amounts of chemical compounds produced on a commercial scale, citric acid leads all other conidial fungus metabolic products, no fungus-produced material has ever stirred the public interest in fungi as much as the antibiotic penicillin. In fact, Turner (1975) characterizes the penicillins as the most important compounds to have been isolated from fungi. The observation of Fleming (1929) that a contaminant *Penicillium* inhibited the growth of the *Staphylococcus* he was culturing attracted little attention at the time, and it was not until about a decade later that the value of this antibiotic became widely apparent. Knowing nothing of the chemical nature of the compound involved, Fleming called it penicillin. The pressures of World War II (for a treatment for contaminated war wounds) unquestionably played a major role in hastening the development of a process to produce this valuable drug.

The history of the penicillin process development and the literature on penicillin, so widely reviewed elsewhere (Coghill, 1944; Florey *et al.*, 1949; Demain, 1966; Chain, 1971), will be discussed only briefly here. Fleming called his contaminant organism *Penicillium rubrum,* but Charles Thom identified it as *Penicillium notatum* (Raper and Alexander, 1945). Following Fleming's original report, Clutterbuck *et al.* (1932) briefly studied penicillin, but it was not until 1939 that a group at Oxford University under the leadership of Florey resumed work on this material. Shortly thereafter two members of the Oxford group were brought to the Northern Regional Research Laboratory in Peoria, Illinois. It was at this laboratory that a discovery was made that resulted in large-scale production of penicillin. By adding corn steeping liquor to the fermentation medium, yields were increased from 1.2 μg/ml up to 24 μg/ml. A search for higher-yielding strains resulted in the isolation (from a cantaloupe) of a strain (NRRL 1951) of *Penicillium chrysogenum* that is the ancestor of today's commercially used strains. By selection from this strain a still higher-yielding strain (NRRL 1951.B25) was obtained. Irradiation of this strain resulted in strain X-1612. Subsequent ultraviolet irradiation of X-1612 resulted in the famous Q-176, which yielded in excess of 540 μg/ml (Raper, 1946). Further strain improvements have since led to strains which yield well over 7000 μg/ml (McCann and Calam, 1972).

It was discovered early in the penicillin investigations that there were several penicillins—all with the same basic structure but differing as to the nature of the side chain R:

The basic structure shown above (6-aminopenicillanic acid, VI) can be obtained from fermentations on media containing no side-chain precursors (Turner, 1975) or by hydrolysis of naturally occurring penicillin G (Carrington, 1971), which is the basis of its production on a commercial scale. Thus, with 6-aminopenicillanic acid available, semisynthetic penicillins can be produced by the introduction of various side chains into position R, and many such compounds have been made (Nayler, 1971).

When the extraordinary value of penicillin was fully realized, investigations were initiated in laboratories all over the world in search of other antibiotics produced by filamentous fungi, conidial as well as others. Surveys of the capacity of fungi to produce antibiotics (Brian, 1951; Broadbent, 1968) seem to indicate that the conidial fungi and Basidiomycetes produce greater numbers of such materials than other fungi. In spite of the thousands of fungi that have been screened, however, no antibiotic of fungal origin comparable in effectiveness to penicillin has ever been discovered. During this period of screening other extremely valuable antibiotics were discovered (streptomycin, aureomycin, terramycin, and so on), but these were formed by Actinomycetes rather than conidial fungi. Miall (1975) reports that another antibiotic of wide use, cephalosporin

C, is synthesized by the conidial fungus *Cephalosporium acremonium,* but that it is used only as the starting material for chemical synthesis of semisynthetic cephalosporins. Brotzu (1948) isolated the original strain of *C. acremonium* that is used, and Abraham and Newton (1961) established the structure of cephalosporin C, showing that it was related to the penicillins.

Unlike the gluconic acid fermentation industry, the penicillin industry does not enjoy the recovery of valuable by-products such as the enzyme glucose aerodehydrogenase, which is recovered from the spent mycelia at the end of the fermentation. With the very great amounts of penicillin produced, large amounts of mycelia are formed and their disposal can sometimes create problems. Pathak and Seshadri (1965) suggested that such mycelia be used as animal food. These workers substituted washed, dried mycelia of *Penicillium chrysogenum* for soybean flour in cooked mice rations and observed weight gains comparable to those in animals fed control rations at the end of 29 days.

G. Griseofulvin

Although griseofulvin (VII) has not received as much attention (nor does it have as wide application) as penicillin in the treatment of infections, it has become an increasingly important fungus-derived antibiotic in recent years. First described as a metabolic product found in small amounts of *Penicillium griseofulvum* (Oxford *et al.,* 1939), it was viewed with interest primarily as an addition to the small list of chlorine-containing metabolities of fungi (e.g., geodin and erdin). Later, Brian *et al.* (1946) isolated a compound from *P. janczewskii* (=*P. nigricans*) which they called the "curling factor," because it caused certain morphological changes in the hyphae of a plant-parasitic fungus, *Botrytis allii.* Grove and McGowan (1947) showed that the curling factor was griseofulvin. According to Turner (1975) it has been isolated from about a dozen other species of *Penicillium* including *P. patulum.* It is also produced by another conidial fungus, *Nigrospora oryzae* (Furuya *et al.,* 1967; Giles *et al.,* 1970).

Griseofulvin is important because it is an effective antibiotic used in the systemic treatment of superficial fungus infections involving hair, skin, and nails. Mutant strains of *P. patulum* are used for the commercial synthesis of griseofulvin according to Turner (1975); this author also presents a brief discussion of the biosynthesis of this antibiotic.

H. Conidial Fungus Enzymes

Even a very cursory examination of a relevant literature reveals that the number of different enzymes produced by fungi is immense. Just how many enzymes are produced by conidial fungi on a truly industrial scale is difficult to determine. Certainly it is possible to purchase small amounts of a great number of enzyme preparations of fungal origin for use in various types of experimental work, but to term this "industrial scale" seems to be stretching the point. Therefore only enzymes or enzyme-containing materials that are produced in large quantities will be discussed in the present context.

1. Amylases

The knowledge of exactly when or by whom alcoholic fermentation was discovered is lost in antiquity, but one estimate puts the discovery back at least 30 centuries. It also seems very likely that alcohol was discovered independently by many different individuals and groups at many different times. The first alcoholic beverages tasted by man were probably accidentally fermented fruit juices or sap from wounded trees. As more knowledge was acquired, it was learned that alcoholic beverages could be made from starchy materials such as cereal grains and potatoes provided that these materials were first manipulated in such a manner that starch was converted to fermentable sugars. The method of such conversion marks a striking difference between the Occident and the Orient. In the West, conversion of starch to sugar for the purpose of making alcohol is usually accomplished by means of the amylases (starch-digesting enzymes) present in dried, germinated small grains (customarily barley), while in the East similar enzymes from fungi are used. There has been one minor exception—in at least one area of Central America salivary enzymes were employed. Squaws, sitting around and chewing corn, expectorated the wet, chewed grain into the pot in which fermentation was to take place. This latter conversion process, efficient though it may be, does not seem to have been widely accepted in an international wave of enthusiasm.

In many countries the consumption of alcoholic beverages has reached staggering proportions, which means that the production of starch-digesting preparations has shown a concomitant increase. About half a century ago Atsuki (1929) reported that 10% of the rice production in Japan was used for the production of rice wine, called *sake*. In China, a zygomycete is used in alcoholic fermentation, but in Japan *Aspergillus oryzae* has long been the organism of choice. As we in the West term dried, sprouted barley "malt," the Japanese term a crude preparation of fungal amylases (as well as many other enzymes) *koji*. The use of *koji* to process food has already been noted in the Section II,B,3.

Apparently *koji* was originally prepared by culturing *A. oryzae* on steamed rice, but Takamine (1913, 1914, 1923) substituted wheat bran for rice and named the molded bran *taka-koji*. Traditionally, *koji* was prepared by placing steamed rice in layers on the floor or in trays, inoculating with *A. oryzae,* and allowing it to incubate until the mass was permeated with mycelia (ca. 48 h). It was then dried as quickly as possible but not at temperature high enough to inactivate the enzymes. Takamine also used a slowly revolving drum with moistened, steamed bran in it instead of incubating the material on floors or in trays. Enzymes could then be extracted by percolating water through the bran and then precipitating the enzymes (principally amylase) by proper chemical means.

Boyer and Underkofler (1945) reported that mold bran was produced in commercial tonnages in a converted government hemp mill at Eagle Grove, Iowa, during World War II. The flow diagram of this plant was shown by Boyer and Underkofler and by Gray (1959). According to Blain (1975), there has been increasing production of fungal enzymes in deep vat cultures over the past 20 yr, and α-amylase of *A. oryzae* origin is used in Western countries. Reed (1966), however, noted the use of enzymes from *A. niger* and *A. awamori* in Russia.

All the enzymes which in the past have been conveniently called amylase do not bring about the same changes in the starch molecule and its constituents. Thus α-amylase hydrolyzes the 1,4-glycosidic links with the formation of dextrins and oligosaccharides, β-amylase hydrolyzes alternate α-1,4-glucan links, and amyloglucosidase converts starch and dextrins to glucose. Weibel *et al.* (1978) has noted that these and other enzymes are used commercially to produce a variety of starch-derived sweeteners. Thus Beckhorn *et al.* (1965) described a process that uses enzymes from *A. oryzae* to produce high-dextrose equivalent syrup. A method for producing glucose and crystalline dextrose using an *A. niger* enzyme has also been described (Kooi and Armbruster, 1967).

2. Pectinolytic Enzymes

The presence of pectins in fruit which is to be converted to jams or jellies has definite advantages, but the presence of such colloidal material in fruit destined for the preparation of fruit juices creates filtration and clarification problems. There are several pectic substances, but basically they consist of linear chains of D-galacturonic acid residues. The pectins may be degraded through the use of widely distributed pectinolytic enzymes (Rombouts and Pilnik, 1972). Blain (1975) has reported that species of *Penicillium* and *Aspergillus* (as well as zygomycete species) have been used as sources of pectinolytic enzymes, and that the enzymes have been obtained from fungi grown on solid media. However, Nyiri (1968) described their production by the submerged culture of *Aspergillus ochraceus.*

3. Proteases

Proteases of microbial origin were inadvertently used in preparing hides for tanning (bating process) long before it was understood that enzymes were involved, and some enzymes of fungal origin were used in this early process. Proteases may be classified on the basis of their pH optima as acid, neutral, or alkaline. Blain (1975) states that commercially produced acid proteases are principally of fungal origin. Actually, all three types can be produced by *Aspergillus oryzae*. Conidial fungi such as *A. oryzae, A. niger,* and *A. saitoi* have been used as sources, but both Zygomycetes and Basidiomycetes have also been used (Keay *et al.,* 1972). Citing Reed (1966), Blain states that more than half of the bread made in the United States is prepared with enzymes of *A. oryzae,* the proteases being more important than the fungal amylase since they reduce the viscosity of the dough. The use of fungal proteases in tenderizing meat was noted in Section II,C.

4. Glucose Aerodehydrogenase

Glucose aerodehydrogenase, the enzyme active in the conversion of glucose to gluconic acid, is recovered as a by-product of gluconic acid fermentation. Being an intracellular enzyme, it is obtained from the spent mycelium after it has been ground and allowed to autolyze. The enzyme is then recovered from the autolysate. It is used for the removal of traces of glucose from material such as eggs in the preparation of dried egg powder. Lockwood (1975) has reported that it is also used for stabilizing color and flavor in beer, canned foods, and soft drinks.

5. Cellulases

Since DeBary's (1886) first description of fungal enzymes capable of dissolving cellulose plant cell walls, a large amount of research has been conducted for both academic and utilitarian purposes. Like the starch-digesting enzymes, all cellulolytic enzymes do not attack the cellulose molecule in the same manner, and they are classified as C_1 enzyme, β-1,4-glucanases, and cellobiases (King and Vassal, 1969). Much of the earlier research was centered on the wood-destroying Basidiomycetes, but more and more effort now seems to be directed toward the end of examining conidial fungi as sources of such enzymes. For example, Reese (1969) investigated *Trichoderma viride,* Selby (1969) isolated C_1 from *Penicillium funiculosum,* and Chahal and Gray (1969b) studied 43 fungi, including 5 conidial fungi.

Blain (1975) notes that American and European manufacturers produce cellulase preparations and that Japan produces about 100 tons annually. Despite the scale of this production, digestion of cellulose by such enzymes is still a laboratory process. This is especially unfortunate, since the capacity for inexpensive production of highly active cellulases would make available the food energy now locked up in the world's most abundant carbohydrate, cellulose. Cellulose digestion on an industrial scale would make enormous quantities of hexose sugar available for direct use as food, for the preparation of a clean liquid fuel, and perhaps for conversion to a host of now unsuspected chemical compounds.

IV. POTENTIAL USES OF CONIDIAL FUNGI

Urgent and frequent warnings concerning the necessity for regulating population growth, issued from a variety of sources for a great many years, still seem to
be largely disregarded. For about two decades the annual world population
increase has varied between 1.7 and 1.9%. These figures seem very low until one
considers that an annual 1.8% increase results in a population that doubles every
40 yr. Stated in a different manner, successively shorter periods of time are
required to increase the world population by an additional 1 billion individuals.
Thus, approximately 2 million yr were required for world population to first
reach 1 billion (1850 A.D.). However, a second billion was added in just 95 yr
(1945 A.D.), a third billion in 17 yr (1962 A.D.), and a fourth in 14 yr (1976
A.D.). According to present trends, Gray's (1962a,b) prediction of a world population of 6 billion in the year 2002 seems a bit conservative.

Rapidly expanding population with concomitant industrial expansion has
created pressures on and competition for most of the resources needed and
desired by humans—breatheable air, drinkable water, arable farmland, fossil
energy sources, food, and thousands of other important materials and objects.
While it would create many inconveniences, it is certainly possible for the human
species to use less of (or even do without) some items now deemed necessities by
an affluent society. However, food is not such an item. It is now becoming more
and more apparent that Malthus's assessment that the major effect of an expanding population would be that of exerting unbearable pressures on the food supply
was correct. Gray (1962a,b) stressed that, although all food would not be in short
supply, there would be a growing scarcity of protein. At that time approximately
two-thirds of the populated areas of the world already suffered protein shortages
of varying degree.

Since there is little evidence that worldwide population regulation is likely to
become a reality in the near future, we must seek ever-increasing supplies of
protein. Since our only source of protein is other living organisms, and since
plants and animals have already been rather thoroughly exploited, the only remaining unexploited sources of protein are microorganisms. Bacteria, algae,
yeast, and filamentous fungi have been advocated for use as food, and arguments
have been presented to demonstrate the advantages of each. Protein from mic-
roorganic sources has been grouped together under the general heading of
single-cell protein (SCP), although strictly speaking the multicellular filamentous
fungi should not be included in this category. A general objection to the use of
the term ''SCP'' will be briefly discussed later.

The concept of employing conidial fungi for the production of edible protein is
not of recent origin. Pringsheim and Lichtenstein (1920) reported that, as an
emergency measure, fungus-processed straw was fed to cattle during World War
I. Ammonium salts were added to moistened straw which was then inoculated

with an *Aspergillus;* as the fungus grew, the protein content of the straw increased from the original 0.9% up to 8.0%. Others who have considered the possibility of using conidial fungi as a source of food are Skinner (1924), Takata (1929), Skinner *et al.* (1933), Woolley *et al.* (1938), Vinson *et al.* (1945), and Chastukhin (1948).

Gray (1962a,b) and his associates directed their attention to the possibilities of employing conidial fungi as sources of protein for a variety of reasons: (1) there are many thousands of different species and strains to choose from; (2) they are ubiquitous and easily isolated in axenic culture; (3) a great many are very fast growing; (4) most of those investigated are capable of utilizing inorganic nitrogen as a sole source of nitrogen; (5) the harvesting of filamentous fungi produced in submerged culture is simpler than the harvesting of unicellular organisms such as bacteria and yeast; and (6) the capacity of many fungi to grow in media of low pH value would partially preclude the growth of many potentially dangerous bacterial contaminants. The program was designed to examine the capacity of conidial fungi to synthesize protein in significant quantities. Refined glucose as well as several crude carbohydrate-containing substances were used to find the most suitable fungus for use on the substrate. For ease of fast determination and comparison, protein percentages were based upon Kjeldahl nitrogen, although it was recognized quite early (Gray *et al.,* 1964) that such values were often slightly higher than values based upon extracted protein. Although a few crude feeding experiments were conducted, protein quality was not considered, and no attention was paid to possible toxicity, although species from which toxins had been reported obviously were avoided in the experimentation.

The entire investigation was focused on determining the total amount of protein that could be produced by conidial fungi. Since it is possible to make a reasonably accurate estimate of the total protein requirements of a population, this figure could then be compared with the protein-producing potential. It was recognized that both protein quality and protein concentration in a particular food are quite important, but that these are considerations that must come *after* the problem of sufficient total protein production is solved. In all probability some conidial fungus proteins are deficient in one or more of the essential amino acids; however, this should not be a deterrent in their use as food. In a parallel situation, the use of corn as an important feed grain did not cease or even lessen when it was learned that corn proteins were deficient in lysine. On the contrary, corn was supplemented with lysine-containing material, and attempts were made to breed corn with a higher lysine content. Foster (1949) stated that mold proteins by themselves are inadequate to satisfy animal requirements—a strange generalization in view of the findings of Vinson *et al.* (1945), who reported normal growth, gestation, and lactation in mice when they were fed *Fusarium lini* mycelia supplemented with thiamine. With respect to the potential alteration of protein quality, a very intriguing possibility is presented by the findings of Stokes and

Gunness (1946). These investigators found that percentages of 10 different amino acids in the protein of *Penicillium notatum* varied with both cultural conditions and medium constituents. Thus the possibility certainly exists that, if the level of a specific amino acid in a fungal protein is low, it might be altered by proper manipulation of culture conditions.

The growth efficiency of conidial fungi in submerged culture varied widely. Thus, in one screening of 175 different isolates on glucose medium (Gray *et al.,* 1964), *Sepedonium* sp. produced 1 g dry wt of mycelia, while using only 1.3 g of glucose, but *Botryosporium* sp. utilized 28.18 g of glucose in forming 1 g of mycelia; obviously the latter species was converting the carbon of glucose to compounds other than those that make up the mycelia. Over 50% of this group of isolates could produce 1 g of mycelia from 3 g or less of sugar. The percentage of protein (based upon extracted protein) varied from 11.1% in *Rhizoctonia* sp. to 32% in *Phoma* sp. This range of protein values is somewhat lower than that reported by Solomons (1975), who listed a range of 19–47% based upon values obtained through use of an α-amino acid nitrogen determination method similar to that of Gehrke and Wall (1971).

Among the crude substrates employed as carbon sources in these investigations were sweet potatoes (Gray and AbouElSeoud, 1966a), fresh manioc roots and dried cassava flour (Gray and AbouElSeoud, 1966b), fresh sugar beet roots and dried beet shreds (Gray and AbouElSeoud, 1966c), brown rice flour (Gray and Karve, 1967), and citrus molasses (Gray and Racle, 1968). By culturing the best-suited conidial fungus (as determined in preliminary screening experiments) on each of these substrates, protein contents (based upon Kjeldahl nitrogen) were substantially increased. For example, the protein content of sweet potatoes was increased more than 4-fold, fresh manioc roots, 5.7-fold; beet roots, 2.6-fold; and brown rice flour, 2.29-fold. Minced, fresh, unpeeled sugar beets provided the best of the crude carbon sources supplied to conidial fungi. Culturing *Cladosporium* sp. on a simple medium consisting of 85 g minced beets, 2 ml corn steep liquor, and 2 g ammonium chloride/liter, 3.48 g of crude protein was produced per 100 g fresh beet root, and growth was complete in 36 h. That a conidial fungus grows with such rapidity is not surprising, since Trinci (1972) reported a doubling of total protein in only 1 h and 6 min for *Geotrichum candidum,* and Solomons and Scammell (1970) reported a doubling time of 2 h and 25 min for *Fusarium graminearum.*

On the basis of these and other experiments, Gray (1966) pointed out that conidial fungus processing of the total world production of eight crops (manioc, sugar beet, sugarcane, paddy rice, corn, yams, potatoes, and sweet potatoes) in 1962 could have resulted in protein production 50% in excess of the total requirements (estimated on the basis of 52.2 lb/person per year) of the world's population in that year.

It was realized quite early in the investigations that to institute large-volume production in submerged culture on a scale great enough to make substantial contributions to the world's protein pool would create demands for vast amounts of fresh water which might not always be available in some areas of the world. Therefore Gray *et al.* (1963) substituted water dipped from the Atlantic Ocean for distilled water in their standard screening medium. Of the 23 fungi tested (20 conidial fungi, 1 zygomycete, 1 ascomycete, 1 basidiomycete) on this medium, 21 exhibited yield increases ranging from 7.6 to 178.5% when compared with yields obtained in media prepared with distilled water. From these results it was inferred that a water shortage need never create a major problem in the development of a process of this type.

In many of the experiments described above, materials which could be used directly as food were used as sources of carbohydrate for fungal growth. Chahal and Gray (1969a,b) attempted to use dried ground wood pulp as a carbon source. Of 43 fungi screened on media containing wood pulp, three of the best four were anamorphic fungi, with *Rhizoctonia* sp. producing the greatest amount of growth and *Trichoderma* sp. containing the highest percentage of protein (17.7%). Compared with many other conidial fungi cultured on more readily utilizable carbohydrates, this percentage of protein is low, and the fiber content of such material is undoubtedly high—too high perhaps for it to be considered for swine feeding, for example. However, this approach shows promise, and a search for other suitable fungi coupled with alterations in cultural procedures might well lead to the production of a material well suited as a source of protein for animal feeding purposes. The advantages of inserting incalculable amounts of waste cellulose into the food chain needs no further comment.

One of the principal objections to using SCP as human food is that it contains high levels of nucleic acids. Chen and Peppler (1978) reported 6–11% for yeast protein and 11–18% for bacteria. These workers cited no figures for filamentous fungi, but apparently these organisms may also have a high nucleic acid content in view of their generally rapid rates of growth. Edozein *et al.* (1970) fed high levels of yeast to young men. On the basis of the results of their trials it was recommended that human intake of SCP nucleic acids should not exceed 2 g/day (Anonymous, 1970). Apparently, high nucleic acid intake results in increased uric acid content in the blood, which may result in gout. High nucleic acid content need not be a deterrent factor in the use of SCP, because many investigators, aware of this potential problem, have developed methods for removing RNA from such material. Solomons (1975) lists only two such methods, but Chen and Peppler (1978) have listed 18 additional methods ranging from the early sodium chloride–sodium acetate treatment of Decker and Dirr (1944) to the sodium hydroxide or aqueous ammonia treatment of Viikari and Linko (1977). Chen and Peppler also noted that a dried yeast of about 1.5% nucleic acid content had been prepared for the trade using the acid treatment of Peppler (1970). Of the many RNA removal methods developed so far, it is probable that at least one would be suitable for use with mycelia of conidial fungi if they were found to contain undesirably high levels of nucleic acids.

Objecting to the use of conidial fungi as food because a few of them are known to produce mycotoxins is sheer nonsense; it merely reflects a modern extension of the early developed distaste for fungi discussed previously in this chapter. The easy and sensible reaction would be to eliminate the toxin producers from consideration—certainly a great many of them have been identified (Mateles and Wogan, 1967; Goldblatt, 1969; Enomoto and Saito, 1972). Without intending to minimize the potential dangers of certain mycotoxins, it does seem that their importance (with respect to the fungi in general) has been grossly overemphasized. The great furore over mycotoxins began with the work on the carcinogenic aflatoxins in the 1960s, and literally hundreds of reports have since appeared in the literature. Butler (1975) notes that aflatoxin has been demonstrated to have carcinogenic action in rats, ducks, trout, and ferrets, but that direct evidence that it causes human liver disease is difficult to obtain. Conversely, Christensen (1975) states that in a number of fatal illnesses aflatoxin was undoubtedly the cause of illness and death.

Many commonly eaten foodstuffs have been reported to contain toxins and other deleterious materials. When Liener (1966) reported that goiterogens, cyanogens, hemaglutinins, and other harmful substances were synthesized in plants such as kidney beans, peanuts, black-eyed peas, and garden peas, Gray (1970) was motivated to suggest that *nothing that grows on this earth is safe to eat*. Judging by the often circumstantial evidence involving compounds suspected of being carcinogenic, to which the general public has been exposed during the last few decades, perhaps the major breakthrough in this area will come in the form of a report of a compound that *does not cause cancer*.

REFERENCES

Abraham, E. P., and Newton, G. G. F. (1961). The structure of cephalosporin C. *Biochem. J.* **79**, 377–393.

Anonymous (1970). "Protein Advisory Group Guideline No. 4, Single Cell Protein." United Nations, New York.

Atsuki, K. (1929). Sake brewing. *J. Soc. Chem. Ind. Jpn.* **32**, 57B–58B.

Batti, M. A. (1966). U.S. Patent 3,290,227.

Batti, M. A. (1967). U.S. Patent 3,335,067.

Beckhorn, E. J., Labbee, M. D., and Underkofler, L. A. (1965). Production and use of microbial enzymes for food processing. *J. Agric. Food Chem.* **13**, 30–34.

Beesch, S. C., and Shull, G. M. (1956). Fermentation. *Ind. Eng. Chem.* **48**, 1585–1603.

Birkinshaw, J. H., Charles, J. H. V., Lilly, C. H., and Raistrick, H. (1931). Studies in the biochemistry of micro-organisms. VII. Kojic acid. *Trans. R. Soc. London, Ser. B* **220**, 127–138.

Blain, J. A. (1975). Industrial enzyme production. *In* "The Filamentous Fungi" (J. E. Smith and D. R. Berry, eds.), Vol. 1, pp. 193–211. Arnold, London.

Boutroux, L. (1880). Chimie physiologique; sur une fermentation nouvelle de glucose. *C.R. Hebd. Seance, Acad. Sci.* **91**, 236–238.

Boyer, J. W., and Underkofler, L. A. (1945). Mold bran aids production of grain alcohol. *Chem. Metall. Eng.* **52**, 110–111.

Brian, P. W. (1951). Antibiotics produced by fungi. *Bot. Rev.* **17**, 357–430.

Brian, P. W., Curtis, P. J., and Heming, H. G. (1946). A substance causing abnormal development of fungal hyphae produced by *Penicillium janczewskii* Zal. *Trans. Br. Mycol. Soc.* **29**, 173–187.

Broadbent, D. (1968). Antibiotics produced by fungi. *PANS, Sect. B* **14**, 120–141.

Brotzu, G. (1948). "Lavori dell'istituto d'Igiene di Cagliari."

Butkewitsch, W. (1923). Uber die citronensaueregarung. *Biochem. Z.* **142**, 195–211.

Butler, W. H. (1975). Mycotoxins. *In* "The Filamentous Fungi" (J. E. Smith and D. R. Berry, eds.), Vol. 1, pp. 320–329. Arnold, London.

Calam, C. T., Oxford, A. E., and Raistrick, H. (1939). Studies in the biochemistry of microorganisms. 63. Itaconic acid, a metabolic product of a strain of *Aspergillus terreus* Thom. *Biochem. J.* **33**, 1488–1495.

Carrington, T. R. (1971). The development of commercial processes for the production of 6-aminopenicillanic acid. *Proc. R. Soc. London, Ser. B* **179**, 321–323.

Chahal, D. S., and Gray, W. D. (1969a). The growth of selected cellulolytic fungi on wood pulp. *In* "Biodeterioration of Materials" (A. H. Walters and J. J. Elphick, eds.), pp. 584–593. Elsevier, Amsterdam.

Chahal, D. S., and Gray, W. D. (1969b). Growth of cellulolytic fungi on wood pulp. I. Screening of cellulolytic fungi for their growth on wood pulp. *Indian Phytopathol.* **22**, 79–91.

Chain, E. (1971). Thirty years of penicillin therapy. *Proc. R. Soc. London, Ser. B* **179**, 293–319.

Chastukhin, V. Y. (1948). "Mass Cultures of Microscopic Fungi" (in Russian). Nat. Reserv. Headquarters Press, Moscow.

Chen, S. L., and Peppler, H. J. (1978). Single-cell proteins in food applications. *Dev. Ind. Microbiol.* **19**, 79–94.

Christensen, C. M. (1975). "Molds, Mushrooms and Mycotoxins." Univ. of Minnesota Press, Minneapolis.

Clutterbuck, P. W., Lovell, R., and Raistrick, H. (1932). Studies in the biochemistry of microorganisms. 26. The formation from glucose by members of the *Penicillium chrysogenum* series of a pigment, an alkali-soluble protein, and penicillin. *Biochem. J.* **26**, 1907–1918.

Cochrane, V. W. (1948). Commercial production of acids by fungi. *Econ. Bot.* **2**, 145–157.

Coghill, R. G. (1944). Penicillin, science's Cinderella. *Chem. & Eng. News* **22**, 588–593.

Currie, J. N. (1914). Flavor of Roquefort cheese. *J. Agric. Res.* **2**, 429–434.

Dattilo-Rubbo, S. (1938). The taxonomy of fungi of blue-veined cheese. *Trans. Br. Mycol. Soc.* **22**, 174–180.

DeBary, A. (1886). Ueber einige Sclerotinien und Sclerotienkrankheiten. *Bot. Z.* **44**, 377–387, 393–404, 409–426.

Decker, P., and Dirr, K. (1944). Uber die Nichteiweiss-stickstoff der Hefe. *Biochem. Z.* **316**, 248–254.

Demain, A. L. (1966). Biosynthesis of penicillins and cephalosporins. *Biosynth. Antibiot.* **1**, 29–94.

Djien, K. S., and Hesseltine, C. W. (1961). Indonesian fermented foods. *Soybean Dig.* **22**, 14–15.

Dyson, G. M. (1928). Mold Food of the Far East. *Pharm. J.* **121**, 375–377.

Edozein, J. C., Udo U. U., Young, V. R., and Scrimshaw, N. S. (1970). Effects of high levels of yeast feeding on uric acid metabolism of young men. *Nature (London)* **228**, 180.

Enomoto, M., and Saito, M. (1972). Carcinogens produced by fungi *Annu. Rev. Microbiol.* **26**, 279–312.

Fleming, A. (1929). On the antibacterial action of cultures of a *Penicillium,* with special references to their uses in the isolation of *B. influenzae. Br. J. Exp. Pathol.* **10,** 226–236.

Florey, H. W., Chain, E., Heatley, N. G., Jennings, M. A., Sanders, A. G., Abraham, E. P., and Florey, M. E. (1949). Penicillin: Historical introduction. *In* "Antibiotics" Vol. II, pp. 631–672. Oxford Univ. Press, London and New York.

Foster, J. W. (1949). "Chemical Activities of Fungi." Academic Press, New York.

Furuya, K., Enokita, R., and Shirasaka, M. (1967). Studies on the antibiotics from fungi. II. A new griseofulvin producer, *Nigrospora oryzae. Annu. Rep. Sankyo Res. Lab.* **19,** 91–95.

Gehrke, C. W., and Wall, L. L. (1971). Automated TNBS method for protein in feed. *J. Assoc. Off. Anal. Chem.* **54,** 187–191.

Giles, D., Hemming, H. G., and Lehan, M. (1970). British Patent 1,186,507.

Girolami, R. L., and Knight, S. G. (1955). Fatty acid oxidation by *Penicillium roqueforti. Appl. Microbiol.* **3,** 264–267.

Goldblatt, L. A. (1969). "Aflatoxin." Academic Press, New York.

Gray, W. D. (1959). "The Relation of Fungi to Human Affairs." Holt, New York.

Gray, W. D. (1962a). Microbial protein for the space age. *Dev. Ind. Microbiol.* **3,** 63–71.

Gray, W. D. (1962b). Fungi as a nutrient source. *In* "Biologistics for Space Systems Symposium," AMRL-TDR-62-116, pp. 356–381. 6570th Aerosp. Med. Res. Lab., Wright-Patterson AFB, Ohio.

Gray, W. D. (1966). Fungi and world protein supply. *Adv. Chem. Ser.* **57,** 261–268.

Gray, W. D. (1970). The use of fungi as food and in food processing. *Crit. Rev. Food Technol.* **1,** 225–329.

Gray, W. D., and AbouElSeoud, M. (1966a). Fungal protein for food and feeds. II. Whole sweet potato as a substrate. *Econ. Bot.* **20,** 119–126.

Gray, W. D., and AbouElSeoud, M. (1966b). Fungal protein for food and feeds. III. Manioc as a potential crude raw material for tropical areas. *Econ. Bot.* **20,** 251–255.

Gray, W. D., and AbouElSeoud, M. (1966c). Fungal protein for food and feeds. IV. Whole sugar beets or beet pulp as a substrate. *Econ. Bot.* **20,** 372–376.

Gray, W. D., and Karve, M. D. (1967). Fungal protein for food and feeds. V. Rice as a source of carbohydrate for the production of fungal protein. *Econ. Bot.* **21,** 110–114.

Gray, W. D., and Racle, F. A. (1968). Citrus molasses as a source of carbohydrate for the fungal synthesis of protein. *Citrus Ind.* **49,** 20–22, 26.

Gray, W. D., Pinto, P. V. C., and Pathak, S. G. (1963). Growth of fungi in sea water medium. *Appl. Microbiol.* **11,** 501–505.

Gray, W. D., Och, F. F., and AbouElSeoud, M. (1964). Fungi imperfecti as a potential source of edible protein. *Dev. Ind. Microbiol.* **5,** 384–389.

Groff, E. H. (1919). Soy sauce manufacturing in Kwangtung, China. *Philipp. J. Sci.* **15,** 307–316.

Grove, S. F., and McGowan, S. C. (1947). Identity of griseofulvin and "curling" factor. *Nature (London)* **160,** 574.

Hammer, B. W., and Bryant, H. W. (1937). A flavor constituent of blue cheese. *Iowa State Coll. J. Sci.* **11,** 281–285.

Hesseltine, C. W. (1965). A millenium of fungi, food, and fermentation. *Mycologia* **57,** 148–197.

Hesseltine, C. W., and Shibasaki, K. (1961). Miso. III. Pure culture fermentation with *Saccharomyces rouxii. Appl. Microbiol.* **9,** 515–518.

Hesseltine, C. W., and Wang, H. L. (1967). Traditional fermented foods. *Biotechnol. Bioeng.* **9,** 275–288.

Hisanga, W., and Nakamura, S. (1966). Japanese Patent 16555/6.

Hollaender, A., Raper, K. B., and Coghill, R. D. (1945). The production and characterization of ultraviolet-induced mutations in *Aspergillus terreus.* I. Production of the mutations. *Am. J. Bot.* **32,** 160–165.

Hope, E. (1927). U.S. Patent 1,644,131.

Ikeda, Y., Takahashi, E., Yokogawa, K., and Yoshimura, Y. (1972). Screening for microorganisms producing gallic acid from Chinese and Tara tannins. *J. Ferment. Technol.* **50**, 361-370.

Jensen, O. (1904). Biologische studien uber den Käsereifungsprozess unter spezieller Berucksichtigung der fluchtigen Fettsauren. *Landwirtsch. Jahrb. Schweiz* **18**(8), 319.

Kane, J. H., Finlay, A. C., and Amann, P. F. (1945). U.S. Patent 2,385,283.

Karmas, E. (1970). "Fresh Meat Processing." Noyes Data Corp., New Jersey.

Karow, E. O., and Waksman, S. A. (1947). Production of citric acid in submerged culture. *Ind. Eng. Chem.* **39**, 821-825.

Katagiri, H., and Kitahara, K. (1933). The formation of kojic acid by *Aspergillus oryzae. Mem. Coll. Agric., Kyoto Imp. Univ.* **26**, 1-29.

Keay, L., Mosely, M. H., Anderson, R. G., O'Connor, R. J., and Wildi, B. (1972). Production and isolation of microbial proteases. *In* "Enzyme Engineering" (L. B. Wingard, ed.), Vol I, pp. 63-92. Wiley (Interscience), New York.

King, K. W., and Vassal, M. I. (1969). Enzymes of the cellulase complex. *In* "Cellulases and Their Applications" (R. F. Gould, ed.), pp. 7-25. Am. Chem. Soc., Washington, D.C.

Kinoshita, H. (1929). Study of the biosynthesis of teichoic acid and mannitol in a new filamentous fungus. *J. Chem. Soc. Jpn.* **50**, 583-593.

Kluyver, A. J., and Perquin, L. H. C. (1933). Uber die Bedingungen de Kojisauerebildung durch *Aspergillus flavus* Link. *Biochem. Z.* **266**, 82-95.

Knudson, L. (1913a). Tannic acid fermentation. I. *J. Biol. Chem.* **14**, 159-184.

Knudson, L. (1913b). Tannic acid fermentation. II. Effect of nutrition on the production of the enzyme tannase. *J. Biol. Chem.* **14**, 185-202.

Kooi, E. R., and Armbruster, F. C. (1967). Production and use of dextrose. *In* "Starch, Chemistry and Technology" (R. L. Whistler and E. F. Paschall, eds.), Vol. 2, pp. 553-568. Academic Press, New York.

Liener, I. E. (1966). Toxic substances associated with seed proteins. *Adv. Chem. Ser.* **57**, 178-194.

Lockwood, L. B. (1947). The production of Chinese soya sauce. *Soybean Dig.* **7**, 10-11.

Lockwood, L. B. (1975). Organic acid production. *In* "The Filamentous Fungi" (J. E. Smith and D. R. Berry, eds.), Vol. 1, pp. 140-159. Arnold, London.

Lockwood, L. B., and Nelson, G. E. N. (1946). Some factors affecting the production of itaconic acid by *Aspergillus terreus* in agitated cultures. *Arch. Biochem.* **10**, 365-374.

Lockwood, L. B., and Reeves, M. D. (1945). Some factors affecting the production of itaconic acid by *Aspergillus terreus. Arch. Biochem.* **6**, 455-469.

Lockwood, L. B., and Smith, A. K. (1950). Fermented soy foods and sauce. *In* "1950-1951 Yearbook of Agriculture," Yearbook Separate No. 2213, pp. 357-361.

McCann, E. P. and Calam, C. T. (1972). The metabolism of *Penicillium chrysogenum* and the production of penicillin using a high yielding strain at different temperatures. *J. Appl. Chem. & Biotechnol.* **22**, 1201-1208.

Mallea, O. (1950). La industria de fermentacion citrica en la Argentina. *Ind. Quim.* **12**, 264-280.

Mateles, R. I., and Wogan, G. N., eds. (1967). "Biochemistry of Some Foodborne Microbial Toxins." MIT Press, Cambridge, Massachusetts.

May, O. E., and Herrick, H. T. (1930). Some minor industrial fermentations. *Ind. Eng. Chem.* **22**, 1172-1176.

May, O. E., Herrick, H. T., Thom, C., and Church, M. B. (1927). The production of gluconic acid by the *Penicillium luteum purpurogenum* group. I. *J. Biol. Chem.* **75**, 417-422.

May, O. E., Moyer, A. J., Wells, P. A., and Herrick, H. T. (1931). The production of kojic acid by *Aspergillus flavus. J. Am. Chem. Soc.* **53**, 774-782.

May, O. E., Ward, G. E., and Herrick, H. T. (1932). *Zentralbl. Bakteriol., Parasitenkd., Infektionskr. Hyg., Abt. 2* **86**, 129-134.

Miall, L. M. (1975). Historical development of the fungal fermentation industry. *In* "The Filamentous Fungi" (J. E. Smith and D. R. Berry, eds.), Vol. 1, pp. 104-121. Arnold, London.

Molliard, M. (1922). Recherches calorimetriques sur l'utilization de l'énergie respiratoire au cours du développement d'une culture de *Sterigmatocystis nigra. C. R. Hebd. Séances Acad. Sci.* **174**, 236–238.

Moyer, A. J., and Coghill, R. D. (1945). The laboratory scale production of itaconic acid by *Aspergillus terreus. Arch. Biochem.* **7**, 167–183.

Moyer, A. J., May, O. E., and Herrick, H. T. (1936). The production of gluconic acid by *Penicillium chrysogenum. Zentralbl. Bakteriol., Parasitenkd. Infektionskr. Hyg., Abt. 2* **93**, 311–324.

Nakano, M. (1959). "Traditional Methods of Food Processing," pp. 1–15. Food Res. Inst., Ministry of Agriculture and Forestry, Tokyo, Japan.

Nayler, J. C. (1971). Structure-activity relationships in semi-synthetic penicillins. *Proc. R. Soc. London, Ser. B* **179**, 357–367.

Nubel, R. D., and Ratajak, E. J. (1962). U.S. Patent 3,044,941.

Nyiri, L. (1968). Manufacture of pectinases. Part 1. *Process Biochem.* **3**, 27–30.

Ochse, J. J. (1931). "Vegetables of the Dutch East Indies." Archipel. Drukkeriji., Buitenzorg, Java.

Oxford, A. E., Raistrick, H., and Simonart, P. (1939). Studies in the biochemistry of microorganisms. 60. Griseofulvin, $C_{17}H_{17}O_6Cl$, a metabolic product of *Penicillium griseofulvum* Dierckx. *Biochem. J.* **33**, 240–248.

Pathak, S. G., and Seshadri, R. (1965). Use of *Penicillium chrysogenum* as animal food. *Appl. Microbiol.* **13**, 262–266.

Patton, S. (1950). The methyl ketones of blue cheese and their relation to flavor. *J. Dairy Sci.* **33**, 680–684.

Peppler, H. J. (1970). Food Yeasts. *In* "The Yeasts" (A. H. Rose and J. S. Harrison, eds.), Vol. 3, pp. 421–462. Academic Press, New York.

Perlman, D. (1949). Mycological production of citric acid. *Econ. Bot.* **3**, 360–374.

Prescott, S. C., and Dunn, C. G. (1940). "Industrial Microbiology." McGraw-Hill, New York.

Pringsheim, H., and Lichtenstein, S. (1920). Versuche zur Anreicherung von Kraftstroh mit Pilzeiweiss. *Cellulose-Chem.* **1**, 29–39.

Ramsbottom, J. (1936). The uses of fungi. *Br. Assoc. Adv. Sci., Annu. Rep.* pp. 189–218.

Raper, K. B. (1946). The development of improved penicillin-producing molds. *Ann. N.Y. Acad. Sci.* **48**, 41–52.

Raper, K. B., and Alexander, D. E. (1945). Penicillin. V. Mycological aspects of penicillin production. *J. Elisha Mitchell Sci. Soc.* **61**, 74–113.

Raper, K. B., Coghill, R. D., and Hollaender, A. (1945). The production and characterization of ultraviolet induced mutations in *Aspergillus terreus*. II. Cultural and morphological characteristics of the mutations. *Am. J. Bot.* **32**, 165–176.

Reed, G. (1966). "Enzymes in Food Processing." Academic Press, New York.

Reese, E. T. (1969). Estimation of exo-β-1-4-glucanase in crude cellulase solutions. *In* "Cellulases and Their Applications" (R. F. Gould, ed.), pp. 26–33. Am. Chem. Soc., Washington, D.C.

Rombouts, F. M. and Pilnik, W. (1972). Research on pectin depolymerases in the sixties, a literature review. *Crit. Rev. Food Tech.* **3**, 1–26.

Saito, K. (1907). Über die Säurebildung bei *Aspergillus oryzae. Bot. Mag. Tokyo.* **21**, 7–11.

Schweiger, L. B. (1961). U.S. Patent 2,970,084.

Scott, R. (1968). Blue veined cheese. *Process Biochem.* **3**, 11–15, 24.

Selby, K. (1969). The purification and properties of the C_1-component of the cellulase complex. *In* "Cellulases and Their Applications" (R. F. Gould, ed.), pp. 34–52. Am. Chem. Soc., Washington, D.C.

Shibasaki, K., and Hesseltine, C. W. (1961a). Miso. I. Preparation of soybeans for fermentation. *J. Biochem. Microbiol. Technol. Eng.* **3**, 161–174.

Shibasaki, K., and Hesseltine, C. W. (1961b). Miso. II. Fermentation. *Dev. Ind. Microbiol.* **2**, 205–214.

Shibasaki, K., and Hesseltine, C. W. (1962). Miso fermentation. *Econ. Bot.* **16**, 180–195.

Skinner, C. S. (1924). The synthesis of aromatic amino acids from inorganic nitrogen by molds and the value of mold proteins in the diet. *J. Bacteriol.* **28**, 95–106.

Skinner, J. T., Petersen, W. H., and Steenbock, H. (1933). Nahwert von schimmel Pilzmycel. *Biochem. Z.* **267**, 169–178.

Smith, A. K. (1958). "Use of United States Soybeans in Japan," ARS-71-12. U.S. Dep. Agric., Washington, D.C.

Smith, A. K. (1963). Foreign uses of soybean protein foods. *Cereal Sci. Today* **8**, 196, 198, 200, 210.

Smith, A. K., Hesseltine, C. W., and Shibasaki, K. (1961). U.S. Patent 2,967,108.

Solomons, G. L. (1975). Submerged culture production of mycelial biomass. *In* "The Filamentous Fungi" (J. E. Smith and D. R. Berry, eds.), Vol. 1, pp. 249–264. Arnold, London.

Solomons, G. L., and Scammell, G. W. (1970). British Patent 1,346,061.

Staley, A. R. (1935). Soy sauce goes American. *Food Indus.* **7**, 66.

Stanton, W. R., and Wallbridge, A. (1969). Fermented food processes. *Process Biochem.* **4**, 45–51.

Starkle, M. (1924). Die methyl Ketone in Oxydativen Abbau einiger Triglyceride (bzw. Fettsauren) durch Schimmelpilze unter berucksichtigung der besonderen Ranziditat des Kokosfettes. *Biochem. Z.* **151**, 371–415.

Steinkraus, K. H., Lee, C. Y., and Buck, P. A. (1965). Soybean fermentation by the ontjom mold *Neurospora. Food Technol.* **19**, 1301–1302.

Stokes, J. L., and Gunness, M. (1946). The effect of cultural conditions on the amino acid content of *Penicillium notatum. J. Bacteriol.* **52**, 195–207.

Takamine, J. (1913). U.S. Patent 1,054,626.

Takamine, J. (1914). Enzymes of *Aspergillus oryzae* and the application of its amylocastic enzyme to the fermentation industry. *J. Ind. Eng. Chem.* **6**, 824–828.

Takamine, J. (1923). U.S. Patent 1,460,828.

Takata, T. (1929). The utilization of microorganisms for human food materials. *J. Soc. Chem. Ind. Jpn.* **32**, 243–244.

Tamiya, H., and Hida, T. (1929). Vergleichen Studien über die Säurebildung die Atmung, die Oxidasereaktion und das Dehydrierungsvermogen von *Aspergillus* arten. *Acta Phytochim.* **4**, 343–361.

Tamura, G., Kirimura, J., Hara, H., and Sugimura, K. (1952). The microbiological determination of amino acids in miso. *J. Agric. Chem. Soc. Jpn.* **26**, 483–485.

Thom, C. (1909). Camembert cheese problems in the United States. *U.S. Dep. Agric., Bur. Anim. Ind., Bull.* **115**.

Thom, C., and Church, M. B. (1926). "The Aspergilli." Williams & Wilkins, Baltimore, Maryland.

Thom, C., and Currie, J. N. (1913). The dominance of roquefort mold in cheese. *J. Biol. Chem.* **15**, 249–258.

Thom, C., and Fisk, W. W. (1925). "The Book of Cheese." Macmillan, New York.

Thom, C., and Raper, K. B. (1945). "A Manual of the Aspergilli." Williams & Wilkins, Baltimore, Maryland.

Trinci, A. P. J. (1972). Culture turbidity as a measure of mould growth. *Trans. Br. Mycol. Soc.* **58**, 467–473.

Turner, W. B. (1975). Commercially important secondary metabolites. *In* "The Filamentous Fungi" (J. E. Smith and D. R. Berry, eds.), Vol. 1, pp. 122–139, Arnold, London.

Van Tieghem, P. (1867a). Chimie vegetale; sur la fermentation gallique. *C. R. Hebd. Séances Acad. Sci.* **65**, 1091–1094.

Van Tieghem, P. (1867b). Recherches pour servir à l'histoire physiologique des mucédinées fermentation gallique. *Ann. Sci. Nat., Bot. Biol. Veg.* [2] **5**(8), 240.

Viikari, L., and Lindo, M. (1977). Reduction of nucleic acid content of SCP. *Process Biochem.* **12**(4), 17–19, 35.

Vinson, L. J., Cerecedo, L. R., Mill, R. P., and Nord, F. F. (1945). The nutritive value of fusaria. *Science* **101**, 388–389.

Ward, G. E. (1967). Production of gluconic acid, glucose oxidase, fructose, and sorbose. *In* "Microbial Technology" (H. J. Peppler, ed.), pp. 200–221. Van Nostrand-Reinhold, Princeton, New Jersey.

Ward, G. E. (1970). Some contributions of the U.S. Department of Agriculture to the fermentation industry. *Adv. Appl. Microbiol.* **13**, 363–382.

Wehmer, C. (1893). Note sur la fermentation citrique. *Bull. Soc. Chim. Fr.* [4] **9**, 728–730.

Weibel, M. K., McMullen, W. H., and Starace, C. A. (1978). Microbial enzymes in the production of nutritive sweeteners from starch. *Dev. Ind. Microbiol.* **19**, 103–116.

Williams, B. E. (1957) U.S. Patent 2,816,836.

Williams, B. E. (1962). U.S. Patent 3,056,679.

Wood, B. J. B., and Yong, F. M. (1975). Oriental food fermentations. *In* "The Filamentous Fungi" (J. E. Smith and D. R. Berry, eds.), Vol. 1, pp. 265–280. Arnold, London.

Woolley, D. W., Berger, J., Peterson, W. H., and Steenbock, H. (1938). Toxicity of *Aspergillus sydowi* and its correction. *J. Nutr.* **16**, 465–476.

Yabuta, T. (1912). On the Koji acid, a new organic acid formed by *Aspergillus oryzae*. *J. Coll. Agric., Tokyo Imp. Univ.* **5**, 51–58.

Yabuta, T. (1924). The constitution of kojic acid, a γ-pyrone dervative formed by *Aspergillus oryzae* from carbohydrates. *J. Chem. Soc.* **125**, 575–587.

Yamada, K. (1965). "Science in Japan," p. 401. Am. Assoc. Adv. Sci., Washington, D.C.

Yokotsuka, T. (1960). Aroma and flavor of Japanese soy sauce. *Adv. Food Res.* **10**, 75–134.

Yokotsuka, T., Sasaki, M., Kikuchi, T., Asao, Y., and Nabuhara, A. (1967). Production of fluorescent compounds other than aflatoxins by Japanese industrial molds. *In* "Biochemistry of Some Foodborne Microbial Toxins" (R. I. Mateles and G. N. Wogan, eds.), pp. 131–152. MIT Press, Cambridge, Massachusetts.

Yong, F. M. (1971). Studies on soy sauce fermentation. M.Sc. Thesis, University of Strathclyde, Glasgow, Scotland (cf. Wood and Yong, 1975).

V

ULTRASTRUCTURE, DEVELOPMENT, PHYSIOLOGY, AND BIOCHEMISTRY

23

Conidiogenesis and Conidiomatal Ontogeny

Garry T. Cole

I. INTRODUCTION

Investigations of developmental aspects of conidia, conidiogenous cells, conidiophores, and conidiomata have been associated primarily with taxonomic studies on dikaryomycotan anamorphic fungi. Attempts to provide a functional classification of the anamorph-subdivision of conidial fungi have, in general, not been very successful. For example, in Saccardo's (1886) classificatory scheme, taxonomic value was given to pigmentation, conidial shape and septation, and conidiophore arrangement. However, these characters often vary even within a single collection and in such instances are of little taxonomic value. As a supplement to comparative morphology, authors have more recently incorporated features of conidiogenesis into classifications, mainly for members of the anamorph-class Hyphomycetes (e.g., Hughes, 1953; Tubaki, 1958, 1963; Subramanian, 1956, 1962a,b, 1965, 1971a,b, 1972a,b, 1973, 1978; Barron, 1968; Kendrick and Carmichael, 1973; Carmichael *et al.*, 1980). The Coelomycetes,

Biology of Conidial Fungi, Vol. 2

with their reproductive cells usually concealed within pycnidia or acervuli, have been more difficult to categorize on the basis of different kinds of conidium and conidiogenous cell development (Sutton, 1973). In view of the absence of sexual-asexual connections for the majority of genera of conidial fungi, in recent years the aim has been to attain a comprehensive scheme for classifying both Hyphomycetes and Coelomycetes based on modern criteria, such as developmental features of conidia and conidiogenous cells, correlated with existing morphological data.

In the case of the Hyphomycetes, a review of developmental and ultrastructural aspects of conidiogenesis has been presented (Cole and Samson, 1979). Conidium ontogeny has been examined in recent taxonomic work on the Coelomycetes (see Nag Raj, Volume 1, Chapter 3, for literature review). Basic similarities exist in the concepts of conidium and conidiogenous cell development in the Hyphomycetes and Coelomycetes, and most illustrated examples of conidiogenesis presented below are characteristic of both anamorph-classes.

II. BLASTIC AND THALLIC DEVELOPMENT

Two basic kinds of conidium ontogeny are recognized, namely, blastic and thallic development. The distinctive features of these ontogenetic processes are largely derived from concepts of hyphal growth, since it is a modified hyphal apex which often gives rise to conidia. Protoplasmic differentiation during hyphal tip growth has been well documented by a number of workers (Grove and Bracker, 1970; Grove et al., 1968, 1970; Bartnicki-Garcia, 1973; Grove, 1978; Najim and Turian, 1979a). Thin sections of hyphal apexes of conidial fungi reveal at least two populations of vesicles (Figs. 1 and 2A and B). The large vesicles contain amorphous material comparable to that found in the hyphal wall. It has been suggested that these vesicles arise from dictyosome-like organelles in the subapical region of the hypha (Girbardt, 1969; Grove and Bracker, 1970; Grove, 1978; Cole and Samson, 1979). Najim, and Turian (1979a) have equilibrated the dense central cluster of microvesicles at the hyphal apex of *Sclerotinia fructigena* Aderh. and Ruhl (also shown in Fig. 1) with the *Spitzenkörper*. The latter was identified on the basis of light microscope examinations (Girbardt, 1957) as a refractile body at growing hyphal tips of many septate fungi. The microvesicles shown in Fig. 1 are 30–50 nm in diameter. Their origin in septate fungi is uncertain. Najim and Turian (1979a) have suggested that they arise from smooth endoplasmic reticulum (ER), perhaps the same dictyosome-like organelles from which the larger vesicles are formed (Cole and Samson, 1979). Microvesicular structures (40–70 nm in diameter) have also been observed in budding yeasts of *Mucor rouxii* (Cal.) Wehm., as well as hyphal tips of aseptate and

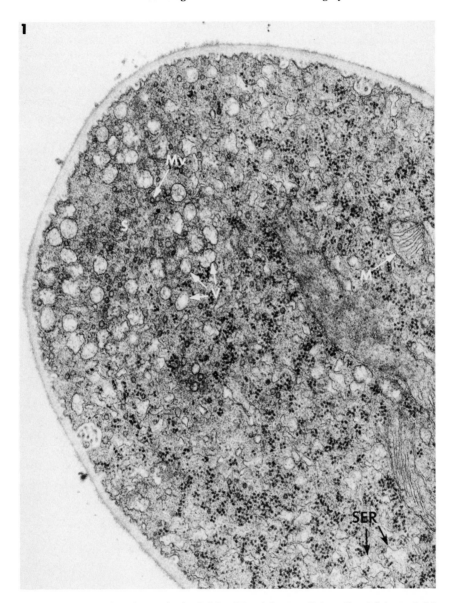

Fig. 1. Thin section of vegetative hyphal tip of *Drechslera sorokiniana* (Sacc.) Subram. & Jain showing polar cluster of large secretory vesicles (V) and microvesicular structures (Mv). Note that the latter are concentrated in a central region devoid of large vesicles and ribosomes. The cluster of microvesicular structures (S) is considered equivalent to Girbardt's (1957) *Spitzenkörper*. M, Mitochondrion; SER, smooth endoplasmic reticulum. ×39,900.

septate fungi (Bracker *et al.,* 1976; Bracker, 1977; Bartnicki-Garcia *et al.,* 1977). These spheroidal organelles have been called chitosomes and are believed to contain the chitin synthase zymogen (Bracker *et al.,* 1976). They are bounded by a tripartite, membranelike shell and may arise from blebbing ER or be released from multivesicular bodies (Grove, 1978). It has been suggested that chitosomes originate by self-assembly of subunits within specialized areas of ER, in the lumen of a macrovesicle (i.e., within the enclosing membrane of a multivesicular body) or in the cytoplasm free of any endomembrane system (Bartnicki-Garcia *et al.,* 1979). Except for a size difference, the microvesicles

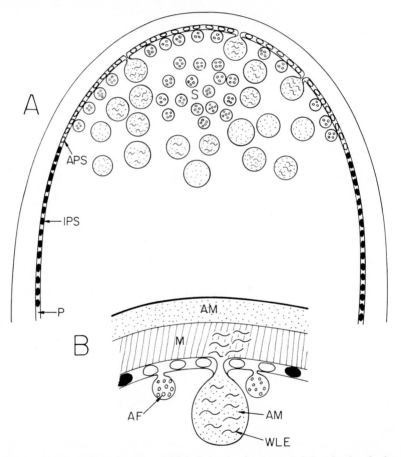

Fig. 2.   The distribution of inactive (solid circles) and active (open circles) molecules of polysaccharide synthases in the plasmalemma of hyphae (A) and blastic conidia (C). An interpretation of intussusception of wall microfibrils in a primary growth region of a hypha or conidium is shown in (B). AF, Activating factor (e.g., protease) for polysaccharide synthase; AM, amorphous component

comprising the *Spitzenkörper* (Figs. 1 and 2A) are morphologically similar to chitosomes. Bartnicki-Garcia *et al*. (1979) postulated that "the chitosome has a key role in cell wall construction, and hence, morphogenesis: it is the microvesicular vehicle by which the cell delivers, to specific sites on the cell surface, individual, organized packets of chitin synthetase, each packet being responsible for the synthesis of one microfibril." However, controversy still exists concerning the experimental evidence for this proposal. Although chitosomes have been clearly demonstrated to synthesize chitin microfibrils *in vitro* (Ruiz-Herrera *et al.*, 1975), their existence *in vivo* has not been unequivocally demonstrated. Farkaš (1979) has raised the possibility that chitosomes may be artifacts, which is "indicated by the fact that molecules of chitin synthase solubilized from the cell

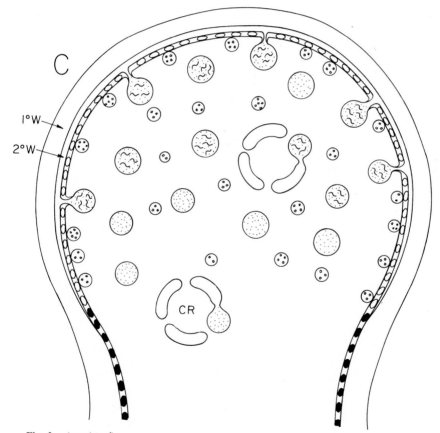

Fig. 2. (*continued*).
of cell wall; APS, active polysaccharide synthase; CR, cisternal ring (dictyosomal equivalent); IPS, inactive polysaccharide synthase; M, microfibrils; P, plasmalemma; S, *Spitzenkörper;* WLE, wall lytic enzymes; 1°W, primary wall; 2°W, secondary wall.

walls of *M. rouxii* by digitonin treatment associate with one another to form vesiculoid structures morphologically and functionally resembling chitosomes." In a recent ultrastructural study by Herth (1979), chitosome-like vesicles were observed in the diatom *Thalassiosira* Cleve, which the author suggests are associated with transport of chitin synthase to the plasmalemma at the site of β-chitin fibril formation. In an examination of chitin synthase activity in subcellular fractions of *Schizophyllum commune* Fr. protoplasts, Vermeulen *et al.* (1979) found that about 30% of the enzyme recovered was associated with the cytoplasmic fraction and occurred in an inactive state. The authors suggest that the "inactive enzyme represents the cytoplasmic transport form of chitin synthase" and that chitisomes are the vehicles of transportation. However, direct cytochemical evidence supporting these proposals is still lacking.

Based primarily on the investigation of septum formation in *Saccharomyces cerevisiae* Hansen, Cabib (1975, 1976) and co-workers (Cabib and Bowers, 1975; Cabib *et al.*, 1973) have presented an alternative concept for the regulation of fungal cell wall biosynthesis. These authors suggest that the polymerization of skeletal polysaccharides, including chitin, is catalyzed by constitutively formed polysaccharide synthases (e.g., chitin synthase) which are uniformly distributed in the plasmalemma. These enzymes may exist either in an active or temporarily inactive (zymogenic) state (Duran and Cabib, 1978; Fig. 2A). Activation and inactivation of these polysaccharide synthases at hyphal tips is presumably influenced by both temporal and spatial control systems. The activating factor of chitin synthase has been identified as a protease (Cabib and Ulane, 1973) and is localized in cytoplasmic vesicles (Cabib *et al.*, 1973; Fig. 2B). Less is known of the inhibition of polysaccharide synthases, although an inhibitory protein acting directly on active chitin synthase has been isolated from the cytoplasm of *M. rouxii* (Lopez-Romero and Ruiz-Herrera, 1976; McMurrough and Bartnicki-Garcia, 1971). Farkaš has suggested that "the activation-inactivation process with polysaccharide synthases seems reversible, and it is assumed that it represents the principal mechanism by which the fungal morphogenesis is regulated." With this in mind, it is tempting to speculate that the protoplasmic differentiation at the hyphal tip and conidium initial is directly associated with temporal and spatial regulation of the biosynthesis of cell wall polymers. The microvesicles may contain both the activating and inhibitory factors for polysaccharide synthases within the plasmalemma. On the other hand the large secretory vesicles probably contain glycoproteins as well as lytic enzymes which are discharged by reverse pinocytosis when the vesicles fuse with the plasmalemma (Fig. 2B). The spatial distribution of these vesicles at the hyphal apex in part explains the polarized nature of hyphal growth (Figs. 1 and 2A). In order for the apex of the hypha to maintain a conical shape during elongation, one can assume that a delicate balance exists between processes of wall synthesis and rigidification on the one hand, and wall lysis on the other (Bartnicki-Garcia, 1973; Fevre, 1979;

Hill and Mullins, 1979; Saunders and Trinci, 1979). Although this is an over-simplification of the events which occur during hyphal wall differentiation and contribute to the maintenance of shape of the hyphal apex, these concepts provide a basis for considering evolved mechanisms of conidium ontogeny. If the rate of wall lysis at the hyphal tip exceeded that of wall synthesis and rigidification, the hyphal apex would swell and could eventually burst. Conversely, too much synthesis and rigidification would result in the arrest of apical elongation. It is suggested that these two variations in the balance between synthesis and lysis play a pivotal role in the development of blastic and thallic conidia.

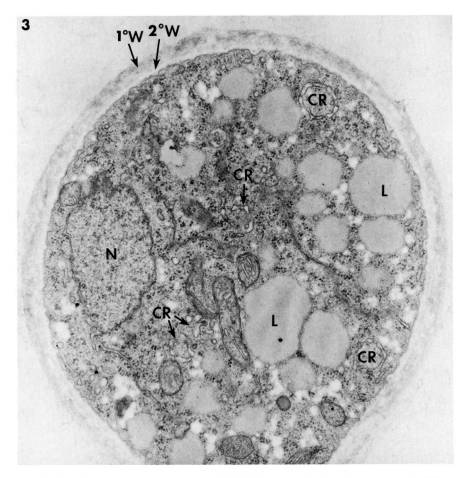

Fig. 3.   Thin section of young conidium of *Scopulariopsis brevicaulis*. 1°W, Primary wall; 2°W, secondary wall; CR, cisternal ring complexes; L, lipid droplet; N, nucleus. ×24,800. (From Cole and Samson, 1979.)

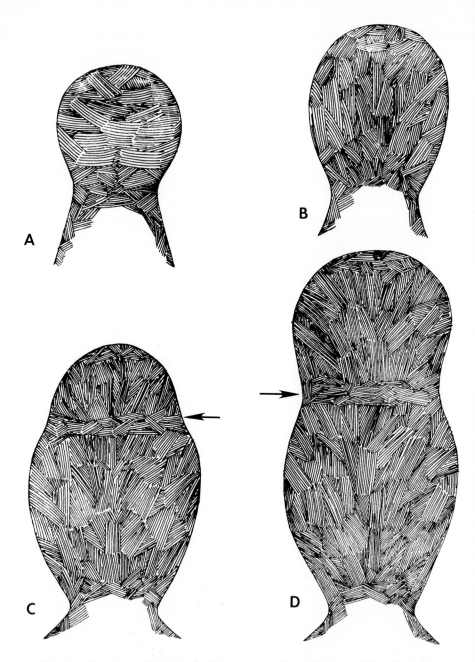

Fig. 4. The changes in rodlet fascicle orientation during blastic conidium formation in *Cladobotryum varium*. A change from mainly diametric expansion (A) to elongation (B) is reflected by a concomitant alteration in rodlet arrangement from a circumferential pattern (A) to an orientation parallel to the long axis of the cell (B). Arrows in (C) and (D) indicate medial constriction of conidium and associated random orientation of rodlet fascicles. (From Cole and Samson, 1979.)

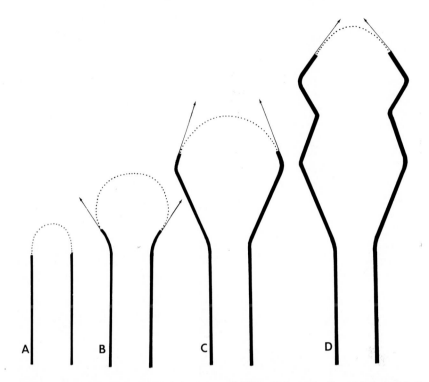

Fig. 5. Interpretation of relationship between wall differentiation and cell growth during blastic conidial development in *Cladobotryum varium*. Solid lines represent set wall layer(s); dotted lines represent unset wall layer(s). Arrows indicate general direction of cell growth. (From Cole and Samson, 1979.)

A blastic conidium develops by the "blowing-out" and *de novo* growth of part of the fertile hypha (Fig. 2C). Protoplasmic differentiation associated with the early stage of blastic conidium initial development is comparable to that of a hyphal tip (Fig. 2A); that is, clusters of vesicles apparently involved in the synthesis and transport of wall precursors are found adjacent to the region of polarized growth (Cook, 1972, 1974). At a later developmental stage, rapid intussusception of new wall material is no longer restricted to the apical dome but involves more of the subapical region of the conidium initial, which then begins to blow out. These processes in turn probably rely on cytoplasmic events, such as the supply of wall precursors and mural enzymes to the conidial wall via the endomembrane system (i.e., ER, dictyosomes, secretory vesicles, and plasmalemma; Fig. 2C). In fact, ultrastructural studies on conidiogenesis in *Neurospora crassa* Shear & Dodge (Turian, 1976; Turian *et al.*, 1973; Ton That and Turian, 1978) and *Sclerotinia fructigena* (Najim and Turian, 1979b) have indi-

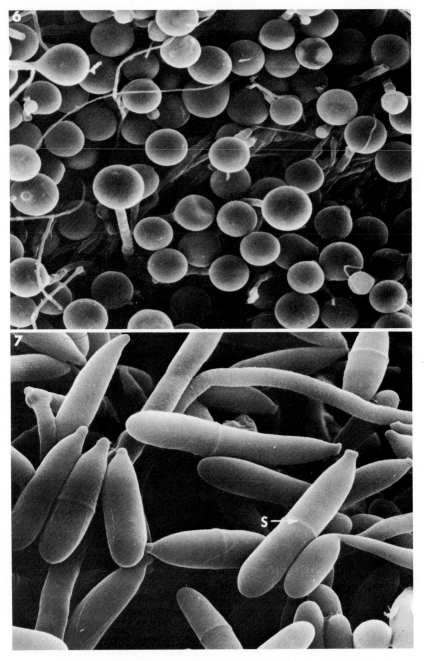

Figs. 6–11. Continued on pages 281 and 282.

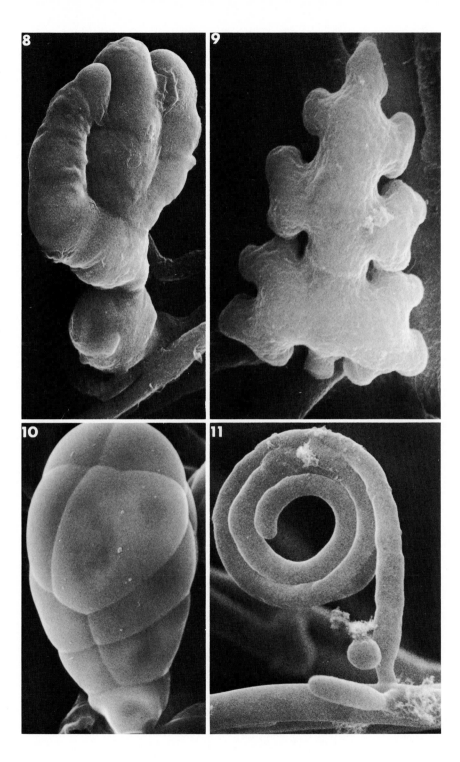

cated that conidium initiation is accompanied by depolarization of apical vesicles and repartition of mitochondria and lipid droplets at the hyphal apex.

The young conidium of *Scopulariopsis brevicaulis* (Sacc.) Bain. in Fig. 3 has assumed a spherical shape. The deposition of new wall material probably involves most of the inner surface of the cell. Wall growth is no longer polarized, and the apical cluster of secretory vesicles is dispersed within the conidium initial. Numerous cisternal rings, suggested to function as dictyosomes (Bracker, 1968; Cole and Aldrich, 1971; Cole and Samson, 1979), are present in the cytoplasm of the young conidium (cf. Fig. 2C). The balance between wall synthesis and plasticization has now shifted in favor of the latter as the initial expands. However, as the new conidium wall thickens, rigidification becomes evident, first at the base and then gradually upward. Concomitantly, the pattern of cell growth shifts from predominantly diametric expansion to elongation. Evidence suggesting that these events do occur during wall differentiation of blastic conidia has been provided by freeze-fracture studies (Cole and Aldrich, 1971; Cole, 1973c; Cole and Samson, 1979). Rodlets, identified as proteinaceous fibrillar components of the outer conidial wall (Hashimoto *et al.*, 1976; Wu-Yuan and Hashimoto, 1977; Cole *et al.*, 1979; Beever *et al.*, 1979), occur in fascicles and appear to change in orientation during conidiogenesis (Fig. 4). It has been suggested that these changes reflect variations in cell wall tensions resulting from alterations in turgor pressure as conidia grow (Cole, 1973c). How these changes in fascicle arrangement occur in unknown. However, rodlet orientation may serve as an indicator of the regions of wall rigidification (random arrangement of rodlet fascicles) and wall plasticization (parallel arrangement of rodlet fascicles) which appear during blastic conidium formation. A diagrammatic interpretation of events associated with wall differentiation during blastic conidium ontogeny in *Cladobotryum varium* Nees is presented in Fig. 4C–D. Kendrick (1971) pointed out that, during the initial stages of blastic development, "wall setting lags behind blowing out by a slight but relatively constant amount. There will thus be more than a hemisphere of unset wall" (Fig. 5A and B), "and setting will thus move along the tangent drawn in the figure. Later, the rate of wall

---

Figs. 6–11.   Variations in blastic conidial shape and septation. Spheroidal, aseptate conidia of *Acrogenospora sphaerocephala* (Berk. & Br.) M. B. Ellis (Fig. 6) contrast with the cylindrical to oblong didymosporous conidia of *Sympodiophora stereicola* G. Arnold (Fig. 7). S, Septum. Palmate and multilobed conidia of *Dictyosporium toruloides* (Corda) Guegen and *Dendrosporium lobatum* Plakidas & Edgerton ex Crane are shown in Figs. 8 and 9, respectively. A muriform conidium of *Monodictys pelagica* (Johnson) Jones and a helicoid conidium of *Helicosporium phragmitis* Höhnel are illustrated in Figs. 10 and 11, respectively Fig. 6: ×399; Fig. 7: ×2200; Fig. 8: ×3240; Fig. 9: ×10,080; Fig. 10: ×3800; Fig. 11: 4620.

Figs. 12 and 13.   Appendages of blastic conidia. Tetraradiate arrangement of appendages formed by the marine fungus *Orbimyces spectabilis* Linder is shown in Fig. 12. The filiform appendages of conidia produced by the coelomycete *Pestalotia olivacea* Guba are shown in Fig. 13. Fig. 12: ×1300; Fig. 13: ×4000.

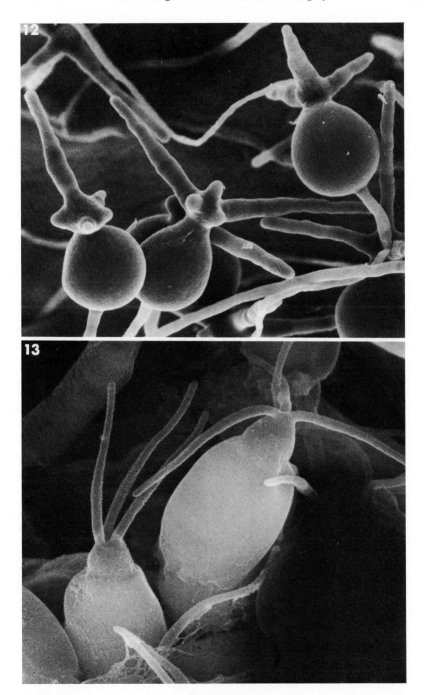

Figs. 12 and 13.

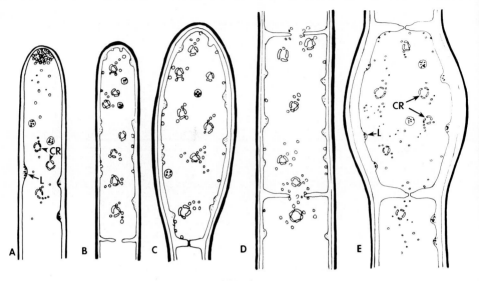

Fig. 14.

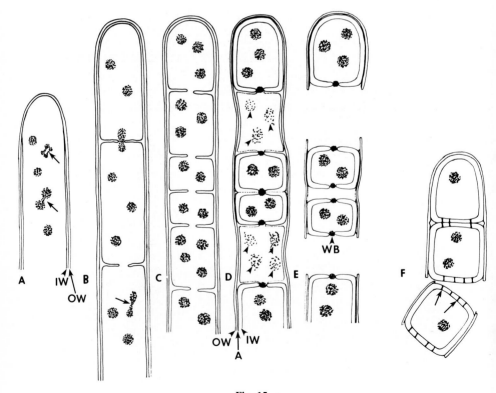

Fig. 15.

setting overtakes the rate of blowing out and exceeds the latter by a slight but relatively constant amount. Now there will be less than a hemisphere of unset wall'' (Fig. 5C), "and setting will move along the tangent drawn in the figure, steadily narrowing the conidium at its apex.'' If the original rates of wall setting and blowing-out are then reestablished, the conidium will assume the shape shown in Fig. 5D, which is comparable to Fig. 4D.

Beginning with a spheroidal initial, a myriad of cell shapes may result from variations in the relative rates of wall rigidification and plasticization during conidium maturation (Figs. 6–13). Conidia may or may not become septate during growth. Septate conidia are categorized as didymosporous (Fig. 7), dictyosporous (Fig. 10), phragmosporous, helicosporous (Fig. 11), or staurosporous (Figs. 36 and 37) (Saccardo, 1886; Kendrick and Nag Raj, 1979). The basal septum of the blastic conidium is frequently differentiated to function in secession of the propagule from the fertile hypha. Aspects of conidial secession will be discussed later. Blastic conidia demonstrate a variety of appendages which arise from the surface of the cell(s). These structures may be functionally significant for insect dispersal, anchorage to the substrate, and/or increased buoyancy in the aerial and aquatic environments (Figs. 12 and 13).

Thallic development occurs by ''conversion'' of an entire fertile hyphal segment into a single aseptate or septate conidium (holothallic conidium; Fig. 14), or by conversion and fragmentation of determinate fertile hyphae into chains of conidia (thallic-arthric conidia; Fig. 15). In contrast to blastic development, initiation of thallic conidium ontogeny is signaled by cessation of apical growth of the fertile hypha. The thallic initial does not show polarized growth. Instead, a segment of the fertile hypha is delimited by one or more cross-walls, the inner cell wall layer thickens, and some enlargement (growth) of the conidium may occur. Conidial wall growth takes place by secondary intussusception, or incorporation of new wall components among existing components of the fertile hypha. Primary intussusception occurs at growing hyphal tips or apexes of blastic conidial initials. Secondary intussusception (Hunsley and Burnett, 1968) occurs in subapical regions of vegetative hyphae and during formation of terminal and intercalary thallic conidia.

Ultrastructural aspects of the cytoplasmic events associated with thallic development are still incompletely known. It is suggested that dispersion of the apical cluster of secretory vesicles occurs at the time of arrest of hyphal tip

---

Fig. 14. Interpretation of protoplasmic and wall differentiation during terminal holothallic conidium formation (A–C) and intercalary holothallic conidium formation (D and E). CR, Cisternal rings; L, lomasome. (From Cole and Samson, 1979.)

Fig. 15. Interpretation of holoarthric (A, B, and F) and enteroarthric (A–E) development emphasizing karyological events and aspects of wall differentiation. Arrows in (A) and (B) point to dividing nuclei. Arrows in (F) locate micropores through septa. Arrowheads in (D) indicate degenerating nuclei. A, Amorphous wall layer of fertile hypha; IW, inner wall layer; OW, outer wall layer; WB, Woronin body. (From Cole and Samson, 1979.)

BLASTIC

Holoblastic

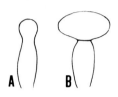  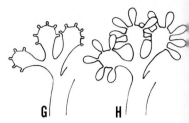

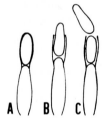 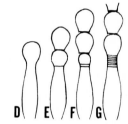 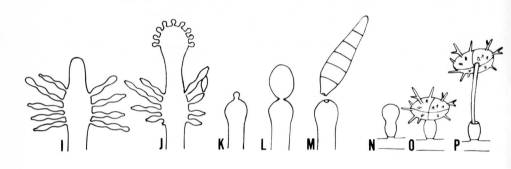

Enteroblastic

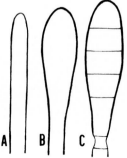 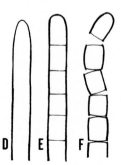 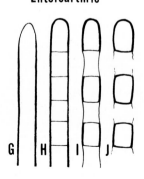

THALLIC

Holothallic   Holoarthric   Enteroarthric

Fig. 16.

growth and that this is followed by the appearance of many dictyosomes and lomasomes in the conidium initial (Fig. 14; Cole and Samson, 1979). Lomasomes, which are multivesicular structures found between the plasmalemma and wall, are considered to be involved in the synthesis and deposition of wall precursors during secondary intussusception (Marchant and Robards, 1968; Heath and Greenwood, 1970; Weisburg and Turian, 1971; Marchant and Moore, 1973; Lin et al., 1975; Hanlin, 1976; Happ et al., 1976). Although our knowledge of the cytological and biochemical aspects of primary intussusception during hyphal growth has expanded considerably in the last few years (see Aronson, this volume, Chapter 28), little information is available on the mechanism(s) of secondary intussusception.

The shape of thallic conidia is much less variable than that of blastic conidia (Figs. 14, 15, and 42–50). This is to be expected because of the restricted modes of thallic development, which are characterized by a limited amount of growth and mainly involve conversion of a preexisting hyphal element into a conidium. Thus most terminal and intercalary holothallic conidia are ellipsoidal (Fig. 14), while thallic-arthric conidia are cylindrical (Fig. 15).

III. SUMMARY OF DIFFERENT MODES OF CONIDIOGENESIS

A summary of the recognized mechanisms of conidiogenesis is presented diagrammatically in Fig. 16. In addition to the two basic modes of conidium ontogeny (i.e., blastic and thallic development), several categories of co-

---

Fig. 16. Summary of modes of conidial and conidiogenous cell development examined in the text. The following symbols are used to denote the three major criteria which distinguish the developmental categories: CW, conidial wall differentiation; OA, order of production and arrangement of conidia on conidiogenous cell; CC, process of conidiogenous cell proliferation. Classification of ontogenetic processes based on modes of conidial wall differentiation (CW) under blastic and thallic development is indicated on the diagram. A list of representative species illustrated in Fig. 16 is presented below. (from Cole and Samson, 1979).

CW: Holoblastic. (A and B) *Nigrospora sphaerica* (Sacc.) Mason; OA: terminal, solitary; CC: determinate. (C–F) *Tritirachium oryzae* (Vincens) de Hoog; OA: terminal, solitary; CC: proliferous (sympodial). (G and H) *Botrytis cinerea;* OA: synchronous, botryose; CC: determinate. (I and J) *Gonatobotryum apiculatum;* OA: 1° conidia synchronous, botryose; 2° conidia asynchronous, catenulate; CC: proliferous (percurrent). (K–M) *Deightoniella torulosa* (Syd.) M. B. Ellis; OA: terminal, solitary; CC: porogenous. (N–P) *Spegazzinia tessarthra* (Berk & Curt.) Sacc.; OA: terminal, solitary; CC: basauxic.

CW: Enteroblastic. (A–C) *Phialophora lagerbergii* (Melin & Nannf.) Conant; OA: basipetal, solitary; CC: phialidic. (D–G) *Scopulariopsis brevicaulis;* OA: basipetal, solitary; CC: proliferous (annellidic). (H–K) *Cladobotryum varium;* OA: basipetal, solitary; CC: retrogressive.

CW: Holothallic. (A–C) *Microsporum gypseum;* OA: terminal, solitary; CC: determinate.

CW: Holoarthric. (D–F) *Geotrichum candidum* Link ex Pers.; OA: random, chains; CC: determinate.

CW: Enteroarthric. (G–J) *Sporendonema purpurascens;* OA: random, chains; CC: determinate.

nidiogenesis are recognized. The latter are based on (1) details of wall differentiation during conidium formation, (2) order of conidium ontogeny, and (3) aspects of conidiogenous cell development. A brief description of each category is presented below.

A. Holoblastic Conidia

1. Determinate Conidiogenous Cells (Fig. 16; Holoblastic, A and B)

Blastic conidia develop by the blowing out of part of the fertile hypha, which may involve all wall layers of the conidiogenous cell. These layers contribute to formation of the conidium wall. Holoblastic development is common in both hyphomycetous and coelomycetous fungi. It seems that increased turgor pressure localized at the tip of the conidiogenous cell combined with lytic enzyme activity associated with the wall encompassing the apex results in enlargement of the fertile region of the cell. Concomitantly an innermost wall layer is synthesized (cf. Fig. 3). Subsequent differentiation of the conidium, which remains attached to the conidiogenous cell, involves continued swelling and elongation. Holoblastic conidia may also undergo septation and appendage formation, resulting in a high degree of variation in form. The conidiogenous cell may cease further extension growth with formation of the terminal, holoblastic conidium.

2. Sympodially Proliferating Conidiogenous Cells (Fig. 16; Holoblastic, C-F)

The conidiogenous cell may instead undergo extension growth from the base and one side of the terminal holoblastic conidium, only to give rise to another conidium at the newly formed apex. Such successive proliferation of the fertile cell is described as sympodial and usually results in formation of a "rachis." Sympodial proliferation is characteristic of members of such anamorph-genera as *Tritirachium* Limber, *Acrodontium* de Hogg, *Beauveria* Vuill., *Phaeoisaria* Höhnel, *Sympodiophora* Arnold, and others. In some species, however, [e.g., *Fonsecaea pedrosoi* Brumpt (Negroni); Fig. 17], the conidiogenous cell apex swells as successive conidia are produced rather than forming a rachis. The conidia are arranged in dense clusters, and the sympodial nature of fertile cell proliferation is not immediately apparent (Cole, 1978). To add to the difficulties of interpretation, an alternative arrangement of successively formed conidia does not necessarily mean that the conidiogenous cell has undergone sympodial pro-

Fig. 17.   Conidia and denticles (D) or abstriction scars at the apex of the sympodially proliferated conidiogenous cell of *Fonsecaea pedrosoi*. ×5720. (From Cole, 1978.)

Figs. 18 and 19.   Stages of basipetal conidium formation at the apex of determinate conidiogenous cells of *Nodulisporium hinnuleum*. 1, First-formed conidium; 2, second-formed conidium. ×13,200.

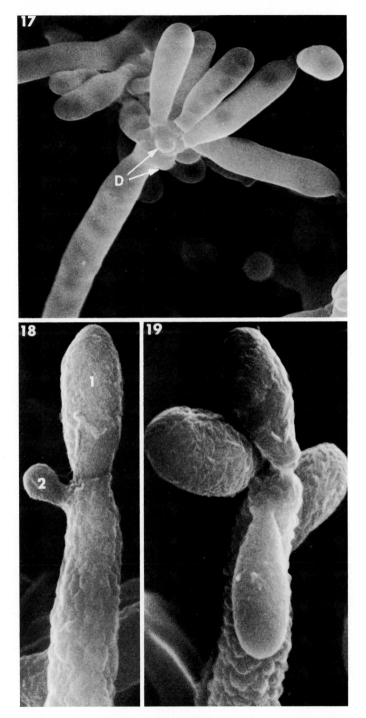

Figs. 17–19.

liferation. For example, the conidia of *Nodulisporium hinnuleum* (Preuss) G. Smith (Figs. 18 and 19) are produced in basipetal succession from determinate conidiogenous cells, a subtle mechanism of conidiogenesis which was elucidated by time-lapse photomicrography (Cole, 1971).

3. Ampullae and Botryose Conidia (Fig. 16; Holoblastic, G–J)

Holoblastic conidia may also form synchronously in botryose clusters from the swollen fertile apexes (ampullae) of conidiogenous cells. The *Chromelosporium* anamorph of *Peziza ostracoderma* Korf produces conidiophores with up to 12 long, divergent fertile hyphae at their apexes, each forming holoblastic conidia synchronously over its entire surface (Fig. 20). Similarly, fertile branches of the conidiophores of *Botrytis cinerea* Pers. ex Fr. (Fig. 21) terminate as ampullae which produce synchronous holoblastic conidia on short denticles. These and other fungi which demonstrate synchronous holoblastic conidiogenesis (e.g., *Nematogonium* Desm., *Gonatobotrys* Corda, *Botryosporium* Schw., *Spiniger* Stalpers) represent potentially fruitful experimental systems for cytologists and wall biochemists to study this morphogenetic process. In each case the ampulla is unicellular. The conidia are not cut off from the fertile cell by a septum until they have matured. The triggering mechanism for synchronous conidium initiation is unknown. However, on the basis of recent concepts of the control of polysaccharide synthase activities in the plasmalemma (Farkaš, 1979), it seems reasonable to assume that localized protoplasmic differentiation (i.e., vesicle aggregation), combined with the activation of wall lytic enzymes and increased turgor pressure within the ampulla, is at least partially responsible for initiating conidiogenesis. Cook (1972, 1974) has shown clusters of vesicles within the ampulla of *Oedocephalum roseum* Cook adjacent to regions which have begun to blow out, initiating synchronous holoblastic conidium formation. The morphology and arrangement of these vesicles are comparable to those found in growing hyphal tips (e.g., Fig. 1).

In *Gonatobotryum apiculatum* (Peck) Hughes, the synchronously formed primary conidia give rise asynchronously to secondary holoblastic conidia (Kendrick *et al.*, 1968; Kendrick and Chang, 1971; Cole, 1973b). The latter in turn proliferate, resulting in the formation of acropetal chains of conidia. The conidiogenous cell itself may subsequently proliferate, growing up through the previously formed ampulla (i.e., percurrent proliferation) only to produce a new ampulla from which 1° and 2° holoblastic conidia arise (Fig. 16, Holoblastic, I and J). Each ampulla serves as a cytoplasmic reservoir providing food material to the tips of the chains. An elaborate kind of septation has evolved between adjacent holoblastic "catenulate" conidia to prevent interruption of this food supply even though one or more aerial chains may be damaged (Cole, 1973b; Cole and Samson, 1979).

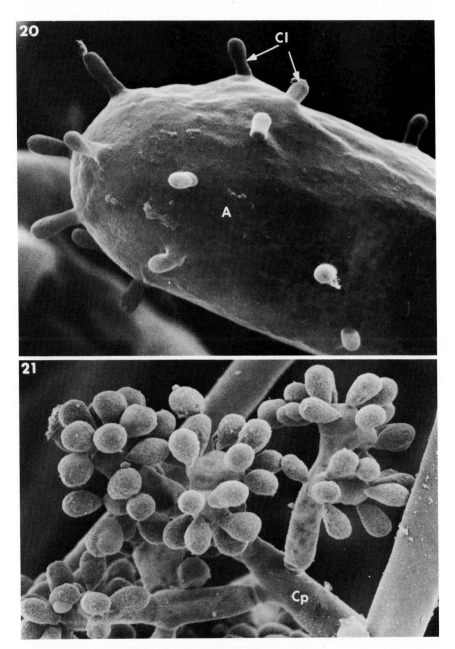

Fig. 20. Cluster of young holoblastic conidia in a synchronized stage of development which have arisen from the ampulla (A) of the anamorph of *Peziza ostracoderma*. CI, Conidium initial. ×5500.

Fig. 21. Botryose, holoblastic conidia which have developed synchronously from the fertile branches of *Botrytis cinerea*. Cp, Conidiophore. ×4500. (From Cole and Samson, 1979.)

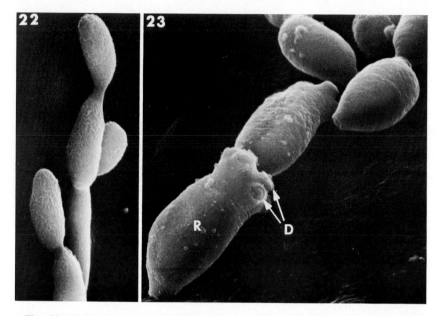

Figs. 22 and 23. Acropetal chains of holoblastic conidia produced by *Cladosporium* sp. Two ramoconidia (R) are shown which have seceded from conidial chains of *Cladosporium* sp. Note the presence of numerous denticles (D) which represent attachment scars of conidia at the base of acropetal chains. Fig. 22: ×3750; Fig. 23: 5625.

4. Acropetal Chains of Conidia (Fig. 16; Holoblastic, I and J)

Formation of acropetal chains of conidia in *Cladosporium* Link ex Fr. does not involve an ampulla but rather direct proliferation of the fertile hyphal apex (Figs. 5, 22, and 23). The fertile hyphae may branch, giving rise to chains of conidia at each new apex or from subapical regions. The first two conidia which form are often morphologically distinct from others of the chain; they are elongate and produce several fertile loci from each of which an acropetal chain of holoblastic conidia develops (Fig. 23). These cells at the base of the chains are called ramoconidia. Ontogenetic variations of this basic theme are demonstrated by *Torula herbarum* (Pers.) Link ex S. F. Gray (Hashmi *et al.,* 1973; D. H. Ellis and Griffiths, 1975a,b) and *Periconia cookei* Mason & M. B. Ellis (M. B.Ellis, 1971b; Cole and Samson, 1979).

5. Porogenous Fertile Cells (Fig. 16; Holoblastic, K–M)

In a special mode of holoblastic development, the conidium first appears as a tiny papilla, restricted to a small area of the conidiogenous cell apex (Fig. 24) or subapical region (Fig. 25). The conidium subsequently elongates (Fig. 27) while remaining attached to the fertile hypha by a narrow isthmus at its base. After

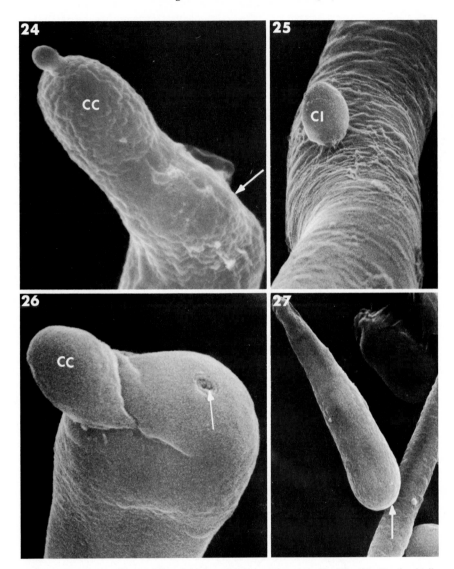

Figs. 24 and 25. Poroconidium initials which have arisen apically (Fig. 24; *Dendryphiella arenaria* Nicot) and laterally (Fig. 25, *Helminthosporium solani* Dur. & Mont.) from conidiogenous cells. Arrow locates point of attachment of previously formed conidium. CC, Conidiogenous cell; CI, conidium initial. Fig. 24: ×9775; Fig. 25: 7630.

Fig. 26. Arrow indicates pore remaining at conidiogenous cell apex of *Ulocladium atrum* after conidial secession. Lateral growth of conidiogenous cell (CC) represents intiation of sympodial proliferation. ×15,000. (From Cole and Samson, 1979.)

Fig. 27. A mature, laterally produced conidium of *Helminthosporium solani*. Arrow indicates isthmus between conidium and conidiogenous cell. ×2160.

conidial secession, a small pore remains in the conidiogenous cell wall (arrow in Fig. 26). A pore is also present in the septal wall at the base of the detached conidium. At the time of secession, the pore is plugged with wall material. However, while the conidium is developing, a narrow channel through the conidiogenous cell wall is present, which allows cytoplasmic streaming and translocation of nutrients into the conidium initial. The channel forms during early stages of conidium formation (i.e., Figs. 24 and 25). There has been some disagreement over the nature of wall relations during this process of conidiogenesis. Some authors support the concept that the channel through the conidiogenous cell wall is first formed by localized lytic enzyme activity and then the inner wall of the fertile hypha blows out through the channel and encompasses the conidium initial (Fig. 28A and B; Campbell, 1968, 1969; M. B. Ellis, 1971a,b; Cole, 1973a). However, thin-section examinations of *Ulo-*

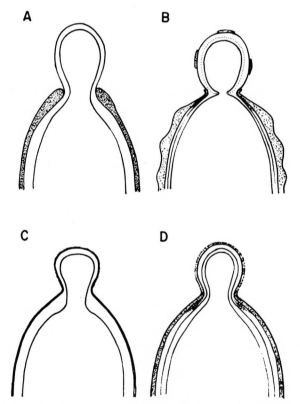

Fig. 28. Various interpretations of wall differentiation during poroconidial development based on thin-section examinations (A, Campbell, 1968, 1969; B, Cole, 1973a; C, Carroll, 1972; Carroll and Carroll, 1974; D, Brotzman *et al.,* 1975). (After Cole and Samson, 1979.)

cladium atrum Preuss (Carroll, 1972; Carroll and Carroll, 1974) and *Helminthosporium maydis* Nisikado & Miyake (Brotzman *et al.*, 1975) have indicated that the conidium initial forms by the blowing out of all wall layers of a restricted region of the conidiogenous cell (i.e., holoblastic development; Fig. 28C) Cole and Samson (1979) have diagrammatically summarized these concepts and concluded that this process of conidiogenesis represents a special mode of holoblastic development. The very early stages of conidium initial formation are holoblastic. However, the outermost wall layer encompassing the developing conidium soon breaks down, while the inner wall layers continue to thicken (e.g., Fig. 28B and D). It seems that the use of features of cell wall differentiation alone for classification of conidial fungi has its limitations. Nevertheless, the "development of solitary conidia at minute, single, or numerous pores in the wall of the conidiophore" (Hughes, 1953) distinguishes this mechanism of conidiogenesis from other processes of conidium ontogeny.

6. Basauxic Proliferation (Fig. 16; Holoblastic, N–P)

A final example of holoblastic conidium formation involves a unique process of conidiogenous cell development; basauxic proliferation (Fig. 29). The first-formed holoblastic conidium is produced terminally at the apex of a fertile cell (Fig. 29A and B). After maturation of the conidium, the outer wall of the fertile cell ruptures below the conidial base as the inner wall layer undergoes extension growth, or basauxic proliferation (Fig. 29C). In the *Arthrinium* anamorph of *Apiospora montagnei* Sacc. (Fig. 29), the filament (conidiogenous cell) which arises from the ruptured cup-shaped cell (conidiophore mother cell) continues to elongate while giving rise to successive, secondary holoblastic conidia just above the ruptured apex of the mother cell (Fig. 29D).

This same mode of conidiogenesis is demonstrated by *Arthrinium phaeospermum* (Corda) M. B. Ellis (Figs. 30 and 31). An additional point of interest in this latter species is the fact that it occurs on plant host tissue both as a hyphomycetous form (i.e., loose, floccose vegetative and reproductive mycelia not enclosed within sporocarps), and a coelomycete (conidia produced within acervuli). The fungus has been found in Texas as both forms growing on bamboo stems (*Phyllostachys aurea*) and carpet grass (*Stenotaphrum secundatum*). Figure 30 shows a median longitudinal section through the acervulus of an infected blade of grass. The cuticle of the host eventually ruptures, allowing the conidia to be released. In Fig. 1, vegetative hyphae, a conidiophore mother cell, a conidiogenous cell, and conidia are shown growing on the surface of the grass cuticle—not within an acervulus. Discovery of the exogenous and endogenous factors which control the alternation of this fungus between a coelomycetous and hyphomycetous mode of development represents a challenging research problem for future investigation.

Basauxic conidiogenous cell proliferation is demonstrated by several additional anamorph-genera, including *Spegazzinia* Sacc., *Pteroconium* Sacc.

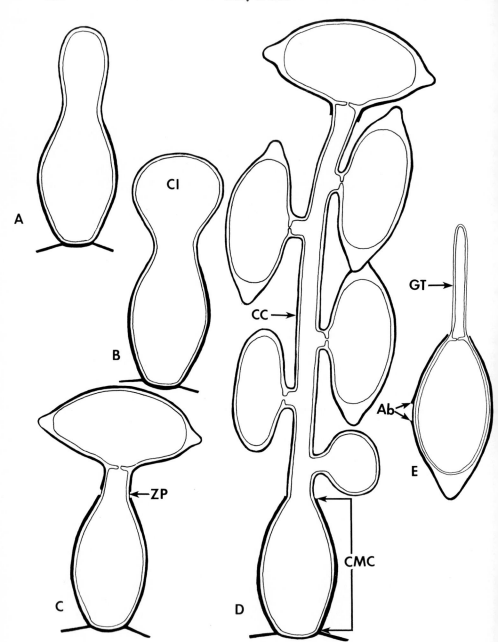

Fig. 29. Interpretations of basauxic conidiogenous cell development, formation of a basipetal succession of conidia, and conidial germination in the *Arthrinium* state of *Apiospora montagnei*. Ab, Abstriction scar; CC, conidiogenous cell; CI, conidium initial; CMC, conidiophore mother cell; GT, germ tube; ZP, zone of proliferation of conidiogenous cell. (From Cole and Samson, 1979.)

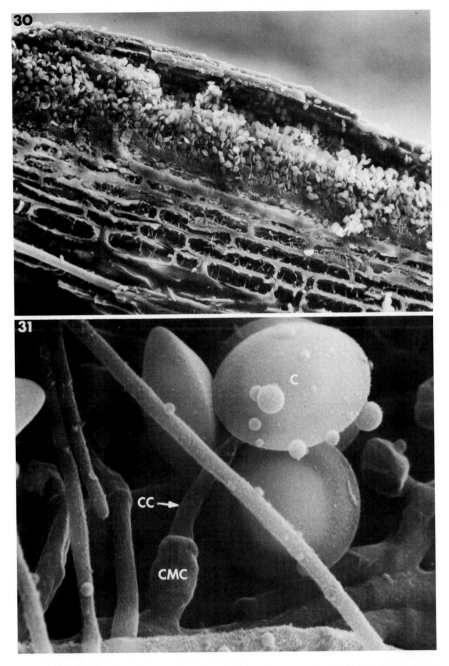

Fig. 30. Section through acervulus of *Arthrinium phaeospermum* within host (carpet grass, *Stenotaphrum secundatum*) showing mass of conidia and conidiogenous cells. ×150.

Fig. 31. A conidiophore mother cell (CMC), basauxically proliferated conidiogenous cell (CC), and conidia (C) produced on agar surface (i.e., hyphomycetous growth) by *Arthrinium phaeospermum*. ×5290.

ex Grove, *Cordella* Speg., *Dictyoarthrinium* Hughes and *Endocalyx* Berk. ex
Br. (Cole and Samson, 1979).

B. Enteroblastic Conidia

1. Phialidic Development (Fig. 16; Enteroblastic, A–C)

Enteroblastic conidial development is a process in which "the outer layer(s) of
the wall of the conidiogenous cell is (are) not involved in the formation of the
conidium wall" (Kendrick, 1971). The phialide is a specialized kind of fertile
cell which produces a basipetal succession of conidia (phialoconidia) from the
conidiogenous locus (Cole and Kendrick, 1969a; Hammill, 1974, 1977b; Cole and
Samson, 1979). The first-formed phialoconidium is holoblastic (Fig. 32A and
F). Initiation of the second conidium involves some extension growth of the inner
phialide wall layer below the base of the primary conidium. As a result, the
first-formed phialoconidium is pushed upward and the outer wall layer of the
phialide ruptures when it reaches its limit of prolongation (Fig. 32B and G). This
ruptured outer wall of the fertile cell is called a collarette (Fig. 32C). Secondary
phialoconidia emerge in basipetal succession through the collarette. Two processes of wall differentiation are recognized during secondary phialoconidium
formation. In one case (Fig. 32C–E), the outer wall layer of each mature secondary conidium ruptures adjacent to its basal septum during formation of the next
conidium initial (arrows in Fig. 32D). Repetition of this process leaves rings of
wall material (i.e., scars of seceded conidia) inside the collarette (Fig. 32E),
which are well illustrated in thin sections of *Trichoderma saturnisporum* Hammill (Hammill, 1974). Under the light microscope, these endogenous lamellae
are visible as thickenings at the base of the collarette (Cole and Samson, 1979).
The phialoconidia produced by this mode of development usually accumulate in
droplets at the phialide apex but may also aggregate in columns [e.g., *Metarrhizium anisopliae* (Metsch.) Sorok.] or become arranged in an imbricate pattern
(e.g., *Mariannaea elegans* (Corda) Samson). In the alternative process of wall
differentiation (Fig. 32H–I), secondary phialoconidia are held together in chains
because of thickened wall connectives and a thin, continuous outer wall layer
between adjacent propagules (arrows in Fig. 32H). Since the outer wall layer of
each mature conidium does not rupture at its base during formation of the next
conidium initial, there are no rings of wall material left within the collarette. This
mechanism of phialoconidium ontogeny is demonstrated by members of *Penicillium* Link ex Fr., *Aspergillus* Mich. ex Fr., *Chalara* (Corda) Rabenh., and
others.

Phialide morphology and arrangement are highly variable (Cole and Samson,
1979). A dense cluster of flask-shaped phialides attached to the fertile vesicle of
Aspergillus fumigatus Fres. is shown in Fig. 33. The single phialide of *A. flavus*

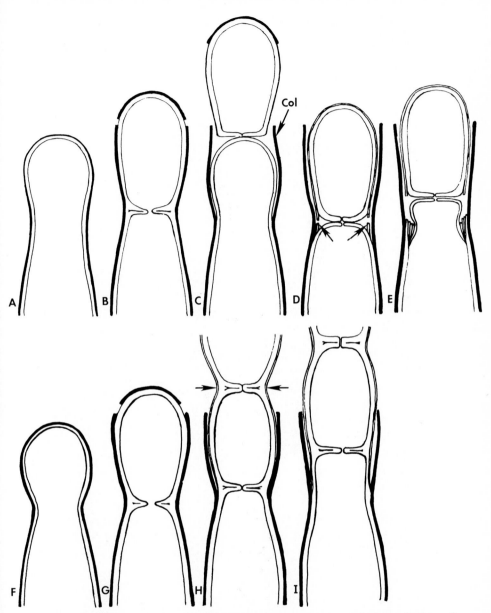

Fig. 32. Interpretations of wall differentiation associated with two basic mechanisms of phialoconidium formation. (A–E) Conidia readily secede from the phialide apex and form droplets or adhere in "false" chains. (F–I) Conidia remain in a chain at the phialide apex because of thick-walled connectives between adjacent cells and/or a thin, continuous outer wall layer encompassing the conidial chain [arrows in (H)]. Note that the first-formed conidium in each case is holoblastic (B and G) and initiation of the second conidium causes the outer phialide wall to rupture, forming the collarette (Col). Arrows in (D) locate the ruptured outer wall of the second conidium which resulted from initiation of the third conidial initial. (From Cole and Samson, 1979.)

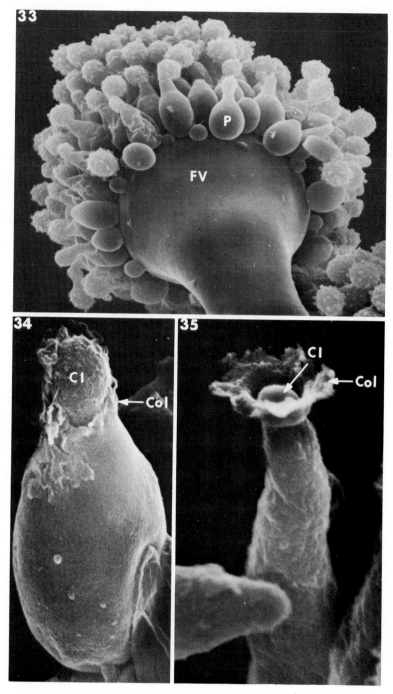

Figs. 33–35.

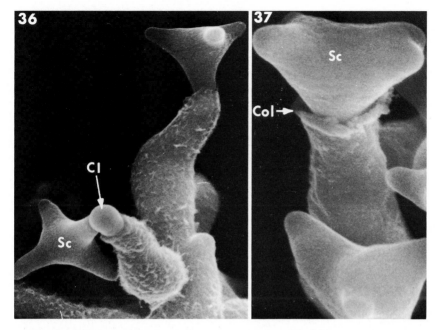

Figs. 36 and 37. Stages of emergence of tetraradiate phialoconidia of *Clavatospora stellatacula*. CI, Conidium initial; Col, collarette; Sc, stauroconidium. Fig. 36: ×5390; Fig. 37: 11,880.

Link ex Fr. in Fig. 34 has ruptured at its apex, revealing the conidium initial. The rather inconspicuous collarette of *Aspergillus* contrasts with the distinctive collarette of *Phialophora richardsiae* (Nannf. apud Melin & Nannf.) Conant (Fig. 35). Phialoconidia also show a high degree of morphological variability. Developmental stages of the stauroconidia of the marine hyphomycete *Clavatospora stellatacula* Kirk are illustrated in Figs. 36 and 37. In spite of these morphological differences, all phialides demonstrate similar features of conidium development; that is, a plurality of conidia are produced in basipetal succession from a localized, endogenous meristem (Cole and Kendrick, 1969a).

Two variations in phialidic development deserve mention: formation of polyphialides and percurrently proliferated phialides. Polyphialides give rise to a succession of conidiogenous loci by sympodial proliferation of the fertile cell from the base and to one side of the previously formed collarette. The

---

Fig. 33. Dense cluster of phialides which have arisen from the fertile vesicle (FV) of *Aspergillus fumigatus*. Phialoconidia are attached to the apex of phialides (P). ×3100.

Fig. 34. Phialoconidium initial (CI) which has partially emerged through the ruptured phialide apex or collarette (Col) of *Aspergillus flavus*. (From Cole and Samson, 1979.) ×7425.

Fig. 35. Flared collarette (Col) and endogenous phialoconidium initial (CI) of *Phialophora richardsiae*. ×14340.

polyphialide therefore has "more than one open end from each of which a basipetal succession of phialospores is produced" (Hughes, 1951) and is formed by such conidial fungi as *Lasiosphaeria hirsuta* (Fries) Ces. & de Not., *Codinaea simplex* Hughes & Kendrick, and *Phialophora geniculata* van Emden. Cole and Samson (1979) have noted that usually one functional conidiogenous locus exists at any time, but in rare cases (e.g., *Polypaecilum insolitum* G. Smith) two or more concurrently active fertile apertures may exist.

Phialides may also proliferate percurrently, as in *Catenularia* Grove ex Sacc., by growing through the previously fertile apex; that is, the phialide undergoes extension growth through its collarette. The phialide then ceases to elongate and gives rise to another basipetal succession of phialoconidia at its new fertile apex. This mode of conidium and conidiogenous cell ontogeny is distinguished from annellidic development, which is described below.

2. Annellidic Development (Fig. 16; Enteroblastic, D–G)

Another specialized kind of conidiogenous cell is the "annellide." As in phialidic development, the first-formed conidium is holoblastic, while the secondary conidia produced in basipetal succession are enteroblastic. However, ontogenetic and morphological differences exist between the annellide and phialide. The first-formed conidium secedes from the annellide as the fertile apex of the conidiogenous cell undergoes percurrent proliferation (Fig. 38A). Development of the septum at the base of this conidium is pivotal in both secession and later proliferation of the fertile cell. Centripetal growth of the inner wall of the conidiogenous cell leads to the formation of a double-layered septum with a medial, electron-translucent zone separating the upper and lower halves of the cross-wall (Fig. 39). This electron-translucent zone is thought to be the site of lytic enzyme activity (Gull, 1978) which eventually results in separation of the two septal layers and thereby leads to secession of the conidium from the conidiogenous cell. This method of conidial secession, referred to as schizolytic (Hughes, 1971a,b; Cole and Samson, 1979), is most common among fungi with blastic conidium ontogeny.

During annellidic development, schizolytic secession occurs concomitantly with percurrent proliferation of the conidiogenous cell. The double-layered septum differentiates into a thicker upper layer, which becomes the base of the conidium, and a thin lower layer, which functions as the new proliferating apex of the fertile cell (Fig. 38A). The pore in the cross-wall at the conidial base is plugged, usually with a Woronin body (Buller, 1933; Carroll, 1966; Cole and Aldrich, 1971; Wergin, 1973). The perforation through the lower septal layer is sealed by continued centripetal growth of the cross-wall (Fig. 38B). The ultrastructure of this latter process has not been clearly defined. The proliferated apex of the annellide (ZP in Fig. 38B) soon differentiates into a second conidium (Fig. 38C–E). This propagule forms enteroblastically in contrast to the holoblastic

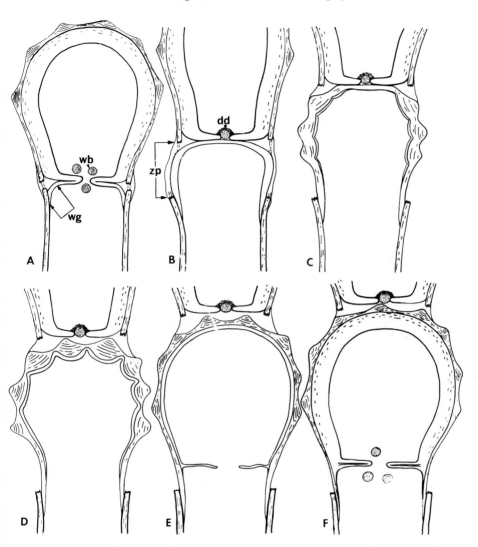

Fig. 38. Diagrammatic interpretation of enteroblastic conidium formation from percurrently proliferating annellides of *Scopulariopsis brevicaulis*. dd, Electron-opaque deposit surrounding septal plug; wb, Woronin body; wg, region of new wall growth; zp, zone of percurrent proliferation. (After Cole and Aldrich, 1971.)

development of the primary conidium (Fig. 38A). The ruptured outer co-nidiogenous cell wall remaining after secession of the first-formed conidium appears as a ring or annellation at the apex of the fertile cell. By this same sequence of events, the new outer wall layer, which is continuous between the secondary

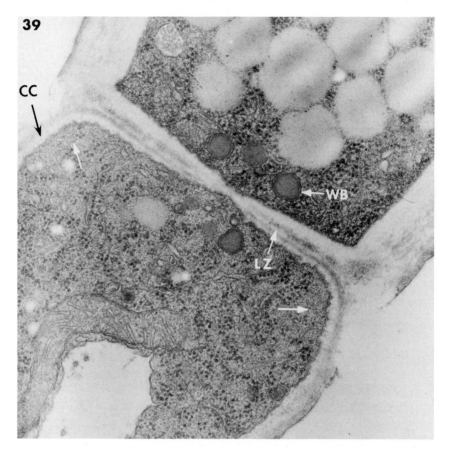

Fig. 39. Septum delimiting conidium from annellide (conidiogenous cell, CC) of *Scopulariopsis brevicaulis*. Arrows indicate cytoplasmic regions of low ribosome density where wall intussusception occurs. LZ, Lytic zone between septal wall layers which have begun to septate centripetally; WB, Woronin body. ×35,650.

conidium and new annellide apex, later ruptures opposite the basal septum of the conidium (Fig. 38F). Because of the earlier percurrent proliferation of the conidiogenous cell, the second annellation appears slightly above the first. Formation of a succession of secondary conidia results in development of an elongated and annellated neck of the fertile cell (Fig. 16; Enteroblastic, G). This mode of conidiogenesis has been illustrated in *Scopulariopsis brevicaulis* by time-lapse light microscopy (Cole and Kendrick, 1969b).

Annellidic development, as described above, has been identified in many conidial fungi, including *Annellophora* Hughes (Hughes, 1953), the *Spilocaea* anamorph of *Venturia inaequalis* (Cooke) Wint. (Reisinger, 1972; Hammill, 1973; Corlett *et al.*, 1976), *Trichurus spiralis* Hasselbring (Hammill, 1977a), and *Exophiala jeanselmei* (Langeron) McGinnis and Padhye (Cole, 1978). However, variations of this developmental process, which have been defined primarily by electron microscope examinations, suggest the convergence of phialidic and annellidic concepts (Cole and Samson, 1979). For example, the formation of distinct annellations within the collarette of *Conioscypha varia* Shearer and *C. lignicola* Höhnel are due to successive percurrent proliferations of the fertile cell, a process similar to annellidic development (Van Emden and Veenbaas-Rijks, 1973; Shearer and Motta, 1973). A comparison of Figs. 32A–E and 38 also demonstrates the basic similarities in wall differentiation between phialidic and annellidic development. Most phialides and annellides are nevertheless distinguishable on the basis of the presence of a collarette in the former and closely spaced annellations in the latter. More clearly differentiated are the percurrently proliferating phialides, which produce a linear arrangement of collarettes resembling annellations except that a plurality of enteroblastic conidia arise through each successively formed collarette.

3. Retrogressive Conidiogenous Cell Development (Fig. 16, Enteroblastic, H–K)

Like phialides and annellides, retrogressive conidiogenous cells give rise to a basipetal succession of conidia from the fertile apex of a conidiogenous cell. However, the fertile hypha in this case gradually shortens as conidium ontogeny proceeds. Two mechanisms of retrogressive conidiogenous cell development are recognized (Cole and Samson, 1979) and are illustrated diagrammatically in Figs. 40 and 41.

The fertile cell of *Cladobotryum varium* looks very much like a phialide (Fig. 40A–D). As in phialoconidiogenesis, the first-formed conidium is holoblastic (Fig. 40A–C), while secondary conidia are enteroblastic and grow through the ruptured outer wall (collarette) of the fertile cell. However, a time-lapse photomicrospic analysis of this fungus has shown that the conidiogenous cell becomes progressively shorter as each conidium matures and secedes (Cole and Kendrick, 1971). Thin-section, scanning electron microscope, and freeze-etch investigations of *C. varium* have revealed the details of this subtle developmental process (Cole and Samson, 1979). The septum at the base of the second conidium forms below the ruptured apex of the fertile cell (Fig. 40E). Unlike the phialide, the outer conidiogenous cell wall is firmly attached to the surface of the conidium wall. As schizolytic conidial secession occurs, a ring of the fertile cell wall remains attached to the base of the conidium (Fig. 40F). As new enteroblastic conidia are formed and secede, the conidiogenous cell shortens (Fig. 40G and H). A similar mechanism of retrogressive development has been illustrated in *Trichothecium roseum* (Pers.) Link ex S. F. Gray (Kendrick and Cole, 1969; Cole and Samson, 1979).

It is not possible to distinguish between vegetative and fertile hyphae of *Basipetospora rubra* Cole & Kendrick with the light microscope until co-

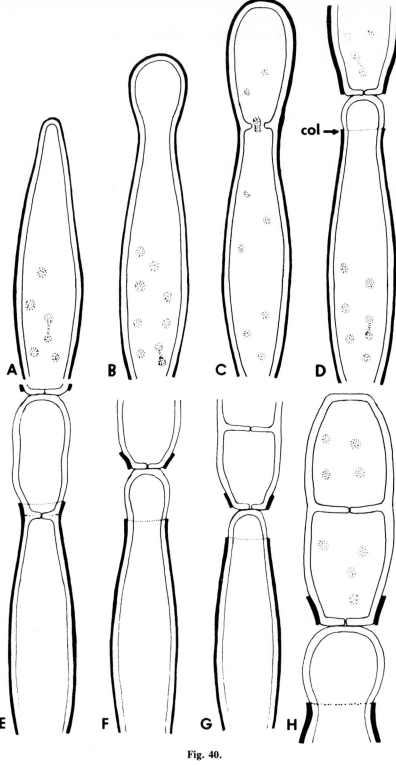

Fig. 40.

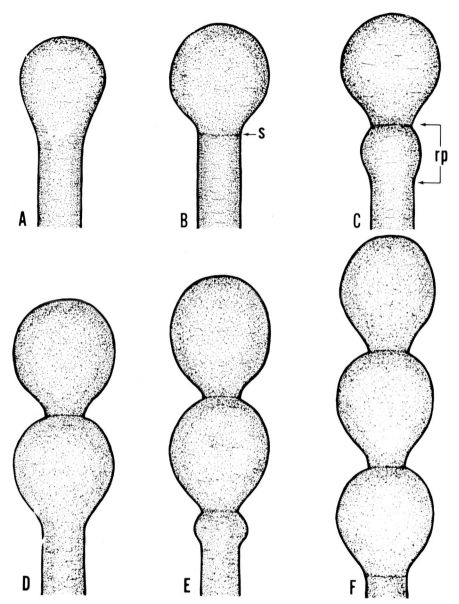

Fig. 41. Diagrammatic interpretation of retrogressive development of the fertile hypha of *Basipetospora rubra*. S, Septum; rp, region of retrogression and proliferation.

Fig. 40. Diagrammatic interpretation of retrogressive conidiogenous cell formation and basipetal conidium ontogeny in *Cladobotryum varium*. col, Collarette. (From Cole and Samson, 1979.)

nidiogenesis has been initiated. The apex of the fertile hypha at first swells (Fig. 41A), and then a cross-wall(s) forms delimiting the terminal, holoblastic conidium (Fig. 41B). Time-lapse photomicrography (Cole and Kendrick, 1968) has shown that the fertile hypha subsequently swells below this septum (rp in Fig. 41C) and a second conidium is formed, still attached to the primary conidium (Fig. 41D). Development of the second conidium primarily involves new growth but also incorporates a portion of the preexisting fertile hyphal wall into the conidial wall. This latter process results in retrogression of the fertile hypha as successive conidia are formed (Fig. 41E and F). Ultrastructural investigations of *B. rubra* (Cole and Samson, 1979) have revealed that the outer hyphal wall remains intact as the basipetal succession of conidia (holoblastic) is produced. In spite of the holoblastic nature of conidium ontogeny in *B. rubra* it is included with other retrogressive forms because of the distinctiveness of this mode of conidiogenous cell development.

C. Holothallic Conidia (Fig. 16; Thallic, A–C)

Thallic development occurs by conversion of a fertile hyphal element into one or more terminal or intercalary conidia (Figs. 14 and 15). Galgóczy (1975) illustrated holothallic conidium formation in *Microsporum gypseum* (Bodin)

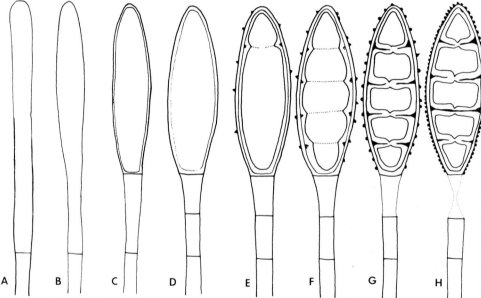

Fig. 42. Diagrammatic interpretation of holothallic conidium formation in *Microsporum gypseum*. (After Galgóczy, 1975.)

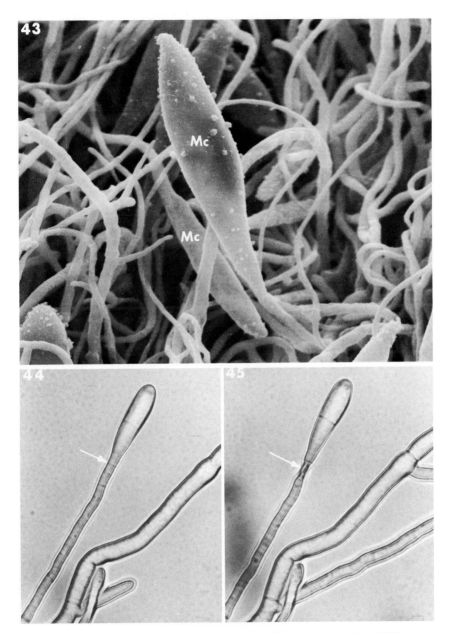

Fig. 43. Echinulate, holothallic macroconidium (Mc) of *Microsporum canis.* ×1150.

Figs. 44 and 45. Stages of rhexolytic secession of a holothallic conidium from the fertile hypha of *Microsporum gypseum*. Arrow in Fig. 44 indicates basal septum of macroconidium. Arrow in Fig. 45 indicates basal cell which has collapsed. $T = 0$ and 10 min, respectively. ×700.

Guiart & Grigorakis (Fig. 42), demonstrating that all wall layers of the fertile hypha become part of the wall of the terminal conidium. However, his diagrammatic interpretation fails to indicate that some extension growth occurs during holothallic development. This growth is probably the result of secondary intussusception (Cole and Samson, 1979). The macroconidia of *M. gypseum* (Figs. 42, 44, and 45) and *M. canis* Bodin (Fig. 43) are septate and verrucose and secede from the fertile hypha by autolysis of the cell adjacent to the conidial base (Figs. 44 and 45). This represents the second method of conidial secession, which has been referred to as a rhexolytic process (Correns, 1899; Hughes, 1971a,b). Some blastic conidia secede by this method (e.g., *Gonatobotryum apiculatum*, anamorph of *Peziza quelepidotia* Korf & O'Donnell, *Epicoccum purpurascens* Ehrenb. ex Schlecht, *Sporotrichum aureum* Link ex Fr.), but the process occurs more commonly among the thallic conidium-forming fungi.

The concept of holothallic development encompasses the definitions of terminal and intercalary chlamydospores (Kendrick, 1971). However, since the term "chlamydospore" is so deeply entrenched in mycological literature and has been used in a variety of connotations, it is not possible to abandon it completely for "holothallic conidium." Of significance is that many holothallic and some blastic conidia do not readily separate from their parent hyphae (i.e., those which undergo rhexolytic secession). This is not a developmental characteristic unique to chlamydospores.

D. Holoarthric and Enteroarthric Conidia (Fig. 16; Thallic, D–J)

The thallic-arthric mode of conidiogenesis occurs by septation and fragmentation of determinate fertile hyphae and has been subdivided into two developmental processes based on mechanisms of wall differentiation. Holoarthric ontogeny involves transformation of the entire fertile hypha into a chain of conidia; the wall of the hypha becomes incorporated into that of the conidia (Fig. 15A, B, and F, and Fig. 46). Enteroarthric conidia differentiate within the fertile hypha (Fig. 47), and the hyphal wall is not necessarily incorporated as part of the conidial walls (Fig. 15A–E). The time-lapse sequence of conidiogenesis in *Sporendonema purpurascens* (Bon.) Mason & Hughes in Figs. 48–50 demonstrates some of the characteristics of this developmental process. Hyphal tip growth, which has accounted for branch elongation (cf. branch located by arrows in Figs.

Fig. 46. Adjacent holoarthric conidia of the anamorph of *Peniophora gigantea*. The conidia have formed by disarticulation of a fertile hypha. ×8600. (From Cole and Samson, 1979.)

Fig. 47. Enteroarthric conidia (C) contained within the parent hyphal wall of *Malbranchea sulfurea*. . Adjacent dark segments represent partially autolyzed cells. The wall of the latter will eventually break down, allowing the endogenous conidia to be released. ×9000.

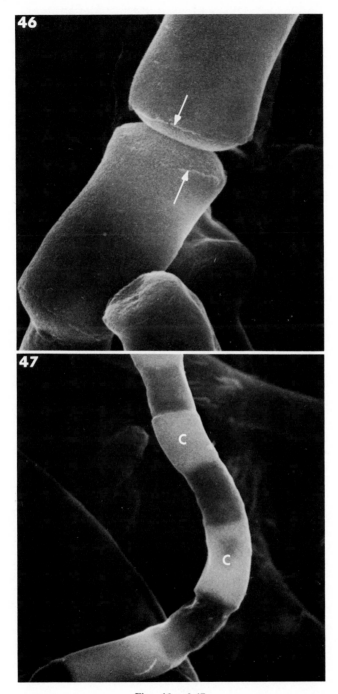

Figs. 46 and 47.

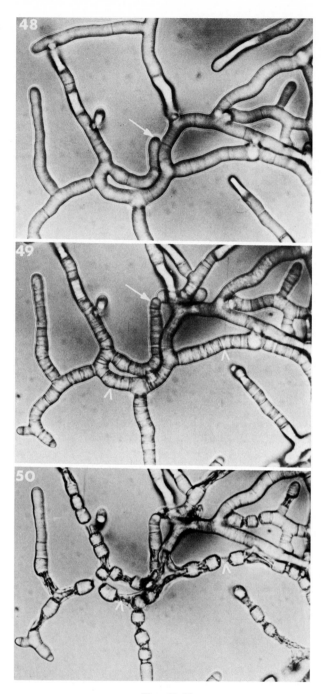

Figs. 48–50.

48 and 49), has ceased in Fig. 50, and multiple septation has occurred along the hyphal branches. Some cells appear optically translucent, as if they have lost some of their cytoplasm (arrowheads in Fig. 49). A comparable developmental stage is shown by the scanning electron micrograph of *Malbranchea sulfurea* (Miehe) Sigler & Carmichael in Fig. 47. Here the light cells are electron-dense and the adjacent dark regions are electron-translucent. The final stage of the time-lapse sequence reveals the dénouement (Fig. 50). The cells indicated by arrowheads in Fig. 49 have collapsed (arrowheads in Fig. 50). Thin sections and freeze fractures of this stage have shown that extensive autolysis occurs in the cells destined to collapse (Cole, 1975). The adjacent intact cells are conidia, encompassed by the fertile hyphal wall. The walls of the collapsed cells eventually rupture, allowing the endogenous conidia to escape (another example of rhexolytic secession). Another mode of enteroarthric development, demonstrated by *Coremiella cubispora* (Berk. & Curt.) Ellis, involves the endogenous formation of conidia without autolysis of intervening cells (Cole and Samson, 1979). Enteroarthric conidial development represents a promising model system for ultrastructural, cytological, and wall biochemical investigations. For example, ultrastructural changes associated with autolysis and conidiogenesis in adjacent cells have not been fully explored. Nuclear division, migration, and compartmentalization during early septation in both holoarthric and enteroarthric forms require further study (Cole, 1975). Biochemical analyses of the changes in cytoplasmic (e.g., phosphatase activity) and wall components during various stages of enteroarthric development may provide some insight into the mechanisms by which conidiogenesis and autolysis are coordinated.

IV. CONIDIOMATAL DEVELOPMENT IN COELOMYCETES AND HYPHOMYCETES

The conidiomata of Coelomycetes "are specialized pseudoparenchymatic structures within which, or on which, conidiogenous cells are produced. They may be separated into three recognizable categories—pycnidia, acervuli, and stromata" (Sutton, 1973). Many morphological variants occur within these categories. In spite of the significance of conidiomatal morphology in coelomycete systematics, no extensive investigation of the developmental and ultrastructural differences between various forms of pycnidia and acervuli have been

---

Figs. 48–50. Time-lapse sequence of enteroarthric conidium formation in *Sporendonema purpurascens*. Hyphal tip growth indicated by arrows in Figs. 48 and 49 has ceased in Fig. 50, and septation of the fertile hyphae has occurred. Progressive autolysis of certain cells (arrowheads in Figs. 49 and 50) and differentiation of endogenous conidia are shown. T = 0, 5, 8, and 12 h, respectively. ×700. (From Institut für Wissenschaftlichen Film, Göttingen; Film C 1302.)

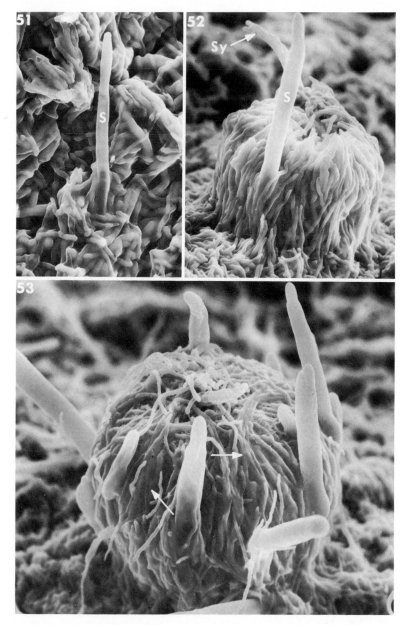

Figs. 51–53. Stages of pycnidial development in *Chaetomella acutiseta* (grown on agar). Note whorl of hyphae at the base of the seta (S) in Fig. 51 and distinct parallel arrangement of filaments adjacent to seta in Fig. 52. Sy, Synnema. Arrows in Fig. 53 indicate mucilaginous material. Fig. 51: ×700; Fig. 52: ×1000; Fig. 53: ×750.

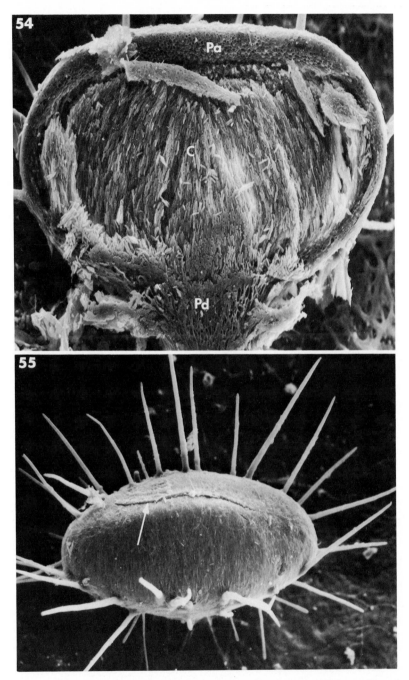

Fig. 54. Cross section of pycnidium of *Chaetomella acutiseta*. Note conidial mass (C) enclosed by pseudoparenchymatous tissue (Pa) which forms peridium and basal pedicel (Pd) of conidioma (on *Quercus texana*). ×214.

Fig. 55. Mature pycnidium of *Chaetomella acutiseta*. Arrow indicates longitudinal slit through which conidia later emerge (on *Quercus texana*). ×151.

performed. Such studies have recently been initiated in my laboratory and some preliminary observations are reported below.

Pycnidial ontogeny in *Chaetomella acutiseta* Sutton & Sarbhoy (Figs. 51–55) has been examined both in pure culture and on a natural substrate (DiCosmo and Cole, 1979, 1980). On potato dextrose agar, initiation of pycnidial development is signaled by the appearance of an erect, thick-walled hypha surrounded by many narrow, thin-walled hyphae at its base (Fig. 51). This erect hypha is destined to become one of the setae of the pycnidium. Additional hyphae aggregate at the base of the erect filament. Although the latter may play a role (thigmotactic? chemotactic?) early in this aggregation process, the seta is soon pushed toward the periphery of the developing pycnidium, and narrow, thin-walled hyphae rise up as a mass of parallel filaments from the agar surface (Fig. 52). Several additional setae later arise from the base of the young pycnidium and penetrate through the edge of the hyphal mass (Fig. 53). The surface hyphae secrete mucilaginous material which binds adjacent filaments together (arrows in Fig. 53). The hyphae at the apex of the young pycnidium interdigitate, forming the roof of the fruiting body. At this stage, a cavity begins to form within which conidiophores bearing phialides and phialoconidia are differentiated. Cavity formation does not appear to involve either lysis or rupture or preexisting hyphae in the central region of the pycnidium, as previously suggested (Dodge, 1930; Stolk, 1963). Instead, the inner hyphae become quickly and apparently synchronously differentiated into conidiophores (DiCosmo and Cole, 1980). As the pycnidium enlarges, the hyphal layers differentiate into an outermost prosenchyma, and an inner, thickened, cellular region referred to as pseudoparenchymatous tissue (Fig. 55; Vuillemin, 1912; Ainsworth, 1971). The two tissues form the encompassing peridium of the conidioma. The pedicel at the base of the pycnidium is composed of elongated cells of the pseudoparenchyma. The locule becomes filled with conidia, and a clear demarcation between the peridial wall, fertile cavity, and sterile pedicel is evident (Fig. 54). On a natural substrate, the surface of the darkly pigmented fructification becomes lustrous as the mucilage deposited by the peridial hyphae dries and forms a sheetlike layer (Fig. 55). When the mature pycnidium absorbs moisture, it swells, and a longitudinal slit or raphe (Stolk, 1963; Sutton and Sarbhoy, 1976) appears on the upper surface of the conidioma (arrow in Fig. 55). Through this slit the conidia are later dispersed.

The conidioma of *Chaetomella acutiseta* demonstrates features of a typical pycnidium. Kendrick and Nag Raj (1979) have defined these as follows: ''1) The

---

Figs. 56–58. Early stages of sporodochial development in the hyphomycetous fungus *Myrothecium leucotrichum* (grown on agar). Arrows in Fig. 56 locate proliferating hypha from which numerous hyphal branches arise. The latter are randomly arranged in Fig. 57 but later become orientated parallel to one another in Fig. 58. S, Seta. Fig. 56: ×4750; Fig. 57: ×755; Fig. 58: ×890.

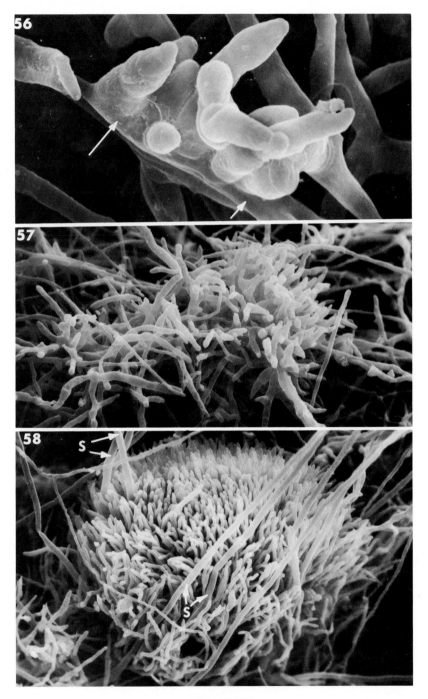

Figs. 56–58.

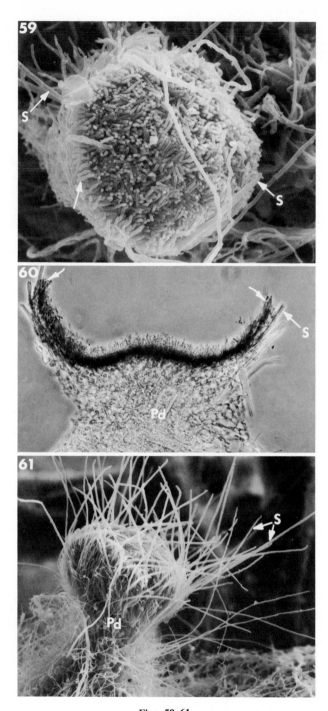

Figs. 59–61.

conidia form more or less completely enclosed by a *fungal* integument. 2) Conidiogenous cells line more or less the *entire* cavity. 3) There is a fairly well-defined and usually restricted ostiole or mouth through which the conidia escape." In contrast, the authors define a typical acervular conidioma as follows: "1) The hymenium develops beneath an integument entirely of *host* origin. 2) Conidiogenous cells are restricted to the *floor* of the cavity. 3) At maturity, there is usually a split of the host integument, and considerable exposure of the relatively flat hymenium. 4) The hymenium layer arises from a more or less well-developed *pseudoparenchymatous* stroma at some level *within* the tissues of the host." This last feature was intended to distinguish the coelomycetous acervular conidioma from the sometimes very similar hyphomycetous sporodochial conidioma which in the authors' opinion "lacks a pseudoparenchymatous basal stroma." However, our ultrastructural studies of conidiomata have instead indicated that certain acervuloid and sporodochial forms are ontogenetically similar. This is illustrated by the hyphomycetous and coelomycetous fungi in Figs. 56–64.

Myrothecium leucotrichum (Pk.) Tulloch, which has been classified as a sporodochial hyphomycete (Rao, 1963; Barron, 1968; M. B. Ellis, 1971b; Tulloch, 1972), produces a fructification in which the conidial mass is supported by a cushionlike arrangement of conidiophores (Ainsworth, 1971). On the basis of comparative developmental and ultrastructural investigation, similarities are demonstrated between the sporodochium of *M. leucotrichum* (Figs. 56–61) and the cupulate conidiomata of the coelomycetous fungi *Stauronema* (Sacc.) Syd (Figs. 62–63) and *Dinemasporium graminum* Lev. The isolate of *M. leucotrichum* was grown on heat-sterilized rabbit dung. It is suggested that formation of the conidioma is initiated from a swollen hyphal segment which proliferates, giving rise to new hyphal branches (Fig. 56). The latter elongate and branch, forming an aggregation of vegetative filaments (Fig. 57). The new hyphal branches which arise from the upper surface of this complex become orientated parallel to one another (Fig. 58). Septate hyphae (setae), which become apically thick-walled, are produced at the perimeter of the conidioma initial and extend above the pulvinate mass of narrow and thin-walled filaments (S in Figs. 58–61). The vegetative hyphae at the base of the young conidioma elongate, forming a pedicel (P in Figs. 60 and 61). The hyphae in the central region of the cushion differentiate into conidiophores bearing phialides which produce chains of cylindrical phialoconidia (Fig. 59). Of significance is that one or more layers of hyphae at the margin of the conidioma grow inward, forming an excipulum

---

Figs. 59–61. Cup-shaped sporodochia of *Myrothecium leucotrichum* (Figs. 59 and 60) which contain a mass of cylindrical conidia. Arrows in Figs. 59 and 60 indicate an excipulum composed of sterile hyphae surrounding the fertile mycelium. Mature sporodochium shown in Fig. 61. Pd, Pedicel; S, seta. Fig. 59: ×600; Fig. 60: ×480; Fig. 61: ×300.

Figs. 62 and 63.

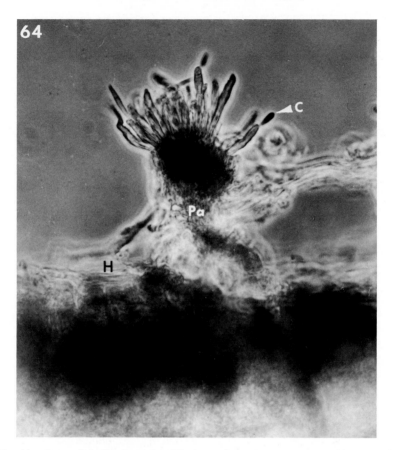

Fig. 64. Cryosection of aggregated conidiophores of *Pseudocercospora* sp. which have arisen from a pseudoparenchymatous stroma immersed in leaf tissue of *Galactia canescens*. C, conidium; H, host tissue; Pa, pseudoparenchymatous tissue. ×1400.

which surrounds the fertile mycelium (arrows in Figs. 59 and 60; Figs. 61–63). Also, the hyphae which form the basal support for the conidioma differentiate into pseudoparenchymatous tissue (Fig. 60). As conidiogenesis proceeds, chains of phialoconidia adhere, producing a thick column of cells.

The cupulate, acervuloid conidioma of *Stauronema* sp. is shown on one of its natural substrates [carpet grass, *Stenotaphrum secundatum* (Walt) O. Ktze.] in Fig. 62. Although initiation of this fruiting body occurs within the host tissue, most of the differentiation of the conidioma occurs on the surface of the epider-

Figs. 62 and 63. Cupulate conidioma (Fig. 62) and appendaged, rough-walled conidia (Fig. 64) of *Stauronema* sp. Fig. 62: ×231; Fig. 63: ×7700.

mis. The conidiogenous cells (phialides) line the inner surfaces of the cupulate fruiting body and produce rough-walled, appendaged conidia (Fig. 63). The setae form an excipulum surrounding and partially covering the fertile mycelium. In a moist chamber, the setae fold back, exposing most of the hymenium. The morphological similarities between the conidiomata of *M. leucotrichum* and of *Stauronema* sp. are evident from a comparison of Figs. 59–62.

In considering ontogenetic and morphological categories of conidiomata, aggregation of conidiophores arising from a stromatal complex composed of pseudoparenchymatous tissue seems to represent the simplest form of fructification in the continuum between Hyphomycetes and Coelomycetes. The aggregated conidiophores of the hyphomycetous fungus *Pseudocercospora* Speg. arise from a stroma immersed in host tissue (Fig. 64). In *Cercospora* Fres., a related hyphomycete, the conidiophores usually emerge through stomata, and "hyphal knots" or "pseudostromata" often form within the substomatal cavities (M. B. Ellis, 1971b). In other related hyphomycetous genera, a basal stroma composed of well-differentiated pseudoparenchymatous tissue is formed (e.g., *Hadrotrichum* Fuckel, *Fusicladium* Bon., *Verrucispora* Shaw & Alcorn, *Cercosporidium* Earle). Present efforts in our studies are directed toward ontogenetic investigations of these and other kinds of conidiomata which cannot be clearly defined either as pycnidia or acervuli.

Sutton (1973) pointed out that "no system has yet been advanced that will satisfactorily accommodate or distinguish the variability and diversity in form shown by coelomycete fructifications." Aspects of pycnidial development, including the formation of a primordium, evolution of a cavity (lysigenous, schizogenous, or both), and the mechanisms of dehiscence have not been clearly illustrated and defined. Concepts of the stroma, pseudostroma, and loculate stroma are vague and require elucidation. Ontogenetic similarities which exist between acervuli and sporodochia produced in culture and on a natural substrate illustrate that certain anamorphs, previously identified as hyphomycetous, may in fact represent forms intermediate between the two major anamorph-classes of Fungi Imperfecti. It is evident that substantial contributions can be made to the taxonomy of Coelomycetes by studying both conidiomatal development and conidiogenesis.

REFERENCES

Ainsworth, G. C. (1971). "Ainsworth and Bisby's Dictionary of the Fungi," 6th ed. Commonw. Mycol. Inst., Kew, Surrey, England.

Barron, G. L. (1968). "The Genera of Hyphomycetes from Soil." Williams & Wilkins, Baltimore, Maryland.

Bartnicki-Garcia, S. (1973). Fundamental aspects of hyphal morphogenesis. *Symp. Soc. Gen. Microbiol.* **23**, 245–267.

Bartnicki-Garcia, S., Bracker, C. E., and Ruiz-Herrera, J. (1977). Reassembly of solubilized chitin synthetase into functional structures resembling chitosomes. *Proc. Int. Mycol. Congr., 2nd, 1977* Abstract, p. 41.

Bartnicki-Garcia, S., Ruiz-Herrera, J., and Bracker, C. E. (1979). Chitosomes and chitin synthesis. *In* "Fungal Walls and Hyphal Growth" (J. H. Burnett and A. P. J. Trinci, eds.), pp. 149–168. Cambridge Univ. Press, London.

Beever, R. E., Redgwell, R. J., and Dempsey, G. P. (1979). Purfication and chemical characterization of the rodlet layer of *Neurospora crassa* conidia. *J. Bacteriol.* **140,** 1063–1070.

Bracker, C. E. (1968). The ultrastructure and development of sporangia in *Gilbertella persicaria*. *Mycologia* **60,** 1016–1067.

Bracker, C. E. (1977). Structure and transformation of chitosomes during chitin microfibril synthesis. *Proc. Int. Mycol. Congr., 2nd, 1977* Abstract, pp. 64.

Bracker, C. E., Ruiz-Herrera, J., and Bartnicki-Garcia, S. (1976). Structure and transformation of chitin synthetase particles (chitosomes) during microfibril synthesis *in vitro*. *Proc. Natl. Acad. Sci. U.S.A.* **73,** 4570–4574.

Brotzman, H. G., Calvert, O. H., Brown, M. F., and White, J. A. (1975). Holoblastic conidiogenesis in *Helminthosporium maydis*. *Can. J. Bot.* **53,** 813–817.

Buller, A. H. R. (1933). "Researches on Fungi," Vol. V. Univ. of Tononto Press, Toronto, Ontario.

Cabib, E. (1975). Molecular aspects of yeast morphogenesis. *Annu. Rev. Microbiol.* **29,** 191–214.

Cabib, E. (1976). The yeast primary septum: A journey into three-dimensional biochemistry. *Trends Biochem. Sci.* **1,** 275–277.

Cabib, E., and Bowers, B. (1975). Timing and function of chitin synthesis in yeast. *J. Bacteriol.* **124,** 1586–1593.

Cabib, E., and Ulane, R. E. (1973). Chitin synthethase activating factor from yeast, a protease. *Biochem. Biophys. Res. Commun.* **50,** 186–191.

Cabib, E., Farkaš, V., Ulane, R. E., and Bowers, B. (1973). Yeast septum formation as a model system for morphogenesis. *In* "Yeast, Mould and Plant Protoplasts" (J. R. Villanueva, I. Garcia-Acha, S. Gascon, and F. Uruburu, eds.), pp. 105–116. Academic Press, New York.

Campbell, R. (1968). An electron microscope study of spore structure and development in *Alternaria brassicicola*. *J. Gen. Microbiol.* **54,** 381–392.

Campbell, R. (1969). Further electron microscope studies of the conidium of *Alternaria brassicicola*. *Arch. Mikrobiol.* **69,** 60–68.

Carmichael, J. W., Kendrick, I. L., Conners, I. L., and Sigler, L. (1980) "The Genera of Hyphomycetes." Univ. of Alberta Press, Edmonton.

Carroll, F. E. (1972). A fine-structural study of conidium initiation in *Stemphylium botryosum* Wallroth. *J. Cell Sci.* **11,** 33–47.

Carroll, F. E., and Carroll, G. C. (1974). The fine structure of conidium initiation in *Ulocladium atrum*. *Can. J. Bot.* **52,** 443–446.

Carroll, G. C. (1966). A study of ascosporogenesis in *Saccobolum kerverni* and *Ascodesmis sphaerospora*. Ph.D. Thesis, University of Texas, Austin.

Cole, G. T. (1971). The sympodula and the sympodioconidium. *In* "Taxonomy of Fungi Imperfecti" (B. Kendrick, ed.), pp. 141–159. Univ. of Toronto Press, Toronto.

Cole, G. T. (1973a). Ultrastructure of conidiogenesis in *Drechslera sorokiniana*. *Can. J. Bot.* **51,** 629–638.

Cole, G. T. (1973b). Ultrastructural aspects of conidiogenesis in *Gonatobotryum apiculatum*. *Can. J. Bot.* **51,** 1677–1684.

Cole, G. T. (1973c). A correlation between rodlet orientation and conidiogenesis in Hyphomycetes. *Can. J. Bot.* **51,** 2413–2422.

Cole, G. T. (1975). The thallic mode of conidiogenesis in the Fungi Imperfecti. *Can. J. Bot.* **53,** 2988–3001.

Cole, G. T. (1978). Conidiogenesis in the black yeasts. *Sci. Publ., Pan Am. Health Organ.* **356**, 66–78.

Cole, G. T., and Aldrich, H. C. (1971). Ultrastructure of conidiogenesis in *Scopulariopsis brevicaulis. Can. J. Bot.* **49**, 745–755.

Cole, G. T., and Kendrick, W. B. (1968). Conidium ontogeny in Hyphomycetes. The imperfect state of *Monascus ruber* and its meristem arthrospores. *Can. J. Bot.* **46**, 987–992.

Cole, G. T., and Kendrick, W. B. (1969a). Conidium ontogeny in Hyphomycetes: The phialides of *Phialophora, Penicillium,* and *Ceratocystis. Can. J. Bot.* **47**, 779–789.

Cole, G. T., and Kendrick, W. B. (1969b). Conidium ontogeny in Hyphomycetes: The annellophores of *Scopulariopsis brevicaulis. Can. J. Bot.* **47**, 925–929.

Cole, G. T., and Kendrick, W. B. (1971). Conidium ontogeny in Hyphomycetes: Development and morphology of *Cladobotryum. Can. J. Bot.* **49**, 595–599.

Cole, G. T., and Samson, R. A. (1979). "Patterns of Development in Conidial Fungi." Pitman, London.

Cole, G. T., Sekiya, T., Kasai, R., Yokoyama, T., and Nozawa, Y. (1979). Surface ultrastructure and chemical composition of the cell walls of conidial fungi. *Exp. Mycol.* **3**, 132–156.

Cook, B. E. (1972). The distribution and role of cytoplasmic membrane bounded vesicles during the development of botryose solitary blastospores of two fungi. *New Phytol.* **71**, 1135–1141.

Cook, B. E. (1974). The development of conidiophores and conidia in the imperfect fungus *Oedocephalum roseum. New Phytol.* **71**, 115–130.

Corlett, M., Chong, J., and Kokko, E. G. (1976). The ultrastructure of the *Spilocaea* state of *Venturia inaequalis in vivo. Can. J. Microbiol.* **22**, 1144–1152.

Correns, C. (1899). "Untersuchungen uber die Vermehrung der Laubmoose durch Brutorgane und Stecklinge." Fischer, Jena.

DiCosmo, F., and Cole, G. T. (1979). "Pycnidial Development and Ultrastructure in *Chaetomella acutiseta* (Coelomycetes)." Can. Bot. Assoc., Ottawa (abstr.).

DiCosmo, F., and Cole, G. T. (1980). Conidiomatal development in *Chaetomella acutiseta* (Coelomycetes). *Can. J. Bot.* **58**, 1127–1137.

Dodge, B. O. (1930). Development of the asexual fructifications of *Chaetomella raphigera* and *Pezizella lythri. Mycologia* **22**, 169–174.

Duran, A., and Cabib, E. (1978). Solubilization and partial purification of yeast chitin synthase. Confirmation of the zymogenic nature of the enzyme. *J. Biol. Chem.* **253**, 4419–4425.

Ellis, D. H., and Griffiths, D. A. (1975a). The fine structure of conidial development in the genus *Torula.* I. *T. herbarum* (Pers.) Link ex S. F. Gray and *T. herbarum* f. *quaternella* Sacc. *Can. J. Microbiol.* **21**, 1661–1675.

Ellis, D. H., and Griffiths, D. A. (1975b). The fine structure of conidial development in the genus *Torula.* II. *T. caligans* (Batista and Upadhyay) M. B. Ellis and *T. terrestris* Misra. *Can. J. Microbiol.* **21**, 1921–1929.

Ellis, M. B. (1971a). Porospores. *In* "Taxonomy of Fungi Imperfecti" (B. Kendrick, ed.), pp. 71–74. Univ. of Toronto Press, Toronto.

Ellis, M. B. (1971b). "Dematiaceous Hyphomycetes." Commonw. Mycol. Inst., Kew, Surrey.

Farcaš, V. (1979). Biosynthesis of cell walls of fungi. *Microbiol. Rev.* **43**, 117–144.

Fevre, M. (1979). Intracellular and cell wall associated $(1\rightarrow3)$ β-glucanases of *Saprolegnia. Mycopathologia* **67**, 89–94.

Galgóczy, J. (1975). Dermatophytes: Conidium ontogeny and classification. *Acta Microbiol. Acad. Sci.* **22**, 105–136.

Girbardt, M. (1957). Der Spitzenkörper von *Polystictus versicolor* (L.). *Planta* **50**, 47–59.

Girbardt, M. (1969). Die Ultrastraktur der Apikalregion von Pilzhyphen. *Protoplasma* **67**, 413–441.

Gooday, G. W. (1978). The enzymology of hyphal growth. *In* "The Filamentous Fungi" (J. E. Smith and D. R. Berry, eds.), pp. 51–72. Wiley, New York.

Grove, S. N. (1978). The cytology of hyphal tip growth. *In* "The Filamentous Fungi" (J. E. Smith and D. R. Berry, eds.), pp. 28–50. Wiley, New York.

Grove, S. N., and Bracker, C. E. (1970). Protoplasmic organization of hyphal tips amongst fungi: Vesicles and Spitzenkörper. *J. Bacteriol.* **104,** 989–1009.

Grove, S. N., Bracker, C. E., and Morré, D. J. (1968). Cytomembrane differentiation in the endoplasmic reticulum-Golgi apparatus-vesicle complex. *Science* **161,** 171–173.

Grove, S. N., Bracker, C. E., and Morré, D. J. (1970). An ultrastructural basis for hyphal tip growth in *Pythium ultimum. Am. J. Bot.* **57,** 245–266.

Gull, K. (1978). Form and function of septa in filamentous fungi. *In* "The Filamentous Fungi" (J. E. Smith and D. R. Berry, eds.), pp. 78–93. Wiley, New York.

Hammill, T. M. (1973). Fine structure of annellophores. IV. *Spilocaea pomi. Trans. Br. Mycol. Soc.* **60,** 65–68.

Hammill, T. M. (1974). Electron microscopy of phialides and conidiogenesis in *Trichoderma saturnisporum. Am. J. Bot.* **61,** 15–24.

Hammill, T. M. (1977a). Transmission electron microscopy of annellides and conidiogenesis in the synnematal hyphomycete *Trichurus spiralis. Can. J. Bot.* **55,** 233–244.

Hammill, T. M. (1977b). Karyology during conidiogenesis in *Gliomastix mororum:* Light microscopy. *Am. J. Bot.* **64,** 1140–1151.

Hanlin, R. T. (1976). Philaide and conidium development in *Aspergillus clavatus. Am. J. Bot.* **63,** 144–155.

Happ, G. M., Happ, C. M., and Barras, S. J. (1976). Bark beetle fungal symbiosis. II. Fine structure of a basidiomycetous ectosymbiont of the southern pine beetle. *Can. J. Bot.* **54,** 1049–1062.

Hashimoto, T., Wu-Yuan, C. D., and Blumenthal, J. J. (1976). Isolation and characterization of the rodlet layer of *Trichophyton mentagrophytes* microconidial wall. *J. Bacteriol.* **127,** 1543–1549.

Hashmi, M. H., Morgan-Jones, G., and Kendrick, W. B. (1973). Conidium ontogeny in Hyphomycetes. The blastoconidia of *Cladosporium herbarum* and *Torula herbarum. Can. J. Bot.* **51,** 1089–1091.

Heath, I. B., and Greenwood, A. D. (1970). The structure and formation of lomasomes. *J. Gen. Microbiol.* **62,** 129–137.

Herth, W. (1979). The site of β-chitin fibril formation in centric diatoms. II. The chitin-forming cytoplasmic structures. *J. Ultrastruct. Res.* **68,** 16–27.

Hill, T. W., and Mullins, J. T. (1979). Hyphal tip growth in *Achlya:* Enzyme activities in mycelium and medium. *Can. J. Bot.* **57,** 2145–2149.

Hughes, S. J. (1951). Studies on microfungi. XI. Some hyphomycetes which produce phialides. *Mycol. Pap.* **45,** 1–36.

Hughes, S. J. (1953). Conidiophores, conidia, and classification. *Can. J. Bot.* **31,** 577–659.

Hughes, S. J. (1971a). Percurrent proliferations in fungi, algae, and mosses. *Can. J. Bot.* **49,** 215–231.

Hughes, S. J. (1971b). On conidia of fungi and gemmae of algae, bryophytes, and pteridophytes. *Can. J. Bot.* **49,** 1319–1339.

Hunsley, D., and Burnett, J. H. (1968). Dimensions of microfibrillar elements in fungal walls. *Nature (London)* **218,** 462–463.

Kendrick, W. B., ed. (1971). "Taxonomy of Fungi Imperfecti." Univ. of Toronto Press, Toronto.

Kendrick, W. B., and Carmichael, J. W. (1973). Hyphomycetes. *In* "The Fungi" (G. C. Ainsworth, A. S. Sussman, and F. K. Sparrow, eds.), Vol. 4, pp. 212–509. Academic Press, New York.

Kendrick, W. B., and Chang, M. G. (1971). Karyology of conidiogenesis in some Hyphomycetes. *In* "Taxonomy of Fungi Imperfecti" (W. B. Kendrick, ed.), pp. 279–291. Univ. of Toronto Press, Toronto.

Kendrick, W. B., and Cole, G. T. (1969). Conidium ontogeny in Hyphomycetes, *Trichothecium roseum* and its meristem arthrospores. *Can. J. Bot.* **47**, 345–350.

Kendrick, W. B., and Nag Raj, T. R. (1979). Morphological terms in Fungi Imperfecti. *In* "The Whole Fungus" (W. B. Kendrick, ed.), Vol. I, pp. 43–61. Natl. Museums of Canada, Ottawa.

Kendrick, W. B., Cole, G. T., and Bhatt, G. C. (1968). Conidium ontogeny in Hyphomycetes. *Gonatobotryum apiculatum* and its botryose blastospores. *Can. J. Bot.* **46**, 591–596.

Lin, L. P., Lee, Y. Y., Fong, J. C., and Hsich, S. I. (1975). Ultrastructure of membrane complex systems in *Agaricus bisporus*. *In* "Fine structure of Fungi" (H. Takeo, ed.), Vol. 2, pp. 136–140. Science Council of Japan, Tokyo.

Lopez-Romero, E., and Ruiz-Herrera, J. (1976). Synthesis of chitin by particulate preparations from *Aspergillus flavus*. *Antonie van Leeuwenhoek* **42**, 261–276.

McMurrough, I., and Bartnicki-Garcia, S. (1971). Properties of a particulate chitin synthetase from *Mucor rouxii*. *J. Biol. Chem.* **246**, 4008–4016.

Marchant, R., and Moore, R. T. (1973). Lomasomes and plasmalemmasomes in fungi. *Protoplasma* **76**, 235–247.

Marchant, R., and Robards, A. W. (1968). Membrane systems associated with the plasmalemma of plant cells. *Ann. Bot. (London)* [N.S.] **32**, 457–471.

Najim, L., and Turian, G. (1979a). Ultrastructure de l'hyphe végétatif de *Sclerotinia fructigena*. *Can. J. Bot.* **57**, 1299–1313.

Najim, L., and Turian, G. (1979b). Conidiogenous loss of structure-functional polarity in the hyphal tips of *Sclerotinia fructigena*. *Eur. J. Cell Biol.* **20**, 24–27.

Rao, P. R. (1963). A new species of *Myrothecium* from soil. *Antonie van Leeuwenhoek* **29**, 180–182.

Reisinger, O. (1972). Contribution à l'étude ultarstructurale de l'appareil sporifère chez quelques Hyphomycètes à paroi melanisée. Thèse Sciences Naturelles, Nancy, France.

Ruiz-Herrera, J., Sing, V. O., van der Voude, V. J., and Bartnicki-Garcia, S. (1975). Microfibril assembly by granules of chitin synthetase. *Proc. Natl. Acad. Sci. U.S.A.* **72**, 2706–2710.

Saccardo, P. A. (1886). Hyphomyceteae. *Syll. Fung.* **4**, 1–87.

Saunders, P. T., and Trinci, A. P. J. (1979). Determination of tip shape in fungal hyphae. *J. Gen. Microbiol.* **110**, 469–473.

Shearer, C. A., and Motta, J. J. (1973). Ultrastructure and conidiogenesis in *Conioscypha* (Hyphomycetes). *Can. J. Bot.* **51**, 1747–1751.

Stolk, A. (1963). The genus *Chaetomella* Fuckel. *Trans. Br. Mycol. Soc.* **46**, 409–425.

Subramanian, C. V. (1956). Hyphomycetes I and II. *J. Indian Bot. Soc.* **35**, 53–91, 446–494.

Subramanian, C. V. (1962a). The classification of the Hyphomycetes. *Bull. Bot. Surv. India* **4**, 249–259.

Subramian, C. V. (1962b). A classification of the Hyphomycetes. *Curr. Sci.* **31**, 409–411.

Subramanian, C. V. (1965). Spore types in the classification of the Hyphomycetes. *Mycopathol. Mycol. Appl.* **26**, 373–384.

Subramanian, C. V. (1971a). Conidial ontogeny in Fungi Imperfecti with special reference to cell wall relationships. *J. Indian Bot. Soc.* **50A**, 51–59.

Subramanian, C. V. (1971b). The phialide. *In* "Taxonomy of Fungi Imperfecti" (B. Kendrick, ed.), pp. 92–119. Univ. of Toronto Press, Toronto.

Subramanian, C. V. (1972a). Conidial chains, their nature and significance in the taxonomy of Hyphomycetes. *Curr. Sci.* **41**, 43–49.

Subramanian, C. V. (1972b). Conidium ontogeny. *Curr. Sci.* **41**, 619–624.

Subramanian, C. V. (1973). Reflections on the conidium. *J. Indian Bot. Soc.* **52**, 1–16.

Subramanian, C. V., ed. (1978). "Proceedings of the International Symposium on Taxonomy of Fungi." Univ. of Madras, Madras, India.

Sutton, B. C. (1973). Coelomycetes. *In* "The Fungi" (G. C. Ainsworth, A. S. Sussman, and F. K. Sparrow, eds.), Vol. 4A, pp. 513–582. Academic Press, New York.

Sutton, B. C., and Sarbhoy, A. K. (1976). Revision of *Chaetomella* and comments upon *Vermiculariopsis* and *Thyriochaetum*. *Trans. Br. Mycol. Soc.* **66**, 297–303.

Ton That, T., and Turian, G. (1978). Ultrastructure of microcycle macroconidiation in *Neurospora crassa*. *Arch. Microbiol.* **116**, 279–288.

Tubaki, K. (1958). Studies on Japanese Hyphomycetes. V. Leaf and stem group with a discussion of the classification of Hyphomycetes and their perfect stages. *J. Hattori Bot. Lab.* **20**, 142–244.

Tubaki, K. (1963). Taxonomic study of Hyphomycetes. *Annu. Rep. Inst. Ferment., Osaka* **1**, 25–54.

Tulloch, M. (1972). The genus *Myrothecium* Tode ex Fr. *Mycol. Pap.* **130**, 1–42.

Turian, G. (1976). Reducing power of hyphal tips and vegetative apical dominance in fungi. *Experientia* **32**, 989–991.

Turian, G., Oulevey, N., and Cortat, M. (1973). Reserches sur la différenciation conidienne de *Neurospora crassa*. V. Ultrastructure de la séquence macroconidiogène. *Ann. Microbiol. (Inst. Pasteur)* **124A**, 443–458.

Van Emden, J. H., and Veenbaas-Rijks, J. W. (1973). *Cylicognone*, a new genus of Hyphomycetes with unusual conidium ontogeny. *Acta Bot. Neerl.* **22**, 637–640.

Vermeulen, C. A., Raeven, M. B. J. M., and Wessels, J. G. H. (1979). Localization of chitin synthase activity in subcellular fractions of *Schizophyllum commune* protoplasts. *J. Gen. Microbiol.* **114**, 87–97.

Vuillemin, P. (1912). "Les champignons: Essai de classification." Doin, Paris.

Weisburg, S. H., and Turian, G. (1971). Ultrastructure of *Aspergillus nidulans* conidia and conidial lomasomes. *Protoplasma* **72**, 55–67.

Wergin, W. P. (1973). Development of Woronin bodies from microbodies in *Fusarium oxysporum* f.sp. *lycopersici*. *Protoplasma* **76**, 249–260.

Wu-Yuan, C. D., and Hashimoto, T. (1977). Architecture and chemistry of microconidial walls of *Trichophyton mentagrophytes*. *J. Bacteriol.* **129**, 1584–1592.

24

Biochemistry of Microcycle Conidiation

J.E. Smith, J.G. Anderson, S.G. Deans, and D.R. Berry

I. INTRODUCTION

The normal developmental program of conidial fungi begins with germination of the conidium to produce a vegetative filamentous growth form characteristic of the organism. Under balanced nutritional and environmental conditions the growth of the fungus will normally remain vegetative. However, should growth conditions change and become unbalanced by nutritional deficiency and/or by critical changes in essential environmental parameters, the vegetative growth pattern may change, resulting in the formation of asexual reproductive morphology, that is, the conidiophore apparatus. Morphologically, the conidiophore may be a simple cell or a system of conidiogenous cells with or without

Biology of Conidial Fungi, Vol. 2

a differentiated supporting structure that orients the developing conidia away from the parent mycelium and the medium (Talbot, 1971; Kendrick, 1971).

At the molecular level conidiation results from the integration of many cellular functions, each having a significant and essential part in the whole process (Smith, 1978). Furthermore, conidiation can only occur as the end result of inherent genetic competence responding to specific chemical and physical environmental messages to the fungus. These in turn promote changes at the cellular level, creating new metabolic patterns (Turian, 1974). Fungal sporulation in general has been extensively reviewed in recent years (Smith and Galbraith, 1971; Smith and Anderson, 1973; Lovett, 1975; Turian, 1974, 1975, 1976; Smith, 1978).

The use of simple and complex fermenter systems in the study of conidiation in fungi has introduced new dimensions to the comprehension of this complex form of differentiation (Smith, 1978). It now is possible to examine conidiation under quite exacting controlled experimental conditions, thus permitting some understanding of how individual environmental parameters such as aeration, temperature, pH, nutrient type, and concentration affect the overall process. In this way most environmental parameters can be considered in a manner not easily available or even possible in natural or surface-growing cultures. However, no one experimental system is perfect and, although the submerged cultivation involved in fermenter systems offers much greater environmental control of the growing organism, it is of course far removed from surface growing conditions (Bull and Bushell, 1976; Smith et al., 1977b; Smith, 1978). It must further be stressed that submerged mycelial cultures will normally be heterogeneous in morphological form (though less so than surface mat cultures), and subsequent biochemical analyses will merely represent a summation of many physiological states. These difficulties may be overcome in part by introducing some degree of synchronous control over morphological development. In this manner synchronous sporulation has been achieved with Aspergillus niger (Anderson and Smith, 1971a,b), Achlya sp. (Griffin and Breuker, 1969), and Penicillium digitatum (Zeidler and Margalith, 1972) by either nutrient changes or cultural manipulation. The medium replacement technique of Anderson and Smith (1971b) permitted a sharp separation between the various developmental phases of conidiation in A. niger and allowed considerable insight into the biochemical changes accompanying the developmental events (Smith and Ng, 1972; Ng et al., 1972; Lloyd et al., 1972; Ng et al., 1973a).

Under batch fermentation conditions nutrient uptake and waste product release occur in a closed system, making it difficult to deduce whether the associated developmental changes result from nutrient limitation or from the limitation of growth rate imposed by the closed environmental conditions. However, by means of chemostat culture it has been possible to grow several fungi at various growth rates and in various metabolic steady states to achieve a fuller understanding of the factors regulating sporulation in filamentous fungi (Righelato *et al.,* 1968; Ng *et al.,* 1973b; Robinson and Smith, 1976; Larmour and Marchant, 1977). Emerging from all these studies is the overriding importance of the growth rate of the fungus in determining the pattern of morphological development.

When the growth of a fungus is being regulated by the inherent genomic characters of the organism rather than by the environment, i.e., when the rate of nutrient uptake into the mycelium and rate of utilization are constant, then the organism can be considered to be showing balanced growth (Bu'Lock, 1975). Growth will be considered unbalanced when the growth rate of the organism is limited by the reduced availability of some essential nutrient or by the influence of other critically relevant environmental factors—temperature, oxygen or carbon dioxide concentration, pH—which may influence metabolism. With filamentous fungi, balanced or autocatalytic growth (Bu'Lock, 1975) will normally produce a continued vegetative filamentous morphology. When limitations result in an unbalanced system, major metabolic alterations (alternative metabolic pathways, excretion of intermediary metabolites, synthesis of storage polymers, and morphological differentiation) occur (Bu'Lock, 1975; Smith, 1978).

Unbalanced growth does not invariably result in conidiation in filamentous fungi. In particular, carbon limitation can often cause autolysis. Thus, although movement away from balanced growth is a prerequisite for fungal conidiation, the manner in which the unbalanced state is achieved is of critical importance. A fuller treatment of this topic is given by Smith (1978). However, in summary, it can be stated that for conidiation to occur in filamentous fungi there must be a check in the normal unlimited pattern of vegetative growth. Such a change can be promoted by limitation of growth rate, and any environmental factor which can influence growth rate may also influence conidiation. The central role of growth rate in determining the nature and extent of conidiation in filamentous fungi cannot be overstated.

This chapter will be concerned primarily with microcycle conidiation in filamentous fungi and in particular with the biochemical and ultrastructural aspects of this novel form of conidiation. Microcycle conidiation has been considered the recapitulation of conidiation following conidial germination without an intervening phase of mycelial growth (Smith *et al.,* 1977a).

II. THE MICROCYCLE AS A MODEL SYSTEM FOR SPORULATION STUDIES

The keen interest in the development of microcycle sporulation systems of filamentous fungi has arisen mainly in laboratories with a long-standing interest in the biochemical analysis of fungal sporulation. One of the major quests of the investigators has been the search for cultivation systems which improve synchrony of development and which can provide fungal material on which meaningful biochemical studies can be carried out.

Surface cultivation techniques have yielded much useful information on the influence of the environment on the sporulation process, together with reliable genetic interpretations. Yet characteristic mycelial mat formation typical of surface cultures precludes any meaningful study of the biochemistry associated with sporulation. Submerged cultivation methods, particularly those involving modern fermentation equipment (Bull and Bushell, 1976), can produce more homogeneous mycelia, although special inducing conditions are usually required to overcome the inhibition of sporulation which characterizes submerged culture development. Even if good induction can be obtained, vegetative growth tends to mask other developmental stages, particularly when sporulation synchrony is poor. As pointed out above, some synchronous control of filamentous fungi (e.g., *Aspergillus niger, Achlya,* and *Penicillium digitatum*) has been achieved using nutrient changes and cultural manipulations. Even these sophisticated techniques still suffer from the major limitation that conidiation is preceded by a period of vegetative filamentous growth. Thus the material harvested for biochemical studies will inevitably be heterogeneous. The fact that vegetative cells greatly outnumber conidiogenous cells severely complicates biochemical interpretations.

The novel microcycle sporulation systems that have been developed undoubtedly offer many advantages over conventional mycelial sporulation, since they tend to eliminate the normal phase of vegetative growth and produce a cultural biomass containing a high proportion of sporulating structures. Another major advantage of the microcycle approach, particularly where the induction system involves thermic variation, is that it imparts much greater synchrony of conidial development than has hitherto been achieved in submerged culture.

For bacteria, microcycle sporulation has been defined as the immediate recapitulation of sporogenesis following spore germination (Vinter and Slepecky, 1965). For filamentous fungi, microcycle conidiation is the direct production of conidiophores and conidia by germinating conidia (Anderson and Smith, 1971a). Thus microcycle conidiation results in the production of a morphologically distinct unit comprising the parent conidium connected to the spore-bearing apparatus (which may have arisen from an abbreviated germ tube) and the second-generation conidia.

Temperature manipulation in appropriate culture media has been found to be an extremely useful method of inducing microcycle conidiation. It is important to realize, however, that microcycle sporulation processes have been described for a range of fungi and other microorganisms in response to environmental conditions other than temperature variation. A brief description of these examples will be given before returning to temperature-induced microcycle conidiation.

Induction of Microcycle Sporulation

1. Induction by Cultural Methods not Involving Temperature Variation

Various reports have appeared in the literature indicating that sporulation can take place immediately following spore germination in such a way that there is an apparent absence or extreme limitation of the vegetative phase. Such phenomena have been variously defined; however, since they fulfill the criteria of the definition given above they have been included here as examples of microcycle sporogenesis.

A variety of cultural conditions have been involved in the induction of microcycle sporulation, but in most instances it appears to arise as a consequence of inhibition of hyphal development. Thus Rotem and Bashi (1969) concluded that the direct formation of secondary conidia in *Alternaria* spp. and *Stemphylium botryosum* was induced by various factors which inhibited the vegetative development of the mother conidium. Similarly, the various conditions which led to conidiation of sporelings of *Geotrichum candidum* were considered by Park and Robinson (1969) to interfere with the metabolic condition characteristic of the somatic phase.

One of the inducing conditions found for *Geotrichum candidum* occurred when germination took place on cellophane over mature cultures of the same or other fungi. There are several other examples which suggest that the inhibitory conditions created by crowding can result in microcycle sporogenesis. In some cases the spores crowd themselves; in others the spores attempt to germinate in the presence of an already established mycelium. Thus spores borne directly on the germ tubes were obtained in *Melanconium fuligineum* by overcrowding of germinating spores (Timmick *et al.*, 1952). Autoinhibitors associated with the conidia were considered by Lingappa and Lingappa (1969) to be responsible for the inhibition of mycelial growth and the preferential development of secondary conidia in *Glomerella cingulata*. An effect of germination inhibitors is also indicated in an observation by Hadley and Harold (1958) in which sporulation from the germ tubes of *Penicillium notatum* conidia occurred when these "germinated" in the medium in which they were produced. More recently, Zeidler and Margalith (1973) have reported microcycle conidiation in *Pencillium digitatum* when conidia were germinated in media containing certain amino acids (glutamate and serine) as the sole nitrogen source. In this study microcycle conidiation was observed to occur from conidia that did not germinate during the early stages of cultivation.

Besides accumulating inhibitory substances, crowded conditions can lead to nutritional deficiency. It is possible that this may have been a factor in some of the above examples. Certainly induction of microcycle conidiation by nutrient limitation has been reported with some fungal spores. So-called iterative germination, in which conidia directly form fertile conidiophores, has been described in *Helminthosporium spiciferum* when the conidia are germinated in water in the absence of an external food supply (Mangenot and Reisinger, 1976). A similar germination pattern was described earlier, under less defined conditions, for *H. sativum* by Boosalis (1962), who termed the phenomenon "precocious sporulation."

An indication that a potential for microcycle sporogenesis exists among diverse types of fungal spores is given by the observation of "lag-phase sporogenesis" in *Blastocladiella emersonii* (Hennessy and Cantino, 1972). Direct sporulation from zoospores of this fungus occurred under starvation conditions in dilute phosphate buffer.

It is interesting also that the phenomenon is not restricted to fungal spores. Caravajal (1947) reported the occurrence of secondary sporulation in a number of *Streptomyces* species when the spores germinated in the medium in which they were produced. Under certain germination conditions Vinter and Slepecky (1965) have described the direct transition of outgrowing bacterial spores to new sporangia without intermediate cell division.

2. Induction by Combined Temperature–Nutrient Manipulation

Combined temperature and nutrient manipulation was first used for the induction of microcycle conidiation in *Aspergillus niger* (Anderson and Smith, 1971a, 1972). The system developed, using a defined sporulation medium and submerged agitated conditions, involved incubation of conidia at a critical supraoptimal temperature (44°C), which selectively allowed conidial enlargement (spherical growth) to occur but inhibited germ tube formation. Subsequent temperature reduction to 30°C allowed the direct production of one or more fertile conidiophores from the enlarged parent conidium (giant cell) in the continuing absence of vegetative filamentous development.

Variations of this system have now been successfully used to induce microcycle conidiation in *Penicillium urticae* (Sekiguchi *et al.,* 1975a,b,c) and *Paecilomyces varioti* (Anderson *et al.,* 1978), and microcycle macroconidiation (Cortat and Turian, 1974) and microcycle microconidiation (Rossier *et al.,* 1977) in *Neurospora crassa.* With the exception of microcycle microconidiation in *N. crassa,* in which plate cultures were used, all these microcycle systems have been developed using submerged agitated conditions. In each case the system comprises two stages. In stage 1, at elevated temperature, spore enlargement and hyphal restriction occur. In stage 2, at the lowered temperature and in the presence of an appropriate nutritional environment, the conidiation phase of the microcycle occurs. The precise temperatures required, the duration of each stage, and the nutritional status of the medium are highly specific for each organism. A summary of the inducing conditions used with each of these species is given in Table I.

The extent to which the temperature must be raised to promote stage 1 (restriction of vegetative development) reflects the temperature tolerance of the species.

TABLE I

Temperature and Nutrient Conditions Used to Induce Microcycle Conidiation in Filamentous Fungi

Species	Reference[a]	Conditions used for stage 1 (conidial enlargement)			Conditions used for stage 2 (conidiophore development)		
		Temperature (°C)	Time (h)	Medium conditions	Temperature (°C)	Time (h)	Medium conditions
Aspergillus niger	(1),(2)	41–44	24–48	Complete growth medium CO_2 promotes stage 1	30	18	Same medium as stage 1 Glutamate promotes conidiation
Penicillium urticae	(3)	37	24	Complete growth medium	35	30	Culture is transferred to nitrogen-poor medium
Neurospora crassa							
Microcycle micro-conidiation	(4)	46	24–48	Nonnutrient (phosphate buffer) medium	25	96	Same medium as stage 1
Microcycle macro-conidiation	(5)	46	15	Complete growth medium	25	12	Same medium as stage 1
Paecilomyces varioti	(6)	48	24	Complete growth medium	40	15	Same medium as stage 1 Glutamate promotes conidiation

[a] References: (1) Anderson and Smith (1971a); (2) Kuboye *et al.* (1976); (3) Sekiguchi *et al.* (1975a); (4) Rossier *et al.* (1977); (5) Cortat and Turian (1974); (6) Anderson *et al.* (1978).

Comparison of the temperature difference between stage 1 and stage 2 used for these different organisms illustrates also the highly specific nature of the inducing conditions. Thus for *Neurospora crassa* the difference in temperature between stage 1 and stage 2 is 21°C, for *Aspergillus niger* 14°C, for *Paecilomyces varioti* 8°C, and for *Penicillium urticae* only 2°C. In general, an approximately similar time is required to achieve microcycle conidiation in each of these systems (between 24 and 48 h after inoculation of conidia) except with microcycle microconidiation in *N. crassa*, which requires 24–48 h for stage 1 and a further 96 h for stage 2.

While temperature manipulation has been a key feature of the microcycle induction process, careful consideration has also been required regarding medium composition. With the exception of the microcycle microconidiation system for *Neurospora crassa*, stage 1 takes place in a medium which would support normal germination and extensive hyphal growth if the temperature were at a permissive level. With *N. crassa*, Rossier *et al.* (1977) could only achieve microcycle microconidiation when both stages of the process occurred in a non-nutrient (phosphate buffer) medium.

With both *Aspergillus niger* (Kuboye *et al.*, 1976) and *Paecilomyces varioti* (Anderson *et al.*, 1978) it has been found that carbon dioxide is also involved with elevated temperature in promoting conidial enlargement and inhibiting germ tube formation. Thus with *A. niger* grown in submerged aerated fermenter culture, complete inhibition of germ tube formation could not be achieved at temperatures below 44°C, whereas in the presence of 5% carbon dioxide in air complete inhibition was obtained at 41°C. Whether or not carbon dioxide (metabolically produced) might play some role in the induction of microcycle conidiation in other species has, as far as we are aware, not been investigated. Carbon dioxide is known to be a powerful inducer of morphogenetic change in filamentous fungi (see Smith and Galbraith, 1971). Its interaction with elevated temperature to induce dimorphic transformations and other spherical cell forms in filamentous fungi has recently been discussed (Anderson and Smith, 1976; Anderson, 1978).

Consideration of the constituents of the media in which stage 2 (conidiation) of the different microcycle systems takes place also raises several points of interest. With *Neurospora crassa*, microcycle microconidiation occurs solely at the expense of endogenous conidial reserves, since both stages occur on nonnutrient medium. Stage 2 of the *Penicillium urticae* microcycle takes place after the temperature shift and when the culture has been transferred from a nutrient-rich medium to a nitrogen-poor medium. With microcycle conidiation in *Aspergillus niger* and *Paecilomyces varioti* and with microcycle macroconidiation in *Neurospora crassa*, nutrient limitation is not imposed during stage 2. These microcycle processes are in some way induced by physiological changes which take place as a result of growth restriction by elevated temperature. In the case of

Aspergillus niger and *Paecilomyces varioti* the incorporation of glutamate into a simple glucose–ammonium medium is also required for optimum expression of microcycle conidiation. Although glutamate is included for convenience in the medium during both stages of the process, this amino acid is not required for spore enlargement and germ tube inhibition during stage 1, but it markedly stimulates conidiation during stage 2. Reference has already been made to the role of glutamate in the induction of microcycle conidiation in *Penicillium digitatum* (Zeidler and Margalith, 1973), although with this organism thermal variation was not required.

A major advantage of the microcycle approach to the study of fungal sporulation is that it usually imparts much greater synchrony of development than is normally achieved in submerged cultivation. This has been reported for all the temperature-induced systems discussed except for microcycle microconidiation in *Neurospora crassa* (Rossier *et al.*, 1977), where lack of synchrony during germination and subsequent development persisted throughout stages 1 and 2. With this exception, temperature manipulation has been found to maintain a separation of the morphological events which comprise each stage. Reasonably synchronous development then proceeds by temperature shift and, as with *Penicillium urticae* (Sekiguchi *et al.*, 1975a), alteration of the medium composition. Synchrony persists because of the severe restriction of vegetative growth resulting from exposure to elevated temperature, and in some cases to the suboptimal nutrient conditions imposed.

Efforts have been made to improve further the degree of synchrony typical of these systems. With microcycle macroconidiation in *Neurospora crassa* synchrony of conidial differentiation in stage 2 has been improved by selective filtration of the heat-treated conidia before the shift down to 25°C (Cortat and Turian, 1974). With *Aspergillus niger* large-scale production of conidia for microcycling is achieved by germinating conidia on cubes of sterile bread (Sansing and Ciegler, 1973). Harvesting the spores after several days' growth yields high numbers of a relatively asynchronous spore population. To further increase the index, a variety of methods have been examined encompassing both induction synchrony and selection synchrony. However, washing the spores repeatedly in sterile distilled water was found to be the most effective method (Deans and Smith, 1978). It has been found that improving the synchrony of giant cell formation greatly facilitates subsequent synchrony in later stages of conidiogenesis. The relationship of this synchronized developmental pattern to time is shown in Fig. 1.

With *Aspergillus niger* (Kuboye *et al.*, 1976; Deans and Smith, 1981a) and *Paecilomyces varioti* (Anderson *et al.*, 1978) it has been possible to move away from the simple flask or tube culture systems used initially to develop the full microcycle system under fermenter conditions. When the microcycle process is carried out within the confines of a single fermentation vessel, not only can more

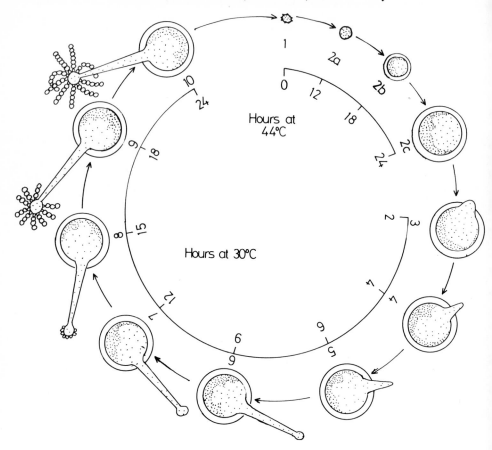

Fig. 1. Developmental pattern of microcycle conidiation in *Aspergillus niger*.

rigorous control on the developmental phases be maintained, but also a sequential sampling from a single culture can be carried out where sufficient biomass is produced for biochemical analysis.

III. MORPHOLOGICAL AND ULTRASTRUCTURAL CHANGES DURING MICROCYCLE CONIDIATION

A. Changes during State 1 (Conidium Enlargement)

Normal germination of most fungal spores comprises an initial phase of spore enlargement followed by a phase of germ tube protrusion and extension. A feature common to most forms of microcycle conidiation is a marked increase in the degree of conidium enlargement during the period of elevated temperature treatment. This abnormal enlargement is most pronounced with conidia of *Aspergillus niger* (Anderson and Smith, 1972). Under normal conditions of germination *A. niger* conidia will undergo on average a 2-fold swelling before producing a germ tube. However, at temperatures from 38° to 43°C the proportion of conidia which produce germ tubes progressively decreases, and at 44°C germ tube formation is completely inhibited. At these temperatures enlargement of the conidia continues, resulting in the formation of large, spherical cells ca. 20–25 μm in diameter. During this process a 30-fold increase in spore surface area occurs. It has been clearly established that this remarkable increase in cell size is a true growth process involving extensive synthesis of macromolecules and a net increase in dry weight (Kuboye *et al.*, 1976; Deans and Smith, 1981a).

It appears that during stage 1 of microcycle conidiation the enlarging conidia are unable to germinate, probably because of the inhibitory effect of elevated temperature on the apical growth process. Growth is not stopped completely but occurs by uniform wall deposition, resulting in spherical cells of variable size and thickness. Only when the cells are removed to a lower temperature can apical outgrowth occur. These observations indicate that the conidium enlargement (spherical growth) phase of germination is less susceptible to high-temperature inhibition than the subsequent germ tube production phase. It has been pointed out that this effect may be involved in the development of the spores of certain pathogenic fungi into enlarged spherical parasitic forms (Anderson and Smith, 1976; Anderson, 1978).

The ultrastructural changes which take place during conidium enlargement at elevated temperatures have been examined in *Penicillium urticae* (Sekiguchi *et al.*, 1975a), *Aspergillus niger* (Smith *et al.*, 1977a), and *Neurospora crassa* (Rossier *et al.*, 1977). Although the conidia are held at temperatures which severely restrict development, no evidence of thermally induced ultrastructural damage has been observed in these cells. With *A. niger* the most dramatic ultrastructural changes observed were a considerable increase in the number of nuclei and other organelles and the production of an extensively thickened, multilayered cell wall (Fig. 2). Also glycogen and lipid droplets not present during early stages of spherical growth began to accumulate after approximately 20 h at 44°C. From the appearance of the cell wall it is clear that isotropic wall deposition continues during spherical growth at elevated temperature. Dictyosome-like structures were observed in the immediate vicinity of the growing wall, and vesicles were seen to bleb off from the dictyosome plates and make their way, by means unknown, to the plasmalemma.

The ultrastructural appearance of the walls of enlarged conidia of both *Penicillium urticae* (Sekiguchi *et al.*, 1975a) and *Neurospora crassa* (Rossier *et al.*, 1977) are consistent with the idea that isotropic wall growth is a characteristic feature of this stage. A particular feature of the heat-treated microconidia of *N.*

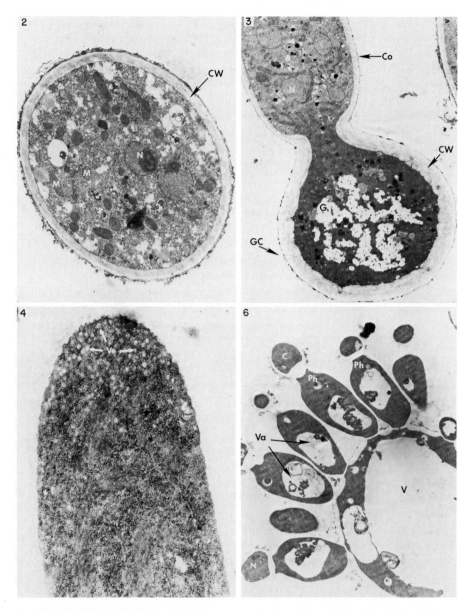

Fig. 2. Giant cell of *Aspergillus niger* produced after 24 h at 44°C. CW, Cell wall; M, mitochondria; N, nucleus.

Fig. 3. Giant cell of *Aspergillus niger* with outgrowing conidiophore. GC, Giant cell; CW, cell wall; G, glycogen; N, nucleus; Co, conidiophore.

crassa, observed by Rossier *et al.* (1977), was the presence of numerous vacuoles with electron-dense inclusions. Progressive vacuolations, observed in the enlarged conidia of *A. niger* and *P. urticae* at later stages in the microcycle process, probably reflect degeneration and cellular aging.

B. Changes during Stage 2 (Conidiophore Development and Spore Production)

With the exception of *Aspergillus niger,* the outgrowths which occur from the enlarged conidia after a temperature shift are morphologically similar to normal germ tubes. With *A. niger* a thick conidiophore stalk is produced (Figs. 3 and 4) which is characteristic of this species, the giant cell apparently performing the function of the *Aspergillus* foot cell (Anderson and Smith, 1972; Smith *et al.,* 1977a).

The germ tubes produced by other species after a temperature decrease exhibit a limited amount of apical extension before conidiogenesis begins. In the case of *Neurospora crassa* microcycle macroconidiation (Rossier *et al.,* 1977) the germ tubes convert to simple conidiophores, since instead of elongating they produce proconidia from the tip by a process of basifugal budding (Fig. 5).

In *N. crassa* microcycle microconidiation (Cortat and Turian, 1974), apical growth of the germ tube stops and is replaced by lateral growth. This gives rise to hooks which finally become phialides bearing microconidia (Fig. 5).

Microcycle conidiogenesis in *Penicillium urticae* (Fig. 5) is accomplished by the formation of a phialide and new conidia from the tips of the germ tubes. Additional phialides are also common as branches near the tip of the germ tube. Although this is a simpler conidiophore than is normally produced in subaerial culture, the ultrastructural changes associated with production of phialoconidia by the microcycle structure are indistinguishable from normal subaerial formation (Sekiguchi *et al.,* 1975c).

Even greater simplification of the conidiogenous apparatus was observed during microcycle conidiation of *Paecilomyces varioti* (Anderson *et al.,* 1978). In this organism short germ tubes were produced which then took on the typical bottle shape of a phialide, and chains of new conidia were produced from the tips of these cells (Fig. 5). Thus in this species microcycle conidiation was expressed as the direct production of phialides from the parent conidium.

Fig. 4.   Apical vesicles at the growing tip of a conidiophore produced from a giant cell of *Aspergillus niger.* Ve, Vesicles.

Fig. 6.   Microcycle conidiation in *Aspergillus niger.* Section through the conidiophore head showing the vesicle, phialides, and new conidia. V, Vesicle; Ph, phialide; C, conidium; N, nucleus; Va, vacuole.

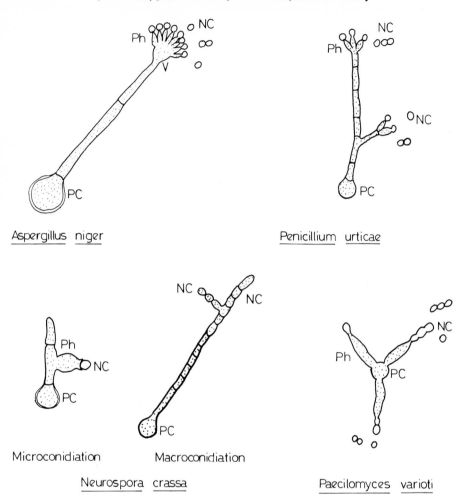

Fig. 5. Appearance of microcycle conidiation in several filamentous fungi induced by temperature and nutrient manipulations (for references see text). The various organisms are not drawn to scale. NC, New conidia; PC, parent conidum (enlarged); Ph, phialide; V, vesicle.

The most complex of the microcycle conidiation structures is that produced by *Aspergillus niger,* where maturation of the conidiophore stalk occurs, forming a vesicle bearing phialides and new conidia. However, in this case the conidiophore also shows some simplification in comparison to the normal subaerial structure. In microcycle conidiation the size of the conidiophore is much reduced and, whereas the normal subaerial conidiophore possesses both metulae and phialides, only phialides are produced on the vesicle of the microcycle structure (Fig. 6). However, the conidia produced by microcycle conidiation are viable and approximately the same size as subaerially produced conidia, although they lack the thick outer cuticle and the black pigmentation characteristic of this species. Pigmentation and surface ornamentations of conidia are features which develop during conidium maturation in an aerial but not in a submerged environment (Mangenot and Reisinger, 1976). The conidia produced by microcycle conidiation of *A. niger* can be separated from the parental culture and used to initiate normal vegetative growth. Alternatively, they can be recycled through the microcycle process, thus making this system an attractive alternative to conventional bulk spore production.

In view of the complexity of the conidiophore in *Aspergillus niger* it seems remarkable that the giant cell frequently produces two and occasionally up to five of these structures during microcycle conidiation (Anderson and Smith, 1972). Even more bizarre is the observation (Deans, 1978) that, when the base of the conidiophore is sheared from the giant cell, it is possible under certain conditions for one or more new conidiophores to grow out from the severed end. This phenomenon occurs by the reversal of growth polarity in that new growth occurs in the direction opposite that which gave rise to the original conidiophore. Conidiation in these double-ended conidiophores is as prolific as that occurring from the normal microcycle structure.

A detailed examination has now been made of the ultrastructural changes associated with the outgrowth and maturation of the conidiophore during microcycle conidiation of *Aspergillus niger* (Deans, 1978). The initiation of outgrowth from the giant cell was marked by an accumulation of apical vesicles at the point of incipient conidiophore emergence. The wall of the new outgrowth, observed to be continuous with the innermost layer of the giant cell, emerged through the outer layers. Apical growth of the conidiophore was established with an organelle distribution pattern in the subapical region, which is typical of tip growth in fungal filaments. Microvesicles were routinely observed at the conidiophore apex, the site of new cellular extension (Fig. 4). Particularly distinct differences in the cellular organization of the giant cell and the emerging conidiophore were the excessively thick walls of the former and the lack of glycogen in the latter (Fig. 3).

Once the conidiophore had developed to the point of autonomy, septum formation effectively sealed off the parent giant cell, and partial autolysis and vacuolation of this cell frequently resulted. In some cases a double septum was found near the giant cell–conidiophore interface. Where more than one conidiophore emerged from the giant cell, septa were again formed to separate these from the parental cell. It is interesting that with both *Penicillium urticae* (Sekiguchi *et al.*, 1975b) and *Neurospora crassa* (Rossier *et al.*, 1977) early septum formation occurred at the junction of the enlarged conidium and the outgrowing germ tube. It is possible that the function of these septa is to separate the developing conidiophore from the degenerating and autolytic contents of the aging parental cells.

IV. BIOCHEMICAL ASPECTS OF MICROCYCLE
CONIDIATION

It is now widely considered that differential gene activation is the principal factor controlling cellular development and will be expressed in quantitative and qualitative changes in the rate of enzyme synthesis. However, when considering enzymes as "markers" of differentiation it is difficult if not impossible to derive from *in vitro* enzyme studies alone (1) that an enzyme does not exist or occur in another form in a different place *in vivo* or (2) the level of enzymes normally present in great excess *in vivo* compared to their substrate (Gustafson and Wright, 1972).

Mechanisms other than the regulation of enzyme synthesis can also profoundly influence developmental patterns. These include stabilization, activation, and modulation of enzyme activity together with allosteric factors; the availability of low-molecular-weight precursors and the flux in the levels of critical intermediates could also be of critical importance in regulating developmental patterns (Killick and Wright, 1974).

The following section will examine some aspects of the biochemical changes occurring during microcycle conidiation of *Aspergillus niger*. For a fuller consideration of the biochemistry of fungal sporulation in general, reference should be made to Lovett (1975) and Smith (1978).

A. The Cell Wall

The fungal cell wall should be considered a dynamic perimeter within which enzymatic activity controls the phenomenon of cellular morphogenesis. Fungal cell walls are chemically and physically very complex, though recent advances in analytic techniques have given meaningful information concerning the macromolecular constituents of the cell wall and their spatial arrangements (for references, see Bartnicki-Garcia, 1973; Rosenberger, 1976; Grove, 1978; Gooday, 1978). However, almost all studies on wall chemistry of filamentous fungi have been limited to vegetative mycelia, and only a very few examples exist where the complete developmental cycle has been analyzed (Zonneveld, 1977).

The polysaccharides involved in wall synthesis do not exist free in nature but involve close interrelationships among enzymes concerned with percursor formation, transport of precursors, and their eventual polymerization into three-dimensional structures with the shape and configuration characteristic of the particular phase of development. Studies with radiolabeled cell wall precursors, notably glucose and *N*-acetylglucosamine, have conclusively shown the site of their final deposition but fail to show the site of synthesis. Electron microscope studies suggest that the cytoplasmic microvesicles do not transport wall fibrillar components. Furthermore, it is now almost certain that the microfibrillar skeleton of the wall is synthesized *in situ* either on the outer surface of the plasma membrane or within the wall fabric. The amorphous matrix material is probably synthesized in the cytoplasm, transported to the wall in vesicles, discharged, and bound to the microfibrillar network.

In the aspergilli the major cell wall polymers are chitin and α- and β-glucans, plus varying amounts of protein and lipid and small amounts of other polysaccharides (for reference, see Deans, 1978). Synthetic and lytic enzymes associated with cell wall biogenesis in aspergilli have activity profiles reflecting their involvement in wall growth (Zonneveld, 1972b, 1977; Moore and Peberdy, 1976; Archer, 1977), and frequently the enzyme levels fluctuate during morphogenetic events in which they participate.

The chemical structure of the cell wall has been determined throughout the microcycle of *Aspergillus niger* with respect to the six major components, i.e., protein, lipid, α- and β-glucan, chitin, and galactomannan polymer (Deans and Smith, 1981a). The major enzymes involved in their synthesis and degradation, i.e., α-glucanase, β-glucan synthase, β-glucanase, chitin synthase, and chitinase, have been subcellularly localized and monitored (Deans and Smith, 1981b).

The major α-glucan found in the giant cell was the (S)-glucan, a straight-chain α-1,3-glucan present in levels up to ca. 73%. Nigeran, an alternating α-1,3- and α-1,4-polymer, constituted ca. 20% throughout. During early stages of giant cell formation α-1,3-glucan was elaborated into the cell wall; in later stages during conidiation there was a net decrease in the levels of this component (Fig. 7). Nigeran levels remained fairly constant throughout, with only a slight increase during the late phase of giant cell formation. Enzyme activity patterns demonstrated that, by the end of giant cell formation, the synthetic enzyme decreased while the lytic enzyme proportionately increased (Fig. 7). Thus there was evidently some reutilization of α-glucan moieties or rechanneling of glucose monomers toward the end of giant cell incubation and into conidiation. This implies in part a carbohydrate storage function for the α-1,3-glucan. However, the fact that the α-glucan did not fall below 18% suggests that it serves in part a defined structural role.

Zonneveld (1972a,b, 1973, 1974) has clearly demonstrated a relationship between α-1,3-glucan and morphogenesis in *Emericella nidulans,* with the α-1,3-glucan functioning as a reserve storage compound during cleistothecium and conidiophore formation. In contrast, Wessels *et al.* (1972), with *Schizophyllum commune,* have shown that α-1,3-glucan is not generally reutilized after being incorporated into cell walls.

The fact that the α-glucanase enzyme complex was located in the wall fraction of the giant cell is consistent with its lytic activity on the storage polymer. The broad pH and temperature range over which this enzyme was active (Deans and Smith, 1981b) may reflect an inbuilt ecological mechanism that allows the en-

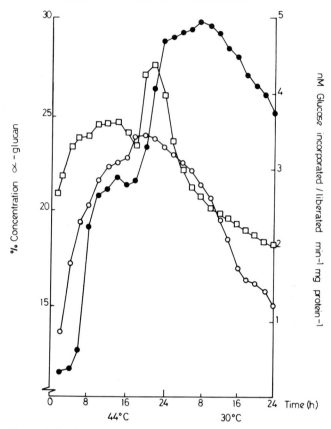

Fig. 7. Changes in levels of α-glucan, α-glucan synthase, and α-glucanase during microcycle conidiation of *Aspergillus niger*. □, α-Glucan; ○, α-glucan synthase; ●, α-glucanase.

zyme to operate under unfavorable conditions. This mechanism enables conidiation to occur, involving the reutilization of glucose wall reserves.

β-Glucans occur widely in fungal cell walls. The most common is (R)-glucan, a polymer with β-1,3 bonding. The precise role of β-glucans within the cell wall matrix is not yet fully understood. It is feasible to consider that the multibranched nature of β-glucan serves as a cementing matrix bonding together the chitin microfibrils and other polymers into a coherent, rigid, layer structure.

β-Glucan and β-glucan synthase increased throughout the microcycle, consistent with the proposed cell wall matrix role of this polymer (Fig. 8). That the activity of the lytic enzyme did not exceed 20% of the synthase enzyme could imply a concerted effort to retain the β-glucans within the cell wall fabric. The low lytic activity could be essential for the insertion of new β-glucan within the matrix of the wall. The process of giant cell formation is one of prolonged cellular growth, not merely wall thickening, hence the need to open up this

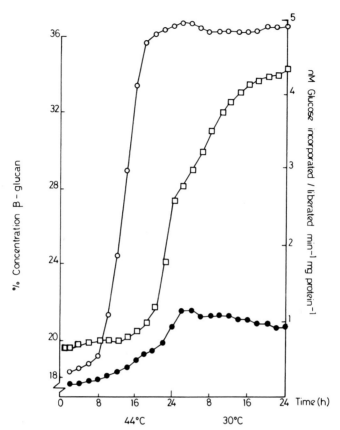

Fig. 8. Changes in levels of β-glucan, β-glucan synthase, and β-glucanase during microcycle conidiation of *Aspergillus niger*. □, β-Glucan; ○, β-glucan synthase; ●, β-glucanase.

matrix material for expansion. Thus β-glucan, in maintaining elevated levels throughout the complete cycle, can be assumed to make a major contribution to the integrity of the cell wall (Fig. 7). β-Glucans have been considered to function in other filamentous fungi [e.g., *Phytophthora palmivora* (Tokunaga and Bartnicki-Garcia, 1971a,b)] and in several yeasts (Manners *et al.*, 1974) as storage polyers. However, in a microcycle system no β-glucan reutilization occurred, either because α-glucans were fulfilling this function or the metabolic enzymes were not being stimulated to degrade β-glucan. Therefore, in the microcycle system of *Aspergillus niger,* β-glucans appear to serve in a constructional capacity in the cell wall matrix.

The microfibrillar nature of chitin within fungal cell walls has led to the proposal that chitin gives both structural rigidity and shape to the organism. Throughout the microcycle, chitin levels increased steadily, achieving a maximum of 34% of total wall material during conidiation (Fig. 9). After the

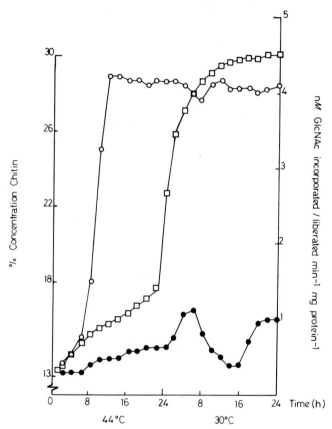

Fig. 9. Changes in levels of chitin, chitin synthase, and chitinase during microcycle conidiation of *Aspergillus niger*. □, Chitin; ○, chitin synthase; ●, chitinase.

initial level of chitin synthase had been established early in giant cell formation the level remained elevated throughout the cycle. In contrast, the generally low activity of chitinase had small peaks of activity at points corresponding to conidophore formation. Thus, as with β-glucan, there was no apparent degradation of chitin during any phase of the microcycle. Chitin appears to be a vital cell wall component contributing structural rigidity to the wall at all phases of development. Any lytic activity directed toward the polymer must be related to the requirement for microfibrillar elongation in order to maintain the structural properties of the cell wall during growth and differentiation.

Several interesting working models for chitin synthase in filamentous fungi and yeasts have been proposed which involve activators, inhibitors, zymogenic forms, and distinct spatial arrangements (Bartnicki-Garcia, 1973; Bartnicki-Garcia *et al.*, 1976; Bracker, 1977; Bracker *et al.*, 1976; Ruis-Herrera, 1977; Cabib and Duran, 1977).

During the microcycle, the chitin synthase complex displayed distinct functional changes (Deans and Smith, 1981b). Chitin synthase isolated from the giant cells demonstrated activation by protease, and it could be stored at low temperatures, features which are generally associated with yeast growth; i.e., a zymogenic form of chitin synthase appeared to exist in the giant cells (Table II). However, on conidiophore outgrowth at 30°C the chitin synthase isolated demonstrated decreased activity on exposure to protease, and its storage requirements were more in line with normal hyphal chitin synthase. Clearly, the difference in duration of activity of the yeast-type (giant cell) chitin synthase compared with that of the shorter-lived hyphal-type (conidiophore) chitin synthase must have profound effects on the type of growth associated with the corresponding enzyme. The prolonged life experienced by the giant cell chitin synthase could partially explain the observed thick cell walls, while the activity present in the conidiophore outgrowths, structures that do not retain the thickened cell walls, supports a different pattern of chitin synthase activity.

The galactomannan polymer fraction of the cell wall probably does not contribute greatly to the wall fabric, since levels seldom exceeded 14% of the total wall material (Fig. 10).

The varying pattern of cell wall protein during the microcycle could be related to the changing activities of specific wall-localized enzymes (Deans and Smith, 1981b). The elevated protein levels during late giant cell formation paralleled the peak activity of the wall-bound enzymes α-glucanase, β-glucan synthase, and chitin synthase, while during conidiation at 30°C a second peak of wall protein was recorded (Fig. 10). However, although wall-linked protein may well contribute structurally to the cell wall filamentous fungi (Mitchell and Taylor, 1969; Bull, 1970), there must also exist the possibility that part of the problem may be contaminating plasma membranes.

TABLE II

The Effect of Protease Activity and Storage on the Activity of Chitin Synthase from Various Stages of Microcycle Conidiation of *Aspergillus niger*

Phase of development	Original activity[a]	Protease addition[b]	Time of storage (h)				
			1.0	2.0	3.0	4.0	5.0
12 h at 44°C	4.20	9.75	6.80	9.75	11.15	11.70	12.20
24 h at 44°C	4.15	10.90	9.85	12.15	13.00	13.20	13.20
12 h at 30°C	4.15	4.35	0.75	0.20	0.05	0	0

[a] Specific activity was measured as nanomoles of GlcNAc incorporated per minute per milligram of protein.

[b] 1 mg protease/ml.

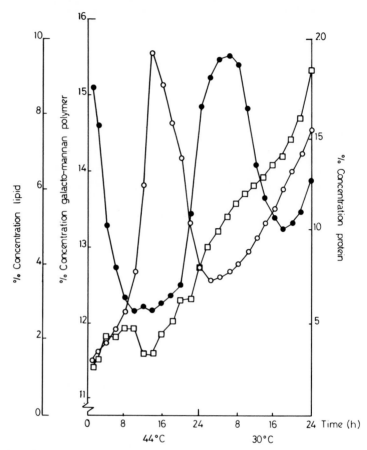

Fig. 10. Changes in levels of a galactomannan polymer, protein, and lipid in wall fractions during microcycle conidiation in *Aspergillus niger*. □, Galactomannan; ○, protein; ●, lipid.

Cell wall lipid levels fell rapidly in the first 15 h of giant cell formation, but this was then followed by an equally rapid accumulation up to the end of giant cell development (Fig. 10). Although wall lipids may represent a convenient storage system, it has been suggested that they also function in elaboration of the main cell wall polysaccharides (Jung and Tanner, 1973; Douglas *et al.*, 1975; Gold and Hahn, 1976).

B. Macromolecular Synthesis

Information concerning the time of production of RNA and protein fractions essential for microcycle conidiation has been obtained using a selection of metabolic inhibitors (Deans and Smith, 1979; Duncan, 1977). Giant cell growth and microcycle conidiation are inhibited by certain inhibitors of protein synthesis, e.g., puromycin, actinomycin D, fluorouracil, 8-azaguanine, and rifampicin when these are present throughout the period of incubation (Deans and Smith, 1979). Spore germination at 30°C was very sensitive to inhibitors of nucleic acid and protein biosynthesis and also to inhibitors of mitochondrial function. The formation of the giant cells at 44°C and conidiophore outgrowth at 30°C were also sensitive to these inhibitors, although inhibition of giant cell enlargement required in general higher concentrations of inhibitors than was required for inhibition of conidiation at 30°C.

Since rifampicin and chloramphenicol are most active against mitochondrial RNA and protein biosynthesis, it implies that mitochondrial function is essential for microcycle conidiation. Preliminary investigations have shown that intact mitochondria can be isolated from the thick-walled giant cells using a Braun MSK rotary homogenizer (Davis and Smith, 1977). Glyoxosomal enzymes were also measured, but evidence for their localization in a microbody or glyoxosome is still inconclusive.

In the *Aspergillus niger* microcycle it was observed in various inhibitor experiments that a minimum giant cell diameter of ca. 15–16 μm was essential for microcycle conidiation (Deans and Smith, 1979). This does not appear to be the case for microcycle conidiation in *Penicillium urticae* (Sekiguchi *et al.*, 1975a).

Addition of actinomycin D, chloramphenicol, or cycloheximide at any time up to 15–16 h inhibited conidiophore outgrowth (Table III); addition of these inhibitors after this time did not result in inhibition of conidiation (Deans and Smith, 1979). Somewhat different results were obtained when actinomycin D was used at 20 μg/ml rather than the 50 μg/ml used in the above-mentioned experiments. When it was added between 0 and 6 h, conidiation was inhibited but not giant cell formation. It was observed that, when the giant cells were subsequently incubated at 30°C, they germinated to produce vegetative hyphae (Duncan, 1977). When actinomycin D was added at later stages, it was found that an increasing number of giant cells gave rise to conidiophores at 30°C and correspondingly fewer produced vegetative hyphae. Thus these studies support the view that any mRNA required for conidiophore production is transcribed during the early stages of giant cell formation. However, whereas actinomycin D at 50 μg/ml inhibits both growth and conidiogenesis, when added at 20 μg/ml it only inhibits conidiogenesis and giant cell growth; the formation of vegetative hyphae can occur. It is possible that actinomycin D at 20 μg/ml only inhibits mRNA formation and that conidiogenesis, but not vegetative growth, requires the formation of new mRNAs. Actinomycin D at 20 μg/ml does not inhibit the incorporation of labeled adenine into bulk RNA in the giant cells of *Aspergillus niger* (Duncan, 1977).

Using the original experimental system of microcycle conidiation (Anderson and Smith, 1971a, 1972) incorporation studies employing [^{14}C]leucine and

TABLE III

Effect of Metabolic Inhibitors on Microcycle Conidiation[a]

	Phases reached in microcycle after 24 h at 30°C		
Time of addition (hours at 44°C)	Actinomycin D 50 μg/ml	Chloramphenicol 60 μg/ml	Cycloheximide 20 μg/ml
Control	10*	10	10
0	2a	2a	2a
2	2a	2a	2a
4	2a	2a	2a
6	2a	2a	2a
8	2b	2b	2b
10	2b	2b	2b
12	2b	4	2b
14	2b	8	4
16	2c	9	6
18	9	10	7
20	9	10	8
22	10	10	8
24	10	10	8

[a] From Deans and Smith, 1978c.
[b] Phases of development as in Fig. 1.

[^{3}H]uridine indicted that the highest rates of RNA and protein biosynthesis occurred during the first 20 h of incubation at 44°C (Duncan *et al.*, 1978). Unfortunately, information on the incorporation of labeled substrates into bulk RNA is not very helpful in determining whether genetic information essential to conidiogenesis is being transcribed. In such experiments the largest fraction of label is incorporated into RNA rather than mRNA. Fractionation of the RNA using acrylamide gel electrophoresis indicated that label had been incorporated into 1.294×10^6 and 0.84×10^6 molecular-weight rRNA and into 4 and 5S RNA fractions. Attempts to isolate mRNA using a Sigma-cell 38 column were less successful; however, a highly labeled RNA fraction with a polydisperse molecular weight was enriched using this technique. Indirect evidence for the presence of mRNA was also obtained by the isolation of polyribosomes from the giant cells (Duncan *et al.*, 1978).

C. Protoplast Formation

By using an induced extracellular enzyme system isolated from culture filtrates of *Trichoderma harzianum* grown on a medium supplemented with purified wall polysaccharides of microcycle *Aspergillus niger,* it has been possible to liberate osmotically sensitive protoplasts both from giant cells (Davis *et al.*, 1977; G. Yahya, personal communication) and from conidiophores (G. Yahya, personal communication). Protoplast release occurred almost entirely with complete digestion of the cell wall. The protoplasts produced from the giant cells and conidiophores regenerated in each case to produce a vegetative mycelium.

REFERENCES

Anderson, J. G. (1978). Light-induced fungal development. *In* "The Filamentous Fungi" (J. E. Smith and D. R. Berry, eds.), Vol. 3, ppl 358–375. Arnold, London.

Anderson, J. G., and Smith, J. E. (1971a). The production of conidiophores and conidia by newly germinated conidia of *Aspergillus niger* (microcycle conidiation). *J. Gen. Microbiol.* **69**, 185–197.

Anderson, J. G., and Smith, J. E. (1971b). Synchronous initiation and maturation of *Aspergillus niger* conidiophores. *Trans. Br. Mycol. Soc.* **56**, 9–29.

Anderson, J. G., and Smith, J. E. (1972). The effects of elevated temperatures on spore swelling and germination in *Aspergillus niger*. *Can. J. Microbiol.* **18**, 289–297.

Anderson, J. G., and Smith, J. E. (1976). Effects of temperature on filamentous fungi. *In* "Inhibition and Inactivation of Vegetative Microbes" (F. A. Skinner and W. G. Hugo, eds.), pp. 191–218. Academic Press, New York.

Anderson, J. G., Aryee, V., and Smith, J. E. (1978). Microcycle conidiation in *Paecilomyces varioti*. *FEMS Microbiol. Lett.* **3**, 57–60.

Archer, D. B. (1977). Chitin biosynthesis in protoplasts and subcellular fractions of *Aspergillus fumigatus*. *Biochem. J.* **164**, 653–658.

Bartnicki-Garcia, S. (1973). Fundamental aspects of hyphal morphogenesis. *Symp. Soc. Gen. Microbiol.* **23**, 245–267.

Bartnicki-Garcia, S., Bracker, C. E., and Ruiz-Herrera, J. (1977). Reassembly of solubilized chitin synthase into functional structures resembling chitosomes. *Proc. Int. Mycol. Congr. 2nd.*

Boosalis, M. G. (1962). Precocious sporulation and longevity of conidia of *Helminthosporium sativum* in soil. *Phytopathology* **52**, 1172–1177.

Bracker, C. E. (1977). Structure and transformation of chitosomes during chitin microfibril synthesis. *Proc. Int. Mycol. Congr. 2nd.*

Bracker, C. E., Ruiz-Herrera, J., and Bartnicki-Garcia, S. (1976). Structure and transformation of chitin synthetase particles (chitosomes) during microfibril synthesis *in vitro*. *Proc. Natl. Acad. Sci. U.S.A.* **73**, 4570–4574.

Bull, A. T. (1970). Chemical composition of wild-type and mutant *Aspergillus nidulans* cell walls: The nature of polysaccharide and melanin constituents. *J. Gen. Microbiol.* **63**, 75–94.

Bull, A. T., and Bushell, M. E. (1976). Environmental control of fungal growth. *In* "The Filamentous Fungi" (J. E. Smith and D. R. Berry, eds.), Vol. 2, pp. 1–31. Arnold, London.

Bu'Lock, J. D. (1975). Secondary metabolism in fungi and its relationship to growth and development. *In* "The Filamentous Fungi" (J. E. Smith and D. R. Berry, eds.), Vol. 1, pp. 33–58. Arnold, London.

Cabib, E., and Duran, A. (1977). Solubilization and properties of yeast chitin synthetase. *Proc. Int. Mycol. Congr. 2nd.*

Caravajal, F. (1947). The production of spores in submerged cultures by some streptomyces. *Mycologia* **39**, 426–440.

Cortat, M., and Turian, G. (1974). Conidiation of *Neurospora crassa* in submerged culture without a mycelial phase. *Arch. Microbiol.* **95**, 305–309.

Davis, B., and Smith, J. E. (1977). Organelle biogenesis in *Aspergillus niger. FEMS Microbiol. Lett.* **1**, 51–53.

Davis, B., d'Avillez-Paixai, M. T., Deans, S. G., and Smith, J. E. (1977). Protoplast formation from giant cells of *Aspergillus niger. Trans. Br. Mycol. Soc.* **60**, 207–212.

Deans, S. G. (1978). Ph.D. Thesis, University of Strathclyde, Glasgow, Scotland.

Deans, S. G., and Smith, J. E. (1979). Influence of metabolic inhibitors on microcycle conidiation of *Aspergillus niger. Trans. Br. Mycol. Soc.* **72**, 201–206.

Deans, S. G., and Smith, J. E. (1981a). Wall composition during microcycle conidiation of *Aspergillus niger* (in preparation).

Deans, S. G., and Smith, J. E. (1981b). Changes in cell wall synthetic and lytic enzymes during microcycle conidiation of *Aspergillus niger* (in preparation).

Douglas, L. J., Atkinson, D. M., Hossack, J. A., and Rose, A. H. (1975). *Proc. Int. Symp. Yeast Other Protoplasts, 4th.* Cave Printing Group, pp. 71–76.

Duncan, D. B. (1977). M.Sc. Thesis, University of Strathclyde, Glasgow, Scotland.

Duncan, D. B., Smith, J. A., and Berry, D. R. (1978). DNA, RNA and protein biosynthesis during microcycle conidiation in *Aspergillus niger. Trans. Br. Mycol. Soc.* **71**, 457–469.

Gold, M. H., and Hahn, H. J. (1976). Role of a mannosyl lipid intermediate in the synthesis of *Neurospora crassa* glycoproteins. *Biochemistry* **15**, 1809–1814.

Gooday, G. W. (1978). The enzymology of hyphal growth. *In* "The Filamentous Fungi" (J. E. Smith and D. R. Berry, eds.), Vol. 3, pp. 51–77. Arnold, London.

Griffin, D. H., and Breuker, C. (1969). Ribonucleic acid synthesis during the differentiation of sporangia in the water mold *Achlya. J. Bacteriol.* **98**, 689–696.

Grove, S. N. (1978). The cytology of hyphal tip growth. *In* "The Filamentous Fungi" (J. E. Smith and D. R. Berry, eds.), Vol. 3, pp. 38–50. Arnold, London.

Gustafson, S. L., and Wright, B. E. (1972). Analyses of approaches used in studying differentiation of the cellular slime moulds. *Crit. Rev. Microbiol.* **1**, 453–478.

Hadley, G., and Harrold, C. E. (1958). The sporulation of *Penicillium notatum* Westling in submerged liquid culture. *J. Exp. Bot.* **9**, 408–417.

Hennessy, S. W., and Cantino, E. C. (1972). Lag phase sporogenesis in *Blastocladiella emersonii*: Induced formation of unispored plantlets. *Mycologia* **64**, 1066–1087.

Jung, P., and Tanner, W. (1973). Identification of lipid intermediate in yeast mannan biosynthesis. *Eur. J. Biochem.* **37**, 1–7.

Kendrick, B. ed. (1971). "Taxonomy of Fungi Imperfecti." Univ. of Toronto Press, Toronto.

Killick, K. A., and Wright, B. E. (1974). Regulation of enzyme activity during differentiation in *Dictyostelium discoideum. Annu. Rev. Microbiol.* **28**, 139–166.

Kuboye, A. O., Anderson, J. G., and Smith, J. E. (1976). Control of autolysis of a spherical cell form of *Aspergillus niger. Trans. Br. Mycol. Soc.* **67**, 27–31.

Larmour, R., and Marchant, R. (1977). The induction of conidiation in *Fusarium culmorum* grown in continuous culture. *J. Gen. Microbiol.* **99**, 59–84.

Lingappa, B. J., and Lingappa, Y. (1969). Role of autoinhibitors in the mycelial growth and dimorphism of *Glomerella cingulata. J. Gen. Microbiol.* **56**, 35–45.

Lloyd, G., Anderson, J. G., Smith, J. E., and Morris, E. O. (1972). The effect of medium nitrogen and of conidiation on esterase synthesis in *Aspergillus niger. Trans. Br. Mycol. Soc.* **59**, 63–70.

Lovett, J. S. (1975). Growth and differentiation in the water mold *Blastocladiella emersonii*: Cytodifferentiation and the role of ribonucleic acid and protein synthesis. *Bacteriol. Rev.* **39**, 345–404.

Mangenot, F., and Reisinger, O. (1976). Form and function of conidia as related to their development. *In* "The Fungal Spore: Form and Function" (D. J. Weber and W. H. Hess, eds.), pp. 789–847. Wiley, New York.

Manners, D. J., Masson, A. J., and Patterson, J. C. (1974). The heterogeneity of glucan preparations from the walls of various yeasts. *J. Gen. Microbiol.* **80**, 411–417.

Mitchell, A., and Taylor, I. F. (1969). Cell wall proteins in *Aspergillus niger* and *Chaetomium globosum*. *J. Gen. Microbiol.* **59**, 103–109.

Moore, P. M., and Peberdy, J. F. (1976). A particulate chitin synthase from *Aspergillus flavus* Link: The properties, location and levels of activity in mycelium and regenerating protoplast preparations. *Can J. Microbiol.* **22**, 915–921.

Ng, A. M. L., Smith, J. E., and McIntosh, A. F. (1973a). Changes in activity of tricarboxylic acid cycle and glyoxylate cycle enzymes during synchronous development of *Aspergillus niger*. *Trans. Br. Mycol. Soc.* **61**, 13–20.

Ng, A. M. L., Smith, J. E., and McIntosh, A. F. (1973b). Conidiation of *Aspergillus niger* in continuous culture. *Arch. Mikrobiol.* **88**, 119–126.

Ng, W. S., Smith, J. E., and Anderson, J. G. (1972). Changes in carbon catabolic pathways during synchronous development of *Aspergillus niger*. *J. Gen. Microbiol.* **71**, 495–504.

Park, D., and Robinson, P. M. (1969). Germination studies with *Geotrichum candidum*. *Trans. Br. Mycol. Soc.* **54**, 83–92.

Righelato, R. C., Trinci, A. P. J., Pirt, S. J., and Peat, A. (1968). The influence of maintenance energy and growth rate on the metabolic activity, morphology and conidiation of *Penicillium chrysogenum*. *J. Gen. Microbiol.* **50**, 399–412.

Robinson, P. M., and Smith, J. H. (1976). Morphogenesis and growth kinetics of *Geotrichum candidum* in continuous culture. *Trans. Br. Mycol. Soc.* **66**, 413–420.

Rosenberger, R. F. (1976). The cell wall. *In* "The Filamentous Fungi" (J. E. Smith and D. R. Berry, eds.), Vol. 2, pp. 328–344. Arnold, London.

Rossier, C., Ton-That, T. C., and Turian, G. (1977). Microcycle microconidiation in *Neurospora crassa*. *Exp. Mycol.* **1**, 52–62.

Roten, J., and Bashi, E. (1969). Induction of sporulation of *Alternaria porri* f.sp. *solani* by inhibition of its vegetative development. *Trans. Br. Mycol. Soc.* **53**, 433–439.

Ruiz-Herrera, J. (1977). *In vitro* synthesis of chitin microfibrils. *Proc. Int. Mycol. Congr., 2nd*, p. 581.

Sansing, G. A., and Ciegler, A. (1973). Mass propagation of conidia from several *Aspergillus* and *Penicillium* species. *Appl. Microbiol.* **26**, 830–831.

Sekiguchi, J., Gaucher, G. M., and Costerton, J. W. (1975a). Microcycle conidiation in *Penicillium urticae:* An ultrastructural investigation of spherical spore growth. *Can. J. Microbiol.* **21**, 2048–2058.

Sekiguchi, J., Gaucher, G. M., and Costerton, J. W. (1975b). Microcycle conidiation in *Penicillium urticae:* An ultrastructural investigation of conidial germination and outgrowth. *Can. J. Microbiol.* **21**, 2059–2068.

Sekiguchi, J., Gaucher, G. M., and Costerton, J. W. (1975c). Microcycle conidiation in *Penicillium urticae:* An ultrastructural investigation of conidiogenesis. *Can. J. Microbiol.* **21**, 2069–2083.

Smith, J. E. (1978). Asexual sporulation in filamentous fungi. *In* "The Filamentous Fungi" (J. E. Smith and D. R. Berry, eds.), Vol. 3, pp. 214–239. Arnold, London.

Smith, J. E., and Anderson, J. G. (1973). Differentiation in the aspergilli. *Symp. Soc. Gen. Microbiol.* **23**, 295–337.

Smith, J. E., and Galbraith, J. C. (1971). Biochemical and physiological aspects of differentiation in the fungi. *Adv. Microb. Physiol.* **5**, 45–134.

Smith, J. E., and Ng, W. S. (1972). Fluorometric determination of glycolytic intermediates and adenylates during sequential changes in replacement culture of *Aspergillus niger*. *Can. J. Microbiol.* **18**, 1657–1664.

Smith, J. E., Anderson, J. G., Deans, S. G., and Davis, B. (1977a). Asexual development in *Aspergillus*. *In* "Genetics and Physiology of *Aspergillus*" (J. E. Smith and J. A. Pateman, eds.), pp. 23–58. Academic Press, New York.

Smith, J. E., Deans, S. G., Anderson, J. G., and Davis, B. (1977b). The nature of fungal sporula-tion. *In* "Biotechnology and Fungal Differentiation" (J. Meyrath and J. D. Bu'Lock, eds.), pp. 17-41. Academic Press, New York.

Talbot, P. H. B. (1971). "Principles of Fungal Taxonomy." Macmillan, New York.

Timnick, M. B., Barnett, H. L., and Lilly, V. C. (1952). The effect of method of inoculation of media on sporulation of *Melanconium fuligineum. Mycologia* **44**, 141-149.

Tokunaga, J., and Bartnicki-Garcia, S. (1971a). Cyst wall formation and endogenous carbohydrate utilization during synchronous encystment in *Phytophthora palmivora* zoospores. *Arch. Mik-robiol.* **79**, 283-292.

Tokunaga, J., and Bartnicki-Garcia, S. (1971b). Structure and differentiation of the cell wall of *Phytophthora palmivora:* Cysts, hyphae and sporangia. *Arch. Mikrobiol.* **79**, 293-310.

Turian, G. (1974). Sporogenesis in fungi. *Annu. Rev. Phytopathol.* **12**, 129-137.

Turian, G. (1975). Differentiation in *Allomyces* and *Neurospora. Trans. Br. Mycol. Soc.* **64**, 367-380.

Turian, G. (1976). Spores in Ascomycetes, their controlled differentiation. *In* "The Fungal Spore: Form and Function" (D. J. Weber and W. M. Hess, eds.), pp. 715-788. Wiley, New York.

Vinter, V., and Slepecky, R. A. (1965). Direct transition of outgrowing bacterial spores to new sporangia without intermediate cell division. *J. Bacteriol.* **90**, 803-807.

Wessels, J. G. H., Kreger, D. R., Marchant, R., Regensberg, B. A., and de Vries, P. M. H. (1972). Chemical and morphological characterization of the hyphal wall surface of the basidiomycete *Schizophyllum commune. Biochim. Biophys. Acta* **273**, 346-358.

Zeidler, G., and Margalith, P. (1972). Synchronized sporulation in *Penicillium digitatum* (Sacc.). *Can. J. Microbiol.* **18**, 1685-1690.

Zeidler, G., and margarlity, P. (1973). Modification of the sporulation cycle in *Penicillium dig-itatum. Can. J. Microbiol.* **19**, 481-483.

Zonneveld, B. J. M. (1972a). A new type of enzyme, an exo-splitting $\alpha(1-3)$ glucanase from non-induced cultures of *Aspergillus nidulans. Biochim. Biophys. Acta* **258**, 541-547.

Zonneveld, B. J. M. (1972b). Morphogenesis in *Aspergillus nidulans:* The significance of $\alpha(1-3)$ glucan of the cell wall and $\alpha(1-3)$ glucanase for cleistothecium development. *Biochim. Biophys. Acta* **273**, 174-187.

Zonneveld, B. J. M. (1973). Inhibitory effect of 2-deoxyglucose on cell wall $\alpha(1-3)$ glucan synthesis and cleistothecium development in *Aspergillus nidulans. Dev. Biol.* **34**, 1-8.

Zonneveld, B. J. M. (1974). $\alpha(1-3)$ glucan synthesis correlated with $\alpha(1-3)$ glucanase synthesis, conidiation and fructification in morphogenetic mutants of *Aspergillus nidulans. J. Gen. Mic-robiol.* **81**, 445-451.

Zonneveld, B. J. M. (1977). Biochemistry and ultrastructure of sexual development in *Aspergillus. In* "Genetics and Physiology of *Aspergillus*" (J. E. Smith and J. A. Pateman, eds.), pp. 59-80. Academic Press, New York.

25

Nuclear Behavior in Conidial Fungi

C.F. Robinow

I. INTRODUCTION

The subject of this chapter is only part of a more inclusive examination of the structure and reproduction of fungal nuclei (Heath, 1978, 1980a). By no means do I intend the following discussion to be interpreted as a review of current knowledge of mitosis in conidial fungi. Instead, it is a brief outline of the major

Biology of Conidial Fungi, Vol. 2

discoveries which have contributed to present concepts of mitosis in these microorganisms. Admittedly, the reader will detect a certain degree of bias in favor of contributions derived from light microscope investigations of nuclear division. Light micrographs of stained preparations are used to illustrate points discussed in the text. On the other hand, the reader will also appreciate, as do I, the immense value of utilizing both light and electron microscope techniques in unraveling the mysteries of nuclear behavior in fungi.

A great deal has been written on the relatively large nuclei in developing asci and basidia, which can be prepared for examination in ways which allow their chromosomes to be counted and each one to be followed in its progress through meiosis. But nuclear images tend to become less distinct when the eye turns from asci and basidia to the supporting tissues and growing mycelia. The student of nuclei investigating vegetative and fertile hyphae produced by conidial fungi has to wrestle with a number of obstacles. Thus nuclei in growing hyphae, as well as those in yeasts, tend to be small, compact, and inconspicuous in living specimens examined with the ordinary compound microscope. They have chromatin that is not always satisfactorily stainable by techniques that work well with meiotic nuclei in developing asci, and they go through processes of mitosis which as a rule do not involve marshalling of chromosomes in the plane of a distinct metaphase plate.

The combined effect of these uninviting properties may account for the fact that, in morphological treatises, surveys, or monographs on the Fungi Imperfecti, nuclei and nuclear detail are usually not mentioned. Nuclei and chromosomes are mentioned in genetic work but are not commonly illustrated. Notable exceptions are the genetic and cytological studies of *Aspergillus nidulans* mutants reported by Morris (1975, 1980), which promise to shed fresh light on fungal mitosis. An absence of discussion of nuclei and chromosomes is also evident in standard treatises on yeasts. If the authors of these renowned and indispensable works and those who make use of them can pursue their studies untroubled by the paucity of data on the number, structure, and behavior of the nuclei of their chosen subjects, then it seems pertinent to ask what gains may be expected from morphological studies on fungal nuclei at this late date in the history of mycology.

It seems to me that there are potential, if modest, rewards of several kinds to be obtained from light microscope studies on the karyology of conidial fungi. There are slight but consistent differences in the organization of resting and dividing nuclei of different anamorphs of ascomycetous (or unknown) affinity, and distinct differences between nuclei of Ascomycetes and Basidiomycetes. The recognition and refinement of these differences seems to be the main value of such nuclear studies for mycology proper. There are other bonuses of a more general biological and tentative nature. The demonstration of countable chromosomes in genera that are or could be the subject of genetic studies is of particular value. Investigations of hyphal and yeast nuclei with clearly visible mitotic spindles and chromosomes may also contribute to our understanding of the currently much discussed mechanism of fungal mitosis and the role of the nuclear envelope in this process (Heath, 1980a,b). Finally, nuclear studies on conidial fungi lead workers to an awareness, desirable in this era of exaggerated importance of the cell concept, that germ tubes and hyphae produced by these evolutionarily advanced members of the kingdom Mycota are actually of coenocytic and not cellular construction. In short, it seems likely that more attention would have been paid to nuclei of imperfect fungi had they been more accessible or, perhaps more accurately, had they been regarded as less inaccessible.

The successful examination of hyphal nuclei has been facilitated by two major advances in technique: the invention of phase-contrast microscopy by Zeeman in the late 1930s, and the publication of the Giemsa nuclear stain procedure by Piekarski (1937). The latter technique involves hydrolysis of fixed cells using hydrochloric acid followed by immersion in Giemsa solution, a differential blood stain previously employed and long established in staining the chromatin of trypanosomes and plasmodia of malaria. Originally used for the detection of sites of DNA in bacteria, Piekarski's hydrochloric acid–Giemsa method was later shown to be applicable to fungi. While phase-contrast microscopy first made it possible to examine nuclei in the living state, the hydrochloric acid–Giemsa technique produced transparent preparations in which stained resting nuclei and mitotic figures stood out brightly against a clear background.

Although acetocarmine and acetoorcein stain chromosomes of plant and animal nuclei well they have not often proved useful in studies of fungal nuclei. Applied directly to cells, as was formerly the custom, acetoorcein stains the contracted chromosomes of hyphal nuclei in mitosis (Figs. 32 and 33) but often does so rather poorly, and the stain is not effectively taken up by either the chromatin or nucleolus of resting nuclei. In staining nuclei whose chromosomes fail to contract during mitosis (e.g., *Saccharomyces*) direct acetoorcein gives no useful results. This stain therefore provides a rather limited view of nuclear behavior in hyphae of anamorphic fungi, and its disappointing performance has partly accounted for the rapid adoption of the hydrochloric acid–Giemsa technique. However, the latter is not the only useful staining technique for hyphal nuclei. The original Feulgen procedure, as well as acetoorcein preceded by hydrolysis (McIntosh, 1954; Elliot, 1960), has also proved a great value. Unfortunately both techniques suffer from limitations which render them less generally useful than hydrochloric acid–Giemsa. The strength of a positive Feulgen reaction depends on the amount of DNA present and, since there is rather little of this nucleic acid in fungal nuclei (e.g., yeast), the Feulgen reaction may be too weak for effective recording by photography. Hydrochloric acid–acetoorcein is as specific in its effects as the Feulgen technique. In many instances, but by no means always, hydrochloric acid–acetoorcein provides more deeply stained and sharply defined images than weakly positive Feulgen preparations. Hydrochloric acid–Giemsa is the most generally useful of the three techniques discussed, but it too has certain shortcomings. The brilliance which makes these preparations visually attractive causes halation and loss of resolution in photographic negatives, as is only too apparent in much published work.

The chemical basis of all this is known only for the Feulgen procedure where the Schiff reagent, basic fuchsin decolorized by sulfur dioxide, is returned to its original mauve color by interaction with aldehyde groups released in the deoxyribose moiety of the DNA molecule by the action of hydrolysis. As regards hydrochloric acid–Giemsa and hydrochloric acid–acetoorcein, one can only say that hydrolysis is known to remove much of the RNA from the cytoplasm and reduce its affinity for basic stains. However, this scarcely accounts for the increased affinity of hydrolyzed chromatin for both Giemsa solution and acetoorcein.

Fungal nuclei may also be stained with acridine orange which at low concentrations is tolerated by growing cells. This stain was used in combination with fluorescent microscopy by Poon and Day (1974) in investigations of mitosis in sporidial "yeasts" of *Ustilago violacea*. The nuclear chromatin of many chemically fixed eukaryotic cells has been revealed against a faintly stained background by the fluorescent nucleophil antibiotic mithramycin (Heath, 1980b). Introduced by Slater (1976), this fluorochrome was effectively employed in studies on yeasts by Conde and Fink (1976). Kŏpecká and Gabriel (1978) have described the remarkable effect of the antibiotic lomofungin. Added to suspensions of living organisms, first washed free of culture medium, lomofungin directly and selectively stains nuclear chromatin of yeasts and mycelial fungi a bright, deep red and may be observed with ordinary transmitted light. Acridine orange may prove useful in studying fine detail of the mitotic process in fungi (Poon and Day, 1974; Poon *et al.,* 1974). Mithramycin and lomofungin provide information on the numbers of nuclei per cell without lengthy preparation and are useful in monitoring the progress of mitosis and meiosis.

In the study of fungal nuclei light microscopy has been most fruitful when it has been employed in combination with thin-sectioning and transmission electron microscopy. It is this kind of approach which ensures lasting value for the work of Aist (1969) and Aist and Williams (1972) on *Fusarium,* and that of Girbardt (1961, 1971, 1978) on the basidiomycete *Polystictus (Coriolus) versicolor.*

II. GENERAL MORPHOLOGY OF FUNGAL NUCLEI

It may seem to the casual observer that if you have seen one fungal nucleus you have seen them all. In fact, distinctions are evident, and on the basis of available data certain generalizations can be made.

A. The Nucleolus

Nuclei in growing hyphae and yeasts invariably have distinct and relatively large nucleoli (Fig. 31) which either persist during mitosis and divide between daughter nuclei together with the genome or are dispersed and reappear at the end of mitosis. As far as I am aware, there is never more than one nucleolus present, and its position is either in the center or at one end of the fungal nucleus. Both kinds of arrangements are encountered in hyphal nuclei of Ascomycetes and conidial fungi. Excentrically located nucleoli are the rule in budding yeasts and Basidiomycetes. The nuclei in conidia are usually much smaller than those in growing hyphae and have correspondingly smaller nucleoli, which may even be invisible under the light microscope. Conidial germination is accompanied by an increase in the dimensions of the nucleus and nucleolus. Nucleoli are readily stained by hematoxylin and acid fuchsin, but only faintly or not at all by acetoorcein. In most instances, nucleoli become inconspicuous after acid hydrolysis. Their affinity for the Giemsa stain after hydrolysis in *Candida ingens* (Figs. 23–27) is unusual.

B. Nucleus-Associated Organelles

The envelope of resting nuclei of members of the Zygomycetes, Oomycetes, Ascomycetes, and Basidiomycetes carries a nucleus-associated organelle (NAO) on its outer or inner surface, from which microtubules emanate at the time of mitosis. NAOs of Ascomycetes and Basidiomycetes are in many instances visible during life when phase-contrast microscopy is used (Girbardt, 1960, 1978; Wilson and Aist, 1967; Aist, 1969). A fresh example is provided by Fig. 34 of the *Chromelosporium* anamorph of *Peziza ostracoderma*.

1. Nucleus-Associated Organelles in Fixed Stained Preparations

The NAOs of different ascomycetes (and their anamorphs) differ in their affinity for stains. It is now clear that those that first attracted attention did so because of their affinity for certain formerly widely used nuclear stains which stained chromosomes equally well. Such NAOs are found at the periphery of nuclei of mildews one of which, *Phyllactinia corylea,* is the subject of a classical, still much quoted karyological study by Harper (1905). Designated "central bodies" by Harper because of their position at the poles of the spindles of mitosis and meiosis the NAOs were later more accurately referred to as "lateral granules" by Colson (1938) in a fresh and equally excellent study of the same species. Two of Colson's drawings have been reproduced in part in Fig. 1 and should be compared with Figs. 9–12 showing lateral granules or, rather, crescents, of nuclei of the *Chromelosporium* anamorph of *Peziza ostracoderma*. It is of interest that just as nucleoli now tend to disappear from textbook illustrations based on electron micrographs of sections of unsuitably preserved materials, so

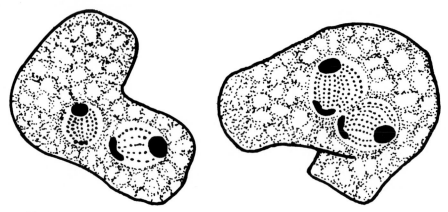

Fig. 1. Nuclei in the oogonium of *Phyllactinia corylea*. Crescent-shaped lateral granules or NAOs occupy the ends of the nuclei opposite the nucleolus. Redrawn from Colson (1938).

also the distinct lateral granules of Colson's explicit illustrations of 1938 were disregarded by Gäumann (1949) in his well-known text, *Die Pilze.* Instead, in Fig. 139 of that Lehrbuch the illustrator replaced Colson's accurately observed lateral granules and nucleoli with a pattern of meaningless stippling. The tenacious retention of hematoxylin by the NAOs of *Phyllactinia* becomes understandable as a reflection of the presence of DNA because the NAOs of another mildew, *Sphaerotheca,* do indeed give a strongly positive Feulgen reaction (C. Robinow, unpublished observations) and NAOs with selective affinity for acetoorcein have been demonstrated, in yet another mildew, *Erisyphe graminis* by McKeen (1972). The strong affinity for acetoorcein (after hydrolysis) found in the NAOs of the *Chromelosporium* anamorph of *Peziza ostracoderma* (Figs. 9–12) already referred to accords well with the cytochemical detection of DNA in the spindle pole bodies (NAOs) of *Ascobolus,* another member of the Pezizales, by Zickler (1973). We have, then, three Erysiphales (*Phyllactinia, Sphaerotheca, Erysiphe*) and two Pezizales (*Peziza* and *Ascobolus*) with NAOs that have marked affinity for nuclear dyes. This is of taxonomic interest because the Erysiphales, though lacking close relatives, are believed to have been derived from the Pezizales (von Arx, 1967). Lack of obvious affinity for nuclear stains and a negative response to the Feulgen procedure probably account for the fact that the NAO of *Saccharomyces* was first identified only recently.

2. Electron Microscopy of Nucleus-Associated Organelles

The fine structure of NAOs and their behavior during mitosis has been studied extensively with the electron microscope. Original observations and critical evaluation of the scattered literature will be found in stimulating reviews by Kubai (1975, 1978), Girbardt (1978), and Heath (1978). The NAOs of Ascomy-

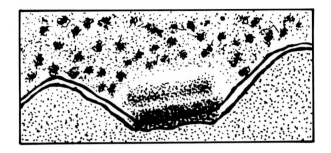

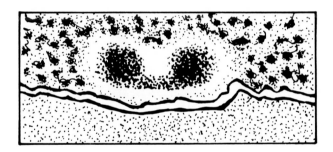

Fig. 2. Nucleus-associated organelles (spindle plaques, spindle pole bodies) of an ascomycetous yeast (top) and a basidiomycetous yeast (bottom). Redrawn and simplified from electron micrographs of sections of *Saccharomyces* (Robinow and Marak, 1966) and *Leucosporidium* (McCully and Robinow, 1972a).

cetes tend to be flat, disc-like, electron-dense structures in contact with or set into large pores of the nuclear envelope. The former kind is found in *Erysiphe* (McKeen, 1972), *Fusarium* (Aist and Williams, 1972), *Ascobolus,* and *Podospora* (Zickler, 1970, 1973), the latter in *Saccharomyes* (Fig. 2a). There is as yet no evidence that the distinct NAOs of *Chromelosporium* (Figs. 9–12) conform to either of these types of construction. In the one published electron micrograph of a favorably oriented slice of a resting *Chromelosporium* nucleus (Hughes and Bisalputra, 1970) a dense segment, resembling in its geometry the NAOs of Figs. 9–12, looks merely like chromatin and lacks the layered structure common to the NAOs of Ascomycetes. It remains to be seen what happens to the fine structure of the NAO of this fungus during the course of mitosis. Of general interest is the discovery that the NAOs of Ascomycetes (e.g., Robinow and Marak, 1966; Peterson *et al.,* 1973; Zickler, 1970, 1973) differ strikingly in shape, fine structure, and mode of reproduction from their counterparts in Basidiomycetes (e.g., Girbardt, 1971, 1978; Girbardt and Hädrich, 1975). This difference is particularly apparent when the NAOs of ascomycetous yeasts are compared with those of basidiomycetous yeasts (McCully and Robinow, 1972a,b; Fig. 2b).

III. MITOSIS

Aspects of mitosis in nuclei of fungal hyphae and yeasts as revealed by the electron microscope have been the subject of several recent review articles (Fuller, 1976; Kubai, 1975, 1978; Girbardt, 1978; Heath, 1978, 1980a). Although these accounts deal with all features of fungal mitosis, a few points seem worth making for the benefit of those whose approach to the subject does not altogether bypass light microscopy.

A. Phase-Contrast Microscopy of Nuclei in Growing Hyphae

Nothing to date has equaled or surpassed the rewarding phase-contrast studies on the behavior of chromosomes and spindles in dividing nuclei of *Fusarium* presented by Aist (1969) or Girbardt's many elegant and meticulous studies on the dynamics of mitosis in Basidiomycetes recently reviewed by him in Heath (1978). An important difference between the results of these two sets of observations illustrates the limitations of phase-contrast microscopy in the study of fungal mitosis. Chromosomes are fleetingly visible during mitosis in *Fusarium* but are never resolved in *Polystictus.* Similarly, the chromosomes of *Penicillium, Aspergillus,* and budding yeasts in general are not visible during nuclear division. In the yeasts, even the electron microscope has not provided us with any significant information on chromosome structure and arrangement. The nuclear envelope is not evident under the light microscope either during life or after staining. Phase-contrast microscopy shows that the nucleus of Ascomycetes remains visible as an object of low density throughout the entire course of mitosis, implying that its contents do not mix with the cytoplasm. In Basidiomycetes, on the other hand, the nucleus in many instances vanishes from sight early in mitosis because its contents apparently become mixed with the cytoplasm. The work on nuclear division in living hyphae established what electron microscopy has borne out; in Ascomycetes mitosis takes place within the intact nuclear envelope, whereas the tortuous process of mitosis in Basidiomycetes is accompanied by more-or-less complete disruption of the envelope.

B. Generalizations from Light Microscopy of Fixed and Stained Preparations: *Trichosporon* as an Example

Fungal nuclei will never rival bean roots or anthers in suitability for demonstrating basic features of mitosis. The mitotic constellations of fungi are often difficult to resolve into separate components. A common pattern of mitosis is illustrated by the pseudomycelial yeast *Trichosporon* (Figs. 3–8), examined at the suggestion of Dr. M. A. Lachance. The granular chromatin of the resting nucleus changes into a more-or-less dense asterisk-like figure, while the nucleolus fades from sight (Fig. 4). The chromosomes, recognized as tiny rodlets, are later seen densely packed along what in optical section appears as a straight ribbon with a long, narrow, translucent central cleft (Fig. 5). The latter is the site of the intranuclear spindle (Fig. 6). For comparison, a corresponding phase in the division of a nucleus of *Aspergillus nidulans* is shown in Fig. 30a and b. This double-track stage of nuclear division merges into telophase (Fig. 7), which is characterized by condensed chromosomes, usually in the shape of two lobes, at either end of a long, thin cord. The latter is faintly stained like chromatin but, as electron microscopy informs us, is also composed of spindle fibers and a membranous sleeve which is provided by the much stretched nuclear envelope. The story concludes with the separation of the chromosomes and the formation of two daughter nuclei which are initially quite small and located some distance in either direction from the original site where mitosis began (Fig. 8). Although a metaphase plate is missing, the behavior of the chromosomes becomes intelligible when they are seen in relation to the spindle which occupies the long, transparent cleft between the two columns of chromosomes in Figs. 7 and 30a and b. The presence of a spindle in dividing nuclei of growing hyphae and budding yeasts was a fairly recent discovery of light microscopy later confirmed by electron microscropy. The double-track orientation of mitotic and meiotic chromosomes on the centrally located spindle which is so characteristic of many fungi was, however, clearly understood and described long ago by Olive (1949). Two passages from his work on meiosis in the basidiomycete *Coleosporium vernoniae* are worth repeating. Describing metaphase I and anaphase I he observed:

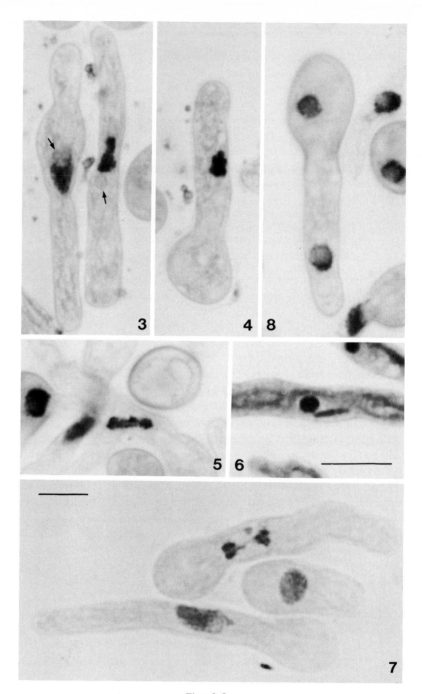

Figs. 3–8.

There is no clearcut metaphase such as is generally distinguishable in higher organisms, as there is no stage at which all the gemini are arranged in a group across the equator of the spindle. . . . Metaphase is of short duration and is quickly followed by anaphase. There is no distinct line of demarcation between the two stages, since some of the chromosomal pairs begin to split apart before others have started . . . some of the chromosomes always reach the poles before others . . . often one or two chromosomes lag considerably behind the others. . . . By mid-anaphase the spindle fibres have generally coalesced into two main strands, so that the chromosomes are found passing to the poles in two lines. As a result, the mass of chromosomes accumulating at each pole frequently has a bilobed appearance.

It could hardly be described better.

The "passing of chromosomes to the poles in two lines" has also been recorded in anaphase II of meiosis in basidia of *Marasmius* (Duncan and Macdonald, 1965) and *Poria* (Setliff *et al.,* 1974). Superficially, there is undoubtedly a close resemblance between these constellations and the stretched-out double-track anaphase figures of mitosis in hyphal nuclei of Ascomycetes and many Hyphomycetes.

The pattern of mitosis represented by *Trichosporon* is also discernible in the illustrations and accounts of mitosis in many other conidial fungi: e.g., in *Ophiostoma* (Bakerspigel, 1961), *Aspergillus* (Robinow and Caten, 1969), *Penicillium* (Crackower and Bauer, 1971), *Cladosporium* (Crackower, 1972), *Fusarium* (Aist and Williams 1972), *Cochliobolus sativus* (Huang and Tinline, 1974), and numerous others.

C. Fine-Structural Details of Mitosis

Electron microscopy of serial sections of dividing *Fusarium* nuclei by Aist and Williams (1972) have shown conclusively that divergent microtubules, which vary considerably in length, are in contact with kinetochores of the chromosomes. Apart from their staggered array over the whole length of the anaphase

---

Figs. 3–8. *Trichosporon* sp. Mercuric chloride–alcohol–acetic acid. Hydrochloric acid–Giemsa stain, except Fig. 6 in which the stain was acid fuchsin. All photographs except Fig. 6 are at the same magnification. Bars in Figs. 6 and 7 represent 5 μm.

Fig. 3. The nucleus of the cell on the left is at rest. An arrow points to its nucleolus. Mitosis, accompanied by increased density of the chromatin, has begun in the nucleus of the cell on the right. An arrow points to the already much reduced nucleolus.

Fig. 4. Nucleus transformed into a densely packed cluster of metaphase chromosomes. The nucleolus has dispersed.

Fig. 5. The double-track stage of anaphase discussed in the text. Compare with Fig. 30a and b showing the same stage in *Aspergillus*.

Fig. 6. The intranuclear spindle at a stage of mitosis corresponding closely to that of the nucleus on the right in Fig. 3.

Fig. 7. Telophase. Typical bilobed appearance of the condensed chromatin of the daughter nuclei shown in the upper cell.

Fig. 8. Daughter nuclei have moved far apart. Cell division is impending.

spindle, *Fusarium* chromosomes are evidently distributed to daughter nuclei in a conventional manner. The same conclusion has been reached for *Saccharomyces* for which details of mitosis had remained obscure until the publication of an admirable piece of quantitative electron microscopy by Peterson and Ris (1976), who counted the microtubules diverging from the inner face of the spindle poles on opposite sides of dividing nuclei and found that their number equaled or was very close to that of the number of linear linkage groups established for this yeast. The microtubules end at the spindle equator in wisps of dense chromatin-like material. Apart from their failure to contract, the behavior of yeast chromosomes at mitosis seems even more ordinary than that of the chromosomes of *Fusarium*. In *Saccharomyces* the kinetochore-bearing segments of the chromosomes (presumably represented by the dense material) are marshaled on a kind of metaphase plate. How the remaining lengths of chromosomes are arranged in the nucleus is not known. The Feulgen reaction suggests that the chromosomes completely fill the nucleus at all times, leaving room only for the nucleolus. Clear evidence of spindle microtubules engaging kinetochores of mitotic chromosomes has also been found in the basidiomycetes *Coprinus* (Thielke, 1974) and *Poria* (Setliff *et al.,* 1974).

The welcomed normalization of mitosis in many fungi does not extend to all of them. For example, in such well studied fungi as *Mucor* and *Neurospora*, the behavior of the chromosomes and their relationship to the spindle remain obscure, and no comprehensive account of mitosis in hyphal nuclei of the latter has yet been provided. There is still some uncertainty about the interrelationship of spindle pole bodies (NAOs), spindle tubules, and chromosomes at the beginning of mitosis. It seems that even at prophase of mitosis the chromosomes in some instances have already made contact with the spindle poles via microtubules growing poleward from the kinetochore. The chromosomes of hyphal nuclei and yeast are perhaps permanently attached to the nuclear envelope close to the NAO and are parted into two sets very early in mitosis during replication of the NAO and subsequent migration of these bodies to opposite sides (not necessarily poles) of the nucleus. This of course raises the possibility that replication of the NAO and the chromosomes may be linked in time and space. Cytogenetic implications of this concept were explored by Day (1972), and observations supporting the idea were presented by Rosenberger and Kessel (1968) who examined the distribution of radioactive nuclei in germ tubes of *A. nidulans* grown from tritium-labeled uninucleate conidia. These authors were led to the important conclusion that "chromatids containing DNA strands of identical age segregate as a unit during mitosis." Although light microscopy of hyphomycete nuclei offers nothing as compelling as the Rosenberger-Kessel experiment, the presence of something that holds members of a set of chromosomes together is suggested by certain regularly encountered anaphase constellations in *Penicillium*. The "flailing arms" revealed by anaphase figures (Fig. 21; Crackower and Bauer, 1971), where all the chromosomes seem anchored at a common center, and the "four-fingered" late-telophase stage of short, straight, often parallel chromosomes seemingly fused together at their poleward ends (Figs. 14 and 21) are examples. In the relatively large fission yeasts *Schizosaccharomyces versatilis* and *S. japonicus,* the light microscopist encounters persuasive images of sets of chromosomes numbering only three in these instances and seemingly tied to spindle poles. Elucidation of this phenomenon will require electron microscopy. However, there is clear evidence that at metaphase of mitosis in basidiomycetes spindle tubules connect spindle pole bodies and kinetochores (Girbardt, 1973; Thielke, 1974; Setliff *et al.,* 1974). This is compatible with but does not prove the idea that such connections may be permanent ones. In at least one instance (e.g., *Saprolegnia*) there is evidence that such connections are established very early during preparation for nuclear division (Heath and Greenwood, 1970). An alternative view, the direct attachment of chromosomes to the nuclear envelope in the region on the NAO, has been put forward by Girbardt (1968, 1971) and Harder (1976a,b). Suggestive as they are, these authors' electron micrographs need corroborative light microscopy, a reversal of the usual state of affairs. The evidence for the attachment of chromosomes directly or indirectly to the nuclear envelope in the region of the NAO is thoroughly discussed in wide-ranging reviews by Kubai (1978) and Heath (1980a) which are recommended to those wishing information about the current state of thought on mitosis in fungi.

IV. SOME FURTHER EXAMPLES OF MITOSIS IN CONIDIAL FUNGI AS REVEALED BY LIGHT MICROSCOPY

The concepts discussed above will now be illustrated using three examples.

A. *Chromelosporium* Anamorph of *Peziza ostracoderma* (Figs. 9–12)

A most striking feature of the resting nucleus of this fungus (examined at the suggestion of Mrs. Susan Legeza) is the clarity of the small peripheral dot, bar, or crescent—the lateral granule of Colson (1938) now called the NAO (Fig. 9). The contrasting appearance of this organelle is due to its affinity for acetoorcein stain, which is as strong as that of the chromosomes during mitosis. It may be unwise to regard these elements as nothing more than NAOs until their fine structure has been elucidated. This matter is dealt with in Section II,B,2.

Three phases of mitosis are illustrated in Figs. 10–12. Metaphase clusters of minute chromosomes (Figs. 10 and 11), an early anaphase (arrow in Fig. 11), and a pair of dense nuclei in telophase (Fig. 12) are revealed as they appear in hydrochloric acid–acetoorcein preparations. The wide separation of the telophase

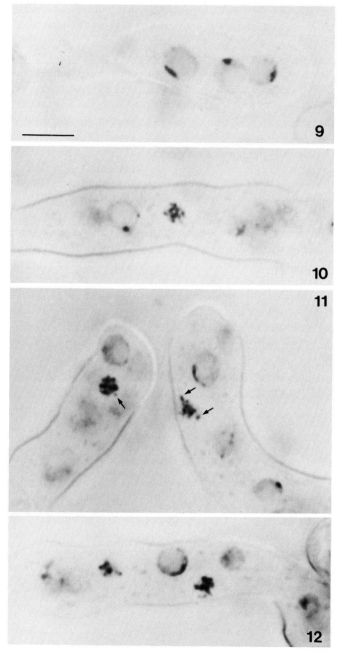

Figs. 9–12.

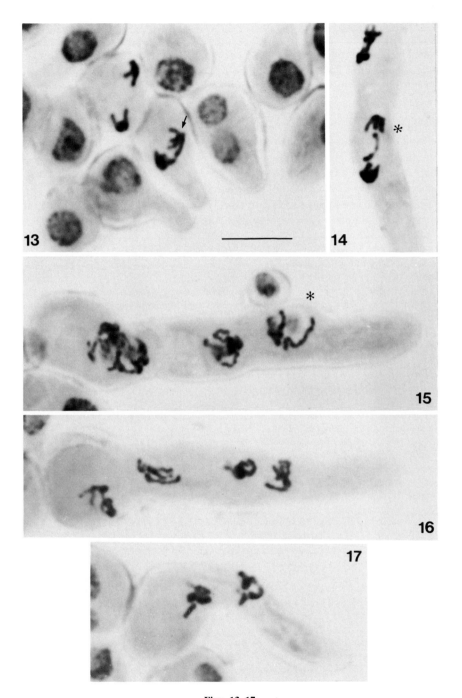

Figs. 13–17.

nuclei in Fig. 12 reflects the tendency of fungal sibling nuclei to move far apart immediately after completion of mitosis. This centrifugal movement (see also Figs. 50 and 51 of *Trichothecium*) may later be reversed.

B. *Penicillium* (Figs. 13–21)

Mitosis is more easily followed in *Penicillium* than in other conidial fungi here examined because its chromosomes are relatively large and there are only few of them, probably not more than four (Figs. 13–21). The chromatin of resting nuclei appears granular or composed of short lengths of winding threads (Figs. 13 and 21). There is a central nucleolus (asterisks in Figs. 15 and 18) which is lost from view early in mitosis. The chromosomes condense and become clearly visible at

---

Figs. 9–12. *Chromelosporium* anamorph of *Peziza ostracoderma*. Mercuric chloride–alcohol–acetic acid. Hydrochloric acid acetoorcein. The magnification is the same for all figures and is indicated by the bar in Fig. 9, which represents 5 μm.

Fig. 9. Resting nuclei with deeply stained NAOs (and/or heterochromatin; see text). The chromatin of the interior of the nuclei is only lightly stained, appearing as a crescent around the relatively large, unstained nucleolus.

Figs. 10–12. Various stages of mitosis. Metaphase clusters are seen in Figs. 10 and 11. Arrows point to putative NAOs (spindle pole bodies). In Fig. 12 members of a pair of telophase nuclei have moved far apart.

Figs. 13–17. *Penicillium* sp. Mercuric chloride–alcohol–acetic acid. Hydrochloric acid–Giemsa. The magnification is the same for all figures and is indicated by the bar in Fig. 13, which represents 5 μm.

Fig 13. Telophase of mitosis in germinating conidia. The arrow points to a constellation of four chromosomes. In the background are many resting nuclei.

Fig. 14. Telophase of a hyphal nucleus. Three chromosomes can be discerned adjacent to the asterisk. An additional chromosome is lagging behind in its poleward movement.

Fig. 15. Prophase. Chromosomes in process of condensation. A nucleolus is still present in the nucleus opposite the asterisk. At the far left the images of two nuclei have been superimposed.

Figs. 16 and 17. Chromosomes preserved at various stages in the process of contraction associated with metaphase.

Figs. 18–21. *Penicillium* sp. Fixation, staining, and magnification as in Figs. 13–17. Bar in Fig. 20 represents 5 μm. Chromosomes preserved in various stages of metaphase contraction.

Fig. 18. Nucleoli are still present at this stage but show stages of reduction in size. One is indicated by an asterisk. A small fragment of nucleolar material is part of the constellation at the top of the figure.

Figs. 19 and 20. Four chromosomes are clearly visible in the nucleus marked with an asterisk. The constellation indicated by an asterisk in Fig. 20 is less readily interpreted. The chromosomes in Fig. 19 were successfully forced apart by pressure exerted on the preparation.

Fig. 21. Telophase of the division of four nuclei (a–b, c–d, e–f, g–h). Daughter nuclei (e) and (f) show flailing arms of chromosomes and are interpreted as sets of four chromosomes (three long, one short) anchored at a common center. The chromosomes of nuclei (g) and (h) are at the same advanced stage of telophase as (a) and (b). Four chromosomes, two short and contracted and two longer and less contracted, can be discerned in nucleus (h). Only three chromosomes are visible in nucleus (g), but the high density of stain at the point of convergence may be due to the presence of a fourth chromosome.

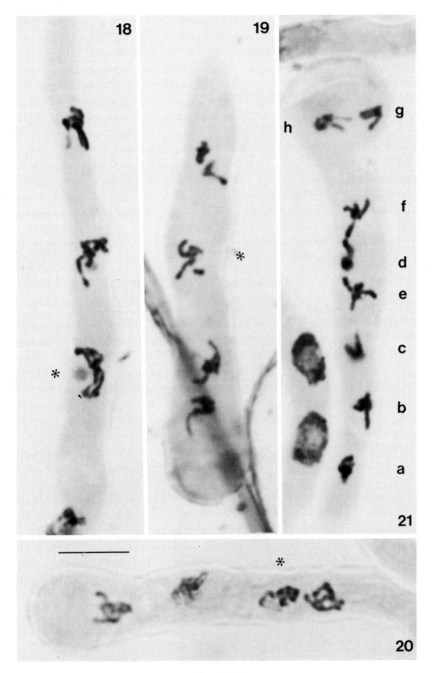

Figs. 18–21.

prophase (Fig. 15). The progress to metaphase is accomplished by further contraction (Figs. 16–20). In the most transparent preparations, not shown here, maximally condensed chromosomes appear to have a hollow center (are doughnut-shaped), which suggests that they have replicated before or during contraction. The chromosomes unfold again as single, slender filaments at anaphase and move apart in the shape of more-or-less coherent sets of three or four "fingers" (Figs. 14, 21 top). Further contraction gives rise to the familiar dense, telophase nuclei (Fig. 21a–b). Small at first but steadily expanding, they change back into the pattern of resting nuclei complete with nucleoli. The course of mitosis in this species of *Penicillium* is representative of the process of nuclear division in many other penicillia I have encountered in more than 25 yr of work with fungal nuclei. However, many details still await elucidation by electron microscopy, foremost among them the relationship between chromosomes and spindle microtubules. Comparison light micrographs of the chromosomes (a) and spindle (b) of a dividing nucleus of *Penicillium* are shown in Fig. 35a and b.

C. *Candida ingens* (Figs. 22–29)

Nuclear division in *Candida ingens* (examined at the suggestion of Dr. W. T. Starmer) is essentially the same as in other yeasts, either ascosporogenous or anamorphic. The time-lapse series in Fig. 22 recalls similar published sequences of other budding yeasts. The nucleus remains visible throughout mitosis, which reflects intactness of the nuclear envelope. The nucleolus divides before constriction of the nucleus begins, about one half of it passing to each daughter nucleus. Chromosomes, however, are never seen. The intranuclear spindle, known to be present from light microscopy of fixed preparations (Fig. 29) is less readily visible in live, budding yeasts than in fission yeasts.

In the fixed and flattened preparations of *C. ingens* shown in Figs. 23–27, the dense, unstructured matter of the nucleolus has retained an unusually strong affinity for the Giemsa stain despite the clearing effects of hydrolysis. The nucleolus is clearly separated from a densely packed aggregate of minute chromosomes. One NAO may be differentiated from the chromosomes in Fig. 23, while NAOs are visible at each end of the elongated constellations in Figs. 23 and 25. The NAOs, if indeed they may be identified as such, are exceptionally large at the poles of the dividing nucleus in Fig. 25. They are clearly distinguishable from nucleolar and chromosomal material in Fig. 26. The nuclei in Fig. 27 have regained the configuration typical of the resting stage.

V. NUCLEAR BEHAVIOR DURING CONIDIOGENESIS

There are several possible interrelationships between nuclei and developing conidia. A conidial initial may receive one of the daughter nuclei from a mitosis

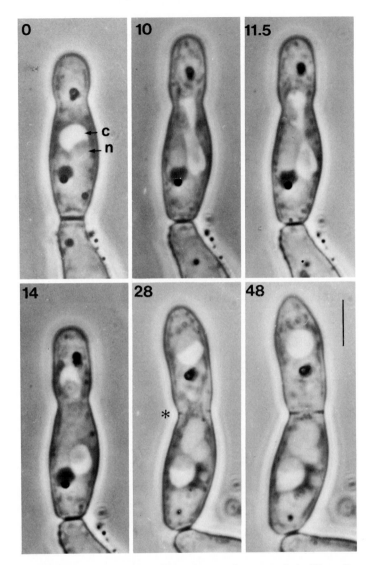

Fig. 22. *Candida ingens.* Time-lapse, phase-contrast microscopy of a budding cell growing in 25% gelatin with 0.5% yeast extract and 2.0% glucose. The magnification is indicated by a bar representing 5 μm. Minutes elapsed are indicated at the upper left in each photograph. Note elongation and constriction of nucleus and nucleolus at 10 and 11.5 min. There is a considerable increase in the volume of both the mother cell and bud after 14 min. The asterisk marks the first appearance of the primary septum. n, Nucleolus; c, Region of the chromosomes.

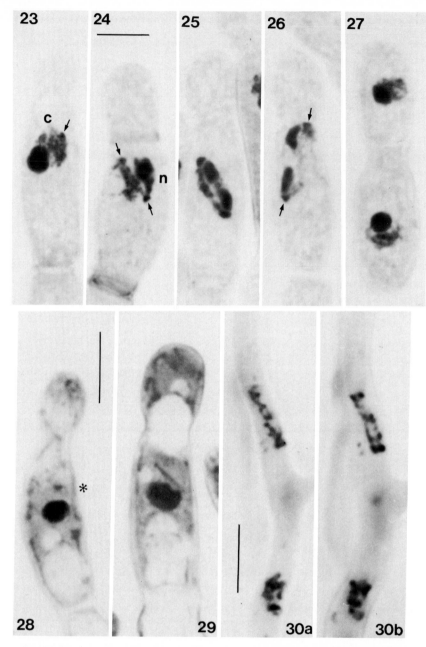

Figs. 23–27. *Candida ingens.* Fixed *in situ* with Formalin–acetic acid–alcohol. Hydrochloric acid–Giemsa stain. The magnification is the same for all figures. The bar above Fig. 24 represents 5 μm. c, Chromosomes; n, nucleolus; arrows point to NAOs (spindle pole bodies). The chromosomes are minute, numerous, and tightly packed.

which takes place at the base of the differentiating cells, the initial may enlarge nearly to the size of a mature conidium before it receives one or more nuclei from a large store accumulated in the supporting structure (i.e., a conidiogenous cell), or the initial may receive a few nuclei which then proceed to multiply within the developing conidium. The first mode of nuclear behavior is demonstrated during conidium ontogeny in *Penicillium* (Zachariah and Metitiri, 1971) and *Aspergillus* (Kozakiewicz, 1978). In the present chapter, this behavior is illustrated in a species of *Penicillium* (Figs. 38 and 39), but regrettably in far less detail than is apparent in the two studies mentioned above. Provision of conidium initials with nuclei generated elsewhere is the mode of behavior encountered in *Gonatobotryum* (Kendrick and Chang, 1971), *Aureobasidium* (Figs. 36 and 37), *Botryosporium* (Figs. 40–42), and *Chromelosporium* (Figs. 43–47). My light micrographs of the last species complement and are in accord with the electron micrographs of the *Chromelosporium* anamorph of *Peziza ostracoderma* presented by Hughes and Bisalputra (1970). The third mode of providing conidia with nuclei, namely, multiplication *in situ*, may be inferred to be that employed by *Helminthosporium* (Hrushovetz, 1956) and has been observed in *Scopulariopsis brevicaulis* by Kendrick and Chang (1971). Another interesting example is provided by *Trichothecium roseum* (Figs. 48–52) in which a few nuclei originally entering the initial through a slender stalk later increase in number by an apparently synchronous wave of mitoses.

VI. METHODOLOGY

A. Phase-Contrast Microscopy of Growing Hyphae

Phase-contrast microscopy, essential for observations of nuclei, is best done with cultures growing on a thin film of agar beneath a coverslip (Girbardt, 1956a). Agar films are obtained by dripping 0.5 ml of molten agar on a slide

---

Figs. 28 and 29. *Candida ingens.* Helly fixation. Acid fuchsin. All magnifications are the same. The bar in Fig. 28 represents 5 μm.

Fig. 28. Resting nucleus in a budding cell. At the base of the nucleus is a deeply stained nucleolus. At the other end of the nucleus (asterisk) is a NAO.

Fig. 29. A stage of mitosis, presumably anaphase, comparable to that in Fig. 24. A spindle traverses the nucleus. It is slightly tilted out of the plane of the optical section, and only the upper end is in focus. The nucleolus is no longer perfectly round. Taken together, Figs. 24 and 29 suggest that anaphase is already well advanced at the time the dividing nucleus is first seen to elongate.

Figs. 30a and b. *Aspergillus nidulans.* Helly fixation. Hydrochloric acid–acetoorcein stain. The bar corresponds to 5 μm. Two focal levels of the typical two-track anaphase constellation of a dividing hyphal nucleus. Compare with Fig. 5 of *Trichosporon*.

using a pipet and then draining some of the agar off by tilting the slide back and forth until about four-fifths of it is evenly coated. The slide is then transferred to a moist chamber to await inoculation. Slightly thicker but equally serviceable films are obtained by dipping a slide into molten agar in a petri dish and then draining it by holding it vertically for a few seconds (Aist, 1969). Agar films can either be inoculated with spores or given a head start with small pieces cut from the edge of growing cultures. To quote Aist (1969):

> Two agar blocks about 3 mm wide were cut from the periphery of a rapidly growing culture and were inverted and placed on the agar slide 2 cm apart. The inoculated agar slides were incubated . . . at 22°C overnight or until about 3 mm of growth were obtained. . . . A razor blade was used to trim the agar film to a thin rectangular block, 10 mm by 5 mm, on which a section of only one of the mycelia was growing. A coverslip was then applied directly to the mycelium and the edges were sealed with paraffin.

An even simpler method of Girbardt's (1956a,b) allows study of the same hypha first in living condition and again after fixation and sectioning for electron microscopy. Here a coverslip is placed directly upon the advancing edge of a slide culture. No agar is cut away, and no seal is applied. A lag period of ½–1 h is allowed for adjustment to the new growth conditions. When observations have been completed, the coverslip is lifted off and the culture flooded with fixative. It is this "target preservation technique" that has enabled Girbardt (1978) to understand the dynamics rather than merely the morphology of mitosis in the basidiomycete *Trametes* (=*Polystictus*).

Fig. 31.   Phase-contrast micrograph of a hypha of *Aspergillus nidulans* (diploid strain) growing on a nutrient 18% gelatin medium. The bar denotes 5 μm. The contrast between the high density of the nucleolus and the uniform low density of the rest of the nuclear contents is characteristic of fungal nuclei in general.

Figs. 32 and 33.   Anaphase and telophase, respectively, of mitosis in *A. nidulans*. Helly fixation. Direct staining with acetoorcein. The affinity of the cytoplasm for the stain is greater than it would have been after hydrolysis. The scale in Fig. 33 denotes 5 μm.

Fig. 34.   Phase-contact micrograph of a hypha of the *Chromelosporium* anamorph of *Peziza ostracoderma* growing on 16% gelatin with yeast extract and glucose. Magnification as in Fig. 31. An arrow points to a dense NAO at the upper pole of the nucleus.

Fig. 35.   Spindle (a) and metaphase (b) chromosomes of the same dividing nucleus of a species of *Penicillium*. Helly fixation. Acid fuchsin and hydrochloric acid–Giemsa, respectively. The bar represents 5 μm.

Figs. 36 and 37.   Two views of hyphae of *Aureobasidium pullulans* showing synchronous mitoses and blastoconidium ontogeny. The conidia are not provided with nuclei at this stage. Neighboring hyphae contain resting nuclei with chromatin confined to a crescent curving around the unstained nucleolus. Helly fixation. Hydrochloric acid acetoorcein. The scale denotes 5 μm.

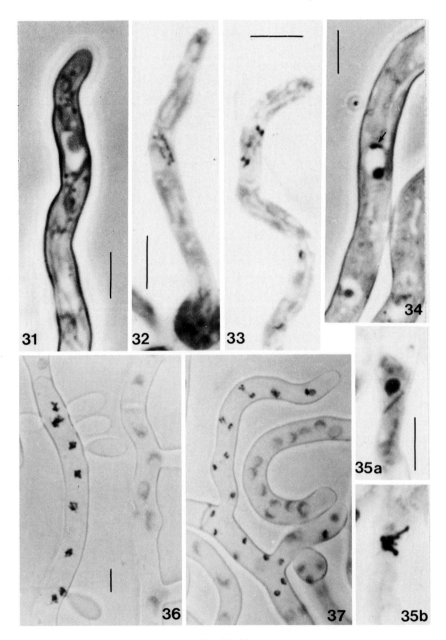

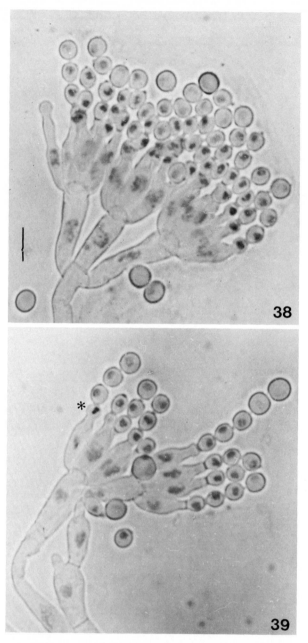

Figs. 38 and 39. Hydrochloric acid–acetoorcein-stained nuclei in metulae, phialides, and conidia of *Penicillium* sp. (same species as shown in Figs. 13–21). The asterisk marks a telophase nucleus that has moved into a conidium initial. The magnification is the same in both figures; the bar in Fig. 38 represents 5 μm. From a coverslip–agar culture. Prefixed in vapors of Formalin–acetic acid–alcohol and postfixed in Newcomer's preservative.

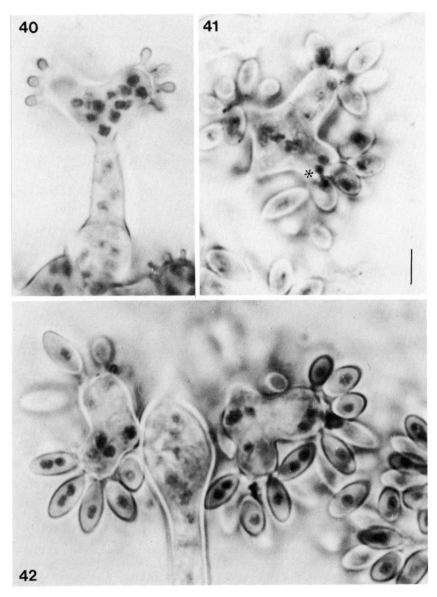

Figs. 40–42. Nuclei in ampullae and developing conidia of a species of *Botryosporium*. In Fig. 41 an asterisk indicates a nucleus preserved in the process of entering a conidium. Hydrochloric acid–acetoorcein. Fixation same as for Figs. 38 and 39. The bar represents 5 μm.

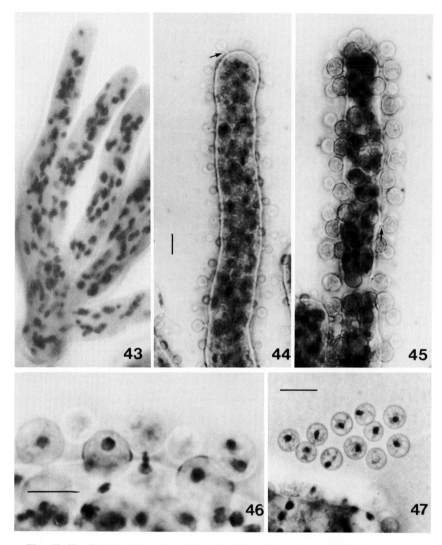

Figs. 43–47. Blastoconidia and ampullae of the *Chromelosporium* anamorph of *Peziza ostraco-derma*. Fixation as in Figs. 38 and 39 of *Penicillium*. Fig. 43. Hydrochloric acid–Giemsa. Figs. 44–47. Hydrochloric acid–acetoorcein. Bars in Figs. 44, 46, and 47 each represent 5 μm. Same magnification for Figs. 43–45. In Fig. 43 the ampullae are crammed with nuclei. Arrows in Figs. 44 and 45 point to conidia arising from ampullae on short stalks (denticles). A nucleus preserved in the process of entering a conidium is shown in Fig. 46. This set of photomicrographs complements the electron micrographs illustrating the same process published by Hughes and Bisulpatra (1970).

Phase-contrast images of hyphae and yeasts examined in watery media tend to be surrounded by halos of light which in work with hyphae are not necessarily inimical to good optical resolution. Girbardt (1965) and Aist (1969), on the other hand, obtained detailed information on the organization of the cytoplasm and nuclei in hyphae of cultures growing in 4% malt extract and 5% maltose agar, respectively. However, the experience of Müller (1956) and Robinow and Marak (1966) has shown that in work with yeasts liquid media should be replaced by 14–25% gelatin whose high refractility greatly reduces the brightness of the disturbing halos.

B. Phase-Contrast Microscopy of Budding Yeasts

Methods for preparing living yeast for phase-contrast microscopy have been described in detail by Robinow and Marak (1966) and Robinow (1975). What follows is based on the latter article with the addition of a diagrammatic outline for clarification (Fig. 53). The steps described below correspond to the numbers in Fig. 53.

Steps 1 and 2. Yeast from an overnight culture is incubated on a suitable agar medium at 30°–35°C for 4–6 h.

Steps 2, 3, and 4. A small loopful of molten gelatin medium containing 0.5% Difco yeast extract and 2% glucose is solidified by placing it in the center of a thin (less than 1.0 mm) slide resting on the lid of a petri dish filled to the brim with ice cold water (step 4). A slab of agar measuring 0.5 × 1.0 mm is next cut from the streaked culture (steps 2 and 3), and the tiny mound of congealed gelatin medium is inoculated using a fine, flexible glass fiber charged with cells from the growth on the agar slab. The cells are first stirred into a slurry with a little water using an inoculation loop (steps 3 and 4). Inoculation can also be done by lightly touching the gelatin with the yeast-bearing surface of a small block cut from the larger block (steps 3 and 4a).

Step 5. A no. 0 coverslip (22 × 30 mm) is placed over the inoculated gelatin, slightly flattening it at the point of contact.

Step 6. The slide with the gelatin and its attached coverslip is then transferred to the lid of a second petri dish filled to overflowing with warm water (about 60°C). As soon as the gelatin begins to spread beneath the cover glass, the slide is returned to the cold dish to arrest further expansion.

Step 7. The assembly is sealed with paraffin wax from a small birthday candle.

Under the light microscope, cells close to the edge of the flattened droplet allow the best resolution. The visibility of intracellular detail improves with time.

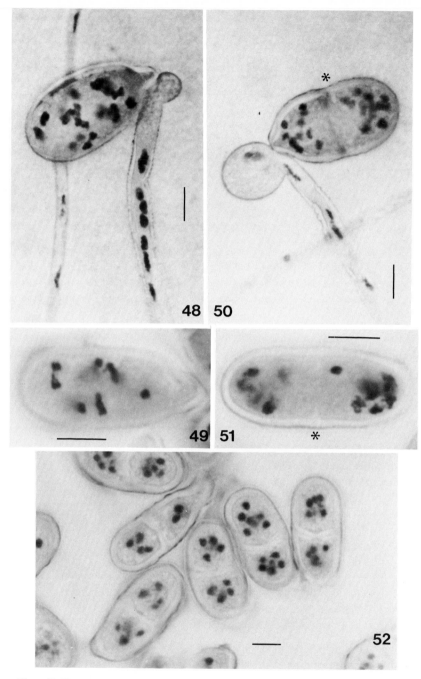

Figs. 48–52. *Trichothecium roseum.* Mercuric chloride–alcohol–acetic acid. Hydrochloric acid–Giemsa. Same magnification for Figs. 48 and 50. Also, Figs. 49 and 51 magnified equally. All

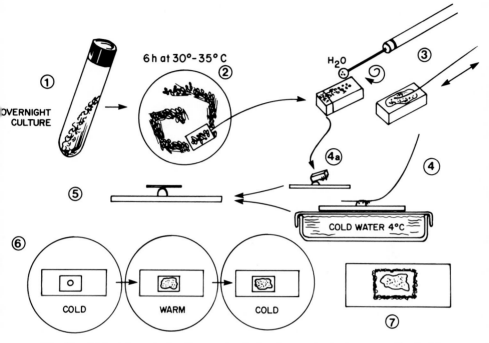

6 h at 30°–35° C

OVERNIGHT CULTURE

H₂O

COLD WATER 4°C

COLD   WARM   COLD

Fig. 53. Method for preparing live yeast cells for phase-contrast microscopy. (See text for explanation.)

C. Fixed and Stained Preparations of Germ Tubes, Mycelia, and Yeast

1. Cultivation and Handling of Germ Tubes and Mycelia

A technique for growing mycelia on the surface of cellophane overlying a nutrient medium was introduced by Fleming and Smith (1944) and has been used in studies on fungal nuclei, chromosomes, and spindles by numerous workers (e.g., Robinow and Caten, 1969; Morris, 1975). This technique merits no further description. Fixed and stained mycelia on cellophane provide excellent overviews of the course of mitosis and are entirely adequate for studies involving counts of nuclei or spindles but cannot, in my experience, be squashed flat enough for optical resolution of mitotic figures because squashing pressure

---

Figs. 48–52 (*continued*).

bars denote 5 μm. The nuclei in the conidia in Figs. 48 and 49 were dividing by mitosis and are distributed at random. In Figs. 50 and 51 the nuclei have completed mitosis and have moved to opposite ends of the as yet undivided conidium. Asterisks mark the beginning of the growth of transverse septa. An overview of mature conidia whose nuclei are again randomly distributed is shown in Fig. 43.

makes hyphae roll about on the cellophane and leaves them distorted. These undesirable aspects can be avoided by growing fungi on glass *beneath* the cellophane using a slight modification of the technique of Clutterbuck and Roper (1966) which is described in detail below. The steps described below correspond to the numbers in Fig. 54.

Step 1. A suspension of conidia in water containing a trace of Tween (soluble oleic acid ester) is centrifuged for several minutes in an ordinary clinical bench centrifuge at 3000 rpm. The pellet is resuspended in distilled water, and the suspension is spun once again.

Step 2. The pellet of washed conidia is resuspended in 3 ml of a mixture of equal parts of a solution of 10% filtered albumin and 0.2% agar in water. The albumin is prepared by shaking 1 pt fresh egg white with 9 pt distilled water in a measuring cylinder. The globulins are allowed to settle, and the supernatant is filtered. The supernatant keeps well in the refrigerator. The agar is melted in a bath of boiling water and kept at a temperature of about 60°C.

Step 3. A loopful of resuspended conidia is spread over a 22 × 22 mm no. 1 coverslip. Four to eight coverslips are prepared and allowed to dry for 15 min at 30°C.

Step 4. A strip of cellophane sterilized by boiling in water is blotted twice between several thicknesses of "bibulous" paper and placed over the dried film of conidia. Very dry cellophane will not make contact with the film of conidia. In this case a small drop of water (one touch of an inoculation loop) is placed on the film of conidia and the cellophane strip then applied. The water will spread beneath it.

Step 5. A slab of agar, 7 × 15 mm, is next placed on top of the cellophane, and the completed assembly, agar facing upward, is placed on a microscope slide which is transferred to a large petri dish containing several layers of glass beads and water. The closed dish is incubated until germ tubes of the desired length have formed. I find that with *Penicillium* .this point is reached after 10 ½ h incubation at 30°C. Cultures may also be started in the evening at 15°C and transferred to 30° or 35°C the following morning.

Step 6. At the appropriate time the slides bearing the agar–cellophane–cover glass sandwiches are removed from the moist chamber. The strip of cellophane is seized by one of the ends protruding from beneath the agar and is peeled off in a single, rapid movement. The coverslip bearing the germ tubes is instantly plunged into freshly prepared fixative standing ready in Columbia staining jars (Arthur H. Thomas Company, Philadelphia, Pennsylvania.)

2. Cultivation and Handling of Yeasts

The preparation of yeast samples for light microscopy has already been described in considerable detail (Robinow and Marak, 1966; Robinow, 1975, 1977), and only a brief summary is provided below.

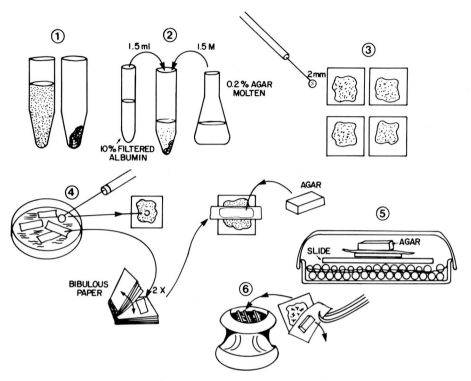

Fig. 54. Method for preparing conidia for nuclear staining and light microscope examination. (See text for explanation.)

Nuclear behavior in yeasts is best studied in young, growing cultures. All stages of nuclear and cell division are encountered in sufficient number in randomly dividing populations. A suspension of yeasts from an overnight slant culture is prepared in sterile distilled water. A petri dish containing a suitable agar medium is flooded with 1 ml of the suspension. The dish, its lid partly displaced, is then propped upright in a 30°–35°C incubator for 20 min. Excess inoculum is removed by aspiration or with pieces of filter paper. The culture, lid replaced, is subsequently incubated for an additional 6–8 h or until a sufficient density of single-layered colonies has been obtained. One of three different procedures may be used for adherence of yeast to a coverslip followed by their fixation.

1. Slabs of agar are cut from the culture and removed from the dish. A coverslip, thinly smeared with fresh egg white, is lowered onto the yeast colonies and tapped lightly on the back with an inoculating needle to ensure good contact. The agar-yeast-coverslip sandwich is turned on its back, the agar block is flicked off, and the coverslip is immediately immersed in fixative.

2. A ridge of yeasts is piled up by scraping the agar surface with the edge of a coverslip. A narrow slab of agar bearing the yeasts is cut from the culture and placed, yeasts down, on one side of a coverslip. With a small wire loop a thin trail of fresh egg white is laid down along the side of the agar slab that faces the exposed surface of the coverslip. The agar slab is then swiftly pushed across the coverslip with the help of a fine scalpel or flattened wire. A thin film of egg white and cells is thus spread over the coverslip and is at once immersed in fixative.

3. Yeast may also be transferred to glass coverslips without disturbing the colonies. This is achieved by growing it on thinly poured agar in petri dishes, inverting agar slabs cut from the culture, and placing them over coverslips coated with a thin, dry film of 8–10% filtered albumin. The intact assemblies are plunged directly into fixative.

3. Fixation and Staining of Germ Tubes, Mycelia, and Yeast

All operations are performed in Columbia dishes. These are Coplin-style slotted vessels of about 12 ml capacity designed to hold four 22 × 22 mm coverslips (or eight, back to back) and are indispensable for work with mycelial fungi and yeasts.

a. Fixatives (as Given by Robinow, 1975)

1. *Modified Helly's.* Stock solution: mercuric chloride, 5 g; potassium dichromate, 3 g; distilled water, 100 ml. Immediately before use add Formalin to make 5–6%. Fix for 10–20 min. Wash and store in 70% ethanol.

2. *Schaudinn's half-strength.* Absolute ethanol, 1 part; saturated solution of mercuric chloride, 2 parts; water, 3 parts. Mix and add glacial acetic to make 2%. Fix for 15 min. Wash and store in 70% ethanol. Recommended for *Penicillium* and *Aspergillus*.

3. *FAA.* Formalin, 5 parts; acetic acid, 5 parts; 95% ethanol, 50 parts; water, 40 parts. Recommended for fixation of films of yeasts on coverslips (in addition to no. 1 and no. 2) and required for fixation of yeasts *through* agar as described in Sections VI, C, 2–3.

b. Staining Procedures (Equally Suitable for Preparations of Mycelial Fungi and Yeasts)

1. *Hydrolysis.* Carried out for 8–10 min in 1 N hydrochloric acid at 60°C. Preparations are taken directly from 70% ethanol to the acid which stands ready in a covered Columbia jar immersed in a water bath. Preparations of yeast are first placed in 1% sodium chloride in water at 60°C for 1 h *before* hydrolysis.

2. *Giemsa.* Stock solution (15–20 drops) is added to 10 ml of Giemsa buffer in a Columbia jar. The buffer is a 10- to 20-fold dilution of Soerensen's 15 M phosphate buffer resulting in a pH of 6.8. Tablets for making ½–1 liter of buffer

are available from biological supply houses. Preparations should be stained for several hours and results checked with a water immersion lens (×40) combined with a good quality ocular lens (×15). Overstaining can be corrected by immersion for a few seconds in distilled water to which two loopfuls of acetic acid have been added. Two open dishes are placed next to the microscope, one containing buffer and the other containing acidified water. The specimen is moved back and forth between the two solutions until contrast between nuclei, chromosomes, and cytoplasm is satisfactory. Further loss of stain and contrast during flattening should be allowed for. The specimen is next mounted on a slide in Giemsa buffer. Excess fluid is removed by gently touching the edges of the coverslip with strips of filter paper. Preparations of fungal germ tubes need to be flattened with the help of a press such as that of Miller and Colaiace (1970). Yeasts tend to become greatly distorted if they are squashed too severely. Mere thumb pressure exerted through a piece of filter paper will often suffice. The flattened preparation is sealed with nail varnish or, better, with Glyceel (Hopkin and Williams, Chadwell Heath, Essex, England) or a similar sealant for wet mounts.

3. *Acetoorcein* (1% synthetic orcein in 60% acetic acid). Four points should be noted: (a) Synthetic orcein from the Fisher Scientific Company is considered the most reliable brand of orcein. (b) Best results are obtained if hydrolysis (8 min at 60°C) and staining (½ h at 30°C) are carried out soon after fixation. (c) Yeasts need not be extracted with 1% sodium chloride before hydrolysis if they are to be stained with acetoorcein. (d) Flattening by mechanical pressure is not recommended. Stained films are mounted over a drop of the stain and gently flattened by pressure through many layers of blotting paper. Results vary among species.

4. *Feulgen–acetocarmine*. Specimens are first subjected to the standard Feulgen staining procedures and poststained with acetocarmine (1% carmine, alum lake) in 45% acetic acid. Recommended by McIntosh (1954), this method has yielded satisfactory results for *Mucor* (Robinow, 1957) and, after fixation in Schaudinn's half-strength solution, has also provided clear images of the chromosomes of *Allomyces* (Robinow, 1962).

5. *Acid fuchsin*. Low concentrations of this dye, 1:40,000–1:60,000 in 1% acetic acid applied for 1 ½–3 ½ min, selectively stain NAOs of resting nuclei and mitotic spindles in mycelial fungi and yeasts (Robinow and Caten, 1969; Robinow, 1975). Close monitoring of the progress of staining with the water immersion lens is essential. Stained films are mounted over 1% acetic acid. Satisfactory results have been obtained for a wide range of species.

4. Fixation of Conidiophores

The following procedure was used for the specimens illustrated in this chapter:

1. Two blocks of yeast extract–glucose agar are placed side by side on a microscope slide but not in contact, inoculated with conidia, and then covered with a 22-mm² flamed coverslip bridging the gap between the agar blocks.

2. The coverslip–agar block cultures are incubated in a moist chamber until a fringe of sporulating structures appears along the edge of the blocks.

3. The culture is initially fixed by placing the slide for 15 min above several layers of glass beads in a petri dish containing 5 ml glacial acetic acid, 5 ml formalin, and 5 ml absolute ethanol.

4. After prefixation the coverslip is removed from the agar blocks and immersed in Schaudinn's half-strength fixative.

5. Fixed conidiophores adhering to the coverslip are then hydrolyzed and stained with Giemsa or acetoorcein.

5. Light Microscopy

The student who has followed these procedures thus far is not ensured of success unless he or she can examine the preparations under a well-equipped and properly illuminated microscope. This is not the place for a discussion of the principles of microscopy, but practical advice will be found in Murray and Robinow (1980).

ACKNOWLEDGMENTS

I wish to thank Mrs. Susan Legeza of the Plant Sciences Department, University of Western Ontario, for help with cultures and much sound mycological advice, Mr. George Sanders for steadfast expert help in the darkroom, and my colleague John Marak for his sustained unfailing technical support in a wide range of matters.

Thanks are due to Bryce Kendrick for initiating the writing of this chapter and defraying some of the expenses of the photographic work.

I am also greatly obliged to Dr. Garry Cole for constructive criticism and seamless carpentry performed on the manuscript.

REFERENCES

Aist, J. R. (1969). The mitotic apparatus in fungi: *Ceratocystis fagacearum* and *Fusarium oxysporum*. *J. Cell Biol.* **40**, 120–135.

Aist, J. R., and Williams, P. H. (1972). Ultrastructure and time course of mitosis in the fungus *Fusarium oxysporum*. *J. Cell Biol.* **55**, 268–389.

Bakerspigel, A. (1961). Vegetative nuclear division in the ascomycete *Ophiostoma fimbriata*. *Cytologia* **26**, 42–49.

Clutterbuck, A. J., and Roper, J. A. (1966). A direct determination of nuclear distribution in heterokaryons of *Aspergillus nidulans*. *Genet. Res.* **7**, 185–194.

Colson, B. (1938). The cytology and development of *Phyllactinia corylea* Lev. *Ann. Bot. (London)* [NS] **2**, 381–402.

Conde, J., and Fink, G. R. (1976). A mutant of *Saccharomyces cerevisiae* defective for nuclear fusion. *Proc. Natl. Acad. Sci. U.S.A.* **73**, 3651–3655.

Crackower, S. (1972). Mitosis in *Cladosporium herbarum*. *Can. J. Microbiol.* **18**, 692–694.

Crackower, S. H. B., and Bauer, H. (1971). Mitosis in *Penicillium chrysogenum* and *Penicillium notatum*. *Can. J. Microbiol.* **17**, 605–608.

Day, A. W. (1972). Genetic implications of current models of somatic nuclear division in fungi. *Can. J. Bot.* **50**, 1337–1347.

Duncan, E. J., and Macdonald, J. A. (1965). Nuclear phenomena in *Marasmius androsaceus* and *M. rotula*. *Trans. Soc. Edinburgh* **66**, 129–141.

Elliot, C. G. (1960). The cytology of *Aspergillus nidulans*. *Genet Res.* **1**, 462–476.

Fleming, A., and Smith, G. (1944). Some methods for the study of moulds. *Trans. Br. Mycol. Soc.* **27**, 13–19.

Fuller, M. S. (1976). Mitosis in fungi. *Int. Rev. Cytol.* **45**, 113–153.

Gaümann, E. (1949). "Die Pilze." Birkhaeuser, Basel.

Girbardt, M. (1956a). Eine Methode zum vergleich Lebender mit fixierten Strukturen bei Pilzen. *Z. Wiss. Mikrosk. Mikrosk. Tech.* **63**, 16–21.

Girbardt, M. (1956b). Eine zielschnitt Methode fur Pilzzellen. *Mikroskopie* **20**, 254–264.

Girbardt, M. (1960a). Licht- und elektronenoptische Unterschungen an *Polystictus versicolor*. **(L.).** *Ber. Dtsch. Bot. Ges.* **73**, 227–240.

Girbardt, M. (1960b). Licht- und elektronenoptische Unterschungen an *Polystictus versicolor*. (L.). VI. *Planta* **55**, 365–380.

Girbardt, M. (1961). Licht- und elektronenoptische Unterschungen an *Polystictus versicolor*. (L.). VII. *Exp. Cell Res.* **23**, 181–194.

Girbardt, M. (1965). Lebendnachweis von *Einzelelementen des endoplasmatischen Retikulums J. Cell Biol.* **27**, 433–440.

Girbardt, M. (1968). Ultrastructure and dynamics of the moving nucleus. *Symp. Soc. Exp. Biol.* **22**, 249–259.

Girbardt, M. (1971). Ultrastructure of the fungal nucleus. II. The kinetochore equivalent (KCE). *J. Cell Sci.* **9**, 453–473.

Girbardt, M. (1973). Die Pilzzelle. *In* "Grundlagen der Cytologie" (G. C. Hirsch, H. Ruska, and P. Sitte, eds.), pp. 441–460. Fischer, Jena.

Girbardt, M. (1978). Historical review and introduction. *In* "Nuclear Division in the Fungi" (I. B. Heath, ed.), pp. 1–20. Academic Press, New York.

Girbardt, M., and Hadrich, H. (1975). Ultrastruktur des Pilzkernes. III.Genase des Kern-assoziierten Organells (NA = "KCE"). *Z. Allg. Mikrobiol.* **15**, 157–173.

Harder, D. E. (1976a). Mitosis and cell division in some cereal rust fungi. I. *Can. J. Bot.* **54**, 981–994.

Harder, D. E. (1976b). Mitosis and cell division in some cereal rust fungi. II. *Can. J. Bot.* **54**, 995–1009.

Harper, R. A. (1905). Sexual reproduction and the organization of the nucleus in certain mildews. *Carnegie Inst. Washington Publ.* **37**, 104.

Heath, I. B. (1978). Experimental studies of mitosis in the fungi. *In* "Nuclear Division in the Fungi" (I. B. Heath, ed.), pp. 38–176. Academic Press, New York.

Heath, I. B. (1980a). Variant Mitoses in Lower Eukaryotes: Indicators of the Evolution of Mitosis? *Int. Rev. Cytol.* **64**, 1–80.

Heath, I. B. (1980b). Fungal mitoses, the significance of variations on a theme. *Mycologia* **74**, 229–250.

Heath, I. B., and Greenwood, A. D. (1970). Centriole replication and nuclear division in *Saprolegnia. J. Gen. Microbiol.* **62**, 139–148.

Hrushovetz, S. B. (1956). Cytological studies of *Helminthosporium sativum*. *Can. J. Bot.* **34**, 321–327.

Huang, H. C., and Tinline, R. D. (1974). Somatic mitosis in haploid and diploid strains of *Cochliobolus sativus*. *Can. J. Bot.* **52**, 1561–1568.

Hughes, G. C., and Bisalputra, A. A. (1970). Ultrastructure of hyphomycetes: Conidium ontogeny in *Peziza ostracoderma*. *Can. J. Bot.* **48**, 361–366.

Kendrick, B., and Chang, M. G. (1971). Karyology of conidiogenesis in some hyphomycetes. *In* "Taxonomy of Fungi Imperfecti" (B. Kendrick, ed.), pp. 279–291. Univ. of Toronto Press, Toronto.

Kôpecká, M., and Gabriel, M. (1978). Staining the nuclei in cells and protoplasts of living yeasts, moulds, and green algae with the antibiotic lomofungin. *Arch. Microbiol.* **119**, 305–311.

Kozakiewicz, Z. (1978). Phialide and conidium development in the aspergilli. *Trans. Br. Mycol. Soc.* **70**, 175–186.

Kubai, D. (1975). The evolution of the mitotic spindle. *Int. Rev. Cytol.* **43**, 167–227.

Kubai, D. (1978). Mitosis and fungal phylogeny. *In* "Nuclear Division in the Fungi" (I. B. Heath, ed.), pp. 177–299. Academic Press, New York.

McCully, E. K., and Robinow, C. F. (1972a). Mitosis in heterobasidiomycetous yeasts. I. *Leucosporidium scottii*. (*Candida scottii*). *J. Cell Sci.* **10**, 857–881.

McCully, E. K., and Robinow, C. F. (1972b). Mitosis in heterobasidiomycetous yeasts. II. *Rhodosporidium* sp. (*Rhodotorula glutinis*) and *Aessosporon salmonicolor*. *J. Cell Sci.* **11**, 1–31.

McCully, E. K., and Robinow, C. F. (1973). Mitosis in *Mucor hiemalis*, a comparative light and electron microscopical study. *Arch. Mikrobiol.* **94**, 133–148.

McIntosh, D. L. (1954). A Feulgen-carmine technique for staining fungus chromosomes. *Stain Technol.* **29**, 29–31.

McKeen, W. E. (1972). Somatic mitosis in *Erysiphe graminis hordei*. *Can. J. Microbiol.* **18**, 1915–1922.

Miller, M. W., and Colaiare, J. D. (1970). Elimination of material that obscures stained chromosomes in squashes of *Vicia faba* root tips. *Stain Technol.* **45**, 81–86.

Morris, N. R. (1975). Mitotic mutants of *Aspergillus nidulans*. *Gent. Res.* **26**, 237–254.

Morris, N. R. (1980). Chromosome structure and the molecular biology of mitosis in eukaryotic microorganisms. *Symp. Soc. Gen. Microbiol.* **30**, 41–75.

Müller, R. (1956). Zur Verbesserung der Phasenkontrastmikroskopie durch Verwendung von medien optimaler Brechungsindices. *Mikroskopie* **11**, 36–46.

Murray, R. G. E., and Robinow, C. F. (1980). Microscopes and microscopy. *In* "A Manual of Methods for General Bacteriology" (P. Gerhardt, ed.). Am. Soc. Microbiol., Washington, D.C. (to be published).

Olive, L. S. (1949). Karyogamy and meiosis in the rust *Coleosporium vernoniae*. *Am. J. Bot.* **36**, 41–54.

Peterson, J. B., and Ris, H. (1976). Electron-microscopic study of the spindle and chromosome movement in the yeast *Saccharomyces cerevisiae*. *J. Cell Sci.* **22**, 219–242.

Peterson, J. B., Gray, R. H., and Ris, H. (1972). Meiotic spindle plaques in *Saccharomyces cerevisiae*. *J. Cell Biol.* **53**, 837–841.

Piekarski, G. (1937). Zytologische Untersuchungen an Paratyphus- und Colibakterien. *Arch. Mikrobiol.* **8**, 428–439.

Poon, N. H., and Day, A. W. (1974). Somatic nuclear division in the sporidia of *Ustilago violacea*. II. Observations on living cells with phase-contrast and fluorescence microscopy. *Can. J. Microbiol.* **20**, 739–746.

Poon, N. H., Martin, J., and Day, A. W. (1974). Conjugation in *Ustilago violacea*. I. Morphology. *Can. J. Microbiol.* **20**(2), 187–191.

Robinow, C. F. (1957). The structure and behaviour of the nuclei in spores and growing hyphae of *Mucorales*. I. *Mucor hiemalis* and *M. fragilis*. *Can. J. Microbiol.* **3**, 771–789.

Robinow, C. F. (1962). Some observations on the mode of division of somatic nuclei of *Mucor* and *Allomyces*. *Arch. Mikrobiol.* **42**, 369–377.

Robinow, C. F. (1975). The preparation of yeasts for light microscopy. *Methods Cell Biol.* **11**, 1–22.

Robinow, C. F. (1977). The number of chromosomes in *Schizosaccharomyces pombe*. Light microscopy of stained preparations. *Genetics* **87**, 491–497.

Robinow, C. F., and Caten, C. E. (1969). Mitosis in *Aspergillus nidulans*. *J. Cell Sci.* **5**, 403–431.

Robinow, C. F., and Marak, J. (1966). A fiber apparatus in the nucleus of the yeast cell. *J. Cell Biol.* **29**, 129–151.

Rosenberger, R. F., and Kessel, M. (1968). Nonrandom sister chromatid segregation and nuclear migration in hyphae of *Aspergillus nidulans*. *J. Bacteriol.* **96**, 1208–1213.

Setliff, E. C., Hoch, H. C., and Patton, R. F. (1974). Studies on nuclear division in basidia of *Poria atemarginata*. *Can. J. Bot.* **52**, 2323–2333.

Slater, M. L. (1976). Rapid nuclear staining method for *Saccharomyces cerevisiae*. *J. Bacteriol.* **126**, 1336–1341.

Thielke, C. H. (1974). Intranucleare Spindeln und Reduktion des Kernvolumens bei der Meiose von *Coprinus radiatus* (Bolt) Fr. *Arch. Microbiol.* **98**, 225–235.

von Arx, J. A. (1967). "Pilzkunde," pp. 1–356. Cramer, Lehre, Germany.

Wilson, C. L., and Aist, J. R. (1967). Motility of fungal nuclei. *Phytopathology* **57**, 769–771.

Zachariah, K., and Metitiri, P. O. (1971). The organization of the penicillus of *Penicillium claviforme* Bainier. *In* "Taxonomy of Fungi Imperfecti" (B. Kendrick, ed.), pp. 120–142. Univ. of Toronto Press, Toronto.

Zickler, D. (1970). Division spindle and centrosomal plaques during mitosis and meiosis in some ascomycetes. *Chromosoma* **30**, 287–304.

Zickler, D. (1973). Evidence for the presence of DNA in the centrosomal plaques of *Ascobolus*. *Histochemie* **34**, 227–238.

26

Viruses of Conidial Fungi

Paul A. Lemke

I. INTRODUCTION

The conidium is a common cell type among fungi, and it is therefore not surprising that many of the fungi known to contain viruses and viruslike particles are indeed conidial fungi (Table I). What is rather surprising in this context is that, among conidial fungi and among anamorphic fungi generally, there are few if any host phenotypes related specifically to the presence of viruses (Lemke, 1976; Hollings, 1978; Saksena and Lemke, 1978).

This is not true in the case of a number of teleomorphic fungi. A virus-specific disease, for example, exists in the basidiomycete *Agaricus brunnescens* (Hollings, 1962; Dieleman-van Zaayen, 1972; Marino *et al.*, 1976; Lemke, 1977a), and killer-immune phenotypes related to specific molecules of viral double-stranded RNA (dsRNA) exist in the yeast *Saccharomyces cerevisiae* (Bevan *et al.*, 1973; Wickner, 1976) and the smut *Ustilago maydis* (Wood and Bozarth, 1973; Koltin and Day, 1976a). In two phytopathogenic teleomorphs, *Endothia parasitica* (Day *et al.*, 1977) and *Thanatephorus cucumeris* (Castanho *et al.*, 1978), the presence of dsRNA (presumably of viral origin) has been correlated

Biology of Conidial Fungi, Vol. 2

TABLE I

Viruses and Viruslike Particles among Fungi

Species[a]	Reference
Conidial fungi	
Acremonium chrysogenum	Day and Ellis (1971),
	Tikchonenko (1978)
Alternaria tenuis	Isaac and Gupta (1964)
Aspergillus flavus	Mackenzie and Adler (1972)
Aspergillus foetidus[b]	Banks *et al.* (1970)
Aspergillus glaucus	Hollings and Stone (1971)
Aspergillus niger[b]	Banks *et al.* (1970)
Candida tropicalis	Nesterova *et al.* (1973)
Candida utilis	Kozlova (1973)
Chalara elegans (=*Thielaviopsis basicola*)	Yamshita *et al.* (1975),
	Bozarth and Goenaga (1977)
Cochliobolus miyabeanus	Yamashita *et al.* (1975)
Colletotrichum atramentarium	Yamashita *et al.* (1975)
Colletotrichum falcatum[c]	Moffitt and Lister (1975)
Colletotrichum graminicola[c]	Moffitt and Lister (1975)
Colletotrichum lindemuthianum[b]	Rawlinson (1973),
	Lecoq and Delhotal (1976)
Coremiella cubispora (=*Briosia cubispora*)	Cole (1975)
Fusarium moniliforme	Bozarth (1972)
Fusarium roseum[b]	Chosson *et al.* (1973),
	Moffitt and Lister (1975)
Gonatobotrys sp.	Spire *et al.* (1972b)
Helminthosporium carbonum[c]	Dunkle (1974b),
	Moffit and Lister (1975)
Helminthosporium maydis[b]	Bozarth *et al.* (1972b),
	Bozarth (1977)
Helminthosporium oryzae	Spire *et al.* (1972a)
Helminthosporium sacchari	Yamashita *et al.* (1975)
Helminthosporium turcicum[c]	Moffitt and Lister (1975)
Helminthosporium victoriae[b]	Bozarth (1972),
	Sanderlin and Gabriel (1976)
Mycogone perniciosa	Lapierre *et al.* (1972)
Penicillium brevicompactum[b]	Wood *et al.* (1971),
	Velikodvorskaya *et al.* (1972),
	Lemke and Ness (1970)
Penicillium chrysogenum[b]	Volkoff *et al.* (1972),
	Banks *et al.* (1969a),
	Bozarth and Wood (1977),
	Tikchonenko (1978)
Penicillium citrinum[b]	Borré *et al.* (1971),
	Benigni *et al.* (1977)
Penicillium claviforme[b]	Lai and Zachariah (1975)
Penicillium cyaneofulvum[b]	Banks *et al.* (1969b)

(*continued*)

TABLE I

Species[a]	Reference
Penicillium funiculosum[b]	Banks *et al.* (1968)
Penicillium multicolor	Mackenzie and Adler (1972)
Penicillium notatum	Hollings and Stone (1971)
Penicillium stoloniferum[b]	Ellis and Kleinschmidt (1967),
	Banks *et al.* (1968)
Penicillium variabile	Borré *et al.* (1971)
Periconia circinata[b]	Dunkle (1974a)
Pyricularia oryzae	Férault *et al.* (1971),
	Yamashita *et al.* (1971)
Rhodotorula glutinis	Kozlova (1973)
Sclerotium cepivorum	Lapierre and Faivre-Amoit (1970)
Stemphylium botryosum	Hollings and Stone (1971)
Verticillium fungicola	Lapierre *et al.* (1973)
Ascomycetes	
Diplocarpon rosae	Bozarth *et al.* (1972a)
Endothia parasitica[c]	Moffitt and Lister (1975),
	Day *et al.* (1977)
Erysiphe graminis	Yamashita *et al.* (1975),
	Nienhaus (1971)
Gaeumannomyces graminis[b]	Rawlinson *et al.* (1973),
	Lapierre *et al.* (1970),
	Moffitt and Lister (1975)
Hypoxylon multiforme	Wyn-Jones and Whalley (1977)
Microsphaera mougeotii	Yamashita *et al.* (1975)
Neurospora crassa	Tuveson *et al.* (1975),
	Tuveson and Sargent (1976),
	Tuveson and Peterson (1972),
	Lechner *et al.* (1972),
	Küntzel *et al.* (1973)
Peziza ostracoderma	Spire (1971),
	Dieleman-van Zaayen (1967)
Saccharomyces carlsbergensis	Volkoff and Walters (1970)
Saccharomyces cerevisiae[b]	Bevan *et al.* (1973)
Saccharomyces ludwigii	Kozlova (1973)
Sphaerotheca lanestris	Nienhaus (1971)
Basidiomycetes	
Agaricus brunescens[b]	Hollings (1962),
	Dieleman-van Zaayen and Temmink (1968),
	Hollings and Stone (1971),
	Dieleman-van Zaayen (1967),
	Marino *et al.* (1976)
Boletus edulis	Huttinga *et al.* (1975)
Coprinus congregatus[b]	Ross (1979)
Coprinus lagopus	Shahriari *et al.* (1973)
Laccaria amethystina	Blattny and Kralik (1968)

(*continued*)

TABLE I (*continued*)

Species[a]	Reference
Laccaria laccata	Blattny and Kralik (1968)
Lentinus edodes[b]	Ushiyama and Nakai (1975),
	Ushiyama and Hashioka (1973),
	Ushiyama *et al.* (1977)
Puccinia allii	Yamashita *et al.* (1975)
Puccinia graminis	Rawlinson and MacLean (1973)
Puccinia horiana	Yamashita *et al.* (1975)
Puccinia malvacearum	Lecoq *et al.* (1974)
Puccinia miscanthi	Yamashita *et al.* (1975)
Puccinia recondita	Yamashita *et al.* (1975)
Puccinia striiformis	Lecoq *et al.* (1974)
Puccinia suaveolens	Yamashita *et al.* (1975)
Puccinia triticina	Yamashita *et al.* (1975)
Schizophyllum commune	Koltin *et al.* (1973)
Thanatephorus cucumeris	Yamashita *et al.* (1975),
	Castanho *et al.* (1978)
Tilletiopsis sp.	Bozarth (1972)
Uromyces alopecuri	Yamashita *et al.* (1975)
Uromyces durus	Yamashita *et al.* (1975)
Uromyces phaseoli	Macdonald and Heath (1978)
Ustilago maydis[b]	Wood and Bozarth (1973)
Lower fungi	
Allomyces arbuscula[b]	Khandjian *et al.* (1975)
Aphelidium sp.	Schnepf *et al.* (1970)
Blastocladiella emersonii	Khandjian *et al.* (1977),
	Cantino and Mills (1979)
Phytophthora infestans	Corbett and Styer (1977)
Pythium butleri[c]	Moffitt and Lister (1975)
Pythium sylvaticum	Brants (1971)
Rhizidiomyces sp.	Wojcik (1977)
Thraustochytrium sp.	Kazama and Schornstein (1972)

[a] Species listed include anamorphic (conidial) and teleomorphic fungi.
[b] Evidence for dsRNA as well as particles has been reported for this species.
[c] Some strains of this species contain dsRNA in the apparent absence of virus particles.

with a cytoplasmically inherited phenotype for hypovirulence. Representative species of teleomorphic fungi therefore have proven to be quite interesting as experimental systems for the study of fungal viruses.

It thus seems, at least on the surface, that the viruses of anamorphic fungi are intrinsically less interesting and unworthy of a separate discussion. The occurrence of virus particles and of viral nucleic acid in these fungi would have no special significance except for the fact that the conidium, as an efficient vehicle for cell transmission, is also an effective mechanism for transmission of an endogenous or heritable virus. This parallel in transmissibility of host cell and virus by such a cell as the conidium may not be a coincidence but rather reflect a considerable degree of coevolution between these fungi and their viruses.

Taken as a group, fungi are prone to extensive vegetative reproduction with the production of many types of propagules, often in great profusion. This is frequently accompanied by a loss or reduction in sexuality. Collectively, the viruses of fungi have thus far proven to be noninfectious as isolated particles and to be transmitted from one cell line to another by cell fusion. Cell-mediated transmission of a virus might lead to considerable loss of integrity in organization and function of a virus, and this is indeed indicated from the physicochemical and organizational details on mycoviruses (Wood, 1973; Lemke, 1976; Saksena and Lemke, 1978). The progressive loss of sexuality among fungi and of infectivity and structure among mycoviruses may very well represent a coadaptation, a compromise in which the conidium as well as other agents for fungal cell transmission has played a significant role.

II. DISCOVERY AND PROPERTIES OF FUNGAL VIRUSES

Interest in an antiviral activity associated with two conidial fungi, *Penicillium stoloniferum* (Kleinschmidt *et al.,* 1968) and *Penicillium funiculosum* (Lampson *et al.,* 1967; Banks *et al.,* 1968), led to early evidence for virus particles and to the discovery of viral dsRNA in fungi. Since these earlier studies, the evidence for viruses in fungi, mainly dsRNA-containing viruses, has grown substantially (Table I).

The dsRNA-containing virus particles described thus far from fungi are small, spherical particles characterized by a rather simple structure (Fig. 1; for review, see references by Bozarth, 1979; Lemke and Nash, 1974; Lemke, 1977b; Wood, 1973). They possess single-shelled or simple capsids, and particle diameters range from 20 to 48 nm. The distribution and complexity of their dsRNA genomes vary considerably, indicating that these viruses, despite their small size, simple structure, and characteristic dsRNA content, may be heterogeneous in evolutionary origin. Most of the dsRNA-containing fungal viruses examined thus far are serologically unrelated, and often serologically unrelated viruses exist within a given host strain. A strain of *Chalara elegans,* for example, has been shown to have at least five such viruses, each containing dsRNA (Bozarth and Goenaga, 1977).

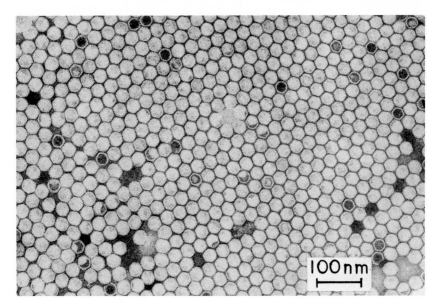

Fig. 1. Electron micrograph of negatively stained virus particles, approximately 36 nm in diameter, from a strain of *Pyricularia oryzae*. These particles contain dsRNA and were purified by sucrose density centrifugation. (Published micrograph of Yamashita and co-workers 1971, reproduced with permission of the Phytopathological Society of Japan.)

Some dsRNA-containing mycoviruses are component viruses (i.e., the genome is segmented and distributed among a population of serologically identical particles); the electrophoretically slow virus of *Penicillium stoloniferum* is an example (Buck and Kempson-Jones, 1973, 1974; Buck, 1979) and has just two dsRNA segments—one sufficiently large (0.94 megadaltons) to encode for its single structural polypeptide (42 kilodaltons) and a second (1.1 megadaltons) sufficiently large to encode for its RNA polymerase (56 kilodaltons). This component virus therefore possesses a minimal genome and phenotype to qualify structurally as a virus. Other mycoviruses have as their genome a single large dsRNA molecule; the 6.3-megadalton dsRNA of a *Helminthosporium maydis* virus is an example and represents the largest dsRNA molecule described to date from any virus (Bozarth, 1977). Still other mycoviruses are intermediate in level of organization. They exhibit varying degrees of physicochemical complexity, especially with regard to the distribution of genome segments and their functions. Viruses associated with the killer systems of *Saccharomyces cerevisiae* (Wickner, 1976; Vodkin *et al.*, 1974; Oliver *et al.*, 1977; Bevan and Mitchell, 1979; Bruenn and Kane, 1978; Hopper *et al.*, 1977; Fried and Fink, 1978) and *Ustilago maydis* (Koltin and Day, 1976a,b; Koltin, 1977; Koltin and Kandel, 1978;

Kandel and Koltin, 1978; Koltin *et al.*, 1978) are rather well characterized in this regard. Essential features such as viral RNA polymerase activity and capsid structure are encoded by a relatively large dsRNA segment, whereas other features, such as specific killer and immune functions, are determined by smaller dsRNA segments. The latter segments are dispensable, as evidenced by the presence of viruses with partial genomes in either wild strains or mutants deficient for killer and immune functions (Bevan *et al.*, 1973; Vodkin *et al.*, 1974; Bevan and Mitchell, 1979; Koltin and Day, 1976b; Koltin, 1977). The serologically identical viruses of *Penicillium chrysogenum* and *Penicillium cyaneoful-vum* vary only insofar as the latter virus contains an extra low-molecular-weight dsRNA of unknown function (Buck and Girvan, 1977). In the killer systems supernumerary dsRNA segments are sometimes found among strains; these segments can encode for suppression of killer function and are apparently not encapsidated into particles (Adler *et al.*, 1976; Vodkin *et al.*, 1974; Koltin, 1977).

In two fungi, *Endothia parasitica* (Day *et al.*, 1977) and *Thanatephorus cucumeris* (Castanho *et al.*, 1978), there exists in certain strains evidence for a dsRNA genome and the apparent absence of particles. In strains of other fungi, and commonly among mycoviruses, empty particles are encountered at high titer (Adler *et al.*, 1976; Bozarth, 1979; Buck and Kempson-Jones, 1973). An extreme case is found in a strain of *Aspergillus flavus* (Fig. 2); particles therein routinely lack nucleic acid and are indeed viruslike (Wood *et al.*, 1974).

All the comparative data available so far on dsRNA-containing fungal viruses indicate that these viruses are greatly reduced relative to dsRNA-containing viruses from nonfungal hosts (Wood, 1973). These data also reveal levels of disorganization among fungal viruses that are not encountered elsewhere. It may only be coincidental, but among mycoviruses greater levels of reduction and disorganization are observed in the viruses of conidial fungi.

One further property of dsRNA-containing fungal viruses deserves mention here and concerns the data thus far available on replication of these viruses. The viruses of *Penicillium stoloniferum* (Buck, 1975, 1978; Buck and Ratti, 1975), *Aspergillus foetidus* (Ratti and Buck, 1975, 1978, 1979), and *Saccharomyces cerevisiae* (Herring and Bevan, 1977; Bevan and Mitchell, 1979) have been studied best in this regard. These data indicate that fungal viruses are heterogeneous for mode of replication and, once again, are unlike the dsRNA-containing viruses of other life forms (Buck, 1979). In *Penicillium stoloniferum* particles called product particles have been identified. These contain two molecules of the same dsRNA segment per particle and indicate that the RNA polymerase associated with them can act as a replicase. This suggests further that replication of this virus, at least in part, may proceed by direct duplication. Such a mechanism for replication is rather unexpected for a virus, since a unit of virus infection is expected to yield many particles and normally this leads to host cell

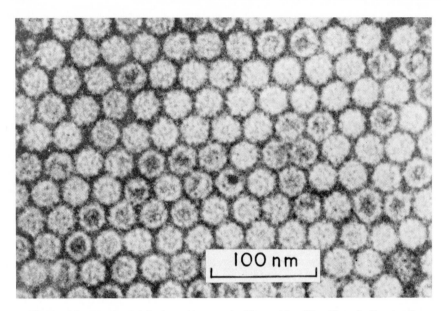

Fig. 2. Electron micrograph of proteinaceous viruslike particles, 27 to 30-nm in diameter, from a strain of *Aspergillus flavus*. Particles are negatively stained and were purified by sucrose density gradient electrophoresis. (Published micrograph of Wood and co-workers, 1974, reproduced from the *Journal of Virology* with permission of the American Society for Microbiology.)

destruction. Replication by duplication or any curtailment of virus replication might well have evolved among fungal viruses as an adaptation for persistent infection of a host system. Such a modified replicative cycle, if successful, would accompany but not necessarily exceed the rate of host cell duplication. Loss of infectivity might accrue in such a situation, provided that virus transmission could accompany and be extended by transmission of the host cell line.

III. TRANSMISSION OF FUNGAL VIRUSES

Genetic elements enter cells as a result of either infection or heredity. Infectious transfer involves extrinsic genetic elements and charactistically proceeds from the environment into cells. Hereditary transfer, on the other hand, involves only cells and accompanies either cell division (serial transmission) or cell fusion (lateral transmission).

Since viruses represent genetic elements extrinsic to the cell, they are expected to be infectious. Fungal viruses, on the basis of their physicochemical and ultrastructural properties, appear to be extrinsic to the cell in evolutionary origin and qualify as viruses in this regard. However, fungal viruses have so far not been shown to be infectious but to be heritable and to be transmitted either serially or laterally (Tables II and III).

The conidium, or at least the macroconidium, as a self-contained agent for transmission of a cell line, qualifies as an agent for serial transmission of any genetic element endogenous to it. The same can be said of other types of asexual spores or of any propagule encountered in fungi. Insofar as conidia can be formed on a heterokaryon or following cytoplasmic exchange between cell lines, the conidium represents as well a potential agent for serial dispersal following lateral transmission of some genetic element. Conidiogenesis may or may not be accompanied on the same mycelium by sexual reproduction. While meiotic spores formed following sexual reproduction may also be effective agents for dispersal, and may perhaps be more regularly associated with lateral transmission of a genetic element, the important thing with regard to a consideration of conidial fungi and their viruses is that the absence of sexuality does not preclude the potential for lateral transmission of a virus. Regardless of sexual competence by the fungus, lateral transmission of an endogenous or heritable virus would be defined by incompatibility mechanisms and the extent of cytoplasmic exchange between mycelia. Fungal species, even sexually competent species, are often very restrictive in this regard, exhibiting a limited potential for outcrossing not only between species but also within a species (Lemke, 1973; Burnett, 1975; Esser and Raper, 1965).

TABLE II

Studies on Serial Transmission of Viruses and Viruslike Particles in Fungi

Species[a]	Reference
Agaricus brunnescens	Schisler et al. (1963)
Aspergillus flavus	Wood et al. (1974)
Aspergillus foetidus	Chang and Tuveson (1973)
Colletotrichum lindemuthianum	Delhotal et al. (1976)
Endothia parasitica	Day et al. (1977)
Gaeumannomyces graminis	Rawlinson et al. (1977)
Lentinus edodes	Ushiyama and Nakai (1975)
Mycogone perniciosa	Albouy and Lapierre (1972)
Penicillium brevicompactum	Sansing et al. (1973)
Penicillium chrysogenum	Lemke et al. (1973)
Penicillium claviforme	Metitiri and Zachariah (1972)
Penicillium cyaneofulvum	Banks et al. (1969b)
Penicillium stoloniferum	Banks et al. (1968)
Puccinia graminis	Rawlinson and MacLean (1973)
Pyricularia oryzae	Boissonnet-Menes and Lecoq (1976)
Uromyces phaseoli	Macdonald and Heath (1978)

[a] Species of conidial fungi are indicated by boldface italics.

TABLE III

Studies on Lateral Transmission of Viruses and Viruslike Particles in Fungi

Species[a]	Reference
Agaricus brunnescens	Gandy (1960),
	Schisler *et al.* (1967)
Aspergillus flavus[b]	Gussack *et al.* (1977)
Aspergillus niger	Lhoas (1970)
Colletotrichum lindemuthianum	Delhotal *et al.* (1976)
Coprinus congregatus	Ross (1979)
Endothia parasitica	Day *et al.* (1977)
Gaeumannomyces graminis	Rawlinson *et al.* (1973)
Penicillium chrysogenum	Lemke and Nash (1974),
	Lemke *et al.* (1976)
Penicillium claviforme	Metitiri and Zachariah (1972)
Penicillium stoloniferum	Lhoas (1971)
Pyricularia oryzae[c]	Boissonnet-Menes and Lecoq (1976)
Saccharomyces cerevisiae	Bevan *et al.* (1973),
	Wickner (1976)
Schizophyllum commune	Koltin *et al.* (1973)
Thanatephorus cucumeris	Castanho *et al.* (1978)
Ustilago maydis	Wood and Bozarth (1973)

[a] Species of conidial fungi indicated by boldface italics.

[b] Negative evidence for transmission between closely related strains of *Aspergillus flavus* and *Aspergillus parasiticus*.

[c] Transmission demonstrated by protoplast fusion.

A. Serial Transmission

A typical mycelium is a cellular network that is asynchronous for both cell division and sporulation. Virus replication in such a mycelium might also be expected to be asynchronous, and the distribution of viruses into spores or into sectors of the mycelium might vary. Studies which bear out these expectations and indicate variation in virus titer have been conducted mainly with species of *Penicillium* (Detroy and Still, 1975, 1976; Detroy *et al.*, 1973; Still *et al.*, 1975; Lemke *et al.*, 1973; Volterra *et .*, 1975; De Marini *et al.*, 1977).

Ultrastructural studies with *Penicillium chrysogenum* have indicated the presence of virus particles either scattered or aggregated in the ground cytoplasm or in membrane-bound vesicles. High titers of particles are reached in older cells. Particles appear to be absent from apical cells, and their presence in conidia is difficult to demonstrate by electron microscopy (Buck, 1979; Yamashita *et al.*, 1973; Volkoff *et al.*, 1972; Metitiri and Zachariah, 1972). Clearly, genetic determinants for synthesis of virus particles, if not the particles per se, exist rather regularly in the conidia of *Penicillium chrysogenum,* as determined by radioimmunoassays indicating the presence of virus particles in an extended series of single-spore colonies derived from a virus-infected strain (Lemke *et al.*, 1973). That the virus of *Penicillium chrysogenum* (strain NRRL 1951) persists is indicated further by the fact that this virus has been retained consistently among descendants of the NRRL strain used for commercial production of penicillin, among descendants derived independently in several laboratories, and following treatment with several mutagenic agents (Banks *et al.*, 1969b; Lemke *et al.*, 1976). Some variation in virus titer is observed among descendants of this strain obtained either spontaneously or following treatment with specific compounds, but there seems to be no correlation between virus titer and penicillin titer among strains (Lemke *et al.*, 1973).

In *Penicillium stoloniferum* (De Marini *et al.*, 1977) and *Penicillium citrinum* (Volterra *et al.*, 1975), qualitative as well as quantitative differences in virus content have been observed among strains derived as single-spore isolates of a given mycelium. These two penicillia have each been shown to contain two viruses; particles are isometric but serologically distinct in the case of *Penicillium stoloniferum*, and in the case of *Penicillium citrinum* particles are anisometric and can be distinguished by electron microscopy. A disparity in the ratio of the two particle types in *Penicillium citrinum* apparently leads to a marked difference in sporulation efficiency (Volterra *et al.*, 1975). The appearance and frequency of asporogenic regions on mycelia of *Penicillium chrysogenum* (Fig. 3) could not be related consistently to a difference in virus titer (Lemke *et al.*, 1973). In some strains of *Penicillium chrysogenum* as well as of *Penicillium citrinum* (Borré *et al.*, 1971; Lemke *et al.*, 1973, 1976) such regions of the mycelium, and not the mycelium generally, exhibit an exceptional lysis (Fig. 4). This lysis is conditional upon growth on a lactose-based medium. Such lysis in these *Penicillium* strains is unrelated to virus-induced lysis following *de novo* infection and is therefore not comparable to the lysis typically induced by phages in bacterial and mammalian cell lines.

B. Lateral Transmission

The vast majority of fungi are mycelial forms and, regardless of their potential for sexual reproduction, exhibit some potential for cell fusion and cytoplasmic exchange over an extended portion of their life cycle. Conidial fungi are no exception and, insofar as vegetative cell fusions occur between strains of these microbes, the possibility of lateral transmission of a noninfectious virus from an infected mycelium to a noninfected one exists.

It has already been stated in this chapter that the fungal viruses studied to date appear to be infectious only by heredity, a feature which in the opinion of some might disqualify them as viruses. Accordingly, some authors have preferred to designate these particles, even dsRNA-containing particles, as *viruslike*. This

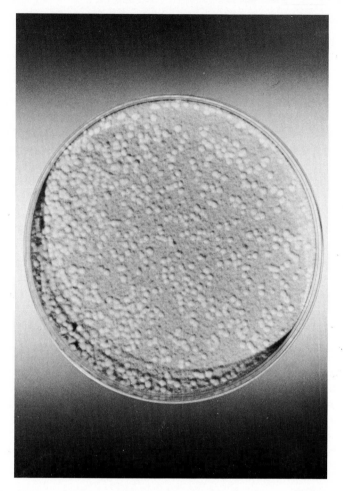

Fig. 3.   Wild-type strain of *Penicillium chrysogenum,* derived as a monoconidial isolate of strain
NRRL 1951 and cultured on a lactose-based medium. White asporogenic patches are produced by
this strain, but lytic plaque formation is not observed. The strain contains a dsRNA virus.

designation simply avoids the issue of their exact nature. These particles qualify
as viruses on the basis of general structure and nucleoprotein composition. When
viewed in the context of fungal biology, any shortcoming with regard to their
infectivity can be reconciled at least on a conceptual basis.

A great many fungi lack sexuality completely or are in nature basically asexual. Among fungi, even sexually competent fungi, inbreeding mechanisms exist
(i.e., vegetative or heterogenic incompatibility, homothallism among some sexually reproducing species, and ecological specializations such as a narrow host

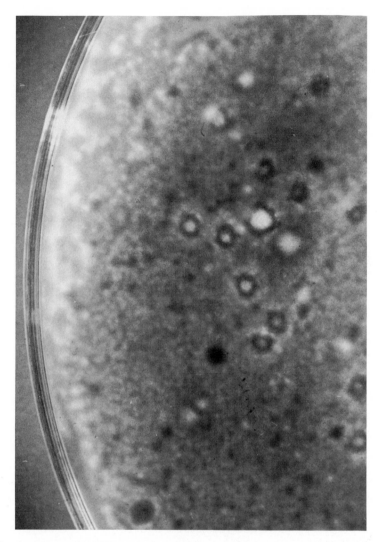

Fig. 4. Mutant strain of *Penicillium chrysogenum*, derived by ultraviolet light treatment from the strain depicted in Fig. 4 and also grown on a lactose-based medium. In this mutant lytic plaques form in association with asporogenic patches. The lysis is conditional upon a nuclear gene mutation and the presence of virus particles at an exceptionally high titer. (For details, see Lemke *et al.*, 1973, and Lemke and Nash, 1974; published photograph reproduced with permission from the *Journal of General Microbiology.*)

range for many plant-pathogenic fungi). Inbreeding sometimes occurs within the species (Esser and Raper, 1965; Lemke, 1973), and the selective pressures for such inbreeding and for loss of sexuality in fungi are largely unexplained.

Inbreeding might be mutually beneficial in a relationship between an endogenous virus and its fungal host. The virus would not be distributed indiscriminately, or to some host system in which it was not well adapted for persistent infection. Alternatively, a noninfected host by inbreeding would be less likely to receive a potentially virulent virus or one poorly adapted to its specific host cell line.

While there are several studies on the lateral transmission of fungal viruses (Table III), not all of them have been conducted with equal attention to detail. In some cases strains have been marked genetically for auxotrophy or with spore color mutations (Lhoas, 1970; Delhotal *et al.*, 1976; Lemke *et al.*, 1976; Boissonnet-Menes and Lecoq, 1976), and the events following cell fusion and heterokaryosis have been well documented. Other studies have used mating-type markers or only morphological criteria (Gandy, 1960; Schisler *et al.*, 1967; Ross, 1979; Rawlinson *et al.*, 1973; Metitiri and Zachariah, 1972), and in most cases the primary assay for infectivity has been electron microscopy. This of course has not been true of studies involving the killer systems of *Saccharomyces cerevisiae* and *Ustilago maydis*. Indeed, these systems offer a considerable advantage in studies on virus transmission. In killer systems and in *Endothia parasitica* and *Thanatephorus cucumeris* lateral transmission of dsRNA molecules has been well documented (Wood and Bozarth, 1973; Bevan *et al.*, 1973; Day *et al.*, 1977; Castanho *et al.*, 1978; Wickner, 1976; Koltin and Day, 1976a,b). These are ideal experimental systems for the study of cytoplasmic inheritance and nucleocytoplasmic interactions in the eukaryotic cell.

IV. EVOLUTIONARY IMPLICATIONS

In this chapter the viruses of fungi, especially the viruses of conidial fungi, have been interpreted as phylogenetically reduced and as having a derived loss of potential for infectivity. The simple structure of these viruses and their adaptation to the fungal host for heritable or cell-mediated transmission are consistent with this interpretation. Serological studies as well as data on differences in genome segmentation indicate further that dsRNA-containing fungal viruses may have evolved more than once from several lines of more complex progenitors. While it is doubtful that the details of this evolution can ever be retraced, the close or endogenous relationship of present-day mycoviruses to present-day fungi has led virologists to reconsider the criteria normally expected of a virus. Mycologists might also take inventory of the consequences of this relationship for the evolution of fungi. With reference to conidial fungi two noteworthy characteristics have appeared rather consistently. First, these fungi exhibit proliferous reproduction that is predominantly, if not exclusively, asexual in nature. Second, these fungi, even those completely lacking in formal sexuality, have some potential for plasmogamy with compatible strains over an extended portion of their life cycle.

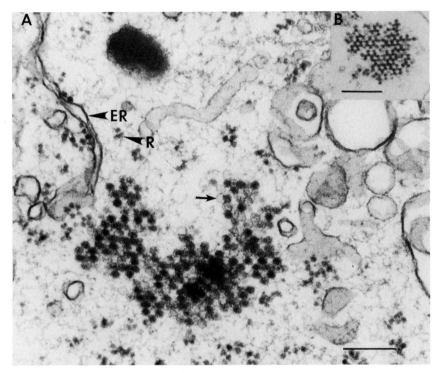

Fig. 5. (A) Virus particles in cytoplasm of degenerated vegetative hypha of *Coremiella cubispora*. Arrow, Empty capsids; R, ribosomes; ER, endoplasmic reticlum. Bar = 200 nm. (B) Lattice arrangement of virus particles in cytoplasm of degenerated conidium. Bar = 300 nm. (From Cole, 1975.)

Selective pressures for both characteristics in the evolution of conidial fungi may well have developed in concert with mycovirus evolution.

Such implications for fungus–virus coevolution will remain speculative in the absence of experimental systems for viral infectivity or systems that might be interpreted as intermediate in this evolution. Such systems are presently unknown among conidial fungi, although there is evidence for the occurrence of viruslike particles (Fig. 5) in an abnormal strain of *Coremiella cubispora*, a strain producing morphogenetically atypical conidia (Cole, 1975). There has also been described in the basidiomycete *Coprinus congregatus* (Ross *et al.*, 1976) an infectious disorder of meiosis. The infectious agent, based on more recent studies (Ross, 1979), may very well be a virus or complex of viruses and seems to involve dsRNA. If this is the case, then such a system, one involving a virus that impairs meiosis and retains infectivity, may very well represent an intermediate condition ancestral to that prevailing at present among conidial fungi and their viruses.

REFERENCES

Adler, J., Wood, H. A., and Bozarth, R. F. (1976). Virus-like particles from killer, neutral, and sensitive strains of *Saccharomyces cerevisiae*. *J. Virol.* **17**, 472–476.

Albouy, J., and Lapierre, H. (1972). Observation en microscopie électronique de souches virosées de *Mycogone perniciosa* agent d'une mole du champignons de couche. *Ann. Phytopathol.* **4**, 353–358.

Banks, G. T., Buck, K. W., Chain, E. B., Himmelweit, F., Marks, J. E., Tyler, J. M., Hollings, M., Last, F. T., and Stone, O. M. (1968). Viruses in fungi and interferon stimulation. *Nature (London)* **218**, 542–545.

Banks, G. T., Buck, K. W., Chain, E. B., Darbyshire, J. E., and Himmelweit, F. (1969a). Virus-like particles in penicillin producing strains of *Penicillium chrysogenum*. *Nature (London)* **222**, 89–90.

Banks, G. T., Buck, K. W., Chain, E. G., Darbyshire, J. E., and Himmelweit, F. (1969b). *Penicillium cyaneo-fulvum* and interferon stimulation. *Nature (London)* **223**, 155–158.

Banks, G. T., Buck, K. W., Chain, E. B., Darbyshire, J. E., Himmelweit, F., Ratti, G., Sharpe, T. J., and Planterose, D. N. (1970). Antiviral activity of double-stranded RNA from a virus isolated from *Aspergillus foetidus*. *Nature (London)* **227**, 505–507.

Benigni, R., Ignazzitto, G., and Volterra, L. (1977). Double-stranded ribonucleic acid in viruses of *Penicillium citrinum*. *Appl. Environ. Microbiol.* **34**, 811–814.

Bevan, E. A., and Mitchell, D. J. (1979). The killer system in yeast. *In* "Viruses and Plasmids in Fungi" (P. A. Lemke, ed.), pp. 161–199. Dekker, New York.

Bevan, E. A., Herring, A. J., and Mitchell, D. J. (1973). Preliminary characterization of two species of dsRNA in yeast and their relationship to killer character. *Nature (London)* **245**, 81–86.

Blattny, C., and Kralik, O. (1968). A virus disease of *Laccaria laccata* (Scop. ex Fr.) Cooke and some other fungi. *Ceska Mykol.* **22**, 161–166.

Boissonnet-Menes, M., and Lecoq, H. (1976). Transmission de virus par fusion de protoplasts chez *Pyricularia oryzae*. *Physiol. Veg.* **14**, 251–157.

Borré, E. L., Morgantine, E., Ortali, V., and Tonolo, A. (1971). Production of lytic plaques of viral origin in *Penicillium*. *Nature (London)* **229**, 568–569.

Bozarth, R. F. (1972). Mycoviruses: A new dimension in microbiology. *In* "Environmental Health Perspective," pp. 23–39. U.S. Department of Health, Education and Welfare, Washington, D.C.

Bozarth, R. F. (1977). Biophysical and biochemical characterization of virus-like particles containing a high molecular weight dsRNA from *Helminthosporium maydis*. *Virology* **80**, 149–157.

Bozarth, R. F. (1979). The physico-chemical properties of mycoviruses. *In* "Virus and Plasmids in Fungi" (P. A. Lemke, ed.), pp. 43–91.

Bozarth, R. F., and Goenaga, A. (1977). Complex of virus-like particles containing double-stranded RNA from *Thielaviopsis basicola*. *J. Virol.* **24**, 846–849.

Bozarth, R. F., and Wood, H. A. (1977). Purification and properties of the virus-like particles of *Penicillium chrysogenum*. *Phytopathology* **61**, 886.

Bozarth, R. F., Wood, H. A., and Goenaga, A. (1972a). Virus-like particles from a culture of *Diplocarpon rosae*. *Phytopathology* **62**, 493.

Bozarth, R. F., Wood, H. A., and Nelson, R. R. (1972b). Virus-like particles in virulent strains of *Helminthosporium maydis*. *Phytopathology* **62**, 748.

Brants, D. H. (1971). Infection of *Phythium sylvaticum in vitro* with tobacco mosaic virus. *Neth. J. Plant Pathol.* **77**, 175–177.

Bruenn, J., and Kane, W. (1978). The relatedness of the double-stranded RNA's present in yeast virus-like particles. *J. Virol.* **26**, 762–772.

Buck, K. W. (1975). Replication of double-stranded RNA in particles of *Penicillium stoloniferum* virus S. *Nucleic Acids Res.* **2**, 1889–1902.

Buck, K. W. (1978). Semi-conservative replication of double-stranded RNA by a virion associated RNA polymerase. *Biochem. Biophys. Res. Commun.* **84**, 639–645.

Buck, K. W. (1979). Replication of double-stranded RNA mycoviruses. *In* "Viruses and Plasmids in Fungi" (P. A. Lemke, ed.), pp. 93–160. Dekker, New York.

Buck, K. W., and Girvan, R. F. (1977). Comparison of the biophysical and biochemical properties of *Penicillium cyaneo-fulvum* virus and *Penicillium chrysogenum* virus. *J. Gen. Virol.* **34**, 145–154.

Buck, K. W., and Kempson-Jones, G. F. (1973). Biophysical properties of *Penicillium stoloniferum* virus S. *J. Gen. Virol.* **18**, 223–235.

Buck, K. W., and Kempson-Jones, G. F. (1974). Capsid polypeptides of 2 viruses isolated from *Penicillium stoloniferum*. *J. Gen. Virol.* **22**, 441–445.

Buck, K. W., and Ratti, G. (1975). A model for the replication of double-stranded ribonucleic acid mycoviruses. *Biochem. Soc. Trans.* **3**, 542–544.

Burnett, J. H. (1975). "Mycogenetics," p. 375. Wiley, New York.

Cantino, E. C., and Mills, G. L. (1979). The gamma particle in *Blastocladiella emersonii:* What is it? *In* "Viruses and Plasmids in Fungi" (P. A. Lemke, ed.), pp. 441–484. Dekker, New York.

Castanho, B., Butler, E. E., and Shepherd, R. J. (1978). The association of double-stranded RNA with *Rhizoctonia* decline. *Phytopathology* **68**, 1515–1519.

Chang, L. T., and Tuveson, R. W. (1973). Genetics of strains of *Aspergillus foetidus* with virus-like particles. *Genetics* **74**, 43–44.

Chosson, J. F., Lapierre, H., Kusiak, C., and Molin, G. (1973). Presence de particules de type viral chez les champignons du genre *Fusarium*. *Ann. Phytopathol.* **5**, 324.

Cole, G. T. (1975). A viral infection of the hyphomycetous fungus, *Briosia cubispora*. *Proc. Can. Microsc. Soc.* **2**, 60–61.

Corbett, M. K., and Styer, E. L. (1977). Intranuclear virus-like particles in *Phytophthora infestans*. *Proc. Am. Phytopathol. Soc.* **3**, 332.

Day, L. E., and Ellis, L. F. (1971). Virus-like particles in *Cephalosporium acremonium*. *Appl. Microbiol.* **22**, 919–920.

Day, P. R., Dodds, J. A., Ellison, J. E., Jaynes, R. A., and Anagnostakis, S. L. (1977). Double-stranded RNA in *Endothia parasitica*. *Phytopathology* **67**, 1393–1396.

Delhotal, P., Legrand-Pernot, F., and Lecoq, H. (1976). Etude des virus de *Colletotrichum lindemuthianum*. II. Transmission des particules virales. *Ann. Phytopathol.* **8**, 437–448.

De Marini, D. M., Kurtzman, C. P., Fennell, D. I., Worden, K. A., and Detroy, R. W. (1977). Transmission of PsV-F and PsV-S mycoviruses during conidiogenesis of *Penicillium stoloniferum*. *J. Gen. Microbiol.* **100**, 59–64.

Detroy, R. W., and Still, P. E. (1975). Fungal metabolites and viral replication in *Penicillium stoloniferum*. *Dev. Ind. Microbiol.* **16**, 145–151.

Detroy, R. W., and Still, P. E. (1976). Patulin inhibition of mycovirus replication in *Penicillium stoloniferum*. *J. Gen. Microbiol.* **92**, 167–174.

Detroy, R. W., Freer, S. N., and Fennell, D. I. (1973). Relationship between the biosynthesis of virus-like particles and mycophenolic acid in *Penicillium brevi-compactum*. *Can. J. Microbiol.* **19**, 1459–1462.

Dieleman-van Zaayen, A. (1967). Virus-like particles in a weed mould growing on mushroom trays. *Nature (London)* **216**, 595–596.

Dieleman-van Zaayen, A. (1972). "Mushroom Virus Disease in the Netherlands: Symptoms, Etiology, Electron Microscopy, Spread and Control." Centre for Agricultural Publishing and Documentation, Wageningen, Netherlands.

Dieleman-van Zaayen, A., and Temmink, J. H. M. (1968). A virus disease of cultivated mushrooms in the Netherlands. *Neth. J. Plant Pathol.* **74**, 48–51.

Dunkle, L. D. (1974a). Double-stranded RNA mycovirus in *Periconia circinata. Physiol. Plant Pathol.* **4**, 107–116.

Dunkle, L. D. (1974b). The relation of virus-like particles to toxin producing fungi in corn and sorghum. *Proc. 28th Annu. Corn. Sorghum Res. Conf.* pp. 72–81.

Ellis, L. F., and Kleinschmidt, W. J. (1967). Virus-like particles of a fraction of statolon, a mould product. *Nature (London)* **215**, 649–650.

Esser, K., and Raper, J. R., eds. (1965). "Incompatibility in Fungi." Springer-Verlag, Berlin, and New York.

Férault, A. C., Spire, D., Rapilly, F., Bertrandy, J., Skajennikoff, M., and Bernaux, P. (1971). Observation de particules virales dans des souches de *Piricularia oryzae* Briosi et Cav. *Ann. Phytopathol.* **3**, 267–269.

Fried, H. M., and Fink, G. R. (1978). Electron microscopic heteroduplex analysis of "killer" double-stranded RNA species from yeast. *Proc. Natl. Acad. Sci. U.S.A.* **75**, 4224–4228.

Gandy, D. (1960). A transmissible disease of cultivated mushrooms ("watery stipe"). *Ann. Appl. Biol.* **48**, 427–430.

Gussack, G., Bennett, J. W., Cavalier, S., and Yatsu, L. (1977). Evidence for parasexual cycle in a strain of *Aspergillus flavus* containing virus-like particles. *Mycopathologia* **61**, 159–165.

Herring, A. J., and Bevan, E. A. (1977). Yeast virus-like particles possess a capsid-associated single-stranded RNA polymerase. *Nature (London)* **268**, 464–466.

Hollings, M. (1962). Viruses associated with a die-back disease of cultivated mushroom. *Nature (London)* **196**, 962–965.

Hollings, M. (1978). Mycoviruses: Viruses that infect fungi. *Adv. Virus Res.* **22**, 1–53.

Hollings, M., and Stone, O. M. (1971). Viruses that infect fungi. *Annu. Rev. Phytopathol.* **9**, 93–118.

Hopper, J. E., Bostian, K. A., Rowe, L. B., and Tipper, D. J. (1977). Translation of the killer-associated virus-like particles of *Saccharomyces cerevisiae. J. Biol. Chem.* **252**, 9010–9017.

Huttinga, H., Wichers, H. J., and Dieleman-van Zaayen, A. (1975). Filamentous and polyhedral virus-like particles in *Boletus edulis. Neth. J. Plant Pathol.* **81**, 102–106.

Isaac, P. K., and Gupta, S. K. (1964). A virus-like infection of *Alternaria tenuis. Proc. Int. Bot. Congr., 10th, 1964* Abstract, pp. 390–391.

Kandel, J., and Koltin, Y. (1978). Killer phenomenon in *Ustilago maydis:* Comparison of the killer proteins. *Exp. Mycol.* **2**, 270–278.

Kazama, F. Y., and Schornstein, K. L. (1972). Herpes-type virus particles associated with a fungus. *Science* **177**, 696–697.

Khandjian, E. W., Roos, U. P., and Turian, G. (1975). Mycovirus d'*Allomyces. Pathol. Microbiol.* **42**, 250–251.

Khandjian, E. W., Turian, G., and Eisen, H. (1977). Characterization of the RNA mycovirus infecting *Allomyces arbuscula. J. Gen. Virol.* **35**, 415–424.

Kleinschmidt, W. J., Ellis, L. F., van Frank, R. M., and Murphy, E. B. (1968). Interferon stimulation by a double-stranded RNA of a mycophage in statolon preparations. *Nature (London)* **220**, 167–168.

Koltin, Y. (1977). Virus-like particles in *Ustilago maydis:* Mutants with partial genomes. *Genetics* **86**, 527–534.

Koltin, Y., and Day, P. R. (1976a). Inheritance of killer phenotypes and double-stranded RNA in *Ustilago maydis. Proc. Natl. Acad. Sci. U.S.A.* **73**, 594–598.

Koltin, Y., and Day, P. R. (1976b). Suppression of the killer phenotype in *Ustilago maydis. Genetics* **82**, 629–637.

Koltin, Y., and Kandel, J. S. (1978). Killer phenomenon in *Ustilago maydis:* The organization of the viral genome. *Genetics* **88**, 267–276.

Koltin, Y., Berick, R., Stamberg, J., and Ben-Shaul, Y. (1973). Virus-like particles and cytoplasmic inheritance of plaques in a higher fungus. *Nature (London), New Biol.* **241**, 108–109.

Koltin, Y., Mayer, I., and Steinlauf, R. (1978). Killer phenomenon in *Ustilago maydis:* Mapping viral functions. *Mol. Gen. Genet.* **166**, 181–186.

Kozlova, T. M. (1973). Virus-like particles in yeast cells. *Mikro-biologiya* **42**, 745–747.

Küntzel, H., Barath, Z., Ali, I., Kind, J., and Althaus, N.-H. (1973). Virus-like particles in an extranuclear mutant of *Neurospora crassa. Proc. Natl. Acad. Sci. U.S.A.* **70**, 574–578.

Lai, H. C., and Zachariah, K. (1975). Detection of virus-like particles in coremia of *Penicillium claviforme. Can. J. Genet. Cytol.* **17**, 525–533.

Lampson, G. P., Tytell, A. A., Field, A. K., Nemes, M. M., and Hilleman, M. R. (1967). Inducers of interferon and host resistance. I. Double-stranded RNA from extracts of *Penicillium funiculosum. Proc. Natl. Acad. Sci. U.S.A.* **58**, 782–789.

Lapierre, H., and Faivre-Amoit, A. (1970). Présence de particules virales chez différentes souches de *Sclerotium cepivorum. Congr. Int. Prot. Plantes, 7th, 1970* pp. 542–543.

Lapierre, H., LeMaire, J.-M., Jouan, B., and Molin, G. (1970). Mise en évidence de particules virales associées à une perte de pathogénicité chez le Piétin-échaudage des céréales, *Ophiobolus graminis Sacc. C. R. Hebd. Seances Acad. Sci., Ser. D* **271**, 1833–1836.

Lapierre, H., Faivre-Amiot, A., Kusiak, C., and Molin G. (1972). Particules de type viral associées au *Mycogone perniciosa Magnus,* agent d'une des môles du champignon de couche. *C. R. Hebd. Séances Acad. Sci., Ser. D* **274**, 1867–1870.

Lapierre, H., Faivre-Amiot, A., and Molin G. (1973). Isolement de particules de type viral associées au *Verticillium fungicola:* Agent d'une môle du champignon de couche. *Ann. Phytopathol.* **5**, 323.

Lechner, J. F., Scott, A., Aaslestad, H. G., Fuscaldo, A. A., and Fuscaldo, K. E. (1972). Analysis of a virus-like particle from *N. crassa. Abstr. Annu. Meet. Am. Soc. Microbiol.* p. 188.

Lecoq, H., and Delhotal, P. (1976). Etude des virus de *Colletotrichum lindemuthianum.* I. Caractérisation des particules virales. *Ann. Phytopathol.* **8**, 307–322.

Lecoq, H., Spire, D., Rapilly, F., and Bertrandy, J. (1974). Mise en évidence de particules de type viral chez les *Puccinia. C. R. Hebd. Seances Acad. Sci., Ser. D* **27**, 1599–1602.

Lemke, P. A. (1973). Isolating mechanisms in fungi: Prezygotic, postzygotic and azygotic. *Persoonia* **7**, 249–260.

Lemke, P. A. (1976). Viruses of eucaryotic microorganisms. *Annu. Rev. Microbiol.* **30**, 105–145.

Lemke, P. A. (1977a). Fungal viruses in agriculture. *In* "Virology in Agriculture" (J. A. Romberger, J. D. Anderson, and R. L. Powell, eds.), pp. 159–175. Allanheld, Osmun & Co., Montclair, New Jersey.

Lemke, P. A. (1977b). Double-stranded RNA viruses among filamentous fungi. *In* "Microbiology 1977" (D. Schlessinger, ed.), pp. 568–570. Am. Soc. Microbiol., Washington, D.C.

Lemke, P. A., and Nash, C. H. (1974). Fungal viruses. *Bacteriol. Rev.* **38**, 29–56.

Lemke, P. A., and Ness, T. M. (1970). Isolation and characterization of a double-stranded ribonucleic acid from *Penicillium chrysogenum. J. Virol.* **6**, 813–891.

Lemke, P. A., Nash, C. H., and Pieper, S. W. (1973). Lytic plaque formation and variation in virus titre among strains of *Penicillium chrysogenum. J. Gen. Microbiol.* **76**, 265–275.

Lemke, P. A., Saksena, K. N., and Nash, C. H. (1976). Viruses of industrial fungi. *In* "Second International Symposium on the Genetics of Industrial Microorganisms" (K. D. MacDonald, ed.), pp. 323–355. Academic Press, New York.

Lhoas, P. (1970). Use of heterokaryosis to infect virus-free strains of *Aspergillus niger. Aspergillus Newsl.* **11**, 8–9.

Lhoas, P. (1971). Transmission of double-stranded RNA viruses to a strain of *Penicillium stoloniferum* through heterokaryosis. *Nature (London)* **230**, 248–249.

Macdonald, J. G., and Heath, M. C. (1978). Rod-shaped and spherical virus-like particles in cowpea rust fungus. *Can. J. Bot.* **56**, 963–975.

Mackenzie, D. W., and Adler, J. P. (1972). Virus-like particles in toxigenic *Aspergilli*. *Abstr. Annu. Meet. Am. Soc. Microbiol.* p. 68.

Marino, R., Saksena, K. N., Schuler, M., Mayfield, J. E., and Lemke, P. A. (1976). Double-stranded RNA from *Agaricus bisporus*. *Appl. Environ. Microbiol.* **32**, 433–438.

Metitiri, P. O., and Zachariah, K. (1972). Virus-like particles and inclusion bodies in penicillus cells of a mutant of *Penicillium*. *J. Ultrastruct. Res.* **40**, 272–283.

Moffitt, E. M., and Lister, R. M. (1975). Application of a serological test for detecting double-stranded RNA mycoviruses. *Phytopathology* **65**, 851–859.

Nesterova, G. F., Kyarner, Y., and Soom, Y. O. (1973). Virus-like particles in *Candida tropicalis*. *Mikrobiologiya* **42**, 162–165.

Nienhaus, E. (1971). Tobacco mosaic virus strains extracted from conidia of powdery mildews. *Virology* **46**, 504–505.

Oliver, S. G., McCready, S. J., Holm, C., Sutherland, P. A., McLaughlin, C. S., and Cox, B. S. (1977). Biochemical and physiological studies of the yeast virus-like particle. *J. Bacteriol.* **130**, 1303–1309.

Ratti, G., and Buck, K. W. (1975). RNA-polymerase activity in double-stranded ribonucleic acid virus particles from *Aspergillus foetidus*. *Biochem. Biophys. Res. Commun.* **66**, 706–711.

Ratti, G., and Buck, K. W. (1978). Semi-conservative transcription in particles of a double-stranded RNA mycovirus. *Nucleic Acids Res.* **5**, 3843–3854.

Ratti, G., and Buck, K. W. (1979). Transcription of double-stranded RNA in virions of *Aspergillus foetidus* virus S. *J. Gen. Virol.* **42**, 59–72.

Rawlinson, C. J. (1973). Virus-like particles in plant pathogenic fungi. *Proc. Int. Congr. Plant Pathol., 2nd, 1973* Abstract, No. 0911.

Rawlinson, C. J., and MacLean, D. J. (1973). Virus-like particles in axenic cultures of *Puccinia graminis tritici*. *Trans. Br. Mycol. Soc.* **61**, 590–593.

Rawlinson, C. J., Hornby, D., Pearson, V., and Carpenter, J. M. (1973). Virus-like particles in the take-all fungus, *Gaeumannomyces graminis*. *Ann. Appl. Biol.* **74**, 209.

Rawlinson, C. J., Muthyabu, G., and Deacon, J. W. (1977). Natural transmission of viruses in *Gaeumannomyces* and *Phialophora*. *Abstr., Int. Mycol. Congr., 2nd, 1977* p. 558.

Ross, I. K. (1979). An infectious disorder of meiosis in *Coprinus congregatus*. *In* "Viruses and Plasmids in Fungi" (P. A. Lemke, ed.), pp. 485–524. Dekker, New York.

Ross, I. K., Pommerville, J. C., and Damm, D. L. (1976). A highly infectious "mycoplasma" that inhibits meiosis in the fungus *Coprinus*. *J. Cell Sci.* **21**, 175–191.

Saksena, K. N., and Lemke, P. A. (1978). Viruses in fungi. *Compr. Virol.* **12**, 103–143.

Sanderlin, R. S., and Ghabrial, S. A. (1976). Physico-chemical properties of virus-like particles from *Helminthosporium victoriae*. *Proc. Am. Soc. Phytopathol.* **3**, 251.

Sansing, G. A., Detroy, R. W., Freer, S. N., and Hesseltine, C. W. (1973). Virus-like particles from conidia of *Penicillium* species. *Appl. Microbiol.* **26**, 914–918.

Schisler, L. C., Sinden, J. W., and Sigel, E. M. (1963). Transmission of a virus disease of mushrooms by infected spores. *Phytopathology* **53**, 888.

Schisler, L. C., Sinden, J. W., and Sigel, E. M. (1967). Etiology, symptomatology, and epidemiology of a virus disease of cultivated mushrooms. *Phytopathology* **57**, 519–526.

Schnepf, E. C., Soeder, J., and Hegewald, E. (1970). Polyhedral virus-like particles lysing the aquatic phycomycete *Aphelidium* sp., a parasite of the green alga, *Scenedesmus armatus*. *Virology* **42**, 482–487.

Shahriari, H., Kirkham, J. B., and Casselton, L. A. (1973). Virus-like particles in the fungus *Coprinus lagopus*. *Heredity* **31**, 428.

Spire, D. (1971). Virus des champignons. *Physiol. Veg.* **9**, 555–567.

Spire, D., Férault, A. C., and Bertrandy, J. (1972a). Observation de particles de type viral dans une souche d'*Helminthosporium oryzae* Br. de H. *Ann. Phytopathol.* **4**, 359–360.

Spire, D., Férault, A. C., Bertrandy, J., Rapilly, F., and Skajennikoff, M. (1972b). Particules de type viral dans un champignon hyperparasite: *Gonatobotrys*. *Ann. Phytopathol.* **4**, 419.

Still, P. E., Detroy, R. W., and Hesseltine, C. W. (1975). *Penicillium stoloniferum* virus: Altered replication in ultraviolet-derived mutants. *J. Gen. Virol.* **27**, 275–281.

Tikchonenko, T. I. (1978). Viruses of fungi capable of replication in bacteria (PB viruses). *Compr. Virol.* **12**, 235–269.

Tuveson, R. W., and Peterson, J. F. (1972). Virus-like particles in certain slow-growing strains of *Neurospora crassa*. *Virology* **47**, 527–531.

Tuveson, R. W., and Sargent, M. L. (1976). Characterization of virus-like particles from slow-growing strains of *Neurospora crassa*. *Genetics* **83**, s77.

Tuveson, R. W., Sargent, M. L., and Bozarth, R. F. (1975). Purification of a small virus-like particle from strains of *Neurospora crassa*. *Annu. Meet. Am. Soc. Microbiol.* Abstract, p. 216.

Ushiyama, R., and Hashioka, Y. (1973). Viruses associated with hymenomycetes. I. Filamentous virus-like particles in the cells of a fruit body of shiitake, *Lentinus edodes*. (Berk.) Sing. *Rep. Tottori Mycol. Inst.* **10**, 797–805.

Ushiyama, R., and Nakai, Y. (1975). Viruses associated with hymenomycetes. II. Presence of polyhedral virus-like particles in shiitake mushrooms, *Lentinus edodes* (Berk.) Sing. *Rep. Tottori Mycol. Inst.* **12**, 53–60.

Ushiyama, R., Nakai, Y., and Ikegami, M. (1977). Evidence for double-stranded RNA from polyhedral virus-like particles in *Lentinus edodes* (Berk.) Sing. *Virology* **77**, 880–883.

Velikodvorskaya, G. A., Bobkova, A. F., Maksimova, T. S., Klimenko, S. M., and Tikchonenko, T. I. (1972). New viruses isolated from the culture of a fungus of *Penicillium* genus. *Byull. Eksp. Biol. Med.* **73**, 90–93.

Vodkin, M., Katterman, F., and Fink, G. R. (1974). Yeast killer mutants with altered double-stranded ribonucleic acid. *J. Bacteriol.* **117**, 681–686.

Volkoff, O., and Walters, T. (1970). Virus-like particles in abnormal cells of *Saccharomyces carlsbergensis*. *Can. J. Gent. Cytol.* **12**, 621–626.

Volkoff, O., Walters, T., and Dejardin, R. A. (1972). An examination of *Penicillium notatum* for the presence of *Penicillium chrysogenum* type virus particles. *Can. J. Microbiol.* **18**, 1352–1353.

Volterra, L., Cassone, A., Tonolo, A., and Bruzzone, M. L. (1975). Presence of two virus-like particles in *Penicillium citrinum*. *Appl. Microbiol.* **30**, 149–151.

Wickner, R. B. (1976). Killer of *Saccharomyces cerevisiae*, a double-stranded ribonucleic acid plasmid. *Bacteriol. Rev.* **40**, 757–773.

Wojcik, V. H. (1977). Isolation and purification of virus-like particles from the aquatic fungus *Rhizidiomyces*. *Abstr., Int. Mycol. Congr., 2nd, 1977* p. 742.

Wood, H. A. (1973). Viruses with double-stranded RNA genomes. *J. Gen. Virol.* **20**, 61–85.

Wood, H. A., and Bozarth, R. F. (1973). Heterokaryon transfer of virus-like particles associated with a cytoplasmically inherited determinant in *Ustilago maydis*. *Phytopathology* **63**, 1019–1021.

Wood, H. A., Bozarth, R. F., and Mislivec, P. B. (1971). Virus-like particles associated with an isolate of *Penicillium brevi-compactum*. *Virology* **44**, 592–598.

Wood, H. A., Bozarth, R. F., Adler, J., and Mackenzie, D. W. (1974). Proteinaceous virus-like particles from an isolate of *Aspergillus flavus*. *J. Virol.* **13**, 532–534.

Wyn-Jones, A. P., and Whalley, J. S. (1977). Virus-like particles in *Hypoxylon multiforme*. *Mycopathologia* **61**, 63–64.

Yamashita, S., Doi, Y., and Yora, K. (1971). A polyhedral virus found in rice blast fungus. *Pyricularia oryzae* Cavara. *Ann. Phytopathol. Soc. Jpn.* **37**, 356–359.

Yamashita, S., Doi, Y., and Yora, K. (1973). Intracellular appearance of *Penicillium chrysogenum* virus. *Virology* **55**, 445–452.

Yamashita, S., Doi, Y., and Yora, K. (1975). Electron microscopic study of several fungal viruses. *Proc. Intersect. Congr. Int. Assoc. Microbiol. Soc., 1975* Vol. 3, pp. 340–350.

27

Physiology of Conidial Fungi

Robert Hall

I. INTRODUCTION

Reliable information on the physiology of conidial fungi has been accumulating since 1869, the year in which Raulin, a student of Pasteur, published his careful work on the nutrition of *Aspergillus niger* (Ainsworth, 1976). A consid-

Biology of Conidial Fungi, Vol. 2

erable body of literature is now available. To keep this review within reasonable bounds, my remarks are confined to conidial fungi that have, or are likely to have, teleomorphs in the Ascomycetes, excluding the Erysiphales. I have also relied heavily on recent literature to illustrate concepts. Examples from earlier literature can be found in texts by Hawker (1950, 1957), Lilly and Barnett (1951), and Cochrane (1958).

Physiology focuses on the whole organism but draws on information ranging from molecular to environmental. Here I consider contributions to our understanding of processes in conidial fungi from three levels of enquiry: the cellular, the organismic, and the environmental.

Figure 1 draws attention to the many differentiated structures produced by conidial fungi. All are produced directly or indirectly from a hypha, and many in turn can produce a hypha. Some structures are constructed principally to absorb nutrients and grow (hyphae), others to move the fungus to new locations (conidia), and others to survive adverse environments (sclerotia, chlamydospores). The type of structure produced depends on the genetic capacity of the organism,

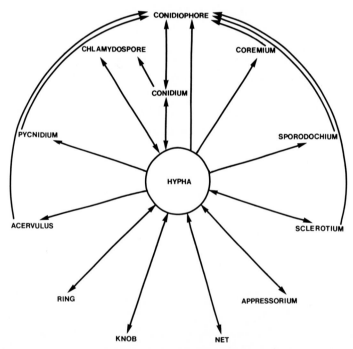

Fig. 1. Paths of differentiation in conidial fungi. As a group, conidial fungi produce many asexually differentiated structures. All of these can develop from the hypha, and most can produce a hypha directly. The hypha is thus seen as the hub of the life cycle and as a subject worthy of intense study. Lines show paths of differentiation, and arrowheads show the direction of development.

its metabolic readiness to proceed to another step in differentiation, and the presence of a permissive or inductive environment. For example, a culture of *Trichoderma viride* has the genetic capacity to produce conidia but does so only after it has reached a certain age and has been exposed to blue light (Gressel and Galun 1967). It is as if the fungal DNA and metabolic machinery sense the environment and select, from the morphological structures available, those best suited to exploit or survive it.

In culture, hyphae, fruiting bodies, conidia, chlamydospores, sclerotia, and so on, may occur together and be organically united. In nature there is a tendency for differentiated structures to lose organic union with one another through death and disintegration of vegetative hyphae. The organism is then best visualized as a set of differentiated structures separated in time or space or both. Continuity of the organism is expressed through its genome, which persists as the fungus moves from one differentiated state to another. For example, macroconidia of *Fusarium solani* may germinate in soil and produce a germ tube. If external nutrients are limiting, the germ tube is likely to produce a chlamydospore, and the conidium and germ tube will disintegrate (Burke, 1965). The fungus has moved through three differentiated states: from conidium, to germ tube, to chlamydospore. At each step it is the same fungus but has different physiological properties. We are challenged to describe both the constancy and the plasticity of the organism. What are the central physiological tendencies and what range of responses can the organism express?

II. CELL PHYSIOLOGY

Cell physiology enables us to explore central tendencies in conidial fungi.

A. Cell Composition and Structure

The cells of conidial fungi, whether differentiated as hyphae, conidia, or some other structure, consist of a rigid wall surrounding a membrane-bound cytoplasm that contains organelles typical of eukaryotic organisms. These organelles include nuclei, mitochondria, ribosomes, endoplasmic reticulum, vacuoles, Golgi bodies, vesicles, microbodies, Woronin bodies, and microtubules (Beckett *et al.*, 1974). The cytoplasm may also contain lipid droplets. Its soluble portion consists largely of enzymes.

The cell may be viewed as a machine that can sustain itself and change its shape and activities. Each component of the cytoplasm is a cog that has specific functions. In conidial fungi, as in all eukaryotic organisms, these functions may be summarized as follows (Bonner and Varner, 1976). The nuclei direct the production of all subcellular components. The chromosomes reproduce DNA and produce mRNA. The nucleolus produces ribosomal subunits. The mitochondria produce ATP in which energy from respirable substrates is stored. Ribosomes are the site of protein synthesis. The plasma membrane regulates movement of solutes into the cell. The endoplasmic reticulum is part of the membrane system of the cell and provides surfaces for chemical reactions. The Golgi apparatus produces vesicles believed to contain enzymes and substrates involved in wall synthesis. Vacuoles of various types contain soluble materials. Those that contain enzymes involved in the lysis of cell components are called lysosomes. Microbodies of various types occur, but their function is unknown. In fungi, Woronin bodies occur close to septa, and plug the pore when the cell is damaged. Microtubules occur in the nuclei and in the cytoplasm. Those in the nuclei are concerned with movement of chromosomes during nuclear division (Beckett *et al.*, 1974). The cytoplasm thus shows structural organization related to the control of cellular processes.

The soluble portion of the cytoplasm is also organized. Many chemicals come into contact with one another in the cell, and thousands of reactions are thermodynamically possible. Order is imposed on this potential chaos by the specific catalytic activities of the enzymes that make up the bulk of the amorphous portion of the cytoplasm and by the location of certain enzymes in specific parts of the cell. Scott and Metzenberg (1970) provided a useful review of the cellular distribution of enzymes in fungi.

The structure and composition of the walls of conidial fungi have been reviewed (see Aronson, this volume, Chapter 28). The cell wall itself has several functions. It determines the size and shape of the cell. It confers resistance to heat, radiation, toxins, and enzymes. It is involved in interhyphal interactions such as anastomosis and the formation of fruiting bodies. It is a reservoir of ions and enzymes and may help to regulate the movement of chemicals between the cell and its environment (Burnett, 1976).

B. Transport

The movement of chemicals into the fungal cell has been reviewed by Jennings (1976). Work on conidial fungi, especially anamorphic *Neurospora crassa, Aspergillus nidulans,* and *Penicillium chrysogenum,* has contributed greatly to this information.

Transport systems show various characteristics which may be used as a guide to the mechanism of uptake. If transport depends on the concentration (more precisely, activity) gradient between the cell and its environment and ceases when the internal concentration of the chemical equals the external concentration, then we may say that transport occurs by *diffusion*. In some cases the kinetics of transport indicate specificity for certain molecular configurations, competition among different chemicals, and a maximum rate of uptake that cannot be increased by further increases in external concentration of the chemical. These characteristics of *specificity, competition,* and *saturation,* respectively, are taken as evidence of a carrier in the membrane that facilitates transport. Thus there may be a system of *facilitated diffusion.* If the chemical being transported reaches a higher concentration in the cell than in the medium, then the cell must do work to achieve this. This *active transport* is characterized by *accumulation* against a concentration gradient and dependence upon *metabolism.* Active transport systems also usually have kinetic properties associated with a carrier. There is evidence that these systems operate in conidial fungi.

It is generally considered that water enters cells by diffusion along a water potential gradient. There is also some evidence of active transport of water. Marchant and White (1966) reported that conidia of *Fusarium culmorum* required a respirable substrate (glucose) in order to take up water.

Potassium (Slayman and Tatum, 1964, 1965a,b), sulfate (Marzluf, 1970; Roberts and Marzluf, 1971), and phosphate (Burns and Beever, 1977) are transported into *N. crassa* by carrier-facilitated active transport, as is choline-*O*-sulfate in *Penicillium notatum* (Bellenger *et al.,* 1968). Two systems for the transport of $H_2PO_4^-$ are known in *N. crassa* (Burns and Beever, 1977).

Nitrate and ammonia enter by diffusion (Nicholas, 1965). Ammonia enters as NH_3 and nitrate as NO_3^-. Uptake of NO_3^- requires exogenous glucose (Macmillan, 1956). In the presence of glucose, NH_3 inhibits the uptake of NO_3^-. For example, *Scopulariopsis brevicaulis* did not take up NO_3^- until external levels of NH_3 had fallen to 0.002 N (Morton and MacMillan, 1954).

The mechanism by which NH_3 represses transport of nitrate is not known. It does not appear to regulate the activity of enzymes that mediate the reduction of nitrate and nitrite. NH_3 also represses transport of amino acids (Whitaker, 1976).

The preferential uptake of NH_3 explains why the use of NH_4NO_3 as the nitrogen source in a poorly buffered medium leads to a decline in pH during growth of the culture (Cochrane, 1958). The culture behaves as an $NH_4^+-H^+$ exchange system; as NH_3 enters the cell, the concentration of H^+ in the medium increases.

Sugars enter by carrier-facilitated diffusion or by active transport (Jennings, 1974). There are several systems involved, and most are highly specific. For example, *A. nidulans* has separate carriers for D-glucose, D-galactose, and D-fructose (Mark and Romano, 1971). There may be more than one transport system for a particular sugar. For example, conidia of *Myrothecium verrucaria* have two systems for transporting trehalose (Mandels and Vitols, 1967), and *N. crassa* has two systems for transporting glucose (Scarborough, 1970).

Amino acids enter fungal cells by carrier-facilitated active transport. Four types of transport systems were recognized by Whitaker (1976). Conidial fungi are represented in each group. There are single systems of broad specificity, multiple systems of broad specificity, systems specific for acidic, neutral, and basic amino acids, and systems specific for single amino acids. Conidial fungi in which one or more of these systems have been observed include anamorphic *N. crassa, A. nidulans, P. chrysogenum, Botrytis fabae* (Whitaker, 1976), and *Arthrobotrys conoides* (Gupta and Pramer, 1970). Whitaker (1976) suggests that systems of narrow specificity operate in complete media and systems of broad specificity in media deficient in carbon, nitrogen, or sulfur.

Proteins which specifically bind amino acids have been extracted from *N. crassa* (Stuart and DeBusk, 1971; Wiley, 1970). Such proteins may act as carriers in amino acid transport. Jennings (1974) described how proteins could complex with hydrophilic solutes and facilitate their movement through the hydrophobic lipid membrane. This has been demonstrated experimentally. A glycoprotein from *N. crassa* increased 1000-fold the rate of movement of L-arginine through a film formed of lipids from *N. crassa* conidia (Stuart and DeBusk, 1973).

Organic acids appear to be transported by carrier-facilitated diffusion. Characteristically they are transported as the undissociated molecule. Therefore they enter the cell most rapidly at a pH lower than the pK_a of the acid. There are very few reports on transport of organic acids in conidial fungi. Lewis and Darnall (1970) described the uptake of caproic acid by *A. niger.*

Two purine transport systems have been reported in *N. crassa;* one transports adenine, hypoxanthine, and guanine, and the second is specific for adenine (Magill and Magill, 1975). The general system occurs in freshly harvested conidia; the specific system develops during germination (Pendyala and Wellman, 1977).

Transport is believed to be controlled largely by the plasmalemma. For example, there is evidence that at least four transport systems are associated with this membrane in *A. nidulans* (Elorza *et al.,* 1969): one for divalent anions, two for purines, and one for sugars. These systems remained functional after removal of the cell wall. There are no reports relating to transport systems in isolated, purified fungal membranes.

The location of transport sites in the thallus is not clearly known. It is likely that transport occurs over the entire surface of conidia and germ tubes. In older hyphae transport may be more rapid at hyphal tips. The enzyme invertase, which is involved in transport of sucrose, was shown to be uniformly distributed along the cell periphery of conidia of *N. crassa* but concentrated at hyphal tips and branching points in the mycelium (Chang and Trevithick, 1970).

C. Metabolism and Metabolites

Chemicals entering the cell become subject to the metabolic activities occurring therein. Metabolism is generally considered to be of two sorts: primary and secondary. Primary metabolism is causally related to growth and development of the organism. It is concerned with the generation and use of energy and molecules to operate the machinery of the cell.

Primary metabolism of fungi is similar to that of other aerobic organisms. The early literature on this subject was reviewed by Foster (1949), while more recent appraisals are to be found in Smith and Berry (1975, 1976, 1978), Burnett (1976), Weber and Hess (1976), and Weete (1974). There is no evidence to suggest that primary metabolism of conidial fungi differs fundamentally from that of other fungi. There is evidence that metabolic patterns change or new chemicals are produced during differentiation. For example, in *N. crassa,* there is higher glycolytic activity in mycelial cultures and higher oxidative metabolism in cultures producing conidia (Turian and Bianchi, 1972). In no case, however, is it possible to point to a critical metabolic event and say that it is the key to the observed morphological change. In fact, several different metabolic patterns appear to be able to generate a common morphological product, such as a conidium (Tan, 1978).

Secondary metabolism produces compounds that have no clear role in maintaining or building the organism. Fungi produce a large number of secondary metabolites. Reviews include those of Turner (1971) and of several authors in volumes edited by Smith and Berry (1975, 1978). Turner (1971) listed over 1000 secondary metabolites produced by 171 genera and 478 species of fungi. Of these, 38 genera (22%) and 217 species (45%) were conidial fungi. Secondary metabolites fall into groups such as organic acids, polysaccharides, lipids, enzymes, hormones, antibiotics, and toxins. Metabolites of industrial importance include penicillins (*P. chrysogenum*), cephalosporin (*Cephalosporium acremonium*), griseofulvin (*Penicillium patulum*), gibberellins (*Fusarium moniliforme*), kojic acid (*Aspergillus oryzae*), gallic acid (*A. niger, Penicillium glaucum*), citric acid (*Penicillium* and *Aspergillus*), gluconic acid (*A. niger* and *Penicillium* sp.), itaconic acid (*Aspergillus itaconicus, Aspergillus terreus*), pectinase (*P. glaucum*), glucose oxidase, and catalase (*A. niger*) (Martin and Demain, 1978).

Most fungal products toxic to animals and man which are referred to as mycotoxins (Purchase, 1974), are also secondary metabolites. Those produced by conidial fungi include aflatoxins (*Aspergillus flavus*), sporidesmins (*Pithomyces chartarum*), patulin (species of *Aspergillus* and *Penicillium*), ochratoxin (*Aspergillus ochraceus* and *Penicillium viridicatum*), and trichothecenes and estrogenic toxins (especially zearalenone) from species of *Fusarium*.

There is evidence that the liberation of metabolites is under cellular control. For example, Yabuki and Fukui (1970) showed that liberation of amylase from the mycelium of *A. oryzae* was controlled by an amylase-binding factor in the cell wall. In media that did not induce the formation of amylase, a factor was present in the wall that prevented amylase from entering the medium. When the medium induced formation of amylase, the binding factor was no longer present in the wall and amylase entered the medium.

Liberation of metabolites may be controlled by the porosity of the wall. In *N. crassa,* invertase exists as heavy and light isozymes. Mutants that liberated small amounts of invertase liberated mostly the light isomer. Mutants that liberated greater amounts of invertase liberated more of the heavy isomer. It was suggested that the wall acted as a sieve and that larger channels occurred in the walls of "leakier" mutants (Trevithick and Metzenberg, 1966). Similarly, Chang and Trevithick (1974) have provided evidence that exoenzymes of *N. crassa* are secreted largely through the porous plastic wall of the hyphal tip.

The ecological importance of extracellular metabolites is considered in Section IV,H.

III. ORGANISM PHYSIOLOGY

In this section we examine how conidial fungi grow and integrate their resources to produce a new generation of reproductive propagules. At this level of enquiry we begin to see a wide range of solutions to common problems.

A. Germination

In conidial fungi germination involves the production of a germ tube from conidia, sclerotia, or fragments of hyphae. The process depends on exogenous supplies of oxygen, water, and possibly one or several nutrients. Morphological events include swelling of the cells and emergence of a germ tube. Cytological events include changes in the number or appearance of organelles and the synthesis of one or more walls inside the original wall. Physiological events include increased rate of respiration, uptake of water, and increased activity of transport systems. Metabolic events include an increase in the activity of metabolic pathways, the repair of metabolic lesions, and the induction of specific enzymes. The general picture is of increased synthetic activity (Madelin 1966, Weber and Hess 1976). Emergence of the germ tube does not coincide with any particular cytological, physiological, or metabolic event. The process of germination of most conidia may fairly be said to be initiated by exogenous water, but the role of water is not known. It does not seem to be simply a question of hydration of the cytoplasm, since the water content of inactive conidia may be similar to that of active hyphae (Hawker and Madelin 1976).

Conidia in a population do not germinate simultaneously. Some produce a germ tube faster than others. The latent period of germination (the time a conidium takes to produce a recognizable germ tube), in a given environment capable of supporting germination, is a property of the conidium which, like other properties, will show some type of distribution in the population. The nature of the distribution has not been analyzed rigorously. However, a plot of the percentage of germinated conidia against time often produces a sigmoid curve. The slope of this curve at any point is a measure of the number of spores producing a germ tube at that time. When the slope of the sigmoid curve is plotted against time, a bell curve is produced that suggests the latent period of germination is distributed normally through the population of conidia. A few conidia germinate rapidly, a few germinate slowly, and most germinate near the mean time for the population.

Germination is usually inhibited when spores are crowded together. This may be due to competition for limited exogenous nutrients, to the presence of endogenous inhibitors of germination or to both. Chemicals that inhibit germination have been identifed in rust urediniospores, but there are no similar reports for ascomycetous conidial fungi. Blakeman (1969) reported that conidia of the *Ascochyta* anamorph of *Mycosphaerella ligulicola* showed self-inhibition of germination in water when produced at 27°C but not when produced at 15°C. Aqueous extracts from both types of conidia could inhibit germination. Conidia produced at lower temperatures were larger. Blakeman suggested the presumed greater nutrient reserves of the larger conidia overcame the effect of their endogenous inhibitors.

B. Growth

Growth is an increase in size, mass, or cell number with time. Studies on conidial fungi have contributed greatly to our understanding of the kinetics of this process in fungi.

Most conidial fungi can be grown in media synthesized in the laboratory. The medium may be used as a liquid or solidified with agar or gelatin. Liquid culture systems are of two basic types: batch culture and continuous-flow culture. Batch culture refers to the common laboratory situation where the medium remains in the vessel during the incubation period. Most substrates are added at the start of incubation, and most products remain in the vessel during the growth period. In this system the fungus continuously alters the medium by withdrawing nutrients and liberating metabolites. The medium may be stationary, stirred, or shaken. In contrast, in a continuous-flow system, such as a chemostat, medium is continuously fed into the culture vessel, and spent medium and some of the fungus are continuously removed. The growth rate, organism concentration, and concentration of all substrates and products can be held constant for prolonged periods.

The manner in which the fungus grows depends to a great extent upon the physical and chemical characteristics of its environment. In a stirred liquid medium the fungus is likely to grow as diffuse, submerged mycelium. In a shaken liquid culture, submerged pellets or diffuse mycelium may form. In stationary medium, liquid or solid, the culture will consist of submerged, surface, and aerial cells. As these cultures grow, they will contain hyphal segments of different physiological ages. They are therefore considered unsuitable for studies on the biochemistry of differentiation. A growing culture consisting of a collection of physiologically similar cells can, however, be approximated in continuous-flow systems.

When growth is measured as radial expansion on an agar medium and plotted against time at frequent intervals, a curve such as that obtained by Trinci (1969) for *A. nidulans* may be obtained. Initially there is no growth; this is the lag period. The colony then begins to grow, and for a short time the growth rate on an arithmetic scale increases with time. When growth in this period is plotted on a logarithmic scale, two phases appear. The first is characterized by a straight line. This is the so-called log phase and is a period of exponential growth. Then follows a period of deceleration during which the rate of change in the growth rate declines (the rate of growth is still increasing in this period). In the fourth, or linear, period the rate of radial growth is constant. The linear phase is usually maintained until the margin of the culture nears the edge of the culture vessel. Most studies on growth in agar media have been confined to this linear phase. Detection of the earlier phases requires microscopic examination and measurements at frequent intervals soon after inoculation.

Race tubes (Ryan tubes) allow radial growth to be followed for longer periods than is possible in conventional petri dishes. They are particularly useful for fungi that grow very quickly. In such tubes linear growth of *N. crassa* was constant over a period of 200 h (Ryan *et al.*, 1943).

The general curve of growth in batch culture is sigmoid. There is an initial period in which no growth occurs (lag phase). This is followed by a phase in which mass increases rapidly and then more slowly until a maximum is reached. The total culture mass may subsequently decline. Mathematical analyses of the period of rapid growth have led to different concepts of growth in batch culture.

Early studies (Lilly and Barnett, 1951) suggested that, when the culture was growing most rapidly, dry weight increased linearly with time. A straight-line relation between the cube root of mycelium weight and the time of incubation was found for the growth of *N. crassa* (Emerson, 1950) and *A. nidulans* (Trinci, 1970). In other cases exponential growth has been observed.

Where growth is unrestricted, we assume it is autocatalytic. The mass produced in unit time depends on the mass at the beginning of the time period. This results in exponential growth. The specific growth rate (μ) is the amount of organism produced by a unit amount of organism in unit time. Its unit is g/g/t or t^{-1}. The formula for exponential growth is

$$\ln x_t = \mu t + \ln x_0 \tag{1}$$

where ln is the natural logarithm and mass x_t after time t is derived from mass x_0. Therefore, during exponential growth a plot of mass on a logarithmic scale against time on an arithmetic scale produces a straight line. Conversely, the production of a straight line by such a plot is taken as evidence of exponential growth. Equation (1) may be rearranged to give

$$\mu t = \ln x_t - \ln x_0 \qquad (2)$$

When t_2 is the time (in hours) required for doubling of the mass, then

$$x_{t_2} = 2x_0 \qquad (3)$$

and

$$\mu t = \ln 2x_0 - \ln x_0 = 0.693 \qquad (4)$$

Therefore doubling time is related to specific growth rate by

$$t_2 = 0.693/\mu \quad \text{hours} \qquad (5)$$

Exponential growth is characterized by a constant doubling time (or generation time) and a constant specific growth rate. If the specific growth rate is increased and growth remains exponential, then the doubling time will be reduced in accordance with Eq. (5).

Exponential growth is now known to occur in all cultures of filamentous fungi for at least a brief period. On agar media, growth has been shown to be exponential when measured as germ tube elongation (*A. nidulans;* Trinci, 1969, 1971b) or as the increase in the total length of hyphae at the colony margin (Trinci, 1971a).

Trinci (1971a) provided a valuable analysis of colony growth on agar media. He noted that no more than the apical 35 μm of the hyphae of *A. nidulans* was growing, but at 37°C the colony expanded radially at a rate of 297 μm/h. The doubling time was about 2 h. He concluded that protoplasm must be increasing in a much longer portion of the hypha than was actually elongating. Zalokar (1959b) had calculated that the terminal 12 mm of hyphae of *N. crassa* must be involved in producing cytoplasm in order to sustain the growth rates observed. Trinci therefore suggested that there was a marginal annulus in which growth was exponential and in which the older portions contributed cytoplasm to the extending hyphal tips. The width of the zone (in micrometers) in conidial fungi was reported to range from 423 (*Geotrichum lactis*) to 10,000 (anamorphic *N. crassa*). The specific growth rate in this zone of exponential growth was estimated to range from 0.426/h (*G. lactis*) to 0.117/h (*A. niger*). Trinci proposed that the rate at which a colony grew radially on an agar medium (*Kr*) was related to its specific growth rate (μ) and the width of the peripheral growth zone (*w*) by

$$Kr = \mu w \qquad (6)$$

The value of *w* did not depend on temperature but was influenced by the concentration of glucose in the medium. Therefore Trinci concluded that radial growth on agar media could be used to measure the effects of temperature, but not of nutrients, on specific growth rate.

Using *P. chrysogenum,* Righelato (1975) confirmed that the width of the marginal annulus growing exponentially was on the order of 0.5–1.2 mm, and that specific growth rates on agar media (0.116–0.348) were similar to those reported for submerged growth in liquid media. A linear increase in colony diameter on agar media is now seen as the natural consequence of an exponential increase in the total hyphal length at the margin of the colony (Righelato, 1975).

Exponential growth has been observed in batch cultures of *N. crassa* (Zalokar, 1959a) and *A. nidulans* (Trinci, 1969). The general situation in batch culture appears to be as follows. When mass of fungal material is plotted on an arithmetic scale against time, the curve is sigmoid. When mass is plotted on a log scale, the curve shows a lag phase, a linear phase (in which growth is occurring exponentially), a deceleration phase, and a decline phase. The exponential phase may be maintained for only three to six generations. In rapidly growing cultures this could occupy less than 24 h.

In continuous-flow systems growth is exponential at steady state. Specific growth rate is deliberately controlled by supplying one substrate (usually the carbon source) at a growth-limiting concentration. The medium is fed at the rate of F liters per hour into a constant-volume culture of V liters. The ratio F/V is known as the dilution rate D. At steady state it equals the specific growth rate μ. By manipulating the substrate concentration the growth rate can be controlled between the maximum the organism is capable of and close to zero.

The organism mass balance is given by

$$dx/dt = \mu x - Dx \qquad (7)$$

At steady state

$$dx/dt = 0 \qquad (8)$$

Therefore

$$\mu = D \qquad (9)$$

As long as the concentration of the limiting substrate is held constant and D is constant, the system will tend toward the steady state. When the specific growth rate (μ) is constant, the doubling time is constant and growth is exponential. Pirt and Callow (1960) used a conidial fungus (*P. chrysogenum*) to show for the first time that these laws of growth, developed with reference to unicellular organisms, can be applied to filamentous fungi growing in a continuous-flow culture.

The minimum doubling time (t_2) observed for filamentous fungi is about 2 h. Solomons (1975) cites specific growth rates of 0.4 for *Neurospora sitophila* ($t_2 = 1.73$ h) and 0.28 ($t_2 = 2.4$ h) for *Fusarium graminearum*. These are representative of the fastest rates obtained in a chemostat. Specific growth rates can be kept low if desired. In penicillin fermentations, where μ may be 0.005, the doubling time is closer to 6 days.

Growth kinetics are greatly influenced by the composition of the medium. For example, specific growth rate is limited by substrate concentration. The equation presented by Monod (1942) was

$$dx/dt = \mu_{max}[s/(k_s + s)]x \qquad (10)$$

where dx/dt is the growth rate, μ_{max} the maximum specific growth rate, k_s the concentration of substrate at which the specific growth rate is half the maximum, s the substrate concentration, and x a measure of growth.

In batch culture k_s is usually very small compared with the initial substrate concentration. We would expect growth to occur at the maximum rate until s approached k_s and then rapidly decline to zero as the substrate became exhausted. Borrow et al. (1961) and many subsequent authors (e.g., Kier et al., 1976) have shown this to be so, especially when the initial concentration of growth-limiting substrate is low. In continuous-flow culture the specific growth rate is controlled by the concentration of the growth-limiting substrate.

The order in which nutrients become limiting influences the shape of the growth curve. For example, Borrow et al. (1961) recognized five phases of growth in batch cultures of F. moniliforme. The first was the balanced phase which covered the period from initiation of growth to the time at which the first nutrient was exhausted. In this period dry weight increased, and rates of nutrient uptake per unit weight of mycelium were constant. For example, "the accumulation of 1 mg dry mycelium was accompanied by the uptake of 3.24 mg glucose, 0.093 mg nitrogen, 0.023 mg phosphorus, 0.0018 mg magnesium and ca. 0.02 mg potassium" (Borrow et al., 1961). If a nutrient other than nitrogen or glucose was the first to become exhausted, there followed a transition phase in which the fungal mass continued to increase but its composition changed. Generally, lipid and carbohydrate concentrations increased. The storage phase was initiated when nitrogen was exhausted from the medium and ended approximately when the mass reached its maximum value. In this period lipid and carbohydrate concentrations in the mycelium continued to increase. During the maintenance phase the mass of mycelium remained constant so long as glucose was available in the medium. Glucose was the only nutrient taken up during this period. The exhaustion of glucose from the medium initiated the terminal phase, characterized by autolysis, release of cell contents into the medium, and reduction in total mass of mycelium.

It is apparent that the form of the growth curve of conidial fungi is determined by many factors, but the more important include the innate capacity for exponential growth, the concentration of limiting substrate, the order in which nutrients are depleted from the medium, and possibly the accumulation of toxic metabolites in the medium.

C. Branching

Formation of a branch requires that a growth zone be initiated in a nongrowing portion of the wall. The mechanism of this is poorly understood. Presumably events such as those that occur at the main apex are involved. Mahadevan and Mahadkar (1970) provided experimental evidence that autolytic enzymes bound to the wall participate in the formation of branches by anamorphic *N. crassa*.

Branching may be controlled by apical dominance exerted by the main axis. Studies on translocation have emphasized the importance of the hyphal tip as a sink for nutrients. In pioneering studies on polar growth and branching Robertson (1958) showed that hyphal apexes of *Fusarium oxysporum* produced branches after tip growth was arrested. In more recent work Collinge *et al.* (1978) found that, when a temperature-sensitive colonial mutant of *N. crassa* was transferred from 25° to 37°C, the hyphae stopped elongating but produced numerous inter- calary septa, and that most of the newly formed compartments produced a branch. These results suggest that branching can occur when the influence of a growing tip is minimized. In addition, they suggest that the formation of branches is related to the formation of septa.

Recognition of exponential growth at colony margins has also prompted con- sideration of the relation of growth to branching. From observations on *A. nidulans,* Katz *et al.* (1972) proposed that branching frequency was proportional to specific growth rate. The pattern of branching may well be determined both by the tendency of the fungus to grow exponentially and by the effect of environ- mental conditions on specific growth rate. However, Clutterbuck (1978) pro- vided evidence against the possibility that branching occurs because tip growth is too slow to accommodate an exponential increase in the mass of cytoplasm. He observed that hyphae of *Neurospora* first branched when the tip was elongating at a rate of 300 μm/h, but that the growth rate of the tips subsequently increased to 3000 μm/h. Neither the growth rate nor acceleration rate was related to branching.

D. Rhythms

Growth or development of many conidial fungi may be rhythmic (Jerebzoff, 1965). Most of these rhythms are caused by periodic fluctuations in environmen- tal conditions and cease immediately or fade away in a constant environment. These are exogenous rhythms. A few conidial fungi, e.g., *Tubercularia vulgaris* and *Penicillium diversum* (Bourret *et al.,* 1969), possess endogenous rhythms, i.e., rhythms that persist in a constant environment. A special type of endoge- nous rhythm is the circadian rhythm, which is defined very precisely. It has a free-running period of about 24 h in constant darkness, has a Q_{10} of about 1 (i.e., the period is not altered by temperature changes), and does not depend on environmental cycles but is entrained (reset) by light or temperature alterations. Circadian rhythms are common in biology but rare in fungi. The *band* and *timex* mutants of *N. crassa* fit these criteria (Lysek, 1978; Sargent *et al.*, 1966).

E. Integration of Resources

There is evidence that different tasks are assigned to different parts of the hypha but that mechanisms exist by which one part of the thallus can communicate with another. There is structural differentiation. Vesicles are concentrated at hyphal tips, Woronin bodies occur in association with septa, vacuoles and lipid bodies become more common in older parts of the mycelium, and certain chemicals occur more commonly in some parts of the hypha than in others. For example, Zalokar (1959b) showed in *N. crassa* that protein and RNA occurred predominantly in the terminal 100 μm of the hypha, whereas glycogen, cytochrome oxidase, and succinic dehydrogenase occurred predominantly in older portions. There is also organized local activity, such as synchronous nuclear division (King and Alexander, 1969), development of septa, movement of Woronin bodies to plug septal pores in damaged hyphae (Beckett *et al.*, 1974), formation of branches and differentiated structures at discrete sites along the hypha, synthesis of wall material at the apex, and storage of energy reserves in older compartments.

At the same time it is clear that there are mechanisms by which one part of the thallus can communicate with another. The septa are perforated, and, except where they are blocked, the thallus can be considered one giant cell with open lines of communication between its various parts. Movement of materials from one part of the thallus to another (translocation) may occur and is clearly of potential importance in the internal regulation of growth and development. We assume that translocation of materials into aerial structures must occur. This view is strengthened by observations of cytoplasmic streaming through the thallus. However, there is virtually no information on translocation in conidial fungi. A report by Howard (1978) showed that *A. niger* was able to translocate radioactive carbon supplied as [^{14}C]glucose. The mechanism of translocation is unknown. Burnett (1976) discussed similarities between translocation and movement of cytoplasm, but attempts to detect movement of cytoplasm in *A. niger* have failed (Howard, 1978; Nishi *et al.*, 1968). Moreover, the mechanism of cytoplasmic streaming is not known.

Another important mechanism of communication within the thallus is anastomosis, in which hyphal tips grow toward each other and become organically united. Anastomosis provides a bridge linking one part of the thallus with another. The implications of this process are manifold, and its mechanism a mystery. In conidial fungi anastomosis provides for the pooling of nutrient and genetic resources. It is presumed to facilitate the production of reproductive structures and to increase flexibility in responding to changing environments.

A third feature that permits cooperation among elements of the thallus is the tendency of some hyphae to associate firmly with one another. Connivence of hyphae is strongly expressed in conidial fungi in structures such as coremia and sporodochia. Connivence may be one expression of a more general tendency of hyphae to associate closely with surfaces. Many studies (e.g., Tsuneda and Skoropad, 1977) show hyphae depressing the surface of the substrate. The mechanisms underlying firm hyphal associations with one another or with non-fungal substrates are not fully understood. We know that, in some cases, the hyphae are attached to the substrate by mucilage (Heitefuss and Williams, 1976).

Hyphae are also notable for their ability to penetrate membranes of various types. They appear to do this by exerting high pressure, by liberating chemicals to dissolve the membrane, or both (Heitefuss and Williams, 1976).

Anastomosis, connivence, ability to attach, and ability to penetrate are fundamental properties of hyphae. Only the last has received detailed experimental investigation. The first three may depend on characteristics of the wall, but no information is available.

The growth and development of conidial fungi are not known to be under hormonal control but may be influenced by self-produced extracellular metabolites. An extracellular metabolite is claimed to influence production of microsclerotia of *Verticillium dahliae* (Brandt and Reese, 1964). Volatile metabolites may influence the fungus that produced them, but most reports refer to the manner in which volatiles affect other fungi (Hutchinson, 1973).

F. Differentiation

The idealized hypha is a cylinder with a hemispherical apical dome. This shape can be generated mathematically if we assume that the rate of wall synthesis (Bartnicki-Garcia, 1973), or the degree of wall plasticity (da Riva Ricci and Kendrick, 1972), is maximal at the top of the dome and declines to zero at the base.

To change the shape of the fungus cell it is necessary to change the shape of the wall. Some authors (e.g., Robertson, 1965; Nickerson, 1963) have emphasized the notions of plasticity and rigidity. The wall retains its shape if rigid but can change its shape if plastic. Nickerson (1963) developed, with reference to yeasts, a concept in which sulfhydryl groups conferred plasticity while disulfide bonds conferred rigidity. The concept has not been applied to conidial fungi. Other writers have related changes in shape to differential rates of wall synthesis (Bartnicki-Garcia, 1968, 1973). This theory suggests that the shape of the fungal cell depends on the particular combination and timing of polar and nonpolar growth the wall has experienced. Polar growth results from the localization of wall synthesis to a small region of the wall. Nonpolar growth occurs when all parts of the wall are synthesized at the same rate. Elongated tubular cells, such as hyphae, are formed by polar growth. Spherical cells are formed by nonpolar growth. Other shapes may be related to these fundamental processes. Chlamydospores, yeast forms of dermatophytes, microsclerotia of *Verticillium dahliae* (Hall and Ly, 1972), and appressoria provide examples of morphogenesis by nonpolar growth. Hyphae, sclerotia of *Botrytis* (Townsend and Willetts, 1954), and elongate conidiophores reflect largely polar growth. The concept that cell shape depends on the site of wall synthesis has been applied to hyphal tips, branching sites, and yeastlike budding. The challenge remains to apply the idea and the available technology to more elaborate structures such as conidiophores and conidia.

Several attempts have been made to relate metabolic patterns to fungal morphogenesis, and *N. crassa* has served as a useful model (Davis, 1976; Scott, 1976). There is considerable information on its metabolism in relation to sporulation (Turian and Bianchi, 1972), and the primary light receptor chemical has been identified (Muñoz and Butler, 1975). However, there is still no clear understanding of the metabolic control of differentiation. Most of our information concerns effects of the environment on growth and development.

IV. ENVIRONMENTAL PHYSIOLOGY

In this section we view a bewildering array of responses to environmental influences. The plasticity of conidial fungi as a group and as individual species becomes apparent.

Recent reviews that contain information relating environmental factors to steps in the life cycle of conidial fungi include Gottlieb (1978) on germination, Ainsworth and Sussman (1965) and Berry (1975) on growth, Hawker (1966), Smith (1978), Rotem *et al.* (1978), and Turian (1974) on sporulation, Emmett and Parbery (1975) on appressoria, and Willetts (1978), Chet and Henis (1975), and Coley-Smith and Cooke (1971) on sclerotia.

Klebs (1928) provided four generalizations which can still be usefully applied to fungi.

1. Growth precedes, and is essential to, reproduction.
2. External conditions determine whether growth or reproduction occurs.
3. If the environment is favorable for growth, reproduction does not occur. Conditions that favor reproduction generally do not favor growth.
4. The range of environmental conditions that permits reproduction is narrower than that permitting growth.

These generalizations emphasize the high sensitivity of fungi to environmental changes. This notion of plasticity is still widely accepted. Booth (1978) went so far as to suggest that the similarities in structure of certain fructifications of conidial fungi were more dependent on their location in the environment than on their genealogical relationships.

Undoubtedly, all environmental factors affect growth and development of fungi. Those about which most is known include nutrition, water, temperature, electromagnetic radiation, pH, and gaseous atmosphere. The environment may also contain specific morphogenetic, stimulatory, or inhibitory chemicals.

A. Nutrition

Germination of most conidia depends on, or is stimulated by, one or more external nutrients (Cochrane, 1958). Washed conidia of very few species will germinate in distilled water. The requirement may be for sugar (anamorphic *Monilinia fructicola*), sugar and nitrogen (*Fusarium roseum*) or, more rarely, several nutrients. Conidia of *Glomerella cingulata* require carbon, nitrogen, phosphorus, and sulfur (Cochrane, 1958). Swelling of conidia of *Penicillium atrovenetum* required exogenous glucose, phosphate, and oxygen. Formation of germ tubes required, in addition, nitrogen (Gottlieb and Tripathi, 1968).

The use of washed conidia to demonstrate requirements for exogenous nutrients may not always be realistic. Many conidia produce, or are surrounded by, a slimy matrix which may provide nutrients for germination. Nevertheless a requirement for exogenous nutrient in the natural habitat is very common; witness the widespread starvation-induced property of fungistasis in soil (Lockwood, 1964; Watson and Ford, 1978).

In a thought-provoking paper Garrett (1973) considered the size of spores in relation to their dependence on exogenous nutrients. Smaller spores are likely to contain fewer nutrient reserves and to depend more on the environment for nutrients to support early growth. However, if the normal habitat provides nutrients to support germination, it is to the advantage of the fungus to produce a large number of small spores rather than a small number of large spores. Smallness may limit invasiveness in a nutrient-poor environment. For example, the conidia of *Botrytis fabae* have nine times the volume of conidia of *B. cinerea*. The presumably greater nutrient reserves of *B. fabae* conidia may explain why they can develop sufficiently in the nutrient-poor environment of a broad-bean leaf surface to establish an infection, while conidia of *B. cinerea* can usually successfully infect only nutritionally richer substrates such as petals and senescent tissues.

Conidial fungi require water, molecular oxygen, an organic source of carbon and energy, a source of nitrogen other than molecular nitrogen, and several other elements. At least 13 elements are essential for growth, namely, oxygen, carbon, hydrogen, nitrogen, phosphorus, potassium, sulfur, magnesium, manganese, iron, zinc, copper, and molybdenum. The first eight are needed in relatively large quantities and are referred to as macronutrients. The latter five are required in small amounts and are called micronutrients. Other elements which may be required in small amounts include gallium, vanadium, and silicon. These are not routinely added even to the most precisely prepared medium. In the laboratory a medium composed of water, a sugar such as glucose, and a nitrogen source such as ammonia, nitrate, or an amino acid will often support growth. Deliberate addition of further essential nutrients is likely to stimulate growth.

Like all fungi, conidial fungi are heterotrophic. This means they require organic compounds as a source of carbon and energy. A few exceptional reports indicate that filamentous fungi can fix carbon dioxide. Mirocha and DeVay (1971) reported that *Fusarium* sp. and *Cephalosporium* sp. not only fixed carbon dioxide but grew on an inorganic salts medium without added carbon. Organic compounds supporting most growth are usually sugars such as D-glucose, D-fructose, and sucrose. Polysaccharides, amino acids, lipids, organic acids, proteins, and hydrocarbons may also be used. A small amount of exogenous carbon may be required to maintain the fungus even when it is not growing. Carter *et al.* (1971) estimated that, at zero growth rate, *A. nidulans* consumed 0.029 g of glucose per gram of fungus biomass per hour.

Nitrogen may be supplied as ammonia, as nitrate, or in organic compounds such as amino acids or proteins. Phosphate and sulfate are convenient and readily used sources of phosphorus and sulfur, respectively. Organic forms can also be used. In fact, organic sulfur is essential for *in vitro* growth of the yeast phase of *Histoplasma capsulatum* (Gilardi, 1965). Potassium, magnesium, manganese, iron, zinc, copper, and molybdenum can all be used as salts.

The efficiency with which nutrient is converted into cell mass (the economic coefficient, EC) may be expressed as grams of substrate required to produce 1 g dry wt of fungus. Gray *et al.* (1964) screened 175 isolates of Fungi Imperfecti and found that 7% were highly efficient (EC of 2 or less) and 45% were moderately efficient (EC between 2 and 3).

That the omission of macronutrients reduces growth is easily shown. Very careful techniques are needed, however, to demonstrate that the micronutrients are essential. Reagent-grade chemicals, which provide the macronutrients, also supply the micronutrients as contaminants. Special procedures are required to produce macronutrients free of micronutrients. The microelement nutrition of *A. niger* has received careful and extensive study, dating from the pioneering work of Raulin (1869, cited in Ainsworth, 1976), in which requirements for iron and zinc were shown. Raulin also suggested that silicon stimulated growth. In a series of papers published between 1919 and 1939, Steinberg (1939) confirmed

Raulin's work and further showed that molybdenum and possibly gallium were essential. Bertrand and Javillier (1911). and Roberg (1928) showed, respectively, that manganese and copper were required for growth.

Growth factors, especially members of the water-soluble B complex of vitamins such as thiamin, biotin, inositol, pyridoxine, nicotinic acid, and pantothenic acid, may be required. A single deficiency for thiamin is the most common. Lilly and Barnett (1951) provided several examples of vitamin requirements among conidial fungi, e.g., thiamin for *Chalara elegans* (*Thielaviopsis basicola*), thiamin, biotin, and inositol for *Melanconium betulinum*, and thiamin, biotin, inositol, and pyridoxine for anamorphic *Ceratocystis ulmi*. However, many conidial fungi, e.g., species of *Fusarium, Aspergillus,* and *Penicillium*, require no added vitamins. These fungi manufacture their own.

A simple synthetic medium recommended for the culture of *Aspergillus* and *Penicillium* is the so-called Czapek–Dox formula. It contains, per liter of medium: $NaNO_3$, 2.0 g; K_2HPO_4, 1.0 g; $MgSO_4 \cdot 7H_2O$, 0.5 g; KCl, 0.5 g; $FeSO_4 \cdot 7H_2O$, 0.1 g; sucrose, 30.0 g; agar (if required), 15 g. The medium is made up to 1 liter with water. The osmotic potential of this medium is about -3 bars. Xerophytic aspergilli, such as members of the *A. glaucus* group, require 200 g of sucrose. This lowers the osmotic potential of the medium to about -13 bars. Dox originally specified KH_2PO_4. However, K_2HPO_4 now used in commercial preparations, places the initial pH of the prepared medium close to neutrality (Tuite, 1969). Essential elements not deliberately added are present as contaminants of one or more of the chemicals added. No growth factors or organic nitrogen are supplied in this medium. If necessary, organic nitrogen may be supplied as peptone or casamino acids. Growth factors may be supplied individually or in a complex such as yeast extract.

Vogel medium N for *Neurospora* is an example of a complex medium in which all required elements and growth factors are deliberately added. It contains, per liter of medium: sucrose, 20 g; Na_3 citrate $\cdot 5H_2O$, 3 g; KH_2PO_4, 5g; NH_4NO_3 2 g; $MgSO_4 \cdot 7H_2O$, 0.2 g; $CaCl_2 \cdot 2H_2O$, 0.1 g; citric acid, 5 mg; $ZnSO_4 \cdot 7H_2O$, 5 mg; $Fe(NH_4)_2 (SO_4)_2 \cdot 6H_2O$, 1 mg; $CuSO_4 \cdot 5H_2O$, 0.25 mg; $MnSO_4 \cdot H_2O$, 0.05 mg; H_3BO_3, 0.05 mg; $Na_2MoO_4 \cdot 2H_2O$, 0.05 mg; biotin, 5 μg. Agar at 15 g/liter is added if desired (Vogel, 1956). Organic nitrogen and other growth factors may be added. This medium supplies, in addition to the 13 essential elements, calcium, chlorine, and boron.

Nutrition has a great effect on morphogenesis. Much of the early work was evaluated by Hawker (1957), while Smith (1978) has provided a more recent review. There is much information but few principles. A period of vegetative growth usually precedes reproduction. Production of a large fruiting body or large numbers of conidia requires a well-nourished mycelium. Yet nutrient concentrations that favor reproduction are usually lower than those optimal for growth. For example, radial growth of *Cercospora nicotianae* was maximal at

50 g sucrose/liter, whereas the greatest number of conidia was produced in a medium containing 5 g sucrose/liter (Stavely and Nimmo, 1968). Starvation or reduction in the food supply often initiates reproduction. For example, transfer of germinated conidia to water or salt solutions stimulated formation of chlamydospores in *Fusarium solani* (Meyers and Cook, 1972). In a medium containing 0.1% glucose, *Geotrichum lactis* grew exponentially until the glucose was exhausted, whereupon growth ceased and arthrospores were formed (Kier *at al.,* 1976). However, in a medium containing 0.5% glucose, arthrospores began to form while the culture was still growing and when less than 15% of glucose and less than 30% of the exogenous ammonium had been consumed.

The order in which nutrients are exhausted in a culture medium may determine whether autolysis or sporulation dominates. Autolysis is likely to dominate if carbon is the first nutrient to become limiting. Sporulation is favored if nitrogen is the first nutrient exhausted (Smith, 1978).

Cessation or physical restriction of growth often stimulates differentiation. Pycnidia and conidia are often produced after the culture reaches the edge of the petri dish. Similarly, conditions that restrict germ tube elongation stimulate the formation of appressoria (Emmett and Parbery, 1975).

Stepwise control of sporulation by manipulation of the nutritional environment has been achieved in continuous-flow culture (Smith, 1978). For example, foot cells and conidiophores of *A. niger* were formed in a low-nitrogen medium in which nitrogen was exhausted before carbon. Transfer of the culture to fresh medium containing nitrogen and citric acid permitted the formation of vesicles and phialides. Subsequent transfer of the culture to a medium containing glucose and nitrate permitted the formation of conidia.

B. Water

Water is essential to growth and reproduction and, in some cases, to spore liberation and dispersal. It also influences the longevity of inactive propagules. Of the fresh weight of the mycelium 85–90% is water (Cochrane, 1958). Conidia may contain less water than the mycelium. Hawker and Madelin (1976) cite conidial moisture contents ranging from 6% (*Penicillium digitatum*) to 88% (*Botrytis fabae*).

The range of water potential that permits fungal activity is generally 0 to -100 bars. This is equivalent to a relative humidity (RH) ranging from 100 to 92%. For example, conidia of *Phyllosticta maydis* germinated maximally at 100% RH (0 bars) and not at all at 94% RH (-80 bars) (Bootsma *et al.,* 1973). To put these water potentials in perspective we may note that the permanent wilting point for plants is commonly taken to be -15 bars. The water potential of most culture media is approximately -1 to -3 bars. However, many conidial fungi grow most rapidly at -10 to -20 bars when the water potential is controlled by solutes

(e.g., sugars, salts) (Congly and Hall, 1976). When the water potential is controlled by the matric potential (Cook *et al.*, 1972) or RH (Bootsma *et al.*, 1973),
maximal growth occurs near 0 bars.

The lowest water potential at which fungi can grow may depend on the chemical used as the osmoticum. *Botrytis cinerea* grew at -54.5 bars when the water
potential was controlled with KNO_3, but not below -27.7 bars when $Ca(NO_3)_2$
was used (Hawkins, 1916).

Although most conidial fungal cannot grow below -100 bars, there are exceptions. Many *Aspergillus* and *Penicillium* species are reported to germinate at a
RH near 85% (-200 bars) (Altman and Dittmer, 1966). Bonner (1948) reported
that the germination of conidia of *A. niger* was greatest at 93% RH (-90 bars).
Raciborski (1905, cited in Lilly and Barnett, 1951) grew *A. glaucus* in a saturated NaCl solution (-380 bars) and a species of *Torula* in saturated LiCl
(-1000 bars). Curran (1971) reported that -450 bars prevented the production
of cleistothecia but permitted the growth and formation of conidia by *A. glaucus*.

The water potential of the fungus is assumed to be in equilibrium with the
water potential of its environment and is the resultant of the osmotic (solute)
potential of the cell sap and the turgor pressure. To permit the fungus to absorb
water osmotically the osmotic potential of the cytoplasm should be more negative
than the water potential of the medium. For example, Adebayo *et al.* (1971)
found that the osmotic potential of the cytoplasm of *Aspergillus wentii* decreased
from -21.5 to -44.3 bars as the water potential of the medium was decreased
from -6.2 to -31.2 bars. However, turgor pressure, calculated as the difference
between measured osmotic potential of the cytoplasm and water potential of the
medium, showed no relation to growth rate. There was no indication that the
higher growth rates observed on drier media were due to higher internal pressures
on the wall.

Since water influences growth so markedly, it is seen to be a major factor
limiting the colonization of a nutrient substrate. Conidial fungi such as *Aspergillus*, *Penicillium*, *Fusarium*, and *Cephalosporium* (*Acremonium*) are common in
moist habitats that supply simple nutrients. The time period over which films of
water persist or RH or soil water potential remains high is of vital importance in
the development of plant diseases caused by conidial fungi (Kozlowski, 1978;
Cook and Papendick, 1972). Similarly, the water potential of stored grains is
perhaps the most important factor determining their susceptibility to colonization
by species of *Aspergillus* (Christensen and Kaufmann, 1965). Seeds with a
moisture content greater than 12% can support growth of *Aspergillus* and
Penicillium (Curran, 1971).

Water also influences the development of reproductive structures. In free
water, *Cristulariella pyramidalis* produced many unbranched pyramidal heads.
In air at 96% RH fewer heads were formed, and they were branched (Latham,
1974). The length of the wet period is important. *Phyllosticta maydis* required
several days of continuous moisture to sporulate on corn leaves (Castor *et al.*,

1977). On the other hand, for a given total wetting period, *Alternaria porri* f.sp. *solani* produced more conidia after intermittent wetting than after continuous wetting of potato leaves (Bashi and Rotem, 1975). Many more examples of the effects of moisture on the sporulation of conidial fungi are provided by Rotem *et al.* (1978).

Release and dispersal of conidia may also be influenced by water. The discharge of spores as a result of desiccation of the conidiophore has been reported for several conidia fungi including *Alternaria, Curvularia,* and *Drechslera* (Ingold, 1971). Discharge by wetting of fruiting bodies or spores is perhaps more common. Pycnidia and acervuli release conidia during wet weather, and many Hyphomycetes possess wettable conidia which are readily dispersed by water. Spores that rely on water for dispersal are often formed in a slime probably largely composed of polysaccharide. For example, the mucilaginous layer around conidia of *Fusarium culmorum* appears to be composed mostly of xylan (Marchant, 1966). On the other hand, many conidia are hard to wet. Some contain surface lipids. (*Alternaria tenuis, Botrytis fabae, N. crassa*), but not all (*Penicillium expansum*) (Hawker and Madelin, 1976).

Storage at high RH values reduces viability of conidia. For example, Rotem (1968) found that at 53°C the longevity of five conidial fungi was maximal at 14.5% RH. *Trichoderma viride* survived best at 0% RH. Survival may depend on the water content of the spore. At the time of harvest from an agar medium the moisture content of conidia of *N. crassa* was 65% (Fahey *et al.,* 1978). Conidia were then stored at room temperature (20° ± 3°C) at various RH values. At 0% RH the water content of the conidia stabilized at about 5% after 1 day, and their germinability remained high (90% of the original value) after 79 days of storage. At 50% RH the water content stabilized at 13% after 2 days, and germinability declined to 60% after 80 days. At 100% RH the water content was maintained near 65%, and germinability declined to zero after 20 days.

C. Temperature

The upper limit for growth of most conidial fungi is near 35°C (Cochrane, 1958). Some grow at temperatures up to 55°C, but temperatures above 60°C are lethal (Baker, 1962). The lower limit for growth is probably near 0°C (Cochrane, 1958). Since most laboratory incubators do not provide temperatures below 3°C, the lower limits for fungi able to grow at refrigerator temperatures generally have not been determined. Conidial fungi can survive freeze-drying and cryogenic storage (−196°C) (Onions, 1971). If cooling procedures are adequate, conidial fungi could probably survive indefinitely at any temperature too low to support metabolism.

Conidial fungi span the range of temperature responses found in fungi. *Fusarium nivale* grows well at temperatures between 0° and 15°C. A few, such as species of *Aspergillus* and anamorphic *Neurospora,* grow rapidly near 35°C.

Anamorphic *Neurospora crassa* grew most rapidly at 36°C but did not grow at all above 44°C (Ryan *et al.*, 1943). Cooney and Emerson (1964) listed several conidial fungi. e.g., *Sporotrichum thermophile*, with characteristics of thermophiles. They did not grow at temperatures less than 20°C, grew most rapidly above 40°C, and required temperatures between 50° and 60°C to prevent growth. Despite these interesting exceptions it appears that most conidial fungi grow at temperatures between 5° and 35°C and make maximum growth at temperatures near 20°–28°C (Lilly and Barnett, 1951; Cochrane, 1958).

Most studies have examined the growth of fungi at constant temperatures. But in nature fungi are exposed to changing temperatures. Burgess and Griffin (1968) found that the growth of *Fusarium graminearum* and *Bipolaris sorokiniana* at fluctuating temperatures could be greater or less than growth at a constant median temperature. Smith (1964) found that fluctuations of small amplitude increased the growth rate of *Macrophomina phaseoli*. Waggoner and Parlange (1974a,b) produced a mathematical model describing spore germination of *Alternaria solani* at changing temperatures.

What are the mechanisms by which temperature affects growth? Ryan *et al.* (1943) found that the rate of growth of anamorphic *N. crassa* increased with temperature in a linear manner from 0.07 mm/h at 4.5°C to a maximum of 5.2 mm/h at 35.7°C. Growth did not occur above 44°C. Rates of growth at a given temperature between 4.5° and 40°C were constant over a period of 200 h. The Q_{10} of the response of *N. crassa* to temperature may be calculated as 2.7 between 10° and 20°C and as 1.6 between 20°C and 30°C. A Q_{10} close to 2 suggests that temperature affects growth mainly through its effects on chemical processes, i.e., through metabolism, rather than through its effects on physical processes such as diffusion (Giese, 1968).

Temperatures optimal for growth and differentiation may differ. *Cercospora nicotianae* grew most rapidly at 26°C but produced the most conidia at 18°C (Stavely and Nimmo, 1969). The production of appressoria by *Colletotrichum trifolii* was more sensitive to high temperatures than the germination of conidia (Miehle and Lukezic, 1972). For example, at 21°C, 96% of conidia germinated and almost all produced appressoria, whereas at 31°C many conidia (56.5%) germinated but none produced appressoria.

In vitro and *in vivo* effects of temperature may differ. Waggoner *et al.* (1972) allowed *Helminthosporium maydis* to produce its mycelium and conidiophores at 23°C in the light. The cultures were then placed in darkness at various temperatures. On filter paper maximum sporulation occurred at about 20°C. Virtually no sporulation occurred at 4° or 30°C. On leaves, maximum sporulation occurred at 25°C, and none occurred below 10° or above 35°C. Conidiophores developed over the range 14°–30°C, with maximum production occurring at 18°C on filter paper and near 30°C on leaves.

A special effect of temperature is the induction of microcycle conidiation. At normal laboratory temperatures (say 15°–30°C) conidia of *A. niger, N. crassa,* and *Penicillium urticae* swell when placed in a germination medium and produce germ tubes. At 37°C, however, the conidia do not produce a germ tube but continue to swell. If the temperature is then reduced to laboratory temperatures, the swollen conidia produce secondary conidia on short germ tubes or conidiophores (Anderson, 1978). Smith *et al.* (this volume, Chapter 24) treat this subject in detail.

Temperature influences survival. Cochrane (1958) has stated that most fungi slowly die if held at a temperature slightly above the maximum for growth. Other factors, especially moisture, interact with temperature to influence survival. Mycelium of *Alternaria porri* f.sp. *solani* stored at 10°C survived 80 months at 40% RH but for only 20 months at 100% RH. At higher temperatures the period of survival decreased but was consistently longer at 20–40% RH than under drier or wetter conditions (Rotem, 1968).

D. Electromagnetic Radiation

It is convenient to classify electromagnetic radiation (EMR) according to wavelength. Far ultraviolet (FUV) covers the range 200–300 nm. Near ultraviolet (NUV) extends from 300 to 380nm. Visible light encompasses wavelengths of 380–720 nm. Radiation with wavelengths less than 300 nm (e.g., X rays, γ rays, FUV) is generally considered mutagenic and germicidal (the maximum emission of energy from germicidal ultraviolet lamps is at 253.7 nm). Wavelengths of extraterrestrial EMR reaching the earth's surface range upward from 290 nm. Wavelengths greater than 320 nm pass through glass.

Very little is known of the effects of light on germination or growth of conidial fungi. In a careful study that separated temperature effects from light effects, Stinson *et al.* (1958) found that visible radiation up to 100 fc had no effect on the growth rate of *Botrytis squamosa*. Conversely, it is well known that light influences the production of conidia. Hedgcock (1906) reported that diurnal daylight induced zonation in cultures of *Cephalosporium* (*Acremonium*). Zonation is now known to occur commonly in fungi.

It is impossible to predict how an untested fungus will react to light. Species within a genus may show different responses. Visible light inhibits the formation of conidia of *Monilinia fructicola* but stimulates their formation in *M. fructigena* (M. P. Hall, 1933). Isolates within a species may differ. Leach (1962b) screened 40 isolates of *Ascochyta pisi* and noted that they ranged in ability to produce pycnidia in darkness from none to high. Similarly, Parbery and Blakeman (1978) found, among 9 isolates of 6 species of *Colletotrichum*, that 8 isolates sporulated in the dark but 1 isolate required exposure to NUV to sporulate. Different reproductive structures may respond differently. *Fusarium solani* produces macroconidia and microconidia. High light intensity favored the production of macroconidia (R. Hall, 1967).

1. Gamma Irradiation

Mohyuddin and Skoropad (1970) showed that 1 Mrad of gamma irradiation completely prevented the growth of species of *Alternaria* and *Fusarium* from wheat seeds. Sporulation of *Alternaria* was completely suppressed by 0.7 Mrad. Generally speaking, 1 Mrad is sufficient to kill all fungi in small amounts of soil, whereas 3 Mrads is required to kill all bacteria and actinomycetes (Jackson *et al.*, 1967).

2. Far Ultraviolet

Far ultraviolet is generally considered mutagenic and germicidal. Norman (1954) showed the 3×10^{16} quanta/cm^2 of FUV at a wavelength of 254 nm killed most uninucleate conidia of *Neurospora*. Conidia with more nuclei required a higher dose of irradiation. The dose required to kill uninucleate conidia may be calculated as equivalent to 23.45×10^3 μW s/cm^2 (conversion formulas are given by Nobel, 1974). However, Leach (1962b) found that, at 237.8 nm, 10×10^3 μW s/cm^2 of irradiation stimulated the production of pycnidia by isolates of *Asochyta pisi* that sporulated poorly in darkness. Cultures exposed to the germicidal wavelength 237.8 nm produced maximum numbers of pycnidia after 100 s of exposure to irradiation of 100 μW/cm? An exposure of 2–3 h completely prevented production of pycnidia. At 313.1 nm maximum numbers of pycnidia were produced after about 1000 s exposure. At each wavelength the numbers of pycnidia increased to a maximum and then declined as the exposure time was increased further. The longer the wavelength, the longer the exposure needed for maximum production of pycnidia. The most sensitive region of the culture was the marginal 1.5–2 mm.

Stimulation of sporulation by low doses of FUV may occur in many conidial fungi. Tan (1978) listed eight species that responded in this way.

3. Near Ultraviolet and Blue Light

Near ultraviolet and blue light often have similar effects. The most effective wavelengths are near 370 nm (NUV) and 450 nm (blue). Light of this quality can reset the circadian rhythm in *N. crassa* and stimulate carotenoid synthesis in *N. crassa* and *Fusarium aquaeductuum*, production of conidiophores and conidia by many fungi including *Trichoderma viride*, *Aspergillus ornatus*, and *Penicillium isariiforme*, and coremium formation and phototropism in *P. isariiforme* (Tan, 1978).

Perhaps the earliest report that NUV stimulated sporulation of fungi was by Stevens (1928). As a result of this and subsequent work, so-called black light lamps which provide NUV are now commonly used together with fluorescent light to stimulate sporulation and facilitate the identification of fungi.

Leach (1962a) exposed many fungi to NUV (310–400 nm) at an intensity of 76 μW/cm^2 for 3–10 days at 21°C. The sporulation of some conidial fungi was initiated or increased compared to that in dark-grown cultures. The list included species of *Alternaria, Fusarium,* and *Phoma.* Different species of *Helminthosporium* responded differently. *Helminthosporium oryzae* formed conidia only when irradiation was followed by a dark period. Sporulation of *H. avenae* was increased by NUV, and *H. sativum* sporulated as well in darkness as after exposure to NUV.

Sclerotia of *Botrytis convoluta* produced conidia abundantly if moistened and then exposed to NUV (Jackson, 1972). Sporulation was not stimulated by visible light.

A reversible NUV–blue light system controls the sporulation of *H. oryzae, Alternaria tomato,* and *Botrytis cinerea.* Near ultraviolet stimulates and blue light inhibits the production of conidia. Conidia develop or not according to the last light regime. Tan (1978) has suggested that a reversible pigment system, which he calls mycochrome, is involved. But no information is available on the action spectrum of the light effects, the absorption spectrum, or the other chemical properties of mycochrome.

Light may inhibit sporulation only above a certain temperature. Near-ultraviolet inhibits the sporulation of *Stemphylium botryosum, Helminthosporium catenarium,* and *Fusarium nivale* above 25°C and of *Alternaria dauci* above 15°C (Leach, 1967). Cool white fluorescent light completely inhibited the sporulation of *Exserohilum* (*Drechslera*) *rostratum* at 34°C. Conidiophores formed but not conidia (Honda and Aragaki, 1978a). Three studies on species of *Alternaria* illustrate the central theme and the differences in detail from one species to another.

Aragaki (1961) showed that *Alternaria tomato* sporulated well in darkness at temperatures between 15° and 30°C. Temperature had no effect on sporulation in the dark. In illuminated cultures sporulation was dramatically reduced at temperatures above 24°C, and no sporulation occurred at temperatures above 26°C. *Alternaria solani, A. passiflorae, A. porri, A. dianthicola, S. botryosum,* and *Helminthosporium turcicum* were said to respond similarly.

Lukens (1963) showed that 1-day-old conidiophores from starved cultures of *Alternaria solani* required 12 h of darkness to produce conidia. Illumination with visible light during the dark period inhibited the production of conidia. However, photoinhibition of sporulation depended on the temperature (Lukens, 1966). Sporulation proceeded uninterrupted in the light at 15°C but was completely prevented in the light at 27°C. Lukens (1965) further showed that the inhibition of sporulation by blue light (84 s of exposure to 450 nm at the sixth hour of darkness) was nullified when blue light irradiation was immediately followed by irradiation with wavelengths of 600–750 nm.

Zimmer and McKeen (1969) examined the effects of temperature and light on the sporulation of *Alternaria dauci.* Near ultraviolet (310–360 nm) was required for the formation of conidiophores. Wavelengths of 370–500 nm and temperatures less than 24°C permitted conidia to develop. Continuous irradiation with wavelengths of 370–510 nm at temperatures above 24°C prevented the formation of conidia. However, a dark period or wavelengths greater than 510 nm permitted conidia to form at high temperatures.

These three studies indicate that in species of *Alternaria* conidiophores develop in the light, but that a combination of blue light and high temperature inhibits the production of conidia.

There is some evidence that NUV and blue light influence the liberation of conidia. Honda and Aragaki (1978b) induced the formation of pycnidia in *Botryodiplodia theobromae* by wavelengths shorter than 520 nm, especially wavelengths of 300 and 310 nm. But conidia were exuded from pycnidia only at wavelengths less than 333 nm.

Near ultraviolet inhibited the production of melanin and microsclerotia by *Verticillium dahliae* (Brandt, 1964). This inhibition could be overcome by the addition of catechol (Brandt, 1965).

Wavelengths longer than 520 nm usually have no direct effect on the formation of conidia, although they may counteract inhibitory effects of shorter wavelengths (Carlile, 1970). For example, red light reverses the blue light inhibition of the production of conidia by *A. solani* and *B. cinerea*. In probably the first record of an infrared effect on fungi, Leach (1975) reported that infrared radiation caused a massive release of conidia of *Drechslera turcica, Cladosporium fulvum, Cercospora* sp., *Stemphylium botryosum,* and *Alternaria tenuis.*

In summary, most of our information on the effects of light on the life cycle of conidial fungi relates to the production of conidia. If there is an effect, it is usually stimulation of sporulation by wavelengths between 300 and 520 nm. In a few conidial fungi these wavelengths inhibit sporulation at moderate to high temperatures.

Very few studies have considered the mechanism by which light exerts its effects. The most thoroughly investigated fungal system is that of *N. crassa* (Muñoz and Butler, 1975). Light is not required for sporulation in this fungus but can reset endogenous rhythms and increase the production of carotenoid pigments. Direct spectrophotometry of the mycelium showed changes in absorbance at 560 nm that indicated photoreduction of a b-type cytochrome. The action spectrum suggested a flavin photoreceptor. In cell-free extracts photoreduction of cytochromes b and c required the addition of flavin mononucleotide or flavin adenine dinucleotide. Therefore the authors concluded that flavin was the primary receptor of light and that photoreduction of cytochrome b was the primary step after light absorption. Flavin photoreceptors have been implicated in other fungi, such as *A. solani* (Lukens, 1963), by matching absorption spectra with action spectra and by noting the effects of flavin compounds in overcoming photoinhibition of sporulation.

The primary event of light absorption by a receptor molecule is believed to set in motion a pattern of metabolic changes that leads to the formation of conidia. Trione and Leach (1969) described "P310", a substance that has a maximum absorption near 310 nm, occurs in NUV-irradiated cultures of *Ascochyta pisi* and stimulates the sporulation of several fungi responding to NUV. The chemical structure of P310 has recently been described by Arpin (cited by Tan, 1978). There is evidence that patterns of nucleic acid metabolism are changed by light. The uracil analogue 5-fluorouracil selectively inhibits the light-induced production of conidia by *Trichoderma viride* and *Botrytis cinerea* and the production of carotenoid pigments in *Verticillium agaricinum* (Tan, 1978). How changes like these exert an influence on the formation of conidia is not known.

E. pH

The pH of the medium is a very important but often neglected environmental factor. It can profoundly affect any activity being studied. Conidial fungi can grow over a wide range of pH. Most tolerate a range from 4 to 9 but grow and sporulate maximally near neutrality (Cochrane, 1958).

The composition of the medium can affect the initial pH and the extent and direction of pH drifts during growth of the fungus. Poorly buffered media containing ammonium salts are likely to become more acidic during growth, while media containing nitrate are likely to become more alkaline. Minimizing pH drifts during growth is a desirable objective that is difficult to achieve. The high concentrations of ions such as phosphate that are required to achieve some measure of pH stability often appreciably influence the biological activity being measured (growth, enzyme activity, and so on). Zwitterionic buffers help to alleviate some of these problems (Child and Knapp, 1973).

F. Gaseous Atmosphere

Tabak and Cooke (1968a) reviewed the earlier literature on effects of gaseous environments on fungi. Most studies show that conidial fungi require molecular oxygen. These microbes do, however, possess pathways for the anaerobic utilization of carbohydrate. This may explain the report by Tabak and Cooke (1968b) that *Geotrichum candidum*, *Fusarium oxysporum*, and *F. solani* grew in an atmosphere of 100% nitrogen. There may also have been traces of oxygen in the atmosphere.

Several reports show that CO_2 is assimilated by conidial fungi. Fixation of CO_2 occurs during the germination of conidia of *Botrytis cinerea* (Kosuge and Dutra, 1963) and is essential to the germination of conidia of *A. niger* (Rippel and

Bortels, 1927; Yanagita, 1957). In *A. niger,* label from $^{14}CO_2$ rapidly appeared in protein, ATP, and nucleic acids (Yanagita, 1963).

The effects of CO_2 on germination and growth depend on the fungus, the concentration of CO_2, and the concentrations of other components of the gaseous environment. The germination of chlamydospores of *Fusarium solani* f.sp. *phaseoli* in soil rose from 25 to 57% as CO_2 concentrations were raised from 0–25% (Bourret *et al.*, 1968). Wells and Uota (1970) examined the effects of gas mixtures on the germination and growth of *Alternaria tenuis, Botrytis cinerea, Cladosporium herbarum,* and *Fusarium roseum.* Neither germination nor growth occurred in an atmosphere of 100% nitrogen. Germination of *C. herbarum* increased as oxygen levels were increased to 21%, but germination of the other three fungi was close to maximal at 0.25% oxygen. At 21% oxygen, addition of CO_2 increased the germination of *F. roseum* but had no effect on *A. tenuis* and reduced the germination of *C. herbarum* and *B. cinerea.* At 1% oxygen, germination of *F. roseum, B. cinerea,* and *C. herbarum* was increased by CO_2 concentrations between 4 and 16%. Growth occurred in an atmosphere of 99.75% nitrogen and 0.25% oxygen and increased as oxygen levels were raised to 4%. Further increases in the concentration of oxygen up to 21% did not increase growth. At 21% oxygen, addition of CO_2 up to 45% progressively inhibited growth. At 2% oxygen, CO_2 increased growth at concentrations between 4 and 16%. It appears that high concentrations of CO_2 generally restrict germination and growth at ambient concentrations of oxygen (21%) but may stimulate them at low oxygen concentrations.

In liquid media, CO_2 effects may operate through the bicarbonate ion. Macauley and Griffin (1969) showed that high concentrations of CO_2 (10–20%) were most inhibitory to several conidial fungi at high pH. Since bicarbonate concentrations would be higher in more alkaline solutions, they concluded that this ion may be involved in the inhibitory effects of CO_2.

Survival in a restrictive atmosphere has been examined. *Aspergillus, Fusarium, Penicillium,* and *Trichoderma* survived in atmospheres permeated by 90% CO_2 for periods of 14–17 days (Stotsky and Goos, 1965, 1966). Species of *Aspergillus* and *Penicillium* survived and grew slowly and abnormally in an atmosphere of 100% CO_2 for 182 days (Calderon and Staffeldt, 1973).

Many other volatiles, e.g., ethanol and ethylene, affect conidial fungi. An overview of the effects of volatile compounds on fungi is provided by Hutchinson (1973).

G. Specific Exogenous Chemical Morphogens

These are likely to be found where there is a specific relationship between the fungus and its habitat or food source. For example, nematode-trapping organs in *Arthrobotrys* species are induced by nitrogenous materials that might be produced by the host nematode (Barron, 1977). Extracts of elm wood promoted the formation of coremia by the *Pesotum* anamorph of the elm parasite *Ceratocystis ulmi* (Hubbes and Pomerleau, 1969). The active ingredients were shown to be fructose and catechin (Taylor *et al.*, 1971). Unidentified metabolites from soil bacteria were claimed to stimulate the production of chlamydospores by the soil-borne *Fusarium solani* (Ford *et al.*, 1970).

H. Ecological Physiology

When we consider the behavior of the fungus in its natural habitat, we are forced to consider the survival value of the physiological properties elucidated in the laboratory. A myriad of questions confront us. To illustrate some of the considerations involved let us consider two questions:

1. Why does light have such a profound effect on the production of conidia? From a teleological point of view we could say that the mycelium needs to stay immersed in the substrate in order to obtain food, but that stimulation of the formation of conidiophores, fruiting bodies, and spores by light and growth of spore-bearing structures toward light would position the spores more favorably for dispersal. However, other external factors can substitute for light. For example, *T. viride* is induced to produce conidia by blue light but also sporulates in continuous darkness after it reaches the edge of the culture vessel (Gressel and Galun, 1967) or if treated with acetylcholine (Gressel *et al.*, 1971). Production of conidia by *Stemphylium solani* normally requires NUV but can be induced by ergosterol or by sterol solvents such as 5% dimethyl sulfoxide and 2% ethanol (Sproston and Setlow, 1968). Furthermore, some fungi that require light to sporulate *in vitro* will sporulate in darkness on their host plants, e.g., *Stemphylium botryosum* f.sp. *lycopersici*, *Helminthosporium graminearum*, *B. cinerea*, and *A. pisi* (Rotem *et al.*, 1978). An extract from leaves can replace the radiation requirement for sporulation of *Dendrophoma obscurans* (Binder and Lilly, 1975). It is clear that quite different external stimuli can lead to a common morphological product, such as a conidium.

2. Why are so many secondary metabolites produced by conidial fungi? Many metabolites are released into the environment. Some are involved in nutrition. If a fungus is growing in a particular location, we may expect to find it capable of producing one or more metabolites that enable it to use the substrate as food. The role of enzymes and organic acids in the colonization of living plants is well known (Heitefuss and Williams, 1976). The relationship of enzymes to the utilization of substrates by saprophytes has also been discussed (Garrett, 1970). It comes as no surprise, for example, to find that the dermatophytic fungus *Trichophyton mentagrophytes,* which grows on hair, produces keratinase, an enzyme which hydrolyzes the hair protein keratin (Yu *et al.*, 1968). Similarly,

Penicillium simplicissimum is able to grow in the unexpected habitat of weathering basalt, partly because it secretes citric acid, which solubilizes nutrients such as iron and magnesium from the rock (Silverman and Muñoz, 1970).

Nutritional abilities have been used to help explain habitat preferences, host specificity of parasites, and the place of saprophytes in succession in organic substrates. Robinson (1967) used nutrition to separate fungi into ecological groups. He distinguished saprophytic fungi, mycorrhizal fungi, and plant-parasitic fungi. To this list we can add animal parasites. Conidial fungi occur in all except the mycorrhizal group.

Extracellular metabolites have many other roles. Some are known to act (and many others may act) as antibiotics and so enable the fungus to obtain and retain possession of a substrate. Some act as adhesives or as nutrient reservoirs. Extracellular slime may absorb water and facilitate spore dispersal. Extracellular hormones such as indoleacetic acid (Buckley and Pugh, 1971) may affect the ability of the fungus to persist on or in host plants. The so-called problem of why fungi produce so many secondary metabolites recedes when their many possible ecological roles are considered.

The details of responses to natural environments are fundamental to the developing science of fungal ecology and are of immense importance in applied areas such as plant pathology. It is necessary to examine fungi in the field, since it is becoming clear that responses observed *in vitro* may not occur *in vivo* (Rotem *et al.*, 1978). This is not the place to offer a digest of the information available on the physiological aspects of fungal ecology. Several overviews are available (Robinson, 1967; Griffin, 1972, W. B. Cooke, 1979). It may suffice to say that, despite an enormous literature on the responses of fungi to controlled environments, we have just begun to investigate the physiology of conidial fungi in their natural habitats.

V. CONCLUSION

Conidial fungi produce a multitude of differentiated structures and grow by many different mechanisms. What do we know of how these processes occur? How do conidia germinate? How do hyphae grow? How do multihyphal structures, fruiting bodies, chlamydospores, conidiophores, conidia, appressoria, and other differentiated structures develop? Partial answers are available from diverse fields of enquiry. But a satisfying integration of metabolic, cellular, organismic, and environmental considerations is not available for a single morphological event, not even for the production of conidia by that much studied fungus *Neurospora crassa* (Turian and Bianchi, 1972). We know a great deal about the biochemistry of tip growth of hyphae but cannot explain related phenomena such as anastomosis and conidiophore development. We can regulate precisely the production of conidia by manipulating the environment but do not have a clear idea of what metabolic events produce the morphological change. We know how to regulate growth and differentiation with great precision in the laboratory but have only hazy notions about how this information applies in nature.

Physiological aspects of the generalized life cycle of conidial fungi may be summarized as follows. Growth and development are the result of genetic capacity, metabolic readiness, and appropriate environmental conditions. Hyphae are produced initially by the process of germination. In most conidial fungi this process starts after detached spores are exposed to free water or high RH and possibly to one or several nutrients. The hypha that develops adopts a tubular form because wall synthesis is confined to the hyphal tip. Nutrients are taken into the spore and mycelium by diffusion or by specific transport systems.

The primary pathways by which nutrients are metabolized in the cell are typical of aerobic, eukaryotic, heterotrophic organisms. Secondary metabolites are also formed. Primary and secondary metabolites are liberated from the cell and many have ecological roles, economic importance to humans, or both.

The thallus grows by tip elongation and branching. Initially growth is rapid and exponential. It slows down as nutrients are exhausted. Growth is followed by the production of differentiated structures and by the decomposition of other parts of the thallus.

The hypha, directly or indirectly, generates all the differentiated structures. Differentiation is a stepwise process highly influenced by the environment. Each type of differentiated structure, each step in its production, and each physiological process is likely to have a unique set of optimal external conditions.

There is much information on the effects of specific environments on growth, differentiation, and metabolism. These details are especially important in areas of applied mycology, such as plant disease epidemiology and industrial fermentation. There are few principles to guide us, and most of the important ones are encapsulated in Klebs's laws. There are even fewer biochemical hypotheses of general application. One of the more important is that the shape of the fungal cell depends on the balance between polar and nonpolar synthesis of the wall.

There are important challenges for the future. The mechanisms of processes that integrate thallus development (branching, anastomosis, connivence, translocation) need to be determined. The physiology of conidial fungi in their natural habitats needs to be explored. Perhaps the greatest challenge is to develop models that integrate environmental and metabolic data into a coherent explanation of growth and development.

REFERENCES

Adebayo, A. A., Harris, R. F., and Gardner, W. R. (1971). Turgor pressure of fungal mycelia. *Trans. Br. Mycol. Soc.* **57**, 145–151.

Ainsworth, G. C. (1976). "Introduction to the History of Mycology." Cambridge Univ. Press, London and New York.

Ainsworth, G. C., and Sussman, A. S. eds. (1965). "The Fungi: An Advanced Treatise," Vol. I. Academic Press, New York.

Altman, P. L., and Dittmer, D. S., eds. (1966). "Environmental Biology." Fed. Am. Soc. Exp. Biol., Baltimore, Maryland.

Anderson, J. G. (1978). Temperature-induced fungal development. In "The Filamentous Fungi" (J. E. Smith and D. R. Berry, eds.), Vol. 3, pp. 358–375. Arnold, London.

Aragaki, M. (1961). Radiation and temperature interaction on the sporulation of Alternaria tomato. Phytopathology 51, 803–805.

Baker, K. F. (1962). Principles of heat treatment of soil and planting material. J. Aust. Inst. Agric. Sci. 28, 118–126.

Barron, G. L. (1977). "The Nematode-Destroying Fungi." Canadian Biological Publications Ltd., Guelph, Ontario.

Bartnicki-Garcia, S. (1968). Cell wall chemistry, morphogenesis, and taxonomy of fungi. Annu. Rev. Microbiol. 22, 87–108.

Bartnicki-Garcia, S. (1973). Fundamental aspects of hyphal morphogenesis. Symp. Soc. Gen. Microbiol. 23, 245–267.

Bashi, E., and Rotem, J. (1975). Sporulation of Stemphylium botryosum f.sp. lycopersici in tomatoes and of Alternari porri f.sp. solani in potatoes under alternating wet-dry regimes. Phytopathology 65, 532–535.

Beckett, A., Heath, I. B., and McLaughlin, D. J. (1974). "An Atlas of Fungal Ultrastructure." Longmans, Green, New York.

Bellenger, N., Nissen, P., Wood, T. C., and Segel, I. H. (1968). Specificity and control of choline-O-sulfate transport in filamentous fungi. J. Bacteriol. 96, 1574–1585.

Berry, D. R. (1975). The environmental control of the physiology of filamentous fungi. In "The Filamentous Fungi" (J. E. Smith and D. R. Berry, eds.), Vol. 1, pp. 16–32. Arnold, London.

Bertrand, G., and Javillier, M. (1911). Influence du manganèse sur le développement de l'Aspergillus niger. C. R. Hebd. Seances Acad. Sci. 152, 225.

Binder, F. L., and Lilly, V. G. (1975). Substitution of the radiation requirement for sporulation by host tissue in Dendrophoma obscurans. Mycologia 67, 1025–1031.

Blakeman, J. P. (1969). Self-inhibition of germination of pycnidiospores of Mycosphaerella ligulicola in relation to the temperature of their formation. J. Gen. Microbiol. 57, 159–167.

Bonner, J., and Varner, J. E., eds. (1976). "Plant Biochemistry," 3rd ed. Academic Press, New York.

Bonner, J. T. (1948). A study of the temperature and humidity requirements of Aspergillus niger. Mycologia 40, 728–738.

Booth, C. (1978). Do you believe in genera? Trans. Br. Mycol. Soc. 71, 1–9.

Bootsma, A., Gillespie, T. J., and Sutton, J. C. (1973). Germination of Phyllosticta maydis conidia in an incubation chamber with control of high relative humidities. Phytopathology 63, 1157–1161.

Borrow, A., Jeffreys, E. G., Kessel, R. H. J., Lloyd, E. C., Lloyd, P. B., and Nixon, I. S. (1961). The metabolism of Gibberella fujikuroi in stirred culture. Can. J. Microbiol. 7, 227–276.

Bourret, J. A., Gold, A. H., and Snyder, W. C. (1968). Effect of carbon dioxide on germination of chlamydospores of Fusarium solani f.sp. phaseoli. Phytopathology 58, 710–711.

Bourret, J. A., Lincoln, R. G., and Carpenter, B. H. (1969). Fungal endogenous rhythms expressed by spiral figures. Science 166, 763–764.

Brandt, W. H. (1964). Morphogenesis in Verticillium: Effect of light and ultraviolet radiation on microsclerotia and melanin. Can. J. Bot. 42, 1017–1023.

Brandt, W. H. (1965). Morphogenesis in Verticillium: Reversal of the near-UV effect by catechol. BioScience 15, 669–670.

Brandt, W. H., and Reese, J. E. (1964). Morphogenesis in Verticillium: A self-produced, diffusible morphogenetic factor. Am. J. Bot. 51, 922–927.

Buckley, N. G., and Pugh, G. J. F. (1971). Auxin production by phylloplane fungi. *Nature (London)* **231**, 332.

Burgess, L. W., and Griffin, D. M. (1968). The influence of diurnal temperature fluctuations on the growth of fungi. *New Phytol.* **67**, 131–137.

Burke, D. W. (1965). *Fusarium* root rot of beans and behavior of the pathogen in different soils. *Phytopathology* **55**, 1122–1126.

Burnett, J. H. (1976). "Fundamentals of Mycology," 2nd ed. Arnold, London.

Burns, D. J. W., and Beever, R. E. (1977). Kinetic characterization of the two phosphate uptake systems in the fungus *Neurospora crassa. J. Bacteriol.* **132**, 511–519.

Calderon, O. H., and Staffeldt, E. E. (1973). Influence of high carbon dioxide concentration on selected fungi. *Dev. Ind. Microbiol.* **14**, 218–228.

Carlile, M. J. (1970). The photoresponses of fungi. *In* "Photobiology of Microorganisms" (P. Halldal, ed.), pp. 309–344. Wiley, New York.

Carter, B. L. A., Bull, A. T., Pirt, S. J., and Rowley, B. I. (1971). Relationship between energy substrate utilization and specific growth rate in *Aspergillus nidulans. J. Bacteriol.* **108**, 309–313.

Castor, L. L., Ayers, J. E., and Nelson, R. R. (1977). Controlled-environment studies of the epidemiology of yellow leaf blight of corn. *Phytopathology* **67**, 85–90.

Chang, P. L. Y., and Trevithick, J. R. (1970). Biochemical and histochemical localization of invertase in *Neurospora crassa* during conidial germination and hyphal growth. *J. Bacteriol.* **102**, 423–429.

Chang, P. L. Y., and Trevithick, J. R. (1974). How important is secretion of exoenzymes through apical cell walls of fungi? *Arch. Microbiol.* **101**, 281–293.

Chet, I., and Henis, Y. (1975). Sclerotial morphogenesis in fungi. *Annu. Rev. Phytopathol.* **13**, 170–192.

Child, J. J., and Knapp, C. (1973). Improved pH control of fungal culture media. *Mycologia* **65**, 1078–1086.

Christensen, C. M., and Kaufman, H. H. (1965). Deterioration of stored grains by fungi. *Annu. Rev. Phytopathol.* **3**, 69–84.

Clutterbuck, A. J. (1978). Hyphal branching and growth acceleration in young *Neurospora* colonies. *Bull. Br. Mycol. Soc.* **12**, 121.

Cochrane, V. W. (1958). "Physiology of Fungi." Wiley, New York.

Coley-Smith, J. R., and Cooke, R. C. (1971). Survival and germination of fungal sclerotia. *Annu. Rev. Phytopathol.* **9**, 65–92.

Collinge, A. J., Fletcher, M. H., and Trinci, A. P. J. (1978). Physiology and cytology of septation and branching in a temperature-sensitive colonial mutant (cot 1) of *Neurospora crassa. Trans. Br. Mycol. Soc.* **71**, 107–120.

Congly, H., and Hall, R. (1976). Effects of osmotic potential on germination of microsclerotia and growth of colonies of *Verticillium dahliae. Can. J. Bot.* **54**, 1214–1220.

Cook, R. J., and Papendick, R. I. (1972). Influence of water potential of soils and plants on root disease. *Annu. Rev. Phytopathol.* **10**, 349–374.

Cook, R. J., Papendick, R. I., and Griffin, D. M. (1972). Growth of two root-rot fungi as affected by osmotic and matric water potentials. *Soil Sci. Soc. Am., Proc.* **36**, 78–82.

Cooke, W. B. (1979). "The Ecology of Fungi." Chem. Rubber Co. Press, Boca Raton, Florida.

Cooney, D. G., and Emerson, R. (1964). "Thermophilic Fungi: An Account of Their Biology, Activities, and Classification." Freeman, San Francisco, California.

Curran, P. M. T. (1971). Sporulation in some members of the *Aspergillus glaucus* group in response to osmotic pressure, illumination and temperature. *Trans. Br. Mycol. Soc.* **47**, 201–211.

de Riva Ricci, D., and Kendrick, B. (1972). Computer modelling of hyphal tip growth in fungi. *Can. J. Bot.* **50**, 2455–2462.

Davis, R. H. (1976). Compartmentation and regulation of fungal metabolism: Genetic approaches. *Annu. Rev. Genet.* **9**, 39–65.

Elorza, M. V., Arst, H. N., Cove, D. J., and Scazzochio, C. (1969). Permeability properties of *Aspergillus nidulans* protoplasts. *J. Bacteriol.* **99**, 113–115.

Emerson, S. (1950). The growth phase in *Neurospora* corresponding to the logarithmic phase in unicellular organisms. *J. Bacteriol.* **60**, 221–223.

Emmett, R. W., and Parbery, D. G. (1975). Appressoria. *Annu. Rev. Phytopathol.* **13**, 147–167.

Fahey, R. C., Mikolajczyk, S. D., and Brody, S. (1978). Correlation of enzymatic activity and thermal resistance with hydration state in ungerminated *Neurospora* conidia. *J. Bacteriol.* **135**, 868–875.

Ford, E. J., Gold, A. H., and Snyder, W. C. (1970). Induction of chlamydospore formation in *Fusarium solani* by soil bacteria. *Phytopathology* **60**, 479–484.

Foster, J. W. (1949). "Chemical Activities of Fungi." Academic Press, New York.

Garrett, S. D. (1970). "Pathogenic Root-Infecting Fungi." Cambridge Univ. Press, London and New York.

Garrett, S. D. (1973). Deployment of reproductive resources by plant pathogenic fungi: An application of E. J. Salisbury's generalization for flowering plants. *Acta Bot. Isl.* **1**, 1–9.

Giese, A. C. (1968). "Cell Physiology," 3rd ed. Saunders, Philadelphia, Pennsylvania.

Gilardi, G. L. (1965). Nutrition of systemic and subcutaneous pathogenic fungi. *Bacteriol. Rev.* **29**, 406–424.

Gottlieb, D. (1978). "The Germination of Fungus Spores." Meadowfield Press Ltd., Shildow, England.

Gottlieb, D., and Tripathi, R. K. (1968). The physiology of swelling phase of spore germination in *Penicillium atrovenetum*. *Mycologia* **60**, 571–590.

Gray, W. D., Och, F. F., and El Seoud, M. A. (1964). Fungi Imperfecti as a potential source of edible protein. *Dev. Ind. Microbiol.* **5**, 384–389.

Gressel, J., and Galun, E. (1967). Morphogenesis in *Trichoderma:* Photoinduction and RNA. *Dev. Biol.* **15**, 575–598.

Gressel, J., Strausbauch, L., and Galun, E. (1971). Photomimetic effect of acetylcholine on morphogenesis in *Trichoderma*. *Nature (London)* **232**, 648–649.

Griffin, D. M. (1972). "Ecology of Soil Fungi." Syracuse Univ. Press, Syracuse, New York.

Gupta, R. K., and Pramer, D. (1970). Amino acid transport by the filamentous fungus *Arthrobotrys conoides*. *J. Bacteriol.* **103**, 120–130.

Hall, M. P. (1933). An analysis of the factors controlling the growth form of certain fungi, with especial reference to *Sclerotinia (Monilia) fructigena*. *Ann. Bot. (London)* **47**, 543–578.

Hall, R. (1967). Proteins and catalase isoenzymes from *Fusarium solani* and their taxonomic significance. *Aust. J. Biol. Sci.* **20**, 419–428.

Hall, R., and Ly, H. (1972). Development and quantitative measurement of microsclerotia of *Verticillium dahliae*. *Can. J. Bot.* **50**, 2097–2102.

Hawker, L. E. (1950). "Physiology of Fungi." Univ. of London Press, London.

Hawker, L. E. (1957). "The Physiology of Reproduction in Fungi." Cambridge Univ. Press, London and New York.

Hawker, L. E. (1966). Environmental influences on reproduction. *In* "The Fungi: An Advanced Treatise" (G. C. Ainsworth and A. S. Sussman, eds.), Vol. 2, pp. 435–469. Academic Press, New York.

Hawker, L. E., and Madelin, M. F. (1976). The dormant spore. *In* "The Fungal Spore: Form and Function" (D. J. Weber and W. M. Hess, eds.), pp. 1–70. Wiley, New York.

Hawkins, L. A. (1916). Growth of parasitic fungi in concentrated solutions. *J. Agric. Res.* **7**, 255–260.

Hedgecock, G. G. (1906). Zonation in artificial cultures of *Cephalosporium* and other fungi. *Rep. Mo. Bot. Gard.* **17**, 115–117.

Heitefuss, R., and Williams, P. H., eds. (1976). "Physiological Plant Pathology." Springer-Verlag, Berlin and New York.

Honda, Y., and Aragaki, M. (1978a). Photosporogenesis in *Exserohilum rostratum:* Influence of temperature and age on conidiophores in the terminal phase. *Mycologia* **70**, 538–546.

Honda, Y., and Aragaki, M. (1978b). Effects of monochromatic radiation on pycnidial formation and exudation of conidia in *Botryodiplodia theobromae. Mycologia* **70**, 605–613.

Howard, A. J. (1978). Translocation in fungi. *Trans. Br. Mycol. Soc.* **70**, 265–269.

Hubbes, M., and Pomerleau, R. (1969). Factors in elm wood responsible for the production of coremia by *Ceratocystis ulmi. Can. J. Bot.* **47**, 1303–1306.

Hutchinson, S. A. (1973). Biological activities of volatile fungal metabolites. *Annu. Rev. Phytopathol.* **11**, 223–246.

Ingold, C. T. (1971). "Fungal Spores: Their Liberation and Dispersal." Oxford Univ. Press (Clarendon), London and New York.

Jackson, N. E., Corey, J. C., Frederick, L. R., and Picken, J. C., Jr. (1967). Gamma irradiation and the microbial population of soils at two water contents. *Soil Sci. Soc. Am. Proc.* **31**, 491–494.

Jackson, R. S. (1972). Environmental factors regulating the production of conidia by sclerotia of *Botrytis convoluta. Can. J. Bot.* **50**, 869–875.

Jennings, D. H. (1974). Sugar transport into fungi: An essay. *Trans. Br. Mycol. Soc.* **62**, 1–24.

Jennings, D. H. (1976). Transport and translocation in filamentous fungi. *In* "The Filamentous Fungi" (J. E. Smith and D. R. Berry, eds.), Vol. 2, pp. 32–64. Arnold, London.

Jerebzoff, S. (1965). Growth rhythms. *In* "The Fungi: An Advanced Treatise" (G. C. Ainsworth and A. S. Sussman, eds.), Vol. 1, pp. 625–645. Academic Press, New York.

Katz, D., Goldstein, D., and Rosenberger, R. F. (1972). Model for branch initiation in *Aspergillus nidulans* based on measurements of growth parameters. *J. Bacteriol.* **109**, 1097–1100.

Kier, I., Allermann, K., Floto, F., Olsen, J., and Sortkjaer, O. (1976). Changes of exponential growth rates in relation to differentiation of *Geotrichum candidum* in submerged culture. *Physiol. Plant.* **38**, 6–12.

King, S. B., and Alexander, L. J. (1969). Nuclear behavior, septation, and hyphal growth of *Alternaria solani. Am. J. Bot.* **56**, 249–253.

Klebs, G. (1928). "Die Bedingungen der Fortpflanzung bei einigen Algen und Pilzen." Fischer, Berlin.

Kosuge, T., and Dutra, T. (1963). Fixation of $^{14}CO_2$ by germinating conidia of *Botrytis cinerea. Phytopathology* **53**, 880 (abstr.).

Kozlowski, T. T., ed. (1978). "Water Deficits and Plant Growth," Vol. 5. Academic Press, New York.

Latham, A. J. (1974). Effect of moisture on conidiophore morphology of *Cristulariella pyramidalis. Phytopathology* **64**, 1255–1257.

Leach, C. M. (1962a). Sporulation of diverse species of fungi under near-ultraviolet radiation. *Can. J. Bot.* **50**, 151–161.

Leach, C. M. (1962b). The quantitative and qualitative relationship of ultraviolet and visible radiation to the induction of reproduction in *Ascochyta pisi. Can. J. Bot.* **40**, 1577–1602.

Leach, C. M. (1967). Interaction of near-ultraviolet light and temperature on sporulation of the fungi *Alternaria, Cercosporella, Fusarium, Helminthosporium,* and *Stemphylium. Can. J. Bot.* **45**, 1999–2016.

Leach, C. M. (1975). Influence of relative humidity and red-infrared radiation on violent spore release by *Drechslera turcica* and other fungi. *Phytopathlogy* **65**, 1303–1312.

Lewis, H. L., and Darnall, D. W. (1970). Fatty acid toxicity and methyl ketone production in *Aspergillus niger. J. Bacteriol.* **101**, 65–71.

Lilly, V. G., and Barnett, H. L. (1951). "Physiology of the Fungi." McGraw-Hill, New York.

Lockwood, J. L. (1964). Soil fungistasis. *Annu. Rev. Phytopathol.* **2**, 341–362.

Lukens, R. J. (1963). Photo-inhibition of sporulation in *Alternaria solani. Am. J. Bot.* **50**, 720–724.

Lukens, R. J. (1965). Reversal by red light of blue light inhibition of sporulation in *Alternaria solani*. *Phytopathology* **55**, 1032.

Lukens, R. J. (1966). Interference of low temperature with the control of tomato early blight through use of nocturnal illumination. *Phytopathology* **56**, 1430–1431.

Lysek, G. (1978). Circadian rhythms. *In* "The Filamentous Fungi" (J. E. Smith and D. R. Berry, eds.), Vol. 3, pp. 376–388. Arnold, London.

Macauley, B. J., and Griffin, D. M. (1969). Effect of carbon dioxide and the bicarbonate ion on the growth of some soil fungi. *Trans. Br. Mycol. Soc.* **53**, 223–228.

MacMillan, A. (1956). The entry of ammonia into fungal cells. *J. Exp. Bot.* **7**, 113–126.

Madelin, M. F., ed. (1966). "The Fungus Spore." Butterworth, London.

Magill, J. M., and Magill, C. W. (1975). Purine base transport in *Neurospora crassa*. *J. Bacteriol.* **124**, 149–154.

Mahadevan, P. R., and Mahadkar, U. R. (1970). Role of enzymes in growth and morphology of *Neurospora crassa:* Cell-wall-bound enzymes and their possible role in branching. *J. Bacteriol.* **101**, 941–947.

Mandels, G. R., and Vitols, R. (1967). Constitutive and induced trehalose mechanisms in the spores of the fungus *Myrothecium verrucaria*. *J. Bacteriol.* **93**, 159–167.

Marchant, R. (1966). Wall structure and spore germination in *Fusarium culmorum*. *Ann. Bot.* (*London*) [N.S.] **30**, 821–830.

Marchant, R., and White, M. F. (1966). Spore swelling and germination in *Fusarium culmorum*. *J. Gen. Microbiol.* **42**, 237–244.

Mark, C. G., and Romano, A. H. (1971). Properties of the hexose transport systems of *Aspergillus nidulans*. *Biochim. Biophys. Acta* **249**, 216–226.

Martin, J. F., and Demain, A. L. (1978). Fungal development and metabolite formation. *In* "The Filamentous Fungi" (J. E. Smith and D. R. Berry, eds.), Vol. 3, pp. 426–450. Arnold, London.

Marzluf, G. A. (1970). Genetic and biochemical studies of distinct sulfate permease species in different developmental stages of *Neurospora crassa*. *Arch. Biochem. Biophys.* **138**, 254–263.

Meyers, J. A., and Cook, R. J. (1972). Induction of chlamydospore formation in *Fusarium solani* by abrupt removal of the organic carbon substrate. *Phytopathology* **62**, 1148–1153.

Miehle, B. R., and Lukezic, F. L. (1972). Studies on conidial germination and appressorium formation by *Colletotrichum trifolii* Bain & Essary. *Can. J. Microbiol.* **18**, 1263–1269.

Mirocha, C. J., and DeVay, J. E. (1971). Growth of fungi on an inorganic medium. *Can. J. Microbiol.* **17**, 1373–1378.

Mohyuddin, M., and Skoropad, W. P. (1970). Effect of ^{60}Co gamma irradiation on the survival of some fungi in single samples of each of three different grades of wheat. *Can. J. Bot.* **48**, 217–219.

Monod, J. (1942). "Recherches sur la croissance des cultures bactériennes." Hermann, Paris.

Morton, A. G., and MacMillan, A. (1954). The assimilation of nitrogen from ammonium salts and nitrate by fungi. *J. Exp. Bot.* **5**, 232–252.

Muñoz, V., and Butler, W. L. (1975). Photoreceptor pigment for blue light in *Neurospora crassa*. *Plant Physiol.* **55**, 421–426.

Nicholas, D. J. D. (1965). Utilization of inorganic nitrogen compounds and amino acids by fungi. *In* "The Fungi: An Advanced Treatise" (G. C. Ainsworth and A. S. Sussman, eds.), Vol. 1, pp. 349–376. Academic Press, New York.

Nickerson, W. J. (1963). Symposium on biochemical bases of morphogenesis in fungi. IV. Molecular bases of form in yeasts. *Bacteriol. Rev.* **27**, 305–324.

Nishi, A., Yanagita, E., and Maruagama, Y. (1968). Cellular events occurring in growing hyphae of *Aspergillus oryzae* as studied by autoradiography. *J. Gen. Appl. Microbiol.* **14**, 171–182.

Nobel, P. S. (1974). "Introduction to Biophysical Plant Physiology." Freeman, San Francisco, California.

Norman, A. (1954). The nuclear role of the ultraviolet inactivation of *Neurospora* conidia. *J. Cell. Comp. Physiol.* **44**, 1–10.

Onions, A. H. S. (1971). Preservation of fungi. *In* "Methods in Microbiology" (C. Booth, ed.), Vol. 4, pp. 113–151. Academic Press, New York.

Parbery, D. G., and Blakeman, J. P. (1978). Effect of substances associated with leaf surfaces on appressorium formation by *Colletotrichum acutatum*. *Trans. Br. Mycol. Soc.* **70**, 7–19.

Pendyala, L., and Wellman, A. M. (1977). Developmental-stage-dependent adenine transport in *Neurospora crassa*. *J. Bacteriol.* **131**, 453–462.

Pirt, S. J., and Callow, D. S. (1960). Studies of the growth of *Penicillium chrysogenum* in continuous flow culture with reference to penicillin production. *J. Appl. Bacteriol.* **23**, 87–98.

Purchase, I. F. H., ed. (1974). "Mycotoxins." Elsevier, New York.

Righelato, R. C. (1975). Growth kinetics of mycelial fungi. *In* "The Filamentous Fungi" (J. E. Smith and D. R. Berry, eds.), Vol. 1, pp. 79–103. Arnold, London.

Rippel, A., and Bortels, H. (1927). Vorlaufige über die allgemeine Bedeuung der Köhlensäure für die Pflanzenzelle (Versuche an *Aspergillus niger*). *Biochem. Z.* **184**, 237–244.

Roberg, M. (1928). Über die Wirkung von Eisen-, Zink- und Kupfersalzen auf Aspergillen. *Zentralbl. Bakteriol. II* **74**, 333–371.

Roberts, K. R., and Marzluf, G. A. (1971). The specific interaction of chromate with the dual sulfate permease systems of *Neurospora crassa*. *Arch. Biochem. Biophys.* **142**, 651–659.

Robertson, N. F. (1958). Observations of the effect of water on the hyphal apices of *Fusarium oxysporum*. *Ann. Bot. (London)* [N.S.] **22**, 159–173.

Robertson, N. F. (1965). The mechanism of cellular extension and branching. *In* "The Fungi: An Advanced Treatise" (G. C. Ainsworth and A. S. Sussman, eds.), Vol. 1, pp. 613–623. Academic Press, New York.

Robinson, R. K. (1967). "Ecology of Fungi." English Universities Press, London.

Rotem, J. (1968). Thermoxerophytic properties of *Alternaria porri* f.sp. *solani*. *Phytopathology* **58**, 1284–1287.

Rotem, J., Cohen, Y., and Bashi, E. (1978). Host and environmental influences on sporulation *in vivo*. *Annu. Rev. Phytopathol.* **16**, 83–101.

Ryan, F. J., Beadle, G. W., and Tatum, E. L. (1943). The tube method of measuring the growth rate of *Neurospora*. *Am. J. Bot.* **30**, 784–799.

Sargent, M. L., Briggs, W. R., and Woodward, D. O. (1966). Circadian nature of a rhythm expressed by an invertaseless strain of *Neurospora crassa*. *Plant Physiol.* **41**, 1343–1349.

Scarborough, G. A. (1970). Sugar transport in *Neurospora crassa*. II. A second glucose transport system. *J. Biol. Chem.* **245**, 3985–3987.

Scott, W. A. (1976). Biochemical genetics of morphogenesis in *Neurospora*. *Annu. Rev. Microbiol.* **30**, 85–104.

Scott, W. A., and Metzenberg, R. L. (1970). Location of aryl sulfatase in conidia and young mycelia of *Neurospora crassa*. *J. Bacteriol.* **104**, 1254–1265.

Silverman, M. P., and Muñoz, E. F. (1970). Fungal attack on rock: Solubilization and altered infrared spectra. *Science* **169**, 985–987.

Slayman, C. W., and Tatum, E. L. (1964). Potassium transport in *Neurospora*. I. Intracellular sodium and potassium concentrations and cation requirements for growth. *Biochim. Biophys. Acta* **88**, 578–592.

Slayman, C. W., and Tatum, E. L. (1965a). Potassium transport in *Neurospora*. II. Measurement of steady-state potassium fluxes. *Biochim. Biophys. Acta* **102**, 149–160.

Slayman, C. W., and Tatum, E. L. (1965b). Potassium transport in *Neurospora*. III. Isolation of a transport mutant. *Biochim. Biophys. Acta* **109**, 184–193.

Smith, J. E. (1978). Asexual sporulation in filamentous fungi. *In* "The Filamentous Fungi" (J. E. Smith and D. R. Berry, eds.), Vol. 3, pp. 214–239. Arnold, London.

Smith, J. E., and Berry, D. R., eds. (1975). "The Filamentous Fungi," Vol. 1. Arnold, London.

Smith, J. E., and Berry, D. R., eds. (1976). "The Filamentous Fungi," Vol. 2. Arnold, London.

Smith, J. E., and Berry, D. R., eds. (1978). "The Filamentous Fungi," Vol. 3. Arnold, London.

Smith, R. S. (1964). Effect of diurnal temperature fluctuations on linear growth rate of *Macrophomina phaseoli* in culture. *Phytopathology* **54**, 849–852.

Solomons, G. L. (1975). Submerged culture production of mycelial biomass. *In* "The Filamentous Fungi" (J. E. Smith and D. R. Berry, eds.), Vol. 1, pp. 249–264. Arnold, London.

Sproston, T., and Setlow, R. B. (1968). Ergosterol and substitutes for the ultraviolet radiation requirement for conidia formation in *Stemphylium solani*. *Mycologia* **60**, 104–114.

Stavely, J. R., and Nimmo, J. A. (1968). Relation of pH and nutrition to growth and sporulation of *Cercospora nicotianae*. *Phytopathology* **58**, 1372–1376.

Stavely, J. R., and Nimmo, J. A. (1969). Effects of temperature upon growth and sporulation of *Cercospora nicotianae*. *Phytopathology* **59**, 496–498.

Steinberg, R. A. (1939). Growth of fungi on synthetic nutrient solutions. *Bot. Rev.* **5**, 327–350.

Stevens, F. L. (1928). Effects of ultra-violet radiation on various fungi. *Bot. Gaz. (Chicago)* **86**, 210–225.

Stinson, R. H., Gage, R. S., and MacNaughton, E. B. (1958). The effect of light and temperature on the growth and respiration of *Botrytis squamosa*. *Can. J. Bot.* **36**, 927–934.

Stotzky, G., and Goos, R. D. (1966). Adaptation of the soil microbiota to high CO_2 and low O_2 *Can. J. Microbiol.* **11**, 853–868.

Stotzky, G., and Goos, R. D. (1966). Adaptation of the soil microbiota to high CO_2 and low O_2 tensions. *Can. J. Microbiol.* **12**, 849–861.

Stuart, W. D., and DeBuck, A. G. (1971). *In vitro* studies of isolated glycoprotein subunits of the amino acid transport system of *Neurospora crassa* conidia. *Arch. Biochem. Biophys.* **144**, 512–518.

Stuart, W. D., and DeBusk, A. G. (1973). Transport of arginine by an *in vitro* system. *Biochem. Biophys. Res. Commun.* **52**, 1046–1050.

Tabak, H. H., and Bridge Cooke, W. (1968a). The effects of gaseous environments on the growth and metabolism of fungi. *Bot. Rev.* **34**, 126–252.

Tabak, H. H., and Bridge Cooke, W. (1968b). Growth and metabolism of fungi in an atmosphere of nitrogen. *Mycologia* **60**, 115–140.

Tan, K. K. (1978). Light-induced fungal development. *In* "The Filamentous Fungi" (J. E. Smith and D. R. Berry, eds.), Vol. 3, pp. 334–357. Arnold, London.

Taylor, P. A., Smalley, E. B., and Strong, F. M. (1971). Synnemata induction in *Ceratocystis ulmi*. *Phytopathology* **61**, 914 (abstr.).

Townsend, B. B., and Willetts, H. J. (1954). The development of sclerotia of certain fungi. *Trans. Br. Mycol. Soc.* **37**, 213–221.

Trevithick, J. R., and Metzenberg, R. L. (1966). Molecular sieving by *Neurospora* cell walls during secretion of invertase isozymes. *J. Bacteriol.* **92**, 1010–1015.

Trinci, A. P. J. (1969). A kinetic study of the growth of *Aspergillus nidulans* and other fungi. *J. Gen. Microbiol.* **57**, 11–24.

Trinci, A. P. J. (1970). Kinetics of apical and lateral branching in *Aspergillus nidulans* and *Geotrichum lactis*. *Trans. Br. Mycol. Soc.* **55**, 17–28.

Trinci, A. P. J. (1971a). Influence of the width of the peripheral growth zone on the radial growth rate of fungal colonies on solid media. *J. Gen. Microbiol.* **67**, 325–344.

Trinci, A. P. J. (1971b). Exponential growth of the germ tubes of fungal spores. *J. Gen. Microbiol.* **67**, 345–348.

Trione, E. J., and Leach, C. M. (1969). Light-induced sporulation and sporogenic substances in fungi. *Phytopathology* **59**, 1077–1083.

Tsuneda, A., and Skoropad, W. P. (1977). The *Alternaria brassicae-Nectria inventa* host-parasite interface. *Can. J. Bot.* **55**, 448–454.

Tuite, J. (1969). "Plant Pathological Methods: Fungi and Bacteria." Burgess, Minneapolis, Minnesota.

Turian, G. (1974). Sporogenesis in fungi. *Annu. Rev. Phytopathol.* **12**, 129–137.

Turian, G., and Bianchi, D. E. (1972). Conidiation in *Neurospora. Bot. Rev.* **38**, 199–154.

Turner, W. B. (1971). "Fungal Metabolites." Academic Press, New York.

Vogel, H. J. (1956). A convenient growth medium for *Neurospora* (medium N). *Microb. Genet. Bull.* **13**, 42.

Waggoner, P. E., and Parlange, J. Y. (1974a). Mathematical model for spore germination at changing temperatures. *Phytopathology* **64**, 605–610.

Waggoner, P. E., and Parlange, J. Y. (1974b). Verification of a model of spore germination at variable moderate temperatures. *Phytopathology* **64**, 1192–1196.

Waggoner, P. E., Horsfall, J. G., and Lukens, R. J. (1972). Epimay: A simulator of southern corn leaf blight. *Conn., Agric. Exp. Stn., New Haven, Bull. Bull.* **729.**

Watson, A. G., and Ford, E. J. (1978). Soil fungistasis—A reappraisal. *Annu. Rev. Phytopathol.* **10**, 327–348.

Weber, D. J., and Hess, W. W., eds. (1976). "The Fungal Spore: Form and Function." Wiley, New York.

Weete, J. D. (1974). "Fungal Lipid Biochemistry." Plenum, New York.

Wells, J. M., and Uota, M. (1970). Germination and growth of five fungi in low-oxygen and high-carbon dioxide atmosphere. *Phytopathology* **60**, 50–53.

Whitaker, A. (1976). Amino acid transport into fungi: An essay. *Trans. Br. Mycol. Soc.* **67**, 365–376.

Wiley, W. R. (1970). Tryptophan transport in *Neurospora crassa:* A tryptophan-binding protein released by cold osmotic shock. *J. Bacteriol.* **103**, 656–662.

Willetts, H. J. (1978). Sclerotium formation. *In* "The Filamentous Fungi" (J. E. Smith and D. R. Berry, eds.), Vol. 3, pp. 197–213. Arnold, London.

Yabuki, M., and Fukui, S. (1970). Presence of binding site for α-amylase and of masking protein for this site on mycelial cell wall of *Aspergillus oryzae. J. Bacteriol.* **104**, 138–144.

Yanagita, T. (1957). Biochemical aspects on the germination of conidiospores of *Aspergillus niger. Arch. Mikrobiol.* **26**, 329–344.

Yanagita, T. (1963). Carbon dioxide fixation in germinating conidiospores of *Aspergillus niger. J. Gen. Appl. Microbiol.* **9**, 343–351.

Yu, R. J., Harmon, S. R., and Blank, F. (1968). Isolation and purification of an extracellular keratinase of *Trichophyton mentagrophytes. J. Bacteriol.* **96**, 1435–1436.

Zalokar, M. (1959a). Enzyme activity and cell differentiation in *Neurospora. Am. J. Bot.* **46**, 555–559.

Zalokar, M. (1959b). Growth and differentiation in *Neurospora* hyphae. *Am. J. Bot.* **46**, 602–610.

Zimmer, R. C., and McKeen, W. E. (1969). Interaction of light and temperature on sporulation of the carrot foliage pathogen *Alternaria dauci. Phytopathology* **59**, 743–749.

28

Cell Wall Chemistry, Ultrastructure, and Metabolism

Jerome M. Aronson

I. INTRODUCTION

It has been noted repeatedly (e.g., Aronson, 1965) that fungal morphology is primarily a reflection of cell wall fabrication, wall growth, and wall modification. The characteristic and stable forms of a variety of reproductive structures and propagules are a direct consequence of developmental events involving some kind of cell wall elaboration. Moreover, the colonization of substrates by vegetative mycelia is dependent upon continuous apical wall synthesis (tip growth). Thus Bartnicki-Garcia (1968) has remarked: "In simplified terms, morphological development of fungi may be reduced to a question of cell wall

Biology of Conidial Fungi, Vol. 2

morphogenesis.'' Without discounting the importance of biochemical reactions and their regulation, the fungal cell wall appears to be the most definitive product of the biochemical processes. Therefore it is not surprising that a variety of investigations of cell walls in fungi have become the subject of numerous reviews. The general nature of fungal walls and the systematic distribution of cellulose and chitin were treated by Aronson (1965). Bartnicki-Garcia (1968) discussed wall chemistry, morphogenesis, and fungal taxonomy and later (1970) considered the phylogenetic implications of cell wall composition. In addition, a specific treatment of wall structure and hyphal morphogenesis has appeared (Bartnicki-Garcia, 1973). LéJohn (1974) also has commented on cell wall chemistry and phylogeny. The chemistry of fungal polysaccharides has been reviewed (Gorin and Spencer, 1968), as have the occurrence and properties of cell wall glycoproteins and peptidoglycans (Gander, 1974). Problems associated with the preparation and analysis of fungal walls were considered by Taylor and Cameron (1973). Most recent among relevant articles are discussions by Grove (1978) on hyphal tip growth, Gooday (1978) on the biochemistry of apical wall synthesis, Rosenberger (1976) on wall chemistry and biosynthesis, and Farkas (1979) on wall biosynthesis. Also, a number of workers, some cited above and elsewhere in this chapter, have cooperated in an edited volume (Burnett and Trinci, 1979) dealing with fungal walls.

The aforementioned reviews will provide the reader with access to most of the relevant literature on fungal walls. The emphasis in this chapter is on recent developments. The coverage is, for the most part, restricted to the conidial fungi, but the discussion in some parts of this text is more comprehensive where the data are of general importance. Although dimorphic pathogenic fungi are mentioned in this chapter, the subject of cell walls as related to dimorphism is discussed in Volume 1, Chapter 5.

II. CONIDIAL WALLS

Since the hyphal continuum of a growing mycelium commonly has its origin in a conidium, it is fitting to focus attention initially on conidial wall structure. Indeed, the wall of the hyphal germ tube, in cases where the matter has been investigated, develops as an extension of a preexisting inner layer of the conidial wall or as an extension of a newly formed layer or layers. Unfortunately, studies on conidial wall ultrastructure considerably outnumber investigations of their chemistry, so that it is generally not possible to attempt a chemical elucidation of the conidial wall layers revealed by transmission electron microscopy (TEM) and freeze-etch methods. Electron microscope studies summarized here are more comprehensively treated in Akai et al. (1976), Mangenot and Reisinger (1976), Smith et al. (1976), and Chapter 23 of this volume.

A. Ultrastructure

Electron density differences noted during TEM observations show that conidial walls generally are found to have two or three layers (Figs. 1–4), although there appear to be exceptional cases where there is but one, and others where there are as many as four (Mangenot and Reisinger, 1976). Among the more than 70 genera enumerated in a review by Mangenot and Reisinger (1976) were some (e.g., *Aspergillus, Fusarium,* and *Penicillium*) in which certain species produced conidia with two wall layers while other congeneric species within a given genus had three-layered walls. Although some difference among species is possible, it is likely that interpretations of electron micrographs by original investigators have contributed to this seeming variability. Moreover, there is evidence that TEM images will vary as a result of the state of the spore (dry or hydrated) (Smith *et al.,* 1976) and/or the fixation method employed (Gull and Trinci, 1971). In spite of the inherent difficulties, Mangenot and Reisinger (1976) have attempted to develop a unifying concept of conidial wall structure. In their view conidial walls are essentially three-layered structures, although all three layers are not always easily discerned. Spores of *Alternaria kikuchiana* (Figs. 1–4), presented here as but one example of conidial wall ultrastructure, are regarded with some justification as having a two-layered wall (Akai *et al.,* 1976). Not included in the Mangenot and Reisinger (1976) review are recent investigations of *Helminthosporium maydis* (Aist *et al.,* 1976), *Trichoderma viride* (Benítez *et al.,* 1976), *Aspergillus clavatus* (Hanlin, 1976), *Pleiochaeta setosa* (Harvey, 1974), *Trichophyton mentagrophytes* (Wu-Yuan and Hashimoto, 1977), *Fusarium culmorum* (Laborda *et al.,* 1974), *F. sulphureum* (Schneider *et al.,* 1977), *F. solani* (van Eck and Schippers, 1976), *F. oxysporum* (Stevenson and Becker, 1979), *Lepteutypa cupressi* and *Monochaetia monochaeta* (Roberts and Swart, 1980), and *Ceratocystis adiposa* (*Chalara* state) (Hawes, 1979, 1980). These recent studies tend to reaffirm the aforementioned general concept of conidial wall ultrastructure. In contrast, Griffiths and Swart (1974) regard conidial walls of *Pestalotia pezizoides* as being single-layers but composed of three zones distinguishable by differences in electron density and melanization. Whether these conidia are fundamentally different from others is questionable, and there is a need for strict definition of terms such as "layer" and "zone."

A frequently encountered problem in studies on conidial wall ultrastructure has been the origin of the wall of the hyphal germ tube in relation to the layers present in the dormant (dehydrated) spore wall (Smith *et al.,* 1976; Akai *et al.,* 1976). As far as fungal spore walls in general are concerned, workers have tended to follow Bartnicki-Garcia (1968) who defined three types (designated types I, II, and III) of vegetative wall formation during germination. The *de novo* formation of a wall on a naked protoplast (type II) occurs only in zoosporic fungi and will not be considered further. In other instances the hyphal germ tube wall develops as an extension of the spore wall or an inner layer of the spore wall

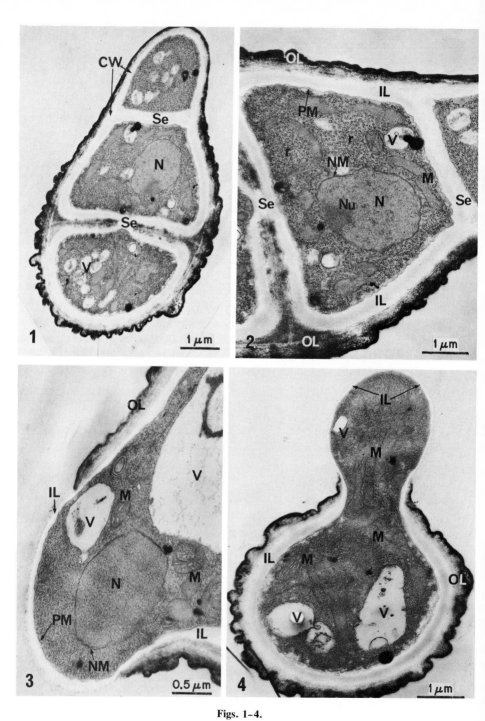

Figs. 1–4.

(type I). In type III spores, a vegetative cell wall forms *de novo* under the spore wall. There is no doubt that spores of type III are characteristic of the Mucorales (Bartnicki-Garcia, 1968). However, in conidial fungi and ascomycetes, both type I and type III conditions have been described (see, for example, review of Akai *et al.*, 1976). A perplexing problem occurs in reference to this classification. In the genus *Aspergillus,* for example, *A. oryzae* and *A. niger* (Akai *et al.*, 1976, and references cited therein) were considered as having type I spores while *A. nidulans* (Border and Trinci, 1970) spores were considered to be of type III. Similarly, Hawker and Hendy (1963) reported that type I spores occurred in *Botrytis cinerea* but, as a result of a reinvestigation of this species, Gull and Trinci (1971) concluded that the conidia were of type III. In the investigations of Border and Trinci (1970) and Gull and Trinci (1971), improved fixation methods not employed by other workers were used, and it was found that the walls of hydrated (germinating) spores increased in thickness and appeared to have additional layers not seen by earlier investigators. Consequently, it was concluded that the newly appearing wall layers in *Botrytis cinerea* spores were products of biochemical synthesis and added to the wall by a process of apposition (Gull and Trinci, 1971). Gull and Trinci assumed that the original spore wall layers were stretched and became thinner during spore swelling and that the dormant spore wall constituents did not swell during hydration. Florance *et al.* (1972) observed similar changes during hydration of spores of *Aspergillus nidulans,* but these workers were of the opinion that the so-called new layers in the wall were manifestations of water inhibition (swelling) and not *de novo* cell wall synthesis. Although in recent discussions some authors (Burnett, 1976; Smith *et al.*, 1976; Stevenson and Becker, 1979; Hawes, 1980) seem convinced that there is extensive *de novo* synthesis of cell wall material, there is still a lack of conclusive evidence of such synthesis, and the possibility that spore wall polymers swell during hydration has been given little consideration except for the earlier statements of Florance *et al.* (1972) and a recent discussion by Ross (1979, p. 225). The exceptional increase in spore wall thickness seen in *Aspergillus niger* when germ tube emergence is inhibited by high temperature (Smith *et al.*, 1976) is not necessarily an indication of a normal tendency for spore wall synthesis since it has been shown that factors arresting normal apical wall synthesis in *A. nidulans* hyphae cause a diversion of wall synthesis to subapical regions (Katz and Rosenberger, 1971a; Sternlicht *et al.*, 1973). In view of the known capacity of some biopolymers, and certain polysaccharides in particular, for

---

Figs. 1–4. Transmission electron micrographs of thin-sectioned conidia of *Alternaria kikuchiana*. Fig. 1. Nongerminating conidium. Fig. 2. Middle cell of conidium shown in Fig. 1. Note the inner (IL) and outer (OL) wall layers. Figs. 3 and 4. Germinating conidia. The outer layer has ruptured, and the germ tube wall is continuous with the inner layer. Se, Septum; CW, cell wall; N, nucleus; V, vacuole; PM, plasma membrane; NM, nuclear membrane; Nu, nucleolus; M, mitochondria; r, ribosomes. (From Akai *et al.*, 1976.)

water imbibition and swelling (Rees, 1977), it would be helpful to know more of the imbibitional and rheological properties of conidial wall components. As far as synthesis of new wall material during germination is concerned, the demonstration, if possible, of wall precursor incorporation into alleged new layers using electron microscopy and autoradiography would provide unequivocal evidence. Alternatively, demonstration of the synthesis of a chemically distinct wall constituent would be a convincing line of evidence, but, on the basis of what is currently known of spore wall chemistry as compared with hyphal walls, it is improbable that this approach would be productive (see Section II,B). Although there are indeed wall layers that become apparent only during germination in *Botrytis cinerea* (Gull and Trinci, 1971), *Aspergillus nidulans* (Border and Trinci, 1970; Florance *et al.*, 1972), *Ceratocystis adiposa* (Hawes, 1980), and *Fusarium oxysporum* (Stevenson and Becker, 1979), this does not appear to be the case with *Alternaria kikuchiana* (Akai *et al.*, 1976) (Figs. 1–4) and, for the present, generalizations appear to be unjustified.

A number of investigations have been concerned with the structural, chemical, and physical properties of conidial surfaces, and a comprehensive treatment of this subject can be found in the review by Hawker and Madelin (1976). Workers have shown particular interest in the rodlet layer occurring at or near the spore surface (Fig. 5). Rodlets were clearly demonstrated by Hess *et al.* (1966) on surfaces of conidia of *Penicillium herquei*. Similar structures were subsequently found in spores of other *Penicillium* species (Hess *et al.*, 1968) and in *Aspergillus* spp. (Hess and Stocks, 1969). Particles measuring 5.0 nm in diameter are combined in rodlets of varying lengths, depending upon the species (Hess *et al.*, 1968; Hess and Stocks, 1969), and parallel rodlets are aggregated into bundles of variable width. The rodlet bundles are interwoven (Fig. 5), giving the wall a texture resembling a basketweave (Hess *et al.*, 1968; Hess and Stocks, 1969). Whether or not rodlets, which appear to be proteinaceous structures (Hashimoto *et al.*, 1976; Cole *et al.*, 1979; Beever *et al.*, 1979), are a consistent feature of conidial walls in all or most deuteromycetous and ascomycetous species is uncertain. Nevertheless, it is clear that they are not exclusive features of spores of aspergilli and penicillia, since they have been observed in conidia of *Epicoccum nigrum, Geotrichum candidum, Gonatobotryum apiculatum, Oidiodendron truncatum, Neurospora crassa, Scopulariopsis brevicaulis,* and *Wallemia sebi* (Hawker and Madelin, 1976; Dempsey and Beever, 1979). In addition, recent studies have extended the range of rodlet observations to include microconidia of *Trichophyton mentagrophytes* (Hashimoto *et al.*, 1976; Wu-Yuan and Hashimoto, 1977). However, rodlets were reported absent in conidia of *Botrytis fabae* (Richmond and Pring, 1971), *Drechslera sorokiniana* (Cole, 1973a), and *Botryodiplodia ricinicola* (Madelin and Ogunsanya, 1979). Whatever the extent to which rodlets occur in spores of conidial fungi, it is apparent that these structures are not found solely in spores. They have been demonstrated in walls of vegetative and fertile hyphae of *Oidiodendron truncatum* (Cole, 1973b),

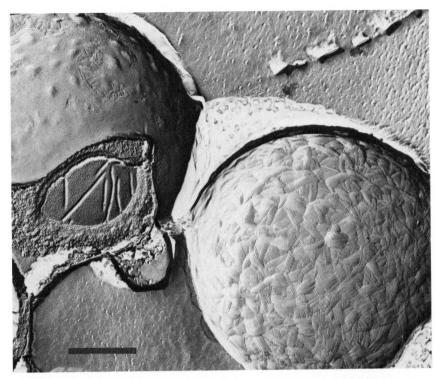

Fig. 5.   Electron micrograph of replicas of freeze-etched *Aspergillus fumigatus* conidia. The rodlet layer is exposed in the spore at the lower right. Rodlets in the spore at the upper left are covered by a thin external layer except for small patches where the external layer has been removed. Grooves occur in a region of exposed plasma membrane. Scale line, 0.5 μm. (From Ghiorse and Edwards, 1973.)

phialides and conidiophores of *Aspergillus fumigatus* (Ghiorse and Edwards, 1973), and apparently in hyphae of *Trichophyton* spp. (Caputo *et al.*, 1977). The presence of a thin layer or film overlying the rodlet layer in conidia of penicillia and aspergilli was suggested by Hess *et al.* (1968) and Hess and Stocks (1969); Fisher and Richmond (1970) found evidence of an external polyphosphate layer on spores of *Penicillium expansum*. The occurrence and composition of this layer varied with the composition of the culture medium and it was easily removed from spore surfaces by washing with water. Ghiorse and Edwards (1973) reported the demonstration of a thin, smooth layer external to the rodlets in *Aspergillus fumigatus* conidia (Fig. 5) but Dempsey and Beever (1979) questioned this interpretation and suggested that the rodlets lie over the smooth layer. There has been some interest in elucidating how the structure and chemistry of conidial surfaces contribute to the water-repellent property of spores in many species (Sassen *et al.*, 1967; Fisher and Richmond, 1969, 1970; Fisher *et al.*,

1972), and Fisher and Richmond (1970) concluded that the rodlet layer, and not the external (polyphosphate) layer, was probably responsible for the hydrophobic nature of *Penicillium expansum* spores. This view has received support from the recent investigation of Beever and Dempsey (1978), who found the rodlet layer to be lacking from easily wettable, but not air dispersable, conidia of a *Neurospora crassa* mutant.

B. Chemistry

Conidial wall chemistry has not been examined in a great many species but, from data that are available, some interesting observations can be made. The summary of data for selected species appearing in Tables I and II indicates that neutral polysaccharides are the principal carbohydrate components. Glucose, galactose, and mannose are the only neutral hexoses that occur consistently in large quantities and, of the three, glucose generally predominates. *Penicillium notatum* appears to differ somewhat in that, in the conidial walls of this species, galactose is the major hexose (Martin *et al.*, 1973) and, in spore walls of

TABLE I

Composition of Conidial Walls of Different Species[a]

Component	*Aspergillus oryzae*[b]	*Penicillium chrysogenum*	*Fusarium culmorum*	*Fusarium sulphureum*
Neutral carbohydrates (total)	27	48.6	60	37.6
Glucose	+	26.3	+++	24.7
Galactose	+	19.4	+	6.7
Mannose	+	2.6	+	6.2
Rhamnose		0.3		
Glucuronic acid				14.5
Chitin	20[c]	11.4[c]	10[c]	12.5[d]
Protein	19	16.9	21	21.6
Lipid			4	1.5
Phosphate	4			
Ash	4		1.5	4.3

[a] Data for *A. oryzae, P. chrysogenum, F. culmorum,* and *F. sulphureum,* respectively, are adapted from Horikoshi and Iida (1964), Rizza and Kornfeld (1969), Laborda *et al.* (1974), and Schneider *et al.* (1977). Data of Rizza and Kornfeld (1969) and Laborda *et al.* (1974) reproduced with the permission of Cambridge Univ. Press; data of Schneider *et al.* (1977) by permission of the National Research Council of Canada. Data are expressed numerically as percent of wall dry weight, as + if identified qualitatively, and as +++ if most abundant.

[b] Walls of this species also contained 4.5% nucleic acids.

[c] Based entirely on analyses for D-glucosamine which is assumed to be derived mainly from chitin.

[d] Chitin identified by x-ray diffraction.

TABLE II

Composition of Walls and Wall Fractions of *Trichophyton mentagrophytes* Microconidia[a]

Component	Walls	Glycoprotein–lipid fraction[b]	Rodlet layer[c]	Insoluble fraction[d]
Neutral carbohydrates (total)	56.1	29.2	7.5	55.6
Glucose	+++	++	+	+
Galactose	tr	tr		
Mannose	+	+	+	
Chitin	16.0	0	0	39.7
Protein	22.6	42.5	83.0	3.0
Lipid	6.5	21.0	0	0
Melanin	0.2	0	2.0	
Phosphate	0.2		0	
Ash	1.7		0	

[a] Reproduced and/or adapted from Hashimoto *et al.* (1976) and Wu-Yuan and Hashimoto (1977). Data are expressed as in Table I; tr indicates trace; multiple + symbols indicate relative proportions.

[b] Extracted from isolated walls using a sodium phosphate buffered (pH 6.5) mixture containing urea, mercaptoethanol, and sodium dodecyl sulfate.

[c] Isolated following sequential treatment of walls with solvents and hydrolytic enzymes.

[d] Insoluble wall residue after extraction of glycoprotein–lipid fraction and hot sodium hydroxide-soluble components.

Trichoderma viride, only glucose was found (Benítez *et al.,* 1976). Rhamnose, occurring only in conidial walls of *P. chrysogenum* (Rizza and Kornfeld, 1969) and *P. notatum* (Martin *et al.,* 1973), is a trace component in these species. Marchant (1966) reported the presence of xylose and arabinose in conidial walls of *Fusarium culmorum* but, in a reinvestigation of this species (Laborda *et al.,* 1974), no pentoses were found, nor were any reported from conidial and chlamydospore walls of *F. sulphureum* (Schneider *et al.,* 1977) and *F. solani* (van Eck, 1978) or from investigated species of other genera (Tables I and II). Walls of *F. sulphureum* conidia contain a sizeable quantity of glucuronic acid, a constituent found also in chlamydospore and hyphal walls of this species (Schneider *et al.,* 1977; Barran *et al.,* 1975). Glucuronic acid was not found in macroconidial and chlamydospore walls of *F. solani* (van Eck, 1978), but it is doubtful that the analytic procedures used in this case could have detected uronic acid, assuming that some was present. As the specific methods required for uronic acid analysis become more routinely used, it is possible that these components will be found in spore walls of other genera. Although conidial walls of *Trichoderma viride* (Benítez *et al.,* 1976) were considered to be devoid of glucosamine, and relatively small quantities of this amino sugar were found in *Pithomyces chartarum* (Sturgeon, 1966) and *Penicillium notatum* (Martin *et al.,* 1973), spore walls of several other species contain appreciable amounts (Tables I and II). Galactosamine does not appear to be a constituent of conidial walls, although it is found in hyphal walls (see Section III,B). It is justifiably assumed that the presence of glucosamine is indicative of some quantity of chitin. It is evident (Tables I and II) that conidial walls invariably contain a large amount of protein, and it is perhaps surprising that the lipid content is relatively low in some cases. Although spore walls of *Neurospora crassa* had a total lipid content approaching 9%, *Alternaria tenuis, Botrytis fabae,* and *Penicillium expansum* conidial walls, respectively, contained only 1.7, 5.1, and 1.1%. No lipid was detected on the spore surface of *P. expansum,* while very small amounts (<0.2%) were found in the other species. In species in which spore surface lipid was detected, the fatty acid and hydrocarbon composition of the surface fraction differed from that of interior parts of the wall (Fisher *et al.,* 1972). Few investigations of spore walls have provided data on melanin content. Isolated conidial walls of *Pithomyces chartarum* (Sturgeon, 1966) contained uncharacterized pigment (presumably melanin) which remained associated with the ethylenediamine-insoluble wall fractions, and Ellis and Griffiths (1974) utilized a number of methods for the qualitative demonstration of melanins occurring in spore walls of *Epicoccum nigrum* and *Humicola grisea.* The quantity of melanin in spore walls undoubtedly varies widely among species. Conidial walls of *Trichoderma viride* contained 21% melanin (Benítez *et al.,* 1976), while only 0.2% was found in *Trichophyton mentagrophytes* (Table II). The value of such comparisons is questionable, since melanin content may vary in response to physical factors (Ellis and Griffiths, 1975) and with the stage of development as influenced by cultural conditions (Bull, 1970a). Furthermore, in the case of *T. mentagrophytes,* the rodlet layer which accounts for approximately 10% of the wall weight contains virtually all the cell wall melanin (Wu-Yuan and Hashimoto, 1977). Thus a component appearing to be of little significance on a total wall weight basis may be of considerable importance if localized in a particular wall layer. Although small quantities of inorganic materials and, in one instance, nucleic acids (Table I) are found in conidial walls, generally little significance has been attributed to these constituents. But the possible occurrence of phosphorus as phosphoglycan(s) should not be overlooked.

As noted previously, it is believed that chitin, an *N*-acetyl-D-glucosamine polymer (Aronson, 1965) is a common constituent of conidial walls. This view is supported by results of cytochemical observations on *Fusarium culmorum* (Marchant, 1966), partial digestion with chitinase of *Trichophyton mentagrophytes* walls (Wu-Yuan and Hashimoto, 1977) and, most definitely, by x-ray diffraction analysis of spore walls of *Fusarium sulphureum* (Schneider *et al.,* 1977; Schneider and Wardrop, 1979). However, it is apparent that some of the glucosamine found in conidial walls is of glycoprotein origin (e.g., Sturgeon, 1964; Page and Stock, 1974a). With respect to neutral glycan(s), currently available information comes only from wall digestion studies using β-D-glucan hydrolases which have, in some instances, been employed as crude mixtures of these plus other enzymes. A purified β-1,3-D-glucanase digested approximately 45–50% of the alkali-insoluble glucan in the microconidial walls of *Trichophyton mentagrophytes* (Wu-Yuan and Hashimoto, 1977), and Horikoshi and Iida (1964) also obtained evidence for the presence of 1,3-linked β-glucan in spore walls of *Aspergillus oryzae*. Using culture filtrates from a variety of hydrolase-producing microorganisms, Benítez *et al.* (1976) deduced that 10% of the conidial wall material of *Trichoderma viride* was 1,3-linked β-glucan and another 35% was 1,6-linked. Although these findings indicate that β1,3- and β-1,6-D-glucans, which seem to be of wide occurrence in fungal walls (Bartnicki-Garcia, 1968), are important spore wall constituents, further elucidation of their chemical structure(s) is needed.

The sizable protein contents of conidial walls are matters of considerable interest. The amino acid compositions of total wall protein and wall protein fractions in selected cases are shown in Table III. It would be most satisfying to be able to derive from such data indications of distinctive properties of conidial wall proteins shared by more than one species, but there seems to be little basis for doing so. Although spore walls and wall fractions all contain relatively large quantities of acidic amino acids, this does not appear to be an unusual feature as far as proteins in general are concerned. The relatively large quantity of cystine in the conidial walls of *Penicillium expansum* suggests the possibility that disulfide cross-linking of polypeptides may be important in this species, but similarly high levels of this sulfur amino acid have not been reported for other investigated penicillia (Rizza and Kornfeld, 1969; Martin *et al.*, 1973). In the case of *Fusarium sulphureum* spore walls, the prevalence of hydroxyamino acids (serine and threonine) along with an unusually large amount of proline may be indicative of specialized wall protein(s), and further study of this species could be interesting. Peptidoglycans containing appreciable amounts of serine and threonine are apparently widespread among filamentous fungi and yeasts (Gander, 1974), and while many of these constituents may be cell wall components, their exact localization for the most part remains to be determined. Treatment of *Microsporum gypseum* macroconidial walls with anhydrous ethylenediamine yielded two glycoproteins (Page and Stock, 1974a). One of these, a water-insoluble polymer (Table III), accounted for 10% of wall dry weight. Aside from the abundance of aspartic and glutamic acids and glycine, this wall constituent has an unusually high leucine content. It is believed (Page and Stock, 1974a) that this glycoprotein is the substrate (*in situ*) for a germination-specific alkaline protease. The other *M. gypseum* glycoprotein, which is water-soluble and accounts for 3.5% of spore wall weight, has a different amino acid composition characterized by approximately 20 mol % proline and a considerably greater hydroxyamino acid content as compared with the water-insoluble component. It does not appear to have any special function in either the formation or germination of mac-

TABLE III

Amino Acid Composition of Conidial Walls and Wall Fractions of Different Species[a]

Amino acid	*Penicillium expansum*	*Fusarium sulphureum*	*Trichophyton mentagrophytes*			*Microsporum gypseum* glycoprotein[c]
			Walls	Glycoprotein–lipid fraction[b]	Rodlet layer[b]	
Asp	9.9	9.2	12.2	8.7	15.7	11.0
Gly	8.1	8.8	6.1	8.0	9.0	10.2
Glu	8.9	9.7	9.5	9.5	8.3	8.5
Lys	2.4	7.1	4.3	4.5	7.7	1.2
Val	5.7	6.4	5.8	4.7	7.5	4.5
Leu	6.0	2.3	6.0	6.0	7.4	15.6
Ala	7.9	6.4	6.4	7.6	7.0	6.5
Thr	8.4	12.6	5.4	5.5	5.1	5.2
Ser	8.7	9.9	5.9	7.8	4.0	7.2
Pro	6.1	11.4	3.8	5.3	3.4	6.9
Ile	3.1	3.2	3.2	3.5	3.2	3.1
Phe	2.4	2.0	2.6	2.9	2.7	7.2
Cys	10.3	3.2	3.0	0.6	1.9	tr
Tyr	2.4	3.6	1.5	2.2	1.3	2.2
His	4.8	1.6	tr	tr	1.3	0.3
Arg	3.9	1.5	1.9	3.8	1.1	0.2
CysA[d]			0.1	0.2	0.1	
Met	1.0	1.2	0.7	0.9	0.1	6.6
NH$_3$			21.6	18.2	11.6	≤2.0

[a] Data are given as mole percentages (or tr, trace) and, for *P. expansum, F. sulphureum, T. menta-grophytes,* and *M. gypseum,* respectively, were taken or calculated from Fisher and Richmond (1970), Schneider *et al.* (1977), Wu-Yuan and Hashimoto (1977), and Page and Stock (1974a). Data of Fisher and Richmond (1970) reproduced with the permission of Cambridge University Press; data of Schneider *et al.* (1977) by permission of the National Research Council of Canada.

[b] See Table II, footnotes *b* and *c,* for a description of the derivation of fractions.

[c] Water-insoluble fraction of an ethylenediamine extract of isolated walls.

[d] Cysteic acid.

roconidia. Among the most significant of recent developments is isolation of the rodlet layer from microconidial walls of *Trichophyton mentagrophytes* (Hashimoto *et al.,* 1976; Wu-Yuan and Hashimoto, 1977) and demonstration of their protein nature (Tables II and III). Following their early freeze-etch investigation of *Penicillium megasporum* conidia, Sassen *et al.* (1967), considering the nonwettable characteristic of the spores, speculated that the rodlets of this species were composed of a cutinlike substance. However, Fisher and Richmond (1970) reported that hydroxy fatty acids characteristic of cutin were absent from conidial walls of penicillia. The latter workers were unable to remove the rodlets (as seen in freeze-etched replicas) by treatments with mercaptoethanol in urea followed by cold, dilute sodium hydroxide but noted that the rodlets were less distinct after alkali treatment. They concluded that some amount of protein had probably been removed from the rodlet layer and found appreciable quantities of protein in the alkaline extract. The findings by Hashimoto *et al.* (1976) that reagents that disrupt hydrogen bonds (urea), reduce protein disulfide linkages (mercaptoethanol), or dissociate oligomeric proteins (sodium dodecyl sulfate) (Lehninger, 1975) have no detectable effect on rodlet structure, and that 1 *N* sodium hydroxide at 100°C is the only treatment causing rapid disintegration, seem to indicate that the rodlets of *T. mentagrophytes* consist of polypeptides with unusual properties. Moreover, the rodlet layer was apparently unaffected by chloroform, ethanol, proteolytic enzymes, lysozyme, β-glucanases, and chitinase (Hashimoto *et al.*, 1976). The amino acid composition of the rodlet layer differs from that of microconidial walls as a whole and from that of the glycoprotein–lipid fraction (Table III). Hashimoto *et al.* (1976) suggested that the relative abundance of glycine, lysine, aspartic acid, and glutamic acid might be indicative of the presence of cross-linkages between amino acids, melanin, or other structural components. While these authors did not speculate on the nature of such linkages, they pointed out that extensive cross-linking of rodlet layer constituents could explain the remarkable resistance of rodlet protein to chemical and enzymatic disruption. The rodlet layers of *Neurospora crassa* macroconidia, recently isolated by Beever *et al.* (1979), contain 91% protein and are also highly resistant to chemical attack.

C. Architecture

By means of TEM and freeze-etch methods following treatments of microconidia and isolated walls with various chemical reagents and enzymes (Hashimoto *et al.*, 1976; Wu-Yuan and Hashimoto, 1977) it has been possible to elucidate the ultrastructural organization of the microconidial walls of *Trichophyton mentagrophytes* and to determine the chemical nature of the structural elements within the wall. Wu-Yuan and Hashimoto (1977) summarized their findings with a diagrammatic reconstruction as shown in Fig. 6. Transmission electron microscopy demonstrated that *T. mentagrophytes* walls consisted of three layers. The outermost layer (Fig. 6), an electron-transparent pellicle, is 15–20 nm in thickness and can be extracted by a phosphate-buffered solution containing urea, mercaptoethanol, and sodium dodecyl sulfate. Observations (TEM) of thin sections show this layer to be absent following the described treatment. This layer also is often removed during freeze-etching procedures. The extracted material, a glycoprotein–lipid complex (Tables II and III), accounts for approximately 25% of the wall weight and gives a single band when studied by disc gel electrophoresis. According to Wu-Yuan and Hashimoto

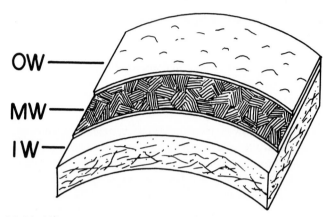

Fig. 6. Model of microconidial wall of *Trichophyton mentagrophytes*. The outer wall (OW) is composed of a glycoprotein–lipid complex. The middle wall (MW) contains proteinaceous rodlets along with polysaccharides and melanin. Chitin microfibrils in an amorphous β-glucan matrix constitute the inner wall (IW). (From Wu-Yuan and Hashimoto, 1977.)

(1977), the glycoprotein–lipid complex may constitute a portion of the inner wall (Fig. 6) also, since distinct microfibrils appeared on the inner wall surface after removal of this fraction. Partial extraction of the outer walls under milder conditions that do not reduce spore viability, results in reduced germination rates, suggesting that the outer wall components have physiological or enzymatic functions in germination (Wu-Yuan and Hashimoto, 1977). The middle wall (Fig. 6) is an electron-dense layer 30–50 nm in thickness (TEM observations). The disappearance of this layer as a result of hot alkali treatment is coincident with loss of the rodlet structure revealed by freeze-etch methods and leads to the conclusion that the rodlets make up a sizable portion of this layer, which accounts for approximately 10% of wall weight. The localization of virtually all the spore wall melanin in isolated rodlet layers (Hashimoto *et al.*, 1976; Wu-Yuan and Hashimoto, 1977) (see Table II) and the high electron density (TEM observations) of the middle wall are consistent with the aforementioned conclusion. The inner wall (Fig. 6) is 200–300 nm in thickness, accounts for approximately 50% of wall weight, and remains essentially intact following treatments that remove the outer and middle walls (Table II). Moreover, the original spore wall shape remains after these treatments. The inner wall glucans are approximately 50% solubilized by digestion with β-1,3-D-glucanase, and hot dilute acid causes rapid hydrolysis. The glucan(s) appears amorphous in the native state but, as a result of hydrolysis of weak linkages (presumably 1,6), the 1,3-linked main glucan chains form the more crystalline hydroglucan (Houwink and Kreger, 1953; Bartnicki-Garcia, 1968; Jelsma and Kreger, 1975) which appears as bundles of microfibrils. These hydroglucan microfibrils are thicker, hence distinguishable from the native fibrils appearing on the inner wall surface following removal of the glycoprotein–lipid complex. The latter are extensively dissociated from the glucan matrix as a result of the β-glucanase treatment mentioned previously. In addition, these fibrils disappear during treatment of the inner wall with chitinase, and N-acetylglucosamine is released. Thus the inner wall appears to be composed of chitin microfibrils in a β-glucan matrix (Wu-Yuan and Hashimoto, 1977). A small amount of protein is associated with inner wall glycans (Table II), but it was not characterized.

Can the architectural design of *T. mentagrophytes* microconidial walls be considered a prototype for conidial walls in other genera? This question will not be answered definitively until more genera have been investigated, but it is of interest to evaluate the currently available information. Sassen *et al.* (1967) presented a graphic interpretation of the conidial wall of *Penicillium megasporum,* which was based on their TEM and freeze-etch observations. Their model depicted an external electron-dense rodlet layer and a thicker internal layer consisting of microfibrils, which they assumed were chitinous, within an amorphous matrix. From these ultrastructural observations, the *P. megasporum* wall appears much like that of *T. mentagrophytes,* except that an outer wall, corresponding to the glycoprotein–lipid layer is lacking. However, as noted previously, earlier investigators of penicillia were aware of a thin, transient outer layer which was assumed (possibly in error) to be polyphosphate. However, as already noted (see Section II,A), Dempsey and Beever (1979) believe that, in some species, the rodlets are not covered with an external wall layer. It seems evident that the concerted use of chemical, biochemical, and electron microscope methods as applied in the investigation of *T. mentagrophytes* provides the means not only for resolving the problem of conidial wall architecture in penicillia but for the elucidation of spore wall structure in other genera as well. In a recent investigation of *Aspergillus niger,* Cole *et al.* (1979) employed a procedure that sheared the walls so that outer and inner conidial wall regions could be isolated and analyzed independently. The outer wall fraction contains much less polysaccharide and considerably more protein than the inner wall fraction. Both outer and inner conidial walls have a sizable lipid content. The rodlets of *A. niger* occur in the outer wall fraction and are similar to those of *T. mentagrophytes* in so far as resistance to various chemical and enzymatic agents is concerned. However, *A. niger* rodlets can be digested by elastase. The investigators of *A. niger* believe that their findings indicate architectural similarity between this species and *T. mentagrophytes.*

D. Wall Alterations during Germination

It was noted previously in this chapter that some workers believe that new cell wall layers are synthesized during spore germination. Unfortunately, little information is available pertaining to changes in the chemical composition of conidial walls during germination. Some work has been reported on chemical differences occurring in *Penicillium notatum* spores in the resting (dormant), swollen, and germinating (germ tube emergence) stages (Martin *et al.*, 1973, 1974). Analyses of intact spores show that they undergo increases in total protein and carbohydrate during the germination process (Martin *et al.*, 1974). Analyses of the composition of isolated walls at various stages indicate that total spore wall protein decreases, as does the total galactose content, and that there is a concomitant increase, in spore wall glucose and hexosamine (Martin *et al.*, 1973). However, microscopic observations of the spores show that outer wall layers, believed to contain appreciable amounts of protein and galactose, are shed during swelling and germ tube emergence. Therefore the observed percentage increases in glucose and hexosamine may be due to the selective loss of other wall components (Martin *et al.*, 1973). An earlier investigation of germinating spores of *Penicillium atrovenetum* (Gottlieb and Tripathi, 1968) reported the incorporation of label from [U-^{14}C]glucose into a spore wall fraction obtained after extensive chemical fractionation of swollen and germinating spores. The total amount of carbon-14 incorporated was quite small, and it is know that chemical preparation of cell walls can produce artifacts. If a minute quantity of cell wall was synthesized, it could have been the product of apical wall synthesis by hyphae, since 40% of *P. atrovenetum* conidia had emergent germ tubes (Gottlieb and Tripathi, 1968). Although the chitin synthetase activity of germinating conidia of *Aspergillus nidulans* increases prior to the microscopic detection of emergent germ tubes, the deposition of hexosamine (chitin) coincides with germ tube growth and not spore swelling (Ryder and Peberdy, 1979). In an investigation of macroconidia of *Microsporum gypseum,* exogenously supplied [U-^{14}C]glucose was incorporated mainly into the intracellular lipid fraction (Dill *et al.*, 1972) during early stages of germination (prior to germ tube emergence). However, the data do not preclude a small amount of wall synthesis, but this possibility was not investigated. Factors involved in wall lysis in *M. gypseum* have been studied more extensively (Page and Stock, 1974a,b). Spores of this species contain alkaline protease, alkaline phosphodiesterase, ethyl esterase, β-1,3-D-glucanase, and chitinase. All enzyme activities were present throughout the germination period, and significant amounts of activity were present in pellet fractions, suggesting that the enzymes may be present in the spore walls. Although the roles of these enzymes in germination require further investigation, the alkaline protease, as noted previously has been implicated in the modification of a water-insoluble glycoprotein accounting for an appreciable amount of spore wall weight. During a 6-h germination period, the protein/hexose ratio in this wall fraction decreased 10-fold (Page and Stock, 1974a). Germ tube emergence is believed to involve partial digestion of the spore wall (Smith *et al.*, 1976; Page and Stock, 1974b), and the chitinase and β-glucanase activities in *M. gypseum* spores are likely to be involved in this process. Page and Stock (1974b) also observed a rapid release of melanin from germinating conidia and, noting that melanin was an inhibitor of chitinase and other cell wall-hydrolyzing enzymes (Kuo and Alexander, 1967; Bull, 1970b), suggested that the melanin release was a prerequisite for the action of the conidial wall glycan hydrolases.

III. HYPHAL WALLS

A. Organization and Functions of the Hyphal Apex

The relationship between the wall of the hyphal germ tube and the inner layers of the conidial wall was considered in the previous section. With the formation of an incipient germ tube, the growth of a fungal hypha becomes strictly polarized and produces a continuum of hyphal segments elongating at the hyphal tip (Bartnicki-Garcia, 1973). Initiation of the germ tube is believed to result from wall synthesis and deposition at the site of germ tube emergence (Smith *et al.*, 1976), and further extension of the hypha by the process of tip growth involves the deposition of new wall material in a restricted region of the hyphal apex. All aspects of tip growth of fungi have been abundantly reviewed (Bartnicki-Garcia, 1973; Smith *et al.*, 1976; Grove, 1978; Trinci, 1978). While earlier workers used external particles or septa and branches as reference points in studies on hyphal growth zones, recent investigators have employed isotope-labeled cell wall precursors combined with autoradiographic techniques. Katz and Rosenberger (1970, 1971b), using the latter approach with *Aspergillus nidulans* mutants, found that tritium-labeled galactose and N-acetylglucosamine were both incorporated into polysaccharides at the hyphal tips. Similarly, Gooday (1971) demonstrated apical incorporation of tritiated glucose and N-acetylglucosamine in hyphae of *Neurospora crassa*. These recent investigations demonstrate that hyphal wall formation associated with hyphal extension involved the preferential deposition or synthesis (or both) of cell wall polysaccharides at the hyphal tip and, in particular, in the apical dome region as opposed to the tubular region behind the dome (e.g., Bartnicki-Garcia, 1973). However, there is actually a decreasing gradient of wall formation extending from the hyphal apex, so that in the nongrowing, tubular zone a limited wall synthesis potential exists (Bartnicki-Garcia, 1973), which may serve to add secondary wall layers, modify existing wall layers or components, or function in branch initiation.

Investigations of germinating spores of *Aspergillus parasiticus* (Grove, 1972) and *Geotrichum candidum* (Steele and Fraser, 1973) show that in conidial fungi a vesicular organization in the cytoplasm of the hyphal apex is present at the time of germ tube emergence. Although concentrations of vesicles have been found in the tips of growing hyphae of fungi representing all major groups, the septate fungi, Zygomycetes, Oomycetes, and Chytridiomycetes, can be delineated on the basis of vesicle distribution patterns (Grove and Bracker, 1970; Roos and

Turian, 1977). The hyphal apex of *Verticillium albo-atrum* (Fig. 7) contains numerous vesicles and is devoid of other cell components except for a small patch of ribosomes (Grove and Bracker, 1970). The vesicle-free region at the tip is believed to correspond to the *Spitzenkörper* (e.g., Girbardt, 1969), a structure observed by light microscopy in the tips of a number of fungi. *Fusarium oxysporum* has a similar apical organization (Grove and Bracker, 1970). In *Aspergillus niger* a similar vesicle distribution occurs, but the *Spitzenkörper* region is occupied by an aggregate of smaller vesicles and small clusters of ribosomes (Grove and Bracker, 1970). In *Neurospora crassa* also, larger apical vesicles surround a region containing mostly minute vesicles, and a similar organization is seen in *Ascodesmis nigricans* (Grove and Bracker, 1970). About 80% of the volume of the extreme apex (the first micrometer) of *Neurospora crassa* hyphae is occupied by vesicles, as opposed to 5% in the base of the extension zone (Collinge and Trinci, 1974).

It is believed that apical vesicles originate in the hyphal endomembrane system (Grove and Bracker, 1970; Grove et al., 1970) located in subapical regions from which they migrate toward the apex. The mechanism of migration is not known, but electrophoresis has been suggested, as has movement as a result of bulk flow along a water potential gradient (Bartnicki-Garcia, 1973; Trinci, 1978). The

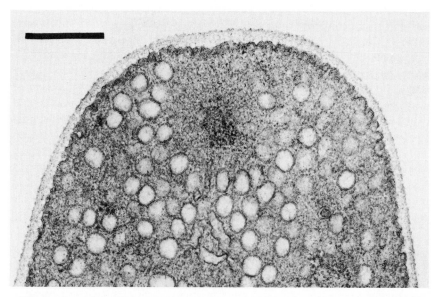

Fig. 7.   Transmission electron micrograph of a thin section of the hyphal apical zone of *Verticillium albo-atrum*. Vesicles are conspicuous, and other cellular structures are virtually absent. The vesicle-free region at the apex corresponds to the *Spitzenkörper*. Scale line, 0.5 μm. (From Grove and Bracker, 1970.)

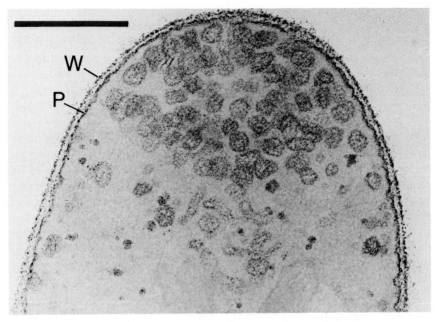

Fig. 8. Transmission electron micrograph of a thin section of the hyphal apical zone of *Hypomyces chlorinus*. The section was treated with a cytochemical procedure that detects some polysaccharides. Electron-dense deposits, indicative of the presence of polysaccharides, occur in the outer part of the cell wall (W), the plasma membrane (P), and the apical veiscles. Compare with Fig. 7. Scale line, 0.5 μm. (From Dargent and Touzé-Soulet, 1976.)

fusion of vesicles with the apical dome plasmalemma appears to provide membrane material to expand the plasmalemma surface and concomitantly discharge materials required for wall formation. In the latter category, polysaccharide synthetases, wall polymer precursors, preformed (polymerized) wall components, and wall-plasticizing enzymes (hydrolases) are thought to be essential in apical wall fabrication, and any or all of these may be transported in vesicles (Bartnicki-Garcia, 1973).

The content of apical vesicles in conidial fungi is largely unknown but, in vesicles of *Saccharomyces cerevisiae* (Matile *et al.,* 1971) and *Achlya ambisexualis* (Nolan and Bal, 1974), respectively, β-1,3-glucanase and cellulase have been detected. Cytochemical methods for polysaccharide localization have been applied to a few species and have provided evidence that apical vesicles contain materials that could serve as wall precursors (Grove, 1978). This appears to be the case in the ascomycete *Hypomyces chlorinus* (Dargent and Touzé-Soulet, 1976) in which the apical wall, plasmalemma, and apical vesicles show similar cytochemical reactivity (Fig. 8). The extent to which wall polymers are preformed as opposed to *in situ* assembly is unknown. However, Katz and

Rosenberger (1970) found evidence of a cytoplasmic polysaccharide fraction having a monosaccharide composition resembling that of the cell wall of *Aspergillus nidulans*. It has been suggested (Bartnicki-Garcia, 1973) that hyphal wall microfibrillar components (chitin) are assembled within the wall itself, while the matrix constituents may be preformed within the cell.

B. Chemical Composition

Whatever the mechanism of hyphal wall deposition, examination of the composition of various species reveals considerable complexity (Table IV). An interesting feature of wall structure in conidial fungi is the marked similarity of the qualitative composition of polysaccharides in conidial and hyphal walls (compare Tables I, II, and IV). The same hexoses are found in either case and, as is true of conidial walls, glucose is the principal neutral sugar in hyphal walls. Pentoses and deoxyhexoses have not been reported frequently, but small quantities of rhamnose, xylose, and arabinose have been found in walls of *Fusicoccum amygdali* (Buck and Obaidah, 1971). Glucuronic acid is becoming a more frequently detected wall component and is probably more prevalent than currently available data (Table IV) suggest. The presence of chitin in hyphal walls of conidial fungi and ascomycetes is well established (Aronson, 1965; Bartnicki-Garcia, 1968), and subsequent investigations confirming earlier work or extending the range of occurrence of chitin include studies on *Trichophyton mentagrophytes* (Kitazima *et al.*, 1972), *Epidermophyton floccosum* (Nozawa *et al.*, 1973), *Aspergillus nidulans* (Bull, 1970a), *Fusarium sulphureum* (Barran *et al.*, 1975), *Pyricularia oryzae* (Nakajima *et al.*, 1970), *Fusicoccum amygdali* (Buck and Obaidah, 1971), *Geotrichum candidum* (Sietsma and Wouters, 1971), *Helminthosporium sativum* (Applegarth and Bozoian, 1969), *Verticillium albo-atrum* (Wang and Bartnicki-Garcia, 1970), and *Penicillium chrysogenum* (Troy and Koffler, 1969).

Bartnicki-Garcia (1968) commented on the small amounts of galactosamine found in walls of *Neurospora crassa* and *N. sitophila* and noted the variability in reports of this sugar in *Aspergillus* spp. While Bull (1970a) reported 10.8% galactosamine in an albino mutant of *A. nidulans* (Table IV), Katz and Rosenberger (1970) found 1.1% in a different mutant. However, this amino sugar is a significant component of walls of *Helminthosporium sativum* (Table IV). In addition, traces have been detected in walls of *Penicillium rubrum* (Unger and Hayes, 1975), and Shah and Knight (1968) found small quantities in walls of *Trichophyton mentagrophytes, Microsporum canis, M. gypseum,* and *Epidermophyton floccosum*. Although variable in occurrence, galactosamine does seem to be a widespread wall component, and Bardalaye and Nordin (1976) have recently characterized a distinct galactosamine-containing polysaccharide from walls of *Aspergillus niger* (see Section III,C,4). The variable content of this

TABLE IV

Composition of Hyphal Walls of Different Species[a]

Component	Trichophyton mentagrophytes	Aspergillus nidulans (albino)	Fusarium sulphureum	Helminthosporium sativum	Verticillium albo-atrum ("yeast" form)[b]
Neutral carbohydrates (total)[c]	47.9	55.1	22.0	41.0	70.9
Glucose	36.2	28.9	12.0	37.0	56.0
Galactose	tr	3.9	3.0	2.0	8.5
Mannose	11.7	2.8	7.0	2.0	6.4
Glucuronic acid		2.6	5.0		1.3
Chitin	30.4	14.3	39.0	8.6	10.0[d]
Galactosamine		10.8		8.3	
Protein	7.8	10.6	7.3	18.0	14.0
Lipid	6.6	9.0	5.5	11.4	3.4
Phosphate	0.1	0.9		1.8	0.2
Ash	2.2		1.8	5.2	

[a] Data for *T. mentagrophytes*, *A. nidulans*, *F. sulphureum*, *H. sativum*, and *V. albo-atrum*, respectively, are adapted from Noguchi et al. (1971), Bull (1970a), Barran et al. (1975), Applegarth and Bozoian (1969), and Wang and Bartnicki-Garcia (1970). Data of Bull (1970a) and Wang and Bartnicki-Garcia (1970) reproduced with permission of Cambridge Univ. Press. Data are expressed as percentages of wall dry weight or as tr (trace).

[b] Cells of the "yeast" form may be conidial; see Wang and Bartnicki-Garcia (1970) for further comments.

[c] Calculated from original sources.

[d] Based on the insoluble glucosamine content; walls also contained 3.3% soluble glucosamine.

amino sugar may be due to nutrient availability and/or the age of cultures (Bardalaye and Nordin, 1976). It can be noted also that galactosamine does not appear to be a constituent of walls of dormant conidia (see Section II,B). Interestingly, the appearance of this sugar in the walls of *Penicillium notatum* was coincident with germ tube emergence, and the quantity increased with further mycelial development (Martin *et al.*, 1973).

All investigated species have appreciable amounts of protein in their hyphal walls (Table IV), and analyses of amino acid composition show a broad assortment of the common protein amino acids (e.g., Unger and Hayes, 1975; Bulman and Chittenden, 1976; Nozawa *et al.*, 1973; references cited in Table IV). Amino acid distributions of whole cell walls generally do not show any unusual features. However, chemical fractionation of walls of *Verticillium albo-atrum* (Wang and Bartnicki-Garcia, 1970) demonstrated that most of the amino acids were solubilized by dilute acid and alkali but that two components, lysine and histidine, were partially bound to the insoluble (chitin) fraction. The protein amino acid distributions of hyphal walls of *Fusarium sulphureum,* as compared with those of spore walls, showed mainly quantitative variations, except for the apparent absence of proline in hyphal walls (Barran *et al.*, 1975; Schneider *et al.*, 1977). Similar results were obtained in studies on *Penicillium notatum* (Martin *et al.*, 1973), but significant qualitative differences were found in investigations of hyphal and conidial walls of *P. chrysogenum* (Rizza and Kornfeld, 1969). Vegetative and synnematal walls of *Ceratocystis ulmi* had the same amino acid assortment, and only quantitative differences were observed (Harris and Taber, 1973).

While there has been some speculation concerning the significance of cell wall lipids, the function(s) of these consistently present constituents are unknown (Bartnicki-Garcia, 1968). The same is true of the inorganic components (ash). The phosphorus frequently found could be indicative of the presence of small quantities of phosphoglycan (Gander, 1974) and/or phospholipid.

Information concerning melanin pigments in hyphal walls is limited. Harris and Taber (1973) observed dense granules, believed to be melanin, during an electron microscope examination of synnematal walls of *Ceratocystis ulmi.* Using electron microscopy, microchemical tests, and infrared spectroscopy, Ellis and Griffiths (1974, 1975) demonstrated melanins in hyphal walls of *Phomopsis* sp. and *Amorphotheca resinae.* In an investigation comparing hyphal wall composition and structure of the wild type with that of an albino mutant of *Drechslera sorokiniana,* the wild-type walls were found to contain 14.3% melanin, while the pigment was lacking in mutant walls (Al-Rikabi and Bonaly, 1977). The most detailed investigation of melanin in hyphal walls has been carried out with *Aspergillus nidulans* (Bull, 1970a). The wild-type walls are similar in composition to those of the albino strain (Table IV), except that approximately 18% of the wall weight is melanin and there is a corresponding reduction in the percentage of carbohydrates. From a number of chemical and physical methods of analysis, it was concluded that the pigment was an indolic melanin. Nitrogen contents of the extracted pigment varied in response to culture conditions and growth medium composition (Bull, 1970a). Bull (1970a,b) has discussed earlier studies of fungal melanin and the evidence for its role in the resistance of fungal walls to lysis.

A considerable number of studies on wall composition were reviewed by Bartnicki-Garcia (1968) and, as noted in the preceding remarks, there has since been a proliferation of investigations of this type. Some additional investigations reporting on wall composition are the following: *Penicillium digitatum* and *P. italicum* (Grisaro *et al.*, 1968); *Aspergillus nidulans* (Zonneveld, 1971); *Aspergillus niger* (Stagg and Feather, 1973); *Epidermophyton floccosum* (Kitajima and Nozawa, 1975); *Hypomyces chlorinus* (Touzé-Soulet and Dargent, 1977); *Cordyceps militaris* (Marks *et al.*, 1969, 1971); *Sporothrix schenckii* (Previato *et al.*, 1979a,b); *Aspergillus fumigatus* (Hearn and Mackenzie, 1979).

C. Characteristics of Hyphal Wall Polymers

It is generally stated axiomatically that cell walls are composed of microfibrils and a nonfibrillar matrix (Aronson, 1965; Bartnicki-Garcia, 1968; Rosenberger, 1976). Chitin generally constitutes the microfibrillar phase and the surrounding matrix, at times described as amorphous (e.g., Aronson, 1965; Bartnicki-Garcia, 1968), is made up of an assortment of glucans, heteroglycans, and peptidoglycans. It is evident that in walls of many fungi, chitin is not the sole fibrillar component, and in the following sections other examples of microfibrillar polymers are discussed. It is appropriate to note also that some matrix polymers may occur, at least in part, in a crystalline state.

1. Chitin and Cellulose

Chitin is a polymer of wide occurrence in fungi (Aronson, 1965; Bartnicki-Garcia, 1968; Rosenberger, 1976). The ribbon (extended helix) conformation of the chains of β-1,4-linked N-acetyl-D-glucosamine units in this polysaccharide is such that a well-ordered, close packing of polymer chains is possible, giving a stable crystalline lattice (Rees, 1977). These regularly arranged chitin molecules are aggregated into bundles forming the microfibrils seen when wall preparations or replicas are viewed in the electron microscope (Fig. 9). Because of its crystalline nature, chitin can be unequivocally identified by x-ray diffraction analysis (e.g., Nozawa *et al.*, 1973; Barran *et al.*, 1975), although the detection of N-acetylglucosamine following digestion with chitinase (e.g., Sietsma and Wouters, 1971; Buck and Obaidah, 1971) has also been employed. The use of older cytochemical and microchemical tests (e.g., Bulman and Chittenden, 1976) as the sole method of detection is becoming less frequent.

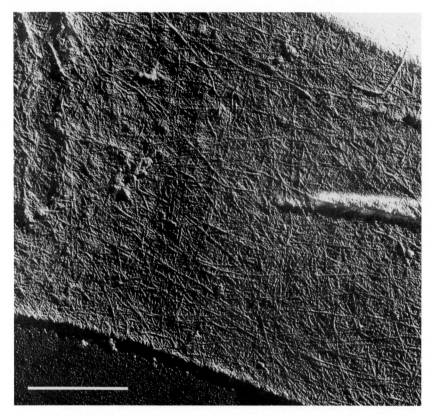

Fig. 9. Transmission electron micrograph of *Neurospora crassa* wall following extensive removal of matrix components by digestion with pronase (a proteinase) and laminarinase (β-1,3D-glucanase). Chitin microfibrils can be seen. Scale line, 1 μm. (Reproduced from Hunsley and Burnett, 1970, with permission of Cambridge Univ. Press.)

Cellulose, a polymer of β-1,4-linked D-glucose units, assumes a ribbon conformation, as does chitin, and the polymer chains are organized into microfibrils (Preston, 1974; Rees, 1977). This polymer has been the subject of considerably more investigation than chitin because of its occurrence in plants and its obvious economic importance (Preston, 1974). Cellulose is of restricted occurrence in fungal walls (Rosenberger, 1976) and, in higher fungi, it is known to occur only in species of the ascomycetes *Europhium* and *Ophiostoma,* where it is regarded as being of taxonomic significance (e.g., Weijman, 1976). Weijman (1976) has utilized the presence or absence of cellulose along with other cell wall characters in addition to morphology to distinguish the ascomycete genera *Cephaloascus, Europhium, Ophiostoma,* and *Ceratocystis* (and their conidial anamorphs, where present).

The fact that chitin and cellulose molecules are extremely long (a single cellulose molecule may be composed of as many as 15,000 glucose units, Preston, 1974) leads to microfibrils of considerable length stabilized by intra- and intermolecular hydrogen bonds. As a result, the fibrils will have high tensile strength (Houwink *et al.*, 1971). Therefore, on theoretical grounds, chitin microfibrils in fungal walls appear to function in absorbing wall strain resulting from turgor-induced stress. Whether or not the chitin microfibrils in growing hyphae have complex functions, like the cellulose fibrils of plants (Preston, 1974), is unknown at present. Whatever the case, their wall-strengthening function is not dispensable. This does not mean that the rheological properties of hyphal walls are derived solely from the chitinous microfibrillar phase. On the contrary, matrix polymers and other fibrillar components are believed to have a significant role also. A stimulating discussion of this aspect of wall structure appears in Rosenberger (1976).

2. β-Glucans

Polymers consisting of β-linked D-glucose units have been repeatedly demonstrated in walls of conidial fungi. Walls of *Aspergillus nidulans* (Bull, 1970a; Zonneveld, 1971), *Fusicoccum amygdali* (Buck and Obaidah, 1971), *Geotrichum candidum* (Sietsma and Wouters, 1971), *Paracoccidioides brasiliensis* (Kanetsuna and Carbonell, 1970), *Pyricularia oryzae* (Nakajima, *et al.*, 1970), *Sporothrix schenkii* (Previato *et al.*, 1979b), *Trichophyton mentagrophytes* (Kitazima *et al.*, 1972), and *Verticillium albo-atrum* (Wang and Bartnicki-Garcia, 1970) contain polymers with β-1,3- and β-1,6-linked glucosyl units. β-Glucans of this type account for 13.8% (Bull, 1970a) and 40% (Troy and Koffler, 1969) of walls of *A. nidulans* and *Penicillium chrysogenum*, respectively, and their occurrence is generally demonstrated by means of digestion of walls or wall fractions with β-1,3-glucanases (Bull, 1970a; Troy and Koffler, 1969; Wang and Bartnicki-Garcia, 1970; Buck and Obaidah, 1971; Kanetsuna and Carbonell, 1971). Although β-glucans are regarded as insoluble wall constituents (e.g., Bartnicki-Garcia, 1968), some workers have reported the occurrence of alkali-soluble wall fractions containing β-1,3- and β-1,6-linkages (Buck and Obaidah, 1971; Kitazima *et al.*, 1972; Kanetsuna and Carbonell, 1970; Previato *et al.*, 1970b). However, in *Fusicoccum amygdali* walls, the soluble β-glucan may constitute a portion of a heteroglycan consisting of a galactomannorhamnan core attached to the β-glucan component (Obaidah and Buck, 1971). Thus there may be considerable heterogeneity in the properties of these wall components. The proportion of 1,3- and 1,6-linkages has not been determined in all cases. In *F. amygdali,* about 50% of the total wall glucan is β-1,3-linked, while approximately 10% of the linkages are β-1,6 (Buck and Obaidah, 1971). There is an approximately 5:1 ratio of 1,3- and 1,6-linkages in the alkali-insoluble glucan fraction of *Pyricularia oryzae* walls also (Nakajima *et al.*,

1972). Using methylation analysis, periodate oxidation, and radiochemical labeling of reducing end groups, Nakajima *et al.* (1972) have investigated the structure of the insoluble glucan from *P. oryzae*. These investigators calculated an average molecular weight of approximately 3.8×10^4, corresponding to an average of 230–240 glucosyl units. It is believed that the polymer is branched, consisting of β-1,3-linked chains with interspersed 1,6-linked units and 1,6-linked side chains (Nakajima *et al.*, 1972).

Insoluble β-glucans from a wide range of fungi are apparently amorphous in the native condition (note that *Aspergillus niger* and *A. awamori* may be exceptions; Bobbitt *et al.*, 1977) but upon treatment with hot, dilute acid are converted to a crystalline, microfibrillar derivative termed "hydroglucan" (e.g., Jelsma and Kreger, 1975; Wang and Bartnicki-Garcia, 1970). This conversion has been observed with walls of *Penicillium notatum* (Kreger, 1954) and *Verticillium albo-atrum* (Wang and Bartnicki-Garcia, 1970). It is believed that dilute acid treatment may cleave sensitive bonds, especially the β-1,6-linkages attaching side chains to the β-1,3-linked glucan chains, and this permits the 1,3-linked chains to aggregate into crystallites (Jelsma and Kreger, 1975). The derived hydroglucan gives an x-ray powder diagram almost identical to that of the β-1,3-glucan paramylon (Jelsma and Kreger, 1975) and, although the hydroglucan may not be a native wall constituent, its preparation and detection can be useful for analytic purposes.

The alkali-insoluble fraction of walls, consisting essentially of chitin and glucan, has been found in at least one instance to contain β-1,4-glucosidic linkages. Wang and Bartnicki-Garcia (1970) found cellobiose in an amount equivalent to approximately 7% of wall weight in a β-glucanase digest of *Verticillium albo-atrum* walls, but x-ray analysis and other methods gave no indication of the presence of any cellulose. Stagg and Feather (1973) found approximately 10–15% of the alkali-insoluble glucan of *Aspergillus niger* to be 1,4-linked and noted that the proportion of β-linkages indicated by nuclear magnetic resonance (nmr) spectrometry was roughly equivalent to the proportion of 1,4-linkages. However, these observations do not demonstrate conclusively that the 1,4-linked glucosyl residues are in the β-configuration. It was noted that a glucan fraction with similar properties occurred in walls of *Penicillium chrysogenum* also (Stagg and Feather, 1973). Similarly, Previato *et al.* (1979b) reported 1,4-linkages in the glucans of *Sporothrix schenkii*. Since cellulose does not occur in these fungi, the nature of the polymer(s) containing the 1,4-linkages remains to be elucidated. One possibility is that these glucosyl units occur in polysaccharides with mixed linkages.

3. α-Glucans

While they are apparently rare in nonseptate fungi, polymers composed primarily or exclusively of α-D-glucose residues appear to be widespread in cell walls of higher fungi (e.g., Rosenberger, 1976). One of the most extensively investigated of these α-glucans is nigeran (mycodextran), a polymer consisting of alternating α-1,3- and α-1,4-glucosidic linkages. This polysaccharide, in addition to the mixed linkages, is characterized by its solubility in hot water and dilute alkali and its high dextrorotation (e.g., Johnston, 1965a). Johnston (1965a) has provided the first evidence that nigeran is a cell wall constituent and reported about 4–6% in walls of *Aspergillus niger*. Similarly, walls of *A. nidulans* contain about 5% nigeran but, in *A. awamori,* the nigeran content may reach 28% under conditions of nitrogen starvation (Bobbitt *et al.,* 1977). Bobbitt and Nordin (1978) surveyed a large number of aspergilli and penicillia representing several of the taxonomic groups (series) in each genus and found nigeran in a great many species of both genera. However, in some taxonomic groupings, the polysaccharide is either absent or of restricted occurrence among the assigned species, and it has been suggested that nigeran distribution is useful as an indicator of systematic relationships. Bobbitt and Nordin (1978) consider nigeran to be a hyphal wall constituent in all species from which it has been isolated. However, it should be noted that investigations employing isolated cell walls have involved few of the many nigeran-producing species of *Aspergillus* and *Penicillium*. Bartnicki-Garcia (1968) noted the small quantities of nigeran found in walls of *A. niger* and expressed the view that it was a dispensable component. Since this polymer is totally lacking in a number of species of *Aspergillus* and *Penicillium* (Bobbitt and Nordin, 1978) its nonessential nature seems clear. Nevertheless, the polysaccharide is of some interest since, in *A. niger* and *A. awamori,* nigeran, along with chitin, is present in a highly crystalline form (Bobbitt *et al.,* 1977). Detailed studies of the enzymatic degradation of crystalline nigeran *in vitro* are being carried out because it is thought that such investigations will contribute to elucidation of the relationship between the fine structure of cell wall polysaccharides and their accessibility to hydrolytic enzymes (J. H. Nordin, personal communication).

Polysaccharides in which α-1,3-linked glucosyl units predominate are more widespread than nigeran. These polymers have been commonly referred to as *S*-glucan (e.g., Rosenberger, 1976), a term originating from investigations of the basidiomycete *Schizophyllum commune* (e.g., Wessels *et al.,* 1972), and recently a similar polysaccharide (referred to as "pseudonigeran") from walls of *Aspergillus niger* has been characterized (Horisberger *et al.,* 1972). These α-1,3-glucans are major components in the walls of most conidial fungi that have been investigated. The more extensively investigated glucans of this type are soluble in dilute alkali, are water-insoluble, and show high dextrorotation. In *Aspergillus niger* (Johnston, 1965b) and *A. nidulans* (Bull, 1970a; Zonneveld, 1971) these glucans account for about 20–25% of the cell wall. Similarly, about one-fourth of the cell wall of *Fusicoccum amygdali* (Buck and Obaidah, 1971) is composed of α-1,3-glucan. Appreciable amounts of similar glucans occur also in walls of the dimorphic, pathogenic fungi *Blastomyces dermatitidus* (Kanetsuna and Carbonell, 1971), *Histoplasma farciminosum* (San-Blas and Carbonell,

1974), and *Paracoccidioides brasiliensis* (Kanetsuna and Carbonell, 1970). The purified, alkali-soluble polymer (pseudonigeran) from *A. niger* was shown by methylation analysis to be composed mainly of 1,3-linkages and has a degree of polymerization (DP) of 330 (Horisberger *et al.*, 1972). The *A. niger* glucan also contains a small proportion (ca. 1.4%) of 1,4-linkages (Johnston, 1965b; Horisberger *et al.*, 1972), as does a similar glucan (DP 230) extracted from the walls of *Fusicoccum amygdali* (Obaidah and Buck, 1971).

In certain *Aspergillus* and *Penicillium* species and in *Fusicoccum amygdali* also, 50% or more of the cell wall material may be alkali-insoluble, and these resistant wall fractions generally consist of a chitin-glucan "core" with smaller and variable quantities of other neutral glycoses (Johnston, 1965a; Buck and Obaidah, 1971; Stagg and Feather, 1973). While it might be assumed that the alkali-insoluble glucan is largely the widely occurring β-1,3- and β-1,6-linked type (Bartnicki-Garcia, 1968), there are now indications that in some instances this may not be the case. From methylation studies and nmr spectroscopy, it appears that well over half of the alkali-insoluble glucan in walls of *A. niger* is α-1,3-linked, and it has been noted that walls of *P. chrysogenum* are similar (Stagg and Feather, 1973). The alkali-insoluble glucan(s) of *A. nidulans* has been only partially characterized (Bull, 1970a; Zonneveld, 1971), but there are indications that insoluble α-1,3-glucan occurs in this species also (Bull, 1970a). Similarly, Buck and Obaidah (1971) concluded that walls of *Fusicoccum amygdali* contained a "backbone" of chitin and α-1,3-glucan. Whether or not α-1,3-glucans are of wide occurrence in the insoluble core of walls of other species is not clear at present. There is no indication of this type of polymer in the walls of the "yeast form" of *Verticillium albo-atrum* (Wang and Bartnicki-Garcia, 1970).

Buck and Obaidah (1971) made the interesting observation that isolated walls of *Fusicoccum amygdali* stained dark blue with iodine (potassium iodide reagent) and were attacked by α-amylase. The products of α-amylase digestion, glucose, maltose, and maltotriose, demonstrated that the walls contained α-1,4-linked glucosyl residues which were calculated to account for about 10% of the total wall glucan. β-Amylase had little effect on these wall components, indicating that there are few, if any, nonreducing ends in the α-1,4-linked regions of the glucan chains and that they must therefore be present as connecting segments between other polymeric units (Obaidah and Buck, 1971). Thus the iodine-staining of the walls is not due to wall contamination with iodine-binding polysaccharides of the starch and glycogen type. *Fusicoccum amygdali* walls are not unique, since at least 18% of the alkali-insoluble glucan of *Aspergillus nidulans* walls is α-1,4-linked (Zonneveld, 1971). Buck and Obadiah (1971) regarded α-1,4-glucan as an important structural component, since it was not extracted by chloral hydrate (a starch solvent), and microscopic observations showed that the walls became thinner after α-amylase treatment. Evidence has been obtained suggesting that α-1,4-linked glucosyl units are covalently associated with α-1,3-glucan and are also linked to a β-glucan-galactomannorhamnan complex (Obadiah and Buck, 1971).

The contribution of α-glucans to the total structure of walls of conidial fungi and their role in the maintenance of structural coherence are matters likely to provoke considerable discussion. The α-1,4-linked glucan components mentioned above are unquestionably significant (at least in *Fusicoccum amygdali* walls), since these constituents seem to be connecting links serving to knit together various polysaccharide fractions to form larger wall complexes. The situation in the case of nigeran is less clear, but it is instructive to give some consideration to the results of Bobbitt *et al.* (1977), since they have a bearing on the discussion (below) of α-1,3-glucans. Based on differential solubility in hot water and dilute alkali, along with susceptibility to mycodextranase digestion, it has been concluded that *Aspergillus niger* and *A. awamori* nigeran exists in three "domains" (configurations), although most of the polymer is localized in the exterior region of the wall. It appears that these domains arise either as a consequence of the location of the polysaccharide (superficial versus buried), the binding of the polymer to other wall components, or both. It has been concluded that the polysaccharide occurs in crystalline form in all three fractions (domains). Electron microscope observations of *A. awamori* have shown no difference in thickness between walls containing 4 and 28% nigeran, but the latter are much more electron-opaque. These observations indicate that, in walls containing the higher nigeran content, there is dense packing of the polymer chains, as would be expected for a highly crystalline wall component. This sizable quantity of crystalline nigeran would be expected to augment the cell wall integrity provided by chitin and other wall constituents but, in view of the dispensable role of nigeran noted previously, the structural effect of the polymer is likely to be incidental. Its actual role is a matter of conjecture, and Bobbitt *et al.* (1977) have discussed some possible functions. These workers have further noted the correlation between α-1,3-glucan metabolism and cleistothecium production in the teleomorph of *A. nidulans* (Zonneveld, 1978) and by analogy have made the interesting suggestion that nigeran is a vestigial storage material in (anamorph) species that have lost their nigeran-degrading enzyme by mutation.

Current evidence suggests the possibility that glucans of the α-1,3-linked type are distributed in wall domains also. As noted above, the S-glucan type (alkali-soluble) seems to be of wide occurrence, and there is evidence of a highly insoluble form as well. In *Schizophyllum commune* S-glucan occurs as crystalline fibrils located toward the outer surface of the wall. The fibrils, being comparatively short, have been referred to as "rodlets" (Wessels *et al.*, 1972), but they are obviously distinct from the protein rodlets (see Section II,B) of conidial walls. The polymer can be recrystallized from alkaline extracts, and microfibrils similar to those found *in situ* are formed (Wessels *et al.*, 1972). The crystalline properties of alkali-soluble α-1,3-glucans of conidial fungi have been observed by x-ray analysis of extracts (Kreger, 1954; Bacon *et al.*, 1968; Obaidah and Buck, 1971; Jelsma and Kreger, 1979), and the *in vitro* formation of short fibrils (rodlets) has been reported for wall material from *Aspergillus niger* (Carbonell *et al.*, 1970) and the yeast phase of *Histoplasma farciminosum* (San-Blas and Carbonell, 1974). The occurrence of α-1,3-glucan fibrils *in situ* has been reported for *H. farciminosum* and *H. capsulatum* (San-Blas and Carbonell, 1974), *Paracoccidioides brasiliensis* (Carbonell *et al.*, 1970), and *Aspergillus nidulans* (Zonneveld, 1978). There is a clear need for x-ray analyses that would give some indication of the configuration of α-1,3-glucan *in situ*, and additional studies on the occurrence of the glucan fibrils could be significant. What is particularly important in my view is the nature of the α-1,3-glucan found in the alkali-insoluble fraction of a number of species. According to Rees (1977), theoretical considerations place α-1,3-glucans in the "ribbon family," a category that includes cellulose, chitin, and other types of polysaccharides conformationally well suited for structural functions in cell walls. Polymers in this category having flat, ribbon conformations (cellulose, chitin, and so on) are considered to have the highest degree of ordered chain packing (Rees, 1977). These theoretical predictions are supported by the occurrence of S-glucan fibrils. In the case of *Aspergillus nidulans*, which as far as is known is exceptional, S-glucan serves as a storage product and is degraded during cleistothecial development of the teleomorph (Zonneveld, 1978). While this is no doubt an important function of the S-glucan in this species, it may tend to obscure the possible role of α-1,3-glucan as a structural polymer in *A. nidulans* walls. Fluctuations in amounts of wall constituents that are correlated with cleistothecial formation occur primarily in the alkali-soluble wall fraction and, when virtually all of the soluble glucan has been utilized, most of the insoluble cell wall fraction remains (Zonneveld, 1973). As noted previously, Bull (1970a) found evidence of alkali-insoluble α-1,3-glucan in *A. nidulans* walls. Thus α-1,3-glucan(s) in *A. nidulans* probably have a dual function.

4. Heteroglycans and Proteoglycans

Hyphal walls generally contain an assortment of heteropolymers occurring in relatively small quantities as compared with chitin and glucans. Some of these components are linked to peptide residues (e.g., Gander, 1974), but very little is known of the covalent structures involved. There have been some earlier investigations of wall peptides in the wild-type and mutant strains of *Neurospora crassa* (Mishra, 1977; Gander, 1974). Five peptide fractions have been obtained from walls of *N. crassa*. They contain a broad assortment of amino acids and are believed to be linked to glycan residues through O-glycosylserine bonds. A number of immunologically active polymers suggested to be peptidogalactomannans (Gander, 1974) and presumed to be wall components in some cases, have been isolated from various dermatophytes, *Aspergillus* spp., and *Penicillium* spp. A phosphogalactomannan from *Penicillium charlesii*, which is one of the best characterized of the numerous heteropolymers (Gander, 1974), occurs extracellularly and not as a cell wall component (Gander and Fang, 1976) and demonstrates the need for caution in deducing the cellular location of these fungal products.

A number of recent investigations have employed isolated cell walls and reveal a remarkable complexity in cell wall heteropolymers. Walls of *Penicillium charlesii* contain an alkali-soluble fraction composed primarily of galactose and glucose along with small amounts of amino acid, phosphate, uronic acid, and ethanolamine residues (Gander and Fang, 1976). A water-soluble polymer representing about 3% of the wall of *Cladosporium trichoides* has been investigated by Miyazaki and Naoi (1976). The polysaccharide is composed of D-galactofuranosyl, D-mannopyranosyl, and D-glucopyranosyl units in a ratio of 1.4:1.4:1.0, along with a trace of L-rhamnose. This polysaccharide is believed to have a branched structure. The main chains consist of galactose and mannose units, each of which carries a short branch chain containing glucose, galactose, or mannose residues. The polymer is associated with protein prior to purification, but the nature of the association was not investigated. Two polysaccharides from walls of *Aspergillus niger* have been recently characterized (Bardalaye and Nordin, 1976, 1977). One is a galactosaminogalactan accounting for about 0.6–0.8% of wall dry weight. This polymer contains approximately 70% D-galactose and 20% D-galactosamine joined by α-1,4-linkages. The polysaccharide is unbranched, has an average DP of 100, and contains a small amount (ca. 6%) of 1,3-linked-D-glucosyl residues. The galactosamine units, which are distributed at random along the chains, are about 20% N-acetylated (Bardalaye and Nordin, 1976). The other *A. niger* polysaccharide, a galactomannan representing about 6% of wall weight (Bardalaye and Nordin, 1977), contains D-galactose and D-mannose in approximately equal amounts along with 12–14% D-glucose. The proposed structure of this wall constituent consists of short chains (two to five residues) of α-1,2-linked mannopyranosyl residues joined together by α-1,6-linkages. Short chains (three to four residues) consisting of α-1,4-linked galactopyranosyl units terminated by nonreducing β-D-galactofuranosyl residues are joined to the mannose chains by 1,2-linkages. The role of glucose in the structure is unknown, and it may be present as a contaminant (Bardalaye and Nordin, 1977).

About 10% of the wall of *Pyricularia oryzae* can be solubilized to yield a proteoheteroglycan fraction (Nakajima *et al.*, 1970). A purified portion of this fraction has been investigated in some detail (Nakajima *et al.*, 1977a). The purified polymer consists of 91% carbohydrate, 9% protein, and a negligible amount of phosphorus. The polysaccharide component contains D-mannose, D-glucose, and D-galactose in a 6:2:1 ratio. There is also a small quantity of D-glucosamine (ca. 0.2%) occurring near the polysaccharide–protein attachment point(s), but the nature of its linkages is apparently unknown. But evidence was obtained for a carbohydrate–protein linkage through O-mannosyl serine or O-mannosyl threonine bonds (Nakajima *et al.*, 1977a). The polysaccharide component consists of main chains of α-1,6-D-mannopyranosyl residues and side chains composed of one to four α-D-mannopyranosyl units joined mainly by 1,2-linkages. Some of the side chains are terminated by D-glucopyranosyl or D-galactofuranosyl residues (Nakajima *et al.*, 1977a,b).

Recent investigations of *Sporothrix schenckii* indicate the occurrence of a neutral rhamnomannan and an acidic rhamnomannan containing D-glucuronic acid (Travassos *et al.*, 1978). Similar rhamnomannans occurring as peptidoglycan complexes appear to be localized primarily on the outer surface of *S. schenckii* walls (Travassos *et al.*, 1977).

D. Biosynthesis of Wall Polymers

Fungi growing on a single-carbon source such as D-glucose can synthesize other monosaccharides required for cell wall synthesis, and the formation of the various monomeric constituents of wall polysaccharides is the first stage of wall formation. Insofar as the neutral hexoses are concerned, they are all interconvertible with D-glucose through well-known biochemical reactions (McCullough *et al.*, 1978; Gander, 1976). The nucleoside diphosphate sugars UDP-glucose, UDP-galactose, and GDP-mannose are considered to be the glycosyl donors in the synthesis of wall polymers containing the corresponding sugar residue. The chitin precursor arises by conversion of glucose 6-phosphate to fructose 6-phosphate, which receives an amino group from glutamine to yield glucosamine 6-phosphate. The latter is N-acetylated and converted to N-acetylglucosamine 1-phosphate which reacts with UTP to yield UDP-N-acetylglucosamine (Gooday, 1978). The origin of sugars occurring less frequently in cell walls or in relatively small quantities has not received a great deal of attention. Among these are glucuronic acid and the pentoses xylose and arabinose. It is generally implied that pentoses arise from glucose 6-phosphate via selected reactions of the pentose phosphate glycolytic pathway (e.g., McCullough *et al.*, 1978). The origin of glucuronic acid is generally ignored, which is not surprising since it has only recently been recognized as a widely occurring cell wall constituent. It is of interest to note that plants, which form cell walls that are chemically and architecturally similar to those of fungi, synthesize glucuronate and pentoses via either of two "pathways" as shown in reaction schemes (1), (2), and (3) (Preiss and Kosuge, 1976).

$$\text{Glucose 6-P} \xrightarrow{+\text{UTP}} \text{UDP-glucose} \xrightarrow{+2\text{NAD}} \text{UDP-glucuronate} \tag{1}$$

$$\text{Glucose-6-P} \rightarrow \rightarrow \text{Myoinositol} \rightarrow \text{Glucuronate} \tag{2}$$

$$\text{UDP-glucuronate} \rightarrow \text{UDP-xylose} \leftrightarrows \text{UDP-arabinose} \tag{3}$$

Any difference in biosynthetic routes is found in the mechanism of glucuronate formation [reactions (1) and (2)]. In either case, the pentoses are derived from UDP-glucuronate [reaction (3)]. It may be of significance that some of the early investigations contributing to elucidation of the inositol pathway in plants [reaction (2)] involved the fungi *Candida utilis, Neurospora crassa,* and *Schwanniomyces occidentalis* (e.g., Gander, 1976), all of which are capable of converting glucose 6-phosphate to myoinositol. The yeast *Schwanniomyces occidentalis* contains the enzyme, myoinositol oxygenase, responsible for the conversion of myoinositol to glucuronate [reaction (2)] (Thonet and Hoffman-Ostenhof, 1966). It may well be that the small amounts of pentoses required for fungal wall synthesis have their origin in the pentose phosphate pathway, but this, as well as the biosynthesis of glucuronate, seems to be in need of clarification.

The formation of nonchitin wall polymers from nucleoside diphosphate glycosyl precursors has not been extensively investigated in fungi generally and in conidial fungi in particular. The following remarks are derived from Gooday's review (1978) which should be consulted for additional details. A β-1,3-glucan and an α-1,4-glucan synthetase from *Neurospora crassa* both use UDP-glucose as the glucosyl donor, as do β-glucan synthetases from *Phytophthora cinnamomi* and *Saccharomyces cerevisiae*. Mannan and mannan–peptidoglycan synthesis appears to involve the transfer of mannosyl residues from GDP-mannose to polyprenol monophosphate mannose. The lipid carrier intermediate then serves as glycosyl donor for polysaccharide synthesis.

Chitin synthesis in fungi has been investigated more extensively, and there is considerably more known concerning the properties of the enzyme, chitin synthetase, which catalyzes the transfer of *N*-acetylglucosamine residues from UDP-*N*-acetylglucosamine to an acceptor (Gooday, 1978) as shown in reaction (4).

$$\text{UDP-GlcNAc} + (\text{GlcNAc})_n \rightarrow (\text{GlcNAc})_{n+1} + \text{UDP} \tag{4}$$

Gooday (1978) has reviewed earlier work on chitin synthetases from a wide range of fungi and compared their kinetic and allosteric properties which, for the most part, are similar. Recent investigations of this enzyme in conidial fungi include studies on *Aspergillus flavus* (Moore and Peberdy, 1976; López-Romero and Ruiz-Herrera, 1976), *A. nidulans* (Ryder and Peberdy, 1977), *A. fumigatus* (Archer, 1977), and *Neurospora crassa* (Bartnicki-Garcia *et al.*, 1978).

Cabib and Farkas (1971) have found that chitin synthetase from *Saccharomyces carlsbergensis* and *S. cerevisiae* occurs in a zymogen (inactive) state which is converted to the active enzyme by an activating factor (presumed to be a protease) from the cells or by treatment with trypsin. The presence of zymogen, activating factor, and an endogenous inhibitor of the activating factor provides a basis for spatial and temporal control of chitin synthesis during septum formation in dividing yeasts (Cabib and Farkas, 1971). A zymogenic form of chitin synthetase occurs in *Aspergillus flavus* (López-Romero and Ruiz-Herrara, 1976), *A. fumigatus* (Archer, 1977), *A. nidulans* (Ryder and Peberdy, 1977), *Neurospora crassa* (Bartnicki-Garcia *et al.*, 1978), and other fungi of diverse taxonomic distribution (Bartnicki-Garcia *et al.*, 1978). Moore and Peberdy (1976) were unable to detect a zymogenic form (active enzyme is present) in *A. flavus*, possibly because of endogenous protease activity released during cell breakage (Bartnicki-Garcia *et al.*, 1978).

The demonstration of enzyme activity capable of catalyzing the synthesis of an insoluble polymer of *N*-acetylglucosamine (chitin) is the first step toward elucidating the mechanism of chitin formation. However, it is widely recognized (e.g., Aronson, 1965; Bartnicki-Garcia *et al.*, 1979) that the synthesis of chitin, in the form in which it naturally occurs in cell walls, involves two concomitant events (processes): (1) biochemical formation of the chitin molecules from *N*-acetylglucosamine monomers, and (2) assembly of the chitin chains into microfibrils. However, Bartnicki-Garcia and co-workers demonstrated the *in vitro* formation of chitin microfibrils (Fig. 10) in a chitin synthetase preparation from *Mucor rouxii* (Ruiz-Herrera and Bartnicki-Garcia, 1974). Subsequently it was shown that small 35- to 100-nm granules called chitosomes (Fig. 11) functioned as "centers" for microfibril formation (Ruiz-Herrera *et al.*, 1975; Bracker *et al.*, 1976), and recently the occurrence of chitosomes in fungi representing major taxa (but not deuteromycetes) was demonstrated (Bartnicki-Garcia *et al.*, 1978). At present, the only species in the anamorph–teleomorph complex known to contain chitosomes is *Neurospora crassa*, but it is believed that they occur in all chitinous fungi (Bartnicki-Garcia *et al.*, 1978). Chitosomal chitin synthetase is zymogenic, and the active enzyme has properties similar to those of the crude enzyme investigated in a number of species (Ruiz-Herrera *et al.*, 1977; Bartnicki-Garcia *et al.*, 1978). Conclusive demonstration of chitosomes *in vivo* has not been accomplished, but multivesicular bodies, seen in thin-sectioned yeast cells of *Mucor rouxii*, have contents resembling thin-sectioned, isolated chitosomes (Bracker *et al.*, 1976). Similar multivesicular structures have been observed in other fungi also (Bartnicki-Garcia *et al.*, 1978). Bracker *et al.* (1976) has suggested that chitosomes are vehicles serving to transport chitin synthetase zymogen to the cell surface. Their actual site of action in filamentous

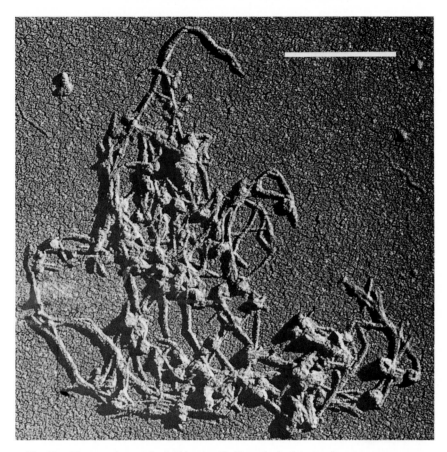

Fig. 10.   Electron micrograph of chitin microfibrils synthesized *in vitro* from an enzyme preparation from *Mucor rouxii*. Compare with "natural" fibrils in Fig. 9. Note the association between the fibrils and granules (chitosomes). Scale line, 0.5 μm. (From Ruiz-Herrera *et al.*, 1975.)

fungi remains to be determined, but in *Saccharomyces cerevisiae* chitin synthetase zymogen is found in the plasma membrane (Duran *et al.*, 1975). It appears significant that a membrane fraction (presumably apical plasma membrane) from protoplasts released from hyphal tips of *Aspergillus fumigatus* has 27 times more active chitin synthetase than preparations from hyphae (Archer, 1977). Since trypsin treatment causes a fourfold increase in chitin synthetase activity in the hyphal preparation, but is without effect on the protoplast membrane preparation (active enzyme rather than zymogen is present), Archer (1977) has suggested that active chitin synthetase in *A. fumigatus* is located in the region of hyphal tips and that an inactive (zymogenic) form of the enzyme occurs in the subapical (tubular) regions of the hyphae. It has been shown (Katz and Rosenberger, 1971a;

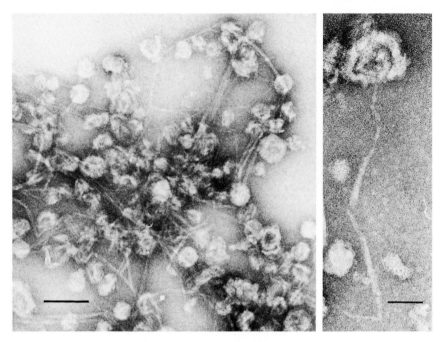

Fig. 11. Electron micrograph of negatively stained *Mucor rouxii* chitosomes and a network of chitin microfibrils synthesized *in vitro* (left). A single chitosome (right) at higher magnification appears to have a shell surrounding the internal contents. A microfibril appears to emanate from the chitosome interior. Left scale line, 100 nm. Right scale line, 40 nm. (From Bracker *et al.*, 1976.)

Sternlicht *et al.*, 1973) that disruption of normal tip growth in *A. nidulans* by osmotic shock or cycloheximide leads to chitin deposition in subapical regions and a noticeable increase in wall thickness. Thus the potential for chitin synthesis occurs over a considerable area. Spatially restricted (apical) synthesis of chitin may depend on controlled delivery, perhaps via apical vesicles, of an activating protease. If it is assumed that chitin synthetase zymogen is indeed located in the plasma membrane, is it possible that zymogen, presumed to be present in subapical regions, "escaped" activation when located in the apical dome region? Alternatively, are there different types of enzyme distinguishable by their location (apical versus subapical)? These are among the numerous questions that remain to be answered.

It should be mentioned that all investigators do not accept the hypothesized *in vivo* function of chitosomes (Bartnicki-Garcia *et al.*, 1979) in chitin microfibril formation. Cabib *et al.* (1979) suggest that fibrils may be products of spontaneous self-assembly once chitin chains (molecules) are produced by chitin synthetase activity. Farkas (1979) believes that chitosomes are too large to be integral structures in the plasma membrane and has suggested that they are artifacts. However, recent studies of freeze-fractured plasma membranes of vascular plants (Mueller and Brown, 1980) and green algae (Giddings *et al.,* 1980) have provided evidence of an intimate association between cellulose microfibril formation and intramembrane particle aggregates. The aggregates consist of hexagonally arranged particles of about 8 nm in diameter and are also found in cytoplasmic vesicles that have been observed to fuse with the plasma membrane. The dimensions of the hexagonal aggregates (Mueller and Brown, 1980; Giddings *et al.,* 1980) are of the same order as those of chitosomes and the latter have been dissociated into enzymically active 16 S particles measuring 7–12 nm (Ruiz-Herrera *et al.,* 1980). In terms of size, these 16 S particles seem similar to the 8-nm particles found in the hexagonal aggregates of plants. Although there is no established relationship between chitosomes and the plasma membrane particles of plants, the *total* body of evidence seems to weigh in favor of concepts involving subcellular particles in microfibril genesis.

E. Architecture

Hyphal wall architecture in anamorphic fungi has been the most neglected of all aspects of wall structure. Among the filamentous members of the anamorph-teleomorph assemblage, only *Neurospora crassa* has been extensively investigated. Hunsley and Burnett (1970) made electron microscope observations of hyphae of *N. crassa* that had been sequentially treated with polysaccharide and protein-hydrolyzing enzymes and concluded that the wall contained a coaxial distribution of polymers. From the original investigation by Hunsley and Burnett (1970), subsequent work by Hunsley and Kay (1976), and a recent discussion by Burnett (1976), it appears that the mature *N. crassa* wall is thought to be made up of four intergrading regions. The outermost region contains α- and β-glucans. The glucans extend into the next region and serve as a matrix for a glycoprotein reticulum. The third layer consists mainly of protein, and the innermost wall layer contains chitin microfibrils intermixed with protein. The wall at the hyphal apex is thought to consist of chitin microfibrils coated externally by an amorphous layer that might be proteinaceous (Hunsley and Kay, 1976). Behind the apex, the wall gradually thickens as additional wall layers are added externally. Some workers (e.g., Trinci, 1978) refer to this process as "secondary wall" formation.

The conclusion that chitin microfibrils occur primarily in the inner region is in accord with numerous observations on fungal walls generally (e.g., Bartnicki-Garcia, 1973). The distribution of protein in *N. crassa* walls is more problematic. Hunsley and Burnett (1970) applied similar methods in studies on wall structure in *Schizophyllum commune* and concluded, in this case also, that the interior of the wall consisted of a chitin–protein region beneath a protein-rich layer. Van der Valk *et al.* (1977) reexamined wall polymer distribution using isolated walls of *S. commune* and were unable to detect a chitin–protein complex or an adjacent discrete layer of protein. The latter investigators found an inner wall layer of chitin microfibrils in a matrix of *R*-glucan, a heteropolymer (peptidoglycan) consisting of β-1,3- and β-1,6-linked glucan (the major component), along with smaller amounts of glucosamine and amino acids. A layer of *S*-glucan occurs outside the chitin–β-glucan zone, and the two regions may intergrade to some extent. The outermost region in the *Schizophyllum* wall consists of a β-glucan mucilage (chemically distinct from *R*-glucan). As noted previously (Section III,B), Wang and Bartnicki-Garcia (1970) found certain amino acids firmly bound to the chitin core in *Verticillium albo-atrum,* but these workers had removed most of the β-glucan from this fraction by converting it to the alkali-soluble hydroglucan. It should be noted also that an earlier investigation of *N. crassa* (Mahadevan and Tatum, 1967) provided evidence of a close association between chitin microfibrils and β-glucan. While there is no basis for assuming a close similarity between *Neurospora* and *Schizophyllum,* there is, however, some uncertainty concerning the former, and it would be wise to exercise caution in attempts to compare wall architecture in conidial fungi with that of the *Neurospora* model. Nevertheless, the Hunsley and Burnett approach to investigations of wall architecture, involving enzymatic and chemical dissection coupled with electron microscopy, has general applicability (see Section II,C). One prerequisite for the use of enzymes in such studies is knowledge of any contaminating enzyme activities that could influence interpretations. Some commercially available enzyme preparations must be carefully screened for contaminating hydrolases (Davis and Domer, 1977).

There is evidence that, in some conidial fungi, α-1,3-glucan occurs in a microfibrillar form. In the yeast-phase cell walls of *Paracoccidioides brasiliensis,* the glucan fibrils occur in the outer region of the wall, while chitin microfibrils occupy the inner layer (Carbonell *et al.,* 1970). A similar distribution of the two types of fibrils occurs in *Histoplasma farciminosum* and *H. capsulatum* (e.g., San-Blas and Carbonell, 1974).

The outermost layer (exolayer) of the walls of *Epidermophyton floccosum,* separated from other wall layers by mechanical means, contains 63% protein, 17% glucosamine, and 11% neutral polysaccharide (Kitajima and Nozawa, 1975). Most of the neutral carbohydrate is mannose, and the separated layer contains no lipid. Walls of this fungus contain an internal layer of chitin microfibrils (Nozawa *et al.,* 1973), but additional features of wall architecture have not been elucidated. The amino acid composition of the exolayer shows a preponderance of aspartic acid, glutamic acid, glycine, alanine, and the hydroxyamino acids serine and threonine, but the nature of protein–polysaccharide linkages is unknown. Analysis of the exolayer composition by gel electrophoresis shows five major components ranging in molecular weight from 4.2 $\times$ 10^4 to 1.5 $\times$ 10^5. The heaviest component is a glycoprotein, and the other four fractions consist only of protein (Kitajima and Nozawa, 1975). Isolated exolayers retain the tubular shape of the hyphae, and electron microscopy shows a reticulate structure reminiscent of the previously mentioned glycoprotein reticulum of *Neurospora crassa* walls. But the *N. crassa* reticulum is believed to be deeply embedded in a glucan matrix (Hunsley and Burnett, 1970), while the *E. floccosum* layer is unquestionably located on the wall surface. In terms of gross composition and resistance to hydrolytic enzymes, the *E. floccosum* exolayer is similar to the rodlet layer found in microconidial walls of *Trichophyton mentagrophytes* (see Section II,C and Table II) but differs in its high glucosamine content and solubility in an sodium dodecyl sulfate–buffer solution (Kitajima and Nozawa, 1975).

Whether or not *E. floccosum* or any other conidial fungus has a proteinaceous rodlet layer in hyphal walls is unknown. Although it was noted previously (Section II,B) that rodlets have been demonstrated in some hyphae, definite conclusions must be deferred until such structures have been chemically characterized.

Little can be said concerning the architecture of septa. In *Neurospora crassa* septa consist of two chitin microfibril layers separated by a central region of unknown composition. A layer of amorphous protein overlying the microfibrils may be present. The microfibrillar layers of the septa are continuous with the chitin inner layer of the lateral hyphal wall, but the amorphous protein coating terminates at the septum–lateral wall junction (Hunsley and Gooday, 1974). Ultrastructural investigations have demonstrated a variety of septal types in conidial fungi (Gull, 1978), and it is possible that other architectural forms exist.

IV. CELL WALLS AND PHYTOPATHOLOGY

Albersheim and Anderson-Prouty (1975) have recently drawn attention to some host–parasite relationships in which host resistance (or susceptibility) to infection may be related to the cell wall structure of certain pathogenic fungi. *Phaseolus vulgaris* (French bean) produces a β-1,3-glucanase capable of lysing the cell walls of *Colletotrichum lindemuthianum*. The fungus produces a protein inhibitor of the glucanase (Albersheim and Valent, 1974). It has been suggested that an excess of enzyme over inhibitor might reduce fungal virulence, while an excess of inhibitor would neutralize the host's defense.

A great many vascular plants produce phytoalexins, which are chemical compounds capable of inhibiting the growth of fungi. Phytoalexins occur in trace amounts in healthy tissues but will accumulate as a result of fungal infection or treatments with fungal metabolites, viruses, bacteria, or a number of chemical substances, as well as mechanical or freezing injury (Kuć and Shain, 1977; Anderson-Prouty and Albersheim, 1975). Dependent on the time and magnitude of phytoalexin accumulation in infected plant tissues, these substances appear to be capable of providing host plants with a mechanism for disease resistance. Among the numerous chemical substances capable of eliciting phytoalexin accumulation are certain polysaccharides found in culture filtrates and cell walls of pathogenic fungi. One such elicitor is a high-molecular-weight polysaccharide containing 1,3- and 1,4-linked glucosyl residues (Anderson-Prouty and Albersheim, 1975). The polymer, whether derived from culture filtrates or cell walls of *Colletotrichum lindemuthianum,* has the same properties. Less than 100 ng of this glucan, applied to cut surfaces of bean cotyledons and hypocotyls, elicits a response similar to that seen when the host is attacked by an incompatible race of the pathogen (Anderson-Prouty and Albersheim, 1975). Glucose-containing polysaccharides with similar elicitor properties have been isolated from other *Colletotrichum* species as well as from cell walls of *Phytophthora megasperma,* and it has been suggested that these observations may be indicative of a general phenomenon whereby plants respond to different fungi by recognition of similar cell wall polymers (Cline *et al.,* 1978).

A number of antifungal agents are known to act through the disruption of normal cell wall formation. Among these are the polyoxins, a group of pyrimidine nucleoside antibiotics widely used in Japan for the control of diseases caused by *Alternaria kikuchiana, Cochliobolus miyabeanus,* and *Pyricularia oryzae* (Misato and Kakiki, 1977; Gooday, 1978). The mode of action of polyoxins involves competitive inhibition of chitin synthetase and results in swelling or bursting of hyphal apexes. Chitin synthetases from polyoxin-resistant strains of *A. kikuchiana* are competitively inhibited when tested *in vitro,* indicating that resistance to the antibiotic is due to some restriction preventing entry of the substance into the cells (Misato and Kakiki, 1977; Gooday, 1978). Nikkomycin, another nucleoside antibiotic with a chemical structure similar to that of polyoxin, inhibits the growth of a number of chitin-producing fungi (Dähn *et al.,* 1976). Kitazin P and some related compounds are organophosphorus fungicides useful in control of the rice blast fungus *Pyricularia oryzae* (Misato and Kakiki, 1977). Kitazin P also is an inhibitor of chitin synthetase (Misato and Kakiki, 1977). Wheat germ agglutininin (WGA), one of several plant lectins, binds strongly to N-acetyl-β-D-glucosamine units as found in chitin. The WGA inhibition of tip growth in *Trichoderma viride* has fostered the speculation that lectins may serve to ward off fungal pathogens (e.g., Gooday, 1978).

Some consideration has been given to the possibility that fungal walls can function in modifying the toxicity of fungicides. Although cell walls are known to bind some antifungal agents, they seem to play a passive role. In some cases, fungicide toxicity occurs after cell wall binding sites have been saturated, while in other instances it is believed that the wall may act as a barrier preventing entry of the fungicide into the cells (Richmond, 1977).

ACKNOWLEDGMENTS

For providing information on their work prior to its publication or for contributing original figures appearing in this chapter, the author thanks S. Bartnicki-Garcia, T. F. Bobbitt, C. E. Bracker, J. H. Burnett, G. T. Cole, R. Dargent, M. R. Edwards, M. Fukutomi, T. Hashimoto, and J. H. Nordin. The statement from Bartnicki-Garcia (1968), quoted in Section I, is reproduced, with permission, from the *Annual Review of Microbiology,* Volume 22, copyright 1968 by Annual Reviews Inc.

REFERENCES

Aist, J. R., Aylor, D. E., and Parlange, J. Y. (1976). Ultrastructure and mechanics of the conidium conidiophore attachment of *Helminthosporium maydis*. *Phytopathology* **66,** 1050–1055.

Akai, S., Fukutomi, M., Kunoh, H., and Shiraishi, M. (1976). Fine structure of the spore wall and germ tube change during germination. *In* "The Fungal Spore: Form and Function" (D. J. Weber and W. M. Hess, eds.), pp. 355–410. Wiley, New York.

Albersheim, P., and Anderson-Prouty, A. J. (1975). Carbohydrates, proteins, cell surfaces, and the biochemistry of pathogenesis. *Annu. Rev. Plant Physiol.* **26,** 31–52.

Albersheim, P., and Valent, B. S. (1974). Host-pathogen interactions. VII. Plant pathogens secrete proteins which inhibit enzymes of the host capable of attacking the pathogen. *Plant Physiol.* **53,** 684–687.

Al-Rikabi, K. H., and Bonaly, R. (1977). Composition chimique et degradation des parois d'une souche sauvage et d'un mutant blanc du champignon *Drechslera sorokiniana*. *Can. J. Microbiol.* **23,** 1508–1517.

Anderson-Prouty, A. J., and Albersheim, P. (1975). Host-pathogen interactions. VIII. Isolation of a pathogen-synthesized fraction rich in glucan that elicits a defense response in the pathogen's host. *Plant Physiol.* **56,** 286–291.

Applegarth, D. A., and Bozoian, G. (1969). The cell wall of *Helminthosporium sativum*. *Arch. Biochem. Biophys.* **134,** 285–289.

Archer, D. B. (1977). Chitin biosynthesis in protoplasts and subcellular fractions of *Aspergillus fumigatus*. *Biochem. J.* **164,** 653–658.

Aronson, J. M. (1965). The cell wall. *In* "The Fungi: An Advanced Treatise" (G. C. Ainsworth and A. S. Sussman, eds.), Vol. 1, pp. 49–76. Academic press, New York.

Bacon, J. S. D., Jones, D., Farmer, V. C., and Webley, D. M. (1968). The occurrence of $\alpha(1\text{-}3)$ glucan in *Cryptococcus, Schizosaccharomyces* and *Polyporus* species, and its hydrolysis by a *Streptomyces* culture filtrate lysing cell walls of *Cryptococcus*. *Biochim. Biophys. Acta* **158,** 313–315.

Bardalaye, P. C., and Nordin, J. H. (1976). Galactosamino-galactan from cell walls of *Aspergillus niger*. *J. Bacteriol.* **125,** 655–669.

Bardalaye, P. C., and Nordin, J. H. (1977). Chemical structure of the galactomannan from the cell wall of *Aspergillus niger*. *J. Biol. Chem.* **252,** 2584–2591.

Barran, L. R., Schneider, E. F., Wood, P. J., Madhosingh, C., and Miller, R. W. (1975). Cell wall of *Fusarium sulphureum*. I. Chemical composition of the hyphal wall. *Biochim. Biophys. Acta* **392,** 148–158.

Bartnicki-Garcia, S. (1968). Cell wall chemistry, morphogenesis and taxonomy of fungi. *Annu. Rev. Microbiol.* **22**, 87–108.

Bartnicki-Garcia, S. (1970). Cell wall composition and other biochemical markers in fungal phylogeny. *In* "Phytochemical Phylogeny" (J. B. Harborne, ed.), pp. 81–103. Academic Press, New York.

Bartnicki-Garcia, S. (1973). Fundamental aspects of hyphal morphogenesis. *Symp. Soc. Gen. Microbiol.* **23**, 245–267.

Bartnicki-Garcia, S., Bracker, C. E., Reyes, E., and Ruiz-Herrera, J. (1978). Isolation of chitosomes from taxonomically diverse fungi and synthesis of chitin microfibrils *in vitro*. *Exp. Mycol.* **2**, 173–192.

Bartnicki-Garcia, S., Ruiz-Herrera, J., and Bracker, C. E. (1979). Chitosomes and chitin synthesis. *In* "Fungal Walls and Hyphal Growth" (J. H. Burnett and A. P.J. Trinci, eds.), pp. 149–168. Cambridge Univ. Press, London and New York.

Beever, R. E., and Dempsey, G. P. (1978). Function of rodlets on the surface of fungal spores. *Nature (London)* **272**, 608–610.

Benítez, T., Villa, T. G., and Garcia Acha, I. (1976). Some chemical and structural features of the conidial wall of *Trichoderma viride*. *Can. J. Microbiol.* **22**, 318–321.

Bobbitt, T. F., and Nordin, J. H. (1978). Hyphal nigeran as a potential phylogenetic marker for *Aspergillus* and *Penicillium* species. *Mycologia* **70**, 1201–1211.

Bobbitt, T. F., Nordin, J. H., Roux, M., Revol, J. F., and Marchessault, R. H. (1977). Distribution and conformation of crystalline nigeran in hyphal walls of *Aspergillus niger* and *Aspergillus awamori*. *J. Bacteriol.* **132**, 691–703.

Border, D. J., and Trinci, A. P. J. (1970). Fine structure of the germination of *Aspergillus nidulans* conidia. *Trans. Br. Mycol. Soc.* **54**, 143–152.

Bracker, C. E., Ruiz-Herrera, J., and Bartnicki-Garcia, S. (1976). Structure and transformation of chitin synthetase particles (chitosomes) during microfibril synthesis *in vitro*. *Proc. Natl. Acad. Sci. U.S.A.* **73**, 4570–4574.

Buck, K. W., and Obaidah, M. A. (1971). The composition of the cell wall of *Fusicoccum amygdali*. *Biochem. J.* **125**, 461–471.

Bull, A. T. (1970a). Chemical composition of wild-type and mutant *Aspergillus nidulans* cell walls: The nature of polysaccharide and melanin constituents. *J. Gen. Microbiol.* **63**, 75–94.

Bull, A. T. (1970b). Inhibition of polysaccharases by melanin: Enzyme inhibition in relation to mycolysis. *Arch. Biochem. Biophys.* **137**, 345–356.

Bulman, R. A., and Chittenden, G. J. F. (1976). The mycelial cell wall of *Penicillium charlesii* G. Smith NRRL 1887. *Biochim. Biophys. Acta* **444**, 202–211.

Burnett, J. H. (1976). "Fundamentals of Mycology," 2nd ed. Arnold, London.

Burnett, J. H., and Trinci, A. P. J., eds. (1979). "Fungal Walls and Hyphal Growth." Cambridge, Univ. Press, London and New York.

Cabib, E., and Farkas, V. (1971). The control of morphogenesis: An enzymatic mechanism for the initiation of septum formation in yeast. *Proc. Natl. Acad. Sci. U.S.A.* **68**, 2052–2056.

Cabib, E., Duran, A., and Bowers, B. (1979). Localized activation of chitin synthetase in the initiation of yeast septum formation. *In* "Fungal Walls and Hyphal Growth" (J. H. Burnett and A. P. J. Trinci, eds.), pp. 189–201. Cambridge Univ. Press, London and New York.

Caputo, R., Innocenti, M., and Shimono, M. (1977). Cell wall and plasma membrane of dermatophytes: A freeze-fracture study. *J. Ultrastruct. Res.* **59**, 149–157.

Carbonell, L. M., Kanetsuna, F., and Gil, F. (1970). Chemical morphology of glucan and chitin in the cell wall of the yeast phase of *Paracoccidioides brasiliensis*. *J. Bacteriol.* **101**, 636–642.

Cline, K., Wade, M., and Albersheim, P. (1978). Host-pathogen interactions. XV. Fungal glucans which elicit phytoalexin accumulation in soybean also elicit the accumulation of phytoalexins in other plants. *Plant Physiol.* **62**, 918–921.

Cole, G. T. (1973a). Ultrastructure of conidiogenesis in *Drechslera sorokiniana*. *Can. J. Bot.* **51**, 629–638.

Cole, G. T. (1973b). A correlation between rodlet orientation and conidiogenesis in Hyphomycetes. *Can. J. Bot.* **51**, 2413–2422.

Cole, G. T., Sekiya, T., Kasai, R., Yokoyama, T., and Nozawa, Y. (1979). Surface ultrastructure and chemical composition of the cell walls of conidial fungi. *Exp. Mycol.* **3**, 78–102.

Collinge, A. J., and Trinci, A. P. J. (1974). Hyphal tips of wild-type and spreading colonial mutants of *Neurospora crassa*. *Arch. Microbiol.* **99**, 353–368.

Dähn, U., Hagenmaier, H., Höhne, H., König, W. A., Wolf, G., and Zähner, H. (1976). Stoffwechselprodukte von Mikroorganismen. 154. Mitteilung: Nikkomycin, ein neuer Hemmstoff der Chitinsynthese bei Pilzen. *Arch. Microbiol.* **107**, 143–160.

Dargent, R., and Touzé-Soulet, J. M. (1976). Sur l'ultrastructure des hyphes d'*Hypomyces chlorinus* Tul. cultive en presence ou en absence de biotine. *Protoplasma* 89, 49–71.

Davis, T. E., and Domer, J. E. (1977). Glycohydrolase contamination of commercial enzymes frequently used in the preparation of fungal cell walls. *Anal. Biochem.* **80**, 593–600.

Dempsey, G. P., and Beever, R. E. (1979). Electron microscopy of the rodlet layer of *Neurospora crassa* conidia. *J. Bacteriol.* **140**, 1050–1062.

Dill, B. C., Leighton, T. J., and Stock, J. J. (1972). Physiological and biochemical changes associated with macroconidial germination in *Microsporum gypseum*. *Appl. Microbiol.* **24**, 977–985.

Durán, A., Bowers, B., and Cabib, E. (1975). Chitin synthetase zymogen is attached to the yeast plasma membrane. *Proc. Natl. Acad. Sci. U.S.A.* **72**, 3952–3955.

Ellis, D. H., and Griffiths, D. A. (1974). The location and analysis of melanins in the cell walls of some soil fungi. *Can. J. Microbiol.* **10**, 1379–1386.

Ellis, D. H., and Griffiths, D. A. (1975). Melanin deposition in the hyphae of a species of *Phomopsis*. *Can. J. Microbiol.* **21**, 442–452.

Farkas, V. (1979). Biosynthesis of cell walls of fungi. *Microbiol. Rev.* **43**, 117–144.

Fisher, D. J., and Richmond, D. V. (1969). The electrokinetic properties of some fungal spores. *J. Gen. Microbiol.* **57**, 51–60.

Fisher, D. J., and Richmond, D. V. (1970). The electrophoretic properties and some surface components of *Penicillium* conidia. *J. Gen. Microbiol.* **64**, 205–214.

Fisher, D. J., Holloway, P. J., and Richmond, D. V. (1972). Fatty acid and hydrocarbon constituents of the surface and wall lipids of some fungal spores. *J. Gen. Microbiol.* **72**, 71–78.

Florance, E. R., Denison, W. C., and Allen, T. C., Jr. (1972). Ultrastructure of dormant and germinating conidia of *Aspergillus nidulans*. *Mycologia* **64**, 115–123.

Gander, J. E. (1974). Fungal cell wall glycoproteins and peptidopolysaccharides. *Annu. Rev. Microbiol.* **28**, 103–119.

Gander, J. E. (1976). Mono- and oligosaccharides. *In* "Plant Biochemistry" (J. Bonner and J. E. Varner, eds.), 3rd ed., pp. 337–380. Academic Press, New York.

Gander, J. E., and Fang, F. (1976). The occurrence of ethanolamine and galactofuranosyl residues attached to *Penicillium charlesii* cell wall saccharides. *Biochem. Biophys. Res. Commun.* **71**, 719–725.

Ghiorse, W. C., and Edwards, M. R. (1973). Ultrastructure of *Aspergillus fumigatus* conidia development and maturation. *Protoplasma* **76**, 49–59.

Giddings, T. H., Jr., Brower, D. L., and Staehelin, L. A. (1980). Visualization of particle complexes in the plasma membrane of *Micrasterias denticulata* associated with the formation of cellulose fibrils in primary and secondary walls. *J. Cell Biol.* **84**, 327–339.

Girbardt, M. (1969). Die Ultrastruktur der Apikalregion von Pilzhyphen. *Protoplasma* **67**, 413–441.

Gooday, G. W. (1971). An autoradiographic study of hyphal growth of some fungi. *J. Gen. Microbiol.* **67**, 125–133.

Gooday, G. W. (1978). The enzymology of hyphal growth. *In* "The Filamentous Fungi" (J. E. Smith and D. R. Berry, eds.), Vol. 3, pp. 51–77. Wiley, New York.

Gorin, P. A. J., and Spencer, J. F. T. (1968). Structural chemistry of fungal polysaccharides. *Adv. Carbohydr. Chem.* **23**, 367–417.

Gottlieb, D., and Tripathi, R. K. (1968). The physiology of swelling phase of spore germination in *Pencillium atrovenetum. Mycologia* **60**, 571–590.

Griffiths, D. A., and Swart, H. J. (1974). Conidial structure in *Pestalotia pezizoides. Trans. Br. Mycol. Soc.* **63**, 169–173.

Grisaro, V., Sharon, N., and Barkai-Golan, R. (1968). The chemical composition of the cell walls of *Penicillium digitatum* Sacc. and *Penicillium italicum* Whem. *J. Gen. Microbiol.* **51**, 145–150.

Grove, S. N. (1972). Apical vesicles in germinating conidia of *Aspergillus parasiticus. Mycologia* **64**, 638–641.

Grove, S. N. (1978). The cytology of hyphal tip growth. *In* "The Filamentous Fungi" (J. E. Smith and D. R. Berry, eds.), Vol. 3, pp. 28–50. Wiley, New York.

Grove, S. N., and Bracker, C. E. (1970). Protoplasmic organization of hyphal tips among fungi: Vesicles and *Spitzenkörper. J. Bacteriol.* **104**, 989–1009.

Grove, S. N., Bracker, C. E., and Morré, D. J. (1970). An ultrastructural basis for hyphal tip growth in *Pythium ultimum. Am. J. Bot.* **57**, 245–266.

Gull, K. (1978). Form and function of septa in filamentous fungi. *In* "The Filamentous Fungi" (J. E. Smith and D. R. Berry, eds.), Vol. 3, pp. 78–93. Wiley, New York.

Gull, K., and Trinci, A. P. J. (1971). Fine structure of spore germination in *Botrytis cinerea. J. Gen. Microbiol.* **68**, 207–220.

Hanlin, R. T. (1976). Phialide and conidium development in *Aspergillus clavatus. Am. J. Bot.* **63**, 144–155.

Harris, J. L., and Taber, W. A. (1973). Compositional studies of the cell walls of the synnema and vegetative hyphae of *Ceratocystis ulmi. Can. J. Bot.* **51**, 1147–1153.

Harvey, I. C. (1974). Light and electron microscope observations of conidiogenesis in *Pleiochaeta setosa. Protoplasma* **82**, 203–221.

Hashimoto, T., Wu-Yuan, C. D., and Blumenthal, H. J. (1976). Isolation and characterization of the rodlet layer of *Trichophyton mentagrophytes* microconidial wall. *J. Bacteriol.* **127**, 1543–1549.

Hawes, C. R. (1979). Conidium ultrastructure and wall architecture in the *Chalara* state of *Ceratocystis adiposa. Trans. Br. Mycol. Soc.* **72**, 177–187.

Hawes, C. R. (1980). Conidium germination in the *Chalara* state of *Ceratocystis adiposa:* A light and electron microscope study. *Trans. Br. Mycol. Soc.* **74**, 321–328.

Hawker, L. E., and Hendy, R. J. (1963). An electron microscope study of germination of conidia of *Botrytis cinerea. J. Gen. Microbiol.* **33**, 43–46.

Hawker, L. E., and Madelin, M. F. (1976). The dormant spore. *In* "The Fungal Spore: Form and Function" (D. J. Weber and W. M. Hess, eds.), pp. 1–70. Wiley, New York.

Hearn, V. M., and Mackenzie, D. W. R. (1979). The preparation and chemical composition of fractions from *Aspergillus fumigatus* wall and protoplasts possessing antigenic activity. *J. Gen. Microbiol.* **112**, 35–44.

Hess, W. M., and Stocks, D. L. (1969). Surface characteristics of *Aspergillus* conidia. *Mycologia* **61**, 560–571.

Hess, W. M., Sassen, M. M. A., and Remsen, C. C. (1966). Surface structure of frozen-etched *Penicillium* conidiospores. *Naturwissenschaften* **53**, 70.

Hess, W. M., Sassen, M. M. A., and Remsen, C. C. (1968). Surface characteristics of *Penicillium* conidia. *Mycologia* **60**, 290–303.

Horikoshi, K., and Iida, S. (1964). Studies of the spore coats of fungi. I. Isolation and composition of the spore coats of *Aspergillus oryzae. Biochim. Biophys. Acta* **83**, 197–203.

Horisberger, M., Lewis, B. A., and Smith, F. (1972). Structure of a (1→3)-α-D-glucan (pseudo-nigeran) of *Aspergillus niger* NNRL 326 cell wall. *Carbohydr. Res.* **23**, 183–188.

Houwink, A. L., and Kreger, D. R. (1953). Observations on the cell wall of yeasts: An electron microscope and X-ray diffraction study. *Antonie van Leeuwenhoek* **19**, 1–24.

Houwink, R., de Decker, H. K., and van den Tempel, M. (1971). Structure of matter in relation to its elastic and plastic behaviour and failure. *In* "Elasticity, Plasticity and Structure of Matter" (R. Houwink and H. K. de Decker, eds.), pp. 12–35. Cambridge Univ. Press, London and New York.

Hunsley, D., and Burnett, J. H. (1970). The ultrastructural architecture of the walls of some hyphal fungi. *J. Gen. Microbiol.* **62**, 203–218.

Hunsley, D., and Gooday, G. W. (1974). The structure and development of septa in *Neurospora crassa. Protoplasma* **82**, 125–146.

Hunsley, D., and Kay, D. (1976). Wall structure of the *Neurospora* hyphal apex: Immunofluorescent localization of wall surface antigens. *J. Gen. Microbiol.* **95**, 233–248.

Jelsma, J., and Kreger, D. R (1975). Ultrastructural observations on (1→3)-β-D-glucan from fungal cell walls. *Carbohydr. Res.* **43**, 200–203.

Jelsma, J., and Kreger, D. R. (1979). Polymorphism in crystalline (1→3)-α-D-glucan from fungal cell walls. *Carbohydr. Res.* **71**, 51–64.

Johnston, I. R. (1965a). The composition of the cell wall of *Aspergillus niger. Biochem. J.* **96**, 651–658.

Johnston, I. R. (1965b). The partial acid hydrolysis of a highly dextrorotatory fragment of the cell wall of *Aspergillus niger:* Isolation of the α-(1→3)-linked dextrin series. *Biochem. J.* **96**, 659–664.

Kanetsuna, F., and Carbonell, L. M. (1970). Cell wall glucans of the yeast and mycelial forms of *Paracoccidioides brasiliensis. J. Bacteriol.* **101**, 675–680.

Kanetsuna, F., and Carbonell, L. M. (1971). Cell wall composition of the yeastlike and mycelial forms of *Blastomyces dermatitidus. J. Bacteriol.* **106**, 946–948.

Katz, D., and Rosenberger, R. F. (1970). The utilisation of galactose by an *Aspergillus nidulans* mutant lacking galactose phosphate-UDP glucose transferase and its relation to cell wall synthesis. *Arch Mikrobiol.* **74**, 41–51.

Katz, D., and Rosenberger, R. F. (1971a). Hyphal wall synthesis in *Aspergillus nidulans:* Effect of protein synthesis inhibition and osmotic shock on chitin insertion and morphogenesis. *J. Bacteriol.* **108**, 184–190.

Katz, D., and Rosenberger, R. F. (1971b). Lysis of an *Aspergillus nidulans* mutant blocked in chitin synthesis and its relation to wall assembly and wall metabolism. *Arch. Mikrobiol.* **80**, 284–292.

Kitajima, Y., and Nozawa, Y. (1975). Isolation, ultrastructure and chemical composition of the outermost layer ("exo-layer") of *Epidermophyton floccosum* cell wall. *Biochim. Biophys. Acta* **394**, 558–568.

Kitazima, Y., Banno, Y., Noguchi, T., Nozawa, Y., and Ito, Y. (1972). Effects of chemical modification of structural polymer upon the cell wall integrity of *Trichophyton. Arch. Biochem. Biophys.* **152**, 811–820.

Kreger, D. R. (1954). Observations on cell walls of yeasts and some other fungi by X-ray diffraction and solubility tests. *Biochim. Biophys. Acta* **13**, 1–9.

Kuć, J., and Shain, L. (1977). Antifungal compounds associated with disease resistance in plants. *In* "Antifungal Compounds" (M. R. Siegel and H. D. Sisler, eds.), Vol. 2, pp. 497–535. Dekker, New York.

Kuo, M. J., and Alexander, M. (1967). Inhibition of the lysis of fungi by melanins. *J. Bacteriol.* **94**, 624–629.

Laborda, F., Garcia Acha, I., Uruburu, F., and Villanueva, J. R. (1974). Structure of conidial walls of *Fusarium culmorum*. *Trans. Br. Mycol. Soc.* **62**, 557–566.

Lehninger, A. L. (1975). "Biochemistry," 2nd ed. Worth, New York.

LéJohn, H. B. (1974). Biochemical parameters of fungal phylogenetics. *Evol. Biol.* **7**, 79–125.

López-Romero, E., and Ruiz-Herrera, J. (1976). Synthesis of chitin by particulate preparations from *Aspergillus flavus*. *Antonie van Leeuwenhoek* **42**, 261–276.

McCullough, W., Payton, M. A., and Roberts, C. F. (1978). Carbon metabolism in *Aspergillus nidulans*. *In* "Genetics and Physiology of *Aspergillus*" (J. E. Smith and J. A. Pateman, eds.), pp. 97–129. Academic Press, New York.

Madelin, M. F., and Ogunsanya, O. C. (1979). The fine structure of conidia of *Botryodiplodia ricinicola* with observations on effects of chilling. *Ann. Bot.* **44**, 417–425.

Mahadevan, P. R., and Tatum, E. L. (1967). Localization of structural polymers in the cell wall of *Neurospora crassa*. *J. Cell Biol.* **35**, 295–302.

Mangenot, F., and Reisinger, O. (1976). Form and function of conidia as related to their development. *In* "The Fungal Spore: Form and Function" (D. J. Weber and W. M. Hess, eds.), pp. 789–846. Wiley, New York.

Marchant, R. (1966). Wall structure and spore germination in *Fusarium culmorum*. *Ann. Bot. (London)* [N.S.] **30**, 821–830.

Marks, D. B., Keller, B. J., and Guarino, A. J. (1969). The composition of the cell wall fraction of the fungus, *Cordyceps militaris*. *Biochim. Biophys. Acta* **183**, 58–64.

Marks, D. B., Keller, B. J., and Guarino, A. J. (1971). Growth of unicellular forms of the fungus, *Cordyceps militaris* and analysis of the chemical composition of their walls. *J. Gen. Microbiol.* **69**, 253–259.

Martin, J. F., Nicolas, G., and Villanueva, J. R. (1973). Chemical changes in the cell walls of conidia of *Penicillium notatum* during germination. *Can. J. Microbiol.* **19**, 789–796.

Martin, J. F., Liras, P., and Villanueva, J. R. (1974). Changes in composition of conidia of *Penicillium notatum* during germination. *Arch. Microbiol.* **97**, 39–50.

Matile, P., Cortat, M., Wiemken, A., and Frey-Wyssling, A. (1971). *Proc. Natl. Acad. Sci. U.S.A.* **68**, 636–640.

Misato, T., and Kakiki, K. (1977). Inhibition of fungal cell wall synthesis and cell membrane function. *In* "Antifungal Compounds" (M. R. Siegel and H. D. Sisler, eds.), Vol. 2, pp. 277–300. Dekker, New York.

Mishra, N. C. (1977). Genetics and biochemistry of morphogenesis in *Neurospora*. *Adv. Genet.* **19**, 341–405.

Miyazaki, T., and Naoi, Y. (1976). Studies on fungal polysaccharides. XIX. Chemical structure of water soluble polysaccharide from the cell wall of *Cladosporium trichoides*. *Chem. Pharm. Bull.* **24**, 1718–1723.

Moore, P. M. and Peberdy, J. F. (1976). A particulate chitin synthase from *Aspergillus flavus* Link: The properties, location, and levels of activity in mycelium and regenerating protoplast preparations. *Can. J. Microbiol.* **22**, 915–921.

Mueller, S. C., and Brown, R. M., Jr. (1980). Evidence for an intramembrane component associated with a cellulose microfibril-synthesizing complex in higher plants. *J. Cell Biol.* **84**, 315–326.

Nakajima, T., Tamari, K., Matsuda, K., Tanaka, H., and Ogasawara, N. (1970). Studies on the cell wall of *Piricularia oryzae*. Part II. The chemical constituents of the cell wall. *Agric. Biol. Chem.* **34**, 553–560.

Nakajima, T., Tamari, K., Matsuda, K., Tanaka, H., and Ogasawara, N. (1972). Studies on the cell wall of *Piricularia oryzae*. Part III. The chemical structure of the β-D-glucan. *Agric. Biol. Chem.* **36**, 11–17.

Nakajima, T., Tamari, K., and Matsuda, K. (1977a). A cell wall proteo-heteroglycan from *Piricularia oryzae:* Isolation and partial structure. *J. Biochem. (Tokyo)* **82**, 1647–1655.

Nakajima, T., Sasaki, H., Sato, M., Tamari, K., and Matsuda, K. (1977b). A cell wall proteo-heteroglycan from *Piricularia oryzae:* Further studies of the structure. *J. Biochem. (Tokyo)* **82,** 1657–1662.

Noguchi, T., Kitazima, Y., Nozawa, Y., and Ito, Y. (1971). Isolation, composition, and structure of cell walls of *Trichophyton mentagrophytes. Arch. Biochem. Biophys.* **146,** 506–512.

Nolan, R. A., and Bal, A. K. (1974). Cellulase localization in hyphae of *Achlya ambisexualis. J. Bacteriol.* **117,** 840–843.

Nozawa, Y., Kitajima, Y., and Ito, Y. (1973). Chemical and ultrastructural studies of isolated cell walls of *Epidermophyton floccosum:* Presence of chitin inferred from X-ray diffraction analysis and electron microscopy. *Biochim. Biophys. Acta* **307,** 92–103.

Obaidah, M. A., and Buck, K. W. (1971). Characterization of two cell-wall polysaccharides from *Fusicoccum amygdali. Biochem. J.* **125,** 473–480.

Page, W. J., and Stock, J. J. (1974a). Changes in *Microsporum gypseum* mycelial wall and spore coat glycoproteins during sporulation and spore germination. *J. Bacteriol.* **119,** 44–49.

Page, W. J., and Stock, J. J. (1974b). Sequential action of cell wall hydrolases in the germination and outgrowth of *Microsporum gypseum* macroconidia. *Can. J. Microbiol.* **20,** 483–489.

Preiss, J., and Kosuge, T. (1976). Regulation of enzyme activity in metabolic pathways. *In* "Plant Biochemistry" (J. Bonner and J. E. Varner, eds.), 3rd ed., pp. 277–336. Academic Press, New York.

Preston, R. D. (1974). "Physical Biology of Plant Cell Walls." Chapman & Hall, London.

Previato, J. O., Gorin, P. A. J., and Travassos, L. R. (1979a). Cell wall composition in different cell types of the dimorphic species *Sporothrix schenckii. Exp. Mycol.* **3,** 83–91.

Previato, J. O., Gorin, P. A. J., Haskins, R. H., and Travassos, L. R. (1979b). Soluble and insoluble glucans from different cell types of the human pathogen *Sporothrix schenckii. Exp. Mycol.* **3,** 92–105.

Rees, D. A. (1977). "Polysaccharide Shapes." Chapman & Hall, London.

Richmond, D. V. (1977). Permeation and migration of fungicides in fungal cells. *In* "Antifungal Compounds" (M. R. Siegel and H. D. Sisler, eds.), Vol. 2, pp. 251–276. Dekker, New York.

Richmond, D. V., and Pring, R. J. (1971). Fine structure of *Botrytis fabae* Sardiña. *Ann. Bot. (London)* [N.S.] **35,** 175–182.

Rizza, V., and Kornfeld, J. M. (1969). Components of conidial and hyphal walls of *Penicillium chrysogenum. J. Gen. Microbiol.* **58,** 307–315.

Roberts, D. C., and Swart, H. J. (1980). Conidium wall structure in *Seiridium* and *Monochaetia. Trans. Br. Mycol. Soc.* **74,** 289–296.

Roos, U.-P., and Turian, G. (1977). Hyphal tip organization in *Alloymces arbuscula. Protoplasma* **93,** 231–247.

Rosenberger, R. F. (1976). The cell wall. *In* "The Filamentous Fungi" (J. E. Smith and D. R. Berry, eds.), Vol. 2, pp. 328–344. Wiley, New York.

Ross, I. K. (1979). "Biology of the Fungi." McGraw-Hill, New York.

Ruiz-Herrera, J., and Bartnicki-Garcia, S. (1974). Synthesis of cell wall microfibrils *in vitro* by a "soluble" chitin synthetase from *Mucor rouxii. Science* **186,** 357–359.

Ruiz-Herrera, J., Sing., V. O., van der Woude, W. J., and Bartnicki-Garcia, S. (1975). Microfibril assembly by granules of chitin synthetase. *Proc. Natl. Acad. Sci. U.S.A.* **72,** 2706–2710.

Ruiz-Herrera, J., Lopez-Romero, E., and Bartnicki-Garcia, S. (1977). Properties of chitin synthetase in isolated chitosomes from yeast cells of *Mucor rouxii. J. Biol. Chem.* **252,** 3338–3343.

Ruiz-Herrera, J., Bartnicki-Garcia, S., and Bracker, C. E. (1980). Dissociation of chitosomes by digitonin into 16S subunits with chitin synthetase activity. *Biochim. Biophys. Acta* **629,** 201–216.

Ryder, N. S., and Pebergy, J. F. (1977). Chitin synthase in *Aspergillus nidulans:* Properties and proteolytic activiation. *J. Gen. Microbiol.* **99,** 69–76.

Ryder, N. S., and Peberdy, J. F. (1979). Chitin synthetase activity and chitin formation in conidia of *Aspergillus nidulans* during germination and the effect of cycloheximide and 5'-fluorouracil. *Exp. Mycol.* **3**, 259–269.

San-Blas, G., and Carbonell, L. M. (1974). Chemical and ultrastructural studies on the cell walls of the yeastlike and mycelial forms of *Histoplasma farciminosum*. *J. Bacteriol.* **119**, 602–611.

Sassen, M. M. A., Remsen, C. C., and Hess, W. M. (1967). Fine structure of *Penicillium megasporum* conidiospores. *Protoplasma* **64**, 75–88.

Schneider, E. F., and Wardrop, A. B. (1979). Ultrastructural studies on the cell walls in *Fusarium sulphureum*. *Can. J. Microbiol.* **25**, 75–85.

Schneider, E. F., Barran, L. R., Wood, P. J., and Siddiqui, I. R. (1977). Cell wall of *Fusarium sulphureum*. II. Chemical composition of the conidial and chlamydospore walls. *Can. J. Microbiol.* **23**, 763–769.

Shah, V. K., and Knight, S. G. (1968). Chemical composition of hyphal walls of dermatophytes. *Arch. Biochem. Biophys.* **127**, 229–234.

Sietsma, J. H., and Wouters, J. T. M. (1971). Cell wall composition and "protoplast" formation of *Geotrichum candidum*. *Arch. Mikrobiol.* **79**, 263–273.

Smith, J. E., Gull, K., Anderson, J. G., and Deans, S. G. (1976). Organelle changes during fungal spore germination. *In* "The Fungal Spore: Form and Function" (D. J. Weber and W. M. Hess, eds.), pp. 301–352. Wiley, New York.

Stagg, C. M., and Feather, M. S. (1973). The characterization of a chitin-associated D-glucan from the cell walls of *Aspergillus niger*. *Biochim. Biophys. Acta* **320**, 64–72.

Steele, S. D., and Fraser, T. W. (1973). Ultrastructural changes during germination of *Geotrichum candidum* arthrospores. *Can. J. Microbiol.* **19**, 1031–1034.

Sternlicht, E., Katz, D., and Rosenberger, R. F. (1973). Subapical wall snythesis and wall thickening induced by cycloheximide in hyphae of *Aspergillus nidulans*. *J. Bacteriol.* **114**, 819–823.

Stevenson, I. L., and Becker, S. A. W. E. (1979). The fine structure of mature and germinating chlamydospores of *Fusarium oxysporum*. *Can. J. Microbiol.* **25**, 808–817.

Sturgeon, R. J. (1964). Components of the cell wall of *Pithomyces chartarum*. *Biochem. J.* **92**, 60P.

Sturgeon, R. J. (1966). Components of the spore wall of *Pithomyces chartarum*. *Nature (London)* **209**, 204.

Taylor, I. E. P., and Cameron, D. S. (1973). Preparation and quantitative analysis of fungal cell walls: Strategy and tactics. *Annu. Rev. Microbiol.* **27**, 243–259.

Thonet, E., and Hoffman-Ostenhof, O. (1966). Über die meso-Inosit-Oxygenase aus dem Sprosspilz *Schwanniomyces occidentalis*. *Monatsh. Chem.* **97**, 107–114.

Touzé-Soulet, J. M., and Dargent, R. (1977). Effets de la carence en biotine sur la composition des parois cellulaires d'*Hypomyces chlorinus*. *Can. J. Bot.* **55**, 227–232.

Travassos, L. R., de Sousa, W., Mendonça-Previato, L., and Lloyd, K. O. (1977). Location and biochemical nature of surface components reacting with concanavalin A in different cell types of *Sporothrix schenckii*. *Exp. Mycol.* **1**, 293–305.

Travassos, L. R., Mendonça-Previato, L., and Gorin, P. A. J. (1978). Heterogeneity of the rhamnomannans from one strain of the human pathogen *Sporothrix schenckii* determined by ^{13}C nuclear magnetic resonance spectroscopy. *Infect. Immun.* **19**, 1107–1109.

Trinci, A. P. J. (1978). Wall and hyphal growth. *Sci. Prog. (Oxford)* **65**, 75–99.

Troy, F. A., and Koffler, H. (1969). The chemistry and molecular architecture of the cell walls of *Penicillium chrysogenum*. *J. Biol. Chem.* **244**, 5563–5576.

Unger, P. D., and Hayes, A. W. (1975). Chemical composition of the hyphal wall of a toxigenic fungus, *Penicillium rubrum* Stoll. *J. Gen. Microbiol.* **91**, 201–206.

van der Valk, P., Marchant, R., and Wessels, J. G. H. (1977). Ultrastructural localization of polysaccharides in the wall and septum of the basidiomycete *Schizophyllum commune*. *Exp. Mycol.* **1**, 69–82.

van Eck, W. H. (1978). Chemistry of cell walls of *Fusarium solani* and the resistance of spores to microbial lysis. *Soil Biol. & Biochem.* **10**, 155–157.

van Eck, W. H., and Schippers, B. (1976). Ultrastructure of developing chlamydospores of *Fusarium solani* and *F. cucurbitae in vitro. Soil. Biol. & Biochem.* **8**, 1–6.

Wang, M. C., and Bartnicki-Garcia, S. (1970). Structure and composition of walls of the yeast form of *Verticillium albo-atrum. J. Gen. Microbiol.* **64**, 41–54.

Weijman, A. C. M. (1976). Cell-wall composition and taxonomy of *Cephaloascus fragrans* and some Ophiostomataceae. *Antonie van Leeuwenhoek* **42**, 315–324.

Wessels, J. G. H., Kreger, D. R., Marchant, R., Regensburg, B. A., and de Vries, O. M. H. (1972). Chemical and morphological characterization of the hyphal wall surface of the basidiomycete *Schizophyllum commune. Biochim. Biophys. Acta* **273**, 346–358.

Wu-Yuan, C. D., and Hashimoto, T. (1977). Architecture and chemistry of microconidial walls of *Trichophyton mentagrophytes. J. Bacteriol.* **129**, 1584–1592.

Zonneveld, B. J. M. (1971). Biochemical analysis of the cell wall of *Aspergillus nidulans. Biochim. Biophys. Acta* **249**, 506–514.

Zonneveld, B. J. M. (1973). Inhibitory effect of 2-deoxyglucose on cell wall α-1,3-glucan synthesis and cleistothecium development in *Aspergillus nidulans. Dev. Biol.* **34**, 1–8.

Zonneveld, B. J. M. (1978). Biochemistry and ultrastructure of sexual development in *Aspergillus. In* "Genetics and Physiology of *Aspergillus*" (J. E. Smith and J. A. Pateman, eds.), pp. 59–80. Academic Press, New York.

VI

GENETICS

29

The Genetics of Conidial Fungi

A.C. Hastie

I. INTRODUCTION

The natural variability of conidial fungi has long been a source of interest and debate among mycologists. Recognition of their variability logically provoked speculation and research because, although sexual reproduction is the conventional cause of variation, conidial fungi are asexual. The origins of our current knowledge of deuteromycete genetics can be traced back directly to the late 1920s. Significant early contributions to a general understanding of the mechanisms underlying genetic variation in Deuteromycetes were made by Brierley (1929, 1931), Hansen and Smith (1932), and Hansen (1938). They were among the first to suggest that mutation and selection alone constituted an unsatisfactory explanation of deuteromycete variation.

While some contemporary mycologists attributed natural variability of conidial fungi to mutation, Brierley sought to explain it in terms of hybridization. He was well aware that the hybridization must be asexual, for he wrote that ''the

Biology of Conidial Fungi, Vol. 2
Copyright © 1981 by Academic Press, Inc.
All rights of reproduction in any form reserved.
ISBN 0-12-179502-0

function of sexuality with its genetic consequences is, in many fungi, taken by hyphal fusion of a somatic nature'' (Brierley, 1931). This statement does not presage precisely the discovery of the parasexual cycle by Pontecorvo (1954), because Brierley went on to say that somatic hyphal fusions would result ''not in heterozygotic but in heterokaryotic and heteroplasmic states.'' Clearly, by excluding heterozygosis Brierley rejected the possibility of nuclear fusion in asexual fungi. However, the ''heterokaryotic'' state describes genetically different nuclei sharing a common cytoplasm, and the ''heteroplasmic'' state refers to genetically different cytoplasmic determinants in the same cytoplasm. Brierley therefore speculated that both nuclear and cytoplasmic genes may be important in Fungi Imperfecti and presaged the discovery of cytoplasmic inheritance in them (Jinks, 1959).

The discovery of ''gigas'' forms of *Penicillium notatum* by Sansome (1946) and the subsequently reported evidence of their diploidy (Sansome 1949) constituted further significant steps in deuteromycete genetics; they prompted Roper's successful search for heterozygous diploids of *Aspergillus nidulans* (Roper, 1952). Pontecorvo *et al.* (1953b) reported the formation of new genotypes at irregular somatic nuclear divisions in these heterozygous diploids. It was soon possible to make formal genetic analyses of *A. nidulans* and related forms (Pontecorvo, 1954; Pontecorvo and Sermonti, 1954). Thus Pontecorvo recognized that there existed in fungi a valid alternative to the sexual cycle, which he called the parasexual cycle (Pontecorvo, 1954). He defined parasexual as genetic recombination in which there is no fine coordination between segregation, recombination, and chromosome reduction as at meiosis (Pontecorvo, 1959).

The parasexual cycle in fungi was first reviewed by Pontecorvo (1956, 1959) and later by Roper (1966). These authors considered parasexuality in *Aspergillus nidulans* in great detail. This process is also described briefly here because it has served as a model for genetic investigations in conidial fungi for over 20 yr. Comprehensive reviews of heterokaryosis (Davis, 1966), cytoplasmic inheritance in fungi (Jinks, 1966), and parasexuality in plant-pathogenic fungi (Tinline and MacNeill, 1969) are also relevant.

II. THE PARASEXUAL CYCLE

A. General Features of the Parasexual Cycle in *Aspergillus nidulans*

The sexual cycle of any organism is essentially a regular alternation of karyogamy and meiosis. Parental genotypes are brought together as gametes, while gene segregation and recombination occur more or less simultaneously, in a highly coordinated sequence of events at meiosis. The organization of a sexual system includes its association with specialized differentiated structures involved in the formation of gametes, karyogamy, meiosis, and the dispersal of meiotic products. This differentiation and coordination inherent in all sexual systems contrasts markedly with the absence of specialized differentiated structures from parasexual systems, in which flexibility and plasticity dominate.

The parasexual cycle is described as it is known in *Aspergillus nidulans,* the species in which it was first discovered (Pontecorvo, 1954). A general plan of the parasexual cycle is illustrated in Fig. 1. Its component stages and the criteria and techniques used to isolate and identify them are described briefly in this section. Subsequent investigations have confirmed that this general scheme operates in numerous other fungi, including those listed in Table I. Variations on this general scheme are considered in later sections.

1. Isolation of Heterokaryons

Standard procedures for demonstrating the parasexual cycle begin with the establishment of presumed homokaryotic cultures and appropriate minimal and complete culture media. Mutagenic treatments are applied to preferably uninucleate conidia from these cultures to cause auxotrophic mutations. These in turn are identified using defined media and characterized by standard techniques

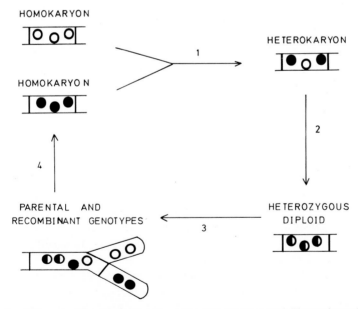

Fig. 1. The parasexual cycle. Stages in the parasexual cycle are named. The numbers refer to the processes joining the stages: 1, hyphal anastomosis; 2, nuclear fusion; 3, mitotic recombination and nondisjunction; 4, nuclear segregation leading to homokaryosis.

TABLE I

Ascomycetes and Deuteromycetes in Which There Is Evidence of Parasexual Recombination

Species	Reference
Ascochyta imperfecta	Sanderson and Srb (1965)
Aspergillus amstelodami	Lewis and Barron (1964)
Aspergillus flavus	Papa (1976)
Aspergillus fumigatus	Berg and Garber (1962)
Aspergillus nidulans	Pontecorvo *et al.* (1953b)
Aspergillus niger	Pontecorvo *et al.* (1953a)
Aspergillus oryzae	Ikeda *et al.* (1957)
Aspergillus soyae	Ishitani *et al.* (1956)
Aspergillus rugulosus	Boam and Roper (1965)
Cephalosporium acremonium	Nüesch *et al.* (1973)
Cephalosporium mycophilum	Tuveson and Coy (1961)
Cochliobolus sativus	Tinline (1962)
Emericellopsis salmosynnemata	Fantini (1962)
Fusarium oxysporum f. *callistephi*	Hoffman (1967)
Fusarium oxysporum f. *cubense*	Buxton (1962)
Fusarium oxysporum f. *pisi*	Buxton (1956)
Humicola sp.	de Bertoldi and Caten (1975)
Penicillium chrysogenum	Pontecorvo and Sermonti (1954)
Penicillium expansum	Barron (1962)
Penicillium italicum	Stromnaes *et al.* (1964)
Pyricularia oryzae	Yamasaki and Niizeki (1965)
Verticillium albo-atrum	Hastie (1962)
Verticillium dahliae	Hastie (1973)

(Pontecorvo, 1949; Holliday, 1960). The ease with which these recessive mutants are detected in wild-type strains implies that they are haploid (Buxton and Hastie, 1962). If mutagen-treated uninucleate conidia were diploid, the recessive mutations induced would not be immediately detectable because they would be heterozygous.

Auxotrophic mutants are essential tools in current techniques for studying parasexuality in the laboratory. Complementary homokaryotic auxotrophs are incubated together as a mixed inoculum on minimal medium. Hyphal fusions occur in the mixed inoculum (step 1 in Fig. 1), and some of these lead to the establishment of heterokaryons. The recessivity of the auxotrophic mutations, together with the complementary nature of the mutants used, ensures that heterokaryons grow on minimal medium although the homokaryotic parents do not. Heterokaryons therefore emerge as sectors growing rapidly from the mixed inoculum. Recognition of heterokaryons of fungi with colored conidia, such as *Aspergillus nidulans,* is further facilitated by arranging that the parent strains have different genotypes with respect to conidial color. Heterokaryons synthesized from such colored variants bear conidiophores with conidia of each parental color if the relevant genes act autonomously.

2. Isolation of Heterozygous Diploids

The second step in the parasexual cycle is the occurrence of nuclear fusions between somatic nuclei in the heterokaryon (step 2 in Fig. 1). Nuclear fusions presumably occur at random between the two haploid nuclear genotypes present, but usually only fusions between genetically different haploid nuclei are detected (Roper, 1952). These nuclear fusions cause the formation of heterozygous diploid nuclei and they, or their mitotic derivatives, ultimately are included in conidia as the heterokaryon grows. A heterokaryon of *Aspergillus nidulans,* or any other fungus with uninucleate conidia, therefore typically bears conidia of three phenotypically distinctive classes: haploid conidia of each parental haploid genotype (e.g., *Ab* and *aB*) and conidia containing a single heterozygous diploid nucleus (e.g., *Aa Bb*). Conidia of each parental type are auxotrophic, but the heterozygous diploid conidia are prototrophic because mutations causing auxotrophy (e.g., *a* and *b*) are always recessive to their wild-type alleles (*A* and *B*). Rare heterozygous diploid conidia borne on a heterokaryon are therefore easily selected from conidia with haploid parental genotypes by plating the mixed conidial suspension on minimal agar medium at a high plating density (Roper, 1952). Only the prototrophic heterozygous conidia grow and form colonies.

The frequency of heterozygous diploid conidia recovered from *Aspergillus* heterokaryons in this way is generally about 1 in 10^6(Pontecorvo, 1959). Proof of their diploidy is obtained routinely by conidial measurements, because diploid conidia have about twice the volume of haploid conidia. Genetic proof of heterozygosity rests upon demonstrating that all recessive mutant alleles introduced into the heterokaryon can be recovered among the segregants, obtained ultimately from one uninucleate conidium. Other criteria of diploidy, such as a doubled DNA content (Heagy and Roper, 1952) and a doubled chromosome number (Robinow and Caten, 1969), are of course unsuitable for routine application.

3. Isolation of New Segregant Genotypes

Relatively stable novel genotypes are formed during the vegetative growth of *Aspergillus* diploids (step 3 in Fig. 1). These are mitotic recombinants, nondisjunctional diploids, and haploids, and they occur in relative proportions of about 100:5:10, respectively, among spontaneous segregants (Pontecorvo, 1959). Two primary processes, namely, mitotic crossing-over and chromosome nondisjunction, are involved in their formation. Mitotic crossing-over in heterozygous diploids, followed by the appropriate chromatid segregation, leads to the formation of mitotic recombinants which are diploid and homozygous for one or more genes on the respective chromosome arm (Stern, 1936; Pontecorvo *et al.*, 1954). Chromosome nondisjunction (or elimination) in dividing diploid nuclei leads to the formation of aneuploid nuclei. Further chromosome nondisjunctions at subsequent nuclear divisions of the primary aneuploid nuclei cause the formation of diverse aneuploids and ultimately stable nondisjunctional diploids or haploids (Fig. 2). These processes have been analyzed in great detail in *Aspergillus* by Kafer (1961), who has shown that mitotic crossing-over and chromosome nondisjunction are entirely independent events, each occurring at low frequencies (about 0.02), so that their coincident occurrence in the same nuclear lineage is exceedingly rare.

Detection of rare, novel segregant genotypes formed during the growth of *Aspergillus* colonies from uninucleate heterozygous diploid conidia requires the use of selective devices (Pontecorvo, 1959). The system originally employed, which is still most commonly used, depends upon arranging that the initial diploid has green conidia (wild-type and dominant) but is heterozygous for one or more genes whose recessive alleles affect conidial color (e.g., *Ww*). Some segregant genotypes formed during growth of a heterozygote have the homozygous recessive genotype (*ww*), and others are haploid and contain only one recessive allele (*w*). In either instance continued growth of the novel genotype results in its inclusion in conidia which are white, and the segregant is revealed as a white sector in an otherwise green colony because of nuclear segregation during hyphal growth (step 4 in Fig. 1). Other unlinked genes causing yellow (*y*) and chartreuse (*cha*) conidia are also available in *A. nidulans* and can be employed similarly because of their autonomous gene action. Analogous selective systems for detecting rare, novel genotypes are based upon recessive mutations causing resistance to toxic substances (Roper and Kafer, 1957) and recessive

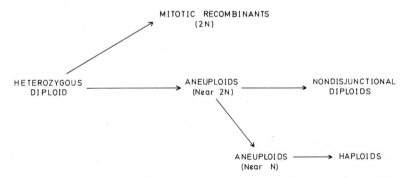

Fig. 2. Origins of relatively stable new genotypes in the parasexual cycle (mitotic recombinants, nondisjunctional diploids, and haploids). Mitotic recombinants are formed following mitotic crossing-over, and formation of the other genotypes involves chromosome nondisjunction or elimination over a series of nuclear divisions.

suppressor mutations which suppress specific auxotrophic requirements (Ponte-corvo, 1959).

Pontecorvo (1959) has described the use of haploid genotypes and mitotic recombinants formed parasexually in mapping the chromosomes of *A. nidulans*. They are of particular value in this species even though a sexual system is also available. The haploid segregants, formed at irregular mitoses, are of special value in assigning unlocated genes to their linkage groups (chromosomes). This property derives directly from the fact that the formation of haploids in the parasexual cycle involves random elimination of chromosomes and no crossing-over. Consequently, a collection of independently formed haploid segregants will reveal either zero recombination or 50% recombination for any pair of segregating genes. In the first instance the gene loci are linked on homologous chromosomes, and the alternative result implies that the genes are located on nonhomologous chromosomes (unlinked). McCully and Forbes (1965) have ex-ploited the unequivocal nature of these recombination fractions in their "master strain technique" for assigning unlocated genes to their linkage groups (chromo-somes). The technique also takes advantage of the discovery by Lhoas (1961) that *p*-fluorophenylalanine incorporated into the growth medium promotes the formation of haploid segregants. Hastie (1970a) has shown that the fungicide benlate also promotes haploidization and can therefore be of value in the same context.

In contrast to the role of haploid segregants in parasexual analyses, mitotic recombinants provide the information required for mapping the positions of genes within linkage groups and also for mapping gene loci relative to their associated centromere. However, there are limitations to the information they provide, which derive directly from the nature and the frequency of mitotic recombination in *A. nidulans*. Because of its rarity and the required use of selective markers to detect it, only mitotic crossovers between the selective gene marker and the corresponding centromere are detected. It follows that a selective marker gene can be used to map only the positions of gene loci between it and its centromere and not the positions of either relatively distal genes or genes located on the other side of the centromere. The extensive chromosome maps which have now been constructed for *A. nidulans* require information from both sexual and parasexual analyses (Dorn, 1967).

B. The Parasexual Cycle in Other Fungi

The discovery of parasexuality in *A. nidulans* prompted the suggestion that a similar system may operate in other anamorphic fungi and provide a means of conducting genetic analyses in them (Pontecorvo, 1956). Numerous researchers have therefore applied the techniques developed with *A. nidulans* to other species over the past 20 yr, and a list of their original references is given in Table I. This list includes only examples for which there is genetic evidence of parasexuality from laboratory studies modeled on the pioneering work with *A. nidulans*. Too few species have been examined to permit a meaningful assessment of the distribution of parasexuality throughout the conidial fungi. Indeed, most of the species listed are important in either industrial mycology or plant pathology. The list of species is therefore selective and reflects a desire among industrial mycologists and plant pathologists to apply genetics to strain improvement and the analysis of host–parasite interactions, respectively. Relatively little progress in either area has as yet resulted directly from the discovery of parasexuality (Roper, 1973; Tinline and MacNeill, 1969).

The investigations listed in Table I all involved the use of auxotrophic mutants, and all the authors ultimately obtained genetic recombinants with respect to these markers among asexual progeny. The systems involved therefore apparently fit the general definition of parasexuality given by Pontecorvo (1959), although there is evidence in some instances of differences in detail from the *Aspergillus* model. However, only one strain, or a few strains of common ancestry, have been studied in some of these fungi, and there is consequently an obvious danger in making generalizations from these observations of limited scope. One cannot legitimately describe a genetic system on the evidence provided by one or a few crosses.

The parasexual cycles of the *Aspergillus* and most of the *Penicillium* species listed in Table I are formally identical to parasexuality in *A. nidulans,* apart from the fact that the asexual species in these genera may produce heterozygous diploids and new segregant genotypes more frequently than *A. nidulans* (Pontecorvo, 1959). More striking differences from the *A. nidulans* system have been reported for other genera. These differences relate to the nature and vigor of heterokaryons, the recovery of heterozygous diploids, and the stability of these diploids.

No heterokaryons of *Humicola* were found by de Bertoldi and Caten (1975), although they recovered heterozygous diploids and therefore suggested that a restricted heterokaryotic phase may occur in mixed cultures of complementary auxotrophs. Hastie (1962) found some heterokaryotic hyphal tips in *Verticillium albo-atrum*, but he has concluded that heterokaryons of *V. albo-atrum* are also transient and unstable (Hastie, 1970b). Similarly, heterokaryons of *Penicillium chrysogenum* and *Cephalosporium acremonium,* composed of complementary auxotrophs, are much less vigorous than the corresponding heterokaryons of *Aspergillus* species grown on minimal medium. The hyphal cells of *Aspergillus* are characteristically multinucleate (Clutterbuck and Roper, 1966), but most hyphal cells of *Verticillium* (Heale *et al.,* 1968; Puhalla and Mayfield, 1974) and *Cephalosporium acremonium* (Nüesch *et al.,* 1973) are uninucleate, and this cytological difference undoubtedly contributes to the relative performances of the respective heterokaryons. The nature of heterokaryosis in *Verticillium dahliae* has been investigated by Puhalla and Mayfield (1974) and is discussed more fully in Section II,C.

The nature and vigor of heterokaryons affects the methods by which heterozygous diploids may be selected from them. The vigorous heterokaryons of *A. nidulans* undergo conidiogenesis freely, and conidia containing a single heterozygous diploid nucleus are recovered from them at a frequency of about 1 in 10^6 (Pontecorvo, 1959). This has been observed repeatedly with other *Aspergillus* species. However, heterozygous diploids of fungi with less vigorous heterokaryons, or no detectable heterokaryotic phase, often appear as rapidly growing prototrophic sectors from mixed inocula of complementary auxotrophs on selective media. Heterozygous diploids have been recovered in this way from *Penicillium chrysogenum* (Pontecorvo and Sermonti, 1954), *Cephalosporium mycophilum* (Tuveson and Coy, 1961), *Cephalosporium acremonium* (Nüesch *et al.*, 1973). *Verticillium albo-atrum* (Ingle and Hastie, 1974), *V. dahliae* (Puhalla and Mayfield, 1974) and *Humicola* sp. (de Bertoldi and Caten, 1975).

A few instances have been reported in which heterokaryons failed to yield heterozygous diploids even after repeated attempts. *Neurospora crassa* is a notable example in which attempts were made both by Case and Giles (1962) and Roper (1966). Similarly, Dutta and Garber (1960) did not find any heterozygous diploid conidia among 10^8 conidia from each of three heterokaryons of *Colletotrichum lagenarium*. Barron (1963) also experienced difficulty in finding spontaneous heterozygous diploid conidia of *Penicillium expansum* but observed them more easily after camphor treatment of heterokaryons. He suggested that the low frequency of spontaneous diploids may be related to the nature and origin of the conidiophores and also to the number of conidia borne on each structure. His argument is explained here because it may well apply to other fungi, especially to pycnidial fungi which may be examined in the future.

Taking *Penicillium expansum* as an example, Barron (1963) points out that most of the conidiophores on heterokaryons are in fact homokaryotic, presumably because they originate from a uninucleate cell. Thus, if the frequency of heterozygous diploid nuclei in a heterokaryon is 1 in 10^6, it will generally be necessary to sample at least 10^6 conidiophores in order to include one which originated from a diploid nucleus. However, because each mature conidiophore of *P. expansum* bears about 10^3 conidia, we can expect 10^9 conidia from these 10^6 conidiophores. We may therefore need to sample about 10^9 conidia from a heterokaryon to find a heterozygous diploid. Transferring this argument to *Colletotrichum lagenarium*, Barron (1963) argues that the respective sample of conidia required may be much larger. The conidia of *C. lagenarium* are borne on acervuli, and each acervulus bears up to 10^5 conidia. If each acervulus originates from a uninucleate cell and the frequency of diploid nuclei in a heterokaryon is 1 in 10^6, it is easily seen that it may be necessary to test about 10^{11} conidia to find one diploid. If these are valid assumptions for *C. lagenarium,* it is perhaps not surprising that Dutta and Garber (1960) found no diploids in samples of 10^8 conidia.

The third feature which apparently distinguishes parasexuality in *Aspergillus* and certain other fungi listed in Table I is the stability of heterozygous diploids. Tuveson and Coy (1961), working with *Cephalosporium mycophilum,* were perhaps the first to report that somatic segregants could be easily found in random samples of conidia from heterozygous diploid strains. They analyzed two prototrophic diploid strains of common ancestry and found that about 22 and 60% of the conidia from them were auxotrophic segregants. Hastie (1962, 1964) reported the same phenomenon in *Verticillium albo-atrum,* and it has also been observed in the related species *V. dahliae* by Hastie (1973) and Puhalla and Mayfield (1974). Hoffman (1967) worked with complementary monoauxotrophic mutants of *Fusarium oxysporum* f. *callistephi* and, although he obtained recombinant auxotrophic segregants, he was unable to find heterozygous diploids. He inferred that an unstable diploid may have been formed in his heterokaryons, but he could not detect it.

Diploid instability has been reported even in *Aspergillus nidulans.* Kafer (1962) found that diploids of *A. nidulans* heterozygous for chromosomal translocations were relatively unstable, producing more frequent segregants than standard diploids. Translocations were indeed found to be widely distributed in laboratory strains of *A. nidulans,* caused apparently by ultraviolet mutagen treatments applied to induce auxotrophic mutants. Analysis of these situations in *A. nidulans* can be supported by information obtained through the sexual cycle, but this advantage is of course unavailable for conidial fungi.

By analogy with the above experience with *Aspergillus* it is possible that the diploid instability found in some conidial fungi is not normal for these species. It may in specific instances be due to heterozygosity for chromosome translocations inadvertently caused by the mutagen treatments necessary to produce auxotrophic mutants. This could well be the case with *Fusarium oxysporum* f. *callistephi,* because other *forma speciales* behave differently (Buxton, 1956, 1962). Diploid instability of *Verticillium* diploids, however, seems to be common. It occurs even when low-intensity chemical mutagen treatments, unlikely to cause simultaneous chromosomal and gene mutations, are used to produce the parent auxotrophs (Ingle, 1972). Excepting the report of Ingram (1968), which apparently describes a special case, diploid instability has been found in every laboratory-synthesized *Verticillium* diploid. These reports cover isolates from different countries, different angiosperm host species, and different species of *Verticillium* (Hastie, 1962, 1964, 1973; Puhalla and Mayfield, 1974; Typas and Heale, 1978).

In most of these examples diploid instability is assessed by the proportion of segregant genotypes recovered in random samples of conidia from segregating diploid cultures. This criterion of diploid stability needs to be applied cautiously, because the frequencies with which segregant genotypes are recovered may not reflect accurately the frequencies with which they are formed, and the latter is the ultimate criterion of stability. An example of *Verticillium albo-atrum* will illustrate the disparity which can exist between the population of nuclei in a mycelium and the conidia borne on it. When diploid conidia are inoculated at low density on very dilute agar media, they form colonies of sparse mycelia with very few conidiophores distributed unevenly in a few clusters. These conidiophores bear almost exclusively haploid conidia (Hastie, 1964). It can nevertheless be shown that most of the mycelium is diploid, because mycelial fragments transferred to richer media produce colonies with diploid conidia.

C. The Parasexual Cycle of *Verticillium*

The parasexual cycle of *Verticillium* was first described by Hastie (1964), and it involves the same sequence of events as parasexuality in *Aspergillus* (Fig. 1). However, it differs significantly from the *Aspergillus* model in two respects. Heterokaryons of *Verticillium* are much less stable, and heterozygous diploids of *Verticillium* apparently form new genotypes more frequently than *Aspergillus* diploids (Hastie, 1962, 1964). These differences are discussed here in some detail because they may be common to other conidial fungi and also because the very frequent genotypic changes in segregating cultures allied to the conidiophore morphology of *Verticillium* provide an opportunity for a novel system of genetic analysis.

1. Heterokaryosis

Davis (1966) and Puhalla and Mayfield (1974) distinguished two types of heterokaryons among higher fungi. They included *Aspergillus nidulans* and *Neurospora crassa* among examples of the first type. The hyphal cells of these fungi are multinucleate, nuclei can migrate from cell to cell, and heterokaryosis regularly extends to apical cells of a colony so that heterokaryons can be regularly propagated using hyphal tips. Puhalla and Mayfield (1974) have investigated heterokaryosis in *Verticillium dahliae* using complementary auxotrophs and have suggested that its heterokaryons are of a distinctive second type characterized by predominantly uninucleate hyphal cells, restricted nuclear migration, and hyphal tips remaining homokaryotic. They have found by microscopic examination that heterokaryosis of this type is confined to binucleate anastomosed cells in which gene complementation can occur and support growth of adjacent homokaryotic cells. Such a heterokaryotic colony is a more-or-less stable mosaic of homokaryotic and isolated heterokaryotic cells.

Heterokaryons of *Verticillium albo-atrum* are likely to have a similar structure. Hastie (1962) and Heale (1966) found that heterokaryons of *V. albo-atrum* could only rarely be propagated by hyphal tips. This could be consistent with the heterokaryon structure described above for *V. dahliae* if the rare heterokaryotic hyphal tips included anastomosed cells. Rigorous proof that heterokaryosis has occurred requires the reisolation of both homokaryotic forms from a single hypha, hyphal tip, or other cytoplasmic unit (Davis, 1966), and this proof is not easily obtained with the *Verticillium* type of heterokaryon.

Maintenance of the heterokaryotic condition in growth from a heterokaryotic cell depends upon a variety of factors. These include the characteristic number of nuclei per cell, the rate of nuclear division relative to cell division, the occurrence and rate of nuclear migration within hyphae, the frequency of hyphal anastomoses, and the continued action of selection in favor of the heterokaryon. Hyphal anastomoses have a particularly important role in *Verticillium* because of the nature of the heterokaryon. Puhalla and Mayfield (1974) found that the number of hyphal anastomoses made by *V. dahliae* was influenced greatly by incubation temperature. Abundant hyphal anastomoses were observed in heterokaryons at 21°C, but few were found at 30°C. These findings correlate with growth of *V. dahliae* heterokaryons, composed of complementary auxotrophs, on minimal medium at 21°C and the absence of growth at 30°C. The authors infer that too few anastomoses are made at 30°C to provide the heterokaryotic cells required to support growth of the homokaryotic areas of the mosaic mycelium.

2. Nuclear Fusion and the Isolation of Diploids

Verticillium dahliae is perhaps the only fungus in which somatic nuclear fusion has been observed cytologically. Most of the anastomosed interhyphal cells seen in heterokaryons of *V. dahliae* by Puhalla and Mayfield (1974) were binucleate. In some instances the two nuclei were on opposite sides of the anastomosis bridge, and in other instances one nucleus had migrated across the bridge. In a few instances the H-shaped anastomosed cell contained a single, relatively large nucleus which was presumably diploid and had resulted from karyogamy. Karyogamy was indeed observed in a single instance using phase-contrast microscopy.

Hastie (1964) described the recovery of heterozygous haploids of *Verticillium albo-atrum* using complementary auxotrophs. Most of its conidia are uninucleate, and the rare binucleate and trinucleate conidia are presumably homokaryotic. In these early experiments the diploids were recovered from mixed cultures of complementary auxotrophs grown at 25°C by plating dense conidial suspensions on minimal medium. This is essentially the technique developed by Roper (1952) with *Aspergillus nidulans*. In later work mixed cultures of complementary auxotrophs were incubated on minimal medium, and heterozygous diploids were recovered as prototrophic sectors emerging from them (Hastie, 1973; Ingle and Hastie, 1974). Ingle and Hastie (1974) examined the effects of seven environmental factors on the frequency with which prototrophic heterozygous diploid sectors emerge from such mixed inocula. They presented evidence that the frequency of sectors reflected the frequency with which diploid nuclei are formed (i.e., diploidization). The nature of both the carbon and nitrogen sources in the selective medium had especially pronounced effects on the frequency of diploidization, and they recommended for future work a minimal medium based on glucose and sodium nitrate. Incubation temperature also had a very marked effect. No diploid sectors were found at either 15° or 20°C, some were found at 22° and 24°C, and many at 26°C. Indeed, on some media nearly all mixed cultures of complementary haploid auxotrophs formed diploid sectors at 26°C.

Hyphal anastomosis, the establishment of at least one heterokaryotic cell, and subsequent nuclear fusion are the essential steps leading to the formation of a heterozygous diploid. The factors which influence the frequency of diploidization may operate on any of these steps. Ingle and Hastie (1974) speculated that the glucose–nitrate stimulation of diploidization may occur because this medium tends to synchronize nuclear cycles, and nuclear fusions may be more probable between synchronous nuclei. Substantial evidence for this hypothesis is still lacking, but it is established that a glucose–nitrate medium tends to promote nuclear synchrony in *Aspergillus nidulans* (Rosenberger and Kessel, 1967).

Ingle and Hastie (1974) rejected the hypothesis that high incubation temperatures increased diploidization by increasing the frequency of hyphal anastomoses, because they observed hyphal anastomoses at all temperatures. However, this conclusion must be treated with caution; the frequency of hyphal anastomoses is difficult to quantify in a meaningful way, and no distinction was made between intrastrain and interstrain anastomoses. Only the latter can be effective anastomoses in this context. Puhalla and Mayfield (1974), working with *V. dahliae,* found fewer hyphal anastomoses at 30°C than at 21°C, although more heterozygous diploid sectors grew from the heterokaryotic cultures at the higher temperature. They suggested that the lower frequency of hyphal anastomoses at 30°C limited the maintenance and growth of the mosaic heterokaryon, and that the higher incubation temperature therefore preferentially favored diploid growth.

These experiments on the effects of temperature on the formation and recovery of heterozygous diploids of *V. albo-atrum* and *V. dahliae* are contradictory. Ingle and Hastie (1974) presented evidence that the frequency with which diploid sectors emerged from mixed cultures reflected the frequency with which diploid nuclei of *V. albo-atrum* were formed, and they concluded that relatively high incubation temperatures promoted the formation of diploid nuclei (diploidization). On the contrary, Puhalla and Mayfield (1974) imply that diploidization is favored by relatively low incubation temperatures because these conditions favor hyphal anastomoses. Resolving the contradiction requires accurate assessment of the frequencies of hyphal anastomoses at different temperatures, and distinguishing between intrastrain and interstrain anastomoses. These are difficult technical problems which merit future research.

3. Formation of Novel Genotypes by *Verticillium* Diploids

New segregant genotypes are formed spontaneously only rarely during vegetative growth of heterozygous diploid strains of *Aspergillus nidulans* and related conidial fungi, and selective gene markers are required to detect and isolate them (see Section II,A). In contrast, spontaneous novel segregant genotypes are easily recovered from heterozygous diploids of *V. albo-atrum* (Hastie, 1962, 1964; Typas and Heale, 1976) and *V. dahliae* (Hastie, 1973; Puhalla and Mayfield, 1974). Segregants are found in appreciable proportions in random samples of the uninucleate conidia taken from young segregating cultures, and Hastie (1964) reported that both mitotic recombinants and new haploid genotypes could be recovered in this way. A 1-week-old segregating culture grown from a single uninucleate heterozygous diploid conidium yields frequent mitotic recombinants, but few or no haploids, in random samples of conidia. However, the relative proportions of mitotic recombinants and haploids in such conidial samples change as the culture ages, so that a 3-week-old segregating culture may yield more than 95% haploids. The higher proportion of haploid conidia from older cultures reflects the fact that haploids are formed very frequently from diploid nuclei, but it also results from more prolific conidiation of haploid mycelium in the segregating colony (Hastie, 1964).

The mechanisms by which new genotypes are formed in these segregating heterozygous strains has been investigated by Hastie (1967, 1968). The novel method used in these investigations makes use of the facility with which *Verticillium* conidia can be removed from the conidiophores by micromanipulation. The conidia of *V. albo-atrum* are formed on verticiliately branched conidiophores, and each conidiogenous cell (phialide) is initially uninucleate. Division of this nucleus is followed immediately by migration of one daughter nucleus into a developing conidium. The other daughter nucleus remains near the base of the phialide and does not divide before formation of the first conidium is completed. This nuclear division and migration cycle is inferred, because cytological observations show that phialides never contain more than two nuclei (Hastie, 1967; Heale *et al.*, 1968), and its repeated occurrence leads to the formation of an indefinite number of uninucleate conidia at the tip of the phialide.

Hastie (1968) has called the population of conidia borne on one phialide a "phialide family." The number of conidia in a phialide family equals the number of nuclear divisions which occur in the phialide, because each conidium is initially uninucleate. Furthermore, because each conidium receives one product of the nuclear division which immediately preceded its formation, the genotypes of the nuclei in a phialide family provide valuable information about the genetic events which occur at phialide nuclear divisions. Micromanipulation can be used to isolate phialide families from segregating heterozygous cultures and separate the individual conidia, and the genotypes of the isolated conidia can be determined by observations on the cultures they produce (Hastie, 1967, 1968). Hastie (1968) has proposed the term "phialide analysis" for genetic analyses which use the information provided by phialide families. Phialide analysis permits the isolation of new genotypes immediately after they are formed, and this analytic technique therefore provides an opportunity to relate the frequency of genotype changes to the frequency of nuclear division in phialides.

 a. Formation of Haploid Segregants.   Three general classes of viable conidia are found in phialide families from segregating cultures. They are diploid, haploid, and aneuploid and are distinguishable mainly by the features of the colonies they produce (Hastie, 1967). Diploid conidia are generally longer than haploid conidia, but their frequency distributions overlap, and the ploidy of an individual conidium often cannot be unequivocally decided from its length. However, diploid conidia form colonies which have on average relatively long conidia (11 μm) and haploid conidia form colonies with shorter conidia (7 μm). Conidia classified as aneuploid produce relatively slow growing colonies which subsequently often form faster growing sectors. These sectors have the properties of euploids, either haploid or diploid.

 Phialide families are classified according to the types of conidia they contain. The family types recovered from young colonies (4 days) of several different heterozygous diploids of *Verticillium albo-atrum* are listed in Table II. The overall percentage of nonviable conidia in these families ranged from 8 to 15%, and similarly high conidial inviability is regularly found wherever conidia from heterozygous diploid strains are micromanipulated. However, only about 1% of the conidia from haploid cultures fail to grow when micromanipulated in the same way. It has therefore been concluded that numerous nonfunctional

TABLE II

Types of Phialide Families Obtained from *Verticillium* Diploids

Family type	Number of families
All conidia diploid	616
All conidia aneuploid	106
All conidia nonviable	26
Conidia diploid or nonviable	251
Conidia aneuploid or nonviable	111
Conidia diploid or aneuploid	303
Conidia diploid, aneuploid, or nonviable	140
Conidia aneuploid or haploid	3
Conidia diploid, aneuploid, or haploid	1
Total	1557

genotypes are formed in diploid strains (Hastie, 1968) and that the occurrence of these is perhaps a regular feature of the formation of haploid nuclei from diploid nuclei. Their occurrence certainly imposes limitations on the information obtainable from phialide families.

The phialide families listed in Table II were isolated from 4-day-old colonies grown on minimal medium from diploid conidia. Genotype changes certainly occur both in the growing mycelium and in the conidiophores. Families containing only one type of conidium (diploid, haploid, or aneuploid) presumably arise when the initial phialide nucleus is of the respective type, and no detectable genotype changes take place during formation of the phialide family (Table II). Other family types include various combinations of diploid, aneuploid, haploid, and nonviable conidia. The absence of families containing only diploid and haploid conidia is regarded as significant, proving that haploid nuclei are not formed directly from diploid nuclei in a one-step process. Some families contain diploids and aneuploids, while others contain haploids and aneuploids. It therefore seems reasonable to postulate that diploid nuclei form haploids through intermediate aneuploid stages, as suggested by other evidence from *Aspergillus nidulans* (Kafer, 1961) and illustrated in Fig. 2. This hypothesis gains additional support from more detailed examination of the aneuploids associated with diploids and haploids in different families. The aneuploids found along with diploids generally produce diploid sectors, and aneuploids associated in families with haploids tend to produce haploid sectors. This suggests that the former aneuploids have near-diploid genomes and that the latter aneuploids have near-haploid genomes.

Phialide analysis therefore provides convincing evidence that diploid nuclei of *Verticillium albo-atrum* form haploids through an intermediate aneuploid phase (Fig. 2), and it is likely that nondisjunctional diploids will occur as a by-product, although they have not been reported in *Verticillium*. The probability of an aneuploid nucleus being formed when a diploid nucleus divides can be estimated only very roughly from the families listed in Table II. There are two obvious reasons for this. First, we know nothing about the ploidy of nonviable conidia, although it is a reasonable assumption that most of them have aneuploid genotypes and are lethal because they are nullisomic. Second, a family containing both diploid and aneuploid conidia may represent either a diploid forming an aneuploid, or vice versa. If only families in which all conidia are viable are considered and it is assumed that all such families containing both diploid and aneuploid conidia are examples of diploid nuclei forming aneuploids, the probability of a diploid nucleus forming an aneuploid is about 0.035 per nuclear division. This is perhaps not significantly different from the corresponding figure of 0.02 estimated for *Aspergillus nidulans* by Kafer (1961), bearing in mind the approximations involved in both instances.

b. Mitotic Recombination in Phialide Families.   The mechanism of mitotic recombination was first described by Stern (1936) to explain results obtained with *Drosophila melanogaster* and has since been confirmed in *A. nidulans* (Roper and Pritchard, 1955). It results in the simultaneous formation of a homozygous dominant (*MM*) and a homozygous recessive (*mm*) diploid genotype following mitosis of a heterozygote (*Mm*) in which there was a mitotic crossover. Consequently, if mitotic recombination occurs in the growth of a phialide family, the family will usually contain conidia with the recessive phenotype as well as conidia with the dominant phenotype (Fig. 3), and these segregating families will be easily detected (Hastie, 1967, 1968).

Examples of families in which mitotic recombination occurred are given in Table III and illustrate the properties of the system. These families were taken from a diploid originally heterozygous with respect to six auxotrophic markers (*tryp an ad/inos hist nic*). All the conidia in each family were viable and diploid, and their genotypes were determined by progeny-testing the cultures they produced.

Family A in Table II contained four conidia, and each was homozygous dominant with respect to the tryptophan (*tryp*) and histidine (*hist*) genes, presumably because of mitotic recombination in the mycelium before the conidiophore was formed. However, the initial phialide nucleus was heterozygous at the aneurin (*an*), adenine (*ad*), inositol (*inos*), and nicotinamide (*nic*) gene loci, because each allele of these genes was recovered in the family as a whole. A mitotic crossover apparently occurred in the first nucleus in the phialide between the *nic* gene and its centromere and led to mitotic recombination at the first nuclear division. Thus the first conidium received the homozygous nic^+/nic^+ nucleus, and the homozygous recessive genotype (nic^-/nic^-) remained in the phialide and generated the three conidia formed later (third family from the top of Fig. 3). The *an, ad,* and *inos* genes remained heterozygous, because apparently no crossover occurred in the relevant chromosome regions. The other families in Table III serve to illustrate that mitotic recombination may occur at any nuclear division in a phialide, and that it may be either the homozygous dominant or recessive genotype which enters the conidium formed first after mitotic recombination. Examples of mitotic recombination causing simultaneous segregation of linked genes in phialide families have been published elsewhere (Hastie, 1967, 1968).

Hastie (1967, 1968) has estimated the probabilities of mitotic recombination at diploid phialide nuclear divisions for two gene loci in *Verticillium albo-atrum*. The probabilities are 0.021 for a gene affecting uracil synthesis (*ur-1*) and 0.036 for the *so-1* gene which affects mycelial pigmentation. The *so-1* gene, because of its autonomous action, also provides an opportunity to estimate the probability of mitotic recombination at the first nuclear division in a germinating conidium.

This estimate turns out to be 0.006 (=3/509), which is much lower than the estimate of 0.036 for phialide nuclear divisions (Hastie, 1968). Nuclear divisions in phialides and at conidial germination are of course successive nuclear divisions, although they occur in very different circumstances, and comparison of these frequencies for *so-1* suggests that mitotic recombination may not be equally frequent at all developmental stages.

Hastie (1967) has also estimated that the probability of mitotic recombination affecting any part of the *V. albo-atrum* genome is about 0.2 at each phialide

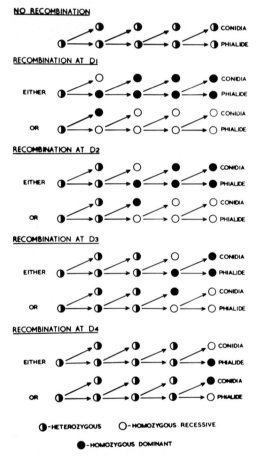

Fig. 3. Possible consequences of mitotic combination in a *Verticillium* phialide. A phialide bearing four conidia is taken as an example, and recombination at each of the four nuclear divisions (D1–D4) is considered. There are two possibilities following each recombination: Either the homozygous recessive or the homozygous dominant segregant nucleus may enter the conidium formed immediately after recombination. (From Hastie, 1967.)

TABLE III

Phialide Families from a *Verticillium albo-atrum* Diploid (*ad an tryp inos nic hist*) Which Show the Consequences of Mitotic Recombination in the Respective Phialides

Family	Number of conidia	Genotype[a]						Division of segregation
		ad	*an*	*tryp*	*inos*	*nic*	*hist*	
A	1	+/−	+/−	+/+	+/−	+/+	+/+	First (*nic*)
	3	+/−	+/−	+/+	+/−	−/−	+/+	
B	1	+/−	+/−	+/+	+/−	+/+	+/+	Second (*ad*)
	1	−/−	+/−	+/+	+/−	+/+	+/+	
	5	+/+	+/−	+/+	+/−	+/+	+/+	
C	1	+/+	+/−	+/+	−/−	+/+	+/−	First (*inos*)
	1	+/+	+/−	+/+	+/+	+/+	+/−	Third (*an*)
	1	+/+	−/−	+/+	+/+	+/+	+/−	
	4	+/+	+/+	+/+	+/+	+/+	+/−	

[a] The gene symbols refer to recessive auxotrophic requirements for adenine (*ad*), aneurin (*an*), tryptophan (*tryp*), inositol (*inos*), nicotinamide (*nic*), and histidine (*hist*). +/+ indicates dominant, +/− indicates heterozygosity, and −/− indicates homozygous recessive for the respective genes.

nuclear division. This rough estimate assumes that a particular gene (*ur-1*) is located at the terminal end of a chromosome arm, that the haploid chromosome number is 4, and also that the centromere to *ur-1* distance represents the average length of a chromosome arm. There is of course no way in which it can be proved that a gene is located at the end of a chromosome, and although there is now some evidence that the haploid chromosome number of *V. albo-atrum* is 4 (Heale *et al.*, 1968), the other assumptions remain unsubstantiated. The estimate for the overall probability of mitotic recombination in a diploid phialide nucleus (0.2) is consequently much less precise and meaningful than estimates for particular gene loci. Nevertheless, this indicates that the general frequency of mitotic recombination per nuclear division in *V. albo-atrum* is likely to be significantly greater than the corresponding frequency (0.02) for *Aspergillus nidulans* estimated by Kafer (1961). These frequencies agree with Pontecorvo's (1959) observation that mitotic recombination seems to be more frequent in anamorphic fungi with no known teleomorphs.

c. Independent Formation of Aneuploids and Mitotic Recombinants. Aneuploids of *V. albo-atrum* are presumably formed from diploids by chromosomal nondisjunction and/or elimination, and there is compelling evidence that mitotic recombinants are formed by mitotic crossing-over (Hastie, 1968). These processes appear to be entirely independent in their occurrence at nuclear divisions, because there is, for example, no evidence that mitotic recombinants tend to be restricted to phialide families which contain both diploid and aneuploid nuclei (conidia). However, both mitotic crossing-over and chromosome nondis-junction are frequent events in diploid *V. albo-atrum,* and it is to be expected that both may readily occur in the same nuclear lineage but at different nuclear divisions.

Haploid conidia, recovered in random samples of conidia from segregating diploid cultures, are therefore quite likely to be derived from a nuclear lineage which included both mitotic crossing-over and haploidization. It is probable that such haploids may be recombinants with respect to either linked or unlinked genes. Consequently, gene recombination fractions calculated from the genotypes of a random sample of haploids of independent ancestry may range from 0 to 50% (Hastie, 1964), and because of frequent mitotic crossing-over it is conceivable that two gene loci on the same pair of homologous chromosomes may show 50% recombination in these samples of haploids. The possibility therefore exists that a recombination fraction of 50% calculated from a sample of independently formed haploid segregants may mean either that the gene loci are on nonhomologous chromosomes or far apart on homologous chromosomes. This ambiguity is a general property of genetic analysis in sexual systems, but it is absent in the use of haploids for linkage detection in the parasexual system of *Aspergillus nidulans* or that of any other fungus with a corresponding low fre-quency of mitotic crossing-over and haploidization.

4. Genetic Analysis in *Verticillium*

Genetic analysis in *Verticillium* may use either random samples of conidia (Hastie, 1964) or phialide families isolated by micromanipulation from the heterozygous diploid strains (Hastie, 1968). Only the former system is described here, because it is generally more efficient in studying gene segregation and recombination, although phialide analysis is particularly useful in mapping the loci of genes within chromosome arms.

Heterozygous diploids appear as sectors from mixed inocula of complemen-tary auxotrophs incubated on minimal medium (Ingle and Hastie, 1974), and 10–20 independent diploid sectors are subcultured on complete medium to obtain a series of segregating cultures. Conidia suspensions are prepared from each of these segregating cultures after 15–20 days' incubation and spread on complete medium at a low plating density. It is assumed that the resulting colonies are formed by single conidia. Over 95% of the colonies obtained are euploid (haploid or diploid); the others grow very slowly, often producing vigorous sectors, are probably aneuploid, and are rejected. Colonies grown from euploid conidia are used to inoculate plates of complete medium (26 colonies per plate) which serve ultimately as templates for subsequent replication on appropriate test media. Thus the phenotypes of the random euploid colonies are determined, and by inference these are the phenotypes of the conidia which formed them.

The haploid conidial phenotypes obtained from a series of 10 segregating cultures of the cross *pyr-1 meth-6 phen-1* × *ur-1 amm-1* are given in Table IV. It is known from previous work (Hastie, 1964) and from the results of phialide analyses that haploid genotypes are formed singly rather than several simultaneously, and it is important that only haploids of independent origin be used for the genetic analysis. Cloning effects must be excluded. All haploids with different genotypes recovered from one segregating culture must be of different origin, but haploids with the same genotype from one segregating culture may be clonal. Therefore only one haploid of specific genotype from each segregating culture is used in the genetic analysis. The column on the right in Table IV gives the minimum number of independent occasions on which each haploid genotype was

TABLE IV

Random Haploid Monoconidial Isolates from 10 Segregating Diploid Cultures of the *Verticillium* Diploid *pyr*-1 *phen*-1 *meth*-6 × *ur*-1 *amm*-1

Genotype[a]	Number of each phenotype from replicate segregating cultures[b]										Independent haploids
pyr							2				1
meth	2			2	1	3	3		3		6
ur	1										1
amm	6	1	4	9	1		2	9			7
pyr phen		1	1	1	1	10	3			13	7
pyr meth	12	2		1	3	2	2				7
pyr inos		1									1
phen meth			1	1				8	20		4
phen ur						1					1
phen amm	1		1		3	1			3		5
amm ur		10	10	4	2	1	4	1		3	8
meth ur		1	1	3	3	1				1	6
pyr amm				5							1
pyr phen meth		2					3	1			3
pyr phen amm	3				1		1	1			4
pyr amm ur		2	2	2	1						3
phen amm ur				2		1					3
meth amm ur		1									1
pyr meth ur					1	1					2
pyr phen amm ur			1								1
pyr phen meth ur				2		1					2
Prototrophic										2	1
Total	24	20	21	26	23	17	26	20	26	19	75

[a] These gene symbols refer to auxotrophic requirements for pyridoxine (*pyr*), methionine (*meth*), uracil (*ur*), ammonium (*amm*), and inositol (*inos*).

[b] A sample of 26 euploid conidia was taken from each segregating culture. Only the haploids in these samples are recorded.

formed over the series of segregating cultures, and these are the numbers used in the analyses of allele segregation and gene recombination given in Tables V, A and B.

For the analysis of allele segregation it is assumed that each haploid nucleus is derived from a diploid ancestor by random loss of chromosomes. Chance then determines which member of a homologous chromosome pair is retained in a haploid. The hypothesis that a given pair of contrasting characters (e.g., $meth^+$ and $meth^-$) is determined by a single pair of alleles is then easily tested by examining the ratio of haploids with each character. The ratio of mutant ($meth^-$) to wild-type ($meth^+$) haploids under these circumstances will not generally be significantly different from 1:1. There is a shortage for each mutant allele, but four of the five auxotrophic requirements segregating in this cross give results

TABLE VA

Allele Segregation among Independent Haploids from the *Verticillium* Diploid *pyr*-1 *phen*-1 *meth*-6 × *ur*-1 *amm*-1[a]

	Genes				
Allele	*pyr*-1	*phen*-1	*meth*-6	*ur*-1	*amm*-1
Mutant allele	32	30	31	28	33
Wild-type allele	43	45	44	47	42

[a] Data are from Table IV.

TABLE VB

Gene Recombination among Independent Haploids from the *Verticillium* Diploid *pyr*-1 *phen*-1 *meth*-6 × *ur*-1 *amm*-1[a]

Genotype				Recombination (%)
meth amm	*meth amm*$^+$	*meth*$^+$ *amm*	*meth*$^+$ *amm*$^+$	
1	30	32	13	18.4
meth phen	*meth phen*$^+$	*meth*$^+$ *phen*	*meth*$^+$ *phen*$^+$	
9	22	21	24	57.4
meth pyr	*meth pyr*$^+$	*meth*$^+$ *pyr*	*meth*$^+$ *pyr*$^+$	
14	17	18	27	44.4
meth ur	*meth ur*$^+$	*meth*$^+$ *ur*	*meth*$^+$ *ur*$^+$	
11	20	17	28	52.0

[a] Data are from Table IV. See footnote to Table IV for an explanation of gene symbols. Only recombination between *meth*-6 and the other four genes is tabulated. The other recombination fractions are all about 50%.

consistent with a 1:1 ratio (Table V). The exceptional requirement is uracil (*ur-1*) which has segregated in a 1:1 ratio in other crosses.

Recombination between the *meth-6* gene and the other segregating genes is also examined in Tables VA and B. Most gene pairs show about 50% recombination and are unlinked; but *meth-6* and *amm-1* show significantly less than 50% recombination and are therefore linked. Haploid recombinants for *meth-6* and *amm-1* presumably had mitotic crossing-over as well as haploidization in their ancestral lineage, and linkage between these two genes has also been detected in other crosses (Hastie, 1978).

This example illustrates how random conidial analysis can be used to investigate the genetic basis of phenotypic differences and detect linkages between genes in *Verticillium*. Numerous gene linkages have been detected in other crosses (Hastie, 1964, and unpublished), and tentative chromosome maps of five linkage groups have been produced (Typas and Heale, 1978).

III. GENETICS AND NATURAL VARIATION IN CONIDIAL FUNGI

The investigations reviewed in the preceding sections are concerned with demonstrating and analyzing components of the parasexual cycle under laboratory conditions. They rely upon using induced auxotrophic mutants to provide the techniques for selecting heterokaryons and heterozygous diploids. All known auxotrophic mutations have chromosomal loci, and they therefore serve as nuclear markers which are essential in proving the occurrence of heterokaryosis and heterozygosity. Given these built-in markers, parasexual phenomena are rather easily investigated. In their absence, as in isolates taken directly from natural populations, one must rely on natural phenotypic differences. These differences may be caused by cytoplasmic (extrachromosomal) rather than nuclear determinants and, even when they are caused by nuclear genes, the phenotypic differences may be unsuitable as a basis for selective techniques. The unavailability of suitable genetic markers has certainly contributed to the lack of progress in studying parasexual recombination in natural cultures of conidial fungi (Hastie, 1970b).

A. Heterokaryosis

1. The Potential Role of Heterokaryosis in Nature

Hansen and Smith (1932) were perhaps the first to appreciate fully the potential significance of heterokaryosis in the biology of filamentous fungi. Its value is at least twofold. First, recessive alleles are not expressed in heterokaryons, and heterokaryosis therefore shelters them from elimination by natural selection.

Davis (1966) has summarized this property of heterokaryons succinctly by describing it as "a possible substitute for heterozygosity in maintaining genetic variability" in populations. He demonstrated the efficiency of this sheltering effect in experiments with *Neurospora crassa* when he showed that, with absolute selection against one component of a heterokaryon, a sequence of about 20 mass conidial transfers was required to remove it entirely from the population (Davis, 1959). The formation of some heterokaryotic conidia by *Neurospora crassa* must have contributed to the persistence of the contraselected genotype. Selection against a given genotype would be more efficient in fungi which make no heterokaryotic conidia, because in such instances the heterokaryon would need to be resynthesized at each transfer. Davis (1959) also showed, both in theory and in practice, that the elimination of one genotype could only happen in such circumstances if the heterokaryon had no selective advantage. Even a small selective advantage favoring the heterokaryon ensures that the heterokaryotic condition persists.

The second significant feature of heterokaryosis in relation to the ecology of fungi results from their capacity to alter nuclear ratios during the growth of heterokaryotic mycelia (Jinks, 1952a). The phenotype of the heterokaryon may therefore adapt progressively to a changing local environment. Such changes in nuclear ratios in heterokaryons are, however, only temporary. They are variations on a theme which is dictated ultimately by the genotypes of the nuclei in the heterokaryon. New themes can only arise by gene mutation and recombination and, without the creation of new genotypes, heterokaryosis has limited evolutionary consequences.

2. Examples of Heterokaryons from Natural Sources

There is no doubt that heterokaryons can have significant advantages over homokaryons both in respect to short-term physiological adaptations and in regard to longer-term evolutionary changes. In the parasexual cycle heterokaryosis is an essential step leading to heterozygosity and genetic recombination. Examples of heterokaryotic conidial fungi isolated from natural sources may be found in the very numerous reports of morphological variation, sectoring, and saltation published during the past 50 yr. However, as has been emphasized by Parmeter *et al.* (1963) and also by Caten and Jinks (1966), rigorous proof of heterokaryosis was not provided in most instances. Sectoring in wild-type isolates may be caused by mutation, or by segregation in either heteroplasmons (see Section III,B,1) or heterokaryons, and many early claims of heterokaryosis did not discriminate between these possibilities. Caten and Jinks (1966) list a few previously published examples of heterokaryosis in homothallic Ascomycetes and conidial fungi. Some of these examples from the latter group are described below, together with a more recent example discovered by Ming *et al.* (1966), but in nearly all these examples no rigorous distinction was made between heterokaryosis and heteroplasmosis (Section III,B,1).

Hansen and Smith (1932) examined 47 cultures of *Botrytis cinerea* collected from various parts of California. Single-conidium isolations from them yielded numerous more-or-less dissimilar morphological types. One wild-type culture was analyzed in great detail through five successive generations of single-conidium isolations. Two morphologically stable monoconidial lines were established (homotypes), while other monoconidial lines were of intermediate morphology (heterotypes) and yielded homotypes at subsequent monoconidial isolations. The authors showed that both the hyphal cells and conidia of *B. cinerea* were multinucleate, and that the segregation patterns they detected were consistent with the homotypes and heterotypes being homokaryotic and heterokaryotic, respectively. They also showed that frequent hyphal anastomoses occurred, and that heterotypes could be synthesized by growing the homotypes together from a mixed inoculum on potato dextrose agar. Their suggestion that the heterotypes were heterokaryons makes the asumption that the genetic difference between the homotypes resides in chromosomal genes rather than extrachromosomal genes. Jinks (1966) has explained that discrimination between chromosomal and extrachromosomal inheritance can be made by a heterokaryon test involving at least one known nuclear marker. Threlfall (1971) obtained auxotrophs and a conidial color mutant in *Botrytis allii,* and this suggests that nuclear markers (auxotrophs) may be obtained in *B. cinerea.* A reexamination of natural variation in this fungus using known nuclear markers is certainly overdue. Hansen (1938) also claimed to have detected heterokaryosis in 29 out of 30 anamorph-genera, but at least some of the examples may have been heteroplasmons.

Jinks (1952b) found several heterokaryons among 16 isolates of *Penicillium* spp. from various natural sources, and he also described a natural heterokaryon of *Penicillium cyclopium* (Jinks, 1952a). These examples included instances in which the heterokaryon was obtained from a point inoculum deposited from the atmosphere onto an exposed plate. Colonization originated presumably from an aggregate of two or more genetically different conidia. It seems therefore that the dispersal of heterokaryons is not dependent entirely on their forming heterokaryotic conidia. Jinks (1959) examined four wild-type isolates of *Penicillium,* employing a heterokaryon test to discriminate between nuclear and cytoplasmic determinants. He found that two of the four isolates were heterokaryons and two were heteroplasmons, and he consequently advised caution before invoking heterokaryosis as the sole cause of the "dual phenomenon" described by Hansen (1938).

Buxton (1954) demonstrated heterokaryosis with respect to color and morphology in *Fusarium oxysporum* f. *gladioli,* but it is not clear whether he actually isolated heterokaryons from infected tissue. Ming *et al.* (1966) reported an especially interesting case of heterokaryosis in isolates of the rice pathogen *Fusarium fujikuroi* obtained from natural material. These isolates subsequently yielded white, red, and purple pigmented homokaryons, and the authors showed that the increased pigmentation was correlated with both increased gibberellin production and pathogenicity. They found that 19 isolates, each from widely separated geographical locations, were heterokaryons composed of any two, or all three, homokaryon types. These authors resynthesized heterokaryons containing any two, and also all three, genotypes on potato dextrose agar without apparently using selective techniques to recover the heterokaryons. This example of natural heterokaryosis merits further investigation both because of the wide distribution of natural heterokaryons and because of the ease with which they appear to be synthesized on artificial media. The facts reported seem to imply that the heterokaryotic mycelium has unidentified selective advantages over homokaryons both on host plants and during saprophytic growth on artificial media. The assumption of Ming *et al.* (1966) that the color variation in their strains is caused by nuclear genes receives support from the more recent work of Sanchez *et al.* (1976). These workers provide compelling evidence that similar color differences in *Fusarium oxysporum* f. *lycopersici* are caused by nuclear genes.

3. Heterokaryon Incompatibility

The capacities of strains of one fungal species to form heterokaryons is restricted by genes causing heterokaryon incompatibility. A heterokaryon incompatibility system in *Aspergillus nidulans* was first demonstrated by Grindle (1963a,b). He obtained many wild-type homokaryons from independent sources, induced conidial color mutants in them, and grew wild-type (green) and mutant (white or yellow) forms together in many pairwise combinations. Genes affecting conidial color in *A. nidulans* have an autonomous action, so that the color of a conidium is determined exclusively by the genotype of the nucleus it contains. Grindle detected heterokaryons by searching for individual conidiophores bearing conidia of both colors where wild-type and mutant colonies merged. Each wild-type isolate formed heterokaryons with conidial color mutants derived from it, but many isolates did not form heterokaryons with color mutants of independent ancestry. Even some paired strains derived ultimately from the same soil sample were incompatible (Grindle, 1963b). Extended studies show that heterokaryon incompatibility depends upon paired strains having the same genotype with respect to the relevant genes, and it has been deduced that homokaryons of *A. nidulans* are compatible only if they have identical alleles of at least five genes (Jinks *et al.*, 1966).

Evidence of a similar heterokaryon incompatibility system has been obtained in other *Aspergillus* spp. with teleomorphs (Jones, 1965) and also in the anamorphic species *A. terreus* and *A. versicolor* (Caten, 1971) and *A. niger* (Brunton,

1975). Caten (1971) has shown that the frequency of heterokaryon incompatibility among isolates from independent sources is about the same in the sexual and asexual species examined. Grindle's method of detecting heterokaryon incompatibility is in principle applicable to any species with colored condidia, particularly if color is controlled by autonomous gene action. This method of detecting heterokaryosis, and therefore heterokaryon incompatibility, has been described by Caten and Jinks (1966) as neutral or nonselective by comparison with forcing heterokaryosis through the use of auxotrophic mutants. They further suggest that information obtained by using such neutral methods will be more relevant to situations in nature.

Caten and Jinks (1966) have discussed the general genetic implications of the occurrence of heterokaryon incompatibility systems for species which have them. The direct effect is to limit heterokaryosis within the species to strains with similar genotypes and thereby promote continued inbreeding. The species will therefore be comprised of inbred subunits or populations between which gene exchange is impeded by heterokaryon incompatibility. These populations will be the units on which evolutionary forces operate, bringing about changes in allele frequencies and creating genetic diversity between heterokaryon-incompatible populations of the same species. Sexual crosses can be made in the laboratory between heterokaryon-incompatible strains of *Aspergillus* (Jinks *et al.,* 1966), and statistical analyses of the ascospore progenies from these crosses have confirmed that they are much more variable than progenies from crosses made within heterokaryon incompatibility groups. Progenies from crosses between strains in different heterokaryon incompatibility groups were also less vigorous than their parents, and these progeny would almost certainly be at a selective disadvantage if intergroup sexual crosses occurred in the wild. However, Butcher *et al.* (1972) inferred from their analyses of wild *Aspergillus nidulans* that some gene flow between heterokaryon-incompatible populations may occur in nature. Such gene exchange presumably takes place through sexual rather than parasexual recombination.

Day (1970) and Caten (1971) suggested that invasive cytoplasmic factors and fungal viruses may be transmitted by hyphal anastomoses and that heterokaryon incompatibility may have value in restricting their spread through species. Lhoas (1971) has subsequently shown transmission of an RNA virus by hyphal anastomosis in *Penicillium stoloniferum,* and Caten (1972) has shown that efficient transmission of the cytoplasmic determinant of "vegetative death" in *Aspergillus amstelodami* occurs only between heterokaryon-compatible strains. Transfer of the determinant was reduced from 100%, in combinations of heterokaryon-compatible strains, to 15% with paired heterokaryon-incompatible strains. Indeed, no transfer of the cytoplasmic determinant was detected between strains differing by more than one incompatibility gene. There is therefore good evidence that heterokaryon incompatibility is an effective barrier to the spread of fungal viruses and other autonomous cytoplasmic factors. These other factors will presumably include natural plasmagenes inherent to the particular fungus. Heterokaryon-compatible strains will thus tend to share the same pool of both plasmagenes and chromosomal genes. Conversely, heterokaryon-incompatible populations will have diverging plasmagene and chromosomal gene pools.

B. Extrachromosomal Inheritance

1. The Heterokaryon Test for Extrachromosomal Inheritance

Heterokaryosis relates to differences between nuclear (chromosomal) genotypes. Extrachromosomal (cytoplasmic) determinants are also known in many organisms, including fungi, and are referred to as plasmagenes (Darlington, 1939). The term "plasmon," introduced to refer to the total extrachromosomal hereditary complement of a cell, is analogous to "genome," which is applied in a nuclear context. The terms "homoplasmon" and "heteroplasmon" are correspondingly analogous to "homozygote" and "heterozygote." A heteroplasmon is a cell in which a given plasmagene occurs in at least two genetically different forms. It seems to be a general feature of eukaryotic cells that each contains numerous copies of each plasmagene. A heteroplasmon may contain variable proportions of mutant and nonmutant plasmagenes, and the ratio of these two forms of the plasmagene may alter as cells grow and multiply.

Some extrachromosomal genes are certainly located in the DNA of mitochondria, but the cytoplasmic loci of most fungal plasmagenes is unknown. Nevertheless, it is generally assumed that there is no mechanical device ensuring equational distribution of plasmagenes to daughter cells, but rather that their distribution to daughter cells is more or less random. Consequently, persistent somatic segregation is a general property of extrachromosomal heredity, and homoplasmons may segregate frequently from heteroplasmons during vegetative growth of conidial fungi (Jinks, 1966).

A heterokaryon test can provide proof of extrachromosomal inheritance in conidial fungi (Jinks, 1959, 1966). The test requires the availability of two homokaryotic haploid strains differing with respect to a trait known to be determined by chromosomal genes, in addition to the trait being investigated. These chromosomal genes serve as nuclear markers. A heterokaryon is synthesized between the two strains, and uninucleate conidia are recovered from it. If the trait under investigation is determined by nuclear genes, all the uninucleate haploid conidia from the heterokaryon will have parental associations of the traits. Alternatively, if the unlocated determinants are extrachromosomal (cytoplasmic), the uninucleate progeny of the heterokaryon will generally include both parentals and recombinants. The latter arise because of reassortment of nuclei and cytoplasm in the heterokaryon. Thus, if we designate the known nuclear differences as

N and n, the unlocated determinants as D and d, and a heterokaryon is synthesized using haploid homokaryons ND and nd, then the recovery of either Nd or nD types among uninucleate conidia from the heterokaryon is proof that the determinants D and d do not have nuclear loci.

Jinks (1963, 1966) has discussed complications which can arise with this heterokaryon test for extrachromosomal inheritance. These difficulties relate primarily to persistent segregation of the unlocated trait and to the possibility that recombinant types have arisen from recombination between nuclear genes rather than reassortment of nuclear (chromosomal) and cytoplasmic (extrachromosomal) genes. In the first instance, a search for frequent persistent segregation of a trait within known haploid homokaryons indicates extrachromosomal inheritance. In the second instance, a heterokaryon test can be designed to incorporate several known nuclear markers, and these can be used to ascertain whether recombination between nuclear genes is occurring in the heterokaryon.

2. Examples of Extrachromosomal Inheritance of Natural Variability in Conidial Fungi and Related Ascomycetes

Jinks (1959) investigated morphological variation in four wild-type isolates of *Penicillium* sp. using the heterokaryon test to determine whether the variation was controlled by chromosomal (nuclear) or extrachromosomal factors. The traits involved were contrasting colony morphology and conidiophore density and resembled the sort of variation referred to as "dual phenomenon" by Hansen (1938). Jinks concluded that chromosomal genes were responsible for the variation within two of the isolates and that cytoplasmic factors exercised control in the others. He therefore urged caution in assuming that heterokaryosis is the only basis of the dual phenomenon in imperfect fungi. Cytoplasmic factors may well be responsible for some of the "duality" found in wild-type isolates of 29 genera of conidial fungi by Hansen (1938). Further investigations along the lines indicated by Jinks (1959) are clearly needed before an accurate assessment can be made of the relative frequency of heteroplasmosis and heterokaryosis as causes of natural variation in these fungi. Many of them are likely to be haploid (Section III,C) and produce mainly uninucleate or homokaryotic conidia. Segregation of cytoplasmic factors is a probable cause of variation in these circumstances.

Both darkly pigmented and hyaline forms of *Verticillium albo-atrum* occur in nature. The pigment of the dark forms is located in the walls of torulose hyphae, which are absent in hyaline forms. The genetic basis of this dimorphism was investigated by Hastie (1962) and more extensively by Typas and Heale (1976). These authors synthesized heterokaryons using complementary auxotrophic mutants isolated from dark and hyaline wild types. These heterokaryons produced more-or-less abundant dark, torulose hyphae, and about 99% of the uninucleate haploid conidia from the heterokaryons also formed colonies with dark hyphae. No reassortment of the auxotrophic nuclear markers occurred during heterokaryosis, but the darkly pigmented conidial progeny included two different phenotypes involving the auxotrophic nuclear markers; dark pigmentation became associated with the auxotrophic nuclear marker from the hyaline parent during heterokaryosis.

This is precisely as expected if a cytoplasmic factor is involved in the inheritance of pigmentation. Typas and Heals (1976) have labeled this factor *hyl* and postulate that it exists in two forms, *hyl$^+$* and *hyl*, with dark pigmentation developing only in strains containing *hyl$^+$*. However, it is possible that *hyl* represents the absence of *hyl$^+$*, hence hyalinity may be caused by the absence of *hyl$^+$* from the cytoplasm of hyaline strains. The existence of this cytoplasmic factor does not of course exclude the possibility that pigmentation can be influenced by chromosomal genes. Indeed, a chromosomal gene affecting mycelial pigmentation of *V. albo-atrum* was shown to be genetically linked to one causing arginine auxotrophy (Hastie, 1968).

Extrachromosomal factors have been implicated in the inheritance of numerous other characters in fungi, including vegetative death, growth rate, absence of conidia (mycelial), and the absence of sexual structures (Jinks, 1966). The latter trait, reported for both *Aspergillus nidulans* and *Aspergillus glaucus,* is especially relevant to the genetics of conidial fungi and any discussion of their origins in particular.

C. Diploidy in Nature

There is general agreement that diploidy offers advantages to diploid species. It permits storage of genetic variability and heterosis and provides an opportunity for genetic recombination. These constitute considerable advantages over haploidy, and this is emphasized by the evolutionary trends evident in the progression from lower to higher organisms. It is therefore perhaps surprising that there is so little evidence of diploidy among wild strains of filamentous higher fungi generally, and conidial fungi in particular (Hastie, 1970b).

Among the higher fungi, yeasts are apparently exceptional in that natural haploids and diploids occur regularly (Phaff *et al.,* 1966). Adams and Hansche (1974), in discussing the relevance of their research on diploidy in *Saccharomyces cerevisiae* to natural populations, have raised the interesting point that, while diploidy is common in terrestial yeasts, all the marine yeasts listed by Van Uden and Fell (1968) are believed to be haploid. Adams and Hansche (1974) suggest that this may be related to nutrient levels in terrestrial and marine environments, because their chemostat experiments with isogenic haploids and diploids indicate that diploids suffer a selective disadvantage when certain nutrients limit growth. It would be interesting to know whether the occurrence of diploid and haploid yeasts in these environments is correlated with the distributions of sexuality and asexuality.

The features which a priori are most likely to distinguish natural diploid conidial fungi from haploids are nuclear cytology and conidial dimensions. Cytological evidence of chromosome number, size, and structure provides the most convincing criterion of ploidy, but the nuclear cytology of vegetative fungal hyphae is notoriously difficult. Strains of *Aspergillus nidulans* (Robinow and Caten, 1969), *Ustilago violacea* (Day, 1972), and *Cochliobolus sativus* (Huang and Tinline, 1974) known to be haploid and diploid from genetic evidence have been compared cytologically. In no instance were the results definitive. The number of chromosomes (or "chromatinic elements") seen in dividing diploid nuclei was generally less than twice the number seen in haploids, presumably because chromosomes from the two haploid genomes may become associated and are often unresolvable in diploid nuclei. Microdensitometry following Feulgen staining has been used to measure the relative DNA contents of nuclei in conidial fungi (Davies and Jones, 1970), but such measurements must be corroborated by other evidence because DNA content alone does not reveal the chromosome number or ploidy.

Conidial dimensions are used routinely to distinguish haploids from diploids synthesized in the laboratory and were of primary importance in recognizing the only two well-documented natural diploid conidial fungi reported. However, conidium size is not by itself an entirely reliable criterion of ploidy. It could also conceivably be influenced by mutations of single genes, so that haploids with relatively large conidia can be envisaged. Furthermore, unstable diploids, such as those of *Verticillium albo-atrum* which rapidly yield spontaneous haploid segregants, may be misclassified as haploids if only older cultures are examined.

Identification of the only natural diploid conidial fungi reported to date depended not only on conidial measurements but also on inferences from genetic evidence. Among the most significant of this genetic evidence was the discovery by Lhoas (1961) that *p*-fluorophenylalanine specifically induced haploidization of laboratory diploids of *Aspergillus niger*. Ingram (1968) established that *Verticillium dahliae* var. *longisporum*, originally isolated by Stark (1961), was a natural diploid, and Nga et al. (1975) reported a natural diploid of *Aspergillus niger*. Diploidy was proved in both instances by showing that (1) the original wild isolate produced segregants with smaller conidia when grown on medium containing *p*-fluorophenylalanine, (2) the original wild strains did not yield auxotrophic mutants when exposed to mutagens, although the small-spored segregants from them did, and (3) the auxotrophic mutants obtained could be used to reconstruct the original large-spored type. These results are consistent with the wild, large-spored strains being diploid, because auxotrophic mutations are always recessive and, even if they were induced in a diploid, dominance would prevent their expression and detection.

Davies and Jones (1970) discovered a diploid "hybrid" of *Cercosporella herpotrichoides* which, although it originated in culture, is mentioned here along with naturally occurring diploids because its origin did not involve the use of auxotrophic markers and associated selection pressures. It originated spontaneously from a mixed culture of two prototrophic wild-type strains. One parent had black pigment and was pathogenic on wheat, while the other was white and pathogenic on barley. Some hyphal tips from a mixed culture produced gray colonies. This hybrid did not conidiate, but plating hyphal fragments confirmed that it was not a heterokaryon. Its diploid nature was established by showing that p-fluorophenylalanine caused it to produce white and black sectors, and also by Feulgen photometry of stained hyphal nuclei. The nuclear DNA contents in the presumed haploid parents (white or black) and diploids (gray) were in the ratio 1:1.94 which, considering the quantities of DNA involved, is remarkably close to the expected 1:2 ratio.

The absence of other reports of wild diploid conidial fungi presumably reflects their scarcity in the field. Conidial size is precisely the sort of descriptive feature recorded by mycologists for many years and is one of the few regularly quantified morphological features of conidial anamorphs. It therefore seems likely that large-spored strains would have been noticed if they had been isolated. Discussion of the reasons for their rarity is perhaps premature, but some speculation on this matter is justified by its importance in relation to an understanding of the biology of conidial fungi and also by a desire to stimulate relevant research.

Caten and Day (1977), after considering effects known in *Schizophyllum commune* and *Ustilago violacea,* have suggested that the rarity of somatic diploids of higher fungi may be caused by karyogamy repressor genes (KR) which specifically suppress somatic nuclear fusions. They have further proposed that rare somatic diploids may form by karyogamy in these fungi when there is a temporary, localized low concentration of repressor. This novel hypothesis has predictive merits and is not invalidated by the rare detection of natural diploids (Ingram, 1968; Nga *et al.,* 1975). Diploid conidial fungi may originate by either spindle failure at a haploid somatic mitosis, so that a homozygous diploid restitution nucleus is formed, or by karyogamy of haploid somatic nuclei. In the first case (spindle failure) the diploid product would certainly be homozygous, but in the second case (karyogamy) the initial diploid product may be either homozygous or heterozygous, depending upon the genotypes involved in the karyogamy. *Verticillium dahliae* var. *longisporum* is apparently homozygous (Ingram, 1968), but the natural diploid of *Aspergillus niger* is certainly heterozygous for genes affecting thiamin synthesis and conidiogenesis (Nga *et al.,* 1975). However, its heterozygosity does not prove that it originated by karyogamy. It may be a descendant of a homozygous diploid that became heterozygous because of the occurrence of gene mutations.

Rare somatic diploidization could be caused by mutation of the proposed repressor gene (KR) to an inactive form (KR⁻) so that the genetic restriction on karyogamy is removed (Caten and Day, 1977). This proposal predicts that rare somatic diploids may have mutant genes (KR⁻) which increase the frequency of karyogamy, and consequently at least some of the haploid segregants obtainable from these diploids would have this mutant gene. Such haploids should show a relatively high frequency of karyogamy in subsequent tests. This was not noticed in the "diploid reconstructions" of either Ingram (1968) or Nga *et al.* (1975), and there is no indication that mutations to KR⁻ have been selected unwittingly in laboratory parasexual studies with *Aspergillus nidulans* extending over the past 25 yr.

IV. CONCLUSIONS

Mutation is the ultimate source of genetic variation, and together with heterokaryosis and heteroplasmosis certainly provides opportunities for variation in wild fungi. Genetic recombination through the parasexual cycle obviously also has a potential value in the general biology of both conidial and other fungi, but the actual existence of parasexuality in wild populations remains largely speculative. To assess the role of the parasexual cycle in nature we need to know whether it occurs, whether incompatibility occurs, the frequencies of the various stages, and how these may be influenced by environmental factors. Our current knowledge is deficient in almost every respect.

The studies on natural heterokaryon incompatibility in members of *Aspergillus* by Jinks and his associates are highly relevant to these requirements and provide the sort of information needed to begin assessing the role of heterokaryosis in this group of fungi. The relevance of these laboratory investigations to natural populations stems largely from the nature of the sensitive technique used for assessing the capacities of strains to form heterokaryons. This technique is certainly of wider application and ought to be more fully exploited in other genera. The relevance of future studies to natural variability will depend upon developing and applying techniques appropriate to other aspects of the parasexual cycle, and perhaps especially the detection of genetic recombination in nature. Ideally this will need clear-cut natural genetic markers, preferably amenable to selective techniques. Genes causing host-range specificity of plant pathogens may be appropriate, and their use was suggested by Pontecorvo (1956). Natural variation in the tolerance of toxic substances may also be similarly exploited.

REFERENCES

Adams, J., and Hansche, P. E. (1974). Population studies in microorganisms. I. Evolution of diploidy in *Saccharomyces cerevisiae*. *Genetics* **76**, 327–338.

Barron, G. L. (1962). The parasexual cycle and linkage relationships in the storage rot fungus *Penicillium expansum*. *Can. J. Bot.* **40**, 1603–1613.

Barron, G. L. (1963). Distribution of nuclei in a heterokaryon of *Penicillium expansum*. *Nature (London)* **200**, 282–283.

Berg, C. M., and Garber, E. D. (1962). A genetic analysis of color mutants of *Aspergillus fumigatus*. *Genetics* **47**, 1139–1146.

Boam, T. B., and Roper, J. A. (1965). Unpublished data, quoted by Roper (1966).

Brierley, W. B. (1929). Variation in fungi and bacteria. *Proc. Int. Congr. Plant Sci. 1926* Vol. 2, pp. 1629–1654.

Brierley, W. B. (1931). Biological races in fungi and their significance in evolution. *Ann. Appl. Biol.* **18**, 420–434.

Brunton, J. (1975). M.Sc. Thesis, University of Dundee, Scotland.

Butcher, A. C., Croft, J. H., and Grindle, M. (1972). Use of genotype-environmental interaction analysis in the study of natural populations of *Aspergillus nidulans*. *Heredity* **29**, 263–283.

Buxton, E. W. (1954). Heterokaryosis and variability in *Fusarium oxysporum* f. *gladioli*. *J. Gen. Microbiol.* **10**, 71–84.

Buxton, E. W. (1956). Heterokaryosis and parasexual recombination in pathogenic strains of *Fusarium oxysporum*. *J. Gen. Microbiol.* **15**, 133–139.

Buxton, E. W. (1962). Parasexual recombination in the banana-wilt *Fusarium*. *Trans. Br. Mycol. Soc.* **45**, 274–279.

Buxton, E. W., and Hastie, A. C. (1962). Spontaneous and ultra-violet irradiation induced mutants of *Verticillium albo-atrum*. *J. Gen. Microbiol.* **28**, 625–632.

Case, M. E., and Giles, N. H. (1962). The problem of mitotic recombination in *Neurospora*. *Neurospora Newsl.* **2**, 6–7.

Caten, C. E. (1971). Heterokaryon incompatibility in imperfect species of *Aspergillus*. *Heredity* **26**, 299–312.

Caten, C. E. (1972). Vegetative incompatibility and cytoplasmic infection in fungi. *J. Gen. Microbiol.* **72**, 221–229.

Caten, C. E., and Day, A. W. (1977). Diploidy in plant pathogenic fungi. *Annu. Rev. Phytopathol.* **15**, 295–318.

Caten, C. E., and Jinks, J. L. (1966). Heterokaryosis: Its significance in wild homothallic Ascomycetes and Fungi Imperfecti. *Trans. Br. Mycol. Soc.* **49**, 81–93.

Clutterbuck, A. J., and Roper, J. A. (1966). A direct determination of nuclear distribution in heterokaryons of *Aspergillus nidulans*. *Genet. Res.* **7**, 185–194.

Darlington, C. D. (1939). "The Evolution of Genetic Systems." Macmillan, New York.

Davies, J. M. L., and Jones, D. G. (1970). The origin of a diploid "hybrid" of *Cercosporella herpotrichoides*. *Heredity* **25**, 137–139.

Davis, R. H. (1959). Asexual selection in *Neurospora crassa*. *Genetics* **46**, 1291–1308.

Davis, R. H. (1966). Heterokaryosis. *In* "The Fungi: An Advanced Treatise" (G. C. Ainsworth and A. S. Sussman, eds.), Vol. 2, pp. 567–588. Academic Press, New York.

Day, A. W. (1972). Genetic implications of current models of somatic nuclear division in fungi. *Can. J. Bot.* **50**, 1337–1347.

Day, P. R. (1970). The significance of genetic mechanisms in soil fungi. *In* "Root Diseases and Soil-Borne Pathogens" (T. A. Toussoun, R. V. Bega, and P. E. Nelson, eds.), pp. 69–74. Univ. of California Press, Berkeley.

de Bertoldi, M., and Caten, C. E. (1975). Isolation and haploidization of heterozygous diploid strains in a species of *Humicola*. *J. Gen. Microbiol.* **91**, 63–74.

Dorn, G. L. (1967). A revised map of the eight linkage groups of *Aspergillus nidulans*. *Genetics* **56**, 619–631.

Dutta, S. K., and Garber, E. D. (1960). Genetics of phytopathogenic fungi. III. An attempt to demonstrate the parasexual cycle in *Colletotrichum lagenarium*. *Bot. Gaz. (Chicago)* **122**, 118–121.

Fantini, A. A. (1962). Genetics and antibiotic production of *Emericellopsis* species. *Genetics* **47**, 161–177.

Grindle, M. (1963a). Heterokaryon compatibility of unrelated strains in the *Aspergillus nidulans* group. *Heredity* **18**, 191–204.

Grindle, M. (1963b). Heterokaryon compatibility of closely related wild isolates of *Aspergillus nidulans*. *Heredity* **18**, 397–405.

Hansen, H. N. (1938). The dual phenomenon in imperfect fungi. *Mycologia* **30**, 442–455.

Hansen, H. N., and Smith, R. E. (1932). The mechanism of variation in the Fungi Imperfecti: *Botrytis cinerea*. *Phytopathology* **22**, 953–964.

Hastie, A. C. (1962). Genetic recombination in the hop-wilt fungus *Verticillium albo-atrum*. *J. Gen. Microbiol.* **27**, 373–382.

Hastie, A. C. (1964). The parasexual cycle in *Verticillium albo-atrum*. *Genet. Res.* **5**, 305–315.

Hastie, A. C. (1967). Mitotic recombination in conidiophores of *Verticillium albo-atrum*. *Nature (London)* **214**, 249–252.

Hastie, A. C. (1968). Phialide analysis of mitotic recombination in *Verticillium*. *Mol. Gen. Genet.* **102**, 232–240.

Hastie, A. C. (1970a). Benlate-induced instability of *Aspergillus* diploids. *Nature (London)* **226**, 771.

Hastie, A. C. (1970b). The genetics of asexual phytopathogenic fungi with special reference to *Verticillium*. *In* "Root Diseases and Soil-Borne Pathogens" (T. A. Toussoun, R. V. Bega, and P. E. Nelson, eds.) pp. 55–62. Univ. of California Press, Berkeley.

Hastie, A. C. (1973). Hybridisation of *Verticillium albo-atrum* and *Verticillium dahliae*. *Trans. Br. Mycol. Soc.* **60**, 511–523.

Hastie, A. C. (1978). Genetic analysis in *Verticillium*. *In* "Pathological Wilting of Plants" (T. S. Sadasivan, C. V. Subramanian, R. Kalyanasundaram, and L. Saraswathi Devi, eds.), pp. 44–58. University of Madras, India.

Heagy, F. C., and Roper, J. A. (1952). Deoxyribose nucleic acid content of haploid and diploid *Aspergillus* conidia. *Nature (London)* **170**, 713.

Heale, J. B. (1966). Heterokaryon synthesis in *Verticillium*. *J. Gen. Microbiol.* **45**, 419–427.

Heale, J. B., Gafoor, A., and Rajasingham, K. C. (1968). Nuclear division in conidia and hyphae of *Verticillium albo-atrum*. *Can. J. Genet. Cytol.* **10**, 321–340.

Hoffman, G. M. (1967). Untersuchungen über die Heterokaryosebildung und den Parasexualcyclus bei *Fusarium oxysporum*. III. Paarungsversuche mit auxotrophen Mutanten von *Fusarium oxysporum* f. *callistephi*. *Arch. Mikrobiol.* **56**, 40–59.

Holliday, R. (1960). A new method for the identification of biochemical mutants of microorganisms. *Nature (London)* **178**, 987.

Huang, H. C., and Tinline, R. D. (1974). Somatic mitosis in haploid and diploid strains of *Cochliobolus sativus*. *Can. J. Bot.* **52**, 1561–1568.

Ikeda, Y., Nakamura, K., Uchida, K., and Ishitani, C. (1957). Two attempts upon improving an industrial strain of *Aspergillus oryzae* through somatic recombination and polyploidisation. *J. Gen. Appl. Microbiol.* **3**, 93–101.

Ingle, M. R. (1972). Ph.D. Thesis, University of Dundee, Scotland.

Ingle, M. R., and Hastie, C. A. (1974). Environmental factors affecting the formation of diploids in *Verticillium albo-atrum*. *Trans. Br. Mycol. Soc.* **62**, 313–321.

Ingram, R. (1968). *Verticillium dahliae* var. *longisporum*, a stable diploid. *Trans. Br. Mycol. Soc.* **51**, 339–341.

Ishitani, C., Ikeda, Y., and Sakaguchi, K. (1956). Hereditary variation and genetic recombination in Koji-molds (*Aspergillus oryzae* and *A. sojae*), VI. Genetic recombination in heterozygous diploids. *J. Gen. Appl. Microbiol.* **2**, 401–430.

Jinks, J. L. (1952a). Heterokaryosis: A system of adaptation in wild fungi. *Proc. R. Soc. London, Ser. B* **140**, 83–99.

Jinks, J. L. (1952b). Heterokaryosis in wild *Penicillium*. *Heredity* **6**, 77–87.

Jinks, J. L. (1959). The genetic basis of "duality" in imperfect fungi. *Heredity* **15**, 525–528.

Jinks, J. L. (1963). Cytoplasmic inheritance in fungi. *In* "Methodology in Basic Genetics" (W. J. Burdette, ed.), pp. 325–353. Holden-Day, San Francisco, California.

Jinks, J. L. (1966). Extranuclear inheritance. *In* "The Fungi: An Advanced Treatise" (G. C. Ainsworth and A. S. Sussman, eds.), Vol. 2, pp. 619–660. Academic Press, New York.

Jinks, J. L., Caten, C. E., Simchen, G., and Croft, J. H. (1966). Heterokaryon incompatibility and variation in wild populations of *Aspergillus nidulans*. *Heredity* **21**, 227–239.

Jones, D. A. (1965). Heterokaryon compatibility in the *Aspergillus glaucus* Link. group. *Heredity* **20**, 49–56.

Kafer, E. (1961). The processes of spontaneous recombination in vegetative nuclei of *Aspergillus nidulans*. *Genetics* **46**, 1581–1609.

Kafer, E. (1962). Translocations in stocks in *Aspergillus nidulans*. *Genetica* **33**, 59–68.

Lewis, L. A., and Barron, G. L. (1964). The pattern of the parasexual cycle in *Aspergillus amstelodami*. *Genet. Res.* **5**, 162–163.

Lhoas, P. (1961). Mitotic haploidisation by treatment of *Aspergillus niger* diploids with para-fluorophenylalanine. *Nature (London)* **190**, 744.

Lhoas, P. (1971). Transmission of a double standard RNA virus to a strain of *Penicillium stoloniferum* through heterokaryosis. *Nature (London)* **230**, 248–249.

McCully, K. S., and Forbes, E. (1965). The use of *p*-fluorophenylalanine with master strains of *Aspergillus nidulans* for assigning genes to linkage groups. *Genet. Res.* **6**, 352–359.

Ming, Y. N., Lin., P. C., and Yu, T. F. (1966). Heterokaryosis in *Fusarium fujikuroi* (Sacc. Wr.) *Sci. Sin.* **15**, 371–378.

Nga, B. H., Teo, S.-P., and Lim. G. (1975). The occurrence in nature of a diploid strain of *Aspergillus niger*. *J. Gen. Microbiol.* **88**, 364–366.

Nüesch, J., Treichler, H. J., and Leirsch, M. (1973). The biosynthesis of cephalosporin C. *In* "The Genetics of Industrial Microorganisms" (Z. Vaněk, Z. Hošťálek, and J. Cudlín, eds.), pp. 309–334. Academia, Prague.

Papa, K. E. (1976). Linkage groups in *Aspergillus flavus*. *Mycologia* **68**, 159–165.

Parmeter, J. R., Jr., Snyder, W. C., and Reichle, R. E. (1963). Heterokaryosis and variability in plant-pathogenic fungi. *Annu. Rev. Phytopathol.* **1**, 51–76.

Phaff, H. J., Miller, M. W., and Mrak, A. M. (1966). "The Life of Yeasts." Harvard Univ. Press, Cambridge, Massachusetts.

Pontecorvo, G. (1949). Auxanographic techniques in biochemical genetics. *J. Gen. Microbiol.* **3**, 122.

Pontecorvo, G. (1954). Mitotic recombination in the genetic systems of filamentous fungi. *Caryologia* **6**, Suppl. 192–200.

Pontecorvo, G. (1956). The parasexual cycle in fungi. *Annu. Rev. Microbiol.* **10**, 393–400.

Pontecorvo, G. (1959). "Trends in Genetic Analysis." Columbia Univ. Press, New York.

Pontecorvo, G., and Sermonti, G. (1954). Recombination without sexual reproduction in *Penicillium chrysogenum*. *Nature (London)* **172**, 126–127.

Pontecorvo, G., Roper, J. A., and Forbes, E. (1953a). Genetic recombination without sexual reproduction in *Aspergillus niger*. *J. Gen. Microbiol.* **8**, 198–210.

Pontecorvo, G., Roper, J. A., Hemmons, L. M., MacDonald, K. D., and Bufton, A. W. J. (1953b). The genetics of *Aspergillus nidulans*. *Adv. Genet.* **5**, 141–238.

Pontecorvo, G., Tarr-Gloor, E., and Forbes, E. (1954). Analysis of mitotic recombination in *Aspergillus nidulans*. *J. Genet.* **52**, 226–237.

Puhalla, J. E., and Mayfield, J. E. (1974). The mechanism of heterokaryotic growth in *Verticillium dahliae*. *Genetics* **76**, 411–422.

Robinow, C. F., and Caten, C. E. (1969). Mitosis in *Aspergillus nidulans*. *J. Cell Sci.* **5**, 403-431.

Roper, J. A. (1952). Production of heterozygous diploids in filamentous fungi. *Experientia* **8**, 14-15.

Roper, J. A. (1966). The parasexual cycle. *In* "The Fungi: An Advanced Treatise" (G. C. Ainsworth and A. S. Sussman, eds.), Vol. I, pp. 589-617. Academic Press, New York.

Roper, J. A. (1973). Mitotic recombination and mitotic nonconformity in fungi. *In* "Genetics of Industrial Microorganisms" (Z. Vanék, Z. Hošťálek, and J. Cudlín, eds.), pp. 81-88, Academia, Prague.

Roper, J. A., and Kafer, E. (1957). Acriflavine resistant mutants of *Aspergillus nidulans*. *J. Gen. Microbiol.* **16**, 660-667.

Roper, J. A. and Pritchard, R. H. (1955). The recovery of the complementary products of mitotic crossing-over. *Nature (London)* **175**, 639.

Rosenberger, R. F., and Kessel, M. (1967). Synchrony of nuclear replication in individual hyphae of *Aspergillus nidulans*. *J. Bacteriol.* **94**, 1464-1469.

Sanchez, L. E., Leary, J. V., and Endo, R. M. (1976). Heterokaryosis in *Fusarium oxysporum* f. *lycopersici*. *J. Gen. Microbiol.* **93**, 219-226.

Sanderson, K. E., and Srb, A. M. (1965). Heterokaryosis and parasexuality in the fungus *Ascochyta imperfecta*. *Am. J. Bot.* **52**, 72-81.

Sansome, E. R. (1946). Induction of "gigas" forms of *Penicillium notatum* by treatment with camphor vapor. *Nature (London)* **157**, 843.

Sansome, E. R. (1949). Spontaneous mutation in standard and "gigas" forms of *Penicillium notatum* strain 1249 B 21. *Trans. Br. Mycol. Soc.* **32**, 305-314.

Stark, C. (1961). Das Auftreten der *Verticillium* Tracheomykosen in Hamburg Gartenbaukulturen: Die Garten-baukulturen. *Gartenbauwissenschaft* **26**, 493-528.

Stern, C. (1936). Somatic crossing-over and segregation in *Drosophila melanogaster*. *Genetics* **21**, 625-730.

Stromnaes, A., Garber, E. D., and Beraha, L. (1964). Genetics of phytopathogenic fungi. IX. Heterokaryosis and the parasexual cycle in *Penicillium italicum* and *Penicillium digitatum*. *Can. J. Bot.* **42**, 423-427.

Threlfall, R. J. (1971). Mutants of *Botrytis allii*. *Microb. Genet. Bull.* **33**, 4.

Tinline, R. D. (1962). *Cochliobolus sativus*. V. Heterokaryosis and parasexuality. *Can. J. Bot.* **40**, 425-437.

Tinline, R. D., and MacNeill, B. H. (1969). Parasexuality in plant pathogenic fungi. *Annu. Rev. Phytopathol.* **7**, 147-170.

Tuveson, R. W., and Coy, D. O. (1961). Heterokaryosis and somatic recombination in *Cephalosporium mycophilum*. *Mycologia* **53**, 244-253.

Typas, M. A., and Heale, J. B. (1976). Heterokaryosis and the role of cytoplasmic inheritance in dark resting structure formation in *Verticillium* spp. *Mol. Gen. Genet.* **146**, 17-26.

Typas, M. A., and Heale, J. B. (1978). Heterozygous diploid analyses via the parasexual cycle, and a cytoplasmic pattern of inheritance in *Verticillium* spp. *Genet. Res.* **31**, 131-144.

Van Uden, N., and Fell, J. W. (1968). Marine yeasts. *Adv. Microbiol. Sea* **1**, 167-202.

Yamasaki, Y., and Niizeki, H. (1965). Studies on variation in the rice blast fungus, *Pyricularia oryzae* Cav. I. Karyological and genetical studies on variation. *Bull. Natl. Inst. Agric. Sci., Ser. D* **13**, 231-273.

VII

TECHNIQUES
FOR INVESTIGATION

30

Isolation, Cultivation, and Maintenance of Conidial Fungi

S.C. Jong

I. ISOLATION

The need for living cultures of conidial fungi is recognized in systematics as well as in the study of problems relating to health, food resources, and the environment. Contemporary work in many areas of mycological research in-

Biology of Conidial Fungi, Vol. 2

volves the isolation and maintenance of fungi in pure culture. For example, in order to prove the connection between an ascomycete or basidiomycete and its conidial state it is essential to obtain cultures derived from both single ascospores or basidiospores and single conidia. Living cultures are required to characterize, identify, and classify conidial fungi. They are used to derive modern taxonomic characters, such as mechanisms of conidiogenesis and conidiomatal ontogeny. Living cultures are also employed in examinations of growth requirements of conidial fungi, the by-products of their metabolism, and their pathogenicity and toxicity.

Many different methods have been developed to isolate fungi. All are selective and designed to prohibit the growth of undesired microorganisms, to enhance the development of the desired fungi, or both. In the case of conidial fungi the best selective method is single-spore isolation (see Kendrick *et al.*, 1979).

A. Direct Isolation

Many terrestrial conidial fungi, such as *Aspergillus, Penicillium,* and *Paecilomyces,* sporulate aerially on the surface of natural substrates. The aerial conidia and conidiophores can be harvested with a sharply pointed microneedle under a binocular dissecting microscope and sown on suitable agar media in petri dishes (Raper and Fennell, 1965). Sometimes it is necessary to incubate freshly collected samples of the fungi in a damp chamber to stimulate the formation of conidia.

The moist chamber may be a jar or petri dish with several layers of sterile paper towels or heavy-grade filter paper placed on the bottom. Usually 1–2 weeks is sufficient for conidial fungi to develop sporulating structures on the surface of incubated substrates. Terrestrial fungi often do not produce conidia on material soaked in water; therefore to keep the chamber moist a small amount of sterile water is added daily near one edge, and the jar or dish is rotated to disperse the water.

Many aquatic conidial fungi have been isolated on submerged leaves, twigs, and test blocks of various plants in both fresh and marine water. Colonized materials collected from the field are washed in distilled water or seawater and placed in sterile petri dishes half full of autoclaved fresh water or seawater. Mycelia and conidia develop on the incubated material within 2–3 days. When released, some conidia remain suspended in the water, while others settle to the bottom of the dish. When a dissecting microscope is used, it is possible to pick up conidia with a needle or a fine capillary pipet (Ingold, 1942; Crane, 1968; Gareth Jones, 1971). For both terrestrial and marine fungi it is important to examine the material in the moist chamber or water frequently so that uncommon or slow-growing species can be discovered and isolated before they are overgrown by more abundant, rapidly growing fungi and bacteria.

B. Direct Inoculation

Conidial fungi can be found in practically any habitat and on all substrates of organic origin. To isolate fungi these materials may be scraped lightly with sterile forceps or microspatulas and sprinkled on the surface of agar plates containing selected media (Waksman, 1916; Crisan, 1964; Riker and Riker, 1936; Ajello *et al.*, 1963). Appropriate substrates include crumbs of soil, small chunks of decaying or submerged wood, a few milliliters of sewage or polluted fresh or marine water, pieces of diseased plant, insect, animal, or human tissue, rotting fruit or spoiled grain, decaying cotton fabric, a pinch of leaf mold, compost, or wet hay, and other organic debris. After a proper period of incubation, hyphae growing out from the material can be transferred to fresh media. Conidia may be picked off using the direct isolation method described above.

Before inoculating the selective media, badly contaminated material may be washed thoroughly with sterile distilled water to eliminate many of the surface contaminants. In some cases the surface layers of material may be removed aseptically to eliminate contaminating microorganisms. The washing technique has been applied to both soil and plant parts. Harley and Waid (1955) have isolated fungi from the surface of living roots by a serial root-washing procedure, while Watson (1960) has obtained a high percentage of *Fusarium* isolates and other genera from washed soil. Serial washing of soil may be performed using an elaborate apparatus designed by Parkinson and Williams (1961), which consists of a Perspex box fitted with stainless steel sieves of graded sizes. An automatic washing apparatus has been described by Hering (1966).

To isolate saprophytic fungi growing beneath the surface of material, or pathogenic fungi from diseased or infected tissues, surface sterilization with chemicals or gases is necessary to eliminate surface contaminants. This sterilization process should be carried out carefully and be limited to destruction of only organisms at the surface. The chemical solutions commonly used include 0.001% mercuric chloride, 0.1% silver nitrate, 0.1% mercuric cyanide, 0.35% sodium or calcium hypochlorite, 1% carbolic acid, 2% potassium permanganate, 3% hydrogen peroxide, 5% Formalin, and 75% ethyl alcohol. The period of immersion is usually 1–5 min, after which the material should be washed in several changes of sterile distilled water before the isolation is attempted. Since fungi appear to possess different susceptibilities to the toxicity of different chemicals, their concentration and length of application vary. Johnson and Sparrow (1961) have isolated caulicolous marine fungi imperfecti by the direct transplant method. Cylindrical segments of invaded material are submerged in 50% hydrogen peroxide for 1 min, washed thoroughly in five to seven changes of sterile seawater, and then placed on an agar medium.

Gases such as formaldehyde, ethylene oxide, and propylene oxide are also used in surface sterilization. There are several types of solid polymers of formaldehyde, but the most common one is paraformaldehyde. At ordinary temperatures it gradually gives off gaseous formaldehyde with a sharp, irritating odor. Ethylene oxide, which has a rather pleasant ethereal odor, is effective against all types of microorganisms. Propylene oxide is similar in properties to ethylene oxide but less active biologically. Booth (1971b) has successfully surface-sterilized woody materials by placing them in a screw-top jar with a small cotton-wool pad moistened with 2 ml of propylene oxide. After 30 min in the closed jar the materials are removed, cut into slices, and then plated out on suitable agar media.

In plant pathology laboratories slowly growing fungus pathogens are generally isolated by placing thin sections of diseased plant parts on agar drops previously prepared on glass slides or coverslips. The latter are inverted and placed on glass rings. Desiccation is prevented by placing the slide or coverslip in a moist chamber. After a suitable period of incubation the sections can be examined with a light microscope. Those showing growth of fungi are transferred to an agar medium. The advantage of hanging agar drops on glass rings is that the tissue sections can be examined with the microscope directly through the glass coverslips without risk of contamination from air, permitting observation of colony and conidial development.

C. Plating Methods

In the study of ecological distribution of various species of soil fungi, Warcup (1950) used a plating technique for isolation. A soil plate is prepared by transferring a small amount of soil with a sterile microspatula to the bottom of a sterile petri dish. A little sterile water may be added to assist in the dispersion of very dry soils that do not break up easily. Approximately 10 ml of melted agar medium held at about 45°C is added, and the soil particles are dispersed by gently shaking and rotating the dish before the agar solidifies. This technique of retaining the intact small particles of soil allows one to culture slowly growing fungi that are usually suppressed in the dilution plate method by heavily sporulating fungi. Using Warcup's method, Parkinson (1957) isolated desired conidial fungi from the rhizosphere, and Eggins and Pugh (1962) obtained cellulolytic fungi from soil.

Johnson and Sparrow (1961) successfully isolated bacteria-free colonies of culm-inhabiting marine fungi by modifying Warcup's technique. About 15 ml of agar medium is poured into a petri dish and allowed to solidify. After a thorough washing with sterile seawater the surface-sterilized culm segments are pushed lightly onto the agar plate. A selected agar medium is then melted, cooled to approximately 45°C, and poured onto the plate sufficiently deep to cover the segments with 1–2 mm of medium. Hyphae often grow out into the medium beyond any adjacent bacterial colonies, which are very much restricted within the agar.

D. Dilution Plate Methods

Dilution plate methods have been used to isolate both fungi and bacteria. Soil, polluted water, infected tissue, wood pulp, or other organic debris is thoroughly suspended in sterile distilled water either mechanically or by hand. A stirrer, mechanical shaker, or Waring blender with the addition of glass beads may be used to facilitate breakdown of the material. The resulting suspension is serially diluted until the desired final dilution is reached. The proper amount of dilution is entirely a matter of choice. Bisby *et al.* (1933) has recommended an average number of 25 colonies per plate, and this number has also been deemed advisable by James and Sutherland (1939) as a result of statistical tests. In order to secure a satisfactory separation of colonies a loopful of the suspension is evenly dispersed over the surface of the agar medium by smearing with a bent sterile glass rod. Each suspension of the dilution series may also be taken up in a loop and streaked repeatedly across the surface of the medium in such a manner that the inoculum is gradually diluted as the streaking proceeds. The surface of the medium should be dried before use to discourage the spread of bacterial colonies.

Many different procedures, with numerous modifications, have been developed for making dilutions in tubes or on plates. A loop or a pipet is used according to the preference of the individual worker, the type of substrate, and the nature of the fungi. The details of these methods have been described by Johnson *et al.* (1959). The conventional method involves introduction of one loopful of the primary suspension into a tube of melted agar medium cooled to about 45°C. The contents of the tube are stirred with a flamed loop transfer needle, and transfer is made directly to a second tube of melted agar medium held at about 45°C. The same manipulation is repeated several times. The contents of each tube are then poured into separate petri dishes. After the agar medium has gelled the plates are inverted to prevent condensation on the cover. Each sample may be plated out in replicates of three to five series, which are then incubated at various temperatures. Likewise, the dilutions can be made directly into melted agar previously poured into petri dishes and cooled to about 45°C (Riker and Riker, 1936).

The dilution method is especially useful and convenient for isolating single conidia and colonies. Various conidial fungi can be selected either by scanning the agar surface for different colonies or by employing a technique combining microscopic examination with removal of the selected conidia. It must be remembered that cultures secured by isolating a single colony are not always free of contaminating bacteria and fungi, because some of the colonies that appear may not be derived from a single conidium or cell. The presence of bacteria and yeasts is particularly difficult to detect, and single-spore isolation of these cultures is desirable.

Single-conidium cultures of fungi may be obtained by a variety of semimechanical cutting devices and mechanical micromanipulators. With semimechanical aids a comparatively thin conidial suspension is spread on or mixed on agar plates by either the streak or dilution method. By inverting these unopened plates on the stage of the microscope, isolated conidia may be observed and located with a dot of ink. The plates are then turned upright, and individual conidia may be obtained by cutting and removing a small disk of agar immediately above the marked area. During observation through a microscope, Hanna (1924) was able to pick up single spores with a fine sewing needle. Booth (1971b) has also transferred *Fusarium* conidia using this technique. Keitt (1915) incised the agar containing the marked single conidium by means of a cylindrical loop. Before the delimited agar disk is lifted out with a flat-tipped needle, it is examined again under a low-power microscope to ascertain that the conidium selected is actually included in the disk and that no other structures are present. Sass (1929) has devised a convex spatula with a hole about 1.5 mm in diameter in the center. When the spatula is pressed down on the agar, a conidium-bearing disk is pushed up through the hole.

Instead of examining the petri dishes through the bottom, La Rue (1920), Lambert (1939), Raper (1963), and Cox (1952) located the conidia by direct microscopic examination of the surface of the agar medium in opened plates. La Rue (1920) designed a special cutting device resembling a microscope objective with a circular blade screwed into the revolving nosepiece. When the cutter is lowered an agar disk bearing a selected conidium is incised and can then be transferred with a flattened needle. Lambert (1939) has devised a "biscuit cutter," about 0.75 mm in diameter or slightly smaller than the field of the low-power objective, which is mounted vertically in a threaded brass plug. This spore isolator may also be screwed onto the oil immersion objective and centered. It is used first to mark off a circle concentric with the microscopic field of the low-power objective. With the latter, an isolated single conidium is then selected. Without disturbing the plate this objective is replaced with the oil immersion objective equipped with the cutter. By carefully lowering the oil immersion objective the agar disk bearing the conidium can be lifted out by the cutter. Raper (1963) has modified a device combining some of the advantages of the La Rue and Lambert methods for rapid isolation of single conidia and single apical cells of vegetative hyphae. Necessary equipment consists of a vertically swinging arm, tipped with a conical cutter and actuated by a cable release, and a supporting frame that can be attached to the barrel of an objective. Cox (1952) has successfully used light as an aid in conidium isolation. A well-isolated conidium is located by adjusting the diaphragm apertures of the microscope lamp and the condenser to reduce the light beam passed through the agar to a ring slightly larger than the conidium itself. The agar surrounding the circumference of the light beam is then cut out by a framed platinum loop and removed with a spatula. This technique is also effective in isolating single hyphal tips of rapidly growing fungi.

E. Baiting Methods

The incorporation of organic matter into soil or water provides a series of habitats which become colonized by fungi present in nature. Special "baiting" techniques have been developed to isolate particular groups of fungi from various habitats. Many conidial fungi have been recovered with the use of various natural substrates or artificial agars (selective media) as bait either in the field or in the laboratory. Root disease organisms in soil have been isolated by baiting the soil sample with suitable plant host material (Garnett, 1944). Paraffinolytic fungi have been obtained by Rynearson and Peterson (1965) by inserting paraffin rods into undisturbed soil. *Microsporum gypseum* has been isolated by White *et al.* (1950) by burying pieces of woolen fabric in the soil. Anastasiou (1962, 1963), Kirk (1969), Shearer (1972), and Anastasiou and Churchland (1969) have retrieved aquatic conidial fungi from wood panels and decaying leaves submerged for 4 weeks to several months in water. Artificial bait such as tubes of agar, agar-coated slides, and agar cylinders has also been used to obtain soil and aquatic fungi (Chesters, 1940, 1948; Thornton, 1952; Bandoni *et al.,* 1975). Detailed coverage of the methods for retrieving aquatic fungi has been provided by Johnson and Sparrow (1961) and Gareth Jones (1971).

To bait conidial fungi in the laboratory the samples collected from different habitats are placed in petri dishes. Various baiting materials, previously sterilized by autoclaving, are added, and the dishes are then incubated. After a week or more, bait infected with the desired conidial fungi can be observed with a binocular dissecting microscope.

Baiting in the laboratory has been particularly useful in recovery of a specific fungus or group of fungi from soil. *Cladosporium resinae,* which occurs in aviation fuel in both bulk storage and aircraft tanks, has been isolated from soil in creosote agar (Christensen *et al.,* 1942) and by baiting with creosote-soaked matchsticks in petri dishes (Parbery, 1967). Predaceous fungi have been obtained by adding soil to agar plates infested with nematodes (Drechsler, 1941; Barron, 1977). Cellulolytic fungi have been recovered by baiting soil and polluted water with agar medium in which very finely powdered cellulose provides the major carbon source (Cooke and Busch, 1957; Eggins and Pugh, 1962). Dermatophytes and other keratinophilic fungi have been collected by baiting soil, sewage, sludge, and polluted water with various keratinized tissues (skin, hair, nails, claws, spines, and so on) of humans and animals. The substrate used most frequently and successfully as bait for this group of fungi is hair. The history of the hair baiting technique has been reviewed by Benedek (1962a). Toma (1929) was the first to employ hair as a culture medium for dermatophytes. Apparently unaware of Toma's work, Karling (1946) independently developed a precise technique using hair as bait for keratinophilic fungi in soil. Vanbreuseghem (1952) later introduced Karling's hair baiting technique into medical mycology.

F. Centrifugation Methods

In isolating nematode-destroying fungi from soil, Barron (1977) used a differential centrifugation technique to separate endoparasites from predators. He realized that endoparasitic fungi persist in soil as hyphae or conidia. Large conidia and other undesired materials can be separated at low centrifugation speeds; the supernatant is then further centrifuged at a higher speed. The resulting pellet containing the small conidia of endoparasites can then be spread over a water agar plate baited with nematodes. Paden (1967) developed a centrifugation technique employing sucrose density gradient columns to separate ascospores from soil material particles, but he also suggested that this method could be used to recover conidial fungi such as *Fusarium* and certain dermatophytes that have readily identifiable propagules.

G. Differential Temperatures

Fungi require certain minimal temperatures in order to grow and carry out their metabolic activities. It is customary to divide fungi into psychrophiles, mesophiles, and thermophiles. The great majority of fungi are mesophiles which grow between 10° and 40°C, and their routine isolation can be performed at room temperature. Psychrophilic fungi, however, are restricted to cold environments in which temperatures rarely rise above 10°C (Deverall, 1968) and are frequently found in refrigerated foods, cold water, and soil in nature. Thermophilic fungi, capable of growing at 50°C or higher and incapable of growing at temperatures below about 20°C (Cooney and Emerson, 1964; Emerson, 1968), are most frequently found in nature where organic materials are decomposing at elevated temperatures. For selective isolation of these two groups of fungi the incubation temperature should be set at 5°C or lower for psychrophiles and 50°C for thermophiles.

H. Selective Media

The value of incorporating selective agents into the medium when isolating specific fungi has been well documented. The details and merits of the various selective media designed for the isolation of pathogenic fungi are reviewed by Tsao (1970). The purpose of using selective media is to inhibit the development of undesired microorganisms, enhance development of the desired fungi, or both. Antibiotics are most commonly used in the selective isolation of fungi from soil and contaminated materials. Penicillin, streptomycin, chlorotetracycline, and chloramphenicol are used to inhibit bacterial growth, while cycloheximide, pimaricin, and nystatin are used in suppression of fungal growth (Thompson, 1945; Georg *et al.*, 1954; Papvizas and Davey, 1961). For example, commercially available BBL mycosal agar, containing 0.4 mg/ml of cycloheximide and 0.05 mg/ml of chloramphenicol, is recommended for the isolation of dermatophytes and most fungi causing systemic mycoses. The incorporated antibiotics inhibit the majority of nonpathogenic fungi and bacteria. Both cycloheximide and chloramphenicol are heat-stable and can be added to the medium before autoclaving. However, cycloheximide also inhibits growth of pathogenic fungi such as *Allescheria boydii, Aspergillus fumigatus, Cryptococcus neoformans, Candida* species, and the yeast forms of certain dimorphic fungi (Georg *et al.*, 1954; McDonough *et al.*, 1960a,b; Emmons *et al.*, 1977). Taplin (1965) and Dolan (1972) therefore recommend the addition of gentamicin which inhibits bacteria without affecting the growth of fungal pathogens. Gentamicin is a basic, stable, water-soluble antibiotic effective against both Gram-positive and Gram-negative bacteria and can be autoclaved. A combination of chloramphenicol at 0.16 mg/ml and gentamicin at 0.5 mg/ml is suggested by Dolan (1971, 1972) for the isolation of pathogenic fungi from clinical specimens.

Acidification of the medium at or below pH 4.0 inhibits the growth of all but acid-tolerant bacteria without seriously disturbing the development of many fungi. Since the pH level of agar media changes when it is autoclaved, the pH is adjusted between the autoclaving and pouring steps. The desired acidity can be achieved by adding a calculated amount of hydrochloric acid, lactic acid, or rose bengal to the medium. To inhibit the growth of acid-tolerant bacteria the addition of a mixture of 2 ppm actinomycin and 50 ppm aureomycin has proved very satisfactory (Beech and Carr, 1955; Beech and Davenport, 1971).

Selective isolation is most effective on media containing the carbon sources most effective for growth of the fungi desired. In medical mycology laboratories Sabouraud's glucose agar is commonly employed in the preliminary isolation of fungi pathogenic to humans and animals (Emmons *et al.*, 1977). Other media often used include a mycosal or mycobiotic agar (Sabouraud's medium plus antibiotics) and a brain–heart infusion—sheep blood (5%) agar (Rippon, 1974). Cooney and Emerson (1964) found yeast starch agar and yeast glucose agar particularly favorable for growth of thermophilic fungi. A neopeptone–dextrose–rose bengal–aureomycin agar gave good results for the recovery of fungi in polluted and sewage waters (Cooke, 1963). A cellulose agar has proved very effective in the detection and isolation of cellulolytic fungi (Eggins and Pugh, 1962; Cooke and Busch, 1957; Dickinson and Pugh, 1965). To isolate fungi from pulp and paper mills Wang (1965) has used Bacto malt extract agar and Bacto potato dextrose agar. Other media vary according to the specific requirements of the desired fungi. Niger seed–creatinine media have been used for the isolation of *Cryptococcus neoformans* (Staib and Seeliger, 1968), an alcohol-containing medium for *Verticillium* species (Nadakavukaren and Horner, 1959), galactose or inulin as a carbon source for *Fusarium* (Park, 1963; Bouhot and Billotte, 1964), a yeast extract-containing medium for *Histoplasma capsulatum* (Smith and Furcolow, 1964), and a V-8 juice–dextrose–yeast extract agar for *Thielaviopsis basicola* (Papavizas and Davey, 1961).

Agar media which undergo specific pigmentation changes during fungal growth are commonly employed in hospital laboratories for rapid diagnosis and primary isolation of certain pathogenic fungi. Since dermatophytes are capable of raising the pH of the medium (Goldfarb and Hermann, 1956), acid ink blue, bromthymol blue, and phenol red have been used as color indicators to differentiate dermatophytes from contaminants (Baxter, 1965; Quaife, 1968; Taplin *et al.*, 1969). Also, phosphomolybdic acid or tetrazolium has been used to differentiate species of *Candida* (Pagano *et al.*, 1958; MacLaren, 1960).

II. CULTIVATION

After the initial isolation of conidial fungi, the cultures may be grown for identification, studies on nutritional requirements and metabolic activity, and so on. Although growth requirements of conidial fungi cover a wide range of conditions, most can readily be cultivated by proper selection of media and other growth factors. Knowledge of the physiological aspects of fungi in general is very helpful in deciding on culture conditions. Useful references on fungal physiology include the books written by Hawker (1950), Lilly and Barnett (1951), and Cochrane (1958) and the current review by Hall (Chapter 27, this volume).

A. Purification

Cultures of fungi contaminated by bacteria, foreign fungi, or mites may be purified using a number of different techniques. Many isolation procedures, dealt with in the preceding section, can be applied in the purification of conidial fungi in culture. When contaminated cultures produce conidia in abundance, single-spore isolation and dilution plate methods are commonly used. Purification may also be achieved by acidification of the medium and incorporation of selective antimicrobial chemical agents. Blank and Tiffney (1936) obtained bacteria-free cultures by growing fungi on ultraviolet-irradiated media. Growth of bacteria is greatly inhibited, while there is little or no effect on development of the fungi.

The ability of many fungi to penetrate the agar medium may be used to separate fungi from bacteria. Brown (1924) has employed this advantage by cutting through the agar with a sterile knife in advance of the growing colony. The agar supporting growth of the mycelium is then removed, inverted, and placed in a sterile petri dish. After a proper period of incubation, mycelia growing through the agar from the undersurface can be cut away to make transfers. Raper (1937) has used a small glass cylinder (van Tieghem cell) to one edge of which are fused three small glass beads 0.3–0.5 mm in diameter. The sterile cylinder is placed in a petri dish so that it rests on the beads. Melted agar is then poured into the dish until it fills the cylinder. The contaminated culture is then inoculated inside the cylinder. As growth proceeds, the fungus grows underneath the free edges of the modified van Tieghem cell while the bacteria remain confined within the cylinder. The hyphae outside the cylinder therefore provide the inoculum for pure cultures. Ark and Dickey (1950) have modified Raper's method by replacing the glass beads with small pellets of modeling clay. The clay secures the cylinder to the bottom of the plate so that the plate can be inverted during incubation. In a third method, the mixed culture is inoculated at the center of an agar plate and a sterile coverslip then pressed down over that region. Many fungi will reach the air by passing through the agar.

Mite infestation of fungus cultures is a common problem (Hansen and Snyder, 1939; Emmons, 1940; Smith, 1960; Dayal et al., 1964). Mites not only eat pure cultures but also contaminate them with other fungi or bacteria. They frequently occur in nature on plant and animal materials and soil particles routinely brought into laboratories for examination and culture. Because mites are usually about 100 μm long and just at the limit of visibility with the unaided eye, they can penetrate tightened cotton plugs and crawl from one culture to another with conidia adhering to their bodies and thus spread contamination. Subden and Threlkeld (1966) have developed a method for eradicating these pests from infested cultures. The absolute control of infested cultures involves two steps. All adults, nymphal instars, and eggs are first killed by freezing the cultures at $-18°C$ for 24 h. This has little or no effect on the viability of conidial fungi. Second, all equipment except tubes, plugs, and media is treated with Kelthane (Rohm and Haas Company, Philadelphia, Pennsylvania).

B. Culture Media

In general, the essential growth substances required by fungi include carbohydrates, proteins, fats, amino acids, vitamins, minerals, air, and water. Although artificial media may contain all the essential elements and compounds that certain fungi need for synthesis of their cell constituents and for operation of their life processes, so far there is no standard medium upon which all fungi will grow. The actual quantity of food material required is usually very small. Some fungi prefer ammonia to nitrate as a nitrogen source, while others require organic nitrogen in forms such as asparagine or amino acids. Some fungi are able to grow normally on a substrate containing no vitamins, while others require the addition of certain vitamins to the growth medium. In order to cultivate these fungi, vitamins such as thiamin, biotin, inositol, pyridoxine, nicotinic acid, and pantothenic acid must be added to the medium. In most cases conidial fungi can be cultivated by supplying only simple sugars such as glucose and sucrose, mineral salts including a source of inorganic nitrogen, air, and water. Many are even able to grow and produce conidia on plain agar media made with tap water. However, a medium conducive to growth may not be suitable for sporulation. As a general rule low-nutrient media are recommended for sporulation.

On the basis of composition, Lilly and Barnett (1951) roughly classified culture media into three categories: natural, semisynthetic, and synthetic. Natural media are based on complex natural materials readily at hand such as cornmeal, V-8 juice, lima beans, carrots, potatoes, onions, prunes, dung, and soil. These are usually used in the form of an extract, infusion, or decoction. Semisynthetic media contain both natural ingredients and defined components. Potato glucose agar, yeast extract–glucose agar, and peptone–glucose agar are examples of this type. Synthetic media are of known composition, and each component and its concentration can be controlled as desired. Thus the primary value of these defined media is for microbiological assays and enzymatic or other biochemical studies.

Culture media may be either solid or liquid. Solid media are either natural substances (e.g., pieces of roots, plugs of potato or carrot, and string beans) or nutrient solutions made into jelly by the addition of agar or gelatin. Agar is a complex polysaccharide extractable with hot water from various marine red algae. It is almost exclusively used at a concentration of about 1.5% for media which are not strongly acid. For media with distinctly acid reactions, 2% or even more agar is required. Agar media do not melt until the temperature exceeds 95°C, and molten media do not resolidify until the temperature falls below 40°C. Gelatin is not used very much because it melts easily at temperatures above 30°C. The advantages of both solid and liquid media have already been discussed by Lilly and Barnett (1951). Agar media are commonly used in preliminary experiments, isolation, identification, and maintenance of cultures. They are particularly valuable in the study of conidial development. Liquid media are primarily used in biochemical work, particularly in studies on metabolic by-products, various metabolite deficiencies, and microbial assays.

A great many different culture media have been used for isolation, identification, propagation, maintenance, morphological and nutritional studies, and biochemical investigations of fungi. A medium is usually designated by the name of the investigator who first used it (e.g., Czapek's medium, Leonian's medium, and Sabouraud's agar) or by its principal ingredients (e.g., glucose–asparagine medium, sucrose–nitrate medium, cornmeal agar). Formulations for some of the most commonly used media and the methods employed in their preparation can be found in the *Mycology Guidebook* (Stevens, 1974), *Methods in Microbiology*, Volume 4 (Booth, 1971c), and other laboratory manuals (Riker and Riker, 1936; Conn, 1957; Johnson *et al.*, 1959; Raper and Fennell, 1965). For this reason formulas for media will not be given consideration here. Furthermore, the American Type Culture Collection (ATTC), which maintains the most diverse collection of reference microorganisms and animal cells in the world, publishes a

Catalogue of Strains every 2 yr. This catalogue includes all the formulations of culture media found satisfactory for growth and maintenance of over 25,000 strains (American Type Culture Collection, 1980).

C. Culture Conditions

There is no doubt that in practically all cases temperature, pH, and light conditions affect the cultivation of fungi. The great majority of conidial fungi grow well at room temperature. However, the optimal temperature for growth may vary among species of a genus and even among strains of a single species. Some fungi which infect plants under snow, or which cause spoilage of refrigerated foods, are tolerant of low temperatures and can grow at or below 0°C (Brooks and Hansford, 1922; Deverall, 1968). Dermatophytes grow best in culture at 25°–35°C, which is below the host body temperature, while fungi causing systemic mycoses frequently require an incubation temperature of 37°C (Rippon, 1974). Thermophilic fungi grow satisfactorily at elevated temperatures of 45°–55°C (Emerson, 1968). Agar media on which thermophiles are grown tend to dry out rapidly, and extra care must be taken to ensure high humidity during the period of incubation. The range of temperatures for conidium formation is usually narrower than for mycelial development.

As reported in the literature compiled by Marsh *et al.* (1959), light influences the growth and sporulation of many conidial fungi. For instance, *Botrytis cinerea* and various species of *Fusarium* were reported to grow better in light than in dark (Paul, 1929; Fikry, 1932). On the contrary, mycelial growth of certain other *Fusarium* species was retarded by continuous exposure to light (Snyder and Hansen, 1941). However, most common fungi seem to grow equally well in light or darkness. Light has either a stimulatory or an inhibitory effect on the sporulation of some conidial fungi (Snyder and Hansen, 1941; Lukens, 1963). *Alternaria brassicae* var. *dauci, A. solani, Helminthosporium oryzae,* and *H. teres* require both light and darkness for sporulation (Leach, 1961; Witsch and Wagner, 1955; Lukens, 1963; Onesirosan and Banttari, 1969). It is common experience that many fungi in culture produce alternating zones of sporulation in response to diurnal changes in light intensity. According to Hawker (1950), many of these zonations are due to the effect of light in increasing the production of conidia, so that zones of intense sporulation alternate with those of sparse or no sporulation. Near-ultraviolet light has been shown to stimulate sporulation in conidial fungi, particularly those in the dematiaceous group. A practical guide to the effects of light on fungi has been provided by Leach (1971). For details, this excellent review article should be consulted.

In general, fungi grow well in slightly acid media, although most conidial fungi will grow normally within a pH range of 4.0–8.0. However, in some fungi a much more restricted pH range is necessary for sporulation than for mycelial growth. The pH of a medium in which a fungus is actively growing can change, depending on the composition and buffering capacity of the medium, the temperature of incubation, and the type of metabolites produced.

III. MAINTENANCE

The primary function of maintenance of cultures is to preserve viability without contamination, variation, mutation, or deterioration. The methods of maintaining fungal cultures have been fully described and discussed by Fennell (1960) and Onions (1971). The traditional method is serial subculture. The frequency of subculturing can be reduced by maintaining the fungi in a reduced metabolic state, which is achieved by covering the culture with distilled water or paraffin oil, drying in soil or on silica gel, or storing at low temperatures. For long-term maintenance freeze-drying (lyophilization) and cryogenic storage in liquid nitrogen are recommended.

A. Periodic Transfer

Like most fungi, the great majority of conidial fungi can be maintained as living cultures if they are transferred periodically. This is general practice for ordinary culture growth in the laboratory. Fungi are grown on test tube slants composed of an agar medium which favors a maximum development of conidia. Conidia are periodically transferred from old cultures to fresh slants of the same medium. The interval between transfers depends largely on the peculiarities of the particular fungi but also on external conditions, especially storage temperature. Transfer of room temperature-maintained cultures every 3 months is sufficiently frequent for most conidial fungi, although certain species may remain viable for many years. To avoid rapid drying of the agar medium and to reduce the metabolic rate of the organism, cultures may be kept in a refrigerator at about 5°C (Fennell *et al.,* 1950; Haynes *et al.,* 1955) or in a freezer at a temperature of about $-10°$ to $-20°$C (Carmichael, 1962). At these low temperatures the viability of most fungi is increased, and the interval between transfers can be longer than at room temperature. Fennell (1960) recommended that the transfer of large pieces of mycelia be done only from young, actively growing marginal areas of fungi that fail to produce specialized propagative cells.

The major disadvantages of the periodic transfer technique are the high possibility of contamination by mites or microorganisms, the risk of deterioration with subsequent loss of morphological characters and desired biochemical responses, and selection of either a mutant strain or a purely vegetative, nonsporulating form.

B. Storage in Sterile Distilled Water

Maintenance of fungal cultures in water alone has proved useful for yeasts and pathogens of man (Castellani, 1939, 1967; Benedek, 1962b; McGinnis *et al.*, 1974) and plants (Person, 1961; Boesewinkel, 1976; Marx and Daniel, 1976). The technique consists of aseptically transferring spores or fragments of hyphae into screw-capped vials, tubes, or bottles half filled with sterile distilled water. The caps are tightened to prevent evaporation of the water. The cultures are then stored either at room or refrigerator temperature. Boesewinkel (1976) has recently reported successful maintenance of 650 plant-pathogenic and saprophytic fungi in sterile distilled water at room temperature for periods up to 7 yr without loss of viability or change in morphological features. To ensure success in applying this technique for storage, McGinnis *et al.* (1974) suggest that care be taken to select actively sporulating isolates and to suspend adequate amounts of spores and pieces of hyphae in water.

C. Storage under Sterile Mineral Oil

Covering young, actively metabolizing cultures on slants of suitable agar media with sterilized mineral or paraffin oil is an effective way of extending the longevity of agar slant cultures of fungi, particularly those that cannot be preserved by other methods. Medicinal-grade mineral oil should be used. The oil can be sterilized by autoclaving at 15 psi for 2 hr. The water is then eliminated by heating in a drying oven at 170°C for 1–2 hr (Fennell, 1960). The culture must be completely covered with a layer of sterile oil at least 1 cm deep to prevent dehydration of the medium and to reduce metabolic activity and growth of the culture (Buell and Weston, 1947). Oil-covered cultures are stored in an upright position at either room or refrigeration temperature. However, they still must be transferred to fresh agar slants and overlaid with sterile mineral oil at appropriate intervals. The transfer process with oil-overlaid cultures is messy.

In conventional methods, cultures are grown in test tubes or McCartney bottles. Elliott (1975) has recently employed a flat-bottomed standard pharmaceutical ampul with an easy-open constriction above the bowl of the ampul and a narrow neck for flame-sealing. Flame-sealed ampuls are particularly suitable for mailing and also eliminate the need for storage in the upright position.

D. Storage in Sterile Soil or Sand

Soil can be used directly as a growth medium or as an absorbent medium to preserve fungal cultures (Weiss, 1957). As a growth medium soil is moistened to a level of about 60% of its maximum water-holding capacity and autoclaved three times for 30 min at 15 psi. The sterile soil is inoculated with a fungal culture and incubated at the optimal temperature. After the soil culture has grown sufficiently it is dried at room temperature and stored in a refrigerator. As an absorbent medium, the soil is first air-dried, then autoclaved, and finally heated in a drying oven until moisture-free. A heavy aqueous suspension of conidia is added to the sterile soil, and the mixture is air-dried and stored in a refrigerator. Fennell (1960) has summarized the advantages of the soil culture method of preservation as follows: (1) longevity of the cultures is usually increased substantially, (2) morphological changes are reduced or eliminated, and (3) uniform inoculum is available for long periods if proper care is taken in handling the cultures.

The soil culture method has been recommended as a means of preserving species of *Aspergillus, Penicillium,* and *Fusarium* (Raper and Fennell, 1965; Raper and Thom, 1949; Booth, 1971a). Cultures of *Aspergillus fischeri, A. sydowi,* and *Penicillium chrysogenum* were successfully maintained in soil by Greene and Fred (1934) for over 2 yr, and *Aspergillus flavus, A. luteus,* and *A. niger* by Guida and DuAmaral (1949–1950) for 5 yr. *Fusarium avenaceum* remained viable in sterile loam soil for 3 yr at room temperature and up to 8 yr under refrigeration at 5°C. Booth (1971a) has reported the most serious objections to this method of maintenance; the considerable time lag before dryness prompts dormancy, induces vegetative strains to overgrow the wild type, or allows saprophytic mutants to overgrow a pathogenic strain.

E. Storage on Silica Gel

Anhydrous silica gel has been used to preserve fungi which produce abundant conidia, such as *Aspergillus nidulans* and *Neurospora crassa* (Perkins, 1962; Barratt *et al.,* 1965; Ogata, 1962). Details of the procedure currently employed in preserving *Neurospora* stock cultures have been given by Perkins (1977). Cotton-plugged, 13×100 mm test tubes filled 65 mm deep with anhydrous silica gel are hot-air sterilized at 180°C for 2 hr and stored at room temperature in a moistureproof box to be ready for use. The cultures to be preserved are grown on 4% agar slants containing appropriate media. About 0.5 ml of sterile water is gently added to a fresh culture, and conidia are suspended by shaking using a vortex-type mixer. About 0.5 ml of sterile nonfat milk is then added and stirred gently, and the entire suspension is pipetted drop by drop over the silica gel in a labeled tube. The tube is vibrated briefly with the mechanical mixer to distribute the inoculum over as many particles as possible throughout the silica gel. The tube is then placed in an ice water bath for 15 min. After 1 day at room temperature the particles appear dry, and the mouth of the tube is sealed by covering it with a 20×20 mm square of Parafilm. The sealed tubes are then stored at 5° or -20°C in moistureproof boxes.

F. Freeze-Drying

Freeze-drying, commonly called lyophilization, is considered an economical and effective means of long-term preservation of the majority of conidial fungi. For this process sporulating cultures are required. Conidia are simultaneously frozen and dried under reduced pressure in a vacuum. Thus the available water in the conidia is removed quickly. In this condition the conidia remain dormant for long periods of time.

A successful lyophilization technique, described by Raper and Alexander (1945), is known as the Northern Regional Research Laboratory (NRRL) method. A heavy suspension of conidia is prepared in sterile bovine serum, and approximately 0.05–0.1 ml of the suspension is added to each of several sterile, cotton-stoppered, Pyrex glass tubes or ampuls that have been properly labeled with glass-marking ink. The tubes are inserted in rubber sleeves on a glass manifold and lowered into a bath of Dry Ice and methyl Cellosolve at a temperature of about −45°C. The conidial suspensions are frozen within seconds, and evacuation is carried out by means of a vacuum pump. When the pellets are dry, the tubes are raised above the bath and evacuation is continued at room temperature for $\frac{1}{2}$ h to ensure thorough desiccation. The tubes are then sealed off under vacuum with a gas–oxygen torch. The freeze-dried preparations are stored in a refrigerator at 5°C.

Haskins and Anastasiou (1953) compared this method with what they refer to as the Prairie Regional Laboratory (PRL) method. The latter method is similar to the double-vial batch method used at the ATCC as described by Weiss (1957). The batch-type technique the ATCC currently employs is a two-step procedure. It involves rapid freezing of conidial suspensions, primary drying at low temperatures, secondary drying at room temperature, and sealing off under vacuum on a manifold. The conidial fungi are generally grown on slants or plates containing an agar medium which will permit abundant sporulation. With the use of aseptic techniques throughout, a heavy suspension containing at least 10^6 conidia/ml is prepared in skim milk (Difco 0032) made as a 20% (w/v) solution and sterilized at 116°C for 20 min. Overheating may cause caramelization of the milk. A 0.2-ml drop of the conidial suspension is pipetted into each sterile, cotton-plugged, 11 × 35 mm freeze-drying shell vial (VWR Scientific, No. 27921-015). The vials are packed upright in a stainless steel container. The latter is placed in the bottom of a mechanical refrigerator at −79°C for 1 h. This container is quickly attached to an Atomo-vac plate (Refrigeration for Science, Inc., 3441 Fifth Street, Oceanside, New York) connected to a dry ethyl Cellosolve-chilled condenser and a vacuum pump. Crushed Dry Ice is placed around the container, and drying proceeds under a vacuum of 20–30 μm of mercury. The Dry Ice completely dissipates within 2 h, and drying continues at ambient temperature overnight. Following drying, the vials are inserted into 14 × 85 mm glass shell ampuls containing a few granules of moisture-indicating silica gel and a small wad of cotton. Two layers of glass fiber paper GF 85 (H. Reeve Angel and Company, Clifton, New Jersey) are packed over the inner vial. The outer ampul is constricted to a narrow capillary above the fiber paper and sealed under a vacuum of at least 50 μm mercury on a manifold.

To rehydrate freeze-dried material, 0.4 ml sterile distilled water from a test tube containing 6 ml water is added to the contents of the vial, mixed well, and transferred back to the test tube. This suspension is allowed to rehydrate for 1 h before being transferred to plates or slants of solid media.

The results of extended tests on freeze-dried fungal cultures report longevity up to 40 months according to Raper and Alexander (1945), up to 15 yr according to Mehrotra and Hesseltine (1958), up to 17 yr according to Hesseltine et al. (1960), up to 18 yr according to Schipper and Bekker-Holtman (1976), and up to 23 yr according to Ellis and Robertson (1968). The ATCC has employed freeze-drying as a means for preserving conidial fungi since 1945.

G. Cryogenic Storage in Liquid Nitrogen

Within the last two decades the successful preservation of living cells and organisms has been achieved through the application of cryogenics. The cryogenic storage temperatures now commonly used are those of liquid nitrogen ($-196°C$) and liquid nitrogen vapor ($-150°C$ and below). The cryoprotective agents are 10% (v/v) glycerol and 5% (v/v) dimethyl sulfoxide (DMSO). Glycerol solutions are sterilized by autoclaving in the manner of ordinary culture media, while DMSO is sterilized by filtration. The ATCC has used this cryogenic technique for the conservation of a wide variety of living fungi since 1960 (Hwang, 1960, 1966, 1968; Hwang et al., 1976). Descriptions of the apparatus and explanations of the procedures of liquid nitrogen refrigeration have been reported by Jong (1978). Both sporulating and nonsporulating fungi can be preserved by this technique. Conidia or mycelia fragments are harvested by flooding the slants or plates with 10% glycerol or 5% DMSO. Approximately a 0.5-ml suspension is pipetted into sterile, cotton-plugged, prescored borosilicate ampuls (T. C. Wheaton Company, Milville, New Jersey) or 2.0-ml screw-top polypropylene vials (Vangard International, Neptune, New Jersey). Strains producing only mycelia are grown on agar plates containing an appropriate medium. Three agar disks containing hyphal tips are placed in an ampul or vial, and 0.4 ml of cryoprotective agent is added. Cultures in broth are fragmented in a sterile Waring blender and suspended in equal parts of 20% glycerol and growth medium, or equal parts of 10% DMSO and growth medium, to give a final concentration of 10% glycerol or 5% DMSO, respectively. Some strains must be concentrated by centrifugation to obtain sufficient material for freezing. Pathogenic strains are not macerated in a mechanical blender because of the hazard of aerosol dispersion. Plastic vials and glass ampuls containing fungi pathogenic to humans and animals are always maintained in the vapor phase of a liquid nitrogen refrigerator and are not sealed. Most nonpathogenic fungi are stored in the liquid phase of a liquid nitrogen refrigerator. It is therefore essential that the ampuls be completely sealed. An improperly sealed ampul immersed in liquid nitrogen will permit entry of the liquid and will explode at the time of thawing as a result of the sudden expansion of nitrogen into gas.

The filled ampuls or vials are placed in labeled aluminum cans (Nasco A545, Fort Atkinson, Wisconsin) and then in boxes. These are placed in the freezing chamber of a programmed freezer (Linde BF3-2, Union Carbide Corporation, New York, New York). The initial cooling is carried out at the rate of $1°C/min$ from room temperature to $-35°C$; subsequent cooling to below $-100°C$ is rapid and uncontrolled. Then the ampuls are immediately transferred and stored in liquid nitrogen at $-196°C$ or liquid nitrogen vapor at about $-150°C$ to $-180°C$. To recover fungal cultures stored in liquid nitrogen the frozen ampuls or vials are thawed rapidly in a $37°C$ water bath until the last trace of ice has disappeared. This usually takes between 40 and 60 s with moderately vigorous agitation. The culture samples may then be aseptically transferred to an appropriate growth medium.

In summary, significant advances have been made in the technology of maintaining viable cultures of fungi. Many of these have been incorporated into the systematic program for storing cultures at the ATCC and other national collections. The overtones of these new techniques are far-reaching. The taxonomist, for instance, may store cultures of original isolates for long periods with little or no fear of mutational changes occurring which affect morphological and/or physiological characters. Isolates of medically important fungi may be maintained at ultralow temperatures without loss of virulence. The cost and space required for storing large numbers of fungi have been significantly reduced by eliminating the need for periodic subculturing. Cultures can be more readily provided to teaching institutions, which is one of the most important contributions made by national fungal collections to the mycological community.

REFERENCES

Ajello, L., Georg, L. K., Kaplan, W., and Kaufman, L. (1963). Laboratory Manual for Medical Mycology,'' Public Health Serv. Publ. 994. National Communicable Disease Center, Atlanta, Georgia.

American Type Culture Collection (1980). ''Catalogue of Strains I,'' 14th ed. ATCC, Rockville, Maryland.

Anastasiou, C. J. (1962). Fungi from Salton Lakes. I. A new species of *Clavariopsis*. *Mycologia* **53**, 11–16.

Anastasiou, C. J. (1963). Fungi from Salton Lakes. II. Ascomycetes and Fungi Imperfecti from the Salton Sea. *Nova Hedwigia* **6**, 243–276.

Anastasiou, C. J., and Churchland, L. M. (1969). Fungi on decaying leaves in marine habitats. *Can. J. Bot.* **47**, 251–257.

Ark, P. A., and Dickey, R. S.(1950). A modification of the Van Tieghem cell for purification of contaminated fungus cultures. *Phytopathology* **40**, 389–390.

Bandoni, R. J., Parsons, J. D., and Redhead, S. A. (1975). Agar "baits" for the collection of aquatic fungi. *Mycologia* **67**, 1020–1024.

Barratt, R. B., Johnson, B., and Ogata, W. N. (1965). Wild-type and mutant stocks of *Aspergillus nidulans*. *Genetics* **52**, 233–246.

Barron, G. L. (1977). "The Nematode-destroying Fungi." Ca. Biol. Publ. Ltd., Ontario.

Baxter, M. (1965). The use of ink blue in the identification of dermatophytes. *J. Invest Dermatol.* **44**, 23–25.

Beech, F. W., and Carr, J. G. (1955). A survey of inhibitory compounds for the separation of yeasts and bacteria in apple juices and ciders. *J. Gen. Microbiol.* **12**, 85–94.

Beech, F. W., and Davenport, R. R. (1971). Isolation, purification and maintenance of yeasts. *In* "Methods in Microbiology" (C. Booth, ed.), Vol. 4, pp. 153–182. Academic Press, New York.

Benedek, T. (1962a). Fragments mycologica. I. Some historical remarks on the development of "hairbaiting" of Toma-Karling-Vanbreuseghem (the ToKaVa-hairbaiting method). *Mycopathol. Mycol. Appl.* **16**, 104–106.

Benedek, T. (1962b). Fragments mycologica. II. On Castellani's "water cultures" and Benedek's "mycotheca" in chlorallactophenol. *Mycopathol. Mycol. Appl.* **17**, 255–260.

Bisby, G. R., James, N., and Timonin, M. (1933). Fungi isolated from Manitoba soil by the plate method. *Can. J. Res.* **8**, 253.

Blank, I. H., and Tiffney, W. N. (1936). The use of ultra-violet irradiated culture media for securing bacteria-free cultures of *Saprolegnia*. *Mycologia* **28**, 324–329.

Boesewinkel, H. J. (1976). Storage of fungal cultures in water. *Trans. Br. Mycol. Soc.* **66**, 183–185.

Booth, C. (1971a). "The Genus *Fusarium*." Commonwealth Mycological Institute, Kew, Surrey, England.

Booth, C. (1971b). Introduction to general methods. *In* "Methods in Microbiology" (C. Booth, ed.). Vol. 4, pp. 1–47. Academic Press, New York.

Booth, C. (1971c). Fungal culture media. *In* "Methods in Microbiology" (C. Booth, ed.), Vol. 4, pp. 49–94. Academic Press, New York.

Bouhot, D., and Billotte, J. M. (1964). Recherches sur l'écologie des champignons parasites dans le sol. II. Choix d'un milieu nutritif pour l'isolement sélectif de *Fusarium oxysporum* et *Fusarium solani* du sol. *Ann. Epiphyt.* **15**, 45–55.

Brooks, F. T., and Hansford, C. G. (1922). Mould growths upon cold-store meat. *Trans. Br. Mycol. Soc.* **8**, 113–142.

Brown, W. (1924). Two mycological methods. I. A simple method of freeing fungal cultures from bacteria. II. A method of isolating single strains of fungi by cutting out a hyphal tip. *Ann. Bot. (London)* **18**, 401–404.

Buell, C. B., and Weston, W. H. (1947). Application of the mineral oil conservation method to maintaining collections of fungus cultures. *Am. J. Bot.* **34**, 555–561.

Carmichael, J. W. (1962). Viability of mold cultures stored at 20°C. *Mycologia* **54**, 432–436.

Castellani, A. (1939). Viability of some pathogenic fungi in distilled water. *J. Trop. Med. Hyg.* **42**, 255.

Castellani, A. (1967). Maintenance and cultivation of the common pathogenic fungi of man in sterile distilled water: Further researches. *J. Trop. Med. Hyg.* **70**, 181–184.

Chesters, C. G. C. (1940). A method of isolating soil fungi. *Trans. Br. Mycol. Soc.* **24**, 352–355.

Chesters, C. G. C. (1948). A contribution to the study of fungi in the soil. *Trans. Br. Mycol. Soc.* **30**, 100–117.

Christensen, C. M., Kaufert, F. H., Schmitz, H., and Allison, J. L. (1942). *Hormodendron resinae* Lindau, an inhabitant of wood impregnated with creosote and coal tar. *Am. J. Bot.* **29**, 552–558.

Cochrane, V. W. (1958). "Physiology of Fungi." Wiley, New York.

Conn, H. J., ed. (1957). "Manual of Microbiological Methods" (by the Society of American Bacteriologists). McGraw-Hill, New York.

Cooke, W. B. (1963). A laboratory guide to fungi in polluted waters, sewage and sewage treatment systems. *U.S. Public Health Serv. Publ.* **999-WP-1,** 1–132.

Cooke, W. B., and Busch, K. A.(1957). Activity of cellulose-decomposing fungi isolated from sewage-polluted water. *Sewage Ind. Wastes* **29,** 210–217.

Cooney, D. G., and Emerson, R. (1964). "Thermophilic Fungi: An Account of Their Biology, Activities, and Classification." Freeman, San Francisco, California.

Cox, R. S. (1952). Reflected light as an aid in the Keitt method of spore and hyphal tip isolations. *Phytopathology* **42,** 118.

Crane, J. L. (1968). Freshwater Hyphomycetes of the northern Appalachian highland including New England and three coastal plain states. *Am. J. Bot.* **55,** 996–1002.

Crisan, E. V. (1964). Isolation and culture of thermophilic fungi. *Contrib. Boyce Thompson Inst.* **22,** 291–302.

Dayal, H. M., Saxena, B. N., and Nigam, S. S. (1964). Eradication of mites from fungal culture collections. *Indian J. Exp. Biol.* **2,** 62–64.

Deverall, B. J. (1968). Psychrophiles. *In* "The Fungi: An Advanced Treatise" (G. C. Ainsworth and A. S. Sussman, eds.), Vol. 3, pp. 129–135. Academic Press, New York.

Dickinson, C. H., and Pugh, G. J. F. (1965). Use of a selective cellulose agar for isolation of soil fungi. *Nature (London)* **207,** 440–441.

Dolan, C. T. (1971). Optimal combination and concentration of antibiotics in media of pathogenic fungi and *Nocardia asteroides. Appl. Microbiol.* **21,** 195–197.

Dolan, C. T. (1972). Effect of gentamicin on growth of yeasts, yeast-like organisms, and *Aspergillus fumigatus. Am. J. Clin. Pathol.* **57,** 30–32.

Drechsler, C. (1941). Predaceous fungi. *Biol. Rev. Cambridge Philos. Soc.* **16,** 265–290.

Eggins, H. O. W., and Pugh, G. J. F. (1962). Isolation of cellulose-decomposing fungi from the soil. *Nature (London)* **193,** 94–95.

Elliott, R. F. (1975). Method for preserving mini-cultures of fungi under mineral oil. *Lab. Pract.* *1975*, 751.

Ellis, J. J., and Robertson, J. A. (1968). Viability of fungus cultures preserved by lyophilization. *Mycologia* **60,** 399–450.

Emerson, R. (1968). Thermophiles. *In* "The Fungi: An Advanced Treatise" (G. C. Ainsworth and A. S. Sussman, eds), Vol. 3, pp. 105–128. Academic Press, New York.

Emmons, C. W. (1940). Fumigation of mite infested cultures of fungi *Mycopathologia* **2,** 320–321.

Emmons, C. W., Binford, C. H., Utz, J. P., and Kwon-Chung, K. J. (1977). "Medical Mycology," 3rd ed. Lea & Febiger, Philadelphia, Pennsylvania.

Fennell, D. I. (1960). Conservation of fungus cultures. *Bot. Rev.* **26,** 79–141.

Fennell, D. I., Raper, K. B., and Flickinger, M. H. (1950). Further observations on the preservation of mold cultures. *Mycologia* **42,** 135–147.

Fikry, A. (1932). Investigations on the wilt disease of Egyptian cotton caused by various *Fusarium* spp. *Ann. Bot. (London)* **46,** 29–70.

Gareth Jones, E. B. (1971). Aquatic fungi. *In* "Methods in Microbiology" (C. Booth, ed.). Vol. 4, pp. 335–365. Academic Press, New York.

Garrett, S. D. (1944). "Root Disease Fungi." Chronica-Botanica, Waltham, Massachusetts.

Georg, L. K., Ajello, L., and Papageorge, C. (1954). Use of cycloheximide in the selective isolation of fungi pathogenic to man. *J. Lab. Clin. Med.* **44,** 422–428.

Goldfarb, J. N., and Hermann, F. (1956). A study of pH changes by molds in culture medium. *J. Invest. Dermatol.* **27**, 193–201.

Greene, H. C., and Fred, E. B. (1934). Maintenance of vigorous mold stock cultures. *Ind. Eng. Chem.* **26**, 1297–1298.

Guida, V. O., and DuAmaral, J. F. (1949–50). Tecnica de conservacao de fungus em areia esteril (Technique of conservation of fungi in sterile sand). *Anq. Inst. Biol., Sao Paulo* **19**, 203–206.

Hanna, W. F. (1924). The dry needle method of making monosporous cultures of Hymenomycetes and other fungi. *Ann. Bot. (London)* **38**, 791–795.

Hansen, H. N., and Snyder, W. C. (1939). Effective control of culture mites by mechanical exclusion. *Science* **89**, 350–351.

Harley, J. L., and Waid, J. S. (1955). A method of studying active mycelia on living roots and other surfaces in the soil. *Trans. Br. Mycol. Soc.* **38**, 104–118.

Haskins, R. H., and Anastasiou, J. (1953). Comparison of the survival of *Aspergillus niger* spores lyophilized by various methods. *Mycologia* **45**, 523–532.

Hawker, L. E. (1950). "Physiology of the Fungi." Univ. of London Press, London.

Haynes, W. C., Wickerham, L. J., and C. W. Hesseltine (1955). Maintenance of cultures of industrially important microorganisms. *Appl. Microbiol.* **3**, 361–368.

Hering, T. F. (1966). An automatic soil-washing apparatus for fungal isolation. *Plant Soil* **25**, 195–200.

Hesseltine, C. W., Bradle, B. J., and Benjamin, C. R. (1960). Further investigations on the preservation of molds. *Mycologia* **52**, 762–774.

Hwang, S. W. (1960). Effects of ultra-low temperatures on the viability of selected fungus strains. *Mycologia* **52**, 527–529.

Hwang, S. W. (1966). Long-term preservation of fungus cultures with liquid nitrogen refrigeration. *Appl. Microbiol.* **14**, 784–788.

Hwang, S. W. (1968). Investigation of ultralow temperature for fungal cultures. I. An evaluation of liquid nitrogen storage for preservation of selected fungal cultures. *Mycologia* **60**, 613–621.

Hwang, S. W., Kwolek, W. F., and Haynes, W. C. (1976). Investigation of ultralow temperature for fungal cultures. III. Viability and growth rate of mycelial cultures following cryogenic storage. *Mycologia* **68**, 377–387.

Ingold, C. T. (1942). Aquatic Hyphomycetes of decaying alder leaves. *Trans. Br. Mycol. Soc.* **25**, 339–417.

James, N., and Sutherland, M. L. (1939). The accuracy of the plating method for estimating the numbers of soil bacteria, actinomycetes and fungi in the dilution method. *Can. J. Res., Sect. C* **17**, 72–82.

Johnson, L. F., Curl, E. A., Bond, J. H., and Fribourg, H. A. (1959). "Methods for Studying Soil Microflora Plant-disease Relationships." Burgess, Minneapolis, Minnesota.

Johnson, T. W., Jr., and Sparrow, F. K., Jr. (1961). "Fungi in Oceans and Estuaries." Cramer, Weinheim.

Jong, S. C. (1978). Conservation of the cultures. *In* "The Biology and Cultivation of Edible Mushrooms" (S. T. Chang and W. A. Hayes, eds.), pp. 119–135. Academic Press, New York.

Karling, J. S. (1946). Keratinophilic chytrids. I. *Am. J. Bot.* **33**, 751–757.

Keitt, G. W. (1915). Simple technique for isolating single-spore strains of certain types of fungi. *Phytopathology* **5**, 266–269.

Kendrick, B., Samuels, G. J., Webster, J. and Luttrell, E. S. (1979). Techniques for establishing connections between anamorph and teleomorph. *In* "The Whole Fungus" (B. Kendrick, ed.), Vol. 2, pp. 635–651. Nat. Mus. Canada, Ottawa.

Kirk, P. W., Jr. (1969). Aquatic Hyphomycetes on wood in an estuary. *Mycologia* **61**, 177–181.

Lambert, E. B. (1939). A spore isolator combining some of the advantages of the La Rue and Keitt methods. *Phytopathology* **29**, 212–214.

La Rue, C. D. (1920). Isolating single spores. *Bot. Gaz. (Chicago)* **70**, 319–320.

Leach, C. M. (1961). The sporulation of *Helminthosporium oryzae* as affected by exposure to near ultraviolet radiation and dark periods. *Can. J. Bot.* **39**, 705–715.

Leach, C. M. (1971). A practical guide to the effects of visible and ultraviolet light on fungi. *In* "Methods in Microbiology" (C. Booth, ed.), Vol. 4, pp. 609–664. Academic Press, New York.

Lilly, V. G., and Barnett, H. L. (1951). "Physiology of the Fungi." McGraw-Hill, New York.

Lukens, R. J. (1963). Photo-inhibition of sporulation in *Alternaria solani*. *Am. J. Bot.* **50**, 720–724.

McDonough, E. S., Ajello, L., Georg, L. K., and Brinkman, S. (1960a). *In vitro* effects of antibiotics on the yeast phase of *Blastomyces dermatitidis* and other fungi. *J. Lab. Clin. Med.* **55**, 116–119.

McDonough, E. S., Georg, L. K., Ajello, L., and Brinkman, S. (1960b). Growth of dimorphic human pathogenic fungi on media containing cycloheximide and chloramphenicol. *Mycopathol. Mycol. Appl.* **13**, 113–120.

McGinnis, M. R., Padhye, A. A., and Ajello, L. (1974). Storage of stock cultures of filamentous fungi, yeasts, and some aerobic actinomycetes in sterile distilled water. *Appl. Microbiol.* **28**, 218–222.

MacLaren, J. A. (1960). The cultivation of pathogenic fungi in a molybdenum medium. *Mycologia* **52**, 148–152.

Marsh, P. B., Taylor, E. E., and Bossler, L. M. (1959). A guide to the literature on certain effects of light on fungi: Reproduction, morphology, pigmentation and phototropic phenomena. *Plant Dis. Rep., Suppl.* **261**, 251–312.

Marx, D. H., and Daniel, W. J. (1976). Maintaining cultures of ectomycorrhizal and plant pathogenic fungi in sterile water cold storage. *Can. J. Microbiol.* **22**, 338–341.

Mehrotra, B. S., and Hesseltine, C. W. (1958). Further evaluation of the lyophil process for the preservation of aspergilli and penicillia. *Appl. Microbiol.* **6**, 179–183.

Nadakavukaren, M. J., and Horner, C. F. (1959). An alcohol agar medium selective for determining *Verticillium* microsclerotia in soil. *Phytopathology* **49**, 527–528.

Ogata, W. N. (1962). Preservation of *Neurospora* stock cultures with anhydrous silica gel. *Neurospora Newsl.* **1**, 13.

Onesirosan, P. T., and Banttari, E. E. (1969). The effect of light and temperature upon sporulation of *Helminthosporium teres* in culture. *Phytopathology* **59**, 906–909.

Onions, A. H. (1971). Preservation of fungi. *In* "Methods in Microbiology" (C. Booth, ed.), Vol. 4, pp. 113–151. Academic Press, New York.

Paden, J. W. (1967). A centrifugation technique for separating ascospores from soil. *Mycopathol. Mycol. Appl.* **33**, 382–384.

Pagano, J., Levin, J. U., and Trejo, W. (1958). Diagnostic medium for differentiation of species of *Candida*. *Antibiot. Annu.* **6**, 137–143.

Papavizas, G. C., and Davey, C. B. (1961). Isolation of *Thielaviopis basicola* from bean rhizosphere. *Phytopathology* **51**, 92–96.

Parbery, D. G. (1967). Isolation of the kerosene fungus, *Cladosporium resinae,* from Australian soil. *Trans. Br. Mycol. Soc.* **50**, 682–685.

Park, D. (1963). The presence of *Fusarium oxysporum* in soils. *Trans. Br. Mycol. Soc.* **46**, 444–448.

Parkinson, D. (1957). New methods for the qualitative and quantitative study of fungi in the rhizosphere. *Symp. Methodes Etud. Microbiol. Sol, 1957.*

Parkinson, D., and Williams, S. T. (1961). A method for isolating fungi from soil microhabitats. *Plant Soil* **14**, 347–355.

Paul, W. R. C. (1929). A comparative morphological and physiological study of a number of strains of *Botrytis cinerea*, with special reference to their virulence. *Trans. Br. Mycol. Soc.* **14**, 118–134.

Perkins, D. D. (1962). Preservation of *Neurospora* stock cultures with anhydrous silica gel. *Can. J. Microbiol.* **8**, 591-594.

Perkins, D. D. (1977). Details for preparing silica gel stocks. *Neurospora Newsl.* **24**, 16-17.

Person, L. H. (1961). A method of maintaining viability and ability to sporulate in isolates of *Ascochyta. Phytopathology* **51**, 797-798.

Quaife, R. A. (1968). Evaluation of ink blue medium with and without cycloheximide, for the isolation of dermatophytes. *J. Med. Lab. Technol.* **25**, 227-232.

Raper, J. R. (1937). A method of freeing fungi from bacterial contamination. *Science* **85**, 342.

Raper, J. R. (1963). Device for the isolation of spores. *J. Bacteriol.* **86**, 342-344.

Raper, K. B., and Alexander, D. F. (1945). Preservation of molds by the lyophil process. *Mycologia* **37**, 499-525.

Raper, K. B., and Fennell, D. I. (1965). "The Genus *Aspergillus.*" Williams & Wilkins, Baltimore, Maryland.

Raper, K. B., and Thom, C. (1949). "A Manual of the Penicillia." Williams & Wilkins, Baltimore, Maryland.

Riker, A. J., and Riker, R. S. (1936). "Introduction to Research on Plant Diseases." John Swift & Co., St. Louis, Missouri.

Rippon, J. W. (1974). "Medical Mycology: The Pathogenic Fungi and the Pathogenic Actinomycetes." Saunders, Philadelphia, Pennsylvania.

Rynearson, T. K., and Peterson, J. L. (1965). Selective isolation of paraffinolytic fungi using a direct soil-baiting method. *Mycologia* **57**, 761-765.

Sass, J. E. (1929). The cytological basis for homothallism and heterothallism in the Agaricaceae. *Am. J. Bot.* **16**, 663-701.

Schipper, M. A. A., and Bekker-Holtman, J. (1976). Viability of lyophilized fungal cultures. *Antonie van Leeuwenhoek* **42**, 325-328.

Shearer, C. A. (1972). Fungi of the Chesapeake Bay and its tributaries. III. The distribution of wood-inhabiting Ascomycetes and Fungi Imperfecti of the Patuxent River. *Am. J. Bot.* **59**, 961-969.

Smith, C. D., and Furcolow, M. L. (1964). Efficiency of three techniques for isolating *Histoplasma capsulatum* from soil, including a new flotation method. *J. Lab. Clin. Med.* **64**, 342-348.

Smith, G. (1960). "An Introduction to Industrial Mycology," 5th ed. Arnold, London.

Snyder, W. C., and Hansen, H. N. (1941). The effect of light on the taxonomic characters in *Fusarium. Mycologia* **33**, 580-591.

Staib, F., and Seeliger, H. P. R. (1968). Zur Selektivzuchtung von *Cryptococcus neoformans. Mykosen* **11**, 267-272.

Stevens, R. B. (ed.) (1974). "Mycology Guidebook" (by Mycology Guidebook Committee, Mycol. Soc. Am.). Univ. of Washington Press, Seattle.

Subden, R. E., and Threlkeld, S. F. H. (1966). Mite control for *Neurospora* labs. *Neurospora Newsl.* **10**, 14.

Taplin, D. (1965). The use of gentamicin in mycology media. *J. Invest. Dermatol.* **45**, 549-550.

Taplin, D., Zaias, N., Rebell, G., and Bland, H. (1969). Isolation and recognition of dermatophytes on a new medium (DTM). *Arch. Dermatol.* **99**, 203-209.

Thompson, L. (1945). Note on a selective medium for fungi. *Proc. Staff Meet. Mayo Clin.* **20**, 248-249.

Thornton, R. H. (1952). The screened immersion plate, a method for isolating soil microorganisms. *Research* **5**, 190-191.

Toma, A. (1929). Sur l'infection des cheveux "*in vitro*" par les champignons des teignes. *Ann. Dermatol. Syphiligr.* **10**, 641-643.

Tsao, P. H. (1970). Selective media for isolation of pathogenic fungi. *Annu. Rev. Phytopathol.* **8**, 157-186.

Vanbreuseghem, R. (1952). Technique biologique pour l'isolement des dermatophytes du sol. *Ann. Soc. Belge Med. Trop.* **32,** 173–178.

Waksman, S. A. (1916). Do fungi live and produce mycelium in the soil? *Science* **44,** 320–322.

Wang, C. J. K. (1965). Fungi of pulp and paper in New York. *State Univ. Coll. For. Syracuse Univ. Tech. Publ.* **87,** 1–115.

Warcup, J. H. (1950). The soil-plate method for isolation of fungi from soil. *Nature (London)* **166,** 117–118.

Watson, R. D. (1960). Soil washing improves the value of the soil dilution and the plate count method of estimating populations of soil fungi. *Phytopathology* **50,** 792–794.

Weiss, F. A. (1957). Maintenance and preservation of cultures. *In* "Manual of Microbiological Methods" (Society of American Bacteriologists, eds.), pp. 99–119. McGraw-Hill, New York.

White, W. L., Mandels, G. R., and Siu, R. G. H. (1950). Fungi in relation to degradation of woolen fabrics. *Mycologia* **42,** 199–223.

Witsch, H., and Wagner, F. (1955). Beobachtungen über den Einfluss des Lichtes auf Mycel und Conidienbildung bei *Alternaria brassicae* var. *dauci. Arch. Mikrobiol.* **22,** 307–312.

31

Techniques for Examining Developmental and Ultrastructural Aspects of Conidial Fungi

Garry T. Cole

I. SPECIAL CULTURING TECHNIQUES

In the preceding chapter, the author outlined general procedures for the isolation, culture, and maintenance of conidial fungi. In addition to these methods of cultivation, several special culturing techniques have been reported which are particularly useful in light and electron microscope investigations. Various kinds of growth chambers have been successfully employed for time-lapse, light microscope analyses of conidial fungi (Duggar, 1909; Sykes and Moore, 1959; Cole and Kendrick, 1968; Cole *et al.*, 1969; Heunert, 1972). The application of a thin-glass culture chamber (Cole and Kendrick, 1968) and the modified plate culture technique (Cole *et al.*, 1969) in studies on conidium and conidiogenous

Biology of Conidial Fungi, Vol. 2
Copyright © 1981 by Academic Press, Inc.
All rights of reproduction in any form reserved.
ISBN 0-12-179502-0

cell ontogeny have been discussed and illustrated by Cole and Samson (1979). With the use of a modification of the thin-glass culture chamber, as well as other growth chambers designed by Heunert (1972), three time-lapse films demonstrating different modes of conidiogenesis were recently completed at the Institut für den Wissenschaftlichen Film (Göttingen, West Germany). The films are entitled, "Conidiogenesis in the Fungi Imperfecti": (1) "Holoblastic Conidia and Poroconidia" (C 1304), (2) "Enteroblastic Conidia" (C 1303), and (3) "Thallic Conidia" (C 1302), and are available from the institute.

Attempts to culture conidial fungi specifically for electron microscope examination have generated special problems. In the case of scanning electron microscopy, it is best that the substrate on which the fungus is growing be smooth and flat so that the mycelium can be easily examined. With this in mind, 12-mm-diameter coverslips were coated on one side with nutrient agar which was then inoculated with the fungus under investigation. This procedure permits examination of the specimen using both light and scanning electron microscopy (Cole, 1975). The coverslip culture is incubated in a closed petri dish. The uncoated surface of the coverslip is placed on a microscope slide which is supported above the surface of the petri dish on glass rods. The bottom of the dish is covered with two pieces of filter paper soaked in 10% glycerol to maintain high relative humidity. The progress of mycelial development can be easily checked using a binocular dissecting microscope without opening the petri dish. After the fungus has begun to sporulate, a larger, thin, uncoated coverslip may be placed over, but not in contact with, the coated coverslip. The former is supported by a ring of petroleum jelly laid down on the surface of the glass slide surrounding the lower coverslip. The mycelium may then be examined using the high-power objectives of a compound light microscope, permitting time-lapse, photomicrographic analyses of conidiogenesis (Cole, 1975). The upper coverslip may be removed at any time, and the lower coverslip culture prepared for the scanning electron microscope (SEM) (Fig. 1). By carefully scoring the agar surface near the area of the culture examined with the light microscope, it is possible to locate the same area in the SEM. This capability of correlating light and electron microscope observations contributes significantly to interpretations of conidiogenesis.

Several variations of this technique have been used in preparing microfungi for ultrastructural examinations. Millipore or Nucleopore filters coated with a thin layer of agar (or uncoated filters) are inoculated with conidial fungi and then incubated on the surface of agar plates in petri dishes. The filter cultures may

---

Fig. 1.   Fragmented fertile hypha (holoarthric conidia) of *Geotrichum candidum* grown on a coverslip culture and prepared for the SEM. 2250. (From Cole and Samson, 1979.)

Fig. 2.   Scanning electron micrograph of vegetative hyphae of *Geotrichum candidum* grown on cellophane which overlaid the surface of agar in a petri dish. 863.

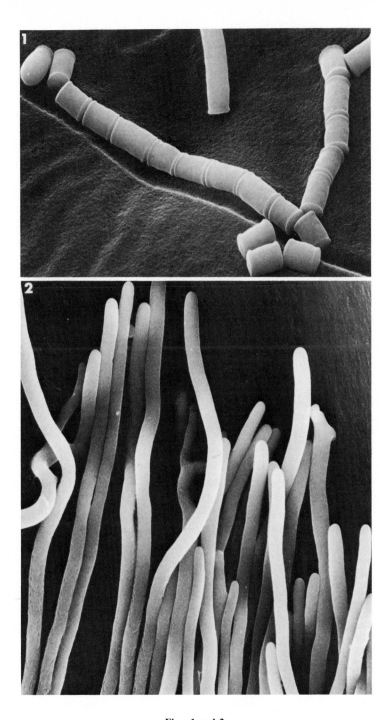

Figs. 1 and 2.

then be prepared for either the SEM or transmission electron microscope (TEM). To ensure that the cells are not lost during preparation, the mycelium on the membrane filters may be covered with 2% agar (Oujezdsky *et al.*, 1973). The filters can be easily sectioned by ultramicrotomy, which is advantageous in TEM preparations. Another successful procedure involves the use of a thin layer of permeable cellophane (DuPont, cellulose xanthate; Kalle Aktiengesellschaft, Wiesbaden, West Germany), which is first cut to cover the bottom of a petri dish, sterilized in boiling water for 3 min, and then placed carefully on the agar surface of a petri dish culture without wrinkling the film. The surface of the cellophane is inoculated and after sufficient mycelial growth may be prepared for either scanning or transmission electron microscopy. This technique has been especially useful in examination of hyphal tips (Fig. 2). The apical regions of the hyphae are monoplanar because they remain in contact with the surface of the cellophane, which increases the ease of obtaining medial longitudinal sections (Grove *et al.*, 1970).

A limitation in the use of coverslip cultures and similar culturing procedures is the high rate of desiccation of the thin layer of agar during incubation. For conidial fungi requiring several days to sporulate, this technique is usually not suitable. The majority of Hyphomycetes prepared for scanning and transmission electron microscopy which were illustrated by Cole and Samson (1979) were grown in conventional petri dish cultures. Small blocks of mycelium and supporting agar (approximately 5 mm sq.) were simply excised from regions of dense sporulation and prepared for the TEM or SEM.

Coelomycetes pose additional problems for developmental and ultrastructural studies. Many do not sporulate in culture, while several pycnidial forms which do produce conidiomata in or on agar have been shown to undergo atypical development when compared to pycnidia grown on natural substrates. Acervular fungi produce their conidiomata only in association with host tissue. Sterilized natural substrates inoculated and maintained in damp chambers have been used successfully in obtaining axenic cultures of Coelomycetes and have proved beneficial in our laboratory for light and electron microscope studies of conidiomatal development (see Cole, this volume, Chapter 23). Many species apparently require alternating light-dark periods during incubation before they will sporulate. *Chaetomella acutiseta* produces abundant pycnidia after 10 days' growth on either steam-sterilized carpet grass (*Stenotaphrum secundatum*) or agar in a petri dish culture incubated at 20°C and subjected to alternating 12-h light and dark periods. Time-lapse light microscope examinations of developing pycnidia on the surface of natural and artificial substrates has been achieved using a fiber optics light source for cool reflected illumination, and Zeiss Luminar epiluminescent objectives (2–14× magnification) attached to a compound microscope. These objectives have excellent flat-field quality. Heat-sterilized rabbit dung has also been demonstrated to be a suitable substrate for the formation of conidiomata by a large number of pycnidial fungi. Intact dung may be prepared for the SEM with minimal disturbance of the delicate conidial fungus on its surface. This simple culturing technique has permitted developmental-ultrastructural examinations of conidiomata produced by several coelomycetous fungi.

II. LIGHT MICROSCOPE EXAMINATIONS

Technology has made great strides forward in development of the compound light microscope, building on the basic principle of Johann Kepler (1571–1630) that combining a convex ocular with a convex objective serves to enlarge the real image of the objective (Gage, 1947). The word "microscope" was coined by Giovanni Faber in 1625 to refer to "an optical instrument capable of producing a magnified image of a small object" (Rochow and Rochow, 1978). The first mass-produced light microscope was designed and manufactured by Anton van Leeuwenhoek (circa 1670) and was capable of resolving about 1.4 μm at 270× magnification (Rooseboom, 1967). Based on the development of achromatic objectives in the eighteenth and early nineteenth centuries, and later investigations under the direction of Ernst Abbe (1840–1905), from which the concept of numerical aperture (NA) and the invention of apochromatic objectives emerged, the modern compound light microscope was constructed. Contemporary flat-field, "high-dry" objectives (NA = 0.5–0.95) permit sufficient working distance to examine thick cells and tissues, while oil immersion objectives (NA = 1.0–1.4) provide the highest resolving power. To obtain maximum resolution from an oil immersion objective, an oil immersion condenser (NA = 1.0 or higher) must be used. The oil should have a refractive index of 1.515 which matches that of glass slides. The thickness of the cover glass should be in the range of 0.17–0.19 mm. For optimal resolution, the high-dry objective should be equipped with a correction collar to accommodate the variation in thickness of commerical cover glasses.

The shortest wavelength of visible light is approximately 400 nm. This in effect limits the resolution of the light microscope to about 0.2 μm. A tungsten light source is commonly employed, and the microscope must be carefully aligned for Koehler illumination before a specimen is examined. In addition, high-intensity halogen light sources and mercury lamps for fluorescence microscopy have been developed during the last 30 yr. This microscopic technique has proved particularly useful to yeast taxonomists and medical mycologists. The process of cell wall growth in *Endomyces magnusii, Schizosaccharomyces octosporus, S. pombe,* and *Saccharomycodes ludwigii* has been successfully examined by Streiblová (1971) and Streiblová and Wolf (1972) using techniques of fluorescence microscopy, thin-sectioning, and time-lapse cinemicrography. Several fluorescent stains (fluorochromes) in combination with appropriate excitation barrier filters have been used successfully in the examination of fungal material (e.g., acridine orange, primulin, calcofluor, fluorescein diacetate, fluorescamine). The value of fluorescence microscopy in medical mycology is most evident from the numerous applications of the fluorescent antibody technique. In this procedure, fluorescein isothiocyanate (FITC) is usually employed as a fluorochrome to tag specific antibodies which in turn react with appropriate antigens. The results of such reactions are used in the identification of common antigenic compounds among fungal pathogens and in serodiagnosis of certain mycoses (Odds, 1979). Most fluorescence microscopes now make use of the incident mode of illumination (i.e., epifluorescence). The exciting radiation is focused on the object from above, and the objective also serves as the condenser, thus eliminating the problem of dissipation of light associated with transmitted illumination. Figure 3 illustrates the combined arrangement of epifluorescence and transmitted light microscopy. In this example, which is from a study on algal chlorophyll fluorescence reported by Wilde and Fliermans (1979), a series of heat, excitation, and barrier filters was used to select for blue incident light (455–490 nm) and red fluorescence emitted from the sample.

Development of phase-contrast objectives was especially significant for

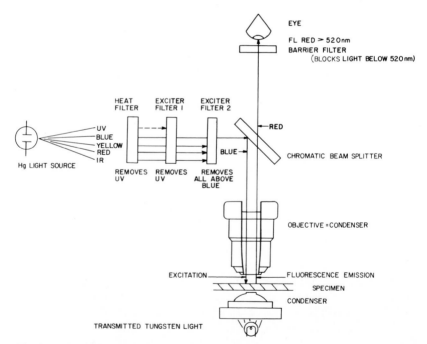

Fig. 3.   Combined epifluorescence and trasmitted-light microscopy. In this example, bright-red chlorophyll fluorescence results from blue excitation of the chlorophyll-containing algal specimen. (After Wilde and Fliermans, 1979.)

mycologists working with hyaline material. Low contrast is not an uncommon problem in the examination of biological specimens "where the watery material differs very little in refractive index from the watery medium in which it is suspended. In such instances, the microscopist can gain the necessary contrast by resorting to phase-contrast, phase-amplitude modulation, . . . dispersion staining" or dark-field microscopy (Rochow and Rochow, 1978). Phase-amplitude modulation is based on the use of a small diaphram or phase plate coated with a thin film of material and inserted in the back focal plane of the objective. By delaying light rays which pass through the phase plate, interference between the delayed rays and direct rays results in enhancement of contrast. A variation of this concept is the basis of interference microscopy. In this case, however, two microscope beams are deliberately caused to interfere with one another, resulting in a pattern of interference fringes which enhances contrast. In the case of Nomarski interference-contrast microscopy (Fig. 4), light emitted from the source is linearly polarized and then passes through the condenser containing a Wollaston prism arranged diagonally to the polarizer. The prism splits the polarized light into two rays. The magnitude of the split is below the resolving power of the microscope, which prevents a double image from being detected. The second Wollaston prism is contained in the rear focal plane of the objective where the light rays are recombined. The light subsequently passes through a polarizing analyzer. Interference depends on the phase difference of the rays of light passing through the analyzer, which in turn depends on the thickness and refractive index of the object. The result is a pseudo-three-dimensional image with good contrast and well-defined boundaries. The halo produced by phase-contrast microscopy is absent. Because the optics of the Nomarski system demonstrate a very limited depth of focus, this technique is most useful in observing monoplanar objects. Nomarski interference-contrast microscopy has been particularly useful in our laboratory for examinations of cryosections of conidiomata and other fungal fructifications. The comparison of photomicrographs produced by bright-field, phase-contrast, and Nomarski interference-contrast microscopy presented in Figs. 5–7 reveals the differences in information provided by these three microscopic techniques.

The Hoffman modulation system (Hoffman and Gross, 1975a,b) is also used to produce three-dimensional images of low-contrast biological material. This system employs a polarizer for the transmitted light, a rotatable holder for a sliding slit with a partial polarizer which fits under the condenser, and a "modulator" disk with dark, gray, and bright areas which fits in the back focal plane of the objective. This modulation of the light beam results in directional oblique illumination which generates a contrast between the image and background, provides shadows and sharp outlines without halos, and produces a good three-dimensional effect.

The stereomicroscope is also a valuable research tool for the mycologist in-

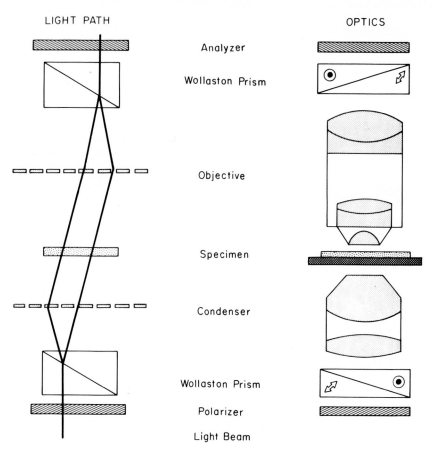

LIGHT PATH   OPTICS

Analyzer

Wollaston Prism

Objective

Specimen

Condenser

Wollaston Prism

Polarizer

Light Beam

Fig. 4.   Diagram of light path and optics of transmitted-light interference-contrast microscopy according to Nomarski.

terested in conidial fungi. For low-magnification viewing (approximately 2–200×), this microscope provides much useful information on such morphological features as the arrangement of conidiophores in hyphomycetous colonies and the arrangement of setae on conidiomata produced by certain of the Coelomycetes. Using a fiber optics illumination system, open culture plates and delicate specimens may be examined for long periods without the problem of rapid desiccation due to heat generated by the tungsten lamps employed with most

Figs. 5–7.   A conidioma of *Graphiola phoenicis* examined by Nomarski interference-contrast (Fig. 5), phase-contrast (Fig. 6), and bright-field microscopy (Fig. 7). The three micrographs were printed under the same conditions. Arrow in Fig. 5 indicates ruptured host (*Phoenix canariensis*) cuticle and epidermis. CC, Conidiogenous cells; E, epidermis of host; HT, host tissues. 300.

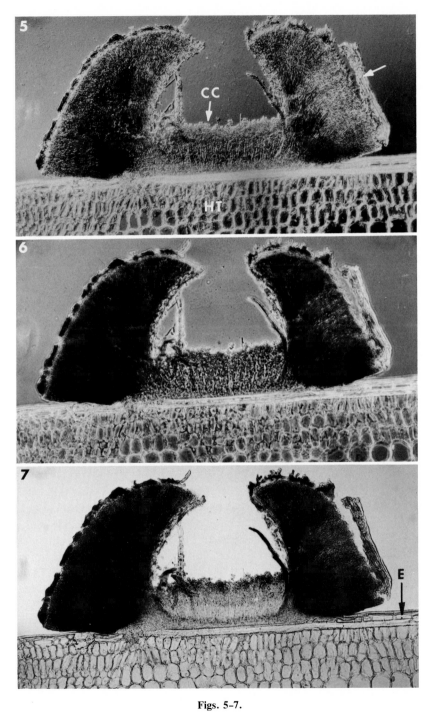

Figs. 5–7.

stereomicroscopes. Achromatic, flat-field objectives (0.8–5×) are available for some of the better quality stereomicroscopes (e.g., Wild-Heerbrugg Model M-5, Zeiss Model SR), combined with a built-in, double-iris diaphragm for maximum depth of field control and a phototube for camera attachment.

Time-lapse light microscope analyses of developing conidial fungi have contributed valuable information on many subtle aspects of conidiogenesis not previously recognized when only fixed cells were examined (e.g., Cole and Kendrick, 1968, 1971; Kendrick and Cole, 1969; Fig. 8A–D). Time-lapse units for photomicroscopes are now provided by several manufacturers at reasonable cost. Making use of a Zeiss Photomicroscope I equipped with automatic exposure and a motorized 35-mm film drive, we designed our own simple time-lapse system (Fig. 9). Several technical requirements of this time-lapse unit are listed below:

1. The exposure cycle, which is normally actuated by pushing the exposure button of the control box, must be automatically activated at predetermined intervals by an electric impulse.

2. Activation of the exposure cycle is dependent on a rotating cam bearing projections which trip a microswitch. Intervals between exposures are determined by arrangement of the projections (the rate of the cam rotation is constant).

3. The light source must be dimmed during intervals to protect the live fungal specimens from excessive light and heat. Heat filters are also positioned in the light path below the microscope condenser.

4. The light source must be returned to a preselected intensity for the duration of each exposure.

5. An override switch is required so that the transformer rheostat can be operated normally, allowing the microscope to be focused and a suitable light intensity selected for film exposure without the necessity of activating the exposure cycle.

Briarty and Ramsey (1978) have described a simple time-lapse unit which makes use of commercially available integrated circuits for controlling delay between successive exposures, duration of operation of the illumination source, and duration and frequency of the exposure. These circuits may be electronically synchronized to most photomicroscopes. In the case of the Zeiss Photomicroscope I, we combined the improvised time-lapse system controlling the 35-mm camera along with a 16-mm cinephotomicrographic unit (Sage Instruments, Inc., White Plains, New York) equipped with a Bolex H16 camera (Fig. 10). The conventional photomicroscope was converted into a time-lapse system with minimal expense and only a basic knowledge of electronics.

Fig. 8. Time-lapse light micrographs of *Trichothecium roseum* showing the retrogressive nature of conidiogenous cell development. Arrows in (B) and (C) indicate a conidium initial. CC, Conidiogenous cell; C, conidium. 5000. (From Kendrick and Cole, 1969.)

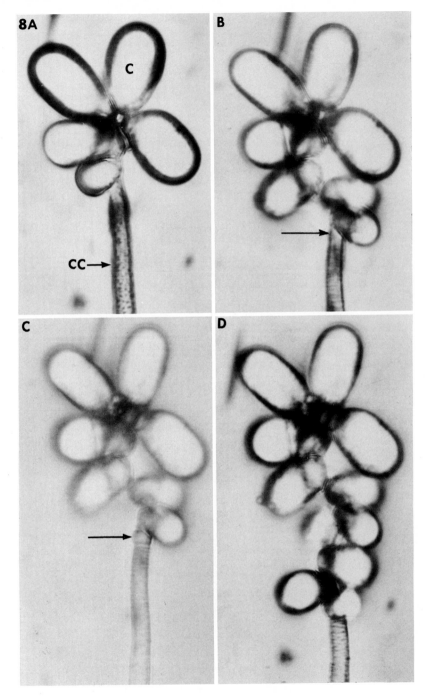

Fig. 8.

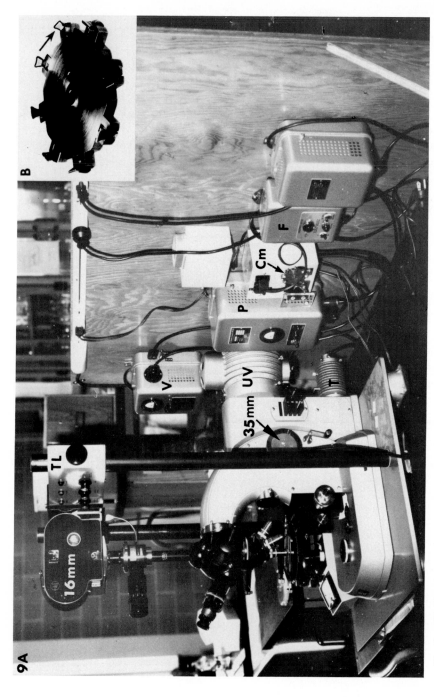

Fig. 9. (A) Time-lapse photomicrographic apparatus. (B) Cam (Cm) used to trigger the 35-mm camera shutter automatically. Arrow in (B) indicates hinged projections used in setting intervals of exposure. (A) shows 16-mm camera, electronic flash power supply (F), ultraviolet light (UV), power supply (P), tungsten lamp housing (T), 16-mm camera electronic time-lapse control system (TL), and voltage regulator (V) for tungsten lamp.

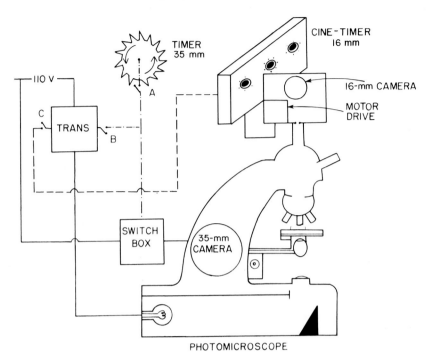

Fig. 10. A simplified circuit diagram showing arrangement of components used in 16- and 35-mm time-lapse photomicrography. When switch B is closed and C is open, the period of bright illumination is controlled by the 35-mm timer. When switch C is closed and B is open, the interval of bright illumination is controlled by the cine-timer. The arrows on the 35-mm timer (Cam) represent the direction of rotation.

Four kinds of 35-mm film are commonly used for photomicrography in our laboratory: Kodak Panatomic-X (ASA 32), Kodak Tri-X (ASA 800), Ilford HP-3 (ASA 800), and Kodak high-contrast copy film (ASA 64). The last has been demonstrated to be most useful for bright-field examinations of dematiaceous fungal cells and has been employed extensively in time-lapse studies on conidiogenesis (Cole and Samson, 1979).

III. ELECTRON MICROSCOPY

"Two events in the 1920s brought about the development of the electron microscope. One was the realization from the de Broglie theory (1924) that particles have wave properties and that very short wavelengths (e.g., 0.005 nm) are associated with an electron beam of high energy. The other event was the demonstration by Busch in 1926–1927 that a suitably shaped magnetic field could be used as a lens in electron microscopes'' (Rochow and Rochow, 1978). In 1934, L. L. Marton built an electron microscope in Brussels and took the first electron micrographs of biological material. This pioneering breakthrough was soon followed (1938–1939) by reports of the construction of electron microscopes, simultaneously and independently, in Berlin by Ruska and von Borries and at the University of Toronto under the supervision of E. F. Burton. However, it was Marton, Hillier, Vance, and Zworykin (inventor of the television picture tube), working at the Radio Corporation of America in 1940, who built the first commercial electron microscope in the U. S. Technology thus played a central role in extending the observational capabilities of the microscopist and, in this case, gave birth to the new field of cell biology. Technological refinement of electron microscopes is still going on, and biologists continue to find new applications and reap the benefits of this engineering research. Several good reviews are available which outline the history and clearly explain the theory of electron microscopy (e.g., Meek, 1970; Wischnitzer, 1970; Dawes, 1971; Oatley, 1972; Wells, 1974; Grimstone, 1977; Rochow and Rochow, 1978). The two principal instruments used for research on biological ultrastructure are the transmission electron microscope (TEM) and the scanning electron microscope (SEM), which are differentiated essentially on the basis of how detail in the specimens is revealed by electrons (Rochow and Rochow, 1978).

A. The Transmission Electron Microscope

In the case of the TEM, the image is formed by electrons which are initially generated by a pointed filament and emitted in a vacuum, pass through the specimen, and finally strike a fluorescent screen at the base of the microscope (Fig. 11). The electrons emitted from the filament are accelerated by the electric potential established between the cathode and anode (50–100 kV for most conventional TEMs). The beam is then condensed by the condenser lenses before it reaches the specimen. Some electrons pass freely through the specimen, while others are changed in velocity and/or direction. The objective lens is used to focus the transmitted beam to an intermediate image, which is subsequently enlarged by the projector lens to form the final image on the fluorescent screen. The screen may be displaced so that the electrons strike and expose a photographic plate. In most TEMs, the camera is maintained under vacuum, necessitating that the film be very dry (prepumped) before it is inserted into the microscope. Even under these conditions, considerable time is often wasted when film is changed, because of the interval required for the vacuum in the film chamber to equilibrate with the rest of the column. New microscopes, however, employ a fiber optics system for film exposure, which allows the film casette to be housed below the fluorescent screen at atmospheric pressure (e.g., Zeiss Model 109).

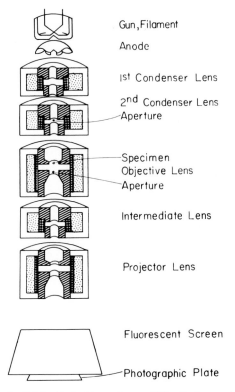

Gun, Filament

Anode

1st Condenser Lens

2nd Condenser Lens
Aperture

Specimen
Objective Lens
Aperture

Intermediate Lens

Projector Lens

Fluorescent Screen

Photographic Plate

Fig. 11.   Scheme of a TEM with double condenser. (After Rochow and Rochow, 1978.)

B. The Scanning Electron Microscope

A significant event in the evolution of electron microscopy was von Ardenne's (1938a,b) innovative addition of scan coils to the TEM. This in effect was the progenitor of the modern scanning or reflection electron microscope. The first SEM constructed by Zworykin *et al.* (1942) was not totally successful because of the poor signal-to-noise ratio which produced a very granular image. Almost 20 yr passed until the development of a scintillator-photomultiplier for electron detection (Everhart and Thornley, 1960), which largely overcame this problem. The first commerically available SEM was subsequently constructed in 1965 at Cambridge University.

The design of the upper column of the SEM (Fig. 12) is comparable to that of the TEM. However, the addition of scan coils causing the beam to sweep back and forth across the surface of the specimen, which is commonly located at the base of the column, is the distinguishing feature of the SEM. When the electron

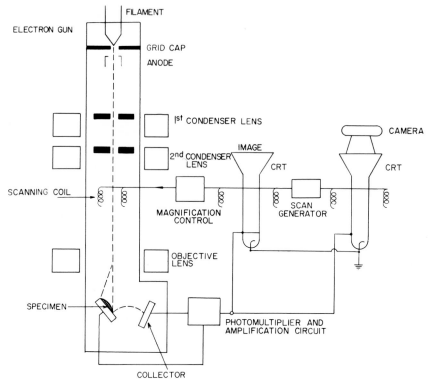

Fig. 12.   Block diagram of a SEM. (After Hayat, 1978.)

beam strikes a point on the specimen, secondary electrons are emitted. The latter
are picked up by the collector and transmitted to the photomultiplier and amplifi-
cation circuit. The amplifier output varies with the intensity of secondary electron emission from the surface of the specimen. In turn, the output of the
amplifier determines the potential of the modulating electrode of the cathode ray
tube (CRT). There are commonly two CRTs, a display and a record unit, which
are electronically synchronized. The current reaching the collector (i.e., secondary electrons) is converted electrically into a voltage signal. Of importance is
that the scan coils in the microscope column and those of the CRTs are powered
by the same scan generator. This means that each point on the specimen corresponds to a point which is reproduced on the CRTs. The voltage signal in effect
varies the brightness of each spot on the CRTs in synchrony with the movement
of the electron beam. Any variation in topography of a region of the specimen
scanned by the electron beam produces a change in secondary-electron emission
and a corresponding variation in the voltage signal, which results in simultaneous
changes in brightness of the matching regions of the display and the record

TABLE I

Comparison of Features of Various Types of Microscopes[a]

Microscope	Resolution (nm)	Magnification range	Depth of focus at ×500 (μm)	Field of focus at ×500 (μm)	Working distance at ×500 (nm)
Compound light microscope	300	10–1500	2	210	0.4
Transmission electron microscope	0.2	55–200,000	800	200	2.0
Scanning electron microscope	5–10	3–100,000	1000	200	20

[a] After Hayat, 1978.

CRTs. The area of the specimen scanned by the incident electron beam is much less than the area scanned on the CRT, which results in a magnified image. Major advantages of the SEM over the TEM and the light microscope are its great depth of focus, long working distance, and high resolving power (5–10 nm), A comparison of these and other features of the three types of microscopes is presented in Table I (Hayat, 1978).

An added feature of the SEM is that signals other than secondary electrons are generated by interaction of the incident electrons and the specimen (Fig. 13). These signals may be monitored with appropriate detectors. Explanations of the nature of all these signals are beyond the scope of this discussion, and the reader is referred to other sources (e.g., Hearle *et al.*, 1972; Oatley, 1972; Murata,

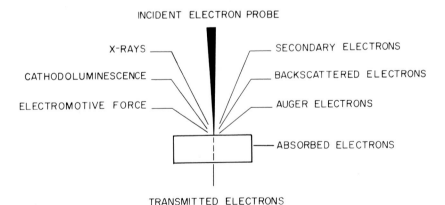

Fig. 13. Kinds of signals produced by the interactions between incident electrons and substances. (After Kimoto, 1972.)

1973; Wells, 1974; Goldstein, 1975; Reimer, 1976; Hayat, 1978). However, some comments on x-ray emission are necessary because of the extensive application of such data in biological research (Marshall, 1975). When electrons of the incident beam interact with the inner shell electrons of atoms comprising the specimen, x-ray photons are usually emitted by a process of ionization. The wavelength of each x-ray emitted is specific for atoms of a given atomic number. Therefore, analysis of the wavelengths of x-rays provides data on the elemental composition of the specimen. By moving the incident beam to various points on the specimen, the elemental composition of different structures may be determined. It is possible to obtain an x-ray profile and morphological image of the same sample simultaneously. The limitation of this technique is that it cannot be applied to elements of very low atomic number. Instead of using a detector which analyzes wavelengths of emitted x rays, the SEM may be equipped with an energy-dispersive spectrometer (EDS). The interaction of incident electrons and constituents atoms of the specimen causes core scattering which results in the emission of continuous radiation. The latter "consists of photons emitted by electrons undergoing deceleration during collisions with atoms. The intensity of the continuum increases with increasing atomic number of the target material" (Hayat, 1978). The elemental composition of the specimen therefore may be analyzed using the EDS on the basis of characteristic energy levels of x rays emitted.

The EDS has been extensively employed in biological applications of x-ray microanalysis and is capable of detecting all elements above fluorine in the periodic table (Hayat, 1978). For example, when plant-pathogenic fungi penetrate the host cuticle and epidermal cell layers, the accumulation of various elements is detected in the cell walls of both the host and pathogen (Fukutomi, 1976). Cucumber cotyledons infected with *Colletotrichum lagenarium* were initially frozen in liquid nitrogen, sectioned, and coated with a thin layer of conductive gold in a vacuum evaporator. The samples were then examined in a SEM equipped with an EDS which revealed silicon, calcium, manganese, and phosphorus at sites of penetration by the fungus. Fukutomi suggested that this technique may be used for further studies on the relationship between the accumulation of certain elements, especially manganese, calcium, and silicon, and fungal parasitism. *In situ* x-ray energy-dispersive analysis of polyphosphate bodies in *Aureobasidium pullulans* was reported by Doonan *et al.* (1979). The fungal cells were unfixed, unstained, and air-dried. The major elemental constituents of these polyphosphate bodies are potassium, magnesium, sulfur, and phosphorus. Schoknecht (1975) used the SEM coupled with x-ray microanalysis as a taxonomic research tool to compare the elemental composition of calcareous deposits in several families of Myxomycetes. In a later study, Schoknecht and Hattingh (1976) employed the same technique to detect elements suspected to be translocated by mycorrhizal fungi from the soil to cortical root cells. Many applications of x-ray microanalysis to mycological and related research areas undoubtedly exist but have not yet been exploited.

C. Recent Developments

High-resolution scanning electron microscopy is now capable of providing structural data which previously could be revealed only by transmission electron microscopy. Panessa-Warren and Broers (1979) described the surface features of a T7 bacteriophage adsorbed to its host bacterium and demonstrated 1- to 2-nm resolution. The authors claimed potential edge sharpness resolution of the secondary-electron-generated image in the range of 0.8–1.0 nm.

This marked improvement in the resolving power of the SEM is largely due to better electron emission guns (i.e., field emission and lanthanum hexaboride cathodes; Crewe *et al.*, 1968; Broers, 1975; Welter, 1975) and ultrahigh vacuum (10^{-8}–10^{-10} torr) in the microscope column. Development of the high-resolution SEM occurred together with the design of the scanning-transmission electron microscope (STEM; Crewe, 1971, 1979; Venables, 1976; Hayat, 1978). In the case of the STEM (Fig. 14), electrons are transmitted through a thin section of the specimen (cf. TEM) and then collected and imaged on a point-to-point basis

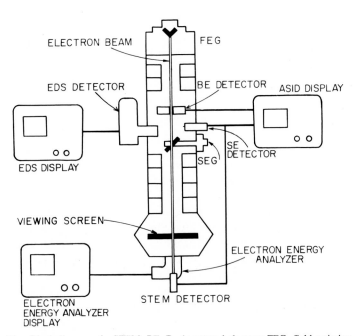

Fig. 14.   Block diagram of a STEM. BE, Backscattered electron; FEG, field emission gun; SE, secondary electron; SEG, side-entry goniometer. (After JEOL News 15(6), 1977.)

on a CRT display (cf. SEM). The electron beam scans the specimen and the transmitted electrons are detected by the collector located at the base of the column. The collector (detector) can analyze the transmitted beam in an electron energy analyzer and simultaneously generate an image on the display CRT of the auxiliary scanning image display (ASID) by essentially the same process as the SEM collector described earlier. The electron beam of a conventional TEM is transmitted through the entire specimen simultaneously, which makes electron energy analysis difficult. The scanning beam of the STEM, on the other hand, provides a means of analyzing the transmitted electrons on a point-to-point basis. The conventional TEM presently requires high-acceleration voltages (100 kV–1MeV) to obtain the best theoretical resolving power (0.3–0.12 nm), which may cause serious electron irradiation damage to biological specimens (Hashimoto, 1979). In contrast, contemporary STEMs obtain a resolution of 0.25 nm using only 40 kV. The micrograph of the STEM image is obtained by photographing the CRT display. However, because the signal originating from the STEM collector may be processed in various ways (e.g., stored, mixed, modulated), Crewe (1979) has suggested that "the ideal way to use the STEM is to store the information in a digital memory and use a mini-computer to extract the required information from the memory. Such a combination could dramatically increase the output of data from the instrument."

The STEM is an analytic electron microscope capable of multiple functions (Ichinokawa, 1979). In addition to the image generated by transmitted electrons, a secondary-electron detector located above the specimen (Fig. 14) is used to generate a topographical image on the ASID. The combination of TEM and SEM capabilities in a single instrument permits observation of both surface and internal ultrastructure of the same cell. Katsumoto *et al.* (1978) examined mouse intraperitoneal macrophages, which were chemically fixed and negatively stained (see Section III,D), using both the transmitted and secondary-electron image modes of the STEM. These authors recognized the value of data derived from these complementary images. A backscattered electron detector conveys information to the ASID on surface topography, elemental composition, and crystal orientation of biological samples. An EDS, also attached to the microscope column, interprets the energy levels of x rays emitted from the specimen, and from these data the elemental composition of the sample is obtained. The analytic STEM is a powerful research tool used for correlating information on shape, structure, and elemental composition of biological specimens.

A final comment relates to development of the specimen stage control (eucentric side-entry goniometer; SEG in Fig. 14). Of particular significance to the success of high-resolution electron microscopy is the design of the specimen stage and its external controls. The SEG is quite resistant to vibration and essentially free of thermal drift. The stage may be tilted to a maximum of 60° and rotated through 360°. This flexibility of movement is useful in obtaining additional information from both transmitted- and secondary-electron images.

For example, precise control of the tilting angle with only slight lateral deviation of the image provides an easy method of obtaining stereoscopic images. The value of ultrastructural data revealed by stereoelectron micrographs is now appreciated (Boyde, 1974, 1975; Chatfield and Pullan, 1974; Boyde and Ross, 1975; Pawley, 1978; Barber and Emerson, 1979).

Electron micrographs contain considerable information which is still largely overlooked. However, application of modern stereological and morphometric methods (Weibel and Bolender, 1973; Bolender *et al.,* 1978) enables workers to quantitate data in their micrographs, such as changes in volume and surface area of organelles, which previously escaped detection. Fawcett (1979) has pointed out that "quantitative digital image analysis makes stereology and morphometry available to microscopists with a reasonable investment of time and without the need for elaborate and costly computer hardware. It is clear that, in the future, electron microscopists must make increasing use of these quantitative methods to give their results greater significance and wider acceptance."

D. Specimen Preparation

Techniques for preparing biological specimens for electron microscopy are almost as variable as the kinds of material examined (Johari and Corvin, 1972). This discussion will concentrate therefore on preparatory methods used in ultrastructural examinations of conidial fungi. Chemical fixatives commonly employed for fungal preparations include glutaraldehyde, osmium tetroxide, acrolein, and formladehyde. These may be used individually or in combination. The chemicals are prepared in a buffer (usually sodium cacodylate or phosphate) which can be used to adjust the pH and balance the fixative osmotically with the cell interior. The latter prevents excessive swelling or shrinkage of cells during fixation. The chemical characteristics of these fixatives and their modes of interaction with cell components have been reviewed by Hayat (1970). The same author has provided a useful list of recipes for the preparation of fixatives and buffers.

1. Thin-Sectioning

Two cardinal rules in preserving live material for thin-sectioning and TEM examination are that the sample should not be exposed to unfavorable growth conditions prior to chemical fixation and that the fixative should penetrate the cells as rapidly as possible. For cultures of conidial fungi, colonies should be flooded with fixative immediately after the lid of the petri dish is removed. Small pieces of the sample may then be excised from the flooded culture and transferred to a vial containing fresh fixative. The same procedure should be used for specimens grown in damp chambers on natural substrates (e.g., coelomycetous fungi).

The preservation of cytological detail in vegetative septate hyphae of anamorphic fungi requires such a rapid fixation procedure (Grove and Bracker, 1970). Any appreciable variation in the microenvironment prior to fixation, especially in the region of the hyphal tip, results in disruption of the vesicular arrangement (i.e., the *Spitzenkörper*; Girbardt, 1957) and disassociation of secretory vesicles and other organelles in the apex of the hypha. As described in Section I, mycelia may be grown on sheets of cellophane, which facilitates preparation of hyphae for thin-section examination. To avoid interruption of tip growth prior to fixation, the culture is quickly flooded with the fixative. Small pieces of the cellophane supporting hyphal tips at the edge of the colony are then transferred to vials containing fresh fixative. Several different fixation and embedding procedures have been used for thin-section examinations of hyphal tips of septate fungi (Girbardt, 1957, 1969; Grove and Bracker, 1970; Collinge and Trinci, 1974; Turian, 1978; Cole and Samson, 1979; Najim and Turian, 1979). A summary of two of these preparatory procedures is presented below. For details of the fixation schedules, the reader is referred to the original reports.

a. From Grove and Bracker (1970)

i. Fixation. 2% glutaraldehyde in 0.1 *M* sodium cacodylate buffer (pH 7.2) for 30 min at room temperature, *or* 1% formaldehyde (prepared from paraformaldehyde) plus 2% glutaraldehyde in 0.1 *M* sodium cacodylate buffer (pH 7.2) for 30 min at room temperature (ca. 24°C).

ii. Postfixation. 1% osmium tetroxide in 0.1 *M* sodium cacodylate buffer (pH 7.2) for 2–4 h at room temperature.

iii. Stain. 0.5% aqueous uranyl acetate for 2–4 h.

iv. Dehydration. Graded ethanol series followed by anhydrous acetone.

v. Embedding. Araldite 605 (Richardson *et al.*, 1960).

vi. Poststain. Sections on collodion-coated grids stained for 10 min with lead citrate (Reynolds, 1963).

b. From Cole and Samson (1979)

i. Fixation. 6% glutaraldehyde and 1% osmium tetroxide, each prepared at 4°C in 0.2 *M* sodium cacodylate buffer (pH 7.1) and mixed in equal volumes, for 2 h on ice and stored in darkness (after Franke *et al.*, 1969; Heintz and Pramer, 1972).

ii. Postfixation. 1% osmium tetroxide in 0.1 *M* sodium cacodylate buffer (pH 7.2) for 2 h on ice in darkness.

iii. Stain. 0.5% aqueous uranyl acetate for 3 h and later in 2% uranyl acetate in 70% ethanol for 2 h during dehydration.

iv. Dehydration. Graded ethanol series followed by anhydrous acetone.

v. Embedding. Spurr's (1969) low-viscosity plastic (intermediate hardness).

vi. Poststain. Sections on Formvar-coated grids stained 20 min with 0.5% aqueous uranyl acetate and 10 min with lead citrate (Reynolds, 1963).

Some differences are evident in the appearance of organelles utilizing these various procedures. However, representation of the degree of cytological preservation is demonstrated by Fig. 1 in Chapter 23 of this volume.

In addition to these conventional fixation procedures, Howard and Aist (1979) have described a freeze-substitution method of preparing hyphal tips for thin-sectioning, which shows good preservation (Figs. 15–19). The mycelium was grown on rectangles of cellulose or polycarbonate membranes which were incubated on the surface of agar in a petri dish. The hyphae were allowed to grow halfway across the supporting substrates. With a pair of forceps, the substrate was then quickly removed from the agar and plunged hyphal side down first into eutectic Freon 22 and then into liquid nitrogen. The specimens were subsequently transferred to 10-ml, glass, screw-capped vials containing precooled (−85°C) anhydrous acetone (electronic grade, dried, and maintained over a molecular sieve which was present in the vials). After 15 days (with one change of acetone during this period), the samples were subdivided and transferred to substitution fluids following one of two different schedules described below.

Schedule A(FS-A): 1% glutaraldehyde plus 1% tannic acid for 24 hr at −85°C and then for 24 hr at −22°C; acetone for 24 hr at −22°C; acetone for 15 min at +22°C.

Schedule B(FS-B): 2% glutaraldehyde . . . plus 1% OsO_4 for 12 hr at −85°C, for 12 hr at −22°C, and then for 45 min at 0°C; a fresh solution of 1% OsO_4 for 2 hr at 0°C; two changes of acetone at +22°C for 15 min each (Howard and Aist, 1979).

Stock solutions of osmium tetroxide were prepared and maintained as described by Feder and Sidman (1958), while 70% glutaraldehyde was distilled to remove water and the distillate (100% glutaraldehyde) diluted using anhydrous acetone and maintained over a molecular sieve at 35°C until used for fixation. The specimens were embedded in Spurr's (1969) plastic, polymerized at 70°C, and thin-sectioned for transmission electron microscopy.

Howard and Aist (1979) have discussed the differences in ultrastructural detail revealed by freeze-substitution and conventional fixation techniques. For example, four wall layers (approximately 60 nm total thickness) are visible at the

---

Fig. 15. Apex of freeze-substituted (FS-B) hyphal tip of *Fusarium acuminatum*. Apical vesicles (V) of two electron densities partially surround a cluster of ribosome-like particles. Note presence of a microtubule (mt) within the apical dome. M, Mitochondrion; SC, smooth cisternae. Area of tip enclosed by a rectangle enlarged in Fig. 16. 30,000. (From Howard and Aist, 1979.)

Fig. 16. The apical cell wall located within the rectangle in Fig. 15 is distinctly four-layered. Note the apical vesicles (V) and smooth contour of the plasma membrane (PM). × 57,000.

Fig. 17. Area from hyphal tip of *Fusarium acuminatum* revealing large apical vesicles (V) with two electron densities, and microvesicles (mv) which appear to be hexagonal in shape. ×130,000. (Contributed by R. J. Howard.)

Fig. 18. One type of fenestrated, smooth membrane cisterna (SC) in a hyphal tip of *Fusarium acuminatum* as seen after freeze substitution. This structure was apparently proliferating (arrows) and giving rise to microvesicles (mv). V, Large apical vesicle. ×62,000. (Contributed by R. J. Howard.)

Fig. 19. Area from freeze-substituted hypha of *Fusarium acuminatum* illustrating a nucleus (N), mitochondria (M) and rough endoplasmic reticulum (RER). × 24,000. (Contributed by R. J. Howard.)

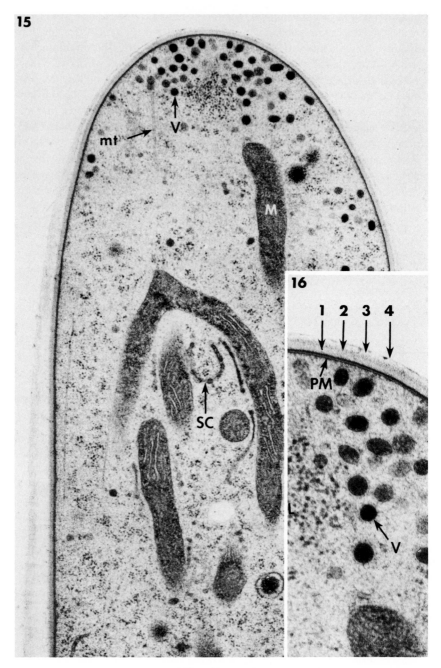

Figs. 15 and 16.

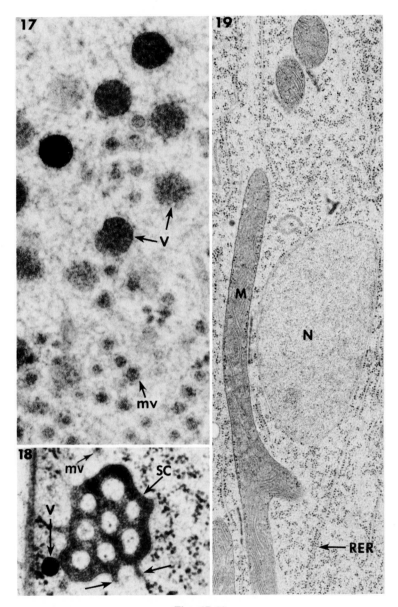

Figs. 17–19.

hyphal tip in Figs. 15 and 16, while only two wall layers are commonly revealed
using conventional fixation procedures (cf. Fig. 1, Chapter 23). The profile of
the plasmalemma in Fig. 16 is smooth, but distinctly undulated in Fig. 1, Chapter
23. The large secretory vesicles shown in Figs. 15–17 are of two electron densities, while the equivalent vesicles in conventionally fixed hyphal tips are
larger and morphologically similar. Microvesicles in freeze-substituted material
are hexagonal in shape and appear to arise from fenestrated smooth membrane
cisternae (Figs. 17 and 18). No such fenestrated membrane complexes have been
associated with microvesicles in conventionally fixed material. Microtubules are
frequently observed at the apex of hyphae prepared by freeze-substitution (mt in
Fig. 15) but are rarely seen after conventional fixation. Although the outermost
mitochondrial membrane is usually absent in freeze-substituted hyphae, the profile of the inner membrane is very smooth and the cristae are evenly spaced (Fig.
19). These features contrast sharply with the undulated profile and unevenly
spaced cristae of conventionally fixed fungal cells. Similar differences in profiles
of nuclei are evident in sections of hyphae prepared by these two techniques.
Although the extraction of lipids from membranes and lipid droplets is one
recognized artifact of the freeze-substitution procedure, cryogenic fixation nevertheless has demonstrated improved preservation of many protoplasmic structures
compared to conventional preparatory techniques.

Fixation procedures successful in the preparation of hyphal tips are usually
effective in preserving cytological detail in conidiogenous cells and young conidia in various stages of development. This is to be expected since "it is often the
hyphal apex that gives rise to the conidia of Hyphomycetes" (Da Riva Ricci and
Kendrick, 1972). However, preparation of mature, thick-walled conidia for
thin-sectioning has presented special difficulties in fixative penetration and plastic embedding. One problem associated with poor penetration of fixatives is the
hydrophobic nature of many conidia. Hammill (1977) added a drop of Tween-20
to the fixative solution (0.8% glutaraldehyde buffered with sodium cacodylate, pH
7.2) to facilitate wetting of hyphal and conidial walls of *Trichurus spiralis*. The
sample was subsequently aspirated (5–10 min) to remove air bubbles and enhance penetration of the fixative. Hawes and Beckett (1977) used the surfactant
Brij 35 (0.1% w/v solution in the fixative) as a wetting agent. Embedding
thick-walled fungal cells has generally involved the use of increasing concentrations (20, 40, 60, 80, 100%) of Spurr's (1969) low-viscosity plastic diluted in
anhydrous acetone (Cole, 1972, 1973a,b, 1975; Harvey, 1974; Brotzman *et al.*,
1975; Ellis and Griffiths, 1975, 1976, 1977; Hanlin, 1976; Murray and Maxwell,
1974; Jones, 1976; Corlett *et al.*, 1976; Hawes and Beckett, 1977). Hammill
(1972, 1977) has demonstrated success in embedding conidial fungi in Araldite
502 (Araldite 502, 15.2 g; DDSA, 11.3 g; DMP-30, 0.375 g). We have found
that gentle but continuous agitation (vials placed on a rotary agitator) improves
plastic infiltration. For the final change in 100% plastic, infiltration is better if

the sample is placed in a vacuum oven set at the polymerization temperature specified for the resin. The sample should remain in the oven under vacuum for 18–24 h. In our laboratory, sections (approximately 60 nm thick) are cut using diamond knives and a Sorvall MT-5000 ultramicrotome.

Instead of embedding fixed and dehydrated specimens in plastic for ultramicrotomy, thin sections may also be obtained from frozen biological tissue. In fact, ultrathin sectioning of frozen cells was used as one of the earliest preparatory techniques in biological electron microscopy (Fernandez-Morán, 1952; Sletyr and Robards, 1977b) Improved contemporary cryoultramicrotomy procedures involve rapid freezing of samples without prior chemical fixation or treatment with cryoprotectants. Sections are cut without flotation in liquid baths and examined in the TEM without staining. The tissue has therefore been exposed to a minimum of solvent and chemical treatment. This feature has been particularly useful in application of the x-ray microanalytic technique discussed in Section III,B. By avoiding exposure of specimens to heavy metals and extraneous ions present in fixatives and buffers, problems of interference by these elements are eliminated and interpretations of x-ray microanalysis are facilitated.

2. Shadowing and Replication

These two techniques evolved early in the development of electron microscopy (Williams and Wyckoff, 1946; Mahl, 1940) as a means of increasing image contrast and providing information on the surface topography of electron-opaque materials. Shadow casting involves the oblique deposition of a thin layer of electron-dense metal (usually platinum–carbon) in a vacuum onto the surface of a dried specimen resting on a support film (Fig. 20). The coated specimen is then examined directly in the TEM. The replication of specimens involves oblique shadow casting (platinum–carbon) followed by deposition of carbon alone at right angles to the surface of the sample (Fig. 21). This second step is important

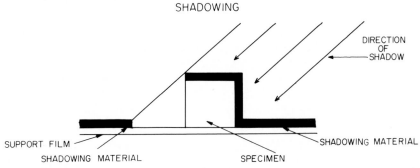

SHADOWING

Fig. 20. Diagrammatic interpretation of the shadowing technique. (After Henderson and Griffiths, 1972.)

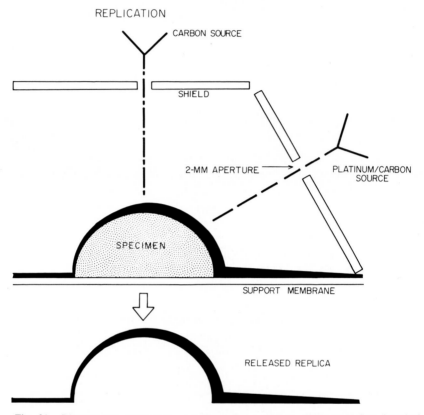

Fig. 21. Diagrammatic interpretation of the replica technique. Platinum is first deposited using a platinum/carbon evaporation source and then carbon alone. The replica is released by placing the coated specimen in a suitable solution which dissolves the biological material.

for structural support of the replica. Unidirectional shadowing at an angle of approximately 45° to the surface of the specimen causes differential deposition of metal across the specimen and reveals structures as raised objects or depressions. The sample to be replicated (e.g., a droplet of cells) is initially dried on the clean, flat surface of a supporting substrate. We use a small piece of freshly cleaved mica. The edge of the coated sample and mica is then gently lowered into a solution which dissolves the biological material but allows the thin metal replica to float to the surface. Chromic acid (a concentrated potassium chromate–sulfuric acid solution) is necessary to dissolve most fungal wall material (i.e., chitin). The replica is subsequently transferred, using a platinum loop, to several small, acid-cleaned dishes containing fresh distilled water (Cole, 1975). The replica is then picked up on a Formvar-coated grid and examined in the TEM. Figures

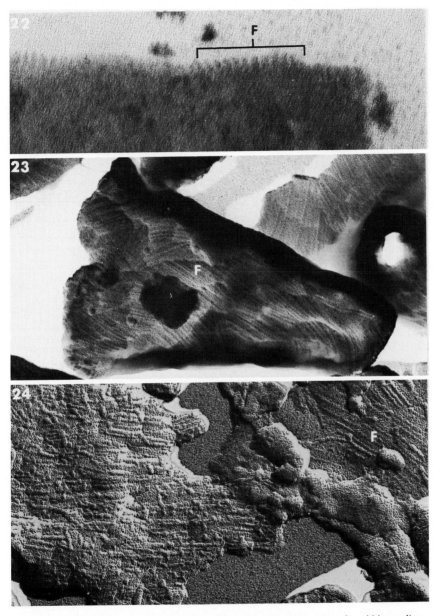

Fig. 22. Thin section of outer conidial wall layer of *Aspergillus niger* after chitinase digestion showing rodlets. F, tangentially sectioned rodlet fascicle. × 90,000. (From Cole *et al.*, 1979.)

Fig. 23. Shadow preparation of outer conidial wall layer of *A. niger* after chitinase digestion. F, Fascicle of rodlets. 50,000.

Fig. 24. Replica of chitinase-digested outer conidial wall layer of *A. niger*. F, Fascicle of rodlets. 66,000. (From Cole *et al.*, 1979.)

22–24 illustrate three different electron microscope images of the same sample—fragments of the isolated outer conidial wall of *Aspergillus niger* (Cole *et al.*, 1979). The surface of the conidial wall is composed of fascicles of fibrous structures called rodlets. Thin sections of electron-opaque wall fragments (Fig. 22) demonstrate poor resolution of the rodlets. On the other hand, shadow casting (Fig. 23) and replication (Fig. 24) clearly show the fascicular arrangement of rodlets and reveal their surface features, respectively.

3. Negative Staining

This technique has been reviewed elsewhere (Haschemeyer and Meyers, 1972; Oliver, 1973) and will be only briefly mentioned here. Samples to be examined by negative staining in the TEM are suspended in a solution of potassium phosphotungstate, and a droplet of this suspension is placed on a grid coated with Formvar. The droplet is carefully removed by touching it with filter paper, and the sample is then allowed to air-dry on the film. The concept of negative staining is that phosphotungstate surrounds particles on the support film as a result of surface tension interactions and penetrates into irregularities on the specimen surface. Under the electron beam, the negatively stained material appears as a light area because of its electron-translucent nature compared to the electron density of the surrounding stain (Fig. 25).

In our laboratory, the following negative staining procedure has proved most useful.

1. With a Pasteur pipet single droplets of stain are placed on a square of Parafilm.

2. A glass rod (or beveled wooden applicator) which has been drawn out to a point is dipped into a suspension containing the material to be stained.

3. A droplet of stain is then pierced with the tip of the liquid-coated rod, thus allowing the sample to flow onto the droplet surface. The droplet should be pierced in the region of maximum curvature where surface tension is reduced.

4. A coated grid is placed film side down on top of the droplet. After the desired staining time has elapsed the grid is picked up with forceps and the excess stain is removed with filter paper. The grid is ready to be examined in the TEM.

4. Preparation of Whole Cells for the Scanning Electron Microscope

Most conidial fungi are susceptible to severe collapse when simply air-dried and then placed in the high-vacuum chamber of an SEM. Although a number of different techniques have been developed for processing soft biological material for scanning electron microscopy (Cole, 1975; Hayat, 1978), the one most successful in preserving conidial fungi is the critical-point drying (CPD) procedure (Anderson, 1951; Lewis and Nemanic, 1973; Kurtzman *et al.*, 1974; Tanaka and Iino, 1974; Bartlett and Burstyn, 1975; Cohen, 1977; Fig. 26). The principal

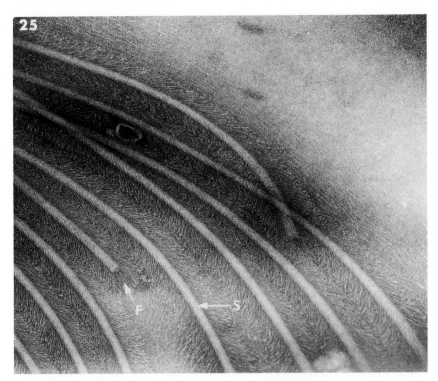

Fig. 25. Group of appendages extending from the bacterial spore of *Clostridium bifermentans.* Each appendage consists of a central shaft (S) from which innumerable filaments (F) of uniform length project. 135,000. (Contributed by L. M. Pope.)

advantage of this technique is that the surface tension forces associated with specimen desiccation are virtually eliminated. Critical-point drying also avoids the solid-phase boundary associated with freeze-drying (Hayat, 1978), which may generate structural artifacts. The critical point is the condition in which a substance in one phase, e.g., as a liquid, has the same density, pressure, and temperature as in another phase, e.g., as a gas. Such a condition is achieved (e.g., for carbon dioxide) by first bleeding the two-phase system into a sealed container under pressure and reduced temperature (less than 20°C for carbon dioxide) and then heating the container, which increases the vapor pressure and vapor density. On the other hand, the density of the liquid decreases as a result of thermal expansion. Under these conditions, a distinct boundary exists between the liquid and vapor phases within the container. Upon continued heating, the critical point is reached (31°C and 72.8 atm for carbon dioxide) when the density of the liquid phase is equal to that of the vapor phase, and the phase boundary disappears. Above the critical temperature and pressure, only the gaseous phase

Fig. 26. Appendaged conidia (C) of *Dinemasporium graminum* prepared by the CPD process. A, Appendage; S, seta. 8680.

exists and the surface tension is zero. At this stage, the pressure within the container can be reduced to atmospheric pressure (while maintaining the temperature) without the appearance of surface tension forces (Hayat, 1978).

In preparation for CPD, colonies of conidial fungi growing on agar or a natural substrate are initially flooded with fixative. Small blocks (approximately 5–10 mm sq.) of the substrate and sporulating mycelium are excised from the flooded culture using a stereomicroscope, transferred to vials containing fresh fixative, and processed according to the fixation procedure outlined above for thin-sectioning, except that the uranyl acetate and embedding steps are omitted. In addition, the final wash in absolute ethanol is followed by washes in a graded series of solutions of amyl acetate in absolute ethanol rather than an acetone series. The samples are finally immersed in absolute amyl acetate before they are transferred to the CPD apparatus.

An abbreviated fixation procedure for conidial fungi has been outlined by Samson et al. (1979). Samples were initially fixed in unbuffered aqueous 2% osmium tetroxide and/or 6% glutaraldehyde at 4°C. Fixation times varied from 1 to 24 h. Treatment in osmium tetroxide alone may be sufficient for the preparation of some material for the SEM. Kelley et al. (1973) has shown that material treated with osmium tetroxide followed by immersion in thiocarbohydrazide results in ligand osmium binding and forms a thin coat over the surface of cells, which provides the specimen with electric conductivity and some resistance to contamination from the electron beam during observation in the SEM. The ligand-mediated osmium binding technique has been proposed as an alternative to deposition of conductive metal on specimens by evaporation or sputter-coating (Malick and Wilson, 1975; Postek and Tucker, 1977). After the osmium-binding treatment, the samples are simply washed with water, dehydrated in 2,2-dimethoxypropane or methoxyethanol, followed by two washes in absolute acetone, and then transferred to the CPD device.

In our laboratory, the specimens are placed in modified BEEM capsules (Nemanic, 1972) before they are transferred to the CPD device (Fig. 27). Care is taken during this step to keep them totally immersed in amyl acetate. Up to five capsules, each containing three to five blocks of substrate and mycelium are then quickly transferred to the wire mesh cylinder and pressure bomb of a Denton DCP-1 CPD apparatus (Fig. 27). The pressure bomb is sealed and, with the bleed valve (V_3 in Fig. 27) closed, liquid carbon dioxide is slowly admitted into the chamber and line by opening the intake valve (V_2 in Fig. 27). The bomb is pre-cooled in an ice bath below the critical temperature, which increases the amount of liquid carbon dioxide inside the chamber. When the pressure gauge of the CPD reaches approximately 900 psi, V_3 is opened slightly, allowing liquid carbon dioxide to flush through the chamber. The samples are flushed for 10 min, displacing some of the amyl acetate with liquid carbon dioxide. The odor of amyl acetate can be easily detected escaping from V_3 at this stage. Amyl acetate should

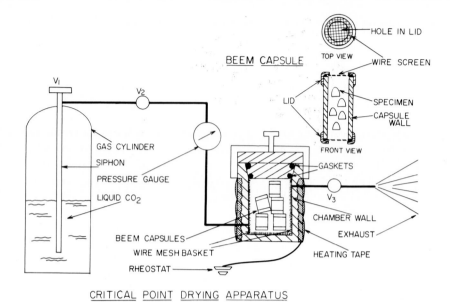

CRITICAL POINT DRYING APPARATUS

Fig. 27. The CPD apparatus and modified Beem capsules used in preparing conidial fungi for the SEM.

be exhausted into a fume hood. Next, V_2 is closed, while V_3 remains open and the pressure is allowed to drop slowly to 200 psi. The chamber is refilled by closing V_3 and slowly reopening V_2. This procedure of flushing, reducing the liquid carbon dioxide level in the chamber, and refilling is repeated five times, or until the odor of amyl acetate is no longer detected. Subsequently, V_3 and V_2 are closed (in that sequence), and the wall of the pressure bomb is heated to 45°C using insulated flexible heating tape (Briscoe Manufacturing Company, Columbus, Ohio) attached to a rheostat (Fig. 27). When the pressure in the chamber reaches 1600 psi, the rheostat is turned off (the chamber remains warm) and V_3 is opened slightly, allowing the chamber pressure to decrease slowly. When the pressure is zero, the dried specimens are removed from the CPD device and transferred directly to a desiccator.

In processing conidial suspensions or wall fragments for the SEM, the sample may be collected on a solvent-resistant filter (e.g., Millipore or Nucleopore) and maintained on this substrate throughout the fixation, dehydration, and CPD steps. Several procedures have been described for handling minute specimens during preparation for the SEM (Marchant, 1973; Ruffolo, 1974; Postek et al., 1974; Newell and Roth, 1975; Rostgaard and Christensen, 1975; Hayat, 1978). An apparatus we have found useful is illustrated in Fig. 28. It consists of a 10-ml interchangeable glass hypodermic syringe, a filter adapter (Millipore Corpora-

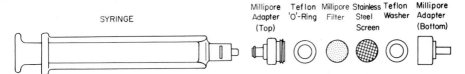

Fig. 28. Apparatus used in the preparation of cell suspensions for the SEM.

tion, Bedford, Massachusetts) and a Nucleopore filter (Nucleopore Corporation, Pleasanton, California). The filter, its pore size (0.03–8.0 μm) selected for the sample to be collected, is placed between the wire screen and a Teflon O-ring inside the adapter. The syringe, without the adapter attached, is first used to aspirate 1–2 ml of the solution containing the suspended sample. The adapter is then connected to the base of the syringe, and the solution is forced through the filter, trapping the suspended material on its surface. The same procedure is repeated for each fixative wash and dehydration solution. The adapter is then carefully dismantled while immersed in absolute amyl acetate, and the filter is transferred directly to a specially constructed Teflon filter holder which fits inside the CPD apparatus. Up to four such holders can be stacked in the cylindrical container of the CPD device.

The specimens must finally be mounted and coated before they can be examined in the SEM. The dried specimens are mounted on aluminum SEM stubs using conductive silver paint. The stubs are then placed on the omnirotary table of a Denton DV 502 high-vacuum evaporator equipped with a cold sputter module. The latter is a DC diode sputtering apparatus designed to deposit thin films of conductive metal (i.e., gold–palladium, 60:40) onto small substrates in an argon atmosphere. An explanation of the sputtering method of coating specimens is presented by Hayat (1978). The advantage of the Denton sputter module is that it permits high deposition rates (approximately 10–15 nm gold/min) with no appreciable substrate heating and no secondary-electron specimen damage.

5. Combined Transmission Electron Microscope and Scanning Electron Microscope Preparations

Innovative techniques have been reported in which the same tissue or cell is prepared for both transmission and scanning electron microscopy. Although these procedures have not yet been applied to ultrastructural studies of conidial fungi, their potential value in such investigations is recognized and is briefly discussed below.

Winborn and Guerrero (1974) described a process in which conventionally fixed and epoxy-embedded tissue was first prepared for the TEM and then subjected to an epoxy solvent, washed in a dehydrant, and critical point-dried in preparation for SEM examination. The material demonstrated excellent preservation in the SEM. Steffens (1978) used sodium methoxide as the epoxy solvent, followed by washes in a graded series of sodium methoxide in methanol–benzene and then in anhydrous acetone prior to CPD. One advantage of this procedure is that thin sections of cells prepared and examined by conventional means in the TEM may be compared to thick sections (approximately 1 μm) of the same tissue or cell subjected to epoxy solvent and examined in the SEM. The thick sections cut from the plastic-embedded block of tissue are mounted on small pieces of a glass slide, exposed to epoxy solvent and dehydrating agents, critical point-dried, coated, and observed in the SEM (Winborn and Guerrero, 1974).

IV. FREEZE-FRACTURE AND FREEZE-ETCH EXAMINATIONS

A. The Technique

As an alternative to fixation, embedding, and thin-sectioning, biological specimens may instead be frozen, fractured, etched, and replicated in preparation for TEM examinations. Although chemical fixation is not a necessary prerequisite for the freezing process, cells are usually pretreated with a cryoprotectant (e.g., 10% glycerol) to avoid excessive ice crystal formation. Prior to fracturing, the material is placed in a vacuum on a cold stage and thus maintained in a frozen state (Fig. 29). By cracking the solidified suspension of cells with a precooled microtome, a fracture plane occurs which largely follows an unpredictable course. This preparatory step may be followed by etching—sublimation of ice from below the plane of fracture, exposing unfractured surfaces of the cells. The sample is subsequently shadowed (both oblique and perpendicular coating) and

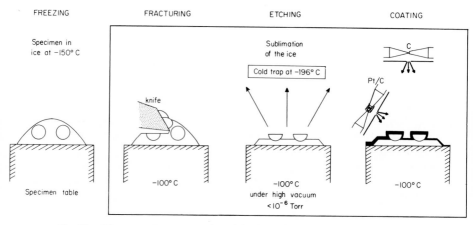

Fig. 29. Diagrammatic representation of the freeze-fracture (freeze-etch) procedure.

then returned to atmospheric pressure. The replica is freed of the biological material in concentrated acid and then washed and mounted on a Formvar-coated grid for examination in the TEM. Figures 30 and 31 are representative images resulting from application of this technique.

The reader is referred to several excellent reviews of the freeze-etch (freeze-fracture) technique for historical background and detailed descriptions of methodology (Koehler, 1972; Benedetti and Favard, 1973; Sletyr and Robards, 1977b; Fineran, 1978). Suffice it to point out that development of this ultrastructural procedure is based on the pioneering work of Hall (1950), Meryman (1957), Meryman and Kafig (1955), Steere (1957), and Haggis (1961). The application of freeze-etching in investigations of biological ultrastructure came of age with the design of the Balzers apparatus (Moor *et al.*, 1961). Improvements in this freeze-etch device over earlier instruments (Steere, 1957) include precise control of the specimen table temperature ($\pm 0.1°C$; fluctuation of temperature within the specimen is less than $0.01°C$), control of the fracturing process (a vertically mounted microtome is mechanically or thermally advanced during the cutting movements), and an efficient reproducible coating procedure. The four principal preparatory steps of this technique—cryofixation, fracturing and etching, coating, and cleaning—are discussed below.

1. Cryofixation

Freezing and freeze-fixation are now commonly used to achieve ultrastructural preservation in biological samples. The improvements in these techniques have been largely derived from research in the field of cryobiology (Franks, 1977). The aim of freezing is to reduce cell metabolism to zero with as little structural damage due to ice crystallization as possible. The latter may be accomplished by using fast coolants or by depressing the freezing point of the specimen with cryoprotectants (antifreeze agents) so that it cools in a shorter time (Fineran, 1978).

In preparing conidia for freeze-etching, sporulating petri dish cultures are flooded with 5% glycerol, which is increased to 10–20% over a 4- to 8-h period, plus 5–10 drops of Teepol (Shell product). The latter serves as a wetting agent for hydrophobic conidia. The glycerol functions as a cryoprotectant, and after gentle agitation the culture is allowed to stand at room temperature for a total of 4–12 h. Glycerination is carried out gradually to allow the cells to absorb and metabolize the glycerol (Richter, 1968a,b; Rottenburg and Richter, 1969). The cryoprotectant is thought to bind the water within the cell, making less water available for freezing. Other cryoprotectants include sugars, alcohols, amides, dimethyl sulfoxide, ethylene glycol, nitrogen oxide, and urea. However, because of its low toxicity and general success in freeze-etch studies on biological samples, glycerol is the most commonly employed antifreeze. Nevertheless, some artifacts have been attributed to glycerol treatment, such as swelling of mitochrondia

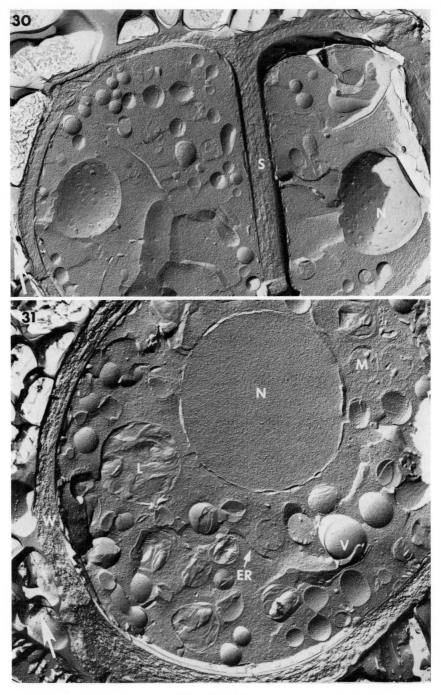

Figs. 30 and 31.

(Bachmann and Schmitt-Fumian, 1971), transformation of endoplasmic reticulum into a mass of vesicles (Moor, 1971), and formation of small vesicles between the cell wall and plasmalemma of fungal spores (Hess and Weber, 1972). Recent investigations of nonpenetrating, polymeric cryoprotective agents (polyvinylpyrrolidone, hydroxyethyl starch, and dextran) have shown that these compounds are very effective in preventing ice crystal formation and generally have a lower toxicity to plant and animal cells than glycerol at similar concentrations (Franks *et al.*, 1977; Echlin *et al.*, 1977; Skaer *et al.*, 1977).

After treatment with glycerol, the sporulating mycelium on the agar surface is gently scraped with an acetone-cleaned razor blade to dislodge conidia from conidiogenous cells. Conidia are then separated from the rest of the mycelium by filtration using a fine-mesh nylon filter (Cole, 1973c) or a Millipore filter with a suitable pore size. The filtrate is centrifuged at 1000x g, and a small droplet of the pellet is transferred to the depression of a Balzers gold–nickel alloy disk and frozen in liquid Freon 22 followed by liquid nitrogen. If the cryoprotectant and coolants have worked properly, the water inside the cell has been solidified into a glassy or vitrified state. Under ideal conditions of vitrification, no ice crystals are produced, but this situation is seldom achieved in practice (Fineran, 1978). As previously pointed out, a rapid rate of freezing in addition to cryoprotective treatment is usually employed to reduce ice crystallization. When conidia are plunged directly into liquid nitrogen, an insulating layer of gas surrounds the sample (Liedenfrost phenomenon), which decreases the rate of heat exchange. Instead, the specimen is quickly placed in a liquid fluorocarbon (i.e., Freon 22) which has a much lower boiling point than liquid nitrogen and does not suffer from the Liedenfrost phenomenon. After 5–10 sec in Freon 22, the specimens are quickly transferred to liquid nitrogen. Several other freezing techniques have been developed, but with only limited success (Fineran, 1978). High-pressure freezing involves lowering the freezing point of water by increasing the atmospheric pressure (Riehle, 1968; Riehle and Hoechli, 1973). Although cryoprotectants are not required for this procedure, thus eliminating artifacts which may be produced by these chemicals, high pressures may cause structural damage to the cells. Spray freezing is another rapid-freezing technique in which a fine jet of cells in suspension (without cryoprotectant) is propelled into a freezing medium (e.g., liquid propane). This technique has demonstrated good preservation of microorganisms (Plattner *et al.*, 1973), but it cannot be used for pieces of tissue. Dempsey and Bullivant (1976a,b) have used a metal block (silver or copper) precooled in liquid nitrogen to freeze specimens. The tissue, without prior treatment with a cryoprotectant, is simply pressed against the surface of the block

---

Figs. 30 and 31. Freeze-fracture and freeze-etch (3 min) replicas of conidia of *Geotrichum candidum*. ER, Endoplasmic reticulum; L, lipid droplet; M, mitochrondria; N, nucleus; S, septum; V, vacuole; W, wall. Arrows indicate direction of shadowing. Fig. 30: 25,200; Fig. 31: 31,500.

which has a much higher thermal capacity and conductivity than liquid nitrogen.

One of the principal aims of cryofixation was to avoid use of chemical fixatives which may generate artifacts during preparation of cells for electron microscopy. The cells prepared by cryofixation are living and in some cases (e.g., yeast) may be revived after thawing (Moor and Mühlethaler, 1963). However, comparative examinations of chemically fixed and cryofixed material have

Fig. 32. Freeze-fracture and deep etch replica of *Cladosporium macrocarpum* conidium. Arrows indicate deep-etch interface between layer of rodlet fascicles and outermost amorphous wall layer. Some rodlets are visible through the amorphous layer at the tips of conidial spines. 79,900. (From Cole *et al.*, 1979.)

yielded comparable results. Glutaraldehyde (1–3%) has been used to reduce the semipermeability of the plasmalemma prior to glycerol treatment and cryofixation (Moor, 1966; Fineran, 1978). This procedure has been particularly important in freeze-etch preparations of delicate tissues and cells of certain plants (e.g., Cole and Wynne, 1974a,b) and animals (e.g., Chailley, 1979), but in general has not proved necessary for conidial fungi. In fact, some yeasts and conidia have been frozen without earlier fixation or treatment with cryoprotectants. For example, conidia of *Cladosporium macrocarpum* were simply suspended in distilled water prior to cryofixation, and yet surface wall fractures revealed good preservation of the wall components (Cole *et al.*, 1979; Fig. 32). However, the cytoplasmic content of these cells was poorly preserved.

2. Fracturing and Etching

In the Balzers freeze-etch device (Model BA 360M), the microtome arm is moved on a pivot (universal joint and gimbal) controlled by an external pulley and wheel located outside the vacuum chamber. The mechanical advance control, also externally mounted, allows the cutting edge of the microtome to be lowered in minute increments (approximately 1 μm accuracy). The knife used in our laboratory is a platinum razor blade which demonstrates good thermal conductivity. The microtome arm and knife are cooled to approximately $-150°C$ by the circulation of liquid nitrogen to and from an enclosed reservoir attached to the microtome. When the knife makes contact with the frozen droplet of cells, a fracture plane is initiated at a level slightly below that traversed by the razor blade (Fineran, 1978). Sletyr and Robards (1977b) interpreted fracturing as a process during which "the cleavage plane through the specimen follows planes of weakness that are not directly related to, nor necessarily spatially close to, the path of the knife . . . " The sample is usually subjected to a series of fractures, the first removing excess suspension medium (cryoprotectant, water) and then one to several additional fractures through cells embedded in the droplet. During this process the temperature of the specimen table is maintained at $-100°C$ ($\pm0.1°C$) while the vacuum is about 10^{-6} torr. The final pass of the knife removes any chips of ice or cellular material which may be present on the fractured surface. The precise pathway of the fracture through the cell suspension is unpredictable. The result is fractures which have passed over the surfaces of cells and organelles, or have scooped out the contents of the cells and/or their components. The replica examined in the TEM therefore reveals three-dimensional aspects of the fractured structures (Figs. 30 and 31).

Numerous devices have been designed which enable one to examine the two new faces exposed when a frozen specimen is fractured (Chalcroft and Bullivant, 1970; Mühlenthaler *et al.*, 1971; Sletyr, 1970a,b; Steere and Moseley, 1969, 1971; Wehrli *et al.*, 1970). The complementary replicas produced after coating reveal morphological features of both fracture faces, which provides valuable information on the structure of cellular components as well as the pathway of the fracture plane. Such comparative analyses have revealed that perfect matching of membrane faces is not always possible. This is due, at least in part, to the plastic flow or deformation of the surrounding membrane material (Sletyr and Robards, 1977a). These deformations may occur during and after the fracture process and cannot be avoided at specimen temperatures down to $-150°C$.

After the specimen has been fractured, it may be shadowed directly, or the fracture face may be etched prior to coating. In performing the etching process, the cold knife (approximately $-150°C$) of the Balzers freeze-etch device is raised well above the specimen, using the external mechanical control, and then brought directly over the fractured surface. Two essential conditions for etching are that the vacuum of the chamber must be higher than the vapor pressure of the ice at the specimen temperature (Dunlop and Robards, 1972) and that the temperature of the microtome arm must be lower than the specimen so that it serves as a cold trap above the fractured surface. To ensure a constant rate of etching, the temperature of the specimen must be highly stable. When these various factors are taken into account, it is possible to calculate the rate of sublimation and thus the etch depth. For example, the conidium in Fig. 32, which was frozen and fractured in distilled water, was etched for 15 min while the specimen temperature was maintained at $-100°C$. The edge of the fracture plane is indicated by arrows. The fascicles of fibrous wall components (rodlets) are revealed by the fracturing process. The amorphous layer (outermost wall component of the conidium) shown below the edge of the fracture plane is revealed by deep etching (etch depth about 1360 nm). In this rather dramatic example, etching has removed ice from the cell surface, unmasking a wall layer which would otherwise have escaped detection. Nonvolatile components of the cell and eutectic mixtures formed by freezing are unchanged by the etching process (Fineran, 1978). For practical purposes, most etching of glycerol-treated conidia is performed for 3–5 min. A novel variation of the etching technique is the combination of complementary freeze-fracturing and freeze-etching (Steere and Erbe, 1979). This procedure should provide valuable information on the effects of etching on the structure of biological specimens.

3. Coating

Development of the platinum–carbon shadowing (Bradley, 1958) and replication techniques (Moor, 1959), discussed briefly in Section III,D, was a significant breakthrough in the evolution of freeze-etching. A coil of platinum wire (0.1 mm in diameter and 6 cm long) is placed over the cylindrical tip (1 mm in diameter) of a carbon rod (Fig. 33). Another carbon rod (spring-loaded) is brought into contact with the tip of the former, and simultaneous evaporation of platinum and carbon is accomplished by application of a current to the rods. Evaporation of carbon alone is performed using a similar arrangement (Fig.

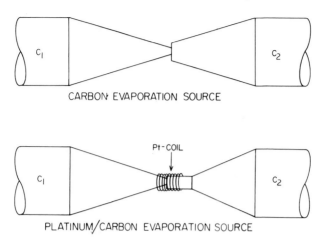

Fig. 33.   Simple carbon and platinum–carbon evaporation sources used for replication in the freeze-etch device. One carbon rod (C_1) is spring-loaded, while the other rod (C_2) is stationary.

33). Although this is the coating procedure used by most freeze-etch workers, certain shortcomings are evident. The amount of platinum evaporated cannot be accurately controlled. Too high a platinum/carbon ratio can lead to recrystallization of platinum within the thin film or replica when subjected to heavy electron bombardment in the TEM (Henderson and Griffiths, 1972). Individual platinum grains are about 2–3 nm in diameter (Sjöstrand, 1979), which in effect sets the limitation of resolving power of cellular substructures revealed by TEM examination of the platinum-carbon replica. On the other hand, high-melting-point metals (e.g., tantalum-tungsten) have smaller crystals and may be used to improve resolution (Moor, 1971, 1973). The electron beam evaporation technique (Bachmann *et al.,* 1969; Zingsheim *et al.,* 1970; Moor, 1971) provides the means for depositing these metals, and the electron beam gun has replaced the conventional platinum-carbon or carbon source in most contemporary freeze-etch devices. Because of the high temperature associated with this evaporation procedure, the electrodes are placed at a considerable distance from the specimen, heat baffles are used within the gun (magnetic baffles are also used to minimize specimen surface damage due to stray electrons), and evaporation is performed only at high vacuum (10^{-6} torr or better). Electron beam evaporation permits extremely fine granular and reproducible films (approximately 3 nm thick) to be laid down over the fractured surfaces of cells. The deposition rates of evaporated metals are constant as long as the operational conditions—vacuum inside the freeze-etch chamber and current and voltage to the electron gun—are exactly the same. However, voltage fluctuations may occur, and accurate reproducibility of film thickness necessitates use of a quartz crystal monitor. This device measures thickness of the film on the basis of the frequency shift of a quartz oscillator exposed to the same incident flux of atoms and molecules as the surface of the specimen.

As previously discussed, unidirectional shadowing results in differential deposition of metal across the surface of the specimen revealing the "humps" and "hollows" of the fractured face of the cells. However, some areas of the specimen surface will not be coated at all and will thus lack structural information (Fineran, 1978). Such loss of detail may be avoided by low-angle rotary shadowing (Branton, 1974) which has been developed for the Balzers freeze-etch device. The value of this coating technique has been demonstrated by studies on intramembrane particles which exhibit tetrameric subunit structures not visible after conventional unidirectional shadowing (Branton and Kirchanski, 1977).

4. Replica Cleaning

The specimen is returned to atmospheric pressure by slowly admitting dry, oil-free nitrogen gas into the vacuum chamber of the freeze-etch device. The Balzers gold–nickel alloy disk supporting the frozen, fractured (etched), coated sample is then removed from the cold stage, and the sample is allowed to thaw in air. For conidial samples the disk is then slowly lowered, at about a 45° angle, into chromic acid contained in a small, glass petri dish. The replica usually floats on the surface of the acid solution, while the disk with the remains of the biological material is removed and cleaned. Care must be taken to keep the top of the replica dry. Conidia and fragments of cellular material (e.g., the chitinous wall) attached to the undersurface of the replica are readily digested by the acid. This cleaning process may be speeded up by heating the dish containing the acid and replica to 50°C. After 15 min, the acid is allowed to cool to room temperature and the replica is then transferred, using a platinum loop, to filtered distilled water contained in an acid-cleaned petri dish. Precautions are taken to avoid contaminants in the water wash, which may adhere to the replica. After transfer to five new water washes (15 min each), the replica is picked up on a Formvar-coated grid, air-dried in a clean environment, and examined in the TEM. This method of replica cleaning has proved satisfactory for freeze-fracture (or freeze-etch) examinations of conidial fungi. Other cleaning techniques have been described by Koehler (1972) and Fineran (1978).

B. Interpretation of Results

Because of unidirectional shadowing during the coating process, a slope facing the platinum–carbon electrode will receive a thicker layer of metal than a slope facing away from the source of vaporization. Fineran (1978) suggested that freeze-etch micrographs should be viewed so that the shadows (electron-transparent regions) extend away from the observer. Viewing micrographs from the opposite direction will often give the impression of reversed relief, which is an optical illusion. On a photographic print (Figs. 30 and 31), the shadow appears as an unexposed (white) umbra adjacent to a coated (black to gray) area. Whether a structure is raised or depressed in the replica may be ascertained by the position of the shadow; "the shadow extends beyond a raised object and remains within the confines of a depressed object" (Fineran, 1978). Freeze-etch micrographs are usually provided with an arrow indicating the direction of shadowing, which facilitates interpretation of the fracture plane (Figs. 30 and 31).

Many variations in freeze-etching nomenclature have appeared in the literature during the last 25 yr, largely as a result of differences in the interpretation of how biological membranes are fractured. It is generally accepted that the fracture plane commonly splits the membrane through its interior, between the two layers of lipid. It has been suggested that hydrophobic bonds holding the lipid bilayer together comprise a zone of weakness through which the fracture preferentially passes (Branton, 1966, 1973). On the exposed interior of the membrane, replicas reveal large quantities of particles protruding through the lipid layer. Although the chemical composition of these particles has not been determined, it is suggested that they are proteinaceous, comprising enzyme complexes associated with the functioning of the membrane (Singer, 1975; Fineran, 1978). In addition, the lipid comprising the bulk of the membrane is usually in a liquid or liquid crystal state (e.g., at physiological temperature) and the particles (globular and amphipathic proteins?) are free to move about in the lipid sea. This concept has been formulated in the fluid-mosaic model of the biomembrane (Singer, 1971, 1974; Singer and Nicolson, 1972; Fig. 34).

Examinations of freeze-fractures of membranes, particularly complementary fractures, reveal an asymmetric distribution of particles between the two fracture faces. Figures 35 and 36 show different fracture faces of the plasmalemma of *Geotrichum candidum*. A much higher concentration of membrane particles is visible in Fig. 35. The grooves and ridges demonstrated in Figs. 35 and 36, respectively, are plasmalemma invaginations commonly revealed by freeze-fractures of fungal cells. A system of nomenclature which differentiates between fracture faces and etched surfaces of cells was formulated by an international group of freeze-etch investigators (Branton *et al.*, 1975). They proposed that, for any membrane which can be split, the half closest to the cytoplasm, nucleoplasm, chloroplast stroma, or mitochondrial matrix be designated the protoplasmic half (PF in Fig. 37), and the half closest to the extracellular space (cell wall), exoplasmic space (interior of endocytic vacuoles, phagosomes, primary and secondary lysosomes, food vacuoles, plant vacuoles, and Golgi vesicles), and endoplasmic space (cisternae of endoplasmic reticulum, cisternae formed between inner and outer nuclear membranes, and cisternae formed by Golgi lamellae) be designated the extracellular, exoplasmic, or endoplasmic half (EF in Fig. 37). The true surface of the membrane revealed by deep etching is designated either ES, if it lies against the extracellular space, or PS, if it lies against the

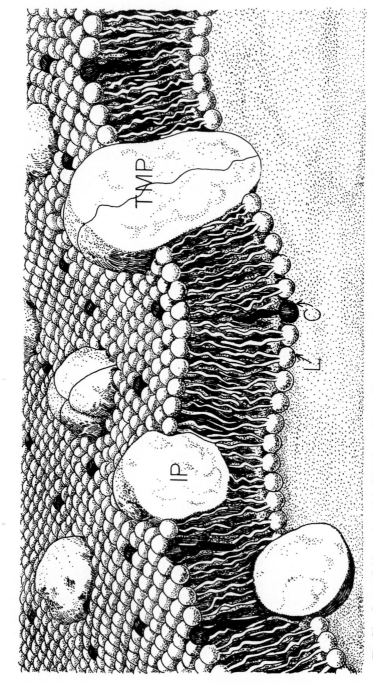

Fig. 34. Current fluid-mosaic model of the biomembrane (Singer and Nicolson, 1972). The proteins are predominantly globular and amphipathic, with their hydrophobic ends embedded in the bilayer of lipids (L) and cholesterol (C). The proteins make up the membrane's active sites; some are simply embedded on one side or the other (integral proteins; IP), while others pass entirely through the bilayer (transmembrane proteins; TMP). Some of the latter presumably contain transport pores. (After Singer, 1975.)

Fig. 35. Continued on page 624

Figs. 35 and 36. Freeze-fracture and freeze-etch of *Geotrichum candidum* conidia. Plasmalemma invaginations are visible on both fracture faces. The concentration of particles on the PF fracture face (Fig. 35) is much higher than on the EF fracture face (Fig. 36). Arrows indicate the direction of shadowing. Fig. 35: 15,000; Fig. 36: 21,600.

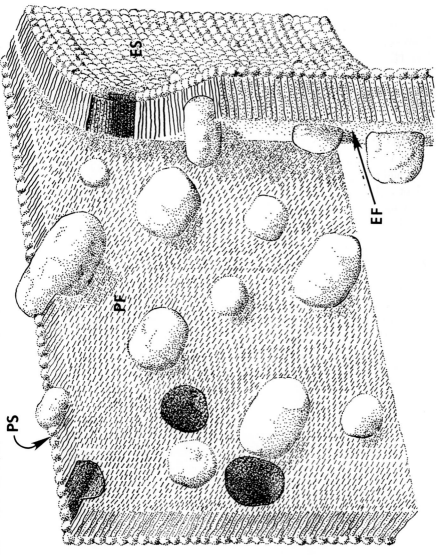

Fig. 37. Schematic of the splitting of a biomembrane through hydrophobic bonds of a lipid bilayer, revealing protein particles on the protoplasmic face (PF) in a higher concentration than on the exoplasmic face (EF). PS and ES represent the true inner and outer surfaces of the membrane, respectively. (After Singer, 1975.)

protoplasm (Fig. 37). In a simplification of this nomenclatural system Branton *et al.* (1975) stated that the surface "designated either PS or ES is the hydrophobic portion of the membrane usually exposed by the etching process" while "the fracture face . . . designated PF or EF . . . is the hydrophobic portion of the fracture process. Thus, a single membrane will possess two true surfaces, PS and ES, and two fracture faces, PF and EF."

However, in spite of the large quantity of experimental data supporting the Singer–Nicolson (1972) model of the biomembrane and the concept that fractures occur through internal hydrophobic regions, Sjöstrand (1979) has presented rather persuasive contradictory evidence. He has pointed out that fractures of certain membranes, such as those of mitochondrial cristae (Sjöstrand and Barajas, 1968; Sjöstrand, 1977), clearly do not conform with current hypothetical models of the biomembrane. Instead, Sjöstrand (1979) has interpreted that fractures of these membranes occur most commonly between the membrane surface and protoplasm, a concept of membrane fracturing originally proposed by Moor and Mühlethaler (1963). Sjöstrand has suggested "that a lipid bilayer can no longer be accepted as a valid analogy of the actual membrane structure." It is evident from Singer's (1975) illustrations of the biomembrane (Figs. 34 and 37) that, in addition to hydrophobic bonds, protein molecules introduce covalent bonds extending across the middle of the membrane, allowing extensive interaction with surrounding lipid molecules. It is unlikely therefore that such a lipoprotein complex could demonstrate the same fracture plane as a pure lipid bilayer. Sjöstrand (1979) directed special attention to the work of Pinto da Silva and Branton (1970) on freeze-fracturing and freeze-etching of labeled red blood cell ghost membranes. Their study is frequently cited as support for the concept that freeze-fractures occur between the lipid layers of biomembranes. After a critical reappraisal of the results of this investigation, Sjöstrand (1979) questioned the authors interpretation of their ultrastructural data. An alternate view advanced by Sjöstrand (1979) supports the possibility that fractures of red blood cell ghosts expose "true membrane surfaces."

Thus it is evident that controversy over the molecular structure of biomembranes is far from over. In the meantime, freeze-etching nomenclature remains in a state of uncertainty. It is clear, however, that this ultrastructural technique, combined with biochemical and biophysical investigations, will continue to contribute significant data which eventually will lead to a better understanding of the molecular structure of biomembranes.

REFERENCES

Anderson, T. F. (1951). Techniques for the preservation of three-dimensional structure in preparing specimens for the electron microscope. *Trans. N.Y. Acad. Sci.* **13**, 130–134.

Bachmann L., and Schmitt-Fumian, W. W. (1971). Improved cryofixation applicable to freeze etching. *Proc. Natl. Acad. Sci. U.S.A.* **68**, 2149–2152.

Bachmann, L., Abermann, R., and Zingsheim, H. P. (1969). Hochauflösende Gefrierätzung. *Histochemie* **20**, 133–142.

Barber, V. C., and Emerson, C. J. (1979). Techniques utilizing real time stereo scanning electron microscopy in the microdissection of biological tissues. *J. Microsc. (Oxford)* **115**, 119–125.

Bartlett, A. A., and Burstyn, H. P. (1975). A review of the physics of critical point drying. *Scanning Electron Microsc.* (O. Johari, ed.), p. 305. IITRI, Chicago, Illinois.

Benedetti, E. L., and Favard, P., eds. (1973). "Freeze-etching: Techniques and Applications." Soc. Fr. Microsc. Electron., Paris.

Bolender, R., Paumgartner, D., Losa, G., Muellener, D., and Weibel, E. (1978). Integrated stereological and biochemical studies on hepatocytic membranes. *J. Cell Biol.* **77**, 565–583.

Boyde, A. (1974). A stereo-plotting device for SEM micrographs; and a real time 3-D system for the SEM. *Scanning Electron Microsc.* (O. Johari, ed.), pp. 93–100. IITRI, Chicago, Illinois.

Boyde, A. (1975). Measurement of specimen height difference and beam tilt angle in anaglyph real time stereo TV SEM systems. *Scanning Electron Microsc.* (O. Johari, ed.), p. 189. IITRI, Chicago, Illinois.

Boyde, A., and Ross, H. F. (1975). Photogammetry and the scanning electron microscope. *Photogammetric Rec.* **8**, 408–457.

Bradley, D. E. (1958). Simultaneous evaporation of platinum and carbon for possible use in higher-resolution shadow-casting for the electron microscope. *Nature (London)* **181**, 875–877.

Branton, D. (1966). Fracture faces of frozen membranes. *Proc. Natl. Acad. Sci. U.S.A.* **55**, 1048–1056.

Branton, D. (1973). The fracture process of freeze-etching. *In* "Freeze-etching: Techniques and Applications" (E. L. Benedetti and P. Favard, eds.), pp. 107–121. Soc. Fr. Microsc. Electron., Paris.

Branton, D. (1974). Interpretation of freeze-etch results. *In* "Electron Microscopy 1974" (J. V. Sanders and D. J. Goodchild, eds.), Vol. 2, p. 28. Aust. Acad. Sci., Canberra.

Branton, D., and Kirchanski, S. (1977). Interpreting the results of freeze-etching. *J. Microsc. (Oxford)* **111**, 117–124.

Branton, D., Bullivant, S., Gilula, N. B., Karnovsky, M. J. K., Moor, H., Mühlethaler, K., Northcote, D. H., Packer, L., Satir, B., Satir, P., Speth, V., Staehelin, L. A., Steere, R. L., and Weinstein, R. S. (1975). Freeze-etching nomenclature. *Science* **190**, 54–56.

Briarty, L. G., and Ramsey, R. L. (1978). A simple time-lapse unit. *J. Microsc. (Oxford)* **114**, 353–355.

Broers, A. N. (1975). Electron sources for scanning electron microscopy. *Scanning Electron Microsc.* (O. Johari, ed.), p. 601. IITRI, Chicago, Illinois.

Brotzman, H. G., Calvert, O. H., Brown, M. F., and White, J. A. (1975). Holoblastic conidiogenesis in *Helminthosporium maydis*. *Can. J. Bot.* **53**, 813–817.

Chailley, B. (1979). Etude par cryofracture des membranes impliquées dans la sécrétion. *Biol. Cell.* **35**, 55–70.

Chalcroft, J. P., and Bullivant, S. (1970). An interpretation of liver cell membrane and junction structure based on observation of freeze-fracture replicas of both sides of the fracture. *J. Cell Biol.* **47**, 49–60.

Chatfield, E. J. and Pullan, H. (1974). Real time 3-D Scanning electron microscopy: its potential and applications. *Cam. Res. and Dev.* (May–June), 17.

Cohen, A. L. (1977). A critical look at critical point drying-theory, practice, and artifacts. *Scanning Electron Microsc.* (O. Johari, ed.), p. 525. IITRI, Chicago, Illinois.

Cole, G. T. (1972). Microfibrils in the cytoplasm of fertile hyphae of the imperfect fungus. *Drechslera sorokiniana*, *J. Ultrastruct. Res.* **41**, 563–571.

Cole, G. T. (1973a). Ultrastructure of conidiogenesis in *Drechslera sorokiniana. Can. J. Bot.* **51**, 629–638.

Cole, G. T. (1973b). Ultrastructural aspects of conidiogenesis in *Gonatobotryum apiculatum. Can. J. Bot.* **51**, 1677–1684.

Cole, G. T. (1973c). A correlation between rodlet orientation and conidiogenesis in Hyphomycetes. *Can. J. Bot.* **51**, 2413–2422.

Cole, G. T. (1975). A preparatory technique for examination of imperfect fungi by scanning electron microscopy. *Cytobios* **12**, 115–121.

Cole, G. T., and Kendrick, W. B. (1968). A thin culture chamber for time-lapse photomicrography of fungi at high magnifications. *Mycologia* **60**, 340–344.

Cole, G. T., and Kendrick, W. B. (1971). Conidium ontogeny in Hyphomycetes: Development and morphology of *Cladobotryum. Can. J. Bot.* **49**, 595–599.

Cole, G. T., and Samson, R. A.(1979). "Patterns of Development in Conidial Fungi." Pitman, London.

Cole, G. T., and Wynne, M. J. (1974a). Endocytosis of *Microcystis aeruginosa* by *Ochromonas danica. J. Phycol.* **4**, 397–410.

Cole, G. T., and Wynne, M. J. (1974b). Nuclear pore arrangement and structure of the Golgi complex in *Ochromonas danica* (Chrysophyceae). *Cytobios* **8**, 161–173.

Cole, G. T., Nag Raj, T. R., and Kendrick, W. B. (1969). A simple technique for time-lapse photomicrography of microfungi in plate culture. *Mycologia* **61**, 726–730.

Cole, G. T., Sekiya, T., Kasai, R., Yokoyama, T., and Nozawa, Y. (1979). Surface ultrastructure and chemical composition of the cell walls of conidial fungi. *Exp. Mycol.* **3**, 132–156.

Collinge, A. J., and Trinci, A. P. J. (1974). Hyphal tips of wild-type and spreading colonial mutants of *Neurospora crassa. Arch. Microbiol.* **99**, 353–368.

Corlett, M., Chong, J., and Kokko, E. G. (1976). The ultrastructure of the *Spilocaea* state of *Venturia inaequalis in vivo. Can. J. Microbiol.* **22**, 1144–1152.

Crewe, A. V. (1971). A high resolution scanning electron microscope. *Sci. Am.* **224**(4), 26–35.

Crewe, A. V. (1979). Development of high resolution STEM and its future. *J. Electron Microsc.* **28**, Suppl., S9–S16.

Crewe, A. V., Eggenberger, D. N., Wall, J., and Welter, L. M. (1968). Electron gun using a field emission source. *Rev. Sci. Instrum.* **39**, 576.

Da Riva Ricci, D., and Kendrick, W. B. (1972). Computer modelling of hyphal tip growth in fungi. *Can. J. Bot.* **50**, 2455–2462.

Dawes, C. J. (1971). "Biological Techniques in Electron Microscopy." Barnes & Noble, New York.

Dempsey, G. P., and Bullivant, S. (1976a). A copper block method for freezing non-cryoprotected tissue to produce ice-crystal-free regions for electron microscopy. I. Evaluation using freeze-substitution. *J. Microsc. (Oxford)* **106**, 251–260.

Dempsey, G. P., and Bullivant, S. (1976b). A copper block method for freezing non-cryoprotected tissue to produce ice-crystal-free regions for electron microscopy. II. Evaluation using freeze fracturing with a cryo-ultramicrotome. *J. Microsc. (Oxford)* **106**, 261–271.

Doonan, B. B., Crang, R. E., Jensen, T. E., and Baxter, M. (1979). *In situ* X-ray energy dispersive microanalysis of polyphosphate bodies in *Aureobasidium pullulans. J. Ultrastruct. Res.* **69**, 232–238.

Duggar, B. M. (1909). "Fungus Diseases of Plants." Ginn & Co., Athenaceum Press, Boston, Massachusetts.

Dunlop, W. F., and Robards, A. W. (1972). Some artifacts of the freeze-etching technique. *J. Ultrastruct. Res.* **40**, 391–400.

Echlin, P., Skaer, H. Le B., Gardiner, B. O. C., Franks, F., and Asquith, M. H. (1977). Polymeric cryoprotectants in the preservation of biological ultrastructure. II. Physiological effects. *J. Microsc. (Oxford)* **110**, 239–255.

Ellis, D. H., and Griffiths, D. A. (1975). The fine structure of conidial development in the genus *Torula*. I. *T. herbarum* (Pers.) Link. ex S. F. Gray and *T. herbarum* f. *quaternella* Sacc. *Can. J. Microbiol.* **21,** 1661–1675.

Ellis, D. H., and Griffiths, D. A. (1976). The fine structure of conidial development in the genus *Torula*. III. *T. graminis* Desm. *Can. J. Microbiol.* **22,** 858–866.

Ellis, D. H., and Griffiths, D. A. (1977). The fine structure of conidiogenesis in *Alysidium resinae* (= *Torula ramosa*). *Can. J. Bot.* **55,** 676–684.

Everhart, T. E., and Thornley, R. M. F. (1960). Wide band detector for microampere low-energy electron currents. *J. Sci. Instrum.* **37,** 246.

Fawcett, D. W. (1979). Electron microscopy in cell biology and its future. *J. Electron Microsc.* **28,** Suppl., S73–S78.

Feder, N., and Sidman, R. L. (1958). Methods and principles of fixation by freeze-substitution. *J. Biophys. Biochem. Cytol.* **4,** 593–600.

Fernandez-Morán, H. (1952). Application of ultra-thin freezing-sectioning technique to the study of cell structures with the electron microscope. *Ark. Fys.* **4,** 471–483.

Fineran, B. (1978). Freeze-etching. *In* "Electron Microscopy and Cytochemistry of Plant Cells" (J. L. Hall, ed.), pp. 279–341. Elsevier, Amsterdam.

Franke, W. W., Krien, S., and Brown, R. M. (1969). Simultaneous glutaraldehyde-osmium tetroxide fixation with postosmication. *Histochemie* **19,** 162–164.

Franks, F. (1977). Biological freezing and cryofixation. *J. Microsc. (Oxford)* **111,** 3–16.

Franks, F., Asquith, M. H., Hammond, C. C., Skaer, H. Le B., and Echlin, P. (1977). Polymeric cryoprotectants in the preservation of biological ultrastructure. I. Low temperature states of aqueous solutions of hydrophilic polymers. *J. Microsc. (Oxford)* **110,** 223–238.

Fukutomi, M. (1976). Utilization of electron probe X-ray microanalyzer to plant pathological fields. *JEOL News* **14,** 2–12.

Gage, S. H. (1947). "The Microscope, "17th ed. Cornell Univ. Press (Comstock), Ithaca, New York.

Girbardt, M. (1957). Der Spitzenkörper von *Polystictus versicolor* (L.). *Planta* **50,** 47–59.

Girbardt, M. (1969). Die Ultrastruktur der Apikalregion von Pilzhyphen. *Protoplasma* **67,** 413–441.

Goldstein, J. I. (1975). Electron beam-specimen interaction. *In* "Practical Scanning Electron Microscopy" (J. I. Goldstein and H. Yakowitz, eds.), p. 69. Plenum, New York.

Grimstone, A. V. (1977). "The Electron Microscope in Biology." Arnold, London.

Grove, S. N., and Bracker, C. E. (1970). Protoplasmic organization of hyphal tips amongst fungi: Vesicles and Spitzenkörper. *J. Bacteriol.* **104,** 989–1009.

Grove, S. N., Bracker, C. E., and Morré, D. J. (1970). An ultrastructural basis for hyphal tip growth in *Pythium ultimum*. *Am. J. Bot.* **57,** 245–266.

Haggis, G. H. (1961). Electron microscope replicas from the surface of a fracture through frozen cells. *J. Biophys. Biochem. Cytol.* **9,** 841–852.

Hall, C. E. (1950). A low temperature replica method for electron microscopy. *J. Appl. Phys.* **21,** 61–63.

Hammill, T. M. (1972). Fine structure of annellophores. III. *Monotosporella sphaerocephala. Can. J. Bot.* **50,** 581–585.

Hammill, T. M. (1977). Transmission electron microscopy of annellides and conidiogenesis in the synnematal hyphomycete *Trichurus spiralis. Can. J. Bot.* **55,** 233–244.

Hanlin, R. T. (1976). Phialide and conidium development in *Aspergillus clavatus. Am. J. Bot.* **63,** 144–145.

Harvey, I. C. (1974). Light and electron microscope observations of conidiogenesis in *Pleiochaeta setosa* (Kirchn.) Hughes. *Protoplasma* **82,** 203–221.

Haschemeyer, R. H., and Myers, R. J. (1972). Negative staining. *In* "Principles and Techniques of Electron Microscopy" (M. A. Hayat, ed.), Vol. 2, pp. 99–147. Van Nostrand-Reinhold, Princeton, New Jersey.

Hashimoto, H. (1979). Development of high resolution electron microscopy in atomic level and its future. *J. Electron Microsc.* **28**, Suppl., S1–S8.

Hawes, C. R. and Beckett, A. (1977). Conidium ontogeny in the *Chalara* state of *Ceratocystis adiposa.* II. Electron microscopy. *Trans. Br. Mycol. Soc.* **68**, 267–276.

Hayat, M. A., ed. (1970). "Principles and Techniques of Electron Microscopy," Vol. 1. Van Nostrand-Reinhold, Princeton, New Jersey.

Hayat, M. A. (1978). "Introduction to Biological Scanning Electron Microscopy." University Park Press, Baltimore, Maryland.

Hearle, J. W. S., Sparrow, J. T., and Cross, P. M., eds. (1972). "The Use of the Scanning Electron Microscope." Pergamon, Oxford.

Heintz, C. E., and Pramer, D. (1972). Ultrastructure of nematode-trapping fungi. *J. Bacteriol.* **110**, 1163–1170.

Henderson, W. J., and Griffiths, K. (1972). Shadow casting and replication. *In* "Principles and Techniques of Electron Microscopy" (M. A. Hayat, ed.), Vol. 2, pp. 151–193. Van Nostrand-Reinhold, Princeton, New Jersey.

Hess, W. M., and Weber, D. J. (1972). Freezing artifacts in frozen-etched *Rhizopus arrhizus* sporangiospores. *Mycologia* **64**, 1164–1166.

Heunert, H.-H. (1972). Microtechnique for the observation of living microorganisms. *Zeiss Inf.* **20**, 40–49.

Hoffman, R., and Gross, L. (1975a). The modulation contrast microscope. *Nature (London)* **254**, 586–588.

Hoffman, R., and Gross, L. (1975b). Modulation contrast microscope. *Appl. Opt.* **14**, 1169–1176.

Howard, R. J., and Aist, J. R. (1979). Hyphal tip cell ultrastructure of the fungus *Fusarium:* Improved preservation by freeze-substitution. *J. Ultrastruct. Res.* **66**, 224–234.

Ichinokawa, T. (1979). Development of analytical electron microscopy and its future. *J. Electron Microsc.* **28**, Suppl., S17–S24.

Johari, O., and Corvin, I., eds. (1972). "Scanning Electron Microscopy/1972." IITRI, Chicago, Illinois.

Jones, J. P. (1976). Ultrastructure of conidium ontogeny in *Phoma pomorum, Microsphaeropsis divoceum,* and *Coniothyrium fuckelii. Can. J. Bot.* **54**, 831–851.

Katsumoto, T., Takayama, H., and Takagi, A. (1978). Comparable observations by means of scanning transmission and scanning electron microscopy of the same cultured cell specimens. *J. Electron Microsc.* **27**, 325–327.

Kelley, R. O., Dekker, R. A. F., and Bluemink, J. G. (1973). Ligand-mediated osmium binding: Its application in coating biological specimens for scanning electron microscopy. *J. Ultrastruct. Res.* **45**, 254–258.

Kendrick, W. B., and Cole, G. T. (1969). Conidium ontogeny in Hyphomycetes: *Trichothecium roseum* and its meristem arthrospores. *Can. J. Bot.* **47**, 345–350.

Kimoto, S. (1972). The scanning microscope as a system. *JEOL News* **10**, 1–30.

Koehler, J. K. (1972). The freeze-etching technique. *In* "Principles and Techniques of Electron Microscopy" (M. A. Hayat, ed.), Vol. 2, pp. 53–98. Van Nostrand-Reinhold, Princeton, New Jersey.

Kurtzman, C. P., Baker, F. L., and Smiley, M. J. (1974). Specimen holder to critical point dry microorganisms for scanning electron microscopy. *Appl. Microbiol.* **28**, 708–712.

Lewis, E. R., and Nemanic, M. K. (1973). Critical point drying techniques. *Scanning Electron Microsc.* (O. Johari, ed.), pp. 767–773. IITRI, Chicago, Illinois.

Mahl, H. (1940). A plaster-cast method for supermicroscopical investigation of metal surfaces. *Metallwirtsch., Metallwiss., Metalltech.* **19**, 488.

Malick, L. E. and Wilson, R. B. (1975). Modified thiocarbohydrazide procedure for scanning electron microscopy: Routine use for normal, pathological or experimental tissues. *Stain Technol.* **50**, 265–269.

Marchant, H. J. (1973). Processing small delicate biological specimens for scanning electron micros-copy. *J. Microsc. (Oxford)* **97**, 369–371.

Marshall, A. T. (1975). X-ray microanalysis of frozen hydrated biological specimens. The effect of charging. *Micron* **5**, 275.

Meek, G. A. (1970). "Practical Electron Microscopy for Biologists." Wiley, New York.

Meryman, H. T. (1957). Physical limitations of the rapid freezing method. *Proc. R. Soc. London, Ser. B* **147**, 452–459.

Meryman, H. T., and Kafig, E. (1955). The study of frozen specimens, ice crystals and ice crystal growth by electron microscopy. *Nav. Med. Res. Inst. Rep.* **13**, 529–544.

Moor, H. (1959). Platin-Kohle-Abdruck-Technik angewandt auf den Feinbau der Milchröhren. *J. Ultrastruct. Res.* **2**, 393–422.

Moor, H. (1966). Use of freeze-etching in the study of biological ultrastructure. *Int. Rev. Exp. Pathol.* **5**, 179–216.

Moor, H. (1971). Recent progress in the freeze-etching technique. *Philos. Trans. R. Soc. London* **261**, 121–131.

Moor, H. (1973). Etching and related problems. *In* "Freeze-etching: Techniques and Applications" (E. L. Benedetti and P. Favard, eds.), pp. 21–30. Soc. Fr. Microsc. Electron., Paris.

Moor, H., and Mühlethaler, K. (1963). Fine structure in frozen-etched yeast cells. *J. Cell Biol.* **17**, 609–628.

Moor, H., Mühlethaler, K., Waldner, H., and Frey-Wyssling, A. (1961). A new freezing ultramic-rotome. *J. Biophys. Biochem. Cytol.* **10**, 1–13.

Mühlethaler, K., Wehrli, E., and Moor, H. (1971). Double fracturing methods for freeze-etching. *Electron Microsc., Proc. Int. Congr., 7th, 1970* Vol. 1, pp. 49–150.

Murata, K. (1973). Monte Carlo calculations on electron scattering and secondary electron produc-tion in the SEM. *Scanning Electron Microsc.* (O. Johari, ed.), p. 267. IITRI, Chicago, Illinois.

Murray, G. M., and Maxwell, D. P. (1974). Ultrastructure of conidium germination of *Cochliobolus carbonus*. *Can. J. Bot.* **52**, 2335–2340.

Najim, L., and Turian, G. (1979). Ultrastructure de l'hyphe végétatif de *Sclerotinia fuctigena*. *Can. J. Bot.* **57**, 1299–1313.

Nemanic, M. K. (1972). Critical point drying, cryofracture, and serial sectioning. *Scanning Electron Microsc.* **5**, 297–304.

Newell, D. G., and Roth, S. (1975). A container for processing small volumes of cell suspensions for critical point drying. *J. Microsc. (Oxford)* **104**, 321–323.

Oatley, C. W. (1972). "The Scanning Electron Microscope." Cambridge Univ. Press, London and New York.

Odds, F. C. (1979). "*Candida* and Candidosis." University Park Press, Baltimore, Maryland.

Oliver, R. M. (1973). Negative stain electron microscopy of protein macromolecules. *In* "Methods in Enzymology" (C. H. W. Hirs and S. N. Timasheff, eds.), Vol. 27, pp. 616–672. Academic Press, New York.

Oujezdsky, K. B., Grove, S. N., and Szaniszlo, P. J. (1973). Morphological and structural changes during the yeast-to-mold conversion of *Phialophora dermatitidis*. *J. Bacteriol.* **113**, 468–477.

Panessa-Warren, B. J., and Broers, A. N. (1979). High resolution SEM of bacterial virus T7. *Ultramicroscopy* **4**, 317–322.

Pawley, J. B. (1978). Design and performance of presently available TV-rate stereo SEM systems. *Scanning Electron Microsc.* (O. Johari, ed.), p. 157. SEM Inc., AMF O'Hare, Illinois.

Pinto Da Silva, P., and Branton, D. (1970). Membrane splitting in freeze-etching. *J. Cell Biol.* **45**, 598–605.

Plattner, H., Schmitt-Fumian, W. W., and Bachman, L. (1973). Cryofixation of single cells by spray-freezing. *In* "Freeze-etching: Techniques and Applications" (E. L. Benedetti and P. Favard, eds.), pp. 81–100. Soc. Fr. Microsc. Electron, Paris.

Postek, M. T., and Tucker, S. C. (1977). Thiocarbohydrazide binding for botanical specimens for scanning electron microscopy: A modification. *J. Microsc. (Oxford)* **110**, 71–74.

Postek, M. T., Kirk, W. L., and Cox, E. R. (1974). A container for the processing of delicate organisms for scanning or transmission electron microscopy. *Trans. Am. Microsc. Soc.* **93**, 265–267.

Reimer, L. (1976). Electron-specimen interactions and applications in SEM and STEM. *Scanning Electron Microsc.* (O. Johari, ed.), pp. 1–8. IITRI, Chicago, Illinois.

Reynolds, E. S. (1963). The use of lead citrate at high pH as an electron-opaque stain in electron microscopy. *J. Cell Biol.* **17**, 208–222.

Richardson, K. C., Jarett, L., and Finke, E. H. (1960). Embedding in epoxy resins for ultrathin sectioning in electron microscopy. *Stain Technol.* **35**, 313–323.

Richter, H. (1968a). Die Reaktion hochpermeabler Pflanzenzellen auf drei Gefrierschutzstoffe (Glyzerin, Äthylenglycol, Dimethysulfoxid). *Protoplasma* **65**, 155–166.

Richter, H. (1968b). Die Gefrierresistenz glyzerinbehandelter *Campanula*-Zellen. *Protoplasma* **66**, 63–78.

Riehle, U. (1968). "Ueber die Vitrifizierung verdünnter wässriger Lösungen," Diss. No. 4271. Eidgen. Technische Hochschule, Zürich, Juris Verlag, Zürich.

Riehle, U., and Hoechli, M. (1973). The theory and technique of high pressure freezing. *In* "Freeze-etching: Techniques and Applications" (E. L. Benedetti and P. Favard, eds.), pp. 31–61. Soc. Fr. Microsc. Electron., Paris.

Rochow, T. G., and Rochow, E. G. (1978). "An Introduction to Microscopy by Means of Light, Electrons, X-rays, or Ultrasound." Plenum, New York.

Rooseboom, M. (1967). The history of the microscope. *Proc. R. Microsc. Soc.* **2**, 266–293.

Rostgaard, J., and Christensen, P. (1975). A multipurpose specimen-carrier for handling small biological objects through critical point drying. *J. Microsc. (Oxford)* **105**, 107–113.

Rottenburg, W., and Richter, H. (1969). Automatische Glyzerinbehandlung Pflanzlicher Dauergewebszellen für die Gefrierätzung. *Mikroskopie* **25**, 313–319.

Ruffolo, J. J. (1974). Critical point drying of protozoan cells and other biological specimens for scanning electron microscopy: Apparatus and methods of specimen preparation. *Trans. Am. Microsc. Soc.* **93**, 124–131.

Samson, R. A., Stalpers, J. A., and Verkerke, W. (1979). A simplified technique to prepare fungal specimens for scanning electron microscopy. *Cytobios* **24**, 7–11.

Schoknecht, J. D. (1975). SEM and X-ray microanalysis of calcareous deposits in myxomycete fructifications. *Trans. Am. Microsc. Soc.* **94**, 216–223.

Schoknecht, J. D., and Hattingh, M. J. (1976). X-ray microanalysis of elements in cells in VA mycorrhizal and nonmycorrhizal onions. *Mycologia* **68**, 296–303.

Singer, S. J. (1971). The molecular organization of biological membranes. *In* "Structure and Function of Biological Membranes" (L. I. Rothfield, ed.), pp. 146–222. Academic Press, New York.

Singer, S. J. (1974). The molecular organization of membranes. *Annu. Rev. Biochem.* **43**, 805–833.

Singer, S. J. (1975). Architecture and topography of biologic membranes. *In* "Cell Membranes: Biochemistry, Cell Biology and Pathology" (G. Weismann and R. Claiborne, eds.), pp. 35–44. HP Publ., New York.

Singer, S. J., and Nicolson, G. L. (1972). The fluid mosaic model of the structure of cell membranes. *Science* **175**, 720–731.

Sjöstrand, F. S. (1977). The arrangement of mitochondrial membranes and a new structural feature of the inner mitochondrial membranes. *J. Ultrastruct. Res.* **59**, 292–319.

Sjöstrand, F. S. (1979). The interpretation of pictures of freeze-fractured biological material. *J. Ultrastruct. Res.* **69**, 378–420.

Sjöstrand, F. S., and Barajas, I. (1968). Effect of modifications in conformation of protein molecules on structure of mitochondrial membranes. *J. Ultrastruct. Res.* **25**, 121–155.

Skaer, H. Le B., Franks, F., Asquith, M. H., and Echlin, P. (1977). Polymeric cryoprotectants in the preservation of biological ultrastructure. III. Morphological aspects. *J. Microsc. (Oxford)* **110**, 257–270.

Sletyr, U. (1970a). Die Grefrierätzung korrespondierender Bruchhälften: Ein neuer Weg zur Aufklärung von Membranstrukturen. *Protoplasma* **70**, 101–117.

Sletyr, U. (1970b). Fracture faces in intact cells and protoplasts of *Bacillus stearothermophilus:* A study by conventional freeze-etching and freeze-etching of corresponding fracture moieties. *Protoplasma* **71**, 295–312.

Sletyr, U. B., and Robards, A. W. (1977a). Plastic deformation during freeze-cleaving: A review. *J. Microsc. (Oxford)* **110**, 1–25.

Sletyr, U. B., and Robards, A. W. (1977b). Freeze-fracturing: A review of methods and results. *J. Microsc. (Oxford)* **111**, 77–100.

Spurr, A. R. (1969). A low-viscosity epoxy resin embedding medium for electron microscopy. *J. Ultrastruct. Res.* **26**, 31–43.

Steere, R. L. (1957). Electron microscopy of structural detail in frozen biological specimens. *J. Biophys. Biochem. Cytol.* **3**, 45–60.

Steere, R. L., and Erbe, E. F. (1979). Complementary freeze-fracture, freeze-etch specimens. *J. Microsc. (Oxford)* **117**, 211–218.

Steere, R. L., and Moseley, M. (1969). New dimensions in freeze-etching. *Proc. Electron Microsc. Soc. Am.* **27**, 202–203.

Steere, R. L., and Moseley, M. (1971). Modified freeze-etch equipment permits simultaneous preparations of 2–10 double replicas. *Electron Microsc., Proc. Congr. 7th, 1970* Vol. 1, pp. 449–450.

Steffens, W. L. (1978). A method for the removal of epoxy resins from tissue in preparation for scanning electron microscopy. *J. Microsc. (Oxford)* **113**, 95–99.

Streiblová, E. (1971). Cell division in yeasts. *Proc. Symp. Int. Congr. Microbiol., 10th, 1970* pp. 131–140.

Streiblová, E., and Wolf, A. (1972). Cell wall growth during the cell cycle of *Schizosaccharomyces pombe. Z. Allg. Mikrobiol.* **12**, 673–684.

Sykes, J. A., and Moore, E. B. (1959). A new chamber for tissue culture. *Proc. Soc. Exp. Biol. Med.* **100**, 125–127.

Tanaka, K., and Iino, A. (1974). Critical point drying method using dry ice. *Stain Technol.* **49**, 203–206.

Turian, G. (1978). The "Spitzenkörper," centre of the reducing power in the growing hyphal apices of two septomycetous fungi. *Experientia* **34**, 1277–1279.

Venables, J. A. (1976). "Developments in Electron Microscopy and Analysis." Academic Press, New York.

von Ardenne, M. (1938a). Das Elektronen-Rastermikroskop: Theoretische Grundlagen. *Z. Phys.* **109**, 553.

von Ardenne, M. (1938b). Das Elektronen-Rastermikroskop: Praktische Ausführung. *Z. Tech. Phys.* **19**, 407.

Wehrli, E., Mühlethaler, K., and Moor, H. (1970). Membrane structure as seen with a double replica method for freeze-fracturing. *Exp. Cell Res.* **59**, 336–339.

Weibel, E. R., and Bolender, R. P. (1973). Stereological techniques for electron microscopic morphometry. *In* "Principles and Techniques of Electron Microscopy" (M. A. Hayat, ed.), Vol. 3, pp. 237–296. Van Nostrand-Reinhold, Princeton, New Jersey.

Wells, O. C. (1974). "Scanning Electron Microscopy." McGraw-Hill, New York.

Welter, L. M. (1975). Application of a field emission source to SEM. *In* "Principles and Techniques of Scanning Electron Microscopy" (M. A. Hayat, ed.), Vol. 3, pp. 195–220. Van Nostrand-Reinhold, Princeton, New Jersey.

Wilde, E. W., and Flierman, C. B. (1979). Fluorescence microscopy for algal studies. *Trans. Am. Microsc. Soc.* **98,** 96–102.

Williams, R. C., and Wyckoff, R. W. G. (1946). Applications of metallic shadow-casting to microscopy. *J. Appl. Phys.* **17,** 23.

Winborn, W. B., and Guerrero, D. L. (1974). The use of a single tissue specimen for both transmission and scanning electron microscopy. *Cytobios* **10,** 83–91.

Wischnitzer, S. (1970). "Introduction to Electron Microscopy." Pergamon, Oxford.

Zingsheim, H. P., Abermann, R., and Bachmann, L. (1970). Apparatus for ultrashadowing of freeze-etched electron microscopic specimens. *J. Phys. E.* **3,** 39–42.

Zworykin, V. K., Hillier, J., and Snyder, R. L. (1942). A scanning electron microscope. *ASTM Bull.* **117,** 15.

Subject Index

A page number set in bold indicates a legend.

Index to Taxa